RNA Technologies

For further volumes:
http://www.springer.com/series/8619

Volker A. Erdmann • Jan Barciszewski
Editors

From Nucleic Acids Sequences to Molecular Medicine

Springer

Editors
Prof. Dr. Volker A. Erdmann
Free University Berlin
Institute of Chemistry/Biochemistry
Thielallee 63
14195 Berlin
Germany
vae@zedat.fu-berlin.de

Prof. Dr. Jan Barciszewski
Polish Academy of Sciences
Institute of Bioorganic Chemistry
Z. Noskowskiego 12/14
61-704 Poznan
Poland
jan.barciszewski@ibch.poznan.pl

ISBN 978-3-642-27425-1 ISBN 978-3-642-27426-8 (eBook)
DOI 10.1007/978-3-642-27426-8
Springer Heidelberg New York Dordrecht London

Library of Congress Control Number: 2012940711

Printed on acid-free paper

Springer is part of Springer Science+Business Media (www.springer.com)

Preface

Nucleic Acids Meet Human Genetic Medicine

With this volume of the series RNA Technologies, we aim to cover current and important developments in RNA chemistry, RNA biology, and the application of RNA molecules in molecular medicine.

Cellular gene expression is based on several levels of information. First the sequence of the canonical adenine, cytosine, guanosine, and thymine bases in DNA encodes for the primary structures of proteins through their specific base triplet arrangements. The "Central Dogma of Molecular Biology" taught us that the DNA is transcribed into RNAs and the protein-coding messenger RNA genes are then translated into proteins by ribosomes. For its function the ribosomes are directly dependent upon noncoding RNAs such as ribosomal RNAs, tRNAs, and small nuclear RNA. It is now evident that RNA, in gene expression, functions far beyond the established roles of these RNAs, for example, in RNA splicing and editing, telomere maintenance, protein secretion, small-molecule sensing, and reaction catalysis. Thus, RNA molecules play a key role in several fundamental cellular processes, serving as a carrier of genetic information and, very importantly, also as its regulator.

In human eukaryotic cells only ca. 1.8% of the genomic DNA is transcribed into mRNAs, while the remaining 98.2% represents the so-called noncoding DNA sequences. As discovered over the last 20 years, a large amount of this noncoding DNA represents actually noncoding RNA. The important role of noncoding RNAs in cellular processes has recently become more and more apparent. How RNA achieves its large number of different functions with a limited assortment of four different building blocks is still a question of great interest, and the answer lies in deciphering the RNA structures at its different levels of complexity. More detailed knowledge about the structural and functional properties of ribonucleic acids, and how they are functioning in the cells, is still required before methods can be developed to use these regulators in solving the problems in molecular medicine. In this book, up-to-date studies are presented, which follow exactly this goal: to understand the structures and functions of RNA molecules and to use this information for applications in cellular biology and molecular medicine.

Over 25 years ago, the central dogma of molecular biology was expanded with the discovery that in addition to proteins, ribonucleic acids can also exhibit enzymatic activities. These RNA enzymes (ribozymes) spurred intense studies into the structural basis of RNA catalysis. All naturally occurring ribozymes catalyze phosphodiester transfer or hydrolysis with exquisite substrate specificity.

Thus, RNA has received increasing interest as a target of chemotherapy and as chemotherapeutic agents to be applied in molecular medicine. The demonstration that synthetic oligonucleotides could be used to interfere with biological information transfer in the 1970s of the last century induced a great interest to a novel type of therapy. Oligomeric nucleic acid-based therapeutics can be subdivided into groups according to their target molecule: antisense oligonucleotides (ODN), ribozymes, microRNAs inhibitors, and short interfering RNAs. These oligonucleotides interact by Watson–Crick base pairing with cellular transcripts leading to their degradation or functional inhibition. First antisense therapies have reached the patients. However, a requirement for the effective application of antisense therapy is based on three fundamental issues, and they are stability, site-specific delivery, and no off-target effects. This is, of course, also true for the employment of the other RNA molecules, such as siRNAs, microRNAs, or different ribozymes. To overcome the stability problems of RNA molecules in the human serum and in the cells, efforts have been directed to chemically modify these ribonucleic acids in such a way that they would become resistant to cellular ribonucleases. But with these modifications one may also increase the unwanted side effects, so that the problems of stability have so far not been optimally solved. Very promising approaches with these goals in mind are also presented in this book.

In the cells oligonucleotides are known to modulate gene expression by binding to proteins as aptamers, and not only that, in addition, the physiological binding to transcription factors to reduce their capacity to mediate gene transcription has also been observed. Very interesting and promising are also antisense molecules that recruit RNase H or RISC factors to mediate RNA hydrolysis. RNA interference, for example, is best known for its various roles in posttranscriptional gene silencing in the cytoplasm.

In this volume the reviewers discuss the general aspects of structure and function of ribonucleic acids as indicated above. Very interesting are the results reported, because they show how new and very promising RNA regulators can be developed for the next generation of therapeutic agents directed at various types of diseases.

In the book, we collected chapters on general aspects of RNA, its stability, and chemical modification. Several reviews highlight recent progress in development and application of antisense technologies for RNA silencing. The therapeutic potential of ribozymes and DNAzymes is also extensively discussed. A current knowledge on microRNA pathways in human diseases is covered in several chapters and will help to round up the current knowledge of RNA technologies and as these technologies may be applied in the field of future molecular medicine.

Berlin, Germany — Volker A. Erdmann
Poznań, Poland — Jan Barciszewski

Contents

RNAissance

Juergen Brosius

Contents

Abstract Half a century ago, at the dawn of molecular biology, few investigators realized the potential of RNA in cellular regulation and function. After a very long lag phase, the quantitative and qualitative evidence for the significance of non-protein-coding RNA (npcRNA) is now overwhelming. In the search for disease-causing gene alterations, RNA genes must be considered equally with protein-coding genes.

Keywords Non-protein-coding RNA • npcRNA • ncRNA • RNA world • Retroposition • npcRNA knockout mice • RNA-encoding disease genes

J. Brosius (✉)
Institute of Experimental Pathology, ZMBE, University of Münster, Von-Esmarch-Street 56, 48149 Münster, Germany
e-mail: RNA.world@uni-muenster.de

V.A. Erdmann and J. Barciszewski (eds.), *From Nucleic Acids Sequences to Molecular Medicine*, RNA Technologies, DOI 10.1007/978-3-642-27426-8_1,

1 From So Simple a Beginning

Unless more or less similar precursor macromolecules or structures were the first self-replicating units and led to RNA via one or more evolutionary transitions—akin to the genetic takeover or major transition from RNA to DNA (Cairns-Smith 1982; Szathmary and Smith 1995)—life might have begun with the fortuitous polymerization of an RNA macromolecule that subsequently could copy itself and later even copied other RNA molecules that arose either de novo by additional polymerizations or by error-prone replication of the primordial replicating RNAs (Brosius 2003). Needless to say, even such a simple scenario is not a trivial feat, given our current conceptions about, for example, environmental conditions and availabilities of precursor molecules at that time (Robertson and Joyce 2011). In any event, RNA was instrumental at very early stages of life and remains so in extant forms. The products of major evolutionary transitions, polypeptides, and later DNA are mere extensions or tools of RNA. For example, proteins enhanced (in complexes with RNA as ribonucleoprotein complexes, RNPs) or completely took over tasks, accomplishing them with higher efficiency and fidelity than RNA alone. With the activity of a protein resembling reverse transcriptase, DNA entered the scene and became the bookkeeper of RNA's hereditary information (Gould 2002); thus, DNA-encoded genes and genomes are not the causative agents (Dawkins 1976). The RNA-centered viewpoint is equally, if not more, suitable to account for altruism and intragenomic conflict (Brosius 2003).

2 Transitions in Slow Motion

An interesting facet of at least some aspects of the aforementioned major transitions is their extreme protraction. Like at the onset of templated protein biosynthesis, evolution is still *trying out* RNA sequences that initially were not earmarked as messenger RNAs or protein exons to translate novel proteins (Guerzoni and McLysaght 2011; Tautz and Domazet-Loso 2011) or novel parts of existing proteins (Baertsch et al. 2008). Likewise, in most genomes of multicellular organisms, the conversion of RNA to DNA via reverse transcription (termed retroposition) is pervasive and unabated (Brosius 1999a), such retromaterial accumulated over long evolutionary periods now constituting the vast majority of genomes (Brosius 1999b, 2005a, b, 2009).

3 RNA First

With the groundbreaking discovery of RNA's enzymatic activities (Guerrier-Takada et al. 1983; Kruger et al. 1982), the RNA world (Gilbert 1986) became more than a mere assumption formulated nearly two decades earlier (Crick 1968;

Orgel 1968; Woese 1967). The 1962 chapter by Alex Rich, however, stands out due to its priority, comprehensiveness, and predictive value (Rich 1962). Since the chapter is not readily available [a scanned copy can be obtained from the author], I am quoting some of the relevant statements. Under the subheading "Did life originate with protein molecules?" Rich concludes: "...it does not really explain this most difficult step, the evolution of nucleic acid-controlled protein synthesis. However, another type of theory can be described as the basis for the origin of life which places the nucleic acids in a more central role." Under the next subheading "Polynucleotides as the origin of living systems" Rich states: "...[it is] imperative for us to reformulate theories about the origin of life in order to place the nucleic acids in proper perspective. As discussed above, the sequence of amino acids in proteins is, in a sense, a derivative of the sequence information encoded in the order of nucleotides in some part of the nucleic acids. Accordingly, it may be more reasonable to consider a theory of the origin of life in which the nucleic acids were developed as the primary agents." And "We postulate that these nucleotide units can be assembled in random chains to make primitive, single-stranded polynucleotides in the absence of any protein catalysis. ... The next step is the development of replicating nucleic acids. We postulate that the primitive polynucleotide chains are able to act as a template or as a somewhat inefficient catalyst for promoting the polymerization of the complementary nucleotide residues to build up an initial two-stranded molecule. This is in a sense the same system as that whereby a single DNA strand is able to make new RNA or DNA strands by the stepwise addition of complementary nucleotides, which are subsequently polymerized together. However, we postulate that this may happen in a primitive environment in the absence of protein catalysts. ...further experiments could be carried out on this type of nonenzymatic polynucleotide catalysis to determine the limits of specificity. What is being pointed out here is an important difference between proteins and nucleic acids. There are significant stereochemical reasons why the polynucleotides can act as their own catalysts for self-replication. However, there are no analogous reasons for believing the polyamino acids have this ability to reproduce themselves."

4 From RNA to RNP World

Furthermore, decades before others (Brosius 2001; Hartman 1975; Ma 2010; Noller 1991; Orgel 1989; Rodin et al. 2011; Tyagi 1981; Walder et al. 1979; Weiner and Maizels 1987; Woese 1980, 2001), Rich envisioned solutions to the major transition of templated protein biosynthesis: "We imagine that the mononucleotide units which are about to be incorporated into polynucleotide chain occasionally have other small molecules attached to them. ... In particular this residue might be an organic acid linked to the hydroxyl group [of the ribose moiety] by an ester linkage. It might also be an amino acid attached through the same linkage. At this stage we theorize this would be the beginning of a prototype system in which the polymerization of a nucleic acid molecule is coupled with the assembly of a series of amino acids. The amino acids might then be subsequently polymerized once they are organized in a linear assembly by their attachment to the adjoining nucleotides" (Rich 1962).

5 Late DNA

Also for the next major transition, the replacement of RNA by DNA as hereditary material, Rich (1962) was right on track by asking: "Why are there two nucleic acids?" He continues: "Despite the great chemical similarities, the molecules nevertheless have quite different functions in the cell. DNA acts as the major carrier of genetic information, while the RNA molecule is used to convert this genetic information into actual protein molecules. Because of the close chemical similarities, we are tempted to ask whether they could have originated historically from a common stem nucleic acid molecule, which then specialized in the course of evolution to produce the two different classes of nucleic acids, which we see today. To discuss this further we should note that the RNA molecule is also able to carry genetic information, as, for example, in the RNA-containing viruses. Thus, it may be reasonable to speculate that the hypothetical stem or parent polynucleotide molecule was initially an RNA-like polymer which was able to convey genetic information as well as organize the amino acids into a specific sequence to make proteins. This implies that the RNA polynucleotide strand had the ability to replicate itself and produce a complement in a manner somewhat similar to that which is found in DNA. It is possible that an enzyme of this type may have been observed already. By this view, DNA may be regarded as a derived macromolecule, which has evolved in a form such that it only carries out part of the primitive nucleic acid function. . . . It will be of considerable interest to study the available simple life forms to see whether some of them may exist with only one type of nucleic acid rather than two types. It is possible that the RNA-containing viruses may be regarded as present-day examples which may have degenerated evolutionarily from such a primitive life form."

6 Rays of Light in the Darkness

Despite such remarkable insight as early as 50 years ago, what followed were at least two to three lost decades, during which, except in their role of messenger RNAs merely providing the link between DNA and protein, RNAs were largely ignored. Before the discovery of RNA's potential catalytical activity and the notion that ribosomal RNA was not only the main functional entity in many ribosome functions (Barta et al. 1984; Noller and Chaires 1972) but also a key player in peptidyl transferase activity (Noller et al. 1992), before small nuclear RNAs were found to be involved in the RNA-splicing mechanism (Lerner et al. 1980), in the 3′

processing of histone mRNAs (Galli et al. 1983), or the maintenance of telomeres (Greider and Blackburn 1989), most scientists in the field thought that non-messenger RNAs were mere fossils from the RNA/RNP worlds, gradually being replaced, where possible, by protein.

Few scientists were open to the idea that during the course of evolution, novel non-protein-coding RNAs were generated analogously to novel proteins and made important contributions to the regulation and function of modern cells. Even fewer laboratories, notably those of Harris Busch, Sheldon Penman, and Volker Erdmann, embarked on the discovery of hitherto unknown non-protein-coding RNAs (npcRNAs), (Erdmann and Wolters 1987; Lorberboum et al. 1986; Prestayko and Busch 1968; Struck et al. 1986; Zieve and Penman 1976). In the mid-1980s, we began to generate cDNA libraries based on npcRNAs that were separated from the large rRNAs and most mRNAs via size fractionation on denaturing polyacrylamide gels and the attachment of oligocytidine tails (DeChiara and Brosius 1987). In so doing, we showed that BC1 RNA, an ~150-nt RNA preferentially expressed in neurons of rodents, was not—as proclaimed by others (Milner et al. 1984; Sutcliffe et al. 1984a, b)—a by-product of the transcription of several thousands of somewhat heterogeneous short interspersed repetitive elements (SINEs) but rather a homogeneous product from one or a few RNA-coding genes. It would be a long and winding road toward elucidating the function of this relatively young npcRNA (see below). In the mid-1990s, incited by the first fully sequenced genomes and the expectation of many others, including that of human, we were concerned that npcRNAs would not receive the attention they deserved (Brosius 1996). Hence, we embarked on generating cDNA libraries derived from small npcRNAs up to 500 nt[1] from various tissues and developmental stages from model organisms of all three domains of life (Filipowicz 2000; Huttenhofer et al. 2001, 2002). The highly significant discovery of hundreds of novel microRNAs (miRNAs) finally opened the floodgates (Ambros 2001; Lagos-Quintana et al. 2001, 2002; Lau et al. 2001; Lee and Ambros 2001; Ruvkun 2001; Ruvkun et al. 2004); no one could ignore the functional significance of npcRNA any longer. Also, phylogenetic analyses of miRNAs and small nucleolar RNAs (snoRNAs) confirmed that RNAs are not necessarily ancient but that novel forms constantly evolve (Berezikov et al. 2006; Schmitz et al. 2008; Zemann et al. 2006). The usual route is by duplication of an RNA, including via retroposition, and

[1] The size cutoff for small RNAs used to be about 500 nt to include SRP RNA (300 nt), 7SK RNA (332 nt), MRP RNA (287 nt), and small nuclear RNAs (depending on species), etc. Large or long RNAs were defined as being in the size range of mRNAs, sometimes even displaying mRNA-like attributes (e.g., processing, polyadenylation) but devoid of a functional open reading frame (Brosius and Tiedge 2004). The first publications on microRNAs (miRNAs) correctly designated them as tiny RNAs to discriminate them from the larger small RNAs (Ambros 2001; Lau et al. 2001; Ruvkun 2001) With time, they were more laxly addressed as small RNAs, and hence, the "new" small RNAs had to be distinguished from the "old" small RNAs, simply by stretching the latter and moving the cutoff to about 200 nt. The fact that numerous small RNAs designated as such in dozens to hundreds of previous publications are now supposed to be found under long RNAs is unfortunate.

subsequent divergence of one of the copies (Volff and Brosius 2007). Once more, Alex Rich can claim priority on the notion of evolution of genes by duplication (Ohno 1970) by stating in 1962: “Here the luxury of surplus, redundant information appears to provide a selective advantage in evolution. If this process goes on several times, eventually the organism ends up with many genetic copies of an original prototype protein molecule. These copies might then evolve along somewhat separate evolutionary lines and give rise to classes of molecules which, though similar, are in fact different in many ways” (Rich 1962).

7 The Avalanche

The age of RNomics began in earnest around the turn of the millennium (Filipowicz 2000), calling up many investigators, including some who were early to sense the importance of RNA (Herbert 2004; Mattick 1994, 2001, 2003). The findings that miRNAs had important functions in regulating the translation of mRNAs were voted “breakthrough of the year” (Couzin 2002) and precipitated a gold rush for more discoveries of not only nearly a thousand miRNA candidates per mammalian genome but of all kinds of transcripts and other RNAs, including piwi-interacting RNAs (piRNAs), promoter-associated and terminator-associated small RNAs, CRISPR RNAs involved in prokaryotic defense against viruses, and long non-protein-coding RNAs, some of these transcripts collectively designated as “dark matter” of the genome (Amaral et al. 2008; Bertone et al. 2004; Birney et al. 2007; Cabili et al. 2011; Carninci et al. 2005; Carrington and Ambros 2003; Cullen 2004; Dinger et al. 2008a; Guttman et al. 2009; Huttenhofer 2006; Kampa et al. 2004; Kapranov et al. 2002, 2007a, b, 2010a, b; Khalil et al. 2009; Kim et al. 2004; Lau et al. 2006; Lipovich et al. 2010; Loewer et al. 2010; Marques and Ponting 2009; Mattick 2004, 2010; Mercer et al. 2010a, b, 2011; Numata et al. 2003; Okazaki et al. 2002; Ozsolak et al. 2010; Pang et al. 2009; Parker et al. 2011; Pasmant et al. 2011b; Ravasi et al. 2006; Rederstorff et al. 2010; Rinn et al. 2003; Saxe and Lin 2011; Taft et al. 2010b; Tang et al. 2002, 2005; Tsai et al. 2010; Voinnet 2002; Wang et al. 2011; Westhof and Filipowicz 2005; Yasuda and Hayashizaki 2008).

In addition, there is a huge body of work on the detection of RNA that is complementary to known or unknown transcripts, so-called antisense RNAs or antisense transcripts (Kampa et al. 2004; Kiyosawa et al. 2003; Raabe et al. 2010; Thomason and Storz 2010). In several cases, it has been shown that antisense transcripts are involved in gene regulation (Faghihi and Wahlestedt 2009; Lapidot and Pilpel 2006; Su et al. 2010; Werner et al. 2009; Werner and Sayer 2009; Werner and Swan 2010). Once more, this mechanism of action was remarkably predicted by Alex Rich half a century ago: “If both strands are active, then the DNA would produce two RNA strands which are complementary to each other. Only one of these might be active in protein synthesis, and the other strand might be a component of the control or regulatory system” (Rich 1962). Likewise, the following quote of Rich’s is highly relevant today: “It may be that further investigation will disclose an

RNA-dependent polymerization mechanism not too unlike that seen in protein synthesis." Interestingly, RNA interference (RNAi) with some involvement of RNA-dependent RNA polymerases is a gene-silencing process involving double-stranded RNA in certain phyla (Baulcombe 2007; Tijsterman et al. 2002).

Additional indirect roles were envisioned for RNA (Brosius 2005). As part of a ribonucleoprotein complex (RNP), RNA can bring two or more proteins, via their capabilities to bind directly or indirectly to RNA, into the same subcellular domain or environment as exemplified by the *Schizosaccharomyces pombe* meiRNA (Watanabe and Yamamoto 1994); some of the thousands of newly discovered npcRNAs might merely assemble various proteins and shuttle them to different cellular compartments. Similar suggestions concerning future insights into RNA functions were advanced recently (Sharp 2009).

8 Is All that Is Transcribed an RNA?

Some investigators cautioned about overinterpreting discoveries, especially with the advent of deep sequencing and the associated danger of capturing more and more degradation products or transcripts that are biologically insignificant (Brosius2005c; Huttenhofer et al. 2005). Also, it had been conceded that the distinction between mRNA and npcRNA can be difficult, especially when the potential open reading frames are short (Dinger et al. 2008b, Ingolia et al. 2011), as there are also mRNAs encoding only short peptides (Sousa et al. 2001). Furthermore, not every cDNA represents a *bona fide*, functional RNA molecule. As predicted, many of the longer RNAs are simply parts of 3′ untranslated regions (3′ UTRs); some of them extending far beyond proximal poly(A) addition signals when alternate, more distal poly(A) addition signals are used. On the other hand, others argue that such transcripts are not mere degradation products but are independently transcribed (Mercer et al. 2010a, 2011). The controversy about long npcRNAs is still raging (Clark et al. 2011; van Bakel et al. 2010).

It is perhaps indicative that not many investigators have taken the risk of depleting genes encoding npcRNAs—apart from those encoding miRNAs (Park et al. 2010)—using the still cumbersome process of generating knockout mice. The faster route of using siRNA in knockdown experiments, especially involving small RNAs, is not always reliable (Ploner et al. 2009).

9 Life Without RNA?

An early and important example of the limited number of RNA knockout mice (apart from miRNA KOs, see above) involved the gene encoding Xist, a very long RNA of ~16.5 kb essential for inactivating the second X chromosome in female mammals. A deletion of several RNA-coding exons led to death in early embryogenesis when the deletion was on the paternal X chromosome (Marahrens et al. 1997).

Another early example was the knockout of the telomerase RNA gene, *Terc*, resulting in viable mice (for at least six generations); however, telomere shortening and chromosomal abnormalities were observed (Blasco et al. 1997; Lee et al. 1998). A further example involved the imprinted oncofetal H19 gene that encodes a ~2.3-kb RNA devoid of open reading frames (Ariel et al. 1997; Leighton et al. 1995a). The deletion is not lethal but affects expression of Insulin Growth Factor 2 (Leighton et al. 1995b), while overexpression of H19 as a transgene is lethal (Brunkow and Tilghman 1991). Somewhat complicating is the fact that recently, it was shown that H19 RNA can serve as a precursor of a miRNA, miR-675 (Cai and Cullen 2007). Deletion of the NEAT1 gene encoding an npcRNA resulted, despite disappearance of paraspeckles in the nucleus, in healthy and fertile mice (Nakagawa et al. 2011).

Lessons from snoRNA deletion mutants in yeast demonstrate that many snoRNAs, albeit highly conserved, generally display rather subtle phenotypes, often manifesting themselves over generations as, for example, disadvantages in growth when compared to the wild type (Badis et al. 2003; Esguerra et al. 2008; King et al. 2003). Examples from our own work underscore the risks and rewards of generating knockout mice of RNA genes. When we embarked on deleting the aforementioned neuronal dendritic BC1 RNA, we were prepared for a subtle phenotype, at best, as the RNA is restricted to the order Rodentia and hence, on an evolutionary scale, relatively young. On the other hand, the RNA-coding gene sequence, its neuronal expression, and dendritic localization are conserved among distantly related rodents corresponding to a separation of about 60 million years (Martignetti and Brosius 1993b; Tiedge et al. 1991). Ablation of the gene in three mouse lines generated from three independent deletions in ES stem cells (Skryabin et al. 2003) showed in several experimental settings in three different laboratories and also under seminatural conditions a mild phenotype, namely reduced exploratory behavior in 129SV × C57BL/6 genetic crosses (Lewejohann et al. 2004). After breeding the mice for ten generations onto a pure C57BL/6 background, the reduced exploratory behavior measured in some of the previous experiments was no longer statistically significant, although a tendency was still observed. Automated tests in the IntelliCage (Endo et al. 2011), however, still revealed a significant reduction in exploration in backcrossed C57BL/6 background mice (Skryabin, Voikar, Wolfer, and J.B., unpublished data). The laboratory of Henri Tiedge found that auditory stress during a relatively small time window (postnatal days 18–21) led to convulsive seizures mediated via group I metabotropic glutamate receptors in two of our $BC1^{-/-}$ mouse lines in the mixed 129SV × C57BL/6 background (Zhong et al. 2009). In our hands, epileptic seizures were not observed in backcrossed C57BL/6 background mice but in mice crossed for at least three generations into the FvB strain. This underscores once more how the genetic background can enhance or nullify the effect of genetic mutations including entire gene deletions (Sibilia and Wagner 1995). The observed phenotype is in line with BC1 RNA functioning as a regulator of translation pathways in neurons, as we suspected for a long time (Brosius and Tiedge 1995; Kindler et al. 2005; Tiedge and Brosius 1996; Tiedge et al. 1991) and then demonstrated in vitro (Kondrashov et al. 2005; Wang et al. 2005, 2002). In anthropoid primates, a possible functional analog (not a homolog), BC200 RNA has been described with similar neuronal expression and dendritic location (Tiedge et al. 1993). The RNA is deregulated in certain tumors and upregulated in Alzheimer's disease patients (Chen et al. 1997; Mus et al. 2007). It will be interesting to examine whether deletions or mutations in the gene encoding BC200 RNA (Martignetti and Brosius 1993a) are associated with certain forms of epilepsies in human patients.

Prader–Willi syndrome (PWS) is a neurodevelopmental disease with manifestations that include infantile hypotonia, usually characterized by neonatal feeding difficulties, followed by hyperphagia leading to obesity, as well as learning disabilities and infertility. Genetically, there are usually large interstitial deletions of 6–6.8 million base pairs at 15q11.2q13. This locus harbors several protein-coding genes and genes encoding small nucleolar RNAs (snoRNAs) that are paternally imprinted (i.e., silenced on the maternal chromosome). The *SNORD116* (*HBII-85*) and *SNORD115* (*HBII-52*) genes are arranged in tandem and transcribed in clusters. As might be expected, several protein-coding regions were knocked out first, but in contrast to a deletion of the imprinting center, which exhibited several phenotypes common to PWS infants, the deleted proteins made no or only minor contributions to the etiology of the disease (Yang et al. 1998). In addition, a microdeletion of the *SNORD115* cluster excluded HBII-52 RNA from a major role in Prader–Willi syndrome (Runte et al. 2005). Targeted deletion of the paternal Snord116 cluster yielded mice that initially failed to thrive (Skryabin et al. 2007). Also, depending on the genetic background of the mice, postnatal lethality ranged from 0% to 15%. In direct contrast to individuals with PWS, the mice remained smaller than wild-type siblings throughout their life and were fertile. An independent deletion of the Snord116 cluster generated in the laboratory of Uta Francke revealed similar phenotypes (Ding et al. 2008). Meanwhile, microdeletions of the *SNORD116* cluster in patients showing the major characteristics of the PWS phenotype were reported (Duker et al. 2010). The phenotypic differences between mice and men are not surprising, as even smaller genetic differences between mouse strains can have major impacts (see above).

An example of an unfavorable risk/reward outcome in generating RNA-encoding gene-depleted mice is *Snora35*, encoding the brain-specific MBI-36 snoRNA of the H/ACA box subtype. The RNA is located within the second huge intron (almost 100 kb) of the serotonin 2c receptor gene (*Htr2c*) and is co-transcribed with the primary *Htr2c* mRNA transcript along with four as yet uncharacterized miRNAs, miR-764, miR-1912, miR-1264, and miR-1298. Our *Snora35* knockout mice (the remote miRNAs remained unaffected) do not thus far appear to display a phenotype different from their wild-type siblings (Boris Skryabin and J.B., unpublished data).

10 RNA in Human Disease and Therapy

Apart from the aforementioned discovery of the catalytic potential of RNA, an ever-growing RNome, and the exciting role of miRNA and siRNA in gene regulation, the fact that mutations in RNA-coding genes are associated with genetic disease is another reason for the heightened attention to RNA. The first report of an RNA as a "disease gene" was, to the best of my knowledge, about MRP RNA, in which mutations cause cartilage–hair hypoplasia (Munoz-Robles et al. 2006; Ridanpaa et al. 2001). MRP RNA is a short RNA polymerase III-transcribed RNA of 267 nt and is related to RNase P, both RNAs functioning in complexes with proteins and processing precursors of rRNA and tRNA, respectively. Mutations can be grouped into two classes: ones that alter the distance of the upstream TATA promoter element to the transcription start site, resulting in lower expression, and others located in the RNA-coding region that do not interfere with the binding of complexed proteins; hence, a direct function of the MRP RNA must be involved, although a mechanism of the functional deficit is still under investigation (Martin and Li 2007; Thiel and Rauch 2011).

While the number of npcRNAs other than miRNAs known to underlie disease is gradually rising (Aguilo et al. 2011; Pasmant et al. 2011a; Qureshi et al. 2010; Taft et al. 2010a; Tsai et al. 2011), changes in the sequences or expressions of miRNAs underlie a multitude of human diseases, ranging from cancer (Farazi et al. 2011; Le Quesne and Caldas 2010; Lichter 2010; Medina et al. 2010; van Kouwenhove et al. 2011), cardiovascular disease (Elton et al. 2011; McManus and Ambros 2011; Small and Olson 2011), metabolic disease (Rottiers and Näär 2012; Shantikumar et al. 2011), hearing loss (Lewis et al. 2009; Mencia et al. 2009; Soukup 2009) to psychiatric disease (Miller and Wahlestedt 2010; Lu et al. 2008; Sayed and Abdellatif 2011).

At the same time, tiny RNAs like siRNA and miRNAs offer avenues toward better diagnosis and eventually treatment of a number of diseases (Boudreau et al. 2011a, b; Bowers et al. 2011; Garzon et al. 2010; Kasinski and Slack 2011; Kim et al. 2011; Lares et al. 2010; McBride et al. 2011; Muntoni and Wood 2011; Rederstorff and Huttenhofer 2010; Seto 2010; Wong 2010) as will be detailed in many chapters of this book.

References

Aguilo F, Zhou MM, Walsh MJ (2011) Long noncoding RNA, polycomb, and the ghosts haunting INK4b-ARF-INK4a expression. Cancer Res 71:5365–5369

Amaral PP, Dinger ME, Mercer TR et al (2008) The eukaryotic genome as an RNA machine. Science 319:1787–1789

Ambros V (2001) microRNAs: tiny regulators with great potential. Cell 107:823–826

Ariel I, Ayesh S, Perlman EJ et al (1997) The product of the imprinted H19 gene is an oncofetal RNA. Mol Pathol 50:34–44

Badis G, Fromont-Racine M, Jacquier A (2003) A snoRNA that guides the two most conserved pseudouridine modifications within rRNA confers a growth advantage in yeast. RNA 9:771–779

Baertsch R, Diekhans M, Kent WJ et al (2008) Retrocopy contributions to the evolution of the human genome. BMC Genomics 9:466

Barta A, Steiner G, Brosius J et al (1984) Identification of a site on 23S ribosomal RNA located at the peptidyl transferase center. Proc Natl Acad Sci USA 81:3607–3611

Baulcombe DC (2007) Molecular biology. Amplified silencing. Science 315:199–200
Berezikov E, Thuemmler F, van Laake LW et al (2006) Diversity of microRNAs in human and chimpanzee brain. Nat Genet 38:1375–1377
Bertone P, Stolc V, Royce TE et al (2004) Global identification of human transcribed sequences with genome tiling arrays. Science 306:2242–2246
Birney E, Stamatoyannopoulos JA, Dutta A et al (2007) Identification and analysis of functional elements in 1% of the human genome by the ENCODE pilot project. Nature 447:799–816
Blasco MA, Lee HW, Hande MP et al (1997) Telomere shortening and tumor formation by mouse cells lacking telomerase RNA. Cell 91:25–34
Boudreau RL, Rodriguez-Lebron E, Davidson BL (2011a) RNAi medicine for the brain: progresses and challenges. Hum Mol Genet 20:R21–R27
Boudreau RL, Spengler RM, Davidson BL (2011b) Rational design of therapeutic siRNAs: minimizing off-targeting potential to improve the safety of RNAi therapy for Huntington's disease. Mol Ther 19:2169–2177
Bowers WJ, Breakefield XO, Sena-Esteves M (2011) Genetic therapy for the nervous system. Hum Mol Genet 20:R28–R41
Brosius J (1996) More haemophilus and mycoplasma genes. Science 271:1302a
Brosius J (1999a) Transmutation of tRNA over time. Nat Genet 22:8–9
Brosius J (1999b) Genomes were forged by massive bombardments with retroelements and retrosequences. Genetica 107:209–238
Brosius J (2001) tRNAs in the spotlight during protein biosynthesis. Trends Biochem Sci 26:653–656
Brosius J (2003) Gene duplication and other evolutionary strategies: from the RNA world to the future. J Struct Funct Genomics 3:1–17
Brosius J (2005a) Echoes from the past—are we still in an RNP world? Cytogenet Genome Res 110:8–24
Brosius J (2005b) Disparity, adaptation, exaptation, bookkeeping, and contingency at the genome level. Paleobiology, 31 (Supplement):1–16
Brosius J (2005c) Waste not, want not – transcript excess in multicellular eukaryotes. Trends Genet 21:287–288
Brosius J (2009) The fragmented gene. Ann N Y Acad Sci 1178:186–193
Brosius J, Tiedge H (1995) Reverse transcriptase: mediator of genomic plasticity. Virus Genes 11:163–179
Brosius J, Tiedge H (2004) RNomenclature. RNA Biol 1:81–83
Brunkow ME, Tilghman SM (1991) Ectopic expression of the H19 gene in mice causes prenatal lethality. Genes Dev 5:1092–1101
Cabili MN, Trapnell C, Goff L et al (2011) Integrative annotation of human large intergenic noncoding RNAs reveals global properties and specific subclasses. Genes Dev 25:1915–1927
Cai X, Cullen BR (2007) The imprinted H19 noncoding RNA is a primary microRNA precursor. RNA 13:313–316
Cairns-Smith AG (1982) Genetic takeover and the mineral origins of life. Cambridge University Press, Cambridge
Carninci P, Kasukawa T, Katayama S et al (2005) The transcriptional landscape of the mammalian genome. Science 309:1559–1563
Carrington JC, Ambros V (2003) Role of microRNAs in plant and animal development. Science 301:336–338
Chen W, Bocker W, Brosius J et al (1997) Expression of neural BC200 RNA in human tumours. J Pathol 183:345–351
Clark MB, Amaral PP, Schlesinger FJ et al (2011) The reality of pervasive transcription. PLoS Biol 9:e1000625, discussion e1001102
Couzin J (2002) Breakthrough of the year. Small RNAs make big splash. Science 298:2296–2297
Crick FH (1968) The origin of the genetic code. J Mol Biol 38:367–379

Cullen BR (2004) Transcription and processing of human microRNA precursors. Mol Cell 16:861–865
Dawkins R (1976) The selfish gene. Oxford University Press, Oxford
DeChiara TM, Brosius J (1987) Neural BC1 RNA: cDNA clones reveal nonrepetitive sequence content. Proc Natl Acad Sci USA 84:2624–2628
Ding F, Li HH, Zhang S et al (2008) SnoRNA Snord116 (Pwcr1/MBII-85) deletion causes growth deficiency and hyperphagia in mice. PLoS One 3:e1709
Dinger ME, Amaral PP, Mercer TR et al (2008a) Long noncoding RNAs in mouse embryonic stem cell pluripotency and differentiation. Genome Res 18:1433–1445
Dinger ME, Pang KC, Mercer TR et al (2008b) Differentiating protein-coding and noncoding RNA: challenges and ambiguities. PLoS Comput Biol 4:e1000176
Duker AL, Ballif BC, Bawle EV et al (2010) Paternally inherited microdeletion at 15q11.2 confirms a significant role for the SNORD116 C/D box snoRNA cluster in Prader–Willi syndrome. Eur J Hum Genet 18:1196–1201
Elton TS, Khan M, Terentyev D (2011) MicroRNAs in cardiovascular disease. F1000 Med Rep 3:10
Endo T, Maekawa F, Voikar V et al (2011) Automated test of behavioral flexibility in mice using a behavioral sequencing task in IntelliCage. Behav Brain Res 221:172–181
Erdmann VA, Wolters J (1987) The Berlin RNA databank. Protein Seq Data Anal 1:127
Esguerra J, Warringer J, Blomberg A (2008) Functional importance of individual rRNA 2′-O-ribose methylations revealed by high-resolution phenotyping. RNA 14:649–656
Faghihi MA, Wahlestedt C (2009) Regulatory roles of natural antisense transcripts. Nat Rev Mol Cell Biol 10:637–643
Farazi TA, Spitzer JI, Morozov P et al (2011) miRNAs in human cancer. J Pathol 223:102–115
Filipowicz W (2000) Imprinted expression of small nucleolar RNAs in brain: time for RNomics. Proc Natl Acad Sci USA 97:14035–14037
Galli G, Hofstetter H, Stunnenberg HG et al (1983) Biochemical complementation with RNA in the Xenopus oocyte: a small RNA is required for the generation of 3′ histone mRNA termini. Cell 34:823–828
Garzon R, Marcucci G, Croce CM (2010) Targeting microRNAs in cancer: rationale, strategies and challenges. Nat Rev Drug Discov 9:775–789
Gilbert W (1986) Origin of life—the RNA world. Nature 319:618
Gould SJ (2002) The structure of evolutionary theory. Belknap, Harvard University Press, Cambridge, MA
Greider CW, Blackburn EH (1989) A telomeric sequence in the RNA of Tetrahymena telomerase required for telomere repeat synthesis. Nature 337:331–337
Guerrier-Takada C, Gardiner K, Marsh T et al (1983) The RNA moiety of ribonuclease P is the catalytic subunit of the enzyme. Cell 35:849–857
Guerzoni D, McLysaght A (2011) De novo origins of human genes. PLoS Genet 7:e1002381
Guttman M, Amit I, Garber M et al (2009) Chromatin signature reveals over a thousand highly conserved large non-coding RNAs in mammals. Nature 458:223–227
Hartman H (1975) Speculations on the evolution of the genetic code. Orig Life 6:423–427
Herbert A (2004) The four Rs of RNA-directed evolution. Nat Genet 36:19–25
Hüttenhofer A, Schattner P, Polacek N (2005) Non-coding RNAs: hope or hype? Trends Genet 21 (5):287–288
Huttenhofer A (2006) RNomics: identification and function of small non-protein-coding RNAs in model organisms. Cold Spring Harb Symp Quant Biol 71:135–140
Huttenhofer A, Kiefmann M, Meier-Ewert S et al (2001) RNomics: an experimental approach that identifies 201 candidates for novel, small, non-messenger RNAs in mouse. EMBO J 20:2943–2953
Huttenhofer A, Brosius J, Bachellerie JP (2002) RNomics: identification and function of small, non-messenger RNAs. Curr Opin Chem Biol 6:835–843

Ingolia NT, Lareau LF, Weissman JS (2011) Ribosome profiling of mouse embryonic stem cells reveals the complexity and dynamics of mammalian proteomes. Cell 147:789–802

Kampa D, Cheng J, Kapranov P et al (2004) Novel RNAs identified from an in-depth analysis of the transcriptome of human chromosomes 21 and 22. Genome Res 14:331–342

Kapranov P, Cawley SE, Drenkow J et al (2002) Large-scale transcriptional activity in chromosomes 21 and 22. Science 296:916–919

Kapranov P, Cheng J, Dike S et al (2007a) RNA maps reveal new RNA classes and a possible function for pervasive transcription. Science 316:1484–1488

Kapranov P, Willingham AT, Gingeras TR (2007b) Genome-wide transcription and the implications for genomic organization. Nat Rev Genet 8:413–423

Kapranov P, Ozsolak F, Kim SW et al (2010a) New class of gene-termini-associated human RNAs suggests a novel RNA copying mechanism. Nature 466:642–646

Kapranov P, St Laurent G, Raz T et al (2010b) The majority of total nuclear-encoded non-ribosomal RNA in a human cell is 'dark matter' un-annotated RNA. BMC Biol 8:149

Kasinski AL, Slack FJ (2011) MicroRNAs en route to the clinic: progress in validating and targeting microRNAs for cancer therapy. Nat Rev Cancer 11:849–864

Khalil AM, Guttman M, Huarte M et al (2009) Many human large intergenic noncoding RNAs associate with chromatin-modifying complexes and affect gene expression. Proc Natl Acad Sci USA 106:11667–11672

Kim J, Krichevsky A, Grad Y et al (2004) Identification of many microRNAs that copurify with polyribosomes in mammalian neurons. Proc Natl Acad Sci USA 101:360–365

Kim M, Kasinski AL, Slack FJ (2011) MicroRNA therapeutics in preclinical cancer models. Lancet Oncol 12:319–321

Kindler S, Wang H, Richter D et al (2005) RNA transport and local control of translation. Annu Rev Cell Dev Biol 21:223–245

King TH, Liu B, McCully RR et al (2003) Ribosome structure and activity are altered in cells lacking snoRNPs that form pseudouridines in the peptidyl transferase center. Mol Cell 11:425–435

Kiyosawa H, Yamanaka I, Osato N et al (2003) Antisense transcripts with FANTOM2 clone set and their implications for gene regulation. Genome Res 13:1324–1334

Kondrashov AV, Kiefmann M, Ebnet K et al (2005) Inhibitory effect of naked neural BC1 RNA or BC200 RNA on eukaryotic in vitro translation systems is reversed by poly(A)-binding protein (PABP). J Mol Biol 353:88–103

Kruger K, Grabowski PJ, Zaug AJ et al (1982) Self-splicing RNA: autoexcision and autocyclization of the ribosomal RNA intervening sequence of Tetrahymena. Cell 31:147–157

Lagos-Quintana M, Rauhut R, Lendeckel W et al (2001) Identification of novel genes coding for small expressed RNAs. Science 294:853–858

Lagos-Quintana M, Rauhut R, Yalcin A et al (2002) Identification of tissue-specific microRNAs from mouse. Curr Biol 12:735–739

Lapidot M, Pilpel Y (2006) Genome-wide natural antisense transcription: coupling its regulation to its different regulatory mechanisms. EMBO Rep 7:1216–1222

Lares MR, Rossi JJ, Ouellet DL (2010) RNAi and small interfering RNAs in human disease therapeutic applications. Trends Biotechnol 28:570–579

Lau NC, Lim LP, Weinstein EG et al (2001) An abundant class of tiny RNAs with probable regulatory roles in *Caenorhabditis elegans*. Science 294:858–862

Lau NC, Seto AG, Kim J et al (2006) Characterization of the piRNA complex from rat testes. Science 313:363–367

Le Quesne J, Caldas C (2010) Micro-RNAs and breast cancer. Mol Oncol 4:230–241

Lee RC, Ambros V (2001) An extensive class of small RNAs in *Caenorhabditis elegans*. Science 294:862–864

Lee HW, Blasco MA, Gottlieb GJ et al (1998) Essential role of mouse telomerase in highly proliferative organs. Nature 392:569–574

Leighton PA, Ingram RS, Eggenschwiler J et al (1995a) Disruption of imprinting caused by deletion of the H19 gene region in mice. Nature 375:34–39
Leighton PA, Saam JR, Ingram RS et al (1995b) An enhancer deletion affects both H19 and Igf2 expression. Genes Dev 9:2079–2089
Lerner MR, Boyle JA, Mount SM et al (1980) Are snRNPs involved in splicing? Nature 283:220–224
Lewejohann L, Skryabin BV, Sachser N et al (2004) Role of a neuronal small non-messenger RNA: behavioural alterations in BC1 RNA-deleted mice. Behav Brain Res 154:273–289
Lewis MA, Quint E, Glazier AM et al (2009) An ENU-induced mutation of miR-96 associated with progressive hearing loss in mice. Nat Genet 41:614–618
Lichter P (2010) All you need is a Mir-acle: the role of nontranslated RNAs in the suppression of B cell chronic lymphocytic leukemia. Cancer Cell 17:3–4
Lipovich L, Johnson R, Lin CY (2010) MacroRNA underdogs in a microRNA world: evolutionary, regulatory, and biomedical significance of mammalian long non-protein-coding RNA. Biochim Biophys Acta 1799:597–615
Loewer S, Cabili MN, Guttman M et al (2010) Large intergenic non-coding RNA-RoR modulates reprogramming of human induced pluripotent stem cells. Nat Genet 42:1113–1117
Lorberboum H, Digweed M, Erdmann VA et al (1986) Small cytoplasmic RNAs from human placental free mRNPs. Structure and their effect on in vitro protein synthesis. Eur J Biochem 155:279–287
Lu M, Zhang Q, Deng M et al (2008) An analysis of human microRNA and disease associations. PLoS One 3:e3420
Ma W (2010) The scenario on the origin of translation in the RNA world: in principle of replication parsimony. Biol Direct 5:65
Marahrens Y, Panning B, Dausman J et al (1997) Xist-deficient mice are defective in dosage compensation but not spermatogenesis. Genes Dev 11:156–166
Marques AC, Ponting CP (2009) Catalogues of mammalian long noncoding RNAs: modest conservation and incompleteness. Genome Biol 10:R124
Martignetti JA, Brosius J (1993a) BC200 RNA: a neural RNA polymerase III product encoded by a monomeric Alu element. Proc Natl Acad Sci USA 90:11563–11567
Martignetti JA, Brosius J (1993b) Neural BC1 RNA as an evolutionary marker: guinea pig remains a rodent. Proc Natl Acad Sci USA 90:9698–9702
Martin AN, Li Y (2007) RNase MRP RNA and human genetic diseases. Cell Res 17:219–226
Mattick JS (1994) Introns: evolution and function. Curr Opin Genet Dev 4:823–831
Mattick JS (2001) Non-coding RNAs: the architects of eukaryotic complexity. EMBO Rep 2:986–991
Mattick JS (2003) Challenging the dogma: the hidden layer of non-protein-coding RNAs in complex organisms. Bioessays 25:930–939
Mattick JS (2004) RNA regulation: a new genetics? Nat Rev Genet 5:316–323
Mattick JS (2010) Linc-ing long noncoding RNAs and enhancer function. Dev Cell 19:485–486
McBride JL, Pitzer MR, Boudreau RL et al (2011) Preclinical safety of RNAi-mediated HTT suppression in the rhesus macaque as a potential therapy for Huntington's disease. Mol Ther 19:2152–2162
McManus DD, Ambros V (2011) Circulating MicroRNAs in cardiovascular disease. Circulation 124:1908–1910
Medina PP, Nolde M, Slack FJ (2010) OncomiR addiction in an in vivo model of microRNA-21-induced pre-B-cell lymphoma. Nature 467:86–90
Mencia A, Modamio-Hoybjor S, Redshaw N et al (2009) Mutations in the seed region of human miR-96 are responsible for nonsyndromic progressive hearing loss. Nat Genet 41:609–613
Mercer TR, Dinger ME, Bracken CP et al (2010a) Regulated post-transcriptional RNA cleavage diversifies the eukaryotic transcriptome. Genome Res 20:1639–1650
Mercer TR, Qureshi IA, Gokhan S et al (2010b) Long noncoding RNAs in neuronal-glial fate specification and oligodendrocyte lineage maturation. BMC Neurosci 11:14

Mercer TR, Wilhelm D, Dinger ME et al (2011) Expression of distinct RNAs from 3′ untranslated regions. Nucleic Acids Res 39:2393–2403
Miller BH, Wahlestedt C (2010) MicroRNA dysregulation in psychiatric disease. Brain Res 1338:89–99
Milner RJ, Bloom FE, Lai C et al (1984) Brain-specific genes have identifier sequences in their introns. Proc Natl Acad Sci USA 81:713–717
Munoz-Robles J, Allende LM, Clemente J et al (2006) A novel RMRP mutation in a Spanish patient with cartilage-hair hypoplasia. Immunobiology 211:753–757
Muntoni F, Wood MJ (2011) Targeting RNA to treat neuromuscular disease. Nat Rev Drug Discov 10:621–637
Mus E, Hof PR, Tiedge H (2007) Dendritic BC200 RNA in aging and in Alzheimer's disease. Proc Natl Acad Sci USA 104:10679–10684
Nakagawa S, Naganuma T, Shioi G, Hirose T (2011) Paraspeckles are subpopulation-specific nuclear bodies that are not essential in mice. J Cell Biol 193:31–39
Noller HF (1991) Ribosomes. Drugs and the RNA world. Nature 353:302–303
Noller HF, Chaires JB (1972) Functional modification of 16S ribosomal RNA by kethoxal. Proc Natl Acad Sci USA 69:3115–3118
Noller HF, Hoffarth V, Zimniak L (1992) Unusual resistance of peptidyl transferase to protein extraction procedures. Science 256:1416–1419
Numata K, Kanai A, Saito R et al (2003) Identification of putative noncoding RNAs among the RIKEN mouse full-length cDNA collection. Genome Res 13:1301–1306
Ohno S (1970) Evolution by gene duplication. Springer, Berlin
Okazaki Y, Furuno M, Kasukawa T et al (2002) Analysis of the mouse transcriptome based on functional annotation of 60,770 full-length cDNAs. Nature 420:563–573
Orgel LE (1968) Evolution of the genetic apparatus. J Mol Biol 38:381–393
Orgel LE (1989) The origin of polynucleotide-directed protein synthesis. J Mol Evol 29:465–474
Ozsolak F, Kapranov P, Foissac S et al (2010) Comprehensive polyadenylation site maps in yeast and human reveal pervasive alternative polyadenylation. Cell 143:1018–1029
Pang KC, Dinger ME, Mercer TR et al (2009) Genome-wide identification of long noncoding RNAs in CD8+ T cells. J Immunol 182:7738–7748
Park CY, Choi YS, McManus MT (2010) Analysis of microRNA knockouts in mice. Hum Mol Genet 19:R169–R175
Parker BJ, Moltke I, Roth A et al (2011) New families of human regulatory RNA structures identified by comparative analysis of vertebrate genomes. Genome Res 21:1929–1943
Pasmant E, Sabbagh A, Masliah-Planchon J et al (2011a) Role of noncoding RNA ANRIL in genesis of plexiform neurofibromas in neurofibromatosis type 1. J Natl Cancer Inst 103:1713–1722
Pasmant E, Sabbagh A, Vidaud M et al (2011b) ANRIL, a long, noncoding RNA, is an unexpected major hotspot in GWAS. FASEB J 25:444–448
Ploner A, Ploner C, Lukasser M et al (2009) Methodological obstacles in knocking down small noncoding RNAs. RNA 15:1797–1804
Prestayko AW, Busch H (1968) Low molecular weight RNA of the chromatin fraction from Novikoff hepatoma and rat liver nuclei. Biochim Biophys Acta 169:327–337
Project AET (2009) Post-transcriptional processing generates a diversity of 5'-modified long and short RNAs. Nature 457:1028–1032, http://www.ncbi.nlm.nih.gov/pubmed?term=processing%20generates%20a%20diversity%20of%205%27-modified%20long%20AND%202009
Qureshi IA, Mattick JS, Mehler MF (2010) Long non-coding RNAs in nervous system function and disease. Brain Res 1338:20–35
Raabe CA, Sanchez CP, Randau G et al (2010) A global view of the nonprotein-coding transcriptome in *Plasmodium falciparum*. Nucleic Acids Res 38:608–617
Ravasi T, Suzuki H, Pang KC et al (2006) Experimental validation of the regulated expression of large numbers of non-coding RNAs from the mouse genome. Genome Res 16:11–19

Rederstorff M, Huttenhofer A (2010) Small non-coding RNAs in disease development and host–pathogen interactions. Curr Opin Mol Ther 12:684–694
Rederstorff M, Bernhart SH, Tanzer A et al (2010) RNPomics: defining the ncRNA transcriptome by cDNA library generation from ribonucleo-protein particles. Nucleic Acids Res 38:e113
Rich A (1962) On the problems of evolution and biochemical information transfer. In: Pullman B, Kasha M (eds) Horizons in biochemistry. Academic, New York, pp 103–126
Ridanpaa M, van Eenennaam H, Pelin K et al (2001) Mutations in the RNA component of RNase MRP cause a pleiotropic human disease, cartilage-hair hypoplasia. Cell 104:195–203
Rinn JL, Euskirchen G, Bertone P et al (2003) The transcriptional activity of human chromosome 22. Genes Dev 17:529–540
Robertson MP, Joyce GF (2011) The origins of the RNA world. In: Atkins JF, Gesteland RF, Cech TR (eds) RNA worlds. Cold Spring Harbor Laboratory, Cold Spring Harbor, NY, pp 21–42
Rodin AS, Szathmary E, Rodin SN (2011) On origin of genetic code and tRNA before translation. Biol Direct 6:14
Rottiers and Näär (2012) MicroRNAs in metabolism and metabolic disorders. Nat Rev Mol Cell Biol 13:239–250
Runte M, Varon R, Horn D et al (2005) Exclusion of the C/D box snoRNA gene cluster HBII-52 from a major role in Prader–Willi syndrome. Hum Genet 116:228–230
Ruvkun G (2001) Molecular biology. Glimpses of a tiny RNA world. Science 294:797–799
Ruvkun G, Wightman B, Ha I (2004) The 20 years it took to recognize the importance of tiny RNAs. Cell 116:S93–S96, 2 p following S96
Saxe JP, Lin H (2011) Small noncoding RNAs in the germline. Cold Spring Harb Perspect Biol 3: a002717
Sayed D, Abdellatif M (2011) MicroRNAs in development and disease. Physiol Rev 91:827–887
Schmitz J, Zemann A, Churakov G et al (2008) Retroposed SNOfall—a mammalian-wide comparison of platypus snoRNAs. Genome Res 18:1005–1010
Seto AG (2010) The road toward microRNA therapeutics. Int J Biochem Cell Biol 42:1298–1305
Shantikumar S, Caporali A, Emanueli C (2011) Role of microRNAs in diabetes and its cardiovascular complications. Cardiovasc Res 93:583–593
Sharp PA (2009) The centrality of RNA. Cell 136:577–580
Sibilia M, Wagner EF (1995) Strain-dependent epithelial defects in mice lacking the EGF receptor. Science 269:234–238
Skryabin BV, Sukonina V, Jordan U et al (2003) Neuronal untranslated BC1 RNA: targeted gene elimination in mice. Mol Cell Biol 23:6435–6441
Skryabin BV, Gubar LV, Seeger B et al (2007) Deletion of the MBII-85 snoRNA gene cluster in mice results in postnatal growth retardation. PLoS Genet 3:e235
Small EM, Olson EN (2011) Pervasive roles of microRNAs in cardiovascular biology. Nature 469:336–342
Soukup GA (2009) Little but loud: small RNAs have a resounding affect on ear development. Brain Res 1277:104–114
Sousa C, Johansson C, Charon C et al (2001) Translational and structural requirements of the early nodulin gene enod40, a short-open reading frame-containing RNA, for elicitation of a cell-specific growth response in the alfalfa root cortex. Mol Cell Biol 21:354–366
Struck JC, Vogel DW, Hartmann RK et al (1986) Isolation and characterization of a novel RNA from *Bacillus stearothermophilus*. Biochem Intl 12:603–611
Su WY, Xiong H, Fang JY (2010) Natural antisense transcripts regulate gene expression in an epigenetic manner. Biochem Biophys Res Commun 396:177–181
Sutcliffe JG, Milner RJ, Gottesfeld JM et al (1984a) Identifier sequences are transcribed specifically in brain. Nature 308:237–241
Sutcliffe JG, Milner RJ, Gottesfeld JM, Reynolds W (1984b) Control of neuronal gene expression. Science 225:1308–1315
Szathmary E, Smith JM (1995) The major evolutionary transitions. Nature 374:227–232

Taft RJ, Pang KC, Mercer TR et al (2010a) Non-coding RNAs: regulators of disease. J Pathol 220:126–139
Taft RJ, Simons C, Nahkuri S et al (2010b) Nuclear-localized tiny RNAs are associated with transcription initiation and splice sites in metazoans. Nat Struct Mol Biol 17:1030–1034
Tang TH, Bachellerie JP, Rozhdestvensky T et al (2002) Identification of 86 candidates for small non-messenger RNAs from the archaeon *Archaeoglobus fulgidus*. Proc Natl Acad Sci USA 99:7536–7541
Tang TH, Polacek N, Zywicki M et al (2005) Identification of novel non-coding RNAs as potential antisense regulators in the archaeon *Sulfolobus solfataricus*. Mol Microbiol 55:469–481
Tautz D, Domazet-Loso T (2011) The evolutionary origin of orphan genes. Nat Rev Genet 12:692–702
Thiel CT, Rauch A (2011) The molecular basis of the cartilage-hair hypoplasia-anauxetic dysplasia spectrum. Best Pract Res Clin Endocrinol Metab 25:131–142
Thomason MK, Storz G (2010) Bacterial antisense RNAs: how many are there, and what are they doing? Annu Rev Genet 44:167–188
Tiedge H, Brosius J (1996) Translational machinery in dendrites of hippocampal neurons in culture. J Neurosci 16:7171–7181
Tiedge H, Fremeau RT, Weinstock PH et al (1991) Dendritic location of neural BC1 RNA. Proc Natl Acad Sci USA 88:2093–2097
Tiedge H, Chen W, Brosius J (1993) Primary structure, neural-specific expression, and dendritic location of human BC200 RNA. J Neurosci 13:2382–2390
Tijsterman M, Ketting RF, Plasterk RH (2002) The genetics of RNA silencing. Annu Rev Genet 36:489–519
Tsai MC, Manor O, Wan Y et al (2010) Long noncoding RNA as modular scaffold of histone modification complexes. Science 329:689–693
Tsai MC, Spitale RC, Chang HY (2011) Long intergenic noncoding RNAs: new links in cancer progression. Cancer Res 71:3–7
Tyagi S (1981) Origins of translation: the hypothesis of permanently attached adaptors. Orig Life 11:343–351
van Bakel H, Nislow C, Blencowe BJ et al (2010) Most "dark matter" transcripts are associated with known genes. PLoS Biol 8:e1000371
van Kouwenhove M, Kedde M, Agami R (2011) MicroRNA regulation by RNA-binding proteins and its implications for cancer. Nat Rev Cancer 11:644–656
Voinnet O (2002) RNA silencing: small RNAs as ubiquitous regulators of gene expression. Curr Opin Plant Biol 5:444–451
Volff JN, Brosius J (2007) Modern genomes with retro-look: retrotransposed elements, retroposition and the origin of new genes. Genome Dyn 3:175–190
Walder JA, Walder RY, Heller MJ et al (1979) Complementary carrier peptide synthesis: general strategy and implications for prebiotic origin of peptide synthesis. Proc Natl Acad Sci USA 76:51–55
Wang H, Iacoangeli A, Popp S et al (2002) Dendritic BC1 RNA: functional role in regulation of translation initiation. J Neurosci 22:10232–10241
Wang H, Iacoangeli A, Lin D et al (2005) Dendritic BC1 RNA in translational control mechanisms. J Cell Biol 171:811–821
Wang KC, Yang YW, Liu B et al (2011) A long noncoding RNA maintains active chromatin to coordinate homeotic gene expression. Nature 472:120–124
Watanabe Y, Yamamoto M (1994) S. pombe mei2+ encodes an RNA-binding protein essential for premeiotic DNA synthesis and meiosis I, which cooperates with a novel RNA species meiRNA. Cell 78:487–498
Weiner AM, Maizels N (1987) tRNA-like structures tag the 3′ ends of genomic RNA molecules for replication: implications for the origin of protein synthesis. Proc Natl Acad Sci USA 84:7383–7387

Werner A, Sayer JA (2009) Naturally occurring antisense RNA: function and mechanisms of action. Curr Opin Nephrol Hypertens 18:343–349

Werner A, Swan D (2010) What are natural antisense transcripts good for? Biochem Soc Trans 38:1144–1149

Werner A, Carlile M, Swan D (2009) What do natural antisense transcripts regulate? RNA Biol 6:43–48

Westhof E, Filipowicz W (2005) From RNAi to epigenomes: how RNA rules the world. Chembiochem 6:441–443

Woese CR (1967) The genetic code: the molecular basis for genetic expression. Harper and Row, New York

Woese CR (1980) Just So Stories and Rube Goldberg machines: speculations on the origin of the protein synthetic machinery. In: Chambliss G, Craven GR, Davies J, Davis K, Kahan L, Nomura M (eds) Ribosomes: structure, function and genetics. University Park Press, Baltimore, pp 357–373

Woese CR (2001) Translation: in retrospect and prospect. RNA 7:1055–1067

Wong JW (2010) MicroRNA-induced silencing of glioma progression. J Neurosci 30:3868–3869

Yang T, Adamson TE, Resnick JL et al (1998) A mouse model for Prader–Willi syndrome imprinting-centre mutations. Nat Genet 19:25–31

Yasuda J, Hayashizaki Y (2008) The RNA continent. Adv Cancer Res 99:77–112

Zemann A, op de Bekke A, Kiefmann M et al (2006) Evolution of small nucleolar RNAs in nematodes. Nucleic Acids Res 34:2676–2685

Zhong J, Chuang SC, Bianchi R et al (2009) BC1 regulation of metabotropic glutamate receptor-mediated neuronal excitability. J Neurosci 29:9977–9986

Zieve G, Penman S (1976) Small RNA species of the HeLa cell: metabolism and subcellular localization. Cell 8:19–31

Nucleic Acids as Therapeutics

Saraswat Pushpendra, Pareek Arvind, and Bhandari Anil

Contents

S. Pushpendra (✉)
Mahatma Gandhi University of Medical Sciences & Technology, RIICO Institutional Area, Sitapura, Jaipur 302 022, India
e-mail: saraswatmgmch@rediffmail.com; saraswatmgumst@gmail.com

P. Arvind
Mahatma Gandhi Institute of Applied Sciences, Tonk Road, Sitapura, Jaipur 302 022, India

B. Anil
Jodhpur National University, Narnadi, Jhanwar Road, Jodhpur 342 001, India

V.A. Erdmann and J. Barciszewski (eds.), *From Nucleic Acids Sequences to Molecular Medicine*, RNA Technologies, DOI 10.1007/978-3-642-27426-8_2,

Abstract Gene therapy is a technique for correcting defective genes responsible for disease development. Nucleic acid-based molecules (deoxyribonucleic acid, complementary deoxyribonucleic acid, complete genes, ribonucleic acid, and oligonucleotides) are utilized as research tools within the broad borders of gene therapy and the emerging field of molecular medicine. Although most of the nucleic acid-based drugs are in early stages of clinical trials, these classes of compounds have emerged in recent years to yield extremely promising candidates for drug therapy to a wide range of diseases, including cancer, infectious diseases, diabetes, cardiovascular, inflammatory, and neurodegenerative diseases, cystic fibrosis, hemophilia, and other genetic disorders. Gene therapy may be classified into two types: somatic and germ line gene therapy. There are many ethical, social, and commercial issues raised by the prospects of treating patients using gene therapy. This chapter summarizes deoxyribonucleic acid-based therapeutics, ribonucleic acid-based therapeutics, and gene transfer technologies. Deoxyribonucleic acid-based therapeutics includes plasmids, oligonucleotides for antisense and antigene applications, deoxyribonucleic acid aptamers, and deoxyribonucleic acidzymes, while ribonucleic acid-based therapeutics includes ribonucleic acid aptamers, ribonucleic acid decoys, antisense ribonucleic acid, ribozymes, small interfering ribonucleic acid, and micro ribonucleic acid. This chapter also includes current status of gene therapy and recent developments in gene therapy research.

Keywords Gene therapy • Nucleic acid therapeutics • DNA-based therapeutics • RNA-based therapeutics • Gene transfer technology • Viral vectors • Nonviral vectors • Liposomes

1 Introduction

Modern drug research aims to discover biologically active molecule(s) that are absolutely specific to the molecular targets responsible for the disease progression. Moreover, there is strong belief that medicine will soon benefit from the development of new therapeutic technologies to directly target human genes. Insertion of new genetic material into the cells of an individual with the intention of producing a therapeutic benefit for the patient is human gene therapy (Anderson 1992; Baltimore 1988; Mizutani et al. 1995), while gene therapy is a technique for correcting defective genes responsible for disease development. Numerous gene therapy strategies are under development, some of which use nucleic acid-based molecules to inhibit gene expression at either the transcriptional or posttranscriptional level (Gewirtz et al. 1998), and this strategy has potential applications, such as in cardiovascular (Mann et al. 1999a, b; Ehsan et al. 2001; Stull and Szoka 1995; Patil and Burgess 2003) and inflammatory disorders (Dean et al. 1994; Yacyshyn et al. 1998), cancer (Mardan et al. 2002), neurological disorders (Shi et al. 2001), and infectious diseases (Macpherson et al. 1999; Welch et al. 1998; Zu Putlitz et al. 1999; Campagno et al. 1999), as well as in organ transplantation (Katz et al. 1997).

A number of human diseases are known to be genetic in origin (e.g., Huntington's chorea and cystic fibrosis), and virtually all diseases, except for some trauma, have a hereditary component (SoRelle 2000). Thus, gene therapy represents an opportunity for the treatment of genetic disorders in humans by modifying their cells genetically (Report 1995).

Despite the obvious advantages that might be gained from human gene therapy (i.e., replacing a defective gene with a normal one), there are many ethical, social, and commercial issues surrounding the technology. The outcome of an error in technology might not be observed for many years. Moreover, it is feared that unpredictable and perhaps irreversible side effects occur in treated individuals. The social implications of such technology include the possibility that patients might suffer from depression as a result of being "genetically altered" or might not be accepted by society in the way that they were before treatment. The commercial implications of such technology are that the insurance companies and other such institutions also would want to access the available information prior to them granting life insurance policies, etc. Hence, it is obvious that a person shown to have a predisposition to a genetic disease could be severely penalized because of a mutation in their DNA.

The possibility of using nucleic acids as drugs for the treatment of genetic diseases is still very much in its infancy. One reason for this is that, unlike monogenetic disorders such as severe combined immune deficiency (SCID), which is caused by a mutation in the adenosine deaminase (ADA) gene, very few diseases are caused by a single gene mutation; most are caused by the mutation of multiple genetic components. For example, cancer usually involves multiple genetic lesions within the same cell and it is unlikely that the nature of every one of these oncogenic mutations is yet known.

Elucidation of the human genome has also provided a major impetus in identifying human genes implicated in diseases, which may eventually lead to the development of nucleic acid based drugs for gene replacement or potential targets for gene ablation (Baker 2001). Moreover, the Human Genome project will help determine genetic markers responsible for patient response to drug therapy, drug interactions, and potential side effects (Van Ommen et al. 1999). Currently, all gene therapy trials approved for human use target somatic cells that will live only as long as the patient, and this is known as the somatic gene therapy. Its purpose is to alleviate disease in the treated individual alone. In contrast, it is also possible to target directly the gametes (sperm and ova) to modify the genetic profile of the subsequent generation of unborn "patients." This gene transfer at an early stage of embryonic development is known as the germ line gene therapy. More than 300 clinical trials involving gene transfer in patients have been approved, and the first nucleic acid drug, an antisense oligonucleotide, fomivirsen (marketed as Vitravene), has been approved by the US Food and Drug Administration (FDA) for the treatment of cytomegalovirus retinitis in immunocompromised patients (Rosenberg et al. 2000).

In 1999, the biggest setback for gene therapy occurred when Jesse Gelsinger, an 18-year-old high-school graduate from Arizona, died as a result of a gene therapy experiment. Gelsinger developed fever and blood clots throughout his body within hours of treatment to correct partial ornithine transcarbamylase (OTC) deficiency, a rare metabolic disease that can cause a dangerous buildup of ammonia in the body, and died 4 days later (Lehrman 1999). The FDA has not yet approved any human gene therapy product for sale.

This chapter summarizes DNA-based therapeutics, RNA-based therapeutics, and gene transfer technologies for the diseases that are known to be genetic in origin. DNA-based therapeutics includes plasmids, oligonucleotides for antisense and antigene applications, DNA aptamers, and DNAzymes, while RNA-based therapeutics includes antisense RNA, ribozymes, RNA decoys, RNA aptamers, small interfering RNA, and microRNA. This chapter also includes the current status of gene therapy and recent developments in gene therapy research.

2 DNA-Based Therapeutics

2.1 Plasmids

Plasmids are high molecular weight, double-stranded DNA constructs containing transgenes, which encode specific proteins. On the molecular level, plasmid DNA molecules can be considered prodrugs that upon cellular internalization employ the DNA transcription and translation apparatus in the cell to biosynthesize the therapeutic entity, the protein (Uherek and Wels 2000). The mechanism of action of plasmid DNA requires that the plasmid molecules gain access into the nucleus after entering the cytoplasm. Nuclear access or lack thereof eventually controls the efficiency of gene expression. In addition to disease treatment, plasmids can be used as DNA vaccines for genetic immunization (Johnston et al. 2002). In the early stages of development, plasmid-based gene therapy was attempted to correct inheritable disorders resulting from a single gene defect. The first federally approved human gene therapy protocol was initiated in 1990 for the treatment of adenosine deaminase deficiency (Anderson 1998). Since then, more than 500 gene therapy protocols have been approved or implemented (Vorburger and Hunt 2002). In 2002, the successful gene-therapy-based cure for SCID was reported (Otsu and Candotti 2002). In 2003, the Chinese drug regulatory agency approved the first gene therapy product for head and neck squamous carcinoma under the trade name Gendicine (Zhaohui 2003). Currently, diseases with complex etiologies such as cancer (Galanis and Russell 2001; Mulherkar 2001) and neurodegenerative disorders such as Alzheimer's disease and Parkinson's disease (Baekelandt et al. 2000) are being targeted. In addition, DNA vaccines for malaria, AIDS, and many other diseases are in development (Bunnel and Morgan 1996). DNA vaccines have also been used to prevent allergic response (Horner et al. 2001).

2.2 Oligonucleotides for Antisense and Antigene Applications

Oligonucleotides are short single-stranded segments of DNA that upon cellular internalization can selectively inhibit the expression of a single protein. For antisense applications, oligonucleotides interact and form a duplex with the mRNA or the pre-mRNA and inhibit its translation or processing, consequently inhibiting protein biosynthesis. For antigene applications, oligonucleotides must enter the cell nucleus, form a triplex with the double-stranded genomic DNA, and inhibit the translation as well as the transcription process of the protein. On the molecular level, numerous mechanisms have been proposed to explain the basis of oligonucleotide action (Crooke 1999a, b; Speedie 2005). For therapeutic purposes, oligonucleotides can be used to selectively block the expression of proteins that are implicated in diseases (Akhtar et al. 2000). With successful antisense inhibition of proteins in animal models, the first antisense drug, fomivirsen sodium (Vitravene, Isis Pharmaceuticals, Carlsbad, CA), was approved for the treatment of cytomegalovirus retinitis in AIDS patients in 1998 (Crooke 1998b). Antisense oligonucleotides such as MG98 and ISIS 5132, designed to inhibit the biosynthesis of DNA methyltransferase and c-raf kinase respectively, are in human clinical trials for cancer (Mardan et al. 2002). Synthetic antisense DNA oligonucleotides and oligonucleotide analogs (Agarawal and Tang 1992), which inhibit the replication of several infectious agents such as hepatitis C virus (Alt et al. 1995), human cytomegalovirus (Azad et al. 1993), human immunodeficiency virus, and papilloma virus (Bordier et al. 1992; Cowsert et al. 1993; Gerviax et al. 1997; Hanecak et al. 1996; Kinchington et al. 1992; Lisziewicz et al. 1992, 1993; Mirabelli et al. 1991; Morvan et al. 1993; Tonkinson and Stein 1993), have also been designed.

2.3 Aptamers

DNA aptamers are double-stranded nucleic acid segments that can directly interact with proteins (Stull and Szoka 1995). Aptamers interfere with the molecular functions of disease-implicated proteins or those that participate in the transcription or translation processes. Aptamers are preferred over antibodies in protein inhibition owing to their specificity, nonimmunogenicity, and stability of pharmaceutical formulation (Jayasena 1999). DNA aptamers have demonstrated promise in intervention of pathogenic protein biosynthesis against HIV-1 integrase enzyme (de Saultrait et al. 2002).

2.4 DNAzymes

DNAzymes are analogs of ribozymes with greater biological stability (Akhtar et al. 2000). The RNA backbone chemistry is replaced by the DNA motifs that confer improved biological stability. DNAzyme directed against vascular endothelial growth factor receptor 2 was confirmed to be capable of tumor suppression by blocking angiogenesis upon intratumoral injections in mice (Zhang et al. 2002).

3 RNA-Based Therapeutics

3.1 RNA Aptamers

RNA aptamers are single-stranded nucleic acid segments that can directly interact with proteins (Stull and Szoka 1995). Aptamers recognize their targets on the basis of shape complementarity (Kaur and Roy 2008). Moreover, their binding specificity and affinity for the target are extremely high and similar to monoclonal antibodies. RNA aptamers have demonstrated promise in intervention of pathogenic protein biosynthesis against HIV-1 transcriptase (Chaloin et al. 2002). Moreover, RNA aptamers that specifically bind and inactivate vascular endothelial growth factor (VEGF) in vitro have been isolated. A clinical study on humans with injection of anti-VEGF aptamers in the eye showed that 80% of the patients retained or improved the eyesight, and they had no side effects (The Eyetech Study Group 2002).

3.2 RNA Decoys

The RNA decoys are designed to provide alternate, competing binding sites for proteins that act as translational activators or mRNA-stabilizing elements (Beelman and Parker 1995; Liebhaber 1997). Decoys can prevent translation or induce instability and, ultimately, destruction of the mRNA. Overexpressed short RNA molecules corresponding to critical *cis*-acting regulatory elements can be used as decoys for *trans*-activating proteins, thus preventing binding of these *trans* activators to their corresponding *cis*-acting elements in the viral genome (Sullenger et al. 1990, 1991). RNA decoys have an advantage over other nucleic acid-based strategies, that is, the decoys are less likely to be affected by variability of the infectious agent because any mutation in the *trans*-activating protein affects not only binding to the decoys but also binding to the endogenous targets. However, there is some question whether RNA decoy strategies will be as benign to the cell physiology as antisense RNAs, since it has been postulated that cellular factors may associate with them (Sullenger et al. 1990). It has previously been demonstrated that a cellular factor termed loop-binding protein is an absolute requirement for Tat-mediated trans-activation in an HIV-infected cell (Marciniak et al. 1990). Finally, RNA decoys do not function by sequestering either the Tat or Rev-protein but by sequestering the cellular factors such as the loop-binding protein.

3.3 Antisense RNA

Antisense drugs are short stretches of deoxyribonucleotide analogs that bind to specific complementary areas of the mRNA by Watson–Crick base pairing to block gene expression in a sequence-specific fashion. Antisense drugs may induce an RNaseH, which cleaves the mRNA at the site of binding, or can physically block translation or other steps in mRNA processing and transport to protein synthesis. The antisense drugs work at an early stage in the production of a disease-causing protein and theoretically can be applied to a number of diseases where the basic pathophysiology involves an overexpression of a given protein molecule (Crook 1998a, Speedie 2005).

Antisense ODNs can base-pair with a gene's transcript and constitute a new technology for the control of gene expression in prokaryotes and eukaryotes, including mammalian cells (Zamecnik and Stephenson 1978). The existence of naturally occurring RNAs and their role in regulating gene expression were shown in the mid-1980s (Simons and Kleckner 1983; Mizuno et al. 1984). Newer antisense oligonucleotides may offer improved pharmacokinetic and safety profiles because of reduced nonspecific interactions (Agarwal and Kandimalla 2000).

A stoichiometric disadvantage of antisense RNA is the high expression required to successfully bind to all target RNA. A major advantage is the lack of immunogenicity of antisense constructs, such that the oligonucleotides and the cells producing them will not be destroyed by the host immune response. The antisense approach to modulating gene expression has been extensively reviewed (Gewirtz et al. 1998; Scanlon et al. 1995; Stein 1998).

3.4 Ribozymes

Antisense RNA alone is not potent enough to produce complete inhibition in vivo. An enzymatic moiety can be included with antisense oligonucleotide, which will cleave the target RNA once the RNA–RNA duplex has formed. These enzymatic RNA strands are called "Ribozymes." They are antisense RNA molecules that are capable of sequence-specific cleaving of RNA molecules (Stull and Szoka 1995). They function by binding to the target moiety through antisense sequence-specific hybridization and inactivating it by cleaving the phosphodiester backbone at a specific site. Thus, they can selectively bind to target mRNAs and form a duplex having highly distorted confirmation that is easily hydrolyzed, and this hydrolysis of mRNA may be used for targeted suppression of specific gene (Mardan et al. 2002). Two types of ribozymes, the hammerhead and hairpin ribozymes (the names are derived from their theoretical secondary structures), have been extensively studied owing to their small size and rapid kinetics (Earnshaw and Gait 1997; Hampel 1998) and for therapeutic applications (Stull and Szoka 1995). Hammerhead ribozymes cleave RNA at the nucleotide sequence U–H (H = A, C, or U) by hydrolysis of a 3′–5′ phosphodiester bond, while hairpin ribozymes utilize the nucleotide sequence C–U–G as their cleavage site (Breaker and Joyce 1994; Burke 1996; Usman et al. 1996). The presence of the RNA backbone in ribozymes makes them easy targets for degradation by RNases, so these molecules are biologically unstable in vivo (Mardan et al. 2002). Ribozymes can be used for knockout gene therapy by targeting overexpressed oncogenes such as the human epidermal growth factor receptor type-2 gene implicated in breast cancer (Aigner et al. 2001) and human papilloma virus infection. The development of ribozymes that colocalize in the same subcellular compartment as their target may further increase their effectiveness (Sullenger and Cech 1993). A significant limitation of the use of ribozymes for gene therapy is that they are susceptible to RNases.

3.5 Small Interfering RNAs

RNA interference (RNAi) is a posttranscriptional mechanism of gene silencing through chromatin remodeling, inhibition of protein translation, or direct mRNA degradation, which is ubiquitous in eukaryotic cells (Caplen 2004; Dorsett and Tuschl 2004; Shankar et al. 2005). Small interfering RNAs (siRNAs) can be used for downregulation of disease-causing genes through RNA interference.

Typically, these are short double-stranded RNA segments with 21–23 nucleotides and are complementary to the mRNA sequence of the protein whose transcription is to be blocked. On administration, siRNA molecules are incorporated into RNA-induced silencing complexes (RISC), which bind to the mRNA of interest and stimulate mRNA degradation mechanisms, such as nuclease activity, that lead to silencing of the particular gene (Bertrand et al. 2002; McMinus and Sharp 2002; Scherr et al. 2003).

Thc structure and functions of RISC have yet to be completely elucidated (Kurreck 2003). Introduction of foreign double-stranded RNAs (dsRNA) can initiate a potent cascade of sequence-specific degradation of endogenous mRNAs that bear homology to the dsRNA trigger (Fire et al. 1998). When dsRNAs are introduced into the cytoplasm, they are processed by the RNase III enzyme called Dicer, which cleaves long dsRNAs into short 21–23 nucleotide duplexes that have symmetric 2–3 nucleotide 3′ overhangs and 5′ phosphate and 3′ hydroxyl groups (Tuschl et al. 1999; Hamilton and Baulcombe 1999). Although it was initially believed that effective RNAi required almost complete sequence homology throughout the length of the mRNA, now it appears that as few as seven contiguous complementary base pairs can direct RNAi-mediated silencing (Jackson and Linsley 2004). The use of siRNA as a therapeutic agent is still in its infancy. siRNAs are being investigated to inhibit HIV (Martinez et al. 2002), hepatitis (Zamore and Aronin 2003), and influenza infection (Ge et al. 2003). Moreover, RNAi technology has been applied to silence the expression of dominant mutant oncogenes, gene amplification, translocations, and viral oncogenes in order to elucidate their function and their interaction with other genes in a number of critical cellular pathways. Since siRNAs do not integrate into the genome and they offer greater safety than plasmid molecules, it is possible to deliver a cocktail of siRNAs targeting multiple disease-causing genes in a single delivery system to control complex diseases such as cancer where several genes are malfunctioning (Wong 2001).

3.6 *MicroRNA*

MicroRNAs (miRNAs) are a class of naturally occurring, small noncoding RNA molecules 21–25 nucleotides in length. These molecules are partially complementary to messenger RNA (mRNA) molecules, and their main function is downregulation of gene expression via translational repression, mRNA cleavage, and deadenylation. MicroRNAs were first described in 1993 (Lee et al. 1993), and the term microRNA was coined in 2001(Ruvkun 2001). Numerous miRNAs have been identified in various organisms hitherto. The central online repository for miRNA nomenclature, sequence data, annotation, and target prediction is miRBase, hosted by the Sanger Institute. MicroRNAs are transcribed by RNA polymerase II called pri-miRNAs that complete with a 5′ cap and poly-A tail (Lee et al. 2004). In the nucleus, pri-miRNAs are processed into pre-miRNAs by the microprocessor complex, which consists of the RNAse III enzyme Drosha (Han et al. 2004) and the double-stranded RNA Pasha/DGCR8 (Denli et al. 2004). These pre-miRNAs are exported by the karyopherin exportin (Exp5) and Ran–GTP complex (Yi et al. 2003). Then Ran GTPase binds with Exp5 to form a nuclear heterotrimer with pre-miRNAs (Yi et al. 2003; Lund et al. 2004). These pre-miRNAs are additionally processed by the RNAse III enzyme Dicer (Bernstein et al. 2001) to generate miRNA. Moreover, Dicer is also responsible for the initiation of the formation of RISC, which is responsible for the gene silencing observed due to miRNA expression and RNA interference (Hammond et al. 2000; Hammond 2005).

4 Gene Transfer Technologies

Gene transfer technologies can be classified into three general types: electrical techniques, mechanical transfection, and vector-assisted delivery systems.

4.1 *Mechanical and Electrical Techniques*

Strategies of introducing naked DNA into cells by mechanical and electrical techniques include microinjection, particle bombardment, the use of pressure, and electroporation. Microinjection is highly efficient since one cell at a time is targeted for DNA transfer, but it is time consuming. Ballistic transfer of gold microparticles may be performed using particle bombardment equipment such as the gene gun.

Electroporation is achieved using high-voltage electrical current to facilitate DNA transfer that results in high cell mortality and is not suitable for clinical use (McAlister et al. 2000; Luo and Saltzman 2000; Regnier et al. 2000; Huang and Viroonchatapan 1999).

4.2 Vector-Assisted Delivery Systems

Vector-assisted DNA/gene delivery systems can be classified into two types based on their origin: biological viral DNA delivery systems and chemical nonviral delivery systems.

4.2.1 Viral Delivery Systems

In Viral Delivery Systems, nonpathogenic attenuated viruses can be used as delivery systems for genes/DNA molecules, especially plasmids (Kamiya et al. 2001; Mah et al. 2002; Lotze and Kost 2002). These viral DNA-delivery vectors include both RNA and DNA viruses. The viruses used as gene therapy vectors can be classified into four types: *Retroviruses* (McTaggart and Al-Rubeai 2002), *Adenoviruses*, *Adeno-associated viruses* (Martin et al. 2002), and *Herpes simplex viruses*. Gene expression using viral vectors has been achieved in tissues such as kidney (Lien and Lie 2002), heart muscle (Chamberlain 2002), eye (Martin et al. 2002), and ovary (Wolf and Jenkins 2002). Moreover, gene therapy using viral systems has made considerable progress for the treatment of a wide range of diseases, such as muscular dystrophy (Chamberlain 2002), AIDS (Lever 1996), and cancer (Zhao et al. 2002). Viruses are used in more than 70% of human clinical gene therapy trials worldwide (Walther and Stein 2000). The only approved gene therapy treatment (Gendicine) delivers the transgene using a recombinant adenoviral vector (Galanis and Russell 2001). DNA delivery using viral vectors has been extensively reviewed (Mah et al. 2002; Lotze and Kost 2002; Hale and Green 2002).

The first-generation retroviral vectors were largely derived from oncoretroviruses, such as the *Moloney Murine Leukemia virus* (MMuLv), and were unable to transfer genes into nondividing cells (Roe et al. 1993; Lewis and Emerman 1994). This limited the potential for their application as a delivery system in gene therapy. The utilization of the lentivirus family of retroviruses has overcome this shortcoming. Lentiviruses, which include Human immunodeficiency virus type 1 (HIV-1), *Bovine immunodeficiency virus* (BIV), *Feline immunodeficiency virus* (FIV), and *Simian immunodeficiency virus* (SIV), are able to transfer genes to nondividing cells (Lewis and Emerman 1994; Naldini et al. 1996). Retroviral vectors used in gene therapy are replication deficient, such that they are unable to replicate in the host cell and can infect only one cell (Mann et al. 1983; Cone and Mulligan 1984). This characteristic, although essential for the safety of viral vectors in gene therapy, imposes restrictions on the amounts of virus that can safely be administered (Kim et al. 1998; Sheridan et al. 2000). Retroviral-mediated delivery of therapeutic DNA has been widely used in clinical gene therapy protocols, including the treatment of cancers, such as melanoma (Fujji et al. 2000) and ovarian cancer (Tait et al. 1999), adenosine deaminase deficiency–severe combined immune deficiency (Bordignon et al. 1995; Onodera et al. 1998), and Gaucher's disease (Dunbar et al. 1998). Retroviral vectors are capable of transfecting high populations (45–95%) of primary human endothelial and smooth muscle cells, a class of cells that are generally extremely difficult to transfer (Garton et al. 2002).

Adenoviruses have been used to deliver therapeutic DNA to patients suffering from metastatic breast, ovarian, and melanoma cancers (Stewart et al. 1999; Alvarez et al. 2000a, b). Indeed, the severe immune response of the host contributes to the limited survival of the adenoviral DNA in the targeted cells and results in a transient expression of the therapeutic gene since the adenoviral DNA is lost over time (Byrnes et al. 1995; Yang et al. 1996a, b; van Ginkel et al. 1997; Michou et al. 1997). First-generation adenoviral vectors were able to accommodate the introduction of therapeutic genes over 7 kb long (but rarely larger) into targeted cells (Bett et al. 1993). However, the generation of gutless adenoviral vectors, which lack all viral genes, has facilitated adenoviral delivery of up to 30 kb of a therapeutic DNA sequence (Clemens et al. 1996; Kochanek et al. 1996; Parks and Grahm 1997; Morsy et al. 1998) with decreased toxicity (Schiedner et al. 1998). Adenoviral-mediated gene transfer in COS-7 cells was significantly higher than that achieved by liposomal delivery systems (Heider et al. 2000).

The use of Adeno-Associated Viral (AAV) vector provides an alternative to adenoviral vectors for gene therapy and a means for long-term gene expression with a reduced risk of adverse reactions upon administration of the vector (Fisher et al. 1997; Jooss et al. 1998). AAV viruses are linear, single-stranded DNA parvoviruses that are not associated with any disease in humans (Rose et al. 1969). In humans, the site of AAV viral DNA integration is on chromosome 19 (Kotin et al. 1990; Samulski et al. 1991). In the engineering of AAV vectors, most of the AAV genome can be replaced with the therapeutic gene (Samulski et al. 1989), which significantly reduces potential adverse responses of the host to viral infection. However, the size of the therapeutic gene is limited to approximately 5 kb (Dong et al. 1996; Hermonat et al. 1997). First-generation adeno-associated viruses had a very small capacity of ~4.7 kb for encapsulation of the plasmid DNA cargo. Recent reports demonstrate efficient production of second-generation adeno-associated viruses with higher encapsulating capabilities (Owens 2002). It has been demonstrated that adenoviruses in formulations may lose their potency after storage in commonly used pharmaceutical vials (Nyberg-Hoffman and Aguilar-Cardova 1999). Herpes simplex virus (HSV) vector is a large and relatively complex enveloped, double-stranded DNA virus that has the capacity to encode large therapeutic genes and, like AAV, can remain latent in infected cells, providing the potential for long-term expression of the therapeutic gene (Carpenter and Stevens 1996). Although able to infect many cell types, HSV vectors currently are limited in their use by vector toxicity (Lowenstein et al. 1994).

4.2.2 Nonviral Delivery Systems

Nonviral delivery systems have the greatest advantage over viral delivery systems—the lack of immune response and ease of formulation and assembly. Commonly used nonviral vectors for delivery of DNA-based therapeutics can be classified into three major types: *Naked DNA delivery systems*, *polymeric delivery systems*, and *liposomal delivery systems* (Mardan et al. 2002; Fattal et al. 1999, 2001; Pedroso de Lima et al. 2001).

Naked DNA can be administered via two possible routes, either by ex vivo delivery or by in vivo delivery. The ex vivo method of naked DNA delivery has been used successfully for the introduction of DNA into endothelial and smooth muscle cells (Nakamura et al. 1998; Mann et al. 1999a, b); its reliance on the culture of harvested cells renders it unsuitable for many cell types. In vivo delivery of naked DNA was first described in 1990 (Wolff et al. 1990). Efficiency of the delivery of naked DNA can be improved when administered in a pressure-mediated fashion (Mann et al. 1999a, b; Liu et al. 1999). Particle bombardment technology enables the localized delivery of DNA readily into skin or muscle (Fynan et al. 1993). Another technique for delivery of naked DNA directly into target cells is electroporation. The successful delivery of DNA by electroporation in vivo has been reported in tissues such as skin and muscle (Wong and Neumann 1982; Neumann et al. 1982; Rols et al. 1998; Rizzuto et al. 1999).

In polymeric delivery systems, cationic polymers are used in gene delivery because they can easily complex with the anionic DNA molecules (Hwang and Davis 2001). The mechanism of action of these polycomplexes is based on the generation of a positively charged complex owing to electrostatic interaction of these cationic polymers with anionic DNA (Luo and Saltzman 2000). Commonly used polymers include polyethylenimine (Lemkine and Demeneix 2001), poly-L-lysine (Lollo et al. 2002), chitosans (Borchard 2001), and dendrimers (Mardan et al. 2002). Agents such as folates, transferrin, antibodies, or sugars such as galactose and mannose can be incorporated for tissue targeting (Mardan et al. 2002). Synthetic polymers such as protective interactive noncondensing polymers (PINC), poly-L-lysine, cationic polymers, and dendrimers offer an alternative to cationic lipids as a vehicle for DNA delivery into target cells (Boussif et al. 1995; Wadhwa et al. 1995; Kukowska-Latello et al. 1996; Tang and Szoka 1997; Mumper et al. 1998). Encapsulation of a DNA molecule or even a therapeutic viral vector within a biodegradable polymer has been demonstrated to permit the controlled release of the DNA in a targeted cell over a period of weeks or months (Naughton et al. 1992; Singh et al. 2000). The inclusion of proteins and peptides in the DNA complex, which are recognized by receptors on targeted cells, has led to an improvement in the efficiency of DNA uptake in several instances (Jenkins et al. 2000). Some polymers have inherent potent pharmacological properties (such as hypercholesterolemia induced by chitosans) that make them extremely unfavorable for human use (LeHoux and Grondin 1993).

Liposomes are one of the most versatile tools for the delivery of DNA therapeutics (Akhtar et al. 2000; Fattal et al. 1999, 2001; Godbey and Mikos 2001). Liposome and drug/lipid complexes have been used for the delivery of the anticancer drugs doxorubicin and daunorubicin (Tejada-Berges et al. 2002). Liposomes can be used as DNA drug delivery systems either by entrapping the DNA-based therapeutics inside the aqueous core or complexing them to the phospholipids lamellae. Liposome can also be used for specialized gene delivery options, such as long circulation half-life and sustained and targeted delivery (Fattal et al. 1999). Numerous studies have demonstrated the use of cationic liposomal formulations for the delivery of different plasmid constructs in a wide range of cells, both in vivo and in vitro (Marshal et al. 1999). The use of cationic lipids to transfer DNA into cells was first described as an in vitro method of DNA delivery (Felgner et al. 1987). Cationic liposomes have also been used in clinical trials to deliver therapeutic DNA (Caplen et al. 1995; Hyde et al. 2000; Noone et al. 2000; Nabel et al. 1993, 1996). Cationic liposomal formulations consist of mixtures of cationic and zwitterionic lipids (Godbey and Mikos 2001; Felgner et al. 1994; Hofland et al. 1996). Proprietary formulations of cationic lipids such as Lipofectamine (Invitrogen, Carlsbad, CA), Effectene (Qiagen, Valencia, CA), and Tranfectam (Promega, Madison, WI) are commercially available (Kang et al. 1999), but most of the kits are useful only for in vitro experimentation. There are reports of improved efficiency of DNA delivery by cationic lipid via the coupling of specific receptor ligands or peptides to DNA/liposome complexes (Jenkins et al. 2000; Ellison et al. 1996; Simoes et al. 1998; Shinmura et al. 2000; Compton et al. 2000). Cytotoxicity of cationic lipids has been established in numerous in vitro (Lappalainen et al. 1994; Patil et al. 2004) and in vivo (Dokka et al. 2000; Filion and Philips 1997; Freimark et al. 1998) studies. Low transfection efficiencies have been attributed to the heterogeneity and instability of cationic lipoplexes (Lee et al. 2001). Another drawback in the use of cationic lipids is their rapid inactivation in the presence of serum (Hofland et al. 1996; Audouy et al. 2000). Some in vivo studies have revealed that the gene transduction responses obtained by cationic liposomes were transient and short-lived (Liu et al. 1997; Wheeler et al. 1996). As an alternative to cationic lipids, the potential of anionic lipids for DNA delivery has been investigated. The safety of anionic lipids has been demonstrated when administered to epithelial lung tissue. In recent years, a few studies using anionic liposomal DNA delivery vectors have been reported. There have been attempts to incorporate anionic liposomes into polymeric delivery systems. However, these vectors have limited applications, mainly because of (1) inefficient entrapment of DNA molecules within anionic liposomes and (2) lack of toxicity data. Lack of further progress of these systems may be attributed, in part, to the poor association between DNA molecules and anionic lipids, caused by electrostatic repulsion between these negatively charged species (Patil et al. 2004, 2005; Dokka et al. 2000; Patil and Rhodes 2000a, b; Fillion et al. 2001; Lakkaraju et al. 2001; Lee and Huang 1997; Guo et al. 2002; Perrie and Gregoriadis 2000). Along with numerous cationic and anionic lipid derivatives, functionalized liposomal formulations serving specific therapeutic objectives have shown promise in gene therapy (Fattal et al. 1999; Venugopalan et al. 2002; Maclean et al. 1997). Specialized liposomal delivery platforms include pH-sensitive liposomes, immunoliposomes, and stealth liposomes. pH-sensitive liposomes can be generated by the inclusion of 1,2-dioleoyl-3-phosphoethanolamine (DOPE) into liposomes composed of acidic lipids such as cholesterylhemisuccinate or oleic acid. At the neutral cellular pH 7, these lipids have the typical bilayer structure; however, upon endosomal compartmentalization, they undergo protonation and collapse into a nonbilayer structure, thereby leading to the disruption and destabilization of the endosomal bilayer, which in turn helps in the rapid release of DNA into the cytoplasm (Venugopalan et al. 2002). Efficient gene delivery of the beta-galactosidase and luciferase reporter plasmids has been obtained using pH-sensitive liposomes in a variety of mammalian cell lines (Legendre and Szoka 1992). A chemical derivative of DOPE, cCitraconyl-DOPE, has been used to deliver DNA-based therapeutics to cancer cells, thereby combining the targeting and the rapid endosome-releasing aspects of specialized liposomal delivery systems (Reddy and Low 2000). A phosphatidylcholine/glycyrrhizin combination was also successful in pH-sensitive gene delivery in mice (Sviridov et al. 2001). Immunoliposomes are sophisticated gene delivery systems that can be used for cell targeting by the incorporation of functionalized antibodies attached to lipid bilayers (Maclean et al. 1997). Immunoliposomes containing an antibody fragment against the human transferrin receptor were successfully used in targeted delivery of tumor-suppressing genes into tumors in vivo (Xu et al. 2002). Tissue-specific gene delivery using immunoliposomes has been achieved in the brain (Shi et al. 2001), embryonic tissue (Khaw et al. 2001), and breast cancer tissue (Krauss et al. 2000). Stealth liposomes are sterically stabilized liposomal formulations that include polyethylene glycol (PEG)-conjugated lipids (Fattal et al. 1999).

Recently, polyethylenimine (PEI), poly(lactic-co-glycolic acid) (PLGA), polypeptides, chitosan, cyclodextrin, dendrimers, and polymers containing different nanoparticles are used in vitro and in vivo with respect to their structure, physicochemical properties, and delivery efficiency as an siRNA delivery vehicle (Kaushik et al. 2011).

5 Current Status of Gene Therapy Research

The FDA has not yet approved any human gene therapy product for sale. Current gene therapy is experimental and has not proven very successful in clinical trials. Little progress has been made since the first gene therapy clinical trial began in 1990. In 1999, gene therapy suffered a major setback with the death of 18-year-old Jesse Gelsinger.

Another major blow came in January 2003 when the FDA placed a temporary halt on all gene therapy trials using retroviral vectors in blood stem cells. FDA took this action after it learned that a second child treated in a French gene therapy trial had developed a leukemia-like condition. Both that child and another who had developed a similar condition in August 2002 had been successfully treated by gene therapy for X-linked severe combined immunodeficiency disease (X-SCID), also known as "bubble baby syndrome."

FDA's Biological Response Modifiers Advisory Committee (BRMAC) met at the end of February 2003 to discuss possible measures that could allow a number of retroviral gene therapy trials for treatment of life-threatening diseases to proceed with appropriate safeguards.

In April of 2003, the FDA eased the ban on gene therapy trials using retroviral vectors in blood stem cells.

6 Recent Developments in Gene Therapy

Month/year	Development in gene therapy	Author(s) name; name of journal(s)/URL	Title of the article	Summary/inference
March 2009	Nanotechnology and gene therapy yields treatment to torpedo cancer	Schatzlein A [http://news.bbc.co.uk/2/hi/health/7935592.stm/ (March 10)]	Nano-treatment to torpedo cancer	School of Pharmacy in London is testing a treatment in mice, which delivers genes wrapped in nanoparticles to cancer cells
April 2008 April 2008	Results of world's first gene therapy for inherited blindness show sight improvement	1. James et al.; N Engl J Med 2. Albert et al.; N Engl J Med	1. Effect of Gene Therapy on Visual Function in Leber's Congenital Amaurosis 2. Safety and Efficacy of Gene Transfer for Leber's Congenital Amaurosis	UK researchers from the UCL Institute of Ophthalmology and Moorefield's Eye Hospital NIHR Biomedical Research Centre have announced results from the world's first clinical trial to test a revolutionary gene therapy treatment for a type of inherited blindness (Leber's congenital amaurosis). The results show that the experimental treatment is safe and can improve sight. The findings are a landmark for gene therapy and could have a significant impact on future treatments for eye diseases
May 2007	This is the first trial to use gene therapy in an operation to treat blindness in humans	Ben H [http://www.reuters.com/article/scienceNews/idUSL0166536200705017?/ (May 1, 2007)]	Doctors test gene therapy to treat blindness	A team of British doctors from Moorefield's Eye Hospital and University College in London conduct the first human gene therapy trials to treat Leber's congenital amaurosis, a type of

(continued)

References

Agarawal S, Tang JY (1992) GEM-91-an antisense oligonucleotide phosphorothioate as a therapeutic agent for AIDS. Antisense Res Dev 2:261–266

Agarwal S, Kandimalla ER (2000) Antisense therapeutics: is it as simple as complementary base recognition? Mol Med Today 6:72–81

Aigner A, Juhl H, Malerczyk C et al (2001) Expression of a truncated 100 kDa HER2 splice variant acts as an endogenous inhibitor of tumor cell proliferation. Oncogene 20:2101–2111

Akhtar S, Hughes MD, Khan A (2000) The delivery of antisense therapeutics. Adv Drug Deliv Rev 44:3–21

Albert MM, Francesca S, Aric AP et al (2008) Safety and efficacy of gene transfer for Leber's congenital amaurosis. N Engl J Med 358:2240–2248

Alt M, Ring R, Hofschneider PH et al (1995) Specific inhibition of hepatitis C viral gene expression by antisense phosphorothioate oligodeoxynucleotides. Hepatology 22:707–717

Alvarez RD, Barns MN, Gomez-Navarro J (2000a) A cancer gene therapy approach utilizing an anti-erbB-2 single-chain antibody-encoding adenovirus (AD21): a phase I trial. Clin Cancer Res 6:3081–3087

Alvarez RD, Gomez-Navarro J, Wang M (2000b) Adenoviral mediated suicide gene therapy for ovarian cancer. Mol Ther 2:524–530

Anderson WF (1992) Human gene therapy. Science 256:808–813

Anderson WF (1998) Human gene therapy. Nature (Lond) 392:25–30

Anil A (2003) Undercover genes slip into the brain. http://www.newscientist.com/section/science-news/. 22 Mar 2003

Audouy S, Molema G, De Leij L et al (2000) Serum as a modulator of lipoplex-mediated gene transfection: dependence of amphiphile cell type and complex stability. J Gene Med 2:465–476

Azad RF, Driver VB, Tanaka K et al (1993) Antiviral activity of a phosphorothioate oligonucleotide complementary to RNA of the human cytomegalovirus major immediate-early region. Antimicrob Agents Chemother 37:1945–1954

Baekelandt V, De Strooper B, Nuttin B et al (2000) Gene therapeutic strategies for neurodegenerative diseases. Curr Opin Mol 2:540–554

Baker BF (2001) The role of antisense oligonucleotides in the wave of genomic information. Nucleosides Nucleotides Nucleic Acids 20:397–399

Baltimore DP (1988) Intracellular immunization. Nature 335:395–397

Beelman CA, Parker R (1995) Degradation of mRNA in eukaryotes. Cell 81:179–183

Ben H (2007) Doctors test gene therapy to treat blindness. http://www.reuters.com/article/scienceNews/idUSL016653620070501?/. 1 May 2007

Bernstein E, Caudy AA, Hammond SM et al (2001) Role for a bidentate ribonuclease in the initiation step of RNA interference. Nature 409:363–366

Bertrand JR, Pottier M, Vekris A et al (2002) Comparison of antisense oligonucleotides and siRNAs in cell culture and in vivo. Biochem Biophys Res Commun 296:1000–1004

Bett AJ, Prevec L, Graham FL (1993) Packaging capacity, and stability of human adenovirus type 5 vectors. J Virol 67:5911–5921

Bob H (2003) Gene therapy may switch off Huntington's. http://www.newscientist.com/section/science-news/. 13 Mar 2003

Borchard G (2001) Chitosans for gene delivery. Adv Drug Deliv Rev 52:145–150

Bordier B, Helena C, Barr PJ et al (1992) In vitro effects of antisense oligonucleotides on human immunodeficiency virus type 1 reverses transcription. Nucleic Acids Res 20:1999–2006

Bordignon C, Notarangilo LD, Nobili N (1995) Gene therapy in peripheral blood lymphocytes and bone marrow for ADA-immunodeficient patients. Science 270:470–475

Boussif O, Lezoualc'h F, Zanta MA (1995) A versatile vector for gene and oligonucleotide transfer into cells in culture and in vivo: polyethylenimin. Proc Natl Acad Sci USA 92:7297–7301

Breaker RR, Joyce GF (1994) DNA enzyme that cleaves RNA. Trends Biotechnol 1:223–229

Bunnel BA, Morgan RA (1996) Gene therapy for HIV infection. Drugs Today 32:09–224

Burke JM (1996) Hairpin ribozyme: current status and future prospects. Biochem Soc Trans 24:608–615

Byrnes AP, Rusby JE, Wood MJ (1995) Adenovirus gene transfer causes inflammation in the brain. Neuroscience 66:1015–1024

Campagno D, Lampe JN, Bourget C et al (1999) Antisense oligonucleotides containing modified bases inhibit in vitro translation of *Leishmania amazonensis* mRNAs by invading the mini-exon hairpin. J Biol Chem 274:8191–8198

Caplen NJ (2004) Gene therapy progress and prospects. Downregulating gene expression: the impact of RNA interference. Gene Ther 11:1241–1248

Caplen NJ, Alton EW, Middleton PG (1995) Liposome-mediated CFTR gene transfer to the nasal epithelium of patients with cystic fibrosis. Nat Med 1:39–46

Carpenter DE, Stevens JG (1996) Long term expression of a foreign gene from a unique position in the latent herpes simplex virus genome. Hum Gene Ther 7:1447–1454

Chaloin L, Lehmann MJ, Sczakiel G et al (2002) Endogenous expression of a high-affinity pseudoknot RNA aptamer suppresses replication of HIV-1. Nucleic Acids Res 30:4001–4008

Chamberlain JS (2002) Gene therapy of muscular dystrophy. Hum Mol Genet 11:2355–2362

Clemens PR, Kochanek S, Sunada Y (1996) In vivo muscle gene transfer of full-length dystrophin with an adenoviral vector that lacks all viral genes. Gene Ther 3:965–972

Compton SH, Mecklenbeck S, Mejia JE (2000) Stable integration of large (>100 kb) PAC constructs in HaCa T keratinocytes using an integrin-targeting peptide delivery system. Gene Ther 7:1600–1607

Cone RD, Mulligan RC (1984) High-efficiency gene transfer into mammalian cells: generation of helper-free recombinant retrovirus with broad mammalian host range. Proc Natl Acad Sci USA 81:6349–6353

Cowsert LM, Fox MC, Zon G et al (1993) In vitro evaluation of phosphorothioate oligonucleotides targeted to the E2 mRNA of papillomavirus: potential treatment for the genital warts. Antimicrob Agents Chemother 37:171–177

Crooke ST (1998a) An overview of progress in antisense therapeutics. Antisense Nucleic Acid Drug Dev 8:115–122

Crooke ST (1998b) Vitravene another piece in the mosaic. Antisense Nucleic Acid Drug Dev 8:vii–v3

Crooke ST (1999a) Molecular mechanisms of antisense drugs. Biochem Biophys Acta 1489:31–43

Crooke ST (1999b) Molecular mechanisms of antisense drugs: human RNaseH. Antisense Nucleic Acid Drug Dev 9:377–379

Danny P (2002) Subtle gene therapy tackles blood disorder. http://www.newscientist.com/section/science-news/. 11 Oct 2002

de Saultrait VR, Lozach P-Y, Altmeyer R et al (2002) DNA aptamers derived from HIV-1 RNaseH inhibitors are strong anti-integrase agents. J Mol Biol 324:195–203

Dean NM, McKay R, Condon TP et al (1994) Inhibition of protein kinase C-α expression in human A549 cells by antisense oligonucleotides inhibits induction of intracellular adhesion molecule 1 (ICAM-1) mRNA by phorbol esters. J Biol Chem 269:16416–16424

Denli AM, Tops BBJ, Plasterk RHA et al (2004) Processing of primary microRNAs by the microprocessor complex. Nature 432:231–235

Dokka S, Toledo D, Shi X et al (2000) Oxygen radical-mediated pulmonary toxicity induced by some cationic liposomes. Pharm Res 17:521–525

Dong JY, Fang PD, Frizzel RA (1996) Quantitative analysis of the packaging capacity of recombinant adeno-associated virus. Hum Gene Ther 7:2101–2112

Dorsett Y, Tuschl T (2004) siRNAs: applications in functional genomics and potential as therapeutics. Nat Rev Drug Discov 3:318–329

Dunbar CE, Kohn DB, Shiffman R (1998) Retroviral transfer of the glucoceribrosidase gene into CD34+ cells from patients with Goucher disease: in vivo detection of transduced cells without myeloblation. Hum Gen Ther 9:2629–2640

Earnshaw DJ, Gait MJ (1997) Progress toward the structure and therapeutic use of the hairpin ribozyme. Antisense Nucleic Acid Drug Dev 7:403–411
Ehsan A, Mann MJ, Dell'Acqua G et al (2001) Long-term stabilization of vein graft wall architecture and prolonged resistance to experimental atherosclerosis after E2F decoy oligonucleotide gene therapy. J Thorac Cardiovasc Surg 121:714–722
Ellison KE, Bisphoric NH, Webster KA (1996) Fusigenic liposome-mediated DNA transfer into cardiac myocytes. J Mol Cell Cardiol 28:1385–1399
Emma Y (2002) 'Miracle' gene therapy trial halted. http://www.newscientist.com/section/science-news/. 3 Oct 2002
Eyetech Study Group (2002) Preclinical and phase 1A clinical evaluation of an anti-VEGF pegylated aptamer (EYE001) for the treatment of exudative age-related macular degeneration. Retina 22:143–152
Fattal E, Dellattre J, Dubernet C et al (1999) Liposomes for the delivery of nucleotides and oligonucleotides. STP Pharma Sci 9:383–390
Fattal E, Dubernet C, Couvrur P (2001) Liposome-based formulations for the delivery of oligonucleotids. STP Pharma Sci 11:31–44
Felgner PL, Gadek TR, Holm M (1987) Lipofection: a highly efficient, lipid-mediated DNA-transfection procedure. Proc Natl Acad Sci USA 84:7413–7417
Felgner JH, Kumar R, Sridhar CN (1994) Enhanced gene delivery and mechanism studies with a novel series of cationic lipid formulations. J Biol Chem 269:2550–2561
Filion MC, Philips NC (1997) Toxicity and immunomodulatory activity of liposomal vectors formulated with cationic lipids toward immune effector cells. Biochim Biophys Acta 1329:345–356
Fillion P, Desjardins A, Sayasith K et al (2001) Encapsulation of DNA in negatively charged liposomes and inhibition of bacterial gene expression with fluid-liposome encapsulated antisense oligonucleotides. Biochim Biophys Acta 1515:44–54
Fire A, Xu S, Montgomery MK et al (1998) Potent and specific genetic interference by double-stranded RNA in *Caenorhabditis elegans*. Nature 391:806–811
Fisher KJ, Jooss K, Alston J (1997) Recombinant adeno-associated virus for muscle directed gene therapy. Nat Med 3:306–312
Freimark BD, Blzinger HP, Florack VJ (1998) Cationic lipids enhance cytokine and cell influx levels in the lung following administration of plasmid: cationic lipid complexes. J Immunol 160:4580–4586
Fujji S, Huang S, Fong TC (2000) Induction of melanoma-associated antigen systemic immunity upon intratumoral delivery of interferon-gamma retroviral vector in melanoma patients. Cancer Gene Ther 7:1220–1230
Fynan EF, Webster RG, Fuller DH (1993) DNA vaccines: protective immunizations by parenteral, mucosal, and gene-gun inoculations. Proc Natl Acad Sci USA 90:11478–11482
Galanis E, Russell S (2001) Cancer gene therapy clinical trials: lessons for the future. Br J Cancer 85:1432–1436
Garton KJ, Ferri N, Raines EW (2002) Efficient expression of exogenous genes in primary vascular cells using IRES-based retroviral vectors. Biotechniques 32:830–843
Ge Q, McManus MT, Nguyen T et al (2003) RNA interference of influenza virus production by directly targeting mRNA for degradation and indirectly inhibiting all viral RNA transcription. Proc Natl Acad Sci USA 100:2718–2723
Gerviax A, Li X, Kraus G et al (1997) Multigene antivirus vectors inhibit divers human immunodeficiency virus type 1 clades. J Virol 71:3048–3053
Gewirtz AM, Sokol DL, Ratajczak MZ (1998) Nucleic acid therapeutics: state of the art and future prospects. Blood 92:712–736
Godbey WT, Mikos AG (2001) Recent progress in gene delivery using non-viral transfer complexes. J Control Release 72:115–125
Guo W, Gosselin MA, Lee RJ (2002) Characterization of a novel diolein-based LPD II vector for gene delivery. J Control Release 83:121–132

Hale SJ, Green NK (2002) Viral approaches to cancer gene therapy. Expert Opin Therapeut Patents 12:369–378
Hamilton AJ, Baulcombe DC (1999) A species of small antisense RNA in posttranscriptional gene silencing in plants. Science 286:950–952
Hammond SM (2005) Dicing and slicing: the core machinery of the RNA interference pathway. FEBS Lett 579:5822–5829
Hammond SM, Bernstein E, Beach D et al (2000) An RNA-directed nuclease mediates post-transcriptional gene silencing in Drosophila cells. Nature 404:293–296
Hampel A (1998) The hairpin ribozyme: discovery, two dimensional model, and development for gene therapy. Prog Nucleic Acid Res Mol Biol 58:1–39
Han J, Lee Y, Yeom K-H et al (2004) The Drosha-DGCR8 complex in primary microRNA processing. Genes Dev 18:3016–3027
Hanecak R, Brown-Driver V, Fox MC et al (1996) Antisense oligonucleotide inhibition of hepatitis C virus gene expression in transformed hepatocytes. J Virol 70:5203–5212
Heider H, Verca HB, Rusconi S et al (2000) Comparison of lipid-mediated, and adenoviral gene transfer in human monocyte-derived macrophages and COS-7 cells. Biotechniques 28:260–270
Hermonat PL, Quirk JG, Bishop BM (1997) The packaging capacity of adeno-associated virus (AAV) and the potential for wild-type-plus AAV gene therapy vectors. FEBS Lett 407:78–84
Hofland HEJ, Shephard L, Sullivan SM (1996) Formation of stable cationic lipid/DNA complexes for gene transfer. Proc Natl Acad Sci USA 93:7305–7309
Horner AA, van Uden JH, Jubeldia JM et al (2001) DNA-based immunotherapeutics for the treatment of allergic disease. Immunol Rev 179:102–118
Huang L, Viroonchatapan E (1999) Introduction. In: Huang L, Huang M-C, Wagner E (eds) Nonviral vectors for gene therapy. Academic Press, San Diego, CA, pp 3–22
Hwang SJ, Davis ME (2001) Cationic polymers for gene delivery: designs for overcoming barriers to systemic administration. Curr Opin Mol Ther 3:183–191
Hyde SC, Southern KW, Gileadi U (2000) Repeat administration of DNA/liposome to the nasal epithelium of patients with cystic fibrosis. Gene Ther 7:1156–1165
Jackson AL, Linsley PS (2004) Noise amidst the silence: off-target effects of siRNAs? Trends Genet 20:521–524
James WB, Alexander JS, Susie SB et al (2008) Effect of gene therapy on visual function in Leber's congenital amaurosis. N Engl J Med 358:2231–2239
Jayasena SD (1999) Aptamers: an emerging class of molecules that rival antibodies in diagnostics. Clin Chem 45:1628–1650
Jenkins RG, Herrick SE, Meng QH (2000) An integrin-targeted non-viral vector for pulmonary gene therapy. Gene Ther 7:393–400
Jennifer FW (2002) Murine gene therapy corrects symptoms of sickle cell disease. Scientist 16:36. http://www.the-scientist.com/2002/3/18
Johnston SA, Talaat AM, McGuire MG (2002) Genetic immunization what's in a name? Arch Med Res 33:325–329
Jooss K, Yang Y, Fisher KJ (1998) Transduction of dendritic cells by DNA viral vectors directs the immune response to transgene products in muscle fibers. J Virol 72:4212–4223
Kamiya H, Tsuchiya H, Yamazaki J et al (2001) Intracellular trafficking and transgene expression of viral and non-viral gene vectors. Adv Drug Deliv Rev 52:153–164
Kang S-H, Zirbes EL, Kole R (1999) Delivery of antisense oligonucleotides and plasmid DNA with various carrier agents. Antisense Nucleic Acid Drug Dev 9:497–505
Katz SM, Tian L, Stepkowski SM et al (1997) Effect of ICAM-1/LFA-1 blockade on pancreatic islet allograft survival function, and early cytokine production. Transplant Proc. 29(1–2):748–749
Kaur G, Roy I (2008) Therapeutic applications of aptamers. Expert Opin Investig Drugs 17:43–60
Kaushik S, Namgung R, Kim WJ (2011) Polymers in small-interfering RNA delivery. Nucleic Acid Ther 21:133–147

Khaw BA, da Silva J, Vural I et al (2001) Intracytoplasmic gene delivery for in vitro transfection with cytoskeleton-specific immunoliposomes. J Control Release 75:199–210
Kim SH, Yu SS, Park JS et al (1998) Construction of retroviral vectors with improved safety, gene expression, and versatility. J Virol 72:994–1004
Kinchington D, Galpin S, Jeroszewski J et al (1992) A comparison of gag, pol, and rev antisense oligodeoxynucleotides as inhibitors of HIV-1. Antiviral Res 17:53–62
Kochanek S, Clemens PR, Mitani K (1996) A new adenoviral vector: replacement of viral coding sequences with 28 kb of DNA independently expressing both full-length dystrophin and beta-galactosidase. Proc Natl Acad Sci USA 93:5731–5736
Kotin RM, Siniscalco M, Samulski RJ (1990) Site-specific integration by adeno-associated virus. Proc Natl Acad Sci USA 87:2211–2215
Krauss WC, Park JW, Kirpotin DB et al (2000) Emerging antibody-based HER2 (ErbB-2/new) therapeutics. Breast Dis 11:113–124
Kukowska-Latello JF, Bielinska AU, Johnson J (1996) Efficient transfer of genetic material into mammalian cells using Starburst polyamidoamine dendrimers. Proc Natl Acad Sci USA 93:4897–4902
Kurreck J (2003) Antisense technologies. Improvement through novel chemical modifications. Eur J Biochem 270:1628–1644
Lakkaraju A, Dubinsky JM, Low WC et al (2001) Neurons are protected from excitotoxic death by p53 antisense oligonuclotides delivered in anionic liposomes. J Biol Chem 276:32000–32007
Lappalainen K, Jaaskelainen I, Syrjanen K et al (1994) Comparison of cell proliferation and toxicity assays using two cationic liposomes. Pharm Res 11:1127–1131
Lee RJ, Huang L (1997) Lipidic vector systems for gene transfer. Crit Rev Ther Drug Carrier Syst 14:173–206
Lee RC, Feinbaum RL, Ambros V (1993) The *C. elegans* heterochronic gene *lin-4* encodes small RNAs with antisense complementarity to *lin-14*. Cell 75:843–854
Lee H, Williams SKR, Allison SD et al (2001) Analysis of self-assembled cationic lipid–DNA gene carrier complexes using flow field-flow fractionation and light scattering. Anal Chem 73:837–8343
Lee Y, Kim M, Han J et al (2004) MicroRNA genes are transcribed by RNA polymerase II. EMBO J 23:4051–4060
Legendre JY, Szoka FC Jr (1992) Delivery of plasmid DNA into mammalian cell lines using pH-sensitive liposomes: comparison with cationic liposomes. Pharm Res 9:1235–1242
LeHoux JG, Grondin F (1993) Some effects of chitosan on liver function in the rat. Endocrinology 132:1078–1084
Lehrman S (1999) Virus treatment questioned after gene therapy death. Nature 401:517–518
Lemkine GF, Demeneix BA (2001) Polyethylenimins for in vivo gene delivery. Curr Opin Mol Ther 3:178–182
Lever AML (1996) Gene therapy in the fight against AIDS. Expert Opin Ther Pat 6:161–167
Lewis PF, Emerman M (1994) Passage through mitosis is required for oncoretroviruses but not for the human immunodeficiency virus. J Virol 68:510–516
Liebhaber SA (1997) mRNA stability and the control of gene expression. Nucleic Acids Symp Ser 36:29–32
Lien Y-HH, Lie L-W (2002) Gene therapy for renal disorders: what are the benefits for the elderly? Drugs Aging 19:553–560
Lisziewicz J, Sun D, Klotman M et al (1992) Specific inhibition of human immunodeficiency virus type 1 replication by antisense oligonucleotides: an in vitro model for treatment. Proc Natl Acad Sci USA 89:11209–11213
Lisziewicz J, Sun D, Metelev V et al (1993) Long term treatment of human immunodeficiency virus infected cells with antisense oligonucleotide phosphorothioates. Proc Natl Acad Sci USA 90:3860–3864
Liu F, Qi H, Huang L et al (1997) Factors controlling the efficiency of cationic lipid-mediated transfection in vivo via intravenous administration. Gene Ther 4:517–523

Liu F, Song Y, Liu D (1999) Hydrodynamics-based transfection in animals by systemic administration of plasmid DNA. Gene Ther 6:1258–1266

Lollo CP, Banaszczyk MG, Mullen PM (2002) Poly-L-lysine based gene delivery systems: synthesis, purification, and application. Methods Mol Med 69:1–13

Lotze MT, Kost TA (2002) Viruses as gene delivery vectors: application to gene function, target validation, and assay development. Cancer Gene Ther 9:692–699

Lowenstein PR, Morrison EE, Bain D (1994) Use of recombinant vectors derived from herpes simplex virus 1 mutant tsK for short-term expression of transgenes encoding cytoplasmic and membrane anchored proteins in post mitotic-polarized cortical neurons and glial cells in vitro. Neuroscience 60:1059–1077

Lund E, Guttinger S, Calado A et al (2004) Nuclear export of microRNA precursors. Science 30:95–98

Luo D, Saltzman WM (2000) Synthetic DNA delivery systems. Nat Biotechnol 18:33–37

Maclean AL, Symonds G, Ward R (1997) Immunoliposomes as targeted delivery vehicles for cancer therapeutics. Int J Oncol 11:325–332

Macpherson JL, Fly JA, Sun LO et al (1999) Ribozymes in gene therapy of HIV-1. Front Biosci 4:D497–D505

Mah C, Byrne BJ, Flotte TR (2002) Virus-based gene delivery systems. Clin Pharmacokinet 41:901–911

Mann R, Mulligan RC, Baltimore D (1983) Construction of a retrovirus packaging mutant and its use to produce helper-free defective retrovirus. Cell 33:153–159

Mann MJ, Gibbons GH, Hutchinson H (1999a) Pressure-mediated oligonucleotide transfection of rat and human cardiovascular tissues. Proc Natl Acad Sci USA 96:6411–6416

Mann MJ, Whittemore AD, Donaldson MC et al (1999b) Ex-vivo gene therapy of human vascular bypass grafts with E2F decoy: the PREVENT single-centre, randomised, controlled trial. Lancet 354:1493–1498

Marciniak RA, Garcia-Blanco MA, Sharp PA (1990) Identification and characterization of a HeLa nuclear protein that specifically binds to the trans-activation-response (TAR) element of human immunodeficiency virus. Proc Natl Acad Sci USA 87:3624–3638

Mardan T, Kopecek J, Kissel T (2002) Prospects for cationic polymers in gene and oligonucleotide therapy against cancer. Adv Drug Deliv Rev 54:715–758

Marshal J, Yew NS, Eastman SJ et al (1999) Cationic lipid-mediated gene delivery to the airways. In: Wagner E (ed) Nonviral vectors for gene therapy. Academic Press, pp 39–68

Martin KR, Klein RL, Quigley HA (2002) Gene delivery to the eye using adeno-associated viral vectors. Methods 28:267–275

Martinez MA, Gutierrez A, Armond-Ugon M et al (2002) Suppression of chemokine receptor expression by RNA interference allows for inhibition of HIV-1 replication. AIDS 16:2385–2390

McAlister DV, Allen MG, Prausnitz MR (2000) Microfabricated microneedles for gene and drug delivery. Annu Rev Biomed Eng 2:289–313

McMinus MT, Sharp PA (2002) Gene silencing in mammals by small interfering RNAs. Nat Rev Genet 3:737–747

McTaggart S, Al-Rubeai M (2002) Retroviral vectors for human gene delivery. Biotechnol Adv 20:1–31

Michou AI, Santoro L, Christ M (1997) Adenovirus-mediated gene transfer: influence of transgene, mouse strain and type of immune response on persistence of transgene expression. Gene Ther 4:473–482

Mirabelli CK, Bennet CF, Anderson K et al (1991) In vitro and in vivo pharmacologic activities of antisense oligonucleotides. Anti-Cancer Drug Des 6:647–661

Mizuno T, Chou MY, Inouye M (1984) A unique mechanism regulating gene expression: translational inhibition by a complementary RNA transcript (micRNA). Proc Natl Acad Sci USA 81:1966–1970

Mizutani TN, Kato M, Hirota K et al (1995) Inhibition of hepatitis C virus replication by antisense oligonucleotids in culture cells. Biochem Biophys Res Commun 212:906–911

Morgan RA, Dudley ME, Wunderlich JR et al (2006) Cancer regression in patients mediated by transfer of genetically engineered lymphocytes. Science 314:126–129. http://www.cancer.gov/newscenter/pressreleases/MelanomaGeneTherapy/. 31 Aug 2006

Morsy MA, Gu M, Motzel S (1998) An adenoviral vector deleted for all viral coding sequences results in enhanced safety and extended expression of a leptin transgene. Proc Natl Acad Sci USA 95:7866–78671

Morvan F, Porumb H, Degols G et al (1993) Comparative evaluation of seven oligonucleotide analogues as potential antisense agents. J Med Chem 36:280–287

Mulherkar R (2001) Gene therapy for cancer. Curr Sci 81:555–560

Mumper RJ, Wang J, Klakamp SL (1998) Protective interactive noncondensing (PINC) polymers for enhanced plasmid distribution and expression in rat skeletal muscle. J Control Release 52:191–203

Nabel GJ, Nabel EG, Yang ZY (1993) Direct gene transfer with DNA–liposome complexes in melanoma: expression, biological activity, and lack of toxicity in humans. Proc Natl Acad Sci USA 90:11307–11311

Nabel GJ, Gordon D, Bishop DK (1996) Immune response in human melanoma after transfer of an allogenic class I major histocompatibility complex gene with DNA–liposome complexes. Proc Natl Acad Sci USA 93:15388–15393

Nakamura M, Devila-Zavala P, Tokuda H (1998) Uptake and gene expression of naked plasmid DNA in cultured brain microvessel endothelial cells. Biochem Biophys Res Commun 245:235–239

Naldini L, Blomer U, Gally P (1996) In vivo gene delivery and stable transduction of nondividing cells by a lentiviral vector. Science 272:263–267

Naughton BA, Dai Y, Sibanda B (1992) Long-term expression of a retrovirally introduced beta-galactosidase gene in rodent cells implanted in vivo using biodegradable polymer meshes. Somat Cell Mol Genet 18:451–462

Neumann E, Schaefer-Ridder M, Wang Y (1982) Gene transfer into lyoma by electroporation in high electric fields. EMBO J 1:841–845

Noone PG, Hohnker KW, Zhou Z (2000) Safety and biological efficacy of a lipid-CFTR complex for gene transfer in the nasal epithelium of adult patients with cystic fibrosis. Mol Ther 1:105–114

Nyberg-Hoffman C, Aguilar-Cardova E (1999) Instability of adenoviral vectors during transport and its implication for clinical studies. Nat Med 5:955–957

Onodera M, Ariga T, Kawamura N (1998) Successful peripheral T-lymphocyte-directed gene transfer for a patient with severe combined immunodeficiency caused by adenosine deaminase deficiency. Blood 91:30–36

Otsu M, Candotti F (2002) Gene therapy in infants with severe combined immunodeficiency. BioDrugs 16:229–239

Owens RA (2002) Second generation adeno-associated virus type 2-based gene therapy systems with the potential for preferential integration into AAVS 1. Curr Gene Ther 2:145–159

Parks RJ, Grahm FL (1997) A helper-dependent system for adenovirus vector production helps define a lower limit for efficient DNA packaging. J Virol 71:3293–3298

Patil SD, Burgess DJ (2003) DNA-based pharmaceuticals: therapeutics for the 21st century. AAPS Newsmagazine 6:27

Patil SD, Rhodes DG (2000a) Conformation of oligodeoxynucleotides associated with anionic liposomes. Nucleic Acids Res 28:4125–4129

Patil SD, Rhodes DG (2000b) Influence of divalent cations on the conformation of phosphorothioate oligodeoxynucleotides: a circular dichroism study. Nucleic Acids Res 28:2439–2445

Patil SD, Rhodes DG, Burgess DJ (2004) Anionic liposomal delivery system for DNA transfection. AAPS J 6:E29

Patil SD, Rhodes DG, Burgess DJ (2005) Biophysical characterization of anionic lipoplexes. Biochim Biophys Acta—Biomembranes 7:E61–E77

Pedroso de Lima MC, Simoes S, Pires P et al (2001) Cationic lipid–DNA complexes in gene delivery: from biophysics to biological applications. Adv Drug Deliv Rev 47:277–294

Perrie Y, Gregoriadis G (2000) Liposome-entrapped plasmid DNA: characterization studies. Biochim Biophys Acta 1475:125–132

Reddy JA, Low PS (2000) Enhanced folate receptor mediated gene therapy using a novel pH-sensitive lipid formulation. J Control Release 64:27–37

Regnier V, Tahiri A, Andre N et al (2000) Electroporation-mediated delivery of 3′-protected phosphodiester oligodeoxynucleotides to the skin. J Control Release 67:337–346

Report (1995) The United Kingdom Health Minister's Gene Therapy Advisory Committee. Guidance on making proposals to conduct gene therapy research on human subjects. Hum Gene Ther 6:335–346

Rizzuto G, Cappelletti M, Maione D (1999) Efficient and regulated erythropoietin production by naked DNA injection and muscle electroporation. Proc Natl Acad Sci USA 96:6417–6422

Roe T, Reynolds TC, Yu G et al (1993) Integration of murine leukemia virus DNA depends on mitosis. EMBO J 12:2099–2108

Rols MP, Delteil C, Golzio M (1998) In vivo electrically mediated protein and gene transfer in murine melanoma. Nat Biotechnol 16:168–171

Rose JA, Berns KI, Hoggan MD (1969) Evidence for a single stranded adenovirus-associated virus genome: formation of a DNA density hybrid on release of viral DNA. Proc Natl Acad Sci USA 64:863–869

Rosenberg SA, Blaese RM, Branner MK et al (2000) Human gene marker/therapy protocols. Hum Gene Ther 11:919–979

Ruvkun G (2001) Molecular biology: glimpses of a tiny RNA world. Science 294:797–799

Samulski RJ, Chang LS, Shenk T (1989) Helper-stocks of recombinant adeno-associated viruses: normal integration does not require viral gene expression. J Virol 63:3822–3828

Samulski RJ, Zhu X, Xiao X (1991) Targeted integration of adeno-associated virus (AAV) into human chromosome 19. EMBO J 10:3941–3950

Scanlon KJ, Ohta Y, Ishida H et al (1995) Oligonucleotide—mediated modulation of mammalian gene expression. FASEB J 9:1288–1296

Schatzlein A (2009) Nano-treatment to torpedo cancer. http://news.bbc.co.uk/2/hi/health/7935592.stm/. 10 Mar 2009

Scherr M, Morgan MA, Eder M (2003) Gene silencing mediated by small interfering RNAs in mammalian cells. Curr Med Chem 10:245–256

Schiedner G, Morral N, Parks RJ (1998) Genomic DNA transfer with a high-capacity adenovirus vector results in improved in vivo gene expression and decreased toxicity. Nat Genet 18:180–183

Scott M (2007) Dual gene therapy suppresses lung cancer in preclinical test. http://www.newswise.com/p/articles/view/526526/. 11 Jan 2007

Shankar P, Manjunath N, Lieberman J (2005) The prospect of silencing disease using RNA interference. J Am Med Assoc 293:1367–1373

Sheridan PL, Bodner M, Lynn A (2000) Generation of retroviral packaging and producer cell lines for large-scale vector production and clinical application: improved safety and high titer. Mol Ther 2:262–275

Shi N, Zhang Y, Zhu C et al (2001) Brain-specific expression of an exogenous gene after IV administration. Proc Natl Acad Sci USA 98:12754–12759

Shinmura K, Morishita R, Aoki M (2000) Catheter-delivered in vivo gene transfer into rat myocardium using the fusigenic liposomal mediated method. Jpn Heart J 41:633–647

Simoes S, Slepushkin V, Gaspar R (1998) Gene delivery by negatively charged ternary complexes of DNA, cationic liposomes and transferrin or fusigenic peptides. Gene Ther 5:955–964

Simons RW, Kleckner N (1983) Translational control of IS10 transposition. Cell 34:683–691

Singh M, Briones M, Ott G (2000) Cationic microparticles: a potent delivery system for DNA vaccines. Proc Natl Acad Sci USA 97:811–816
SoRelle R (2000) Who owns your DNA? Who will own it? Circulation 101:e67–e68
Speedie M (2005) Antisense therapeutic agents. In: Williams DA, Lemke TL (eds) Foye's principles of medicinal chemistry, 5th edn. Lippincott, Philadelphia, PA
Stein CA (1998) How to design an antisense oligodeoxynucleotides experiment: a consensus approach. Antisense Nucleic Acid Drug Dev 8:129–132
Stewart AK, Lassam NJ, Quirt IC (1999) Adenovector-mediated gene delivery of interleukin-2 in metastatic breast cancer and melanoma: results of a phase 1 clinical trial. Gene Ther 6:350–363
Stull RA, Szoka FC Jr (1995) Antigene, ribozyme and aptamer nucleic acid drugs: progress and prospects. Pharm Res 12:463–465
Sullenger BA, Cech TR (1993) Tethering ribozymes to a retroviral packaging signal for destruction of viral RNA. Science 262:1566–1569
Sullenger BA, Gallardo HF, Ungers GE et al (1990) Overexpression of TAR sequences renders cells resistant to human immunodeficiency virus replication. Cell 63:601–608
Sullenger BA, Gallardo HF, Ungers GE et al (1991) Analysis of trans-acting response decoy RNA-mediated inhibition of human immunodeficiency virus type 1 transactivation. J Virol 65:6811–6816
Sviridov YV, Zhadanov RI, Podobed OV et al (2001) The LacZ gene transfer into L929 cells and [14 C]-DNA tissue distribution following intraperitoneal administration of new pH-sensitive lipoplexes in mice. Cytobios 106:7–14
Sylvia PW (2002) DNA nanoballs boost gene therapy. http://www.newscientist.com/section/science-news/. 12 May 2002
Tait DL, Obermiller PS, Hatmaker AR (1999) Ovarian cancer BRCA1 gene therapy: phase I and II trial differences in immune response and vector stability. Clin Cancer Res 5:1708–1714
Tang MX, Szoka FC (1997) The influence of polymer structure on the interactions of cationic polymers with DNA and morphology of the resulting complexes. Gene Ther 4:823–832
Tejada-Berges T, Granai CO, Gordinier M et al (2002) Caelyx/Doxil for the treatment of metastatic ovarian and breast cancer. Expert Rev Anticancer Ther 2:143–150
Tonkinson JL, Stein CA (1993) Antisense nucleic acids-prospects for antiviral intervention. Antiviral Chem Chemother 4:193–199
Tuschl T, Zamore PD, Lehmann R et al (1999) Targeted mRNA degradation by double-stranded RNA in vitro. Genes Dev 13:3191–3197
Uherek C, Wels W (2000) DNA-carrier proteins for targeted gene delivery. Adv Drug Deliv Rev 44:153–166
Usman N, Beigelman L, McSwiggen JA (1996) Hammerhead ribozyme engineering. Curr Opin Struct Biol 1:527–533
Van Ginkel FW, Mc Ghee JR, Liu C (1997) Adenoviral gene delivery elicits distinct pulmonary-associated T helper cell responses to the vector and to its transgene. J Immunol 159:685–693
Van Ommen GJB, Bakker E, Den Dunnen JT (1999) The human genome project and the future of diagnostics, treatment, and prevention. Lancet 354:5–10
Venugopalan P, Jain S, Sankar S et al (2002) pH sensitive liposomes: mechanism of triggered release to drug and gene delivery prospects. Pharmazai 57:659–671
Vorburger SA, Hunt KK (2002) Adenoviral gene therapy. Oncologist 7:46–59
Wadhwa MS, Knoell DL, Young AP (1995) Targeted gene delivery with a low molecular weight glycopeptide carrier. Bioconjug Chem 6:283–291
Walther W, Stein U (2000) Viral vectors for gene transfer: a review of their use in the treatment of human diseases. Drugs 60:249–271
Welch PJ, Yei S, Barber JR (1998) Ribozyme gene therapy for hepatitis C virus infection. Clin Diagn Virol 10:163–171
Wheeler CJ, Felgner PL, Tsai YJ (1996) A novel cationic lipid greatly enhances plasmid DNA delivery and expression in mouse lung. Proc Natl Acad Sci USA 93:11454–11459
Wolf JK, Jenkins AD (2002) Gene therapy for ovarian cancer. Int J Oncol 21:461–468

Wolff JA, Malone RW, Williams P (1990) Direct gene transfer into mouse muscle in vivo. Science 247:1465–1468
Wong IH (2001) Methylation profiling of human cancers in blood: molecular monitoring and prognostication. Int J Oncol 19:1319–1324
Wong TK, Neumann E (1982) Electric field mediated gene transfer. Biochem Biophys Res Commun 107:584–587
Xu L, Huang C-C, Huang W (2002) Systemic tumor-targeted gene delivery by anti-transferrin receptor scFv-immunoliposomes. Mol Cancer Ther 1:337–346
Yacyshyn BR, Bowen-Yacyshyn MB, Jewell L et al (1998) A placebo-controlled trial of ICAM-1 antisense oligonucleotide in the treatment of Crohn's disease. Gastroenterology 114:1133–1142
Yang Y, Jooss KU, Su Q (1996a) Immune responses to viral antigens versus transgene product in the elimination of recombinant adenovirus-infected hepatocytes in vivo. Gene Ther 3:137–144
Yang Y, Su Q, Wilson JM (1996b) Role of viral antigens in destructive cellular immune responses to adenovirus vector-transduced cells in mouse lungs. J Virol 70:7209–7212
Yi R, YiQ IGM et al (2003) Exportin-5 mediates the nuclear export of pre-microRNAs and short hairpin RNAs. Genes Dev 17:3011–3016
Zamecnik PC, Stephenson ML (1978) Inhibition of Rous sarcoma virus replication and cell transformation by a specific oligodeoxynucleotides. Proc Natl Acad Sci USA 75:280–284
Zamore PD, Aronin N (2003) siRNAs knock down hepatitis. Nat Med 9:266–267
Zhang L, Gasper WA, Stass SA et al (2002) Angiogenic inhibition mediated by a DNAzyme that target vascular endothelial growth factor receptor 2. Cancer Res 62:5463–5469
Zhao W, Kobayashi M, Hosokawa M et al (2002) Adenoviral vectors for cancer gene therapy. Curr Genomics 3:163–180
Zhaohui P (2003) China OKs gene therapy drug. Genetic Engineering News 6. http://www.p53therapy.com/news/. 19 Nov 2003
Zu Putlitz J, Yu Q, Burke JM et al (1999) Combinatorial screening and intracellular antiviral activity of hairpin ribozymes directed against hepatitis B virus. J Virol 73:5381–5387

Thio Effects as a Tool for Mechanistic Studies of the Cleavage of RNA Phosphodiester Bonds: The Chemical Basis

Mikko Ora, Tuomas Lönnberg, and Harri Lönnberg

Contents

Abstract Replacement of one of the phosphorus-bound oxygen atoms with sulfur has extensively been used for elucidation of mechanistic details of the cleavage of RNA phosphodiester bonds by ribozymes. Since sulfur atom is larger, less electronegative, and more readily polarizable than oxygen, this substitution affects in many ways metal ion binding and the ease of formation and breakdown of the phosphorane intermediate/transition state obtained by the attack of the entering hydroxyl group on the phosphorus atom. The factors that may be altered by thio substitution include the geometry of the phosphorane intermediate, relative apicophilicities of the ligands, the leaving group property, hydrogen bonding, solvation, and the affinity to metal ions. Experimental studies and theoretical calculations on various model systems have been undertaken to obtain a solid

M. Ora • T. Lönnberg • H. Lönnberg (✉)
Department of Chemistry, University of Turku, 20014 Turku, Finland
e-mail: harlon@utu.fi

V.A. Erdmann and J. Barciszewski (eds.), *From Nucleic Acids Sequences to Molecular Medicine*, RNA Technologies, DOI 10.1007/978-3-642-27426-8_3,

chemical basis for the mechanistic interpretations based on thio effects in ribozyme catalysis. The results of such studies are surveyed in this chapter.

Keywords Hydrolysis • Kinetics • Phosphorothioates • RNA • Thio effects

1 Introduction

Thio-substituted oligonucleotides have extensively been used as tools in mechanistic studies of ribonucleases and ribozymes to identify the stereochemistry and rate-limiting stage of the cleavage reaction, to discern dissociative and associative transition state structures, and to determine the binding sites of catalytically important metal ions. Replacement of one of the nonbridging oxygen atoms of the scissile phosphodiester linkage with sulfur, for example, allows the determination of the stereochemistry of the bond cleavage. The observed inversion of configuration around the phosphorus atom has lent considerable support for direct in-line attack with development of a pentacoordinate transition state or intermediate as a common mechanistic feature of the action of protein ribonucleases (Saenger et al. 1974) and small (Slim and Gait 1991; Koizumi and Ohtsuka 1991) and large (Rajagopal et al. 1989; Padgett et al. 1994) ribozymes. Bridging 5′-thiosubstitution has, in turn, been exploited to obtain evidence for the cleavage of the PO5′ bond in the rate-limiting step of the hammerhead-ribozyme-catalyzed reactions (Zhou et al. 1996, 1997). Comparisons of the magnitude kinetic thio effect in enzymatic and nonenzymatic reactions have been utilized in distinguishing between the alternative mechanisms proposed for the action of RNase A (Herschlag 1994; Hollfelder and Herschlag 1995). The most common applications of thio-substituted oligonucleotides, however, concern the so-called rescue effect. Ribozymes often require Mg^{2+} for full catalytic activity. Since Mg^{2+} prefers oxygen over sulfur as a donor atom, the rate retardation caused by replacement of a particular oxygen atom, a bridging or nonbridging one, with sulfur suggests that the interaction of Mg^{2+} with this particular oxygen atom is essential for the cleavage reaction. Restore of the diminished catalytic activity by a softer Lewis acid, usually by Mn^{2+}, Zn^{2+}, or Cd^{2+} that are more thiophilic than Mg^{2+}, is known as a rescue effect and regarded as a strong additional evidence for correct identification of the binding site. Such rescue effects naturally provide relevant information on the metal ion binding only on condition that the thio substitution does not markedly alter the conformation of the biologically active state. Although examples exist that this is not always the case (Smith and Nikonowicz 2000; Suzumura et al. 2004), the significance of rescue effects as a mechanistic tool cannot be denied. Examples of marked rate retardation that have been restored by thiophilic metal ions are numerous, including studies with hammerhead ribozymes (Knöll et al. 1997; Peracchi et al. 1997; Scott and Uhlenbeck 1999; Wang et al. 1999; Derrick et al. 2000; Osborne et al. 2005, 2009), group I (Herschlag et al. 1991; Piccirilli et al. 1993; Weinstein et al. 1997; Sjögren et al. 1997; Shan et al. 1999, 2001; Shan and Herschlag 1999, 2000; Yoshida et al. 1999, 2000; Szewczak et al. 2002; Forconi et al. 2008) and group II (Sontheimer et al. 1999; Gordon et al. 2000a, 2007) introns, P5 ribozyme of RNase P (Warnecke et al.

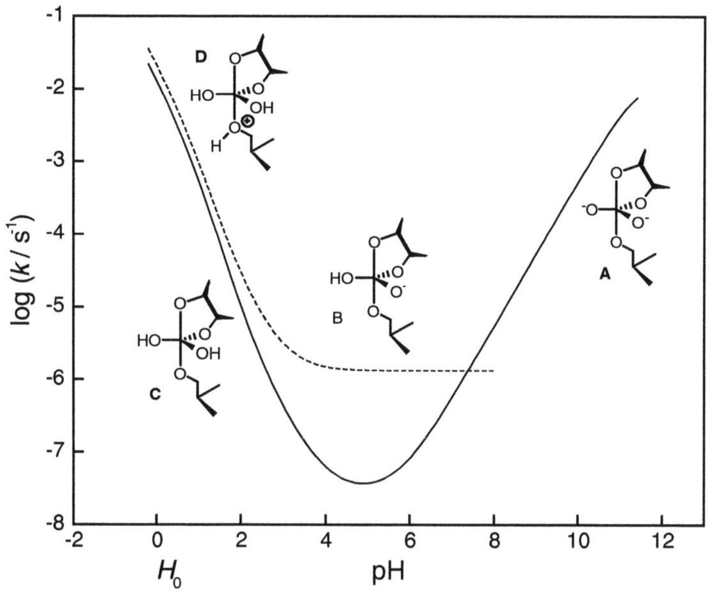

Fig. 1 pH-rate profile for the cleavage (*solid line*) and isomerization (*dotted line*) of the phosphodiester linkage of 3′,5′-UpU at 90°C (Järvinen et al. 1991). H_0 refers to Hammett acidity function (extension of pH scale)

1996, 1999; Chen et al. 1997; Pfeiffer et al. 2000; Christian et al. 2000, 2002), and spliceosome (Sontheimer et al. 1997; Gordon et al. 2000b; Yean et al. 2000). The results of these studies have recently been reviewed (Frederiksen and Piccirilli 2009; Lönnberg 2011b). This chapter tends to summarize the results of experimental and theoretical studies on various model systems undertaken to obtain a solid chemical basis for mechanistic interpretations of thio effects in ribozyme or ribonuclease catalysis.

2 The Effect of Nonbridging Thio Substitution on Metal-Ion-Independent Cleavage and Isomerization of RNA Phosphodiester Linkages

As known from early 1950s (Brown and Todd 1955), the internucleosidic phosphodiester bonds of RNA are hydrolytically much more labile than the corresponding linkages in DNA, owing to the presence of the 2′-OH that serves as an intramolecular nucleophile attacking the phosphorus atom of the neighboring phosphate moiety. Depending on pH, the reaction proceeds via a monocationic, neutral, monoanionic, or dianionic intermediate (Oivanen et al. 1998; Lönnberg and Lönnberg 2005; Lönnberg 2011a), as indicated in Fig. 1. The mechanisms of

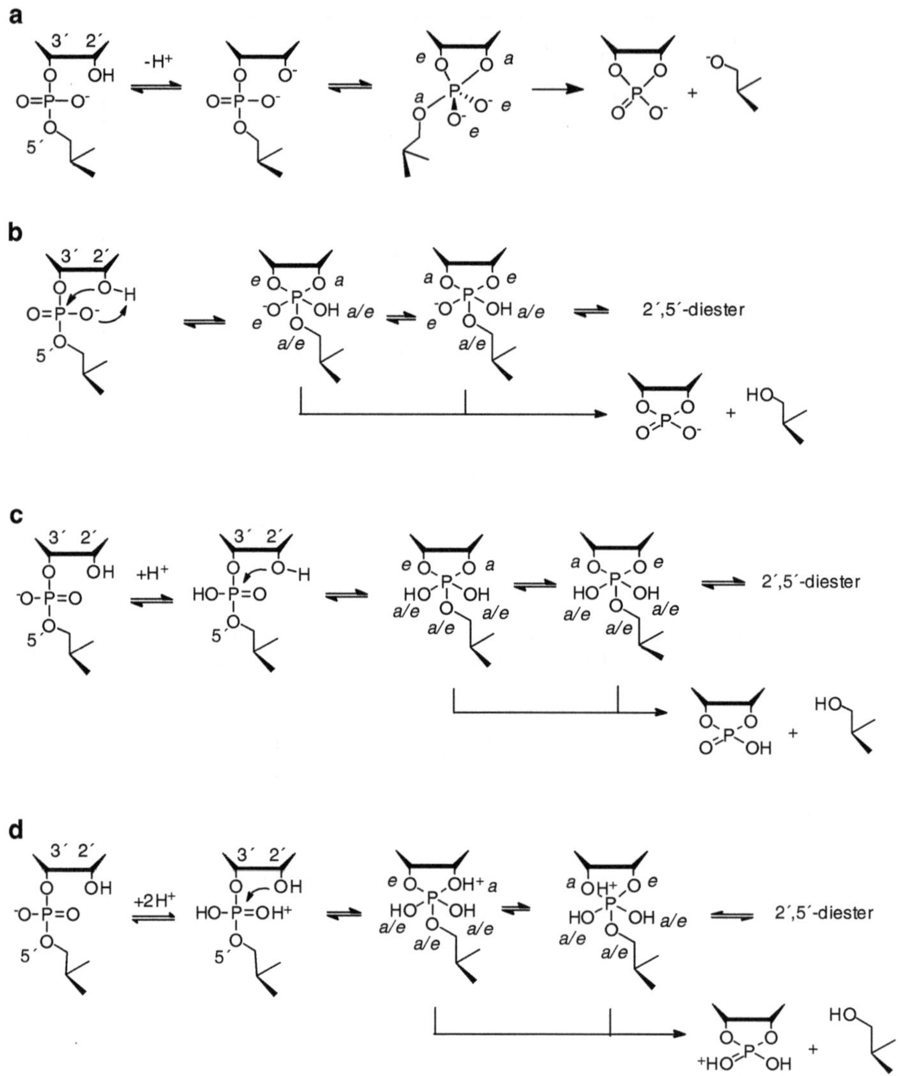

Scheme 1

these reactions are depicted in Scheme 1. The susceptibility of each reaction to thio substitution is different, and hence, the overall thio effect is pH dependent.

At pH $>$ 8, the predominant reaction is the attack of $2'$-O^- on the monoanionic phosphodiester linkage, followed by rate-limiting breakdown of the dianionic phosphorane by departure of the $5'$-linked nucleoside as an oxyanion (reaction A in Scheme 1) (Lönnberg et al. 2004). Studies with R_P (**1a**) and S_P (**1b**) phosphoromonothioate analogues of $3',5'$-UpU (Fig. 2) have indicated that the reaction proceeds with 100% inversion (Oivanen et al. 1995). The thio effect on the rate

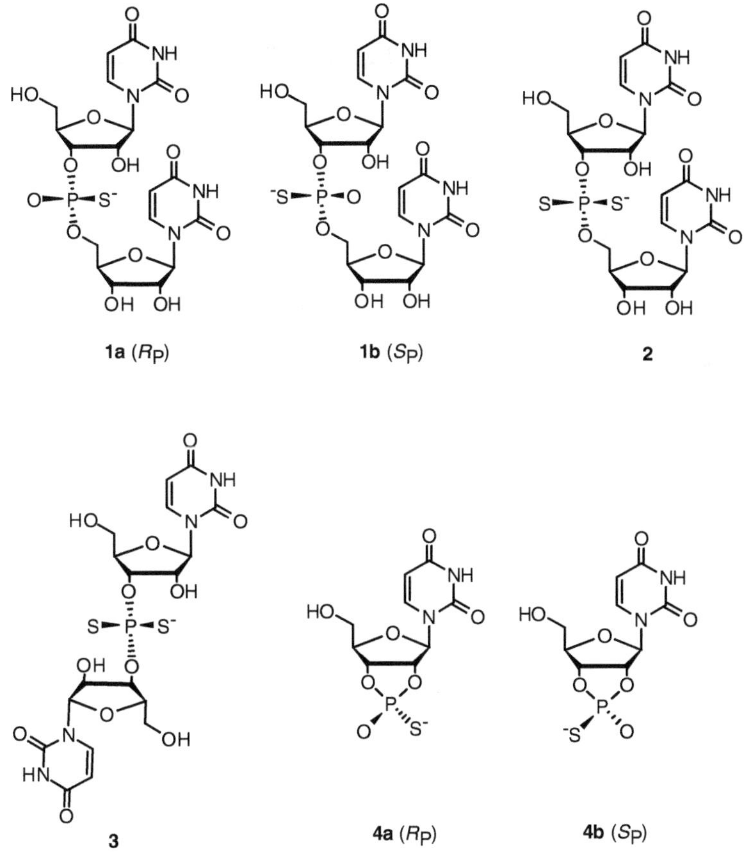

Fig. 2 Structures of ribonucleoside phosphorothioate diesters bearing a nonbridging sulfur atom

of this reaction is almost negligible, k_O/k_S being 1.3 and 0.8 for the R_P and S_P diastereomer, respectively (Oivanen et al. 1995; Almer and Strömberg 1991) (Table 1). According to semiempirical hybrid quantum mechanical/molecular mechanical (QM/MM) calculations (Gregersen et al. 2004) and density-functional theory (DFT) studies with explicit and implicit (continuum) solvation models (Liu et al. 2005, 2006), sulfur atom, being softer than oxygen, stabilizes the dianionic phosphorothioate intermediate. This electronic stabilization does not, however, result in any marked rate acceleration since it is counterbalanced by less efficient solvation of the thio-substituted species, leaving the free energy barriers for the formation and breakdown of the intermediate almost unchanged. Evidently, the transition state structures with phosphodiesters and phosphorothioates are quite similar, since both reactions exhibit almost identical susceptibility to the basicity of the leaving group. The β_{lg} values for the cleavage of uridine 3′-(alkyl phosphorothioate) R_P and S_P diastereomers are -1.23 ± 0.03 and -1.24 ± 0.03, respectively (Ora and Hanski 2011), and -1.28 ± 0.05 for their phosphate counterpart

Table 1 Thio effects, k_O/k_S, for the cleavage of nucleoside phosphoesters

Reaction	R_PS	S_PS	R and S_PS	$3'S$	$5'S$
Cyclization of 3′,5′-diesters to 2′,3′-cyclic esters					
OH^--catalyzed (reaction A)[a]	1.3	0.8	2.8	10^{-2}–10^{-3}	10^{-4}–10^{-5}
pH independent (reaction B)[a]	0.1	0.3	0.2		
H^+-catalyzed (reaction C)[a]	0.9	0.4	0.6		
H^+-catalyzed (reaction D)[a]	22	45			
Metal-ion-catalyzed[b] Mg^{2+}	1.5	0.7			
Mn^{2+}	1.6	1.2			
Zn^{2+}	0.08	0.2			
Cd^{2+}	0.005	0.006			
Isomerization of 3′,5′- and 2′,5′-diesters					
pH independent (reaction B)[a]	5	7	8		
H^+-catalyzed (reaction C)[a]	9	9	4		
H^+-catalyzed (reaction D)[a]	80	60			
Hydrolysis of 2′,3′-cyclic esters					
OH^--catalyzed	2.1	1.4		0.02	
pH independent	3.4	1.7			
H^+-catalyzed	16	11			
Reactions of 3′-(dimethyl phosphate)					
OH^--catalyzed cyclization	4[c]				
pH-independent cyclization	≈1				
OH^--catalyzed isomerization	4[c]			0.04	
Hydrolysis of 3′-monophosphate					
pH-independent reaction	0.004[c]				

[a]Scheme 1
[b]Scheme 2
[c]Achiral compound

(Kosonen et al. 1997). With 3′-(aryl phosphorothioate)s and phosphates, the susceptibility is again almost identical, −0.55 ± 0.05 (Almer and Strömberg 1996) and −0.54 ± 0.05 (Davis et al. 1988), respectively, but much lower than with the alkyl esters, suggesting that with aryl esters, the formation of the phosphorane intermediate rather than its breakdown is rate limiting (Lönnberg et al. 2004). The cleavage of aryl ester is additionally subject to general base catalysis. In other words, deprotonation of the attacking 2′-OH is partly rate limiting, taking place in concert with formation of the phosphorane intermediate. With these reactions, the thio effect is slightly more marked than with the specific base catalysis discussed above. The k_{PO}/k_{PS} values reported for uridine 3′-(aryl phosphorothioate)s fall in the range 1.2–2.6 and 1.7–3.6 with the R_P and S_P diastereomers, respectively (Almer and Strömberg 1996). On using imidazole as a catalyst, the $\beta_{lg} = -0.63 \pm 0.04$ is slightly more negative than with the specific acid catalysis, which is also the case with aryl phosphates ($\beta_{lg} = -0.59 \pm 0.08$).

Replacing the remaining nonbridging oxygen of phosphoromonothioates with sulfur has only a modest effect on the rate of the hydroxide-ion-catalyzed cleavage. The cleavage rate of the dithioate analogue of 3′,5′-UpU (**2**) is 36% of that of the parent compound, i.e., $k_O/k_S = 2.8$ (Ora et al. 2000). With the dithio analogue of 3′,3′-UpU (**3**), the thio effect is even smaller, $k_O/k_S = 1.5$ (Lönnberg et al. 2007). DFT calculations suggest that the phosphorane intermediate is slightly more stable than with phosphates or even with monothioates, but the resulting rate enhancement is compensated by increased barrier for the rate-limiting breakdown of the intermediate, and hence, the overall effect on rate remains small (Liu et al. 2006). In summary, as far as reaction A in Scheme 1 is concerned, nonbridging thio substitution exerts only a modest effect on the rate and transition state structure of RNA phosphodiester cleavage.

The situation is much more complicated under slightly acidic conditions (pH 5–7), where the cleavage takes place via a monoanionic phosphorane intermediate (reaction B). The monoanionic phosphorane is sufficiently stable to pseudorotate (Westheimer 1968), and this allows mutual isomerization of 3′,5′- and 2′,5′-phosphrothioate diesters (Oivanen et al. 1995). As expected on the basis of Westheimer's rules, the reaction proceeds by retention of configuration around phosphorus. It is also worth noting that the existence of a sufficiently stable phosphorane intermediate allows desulfurization. Upon formation of the thiophosphorane, the sulfur ligand must initially adopt an equatorial (*e*) position. The two apical positions (*a*) are occupied by the entering 2′-O and, most likely, the departing 5′O, which is more electronegative than the sulfur ligand. The proton of the attacking 2′-OH is evidently transferred, concerted with the nucleophilic attack, to the nonbridging oxygen rather than to sulfur, but it appears anyway to be equilibrated between the oxy- and thioanions. Since HS^- is five orders of magnitude less basic than HO^-, it is a much better leaving group and may depart instantly when being brought to an apical position by pseudorotation. Under neutral conditions, desulfurization, in fact, is the predominant reaction, representing 80% of the total disappearance of 3′,5′-Up(s)U, although no desulfurization takes place in aqueous alkali. The fast desulfurization makes accurate determination of the thio effects difficult. Thio substitution anyway seems to accelerate also the cleavage reaction, the values for k_O/k_S being 0.1 (R_P), 0.3 (S_P), and 0.2 (dithioate) (Oivanen et al. 1995; Ora et al. 2000). The isomerization is, in turn, markedly retarded, possibly due to the efficiently competing desulfurization, the values for k_O/k_S being 5 (R_P), 7 (S_P), and 8 (dithioate). According to DFT calculations, thio substitution does not appreciably affect the stability of the monoanionic phosphorane, but it influences, especially in the case of dithioate, on the timing of the proton transfer between the entering/departing nucleophile and the phosphorane oxy/thio anion compared to the formation/rupture of covalent bonds (Liu et al. 2005).

Desulfurization also predominates under more acidic conditions (pH 2–4), where the cleavage and isomerization mainly take place via a neutral phosphorane intermediate (reaction C). With the diasteromeric phosphoromonothioate analogues of 3′,5′-UpU, the thio effects are 0.9 (R_P), 0.4 (S_P), 0.6 (dithioate) for the cleavage and 9 (R_P and S_P) and 4 (dithioate) for the isomerization (Oivanen et al. 1995; Ora et al. 2000). On going to very acidic solutions (pH < 2), where RNA phosphodiester bonds are cleaved and isomerized via a monocationic intermediate (reaction D), the thio effects become much more prominent. The reactions of phosphoromonothioates are much slower than with phosphodiesters. The thio effects are 22 (R_P) and 45 (S_P) for the cleavage, and 80 (R_P) and 60 (S_P) for the isomerization (Oivanen et al. 1995). The marked retardation of the reactions via a monocationic phosphorane means that the cleavage and isomerization of phosphorothioates remain first order in hydronium ion up to pH 1, indicating that reaction C takes place under conditions where phosphodiesters utilize mechanism D. Another striking difference is that the isomerization is with phosphorothioates almost one order of magnitude slower than the cleavage, whereas phosphodiesters are isomerized even slightly faster than cleaved. The only reaction observed for dithioates at $H_0 < 0$ is cleavage (Ora et al. 2000). According to QM/MM calculations, the pseudorotation barriers of thiophosphoranes are comparable to those of the corresponding oxyphosphoranes. The thiophosphoranes, however, have longer lifetimes, and this may lead to differences in concentration ratios of migration/cleavage products between phosphates and phosphorothioates (Lopez et al. 2004).

3 The Effect of Nonbridging Thio Substitution on Metal-Ion-Dependent Cleavage of RNA Phosphodiester Linkages

The cleavage of RNA phosphodiester bonds is also catalyzed metal ion complexes (Lönnberg 2011a). Most likely, the phosphorane intermediate formed in a rapid pre-equilibrium stage is stabilized by a metal ion, and the aqua ligand of the phosphate bound metal ion then facilitates the rate-limiting departure of the 5′-linked nucleoside by donating a proton to the leaving oxygen atom (Scheme 2). Since sulfur is a considerably softer donor atom than oxygen, thio substitution has a marked effect on the affinity of various metal ions to the dianionic phosphorane intermediate and, hence, to the catalytic efficiency. Potentiometric studies with adenosine 5′-monophosphate and phosphoromonothioate dianion have indicated that thio substitution decreases the affinity of Mg^{2+} to dianionic phosphate group, the difference between the stability constants being 0.34 log units (Sigel et al. 1997). By contrast, Mn^{2+}, Zn^{2+}, and Cd^{2+} are more thiophilic: thio substitution increases the affinity by 0.2, 0.7, and 2.4 log units, respectively, the percentage of the sulfur-coordinated species being 30%, 80%, and 100%. Consistent with these findings, Mg^{2+}, Mn^{2+}, Zn^{2+}, and Cd^{2+} have been shown to accelerate the cleavage of 3′,5′-Up(s)U at 5 mmol L^{-1} concentration (pH 5.6 at 90°C) by a factor of 1.7, 6, 410, and 3,600, respectively (Ora et al. 1998). The acceleration was similar with both diastereomers. The respective accelerations for the cleavage of 3′,5′-UpU under the same conditions are 2.0, 12, 81, and 38 (Kuusela and Lönnberg 1993). Accordingly, the thio effect k_O/k_S is 1.2, 2.0, 0.2, 0.011 for Mg^{2+}, Mn^{2+}, Zn^{2+}, and Cd^{2+}. These results suggest that a really sizable rescue effect could be expected only with Cd^{2+} and to a lesser extent with Zn^{2+}. It is noteworthy that Mn^{2+}, though more

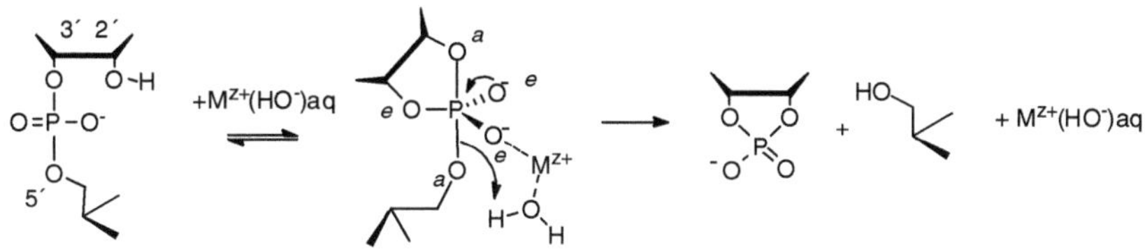

Scheme 2

thiophilic than Mg^{2+}, still appears to favor oxygen over sulfur. Another interesting thing is that Zn^{2+} has been shown to catalyze the isomerization and desulfurization of 3′,5′-Up(s)U, although much less efficiently than the cleavage, isomerization by a factor of 6.4 and desulfurization by a factor of 7.7 (Ora et al. 1998).

4 The Effect of Thio Substitution on the Hydrolysis Ribonucleoside 2′,3′-Cyclic Phosphates

The initial product of the cleavage of RNA phosphodiester bonds is a 2′,3′-cyclic monophosphate. As far as catalysis by Brönsted acids and bases is concerned, the subsequent hydrolysis to an approximately equimolar mixture of 2′- and 3′-monophosphates is so rapid that the appearance of the cyclic phosphodiester can hardly be detected (Järvinen et al. 1991). With ribozyme-catalyzed reactions, the situation is, however, different, and the effect of nonbridging thio substitution on the hydrolysis of 2′,3′-cyclic phosphate is of interest. In basic solutions (pH $>$ 9), where the reaction of 2′,3′-cUMP(S) (**4a,b**) proceeds by an attack on hydroxide ion on the phosphorothioate monoanion, the thio effect is modest, as with the corresponding reaction of internucleosidic phosphodiester linkage. The value of k_O/k_S is 2.1 and 1.4 with the R_P and S_P diastereomer, respectively (Ora et al. 1996a; Oivanen and Lönnberg 1989). The thio effects are considerably increased with the decreasing pH, again analogously to the reactions of RNA phosphodiester bonds, and desulfurization gradually becomes the predominant reaction. For the hydrolysis via monoanionic thiophosphorane, obtained by the attack of water on the monoanionic cyclic thiophosphate, the thio effects are 3.4 (R_P) and 1.7 (S_P). When the attack of water is preceded by pre-equilibrium protonation of the thiophosphate, the thio substitution is clearly rate retarding, k_O/k_S being 16 for the R_P and 11 for the S_P diastereomer. In spite of concurrent desulfurization, intermediary accumulation of cyclic phosphorothioate is observed under acidic conditions. It is also worth noting that the (R_P)-(**4a**) and (S_P)-(**4b**) 2′,3′-cyclic phosphorothioates, unlike their oxygen counterpart, undergo depyrimidination (up to 40%) concurrent with cleavage at pH 6–8 (Ora et al. 1996a).

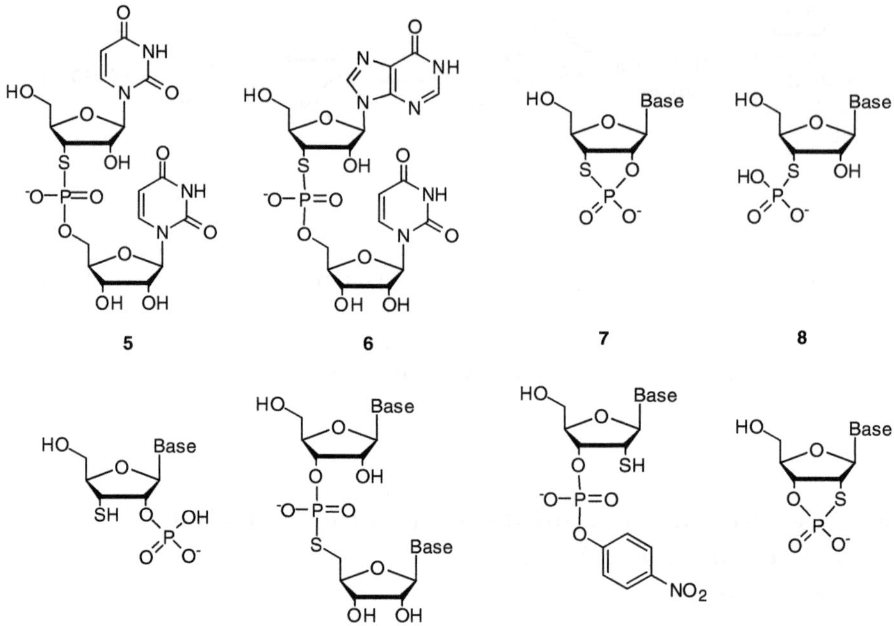

5 6 7 8 9 10 11 12

Fig. 3 Structures of ribonucleoside phosphorothioates containing a bridging sulfur atom and nucleoside phosphates bearing an unsubstituted 2′/3′-mercapto function

5 The Effect of Bridging Thio Substitution on the Cleavage and Isomerization of RNA Phosphodiester Linkage

The effect of 3′-thio substitution on the rate of the cleavage of RNA phosphodiester bonds is somewhat unexpected. Studies with 3′-deoxy-3′-thiouridylyl(3′,5′)uridine (**5**; Liu and Reese 1996) and 3′-deoxy-3′-thioinosylyl(3′,5′)uridine (**6**; Weinstein et al. 1996; Elzagheid et al. 1999) have shown that the hydroxide-ion-catalyzed cleavage is from two to three orders of magnitude faster than with 3′,5′-UpU (Fig. 3). Although sulfur is less electronegative element than oxygen and, hence, reduces the electron density on phosphorus less than oxygen, its higher polarizability compared to oxygen seems to play a more decisive role. According to QM/MM calculations with dynamics simulations in an explicit 20 Å sphere of water molecules (TIP3P model) surrounded by a solvent boundary potential, the 3′-thio substitution slightly stabilizes the thiophosphorane intermediate and lowers the barrier of the rate-limiting breakdown of the intermediate (Gregersen et al. 2004). As mentioned above, the observed thio effects consist of electronic and solvation contributions that tend to cancel one another. Evidently, the solvation effects are somewhat less important with the 3′-thio substitution compared to nonbridging thio substitution.

2′,3′S-Cyclic phosphorothioate (**7**) is clearly accumulated during the hydroxide-ion-catalyzed cleavage of dinucleoside(3′S,5′)monophospates, reaching its maximum 40% level at around one half-life of the cleavage of the starting material (Liu and Reese 1996; Elzagheid et al. 1999). In other words, the cleavage of 2′,3′-cyclic monophosphate is not accelerated by 3′-thio substitution as much as the cleavage of 3′,5′-phosphodiester. The cyclic phosphorothioate is cleaved entirely by PO2′ bond fission (**8**), evidently for the reason that the sulfur atom adopts an equatorial position upon formation of the thiophosphorane intermediate by the attack of hydroxide ion. The reaction is 60 times as fast as with the 2′,3′-cyclic monophosphate (Elzagheid et al. 1999).

Over a pH range 3–5, pH-independent isomerization of the 3′,5′-phosphorothioate linkage to the 2′,5′-phosphodiester linkage predominates, indicating that the monoanionic thiophosphorane is able to pseudorotate, bringing the 3′-sulfur into an apical position (Elzagheid et al. 1999). The reaction is 50 times as fast as with the oxygen counterparts. The acid-catalyzed cleavage and isomerization, proceeding via a neutral and possibly monocationic phosphorane, exhibit only modest thio effects, k_O/k_S being less than unity for the cleavage and greater than unity for the isomerization. The cyclic phosphorothioate is not accumulated under acidic conditions, and the final product is 3′-deoxy-3′-thionucleoside 2′-monophosphate (**9**), in striking contrast to the situation in aqueous alkali. Evidently, the 3′S is able to adopt an apical position within the neutral thiophosphorane obtained by the attack of water on neutral cyclic phosphorothioate formed as the initial product (Elzagheid et al. 1999). Interestingly, the product mixture obtained in glacial acetic acid is quite different, resembling that observed in aqueous alkali (Liu and Reese 2000).

The effect of 5′-thio substitution is expectedly even more dramatic. Thiols are five orders of magnitude more acidic than alcohols and, hence, according to the Hammond principles, better leaving groups. Diribonucleoside 3′,5′S-phosphoromonothioates (**10**) are cyclized under neutral and basic conditions to 2′,3′-cyclic monophosphate from four to five orders of magnitude faster than their oxygen counterparts (Liu and Reese 1995; Thomson et al. 1996). The small kinetic leaving group isotopic effect, $^{34}k_{lg} = 1.0009 + 0.0001$, obtained for the hydrolysis of O-(2-hydroxypropyl) S-(3-nitrobenzyl) phosphorothioate, a simple RNA model, suggests a very early transition state, where the cleavage of the PS bond is not appreciably advanced (Iyer and Hengge 2008). QM/MM calculations do not provide any indication of even a marginally stable intermediate, but suggest a concerted displacement of the 5′-sulfide anion by the 2′-oxyanion (Gregersen et al. 2004).

Cleavage of diribonucleoside 3′,5′S-phosphorothioates is also catalyzed by metal ions. The relative rates obtained at 10 mmol L^{-1} concentration of Mg^{2+}, Zn^{2+} and Cd^{2+} (pH 7.5) are 2.3-, 13-, and 8.3-fold compared to the rate in the absence of metal ions (Thomson et al. 1996). The catalysis is, hence, less efficient than with 3′,5′-UpU, for which relative rates 2.0, 81, and 38 have been reported at 5 mmol L^{-1} metal ion concentration (Kuusela and Lönnberg 1993). The fact that the rate-accelerating effect of the thiophilic metal ions, Zn^{2+} and Cd^{2+}, is so modest lends some support for the suggestion that the formation rather than the breakdown of the thiophosphorane intermediate is rate limiting. One should bear in mind that the non-bridging thio substitution accelerates the cleavage by a factor of 3,600 at $[Cd^{2+}] = 5.0$ mmol L^{-1} (Ora et al. 1998).

While sulfide ions attack more readily than oxyanions on carbon electrophiles, the opposite is true when attack on phosphorus atom is concerned. Accordingly, 2′-deoxy-2′-thionucleoside 3′-phosphodiesters (**11**) react by cyclization to 2′*S*,3′-cyclic phosphorothioates (**12**) only when the leaving group is exceptionally good, such as a 4-nitrophenyl group (Dantzman and Kiessling 1996). The attack of the 2′-sulfide ion has been estimated to be 10^7 times slower than the attack of 2′-oxyanion. The 2′,3′-cyclic phosphorothioate obtained is then hydrolyzed to 2′-phosphorothioate, suggesting that the thiophosphorane intermediate, even if marginally stable, is unable to pseudorotate. When cleavage of 2′-deoxy-2′-thiouridylyl(3′,5′)uridine was attempted under acidic conditions, only oxidation to a 2′,2′-disulfide took place (Reese et al. 1994). No cleavage or migration of the phosphodiester ligand could be detected. Under alkaline conditions, the *N*-glycosidic linkage was cleaved by an intramolecular attack of 2′-sulfide ion on the anomeric carbon atom.

6 Thio Effects on Reactions of Mono- and Triester Congeners of Ribonucleoside 3′-Phosphodiesters

The effects of thio substitution on the hydrolysis of the mono- and triesters of phosphoric acid are very different. Monoesters prefer a dissociative mechanism through a transition state where the PO bond fission is more advanced than the bond formation to the entering nucleophile (Hengge 1998). The thio substitution is clearly rate enhancing, the typical values of k_O/k_S being around 0.001 and 0.1 for the hydrolysis of the mono- and dianions, respectively (Catrina and Hengge 1999). Phosphotriesters, in turn, are hydrolyzed by an associative mechanism where the nucleophilic attack precedes the PO bond cleavage (Thatcher and Kluger 1989; Ba-Saif et al. 1990; Purcell and Hengge 2005). The thio effects, k_O/k_S, fall in the range 10–160 (Hengge 1998; Thatcher and Kluger 1989; Cox and Ramsey 1964). Accordingly, the magnitude of the thio effect seems to correlate with the relative timing of the entering and departure of the nucleophile and, hence, shed light in the transition state structures of enzyme-catalyzed reactions. For this reason, the thio effects observed for the cleavage of the mono- and triester congeners of ribonucleoside 3′-phosphodiesters are of interest.

Uridine 2′-, 3′-, and 5′-thiophosphates are all hydrolyzed to uridine under conditions where the monoanion is the predominant ionic form, i.e., at pH 2–5 (Ora et al. 1996b). The thio effect, k_O/k_S, is 0.003, 0.004, and 0.004 with the 2′-, 3′-, and 5′-phosphate, respectively (Ora et al. 1996b; Oivanen and Lönnberg 1990). In contrast to their oxygen counterparts, no thiophosphate migration takes place under these conditions. Neighboring hydroxyl function seems to participate in the

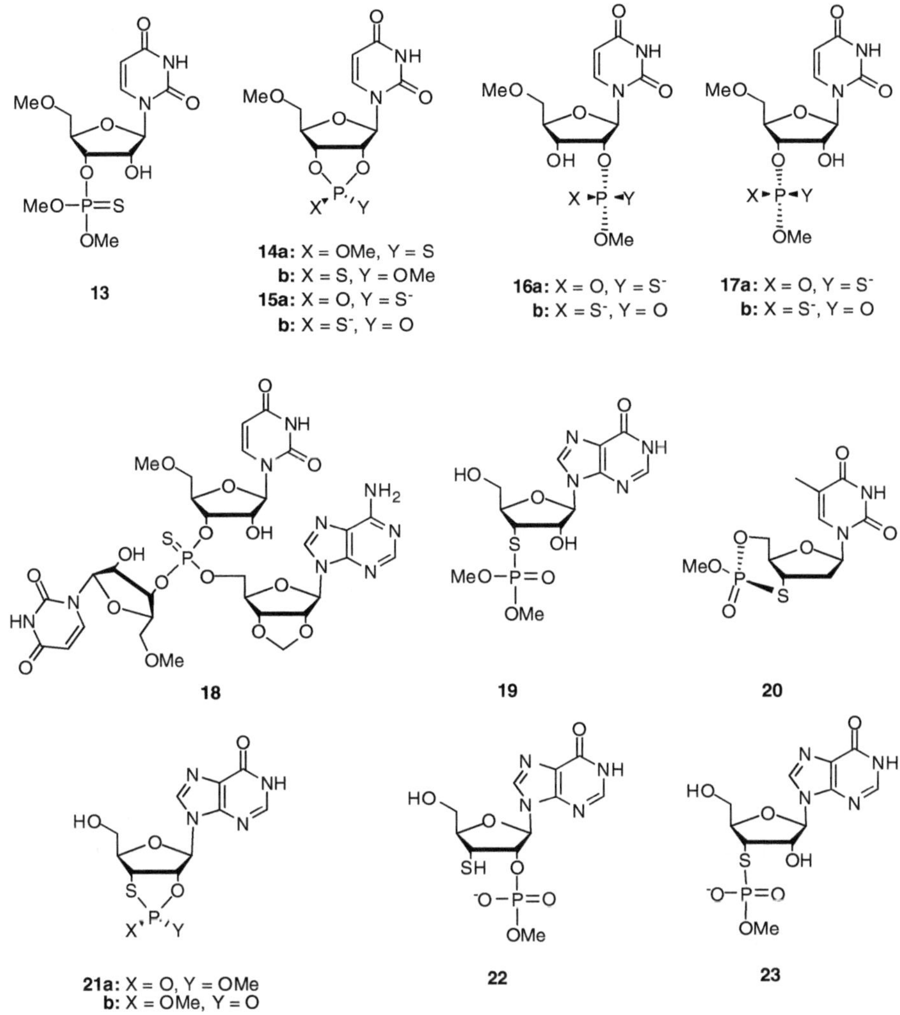

Fig. 4 Structures of nucleoside phosphorothioate triesters and their hydrolysis products

reaction only at pH < 1. Under so acidic conditions, 2′- and 3′-phosphorothioates undergo desulfurization to a mixture of 2′- and 3′-phosphates, probably via intermediary formation of a 2′,3′-cyclic monophosphate.

5′-Protected uridine 3′-(dimethyl phosphorothioate) (**13**), bearing a nonbridging sulfur atom, has been shown to undergo (Ora et al. 1997), analogously to its oxygen counterpart (Kosonen and Lönnberg 1995), two reactions over the entire pH range, viz., isomerization to the 2′-(dimethyl phosphoromonothioate) and hydrolysis to a mixture of diastereomeric 2′,3′-cyclic phosphoromonothioates (**15a,b**) and 2′-(**16a,b**) and 3′-(**17a,b**) (monomethyl phosphorothioates), evidently via a 2′,3′-cyclic phosphorothioate triester intermediate (**14a,b**) (Fig. 4). No desulfurization to 2′- or

3′-phosphate triesters occurs. The rate of the buffer-independent hydrolysis is pH independent over a wide pH range from pH 0 to 7, becoming hydronium and hydroxide ion catalyzed under more acidic and basic conditions, respectively. This pH-independent hydrolysis is approximately as fast as the hydrolysis of the corresponding phosphate triesters, whereas the hydroxide-ion-catalyzed reaction exhibits a marked thio effect, $k_O/k_S = 4$. The thio effect is thus larger than with the corresponding 3′-phosphorothioate diesters, but considerably smaller than the values reported for the alkaline hydrolysis of simple phosphotriesters, proceeding by an intermolecular attack of hydroxide ion. The isomerization is pH independent only over a narrow pH around pH 1 and turns hydroxide ion catalyzed already at pH > 2. The thio effects are comparable to those of the hydrolysis reaction.

A much greater thio effect has been observed for the hydrolysis of 3′,3′,5′-phosphoromonothioate triesters (**18**). For the hydroxide-ion-catalyzed reaction, which takes place by displacement of one of the nucleosides by the 2′-oxyanion, $k_O/k_S = 19$ (Lönnberg et al. 2007, 2005; Lönnberg and Korhonen 2005). In other words, the thio effect is almost fivefold compared to that referring to hydrolysis of uridine 3′-(dimethyl phosphoromonothioate), although both reactions proceed via rather similar phosphorane intermediates, incorporating a dioxaphospholane ring derived from uridine. The phosphate migration between the 2′- and 3′-hydroxy functions is too fast to allow accurate determination of the thio effect, but it appears to be greater than with the hydrolysis, approximately two orders of magnitude, consistent with the general trend that nonbridging thiosubstitution retards phosphate migration. The thio effect, k_O/k_S, for the pH-independent hydrolysis proceeding via a neutral phosphorane intermediate is 4, again greater than with the dimethyl analogue.

With 3′-deoxy-3′-thiouridine 3′-(dimethyl phosphorothioate) (**19**), the only reaction taking place at pH > 3 is the hydroxide-ion-catalyzed isomerization to the corresponding 2′-dimethylphosphate (Elzagheid et al. 2000a). This bridging thio substitution results is a 25-fold rate acceleration, which is about half of the 50-fold rate enhancement observed on replacing the 3′-oxygen of the *cis*-methyl ester of thymidine 3′,5′-cyclic monophosphate with sulfur (**20**; Elzagheid et al. 2000b). On going to more acidic solutions, hydrolysis to a mixture of 2′-(monomethyl phosphate) (**22**) and 3′-(monomethyl phosphorothioate) (**23**) gradually becomes the predominant reaction, the latter product being favored by the increasing hydronium ion concentration. Most likely, an attack of 2′-OH on the protonated triester monocation and concomitant departure of methanol gives protonated 2′,3′-cyclic phosphorothioate triester (**21a**,**b**) as an initial product, and this undergoes rather PO than PS bond cleavage. For comparison, studies with *O*-methyl *O*,*S*-ethylene phosphorothioate have shown that cleavage of the PS bond predominates at pH > 2, whereas 90% of the products are formed by rupture of the endocyclic PO at pH 0 (Gay and Hamer 1972). Evidently, oxygen rather than sulfur becomes protonated and is, hence, able to depart via an apical position more readily than sulfur.

In summary, while thio effects offer a useful tool for the mechanistic studies of ribozymes, extrapolation of the observations obtained with small molecular models to ribozyme catalysis is not always straightforward. It is important to bear in mind that replacing a phosphorus-bound oxygen atom with sulfur exerts two opposite effects on the reaction kinetics, viz., electronic stabilization of intermediates and transition states by better polarizability of sulfur compensated by less efficient stabilization of the thio-substituted species by solvation. In particular, the role of solvation may considerably be altered on going from catalysis by solvent-derived species to more structured microenvironment within ribozymes. Another source of misinterpretations originates from the use of thiophilic metal ions in rescue effect studies. The binding properties of thiophilic metal ions differ in many ways from those of Mg^{2+}, not only by higher affinity to sulfur, and this may cause significant structural reorganization of the catalytic center. In spite of these limitations, thio substitutions still constitute a most versatile tool for mechanistic studies of RNA cleavage.

References

Almer H, Strömberg R (1991) Intramolecular transesterification in thiophosphate-analogues of an RNA-dimer. Tetrahedron Lett 32:3723–3726

Almer H, Strömberg R (1996) Base catalysis and leaving group dependence in intramolecular alcoholysis of uridine 3′-(Aryl phosphorothioate)s. J Am Chem Soc 118:7921–7928

Ba-Saif SA, Waring MA, Williams A (1990) Single transition state in the transfer of a neutral phosphoryl group between phenoxide ion nucleophiles in aqueous solution. J Am Chem Soc 112:8115–8120

Brown DM, Todd AR (1955) Nucleic acids. Annu Rev Biochem 24:311–338

Catrina IE, Hengge AC (1999) Comparisons of phosphorothioate and phosphate monoester transfer reactions: activation parameters, solvent effects, and the effect of metal ions. J Am Chem Soc 121:2156–2163

Chen Y, Li X, Gegenheimer P (1997) Ribonuclease P catalysis requires Mg^{2+} coordinated to the *pro-*R_P oxygen of the scissile bond. Biochemistry 36:2425–2438

Christian EL, Kaye NM, Harris ME (2000) Helix P4 is a divalent metal ion binding site in the conserved core of the ribonuclease P ribozyme. RNA 6:511–519

Christian EL, Kaye NM, Harris ME (2002) Evidence for a polynuclear metal ion binding site in the catalytic domain of ribonuclease P RNA. EMBO J 21:2253–2262

Cox JR Jr, Ramsey OB (1964) Mechanisms of nucleophilic substitution in phosphate esters. Chem Rev 64:317–351

Dantzman CL, Kiessling LL (1996) Reactivity of a 2′-thio nucleotide analog. J Am Chem Soc 118:11715–11719

Davis AM, Hall AD, Williams A (1988) Charge description of base-catalyzed alcoholysis of aryl phosphodiesters: a ribonuclease model. J Am Chem Soc 110:5105–5108

Derrick WB, Greef CH, Caruthers MH et al (2000) Hammerhead cleavage of the phosphorodithioate linkage. Biochemistry 39:4947–4954

Elzagheid MI, Oivanen M, Klika KD et al (1999) Hydrolytic reactions of 3′-deoxy-3′-thioinosylyl (3′,5′)uridine: an RNA dinucleotide containing a 3′-*S*-phosphorothiolate linkage. Nucleosides Nucleotides 18:2093–2108

Elzagheid MI, Mäki E, Kaukinen U et al (2000a) Preparation, hydrolysis and intramolecular transesterification of 3′-deoxy-3′-thioinosine 3′-*S*-dimethylphosphorothiolate. Nucleosides Nucleotides Nucleic Acids 19:827–838

Elzagheid MI, Mattila K, Oivanen M et al (2000b) Hydrolytic reactions of the *cis*-methyl ester of 3′-deoxy-3′-thiotymidine 3′,5′-cyclic phosphorothiolate. Eur J Org Chem: 1987–1991
Forconi M, Lee J, Lee JK et al (2008) Functional identification of ligands for a catalytic metal ion in group I introns. Biochemistry 74:6883–6894
Frederiksen JK, Piccirilli JA (2009) Identification of catalytic metal ion ligands in ribozymes. Methods 49:148–166
Gay DC, Hamer NK (1972) Solvolysis of methyl *OS*-ethylene phosphorothioate. J Chem Soc Perkin Trans 2:929–932
Gordon PM, Sontheimer EJ, Piccirilli AJ (2000a) Kinetic characterization of the second step of group II intron splicing: role of metal ions and the cleavage site 2′-OH in catalysis. Biochemistry 39:12939–12952
Gordon PM, Sontheimer EJ, Piccirilli JA (2000b) Metal ion catalysis during the exon-ligation step of nuclear pre-mRNA splicing: extending the parallels between the spliceosome and group II introns. RNA 6:199–205
Gordon PM, Fong R, Piccirilli JA (2007) A second divalent metal ion in the group II intron reaction center. Chem Biol 14:607–612
Gregersen BA, Lopez X, York DM (2004) Hybrid QM/MM study of thio effects in transphosphorylation reactions: the role of solvation. J Am Chem Soc 126:7504–7513
Hengge AC (1998) Transfer of the PO_3^{2-} group. In: Sinnott M (ed) Comprehensive biological catalysis: a mechanistic reference, vol 1. Academic, New York, pp 517–542
Herschlag D (1994) Ribonuclease revisited: catalysis via the classical general acid-base mechanism or a triester-like mechanism? J Am Chem Soc 116:11631–11635
Herschlag D, Piccirilli JA, Cech TR (1991) Ribozyme-catalyzed and nonenzymic reactions of phosphate diesters: rate effects upon substitution of sulfur for a nonbridging phosphoryl oxygen atom. Biochemistry 30:4844–4854
Hollfelder F, Herschlag D (1995) The nature of the transition state for enzyme-catalyzed phosphoryl transfer. Hydrolysis of O-aryl phosphorothioates by alkaline phosphatase. Biochemistry 34:12255–12264
Iyer S, Hengge A (2008) The effects of sulfur substitutions for the nucleophile and bridging oxygen atoms in reactions of hydroxyalkyl phosphate esters. J Org Chem 73:4819–4829
Järvinen P, Oivanen M, Lönnberg H (1991) Interconversion and phosphoester hydrolysis of 2′,5′- and 3′,5′-dinucleoside monophosphates: kinetics and mechanisms. J Org Chem 56:5396–5401
Knöll R, Bald R, Fürste JP (1997) Complete identification of nonbridging phosphate oxygens involved in hammerhead cleavage. RNA 3:132–140
Koizumi M, Ohtsuka E (1991) Effects of phosphorothioate and 2-amino groups in hammerhead ribozymes on cleavage rates and Mg^{2+} binding. Biochemistry 30:5145–5150
Kosonen M, Lönnberg H (1995) General and specific acid/base catalysis of the hydrolysis and interconversion of ribonucleoside 2′- and 3′-phosphotriesters: kinetics and mechanisms of the reactions of 5′-*O*-pivaloyluridine 2′- and 3′-dimethylphosphates. J Chem Soc Perkin Trans 2:1203–1209
Kosonen M, Yousefi-Salakdeh E, Strömberg R et al (1997) Mutual isomerization of uridine 2′- and 3′-alkylphosphates and cleavage to a 2′,3′-cyclic phosphate: the effect of the alkyl group on the hydronium- and hydroxide-ion-catalyzed reactions. J Chem Soc Perkin Trans 2:2661–2666
Kuusela S, Lönnberg H (1993) Metal ions that promote the hydrolysis of nucleoside phosphoesters do not enhance intramolecular phosphate migration. J Phys Org Chem 6:347–356
Liu X, Reese CB (1995) Uridylyl-(3′ → 5′)-(5′-thiouridine). An exceptionally base-labile di-ribonucleoside phosphate analogue. Tetrahedron Lett 36:3413–3416
Liu X, Reese CB (1996) 3′-Thiouridylyl-(3′ → 5′)-uridine. Tetrahedron Lett 37:925–928
Liu X, Reese CB (2000) Preparation and cleavage reactions of 3′-thiouridylyl-(3′ → 5′)-uridine. J Chem Soc Perkin Trans 1:2227–2236
Liu Y, Gregersen BA, Lopez X et al (2005) Density functional study of the in-line mechanism of methanolysis of cyclic phosphate and thiophosphate esters in solution: insight into thio effects in RNA transesterification. J Phys Chem B 109:19987–20003

Liu Y, Gregersen BA, Hengge A et al (2006) Transesterification thio effects of phosphate diesters: free energy barriers and kinetic and equilibrium isotope effects from density-functional theory. Biochemistry 45:10043–10053

Lönnberg H (2011a) Cleavage of RNA phosphodiester bonds by small molecular entities: a mechanistic insight. Org Biomol Chem 9:1687–1703

Lönnberg T (2011b) Understanding catalysis of phosphate-transfer reactions by large ribozymes. Chem Eur J 17:7140–7153

Lönnberg T, Korhonen J (2005) Hydrolysis of 2′,3′-*O*-methyleneadenosin-5′-yl bis-5′-*O*-methyluridin-3′-yl phosphate: the 2′-hydroxy group stabilizes the phosphorane intermediate, not the departing 3′-oxyanion, by hydrogen bonding. J Am Chem Soc 127:7752–7758

Lönnberg T, Lönnberg H (2005) Chemical models for ribozyme action. Curr Opin Chem Biol 9:665–673

Lönnberg H, Strömberg R, Williams A (2004) Compelling evidence for a stepwise mechanism of the alkaline cyclisation of uridine 3′-phosphate esters. Org Biomol Chem 2:2165–2167

Lönnberg T, Kiiski J, Mikkola S (2005) Hydrolytic stability of 2′,3′-*O*-methyleneadenos-5′-yl 2′,5′-di-*O*-methylurid-3′-yl 5′-*O*-methylurid-3′(2′)-yl phosphate: implications to feasibility of existence of phosphate-branched RNA under physiological conditions. Org Biomol Chem 3:1089–1096

Lönnberg T, Ora M, Virtanen S et al (2007) Thio effects on the departure of the 3′-linked ribonucleoside from diribonucleoside 3′,3′-phosphorodithioate diesters and triribonucleoside 3′,3′,5′-phosphoromonothioate triesters: implications to ribozyme catalysis. Chem Eur J 13:4614–4627

Lopez CS, Faza ON, Gregersen BA et al (2004) Pseudorotation of natural and chemically modified biological phosphoranes: implications for RNA catalysis. ChemPhysChem 5:1045–1049

Oivanen M, Lönnberg H (1989) Kinetics and mechanisms for reactions of adenosine 2′- and 3′-monophosphates in aqueous acid: competition between phosphate migration, dephosphorylation, and depurination. J Org Chem 54:2556–2560

Oivanen M, Lönnberg H (1990) Kinetics of reactions of pyrimidine nucleoside 2′- and 3′-monophosphates under acidic and neutral conditions: concurrent phosphate migration, dephosphorylation and deamination. Acta Chem Scand B 44:239–242

Oivanen M, Ora M, Almer H et al (1995) Hydrolytic reactions of the diastereomeric phosphoromonothioate analogs of uridylyl(3′,5′)uridine: kinetics and mechanism for desulfurization, phosphoester hydrolysis and transesterification to the 2′,5′-isomers. J Org Chem 60:5620–5627

Oivanen M, Kuusela S, Lönnberg H (1998) Kinetics and mechanisms for the cleavage and isomerization of the phosphodiester bonds of RNA by Brönsted acids and bases. Chem Rev 98:961–990

Ora M, Hanski A (2011) Stepwise mechanism of hydroxide-ion catalyzed cyclization of uridine 3′-thiophosphate esters. Helv Chim Acta 94:1563–1574

Ora M, Oivanen M, Lönnberg H (1996a) Hydrolysis and desulfurization of the diastereomeric phosphoromonothioate analogs of uridine 2′,3′-cyclic monophosphate. J Org Chem 61:3951–3955

Ora M, Oivanen M, Lönnberg H (1996b) Hydrolytic dethiophosphorylation and desulfurization of the monothioate analogues of uridine monophosphates under acidic conditions. J Chem Soc Perkin Trans 2:771–774

Ora M, Oivanen M, Lönnberg H (1997) Phosphoester hydrolysis and intramolecular transesterification of ribonucleoside 2′- and 3′-phosphoromonothioate triesters: kinetics and mechanisms for the reactions of 5′-*O*-methyluridine 2′- and 3′-dimethylphosphoromonothioates. J Org Chem 62:3246–3253

Ora M, Peltomäki M, Oivanen M et al (1998) Metal-ion promoted cleavage, isomerization, and desulfurization of the diastereomeric phosphoromonothioate analogues of uridylyl(3′,5′)uridine. J Org Chem 63:2939–2947

Ora M, Järvi J, Oivanen M et al (2000) Hydrolytic reactions of the phosphorodithioate analogue of uridylyl(3′,5′)uridine: kinetics and mechanisms for the cleavage, desulfurization, and isomerization of the internucleosidic linkage. J Org Chem 65:2651–2657
Osborne EM, Schaak JE, DeRose VJ (2005) Characterization of a native hammerhead ribozyme derived from schistosomes. RNA 11:187–196
Osborne EM, Ward WL, Rühle MZ et al (2009) The identity of the nucleophile substitution may influence metal interactions with the cleavage site of the minimal hammerhead ribozyme. Biochemistry 48:10654–10664
Padgett RA, Podar M, Boulanger SC et al (1994) The stereochemical course of group II intron self-splicing. Science 266:1685–1688
Peracchi A, Beigelman L, Scott EC et al (1997) Involvement of a specific metal ion in the transition of the hammerhead ribozyme to its catalytic conformation. J Biol Chem 272:26822–26826
Pfeiffer T, Tekos A, Warnecke JM et al (2000) Effects of phosphorothioate modifications on precursor tRNA processing by eukaryotic RNase P enzymes. J Mol Chem 298:559–565
Piccirilli JA, Vyle JS, Caruthers MH et al (1993) Metal ion catalysis in the *Tetrahymena* ribozyme reaction. Nature 361:85–88
Purcell J, Hengge AC (2005) The thermodynamics of phosphate versus phosphorothioate ester hydrolysis. J Org Chem 70:8437–8442
Rajagopal J, Doudna JA, Szostak JW (1989) Stereochemical course of catalysis by the *Tetrahymena* ribozyme. Science 244:692–694
Reese CB, Simons C, Pei-Zhuo Z (1994) The synthesis of 2′-thiouridylyl-(3′ → 5′)-uridine. J Chem Soc Chem Commun N15:1809–1810
Saenger W, Suck D, Eckstein F (1974) On the mechanism of ribonuclease A. Eur J Biochem 46:559–567
Scott EC, Uhlenbeck OC (1999) A re-investigation of the thio effect at the hammerhead cleavage site. Nucleic Acids Res 27:479–484
Shan SO, Herschlag D (1999) Probing the role of metal ions in RNA catalysis: kinetic and thermodynamic characterization of a metal ion interaction with the 2′-moiety of the guanosine nucleophile in the *Tetrahymena* group I ribozyme. Biochemistry 38:10958–10975
Shan SO, Herschlag D (2000) An unconventional origin of metal-ion rescue and inhibition in the *Tetrahymena* group I ribozyme reaction. RNA 6:795–813
Shan SO, Yoshida A, Sun S et al (1999) Three metal ions at the active site of the *Tetrahymena* group I ribozyme. Proc Natl Acad Sci USA 96:12299–12304
Shan SO, Kravchuk AV, Piccirilli JA et al (2001) Defining the catalytic metal ion interactions in the *Tetrahymena* ribozyme reaction. Biochemistry 40:5161–5171
Sigel RKO, Song B, Sigel H (1997) Stabilities and structures of metal ion complexes of adenosine 5′-*O*-thiomonophosphate ($AMPS^{2-}$) in comparison with those of its parent nucleotide (AMP^{2-}) in aqueous solution. J Am Chem Soc 119:744–755
Sjögren AS, Pettersson E, Sjöberg BM et al (1997) Metal ion interaction with cosubstrate in self-splicing of group I introns. Nucleic Acids Res 25:648–653
Slim G, Gait MJ (1991) Configurationally defined phosphorothioate-containing oligoribonucleotides in the study of the mechanism of cleavage of hammerhead ribozymes. Nucleic Acids Res 19:1183–1188
Smith JS, Nikonowicz EP (2000) Phosphorothioate substitution can substantially alter RNA conformation. Biochemistry 39:5642–5652
Sontheimer EJ, Sun S, Piccirilli JA (1997) Metal ion catalysis during splicing of premessenger RNA. Nature 388:801–805
Sontheimer EJ, Gordon PM, Piccirilli JA (1999) Metal ion catalysis during group II intron self-splicing: parallels with the spliceosome. Genes Dev 13:1729–1741
Suzumura K-I, Takagi Y, Orita M et al (2004) NMR-based reappraisal of the coordination of a metal ion at the pro-*R*p oxygen of the A9/G10.1 site in a hammerhead ribozyme. J Am Chem Soc 126:15504–15511

Szewczak AA, Kosek AB, Piccirilli JA et al (2002) Identification of an active site ligand for a group I ribozyme catalytic metal ion. Biochemistry 41:2516–2525

Thatcher GRJ, Kluger R (1989) Mechanism and catalysis of nucleophilic-substitution in phosphate-esters. Adv Phys Org Chem 25:99–265

Thomson JB, Patel BK, Jimenez V et al (1996) Synthesis and properties of diuridine phosphate analogues containing thio and amino modifications. J Org Chem 61:6273–6281

Wang S, Karbstein K, Peracchi A et al (1999) Identification of the hammerhead ribozyme metal ion binding site responsible for rescue of the deleterious effect of a cleavage site phosphorothioate. Biochemistry 38:14363–14378

Warnecke JM, Furste JP, Hardt WD et al (1996) Ribonuclease P (RNase P) RNA is converted to a Cd(2+)-ribozyme by a single *R*p-phosphorothioate modification in the precursor tRNA at the RNase P cleavage site. Proc Natl Acad Sci USA 93:8924–8928

Warnecke JM, Held R, Busch S et al (1999) Role of metal ions in the hydrolysis reaction catalyzed by RNase P RNA from *Bacillus subtilis*. J Mol Biol 290:433–445

Weinstein LB, Earnshaw DJ, Cosstick R et al (1996) Synthesis and characterization of an RNA dinucleotide containing a 3′-*S*-phosphorothiolate linkage. J Am Chem Soc 118:10341–10350

Weinstein LB, Jones BCNM, Cosstick R et al (1997) A second catalytic metal ion in a group I ribozyme. Nature 388:805–808

Westheimer F (1968) Pseudo-rotation in the hydrolysis of phosphate esters. Acc Chem Res 1:70–78

Yean S-L, Wuenschell G, Termini J et al (2000) Metal-ion coordination by U6 small nuclear RNA contributes to catalysis in the spliceosome. Nature 408:881–884

Yoshida A, Sun S, Piccirilli JA (1999) A new metal ion interaction in the *Tetrahymena* ribozyme reaction revealed by double sulfur substitution. Nat Struct Biol 6:318–321

Yoshida A, Shan S, Herschlag D et al (2000) The role of the cleavage site 2′-hydroxyl in the *Tetrahymena* group I ribozyme reaction. Chem Biol 7:85–96

Zhou DM, Usman N, Wincott FE et al (1996) Evidence for the rate-limiting departure of the 5′-oxygen in nonenzymatic and hammerhead ribozyme-catalyzed reactions. J Am Chem Soc 118:5862–5866

Zhou DM, Zhang L-H, Taira K (1997) Explanation by the double-metal-ion mechanism of catalysis for the differential metal ion effects on the cleavage rates of 5′-oxy and 5′-thio substrates by a hammerhead ribozyme. Proc Natl Acad Sci USA 94:14343–14348

Off-Target Effects and Safety Aspects of Phosphorothioate Oligonucleotides

Martina Stessl, Christian R. Noe, and Johannes Winkler

Contents

Abstract In addition to the intended sequence-specific downregulation of the gene target, antisense and siRNA oligonucleotides can trigger off-target effects that are often caused by a particular chemical modification. The phosphorothioate backbone modification is dominating the clinical development of antisense agents, but reports about unspecific effects with clinical relevance, such as influence on blood clotting, activation of the complement system, and increasing levels of transaminases, shed doubt on its therapeutic value. In vitro mechanistic experiments have shown a class-related interaction of phosphorothioates with cellular pathways involved in apoptosis regulation, endoplasmatic reticulum stress, and glycolysis. Although clinical safety data indicate that severe adverse effects of

M. Stessl • C.R. Noe • J. Winkler (✉)
Department of Medicinal Chemistry, University of Vienna, Althanstrasse 14, 1090 Vienna, Austria
e-mail: johannes.winkler@univie.ac.at

V.A. Erdmann and J. Barciszewski (eds.), *From Nucleic Acids Sequences to Molecular Medicine*, RNA Technologies, DOI 10.1007/978-3-642-27426-8_4,

phosphorothioates are not dose limiting, concerns of unspecific effects, in particular hepatotoxicity, after long-term treatment remain.

Keywords Antisense • siRNA • Toxicology • Chemical modifications

1 Introduction

1.1 Oligonucleotide Therapeutics

When the concept of therapeutic gene silencing with antisense oligonucleotides was conceived three decades ago, unprecedented specificity—often referred to as a magic bullet—was thought to be possible by sequence-specific gene downregulation. Only after clinical developments were under way, it became increasingly clear that, similar to all therapeutic agents, relevant unspecific interactions with proteins or genes other than the intended target may well occur (Wagner 1994; Yamamoto et al. 2011). In retrospect, it seems that during the preclinical development for this new class of biotherapeutics, the possibility of off-target effects was not sufficiently taken into account, and studies of specificity may have been neglected to a certain degree. In particular, the properties conferred to oligonucleotides by the necessary chemical modifications have to be considered when choosing clinical development candidates.

There are several gene expression inhibition mechanisms described for antisense oligonucleotides (ASOs) (Bennett and Swayze 2010; Noe and Kaufhold 2000); the most relevant, however, is the degradation of the mRNA by RNase H. This endonuclease is located in the cytoplasm and in the nucleus and becomes activated at the site of ASO binding to mRNA. During the enzymatic attack, the ASO is undamaged functioning in a catalytic manner (Opalinska and Gewirtz 2002). Besides mRNA degradation, ASOs can inhibit gene functions by a steric block mechanism. This has been recently exploited for correcting aberrant splicing patterns by specifically hybridizing to pre-mRNA and modulating splicing (Bauman et al. 2009). These so-called splice-switching oligonucleotides should not activate RNase H, and therefore, chemical modifications which impede RNase H recognition such as phosphorodiamidate morpholino oligonucleotide (PMO) (Kinali et al. 2009) and locked nucleic acids (LNAs) (Guterstam et al. 2008) are preferred. The understanding of the antisense mechanism has led to the design of different chemical modifications of the phosphate backbone, the sugar components, or the nucleobases in order to improve the most important features of ASOs, such as binding affinity, base-pairing specificity, nuclease resistance, support of endonucleolytic mRNA cleavage, lipophilicity, and solubility, as well as pharmacological properties (Shukla et al. 2010; Urban and Noe 2003; Winkler et al. 2004, 2010).

Phosphorothioates, which are referred to as first-generation compounds contain a non-bridging sulfur substitution in the backbone phosphate linkage (Fig. 1). Second-generation ASOs are based on 2′-*O*-alkyl derivatives which have exhibited some superior in vitro characteristics. Prominent representatives are 2′-*O*-methoxyethyl

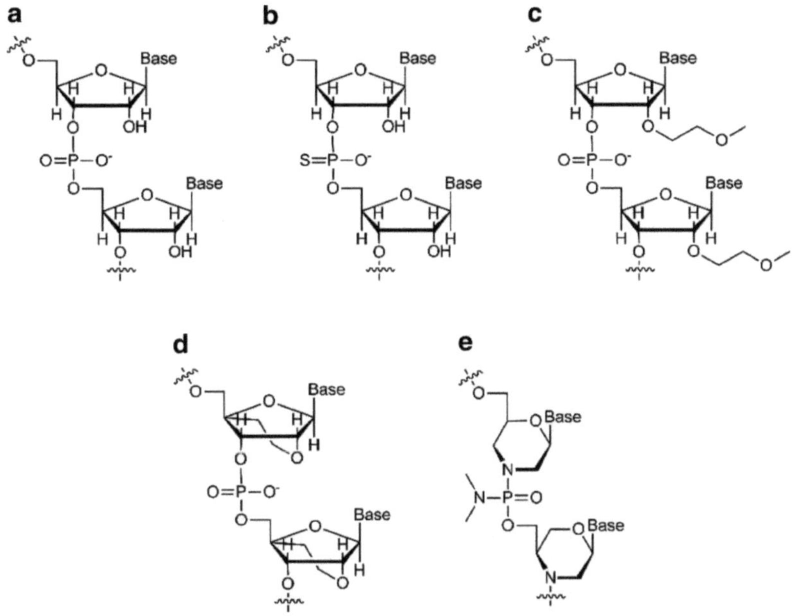

Fig. 1 Chemically modified antisense oligonucleotides used in clinical trials. (**a**) Wild-type DNA, (**b**) phosphorothioate (PS), (**c**) 2′-*O*-methoxy-ethyl-modified oligonucleotides (2′-MOE), (**d**) locked nucleic acids (LNAs), (**e**) phosphorodiamidate morpholino oligomers (morpholinos, PMOs)

(2′-MOE) derivatives showing increased resistance to plasma and tissue breakdown and improvements in the affinity with which they bind RNA (Prakash et al. 2008). Subsequently, more rigorous chemical modifications have been introduced in order to increase stability and target affinity, such as zwitterionic compounds (Winkler et al. 2008, 2009), PMOs (Gerety and Wilkinson 2011; Summerton 2007), and LNAs (Campbell and Wengel 2011; Jepsen and Wengel 2004) as the most promising therapeutic derivatives (Fig. 1). Because 2′-*O*-modified oligonucleotides possess RNA-like characteristics, they lose the capability for RNase H activation. Consequently, gapmer antisense agents were developed, exhibiting 2′-*O*-alkylated nucleosides only at the flanking regions with a stretch of at least eight 2′-deoxynucleosides linked by phosphorothioates in the center. For therapeutic applications, all 2′-*O*-alkylated antisense oligonucleotides and LNAs are partially phosphorothiolated (Table 1). The main reasons why phosphorothioates are being used in therapeutic oligonucleotides are the increase in stability and their uptake-enhancing properties. Recently, the unassisted cellular uptake has been investigated more closely, and the term "gymnotic delivery" has been created (Stein et al. 2010). It is a widespread but not universal phenomenon that phosphorothioates, unlike phosphodiester oligonucleotides, are able to cross cellular membranes and cause transient gene silencing in vitro after administration of relatively high

Table 1 Selected antisense agents in clinical development

ASO name	Target molecule	Chemical modification	Developer	Clinical status	Remarks	References
Oblimersen	Bcl-2	PS	Genta	Phase III	Tested against several tumor types	Bedikian et al. (2006, 2011)
Trabedersen	TGF-β2	PS	Antisense Pharma	Phase III	Application directly into brain tissue	Bogdahn et al. (2011)
Alicaforsen	ICAM-1	PS	ISIS/Atlantic	Phase II		Gewirtz and Sitaraman (2005)
Aprinocarsen	Protein kinase C-α	PS	ISIS/Lilly	Phase II		Paz-Ares et al. (2006)
Mipomersen	Apolipoprotein B	PS/2′-MOE	ISIS/Genzyme	Phase III	Drug application submitted for Europe (2011) and USA (2012)	Patel and Hegele (2010)
Custirsen	Clusterin	PS/2′-MOE	ISIS/Oncogenex	Phase III		Saad et al. (2011)
Miravirsen	miR-122	PS/LNA	Santaris	Phase II	Against hepatitis C	Lanford et al. (2010)
Eteplirsen	Dystrophin	PMO	AVI BioPharma	Phase II		Kinali et al. (2009)

PS phosphorothioate, *MOE* methoxyethyl, *PMO* phosphorodiamidate morpholino oligomer

concentrations. The effect is dependent on cell line and gene target and is a rather slow process. The exact mechanism and requirements are still unclear.

1.2 Clinical Development

The therapeutic spectrum of antisense agents actually tested in clinical phases includes cardiovascular disease, inflammation, infection, and, above all, cancer. Diseases that are characterized by overexpression or inappropriate expression of specific genes, or genes that are expressed by invading microorganisms, are candidates for this therapeutic approach. Malignant diseases, in particular, are attractive candidates, since conventional cancer therapies are highly toxic and affect also normal cells (Opalinska and Gewirtz 2002). Nevertheless, many clinical studies in the field of oncology failed to reach their endpoints, and so far, the antisense concept could not fulfill all expectations (Agarwala et al. 2009; Gekeler et al. 2006; Gjertsen et al. 2007; Pisano et al. 2008). Consequently, no systemically administered antisense oligonucleotide has yet succeeded in clinical trials. All antisense agents currently tested in the clinics feature the phosphorothioate backbone modification with the exception of PMOs. The most advanced substances, such as oblimersen, aprinocarsen, and trabedersen, belong to the first generation, while newer developments of the second generation like mipomersen have additional 2′-*O*-modifications.

2 Phosphorothioate Antisense Oligonucleotides

2.1 Chemical Background

Phosphorothioates (PSs) are more hydrophobic and acidic than wild-type DNA. The negative charge of a PS bond is polarized toward the sulfur atom, while the negative charge of a phosphodiester bond is shared equally by two non-bridging oxygens (Yamamoto et al. 2011). The mentioned substitution generates chirality at the phosphorus center (Rp and Sp, respectively), meaning that each PS internucleotide linkage produces 2^{n-1} stereoisomers of an n-mer PS (Krieg et al. 2003). Although the diastereomers are not equally effective, the differences do not warrant the much more complicated diastereoselective synthesis. Initial publications on the use of PS oligonucleotides (Agrawal et al. 1988; Matsukura et al. 1987) reported a significantly higher resistance of these compounds to nucleases compared to unmodified ASOs. PS oligonucleotides show reasonable stability of duplexes by Watson–Crick base pairs, acceptable resistance against nuclease degradation, and are able to activate RNase H—all essential properties for efficacious ASOs (Eckstein 2000). PSs are not ideal regarding the mentioned parameters and on the downside exhibit a high affinity for certain proteins, thus leading to non-sequence-specific adverse effects. Nevertheless, they dominate the clinical development either with or without additional 2′-*O*-modifications (Table 1).

2.2 Clinical Trials

Phosphorothioate are large molecules which usually harbor many solvent-exposed negative charges and as such cannot readily pass cellular membranes. PSs have been described to be taken to a certain extent up by receptor-mediated endocytosis at lower concentrations, but via fluid-phase endocytosis at higher concentrations (Mahato et al. 2005). The uptake appears to be an active, highly variable process dependent on cell type, time, concentration, energy, temperature, saturation, sequence, and type of the oligonucleotide, and its mechanism is not fully understood. The ASO needs to be released from endosomes after endocytosis, travel to the cell surface, cross the plasma membrane, and migrate to the cyto- or nucleoplasm in order to reach the target mRNA. Keeping these challenges in mind, it is not surprising that only one compound (fomivirsen, Vitravene®) has been approved (Grillone and Lanz 2001). For the treatment of cytomegalovirus (CMV) retinitis, fomivirsen is administered once monthly by intravitreal injection, providing local therapy for retinal infection, thereby preventing systemic toxicity. Examinations of ocular pharmacokinetics in rabbits and monkeys have revealed a favorable tissue distribution pattern and clearance rate after intravitreal injection (Henry et al. 2001). However, the application route and indication, as well as the fact that it is not the therapy of first choice, mean that it is by no means an example of a breakthrough new therapeutic principle.

Other phosphorothioates are in late-stage clinical development, such as oblimersen (G3139, Genasense®) (Bedikian et al. 2006, 2011), alicaforsen (ISIS 2302) (Gewirtz and Sitaraman 2005), mipomersen (Patel and Hegele 2010), and trabedersen (AP 12009) (Bogdahn et al. 2011), but none could meet the requirements for approval yet. The 18-mer oblimersen which is directed against the first six codons of Bcl-2, a key protein in the cellular apoptotic machinery, has undergone a variety of clinical phase II and III studies in various cancer types. Bcl-2 is critical for tumor cell survival, and its effective knockdown positively influences the resistance of human malignancies toward standard chemotherapy (Kang and Reynolds 2009). It has been tested mainly in combination with chemotherapeutic agents, based on data supporting a relevant role for Bcl-2 in cell survival and enhanced tumor response. An initial large phase III study on patients with advanced melanoma in combination with dacarbazine failed to reach its primary endpoint, overall survival (Bedikian et al. 2006). The recently concluded AGENDA trial (Agarwala et al. 2009) was designed after detailed analysis indicated a dependence of treatment susceptibility with serum lactate dehydrogenase (LDH) levels, but also ended in disappointment (Bedikian et al. 2011), and Genta has announced to conclude their efforts in developing oblimersen for the treatment of advanced melanoma.

Mipomersen (ISIS 301012) is a second-generation ASO targeted against apolipoprotein B and has been developed for hypercholesterolemia. Mipomersen takes advantage of the preferential accumulation of phosphorothioates in the liver. Four separate phase III trials have resulted in significant reduction of LDL-C of between 25% and 37%, as well as reductions in apolipoprotein B, and total cholesterol. Based on these results, Genzyme has filed for approval in 2011 in the European Union and 2012 in the USA. Adverse reactions included injection site reactions, elevation of liver transaminases, and accumulation of fat in the liver (Akdim et al. 2010).

2.3 *Pharmacokinetics and Toxicology*

Typical administration routes of PS oligonucleotides are intravenous injection and infusion. However, over the last decades, also topical application to the skin, enema, inhalation, and subcutaneous injection have been reported. Research efforts are advancing even oral formulations of oligonucleotides toward the clinic (Akhtar 2009). Pharmacokinetic properties of phosphorothioate oligonucleotides in humans are dose dependent and characterized by high plasma protein binding. High serum protein binding rates over 90% have been reported for PS oligonucleotides, mainly to albumin and α_2-macroglobulin. The affinity of single-stranded phosphorothioates to plasma proteins is higher than that of uncharged oligonucleotides such as PMOs and PNAs, but also than that of phosphodiesters (Bennett and Swayze 2010; Geary et al. 2001; Watanabe et al. 2006). This fact accounts for a long circulation time of phosphorothioates and prevents elimination by renal excretion. In addition, plasma protein binding may play an important role in tissue distribution by facilitating cellular uptake (Bennett and Swayze 2010). Phosphorothioates are distributed to nearly all organs and tissues, with the highest doses found in liver, kidney, and spleen (Levin 1999). Other organs accumulate oligonucleotides to a lower extent with the exception of the brain and testes, where endothelial barriers prevent distribution. Plasma half-life ranges between 30 and 90 min and is dependent on the dose (Geary 2009; Yu et al. 2001). The metabolism of PS oligonucleotides is mainly driven by exonucleases starting at the 3′ end. The parent compound is found as the major substance in serum. Pharmacokinetics and tissue distribution are significantly impacted by using distinct formulations, and the generation and characterization of many delivery systems based on liposomes (Wang et al. 2010), SNALPs (stable nucleic acid lipid particles) (Whitehead et al. 2009), or nanoparticles (Winkler 2011) have been reported.

In animal models and clinical trials, toxicological effects generally occur at high doses well over the therapeutically active doses and are of transient nature. They can be either sequence dependent or sequence independent (Bennett and Swayze 2010). Class-related toxicological effects after systemic administration are related to sequence-independent, nonspecific interactions with serum proteins and include complement activation, prolongation of the activated partial thromboplastin time (aPTT), proinflammatory action, and reactions at the injection site (Levin et al. 2001). Interaction with factor H, a circulating negative regulatory protein of the complement cascade, facilitates activation of the alternative pathway of the complement system. As a result, cardiovascular effects such as hypotension and bradycardia can occur, and some cases of lethal hemodynamic changes and cardiovascular collapse have been observed in rodents and monkeys (Henry et al. 2002). A second adverse effect observed in animals and humans is the transient prolongation of clotting time (aPTT), caused by inhibition of the intrinsic tenase complex (Sheehan and Lan 1998).

The liver is accumulating a large part of the administered phosphorothioate dose and is the predominant target organ for toxicity. After high-dose injections, reversible hepatic changes like Kupffer cell hypertrophy in rodents and monkeys were reported (Levin et al. 2001). Hepatic dysfunction, hepatocellular degeneration, and single cell necrosis, accompanied by increases of serum transaminases, were found in rare cases in animal models. A transient dose-dependent elevation of serum transaminases aspartate aminotransferase (AST) and alanine aminotransferase (ALT) is frequently observed in clinical trials with phosphorothioate antisense agents, such as mipomersen, an apolipoprotein B synthesis inhibitor (Akdim et al. 2010; Goel et al. 2006). ALT was increased over threefold the upper limit of normal in 8–20% of patients in phase III trials of mipomersen. Triglyceride accumulation in the liver was also reported (Patel and Hegele 2010) and raises concerns for possible development of steatohepatitis. Although there is no evidence of clinically relevant hepatotoxicity after short-term application, concerns for long-term treatment remain (Kling 2010; Visser et al. 2010).

Induction of immune responses is the most relevant sequence-dependent side effect. Toll-like receptor 9 has been identified to recognize specific CpG dinucleotide motifs of antisense oligonucleotides, leading to the rapid activation of cell signaling pathways, including mitogen-activated protein kinases and NF-κB (Gekeler et al. 2006; Krieg et al. 2003). CpG-containing ASOs have been shown to trigger humoral immunity by inducing B-cell activation, proliferation, and interleukin-6 and IgM secretion (Jurk et al. 2006).

3 Specificity of Phosphorothioates

3.1 Off-Target Effects

Despite the promising concept of the antisense strategy, highly specific inhibition of mRNA translation of the target protein, an increasing number of reports about off-target effects of these compounds have been published (Anderson et al. 2006; Lai et al. 2006; Pisano et al. 2008; Stein and Colombini 2008; Stessl et al. 2009; Winkler et al. 2010). Sequence-dependent unspecific effects can be caused by regulatory mechanisms in reply to the actual reduction in target protein expression itself or by antisense hybridization with related or nearly homologous sequences on nontarget mRNAs or miRNAs. Careful sequence choice and evaluation can usually eliminate nonspecific RNA hybridization (Bennett and Swayze 2010). A 20-mer oligonucleotide will usually hybridize to only one sequence in the genome with perfect complementarity. Three mismatches are believed to be sufficient to prevent hybridization to other, untargeted nucleic acid sequences, which can easily be verified by bioinformatical analysis. Sequence-independent or class-related effects of antisense oligonucleotides are caused in rare cases by alterations of endogenous metabolic pathways induced by metabolites of oligonucleotides. Finally and most importantly off-target effects can be based on unspecific interactions of antisense agents with proteins (Levin et al. 2001). In this context, the term "off-target effects" will be used for molecular effects based on interactions among gene knockdown agent and extracellular cell-surface or intracellular proteins.

3.2 Molecular Mechanisms of Off-Target Effects in Vitro

Phosphorothioates have been reported to strongly interact with some heparin-binding proteins, e.g., fibroblast growth factor (FGF), vascular endothelial growth factor (VEGF), or collagen I (Guvakova et al. 1995; Rockwell et al. 1997; Stein et al. 2009), and to induce unspecific apoptosis (Lai et al. 2005; Winkler et al. 2010). Some off-target effects may be caused by CpG motifs (Gekeler et al. 2006), but most can be attributed to the phosphorothioate chemistry (Stessl et al. 2009; Tan et al. 2007; Winkler et al. 2010).

Specific deoxycytidylyl-deoxyguanosine dinucleotide (CpG) motifs are present at relatively high frequency in bacterial DNA and in many viruses. In contrast to bacterial DNA, CpG motifs of vertebrates normally are methylated, but it has been shown that short oligonucleotides are capable of mimicking the toxicity immune-stimulatory effects of unmethylated bacterial sequences by interaction with TLR9 (Jurk et al. 2006) which is expressed especially in B and dendritic cells. Phosphorothioate oligonucleotides containing these CpG dinucleotides can trigger humoral immunity by inducing B-cell activation, proliferation, interleukin-6, and IgM secretion. CpG oligonucleotides can also directly activate monocytes and macrophages, leading to a secretion of cytokines, chemokines, and death receptor ligands and subsequently triggering lytic activity of natural killer cells (Krieg et al. 2003). Due to their immune-stimulatory effects as TLR9 agonists, a number of CpG-containing oligonucleotides have already entered the clinics for the potential treatment of cancer, asthma, and allergies and as vaccine adjuvants for infectious diseases (Zent et al. 2011). HIV-infected patients benefitted from enhanced vaccination efficiency, and antitumor activity could already be proven in phase I and II trials using CpG oligonucleotides as tumor vaccine adjuvant, monotherapy, or in combination with chemotherapy (Jurk and Vollmer 2007).

Unspecific effects of phosphorothioates caused by the backbone chemistry are still not well understood. The disappointing clinical results of oblimersen may be partially attributed to poor target selection (Pisano et al. 2008). Increased apoptotic resistance of cancer cell is likely not exclusively due to Bcl-2 upregulation, but rather caused by changes in complex cellular pathways. In addition, doubts about the molecular mechanisms and specificity of oblimersen have surfaced (Lai et al. 2005). A microarray gene profiling study investigating the phosphorothioate oblimersen and an siRNA-based gene knockdown in a side-by-side comparative analysis revealed that these two agents both effectively downregulated *bcl-2*, but were not equivalent as apoptotic/cytostatic agents and induced disparate phenotypes in PC3 and LNCaP prostate cancer cells (Anderson et al. 2006). Genes differentially regulated by oblimersen included *RASSF1* (belonging to a family of growth-inhibitory proteins capable of inducing cell cycle arrest, upregulated), *MAD* (part of a family of transcriptional repressors which inhibit Myc-directed cell growth and differentiation, upregulated), *DMTF1* (reported to induce growth arrest, upregulated), or *CTNNBL1* (a protein with proapoptotic C-terminus, which was observed to be downregulated). The different effects on cell proliferation and apoptosis together with the observed unique molecular signature of phosphorothioate, absent in all relevant controls and siRNA-based samples, indicated that the mechanism of action of the phosphorothioate is not exclusively the result of its action on Bcl-2 (Anderson et al. 2006).

Off-target effects observed on the genomic level are not necessarily the same on the proteome due to splice variants or posttranslational modifications. A comparison of the effects of oblimersen and a *bcl-2*-targeted siRNA on the protein level showed that the phosphorothioate significantly influenced 22 different proteins in addition to the target on the cytosolic proteome of human melanoma cells, whereas the siRNA sequence did not (Stessl et al. 2009). Both antisense strategies, however, did effectively downregulate the target protein to the same extent. Oblimersen downregulated proteins involved in apoptosis resistance, chaperones with a role in endoplasmatic reticulum stress (ER stress), and cytoskeletal proteins. Among the proteins with relevance to the therapeutic aim of oblimersen were calumenin, cofilin-1, galectin-1, glucose-regulated protein 78 (GRP78), heat shock protein 60 (HSP60), nucleophosmin, peroxiredoxins (PRDX) 1 and 6, and T-complex proteins (TCP) β and ε, as well as cytoskeleton proteins actin and tropomyosins (TPMs) 1 and 2 (Stessl et al. 2009). Other phosphorothioates, including alicaforsen, targeting ICAM-1, which plays an important role in inflammation processes and is not involved in the apoptosis cascade, and a nontargeted oligonucleotide, also significantly downregulated most of those proteins (with the exception of cytoskeleton proteins), while 2′-*O*-methylated gapmers with phosphodiester linkages did not (Winkler et al. 2010). Likewise, phosphorothiolated siRNA, unlike wild-type siRNA, showed similar effects on the proteome (Winkler et al. 2010). The reduction of apoptosis regulating proteins was associated with increased caspase activity.

Surprisingly, a number of glycolytic enzymes, namely, enolase-1 (ENOA1), phosphoglycerate mutase (PGAM), phosphoglycerate kinase (PGM), and triosephosphateisomerase (TPIS), were also found to be differentially expressed after phosphorothioate treatment (Stessl et al. 2009; Winkler et al. 2010). L-Lactate production was decreased after phosphorothioate treatment, verifying a glycolysis reversing effect. Cancer cells, unlike normal tissue, heavily rely on increased glycolysis even under abundant oxygen tension, which is widely known as the Warburg effect (Pedersen 2007). The findings indicate that the Warburg effect is partially reversed after treatment with the phosphorothioate (Winkler et al. 2010).

Independently of their target, all phosphorothioate sequences caused a significant reduction of a number of cancer markers linked to apoptosis regulation and endoplasmatic reticulum (ER) stress. Off-target effects observed after oblimersen treatment were slightly more pronounced than those of other phosphorothioates, indicating that the Bcl-2 reduction has a synergistic effect, possibly by a cross-link of Bcl-2 and ER chaperones. Phosphorothioate oligonucleotides were found to be potent and specific inhibitors of the voltage-dependent anion channel (VDAC) (Lai et al. 2006; Stein and Colombini 2008; Tan et al. 2007). This channel protein has been widely implicated in the initiation of the mitochondrially mediated intrinsic apoptosis pathway and leads to permeabilization of the mitochondrial outer membrane to small proteins and cytochrome c, followed by an activation of caspase-dependent apoptosis. VDAC is the binding site of hexokinase II, the glycolysis initiating enzyme that is pivotal in the Warburg effect in cancer cells. The interaction of a phosphorothioate might lead to an inhibition of the docking of hexokinase II to VDAC, resulting in a lower amount of ATP being transferred from the mitochondria for use in glycolysis.

The mechanistic studies signify that phosphorothioate-mediated unspecific effects are of multifactorial origin and seem to involve chaperones in the endoplasmatic reticulum and binding to the mitochondrial anion channel VDAC.

3.3 *In Vivo Correlation*

In the clinic, side effects after therapeutically active doses have been transient and rather mild. Many safety events such as increase of clotting time and complement activation take place in the bloodstream. Whether the data regarding off-target effects gathered in cell culture experiments have clinical relevance remains to be examined. In cell culture, oligonucleotides are applied together with a transfection enhancing agent. Unassisted cellular uptake in vivo likely leads to lower intracellular concentrations, and the uptake mechanism may play a role for the off-target effects. Clinical data indicate mild and transient toxic effects of phosphorothioates at sites of highest concentrations, namely, the injection site and the liver, reflected by a regularly observed increase in transaminase levels. This supports the concept of an unspecific apoptosis-inducing effect observed at tissues with high local phosphorothioate concentrations. As with other adverse effects, increase in transaminase levels was transient and manageable, so it is not necessarily a reason for major concern. However, clinical trials are being conducted over a limited time period and with a small patient cohort, and long-term effects in a more diverse set of patients will have to be monitored carefully (Kling 2010; Patel and Hegele 2010)

4 Off-Target Effects of Other Modifications

Second-generation antisense agents generally show similar toxicological profiles in preclinical and clinical settings, but often to a lower extent (Levin et al. 2001). This is likely due to the reduced number of phosphorothioate linkages and a higher stability conferred to by the flanking 2′-*O*-substitutions, which means less frequent injections. Hematologic toxicity and increases in serum transaminase levels were observed in clinical safety trials (Goel et al. 2006). In vitro, an apoptosis-inducing effect, which is dependent on the sequence, was reported (Drygin et al. 2004).

High doses of PS oligonucleotides with flanking LNA nucleotides were reported to induce hepatotoxicity in mice, indicated by substantial increase in serum transamines, and liver weight (Swayze et al. 2007). Signs of apoptosis were found in liver sections. In a different study, a number of genes were found to be differentially regulated in the liver of test animals (Mook et al. 2010). A correspondence between the specific activity and off-target effects was reported. Lower dosage seems not to induce hepatotoxicity, and safety studies in mice and monkeys concluded that LNA–PS hybrids can be safely applied in therapeutic settings (Gupta et al. 2010; Lanford et al. 2010).

Morpholino antisense oligonucleotides (PMOs) are widely being used in developmental studies with zebrafish (Bill et al. 2009), partly because siRNAs exhibit widespread off-target activity (Eisen and Smith 2008). Up to 20% of PMOs are estimated to induce unspecific apoptosis in the zebrafish embryo. This effect appears to be sequence dependent and is mediated by activation of the tumor suppressor protein p53. Consequently, phenotypes with increased cell death have to be examined with great care and proper control experiments (Gerety and Wilkinson 2011). In the clinic, PMOs are well tolerated up to the defined maximum dose showing mild accumulation in the kidney, but no indications of functional consequences (Cirak et al. 2011; Sazani et al. 2011). A trial using eteplirsen (AVI-4658) in an intramuscular injection for Duchenne muscular dystrophy has been initiated (Kinali et al. 2009). Because of the lack of an anionic backbone, morpholino oligonucleotides do not show protein interaction, and neither immune stimulation nor prolongation of clotting times nor complement activation has been encountered (Summerton 2007).

Phosphorothioate siRNA induced a similar phenotype in melanoma cells as antisense phosphorothioates (Winkler et al. 2010), indicating that the unspecific effects are analogous regardless of the nucleotide chemistry. Because the molecular effector of siRNA, the RNA-induced silencing complex (RISC), is sensitive to backbone modifications, only a limited number of phosphorothioate linkages are usually engineered into siRNAs for therapeutic purposes (Morrissey et al. 2005; Soutschek et al. 2004). A direct and relevant effect of phosphorothioate-induced off-target effects on siRNA therapeutics is therefore unlikely.

The main concern in the therapeutical development of siRNA-based agents is the activation of an immune response. Similar to the recognition of CpG motifs by TLR9, siRNAs are ligands for Toll-like receptors 7 and 8, expressed exclusively in endosomes (Sioud 2010). As a result, several hundred genes are affected by TLR activation, including interferons and proinflammatory cytokines, leading to significant unspecific effects. It is unclear which structural motif is responsible for TLR activation, but the innate immune response can be abrogated by either chemical modifications at the 2′-position (Fluiter et al. 2009) or by applying delivery systems which prevent localization of free siRNAs in endosomal compartments.

In addition, the RNAi mechanism harbors a higher risk for sequence-dependent off-target effects, because complementarity in the seed region of six to eight nucleotides can be sufficient for activating gene regulation (Burchard et al. 2009; Caffrey et al. 2011). This mechanism is exploited by naturally occurring miRNAs, which can interact with hundreds of different mRNAs and result in profound alterations of gene expression patterns (Huntzinger and Izaurralde 2011).

5 Summary

Clinical experience and molecular studies in vitro have changed the perception of the high specificity of antisense phosphorothioate agents. A number of relevant interactions with targets other than the intended gene have been shown to occur. For the development of highly specific therapeutic oligonucleotides, regardless if they are antisense or siRNA, the chemical modifications have to be selected carefully, and off-target effects should be excluded or minimized preferentially before initiating clinical testing. Newly available genomic and proteomic analyses in relevant preclinical models can be useful in the determination of off-target effects and the prediction of adverse drug reactions.

On the other hand, the elucidation of off-target effects of phosphorothioates may even open up new avenues for potential therapeutic approaches. The paradigm of one disease—one gene has been displaced by a systemic view on cellular pathways and regulatory mechanisms in diseases like cancer or autoimmune disorders. Pharmaceutical development is beginning to aim at several molecular targets in order to achieve synergistic effects and prevent evading mechanisms by shifting toward equivalent pathways. This can be reached either by designing therapeutic agents with multiple molecular targets or by combining highly specific agents. If the obstacles in the application of antisense and siRNA oligonucleotides can be overcome, nucleic acid-based therapeutics promise a tailor-made approach for efficient therapeutic combinations.

References

Agarwala SS, Keilholz U, Gilles E et al (2009) LDH correlation with survival in advanced melanoma from two large, randomised trials (oblimersen GM301 and EORTC 18951). Eur J Cancer 45:1807–1814

Agrawal S, Goodchild J, Civeira MP et al (1988) Oligodeoxynucleoside phosphoramidates and phosphorothioates as inhibitors of human immunodeficiency virus. Proc Natl Acad Sci USA 85:7079–7083

Akdim F, Stroes ESG, Sijbrands EJG et al (2010) Efficacy and safety of mipomersen, an antisense inhibitor of apolipoprotein B, in hypercholesterolemic subjects receiving stable statin therapy. J Am Coll Cardiol 55:1611–1618

Akhtar S (2009) Oral delivery of siRNA and antisense oligonucleotides. J Drug Target 17:491–495

Anderson EM, Miller P, Ilsley D et al (2006) Gene profiling study of G3139- and Bcl-2-targeting siRNAs identifies a unique G3139 molecular signature. Cancer Gene Ther 13:406–414

Bauman J, Jearawiriyapaisarn N, Kole R (2009) Therapeutic potential of splice-switching oligonucleotides. Oligonucleotides 19:1–13

Bedikian AY, Millward M, Pehamberger H et al (2006) Bcl-2 antisense (oblimersen sodium) plus dacarbazine in patients with advanced melanoma: the oblimersen melanoma study group. J Clin Oncol 24:4738–4745

Bedikian AY, Lebbé C, Robert C et al (2011) Survival in a phase III, randomized, double-blind study of dacarbazine with or without oblimersen (Bcl-2 antisense) in patients with advanced melanoma and low-normal serum lactate dehydrogenase (LDH; AGENDA). J Clin Oncol 29:8531

Bennett CF, Swayze EE (2010) RNA targeting therapeutics: molecular mechanisms of antisense oligonucleotides as a therapeutic platform. Annu Rev Pharmacol Toxicol 50:259–293

Bill BR, Petzold AM, Clark KJ et al (2009) A primer for morpholino use in zebrafish. Zebrafish 6:69–77

Bogdahn U, Hau P, Stockhammer G et al (2011) Targeted therapy for high-grade glioma with the TGF-beta2 inhibitor trabedersen: results of a randomized and controlled phase IIb study. Neuro Oncol 13:132–142

Burchard J, Jackson AL, Malkov V et al (2009) MicroRNA-like off-target transcript regulation by siRNAs is species specific. RNA 15:308–315

Caffrey DR, Zhao J, Song Z et al (2011) SiRNA off-target effects can be reduced at concentrations that match their individual potency. PLoS One 6:e21503

Campbell MA, Wengel J (2011) Locked vs. unlocked nucleic acids (LNA vs. UNA): contrasting structures work towards common therapeutic goals. Chem Soc Rev 40(12):5680–5689

Cirak S, Arechavala-Gomeza V, Guglieri M et al (2011) Exon skipping and dystrophin restoration in patients with duchenne muscular dystrophy after systemic phosphorodiamidate morpholino oligomer treatment: an open-label, phase 2, dose-escalation study. Lancet 378:595–605

Drygin D, Barone S, Bennett CF (2004) Sequence-dependent cytotoxicity of second-generation oligonucleotides. Nucleic Acids Res 32:6585–6594

Eckstein F (2000) Phosphorothioate oligodeoxynucleotides: what is their origin and what is unique about them? Antisense Nucleic Acid Drug Dev 10:117–121

Eisen JS, Smith JC (2008) Controlling morpholino experiments: don't stop making antisense. Development 135:1735–1743

Fluiter K, Mook OR, Baas F (2009) The therapeutic potential of LNA-modified siRNAs: reduction of off-target effects by chemical modification of the siRNA sequence. Methods Mol Biol 487:189–203

Geary RS (2009) Antisense oligonucleotide pharmacokinetics and metabolism. Expert Opin Drug Metab Toxicol 5:381–391

Geary RS, Watanabe TA, Truong L et al (2001) Pharmacokinetic properties of 2′-O-(2-methoxyethyl)-modified oligonucleotide analogs in rats. J Pharmacol Exp Ther 296:890–897

Gekeler V, Gimmnich P, Hofmann HP et al (2006) G3139 and other CpG-containing immunostimulatory phosphorothioate oligodeoxynucleotides are potent suppressors of the growth of human tumor xenografts in nude mice. Oligonucleotides 16:83–93

Gerety SS, Wilkinson DG (2011) Morpholino artifacts provide pitfalls and reveal a novel role for pro-apoptotic genes in hindbrain boundary development. Dev Biol 350:279–289

Gewirtz AT, Sitaraman SV (2005) Technology evaluation: alicaforsen (Isis). Curr Opin Mol Ther 7:273–281

Gjertsen BT, Bredholt T, Anensen N et al (2007) Bcl-2 antisense in the treatment of human malignancies: a delusion in targeted therapy. Curr Pharm Biotechnol 8:373–381

Goel S, Desai K, Macapinlac M et al (2006) A phase I safety and dose escalation trial of docetaxel combined with GEM231, a second generation antisense oligonucleotide targeting protein kinase A R1alpha in patients with advanced solid cancers. Invest New Drugs 24:125–134

Grillone LR, Lanz R (2001) Fomivirsen. Drugs Today 37:245–255

Gupta N, Fisker N, Asselin M-C et al (2010) A locked nucleic acid antisense oligonucleotide (LNA) silences PCSK9 and enhances LDLR expression in vitro and in vivo. PLoS One 5:e10682

Guterstam P, Lindgren M, Johansson H et al (2008) Splice-switching efficiency and specificity for oligonucleotides with locked nucleic acid monomers. Biochem J 412:307–313

Guvakova MA, Yakubov LA, Vlodavsky I et al (1995) Phosphorothioate oligodeoxynucleotides bind to basic fibroblast growth factor, inhibit its binding to cell surface receptors, and remove it from low affinity binding sites on extracellular matrix. J Biol Chem 270:2620–2627

Henry SP, Miner RC, Drew WL et al (2001) Antiviral activity and ocular kinetics of antisense oligonucleotides designed to inhibit CMV replication. Invest Ophthalmol Vis Sci 42:2646–2651

Henry SP, Beattie G, Yeh G et al (2002) Complement activation is responsible for acute toxicities in rhesus monkeys treated with a phosphorothioate oligodeoxynucleotide. Int Immunopharmacol 2:1657–1666

Huntzinger E, Izaurralde E (2011) Gene silencing by microRNAs: contributions of translational repression and mRNA decay. Nat Rev Genet 12:99–110

Jepsen JS, Wengel J (2004) LNA-antisense rivals siRNA for gene silencing. Curr Opin Drug Discov Dev 7:188–194

Jurk M, Vollmer J (2007) Therapeutic applications of synthetic CpG oligodeoxynucleotides as TLR9 agonists for immune modulation. BioDrugs 21:387–401

Jurk M, Kritzler A, Debelak H et al (2006) Structure-activity relationship studies on the immune stimulatory effects of base-modified CpG toll-like receptor 9 agonists. ChemMedChem 1:1007–1014

Kang MH, Reynolds CP (2009) Bcl-2 inhibitors: targeting mitochondrial apoptotic pathways in cancer therapy. Clin Cancer Res 15:1126–1132

Kinali M, Arechavala-Gomeza V, Feng L et al (2009) Local restoration of dystrophin expression with the morpholino oligomer AVI-4658 in Duchenne muscular dystrophy: a single-blind, placebo-controlled, dose-escalation, proof-of-concept study. Lancet Neurol 8:918–928

Kling J (2010) Safety signal dampens reception for mipomersen antisense. Nat Biotechnol 28:295–297

Krieg AM, Guga P, Stec W (2003) P-chirality-dependent immune activation by phosphorothioate CpG oligodeoxynucleotides. Oligonucleotides 13:491–499

Lai JC, Benimetskaya L, Khvorova A et al (2005) Phosphorothioate oligodeoxynucleotides and G3139 induce apoptosis in 518A2 melanoma cells. Mol Cancer Ther 4:305–315

Lai JC, Tan W, Benimetskaya L et al (2006) A pharmacologic target of G3139 in melanoma cells may be the mitochondrial VDAC. Proc Natl Acad Sci USA 103:7494–7499

Lanford RE, Hildebrandt-Eriksen ES et al (2010) Therapeutic silencing of microRNA-122 in primates with chronic hepatitis C virus infection. Science 327:198–201

Levin AA (1999) A review of the issues in the pharmacokinetics and toxicology of phosphorothioate antisense oligonucleotides. Biochim Biophys Acta 1489:69–84

Levin AA, Henry SP, Monteith D et al (2001) Toxicity of antisense oligonucleotides. In: Crooke ST (ed) Antisense drug technology: principles, strategies and applications. Dekker, New York, NY, pp 201–267

Mahato RI, Cheng K, Guntaka RV (2005) Modulation of gene expression by antisense and antigene oligodeoxynucleotides and small interfering RNA. Expert Opin Drug Deliv 2:3–28

Matsukura M, Shinozuka K, Zon G et al (1987) Phosphorothioate analogs of oligodeoxynucleotides: inhibitors of replication and cytopathic effects of human immunodeficiency virus. Proc Natl Acad Sci USA 84:7706–7710

Mook O, Vreijling J, Wengel SL et al (2010) In vivo efficacy and off-target effects of locked nucleic acid (LNA) and unlocked nucleic acid (UNA) modified siRNA and small internally segmented interfering RNA (sisiRNA) in mice bearing human tumor xenografts. Artif DNA PNA XNA 1:36–44

Morrissey DV, Lockridge JA, Shaw L et al (2005) Potent and persistent in vivo anti-HBV activity of chemically modified siRNAs. Nat Biotechnol 23:1002–1007

Noe CR, Kaufhold L (2000) Chemistry of antisense oligonucleotides. Methods Princ Med Chem 7:261–347

Opalinska JB, Gewirtz AM (2002) Nucleic-acid therapeutics: basic principles and recent applications. Nat Rev Drug Discov 1:503–514

Patel N, Hegele RA (2010) Mipomersen as a potential adjunctive therapy for hypercholesterolemia. Expert Opin Pharmacother 11:2569–2572

Paz-Ares L, Douillard JY, Koralewski P et al (2006) Phase III study of gemcitabine and cisplatin with or without aprinocarsen, a protein kinase C-alpha antisense oligonucleotide, in patients with advanced-stage non-small-cell lung cancer. J Clin Oncol 24:1428–1434

Pedersen P (2007) Warburg, me and hexokinase 2: multiple discoveries of key molecular events underlying one of cancers' most common phenotypes, the "Warburg effect," i.e., elevated glycolysis in the presence of oxygen. J Bioenerg Biomembr 39:211–222

Pisano M, Baldinu P, Sini MC et al (2008) Targeting Bcl-2 protein in treatment of melanoma still requires further clarifications. Ann Oncol 19:2092–2093

Prakash TP, Kawasaki AM, Wancewicz EV et al (2008) Comparing in vitro and in vivo activity of 2′-O-[2-(methylamino)-2-oxoethyl]- and 2′-O-methoxyethyl-modified antisense oligonucleotides. J Med Chem 51:2766–2776

Rockwell P, O'Connor WJ, King K et al (1997) Cell-surface perturbations of the epidermal growth factor and vascular endothelial growth factor receptors by phosphorothioate oligodeoxynucleotides. Proc Natl Acad Sci USA 94:6523–6528

Saad F, Hotte S, North S et al (2011) Randomized phase II trial of custirsen (OGX-011) in combination with docetaxel or mitoxantrone as second-line therapy in patients with metastatic castrate-resistant prostate cancer progressing after first-line docetaxel: CUOG trial P-06c. Clin Cancer Res 17:5765–5773

Sazani P, Van Ness KP, Weller DL et al (2011) Repeat-dose toxicology evaluation in cynomolgus monkeys of AVI-4658, a phosphorodiamidate morpholino oligomer (PMO) drug for the treatment of duchenne muscular dystrophy. Int J Toxicol 30:313–321

Sheehan JP, Lan H-C (1998) Phosphorothioate oligonucleotides inhibit the intrinsic tenase complex. Blood 92:1617–1625

Shukla S, Sumaria CS, Pradeepkumar PI (2010) Exploring chemical modifications for siRNA therapeutics: a structural and functional outlook. ChemMedChem 5:328–349

Sioud M (2010) Recent advances in small interfering RNA sensing by the immune system. N Biotechnol 27:236–242

Soutschek J, Akinc A, Bramlage B et al (2004) Therapeutic silencing of an endogenous gene by systemic administration of modified siRNAs. Nature 432:173–178

Stein CA, Colombini M (2008) Specific VDAC inhibitors: phosphorothioate oligonucleotides. J Bioenerg Biomembr 40:157–162

Stein CA, Wu S, Voskresenskiy AM et al (2009) G3139, An anti-Bcl-2 antisense oligomer that binds heparin-binding growth factors and collagen I, alters in vitro endothelial cell growth and tubular morphogenesis. Clin Cancer Res 15:2797–2807

Stein CA, Hansen JB, Lai J et al (2010) Efficient gene silencing by delivery of locked nucleic acid antisense oligonucleotides, unassisted by transfection reagents. Nucleic Acids Res 38:e3

Stessl M, Marchetti-Deschmann M, Winkler J et al (2009) A proteomic study reveals unspecific apoptosis induction and reduction of glycolytic enzymes by the phosphorothioate antisense oligonucleotide oblimersen in human melanoma cells. J Proteomics 72:1019–1030

Summerton JE (2007) Morpholino, siRNA, and S-DNA compared: impact of structure and mechanism of action on off-target effects and sequence specificity. Curr Top Med Chem 7:651–660

Swayze EE, Siwkowski AM, Wancewicz EV et al (2007) Antisense oligonucleotides containing locked nucleic acid improve potency but cause significant hepatotoxicity in animals. Nucleic Acids Res 35:687–700

Tan W, Loke Y-H, Stein CA et al (2007) Phosphorothioate oligonucleotides block the VDAC channel. Biophys J 93:1184–1191

Urban E, Noe CR (2003) Structural modifications of antisense oligonucleotides. Farmaco 58:243–258

Visser ME, Kastelein JJ, Stroes ES (2010) Apolipoprotein B synthesis inhibition: results from clinical trials. Curr Opin Lipidol 21:319–323

Wagner RW (1994) Gene inhibition using antisense oligodeoxynucleotides. Nature 372:333–335. doi:10.1038/372333a0

Wang J, Lu Z, Wientjes M et al (2010) Delivery of siRNA therapeutics: barriers and carriers. AAPS J 12:492–503

Watanabe TA, Geary RS, Levin AA (2006) Plasma protein binding of an antisense oligonucleotide targeting human ICAM-1 (ISIS 2302). Oligonucleotides 16:169–180

Whitehead KA, Langer R, Anderson DG (2009) Knocking down barriers: advances in siRNA delivery. Nat Rev Drug Discov 8:129–138

Winkler J (2011) Nanomedicines based on recombinant fusion proteins for targeting therapeutic siRNA oligonucleotides. Ther Deliv 2:891–905

Winkler J, Urban E, Losert D et al (2004) A novel concept for ligand attachment to oligonucleotides via a 2′-succinyl linker. Nucleic Acids Res 32:710–718

Winkler J, Gilbert M, Kocourková A et al (2008) 2′-O-lysylaminohexyl oligonucleotides: modifications for antisense and siRNA. ChemMedChem 3:102–110

Winkler J, Saadat K, Díaz Gavilán M et al (2009) Oligonucleotide polyamine conjugates: influence of length and position of 2′-attached polyamines on duplex stability and antisense effect. Eur J Med Chem 44:670–677

Winkler J, Stessl M, Amartey J et al (2010) Off-target effects related to the phosphorothioate modification of nucleic acids. ChemMedChem 5:1344–1352

Yamamoto T, Nakatani M, Narukawa K et al (2011) Antisense drug discovery and development. Future Med Chem 3:339–365

Yu RZ, Geary RS, Leeds JM et al (2001) Comparison of pharmacokinetics and tissue disposition of an antisense phosphorothioate oligonucleotide targeting human Ha-ras mRNA in mouse and monkey. J Pharm Sci 90:182–193

Zent CS, Smith BJ, Ballas ZK et al (2011) A phase I clinical trial of CpG oligonucleotide 7909 (PF-03512676) in patients with previously treated chronic lymphocytic leukemia. Leuk Lymphoma 53(2):211–217

Oligonucleotide Conjugates: Rationale, Synthesis, and Applications

Yashveer Singh, Pierre Murat, Nicolas Spinelli, and Eric Defrancq

Contents

Y. Singh (✉)
Department of Pharmaceutics, Ernest Mario School of Pharmacy, Rutgers, The State University of New Jersey, 160 Frelinghuysen Road, Piscataway, NJ 08854-8022, USA
e-mail: yashveer@rci.rutgers.edu

P. Murat
Department of Chemistry, The University of Cambridge, Cambridge CB2 1EW, UK

N. Spinelli • E. Defrancq (✉)
Département de Chimie Moléculaire, UMR CNRS 5250, Université Joseph Fourier, BP 53-38041, Grenoble cedex 9, France
e-mail: Eric.Defrancq@ujf-grenoble.fr

V.A. Erdmann and J. Barciszewski (eds.), *From Nucleic Acids Sequences to Molecular Medicine*, RNA Technologies, DOI 10.1007/978-3-642-27426-8_5,

Abstract Synthetic oligonucleotides can cause specific inhibition of gene expression by a variety of mechanisms (e.g., antigene, antisense, siRNA). Such oligonucleotides can also cause catalytic cleavage of the target sequence (e.g., ribozyme, DNAzyme) and selectively bind to the target molecules (e.g., aptamers). However, oligonucleotides possess unfavorable pharmacokinetic and pharmacodynamic properties, like extremely short plasma half-life due to the degradation by nucleases, low cellular uptake, and poor target specificity. These unfavorable properties can be improved by either incorporating structural modifications in oligonucleotide or by conjugating (covalently linking) molecules with relevant biological properties (e.g., peptides and proteins, carbohydrates, antibodies, enzymes, polymers, drugs, fluorophores) to oligonucleotide. The conjugate design usually aims to improve the poor pharmacokinetic and pharmacodynamic properties of the unmodified oligonucleotide and, in some cases, to impart new properties to the oligonucleotide. Over the years, a very large number of diverse oligonucleotide conjugates have been developed and evaluated for diagnostic, therapeutic, and nanotechnology-based applications. The aim of this chapter is to describe the underlying reasons and challenges in oligonucleotide conjugate design, provide an overview of chemical approaches available for their synthesis, and highlight some of their recent applications.

Keywords Antisense • Aptamer • Biomaterial • Conjugate • Diagnostics • Oligonucleotide • siRNA • Therapeutics

Abbreviations

AMD	Age-related macular degeneration
COC	Carbohydrate–oligonucleotide conjugate
CMV	Cytomegalovirus
CPG	Controlled pore glass
CPP	Cell-penetrating peptide
DCC	Dicyclohexyl carbodiimide
Dde	1-(4,4-Dimethyl-2,6-dioxacyclohexylidene)ethyl
DMT	4,4′-Dimethoxy trityl
DNA	Deoxyribonucleic acid
DTPA	Dithiol phosphoramidite
DTT	Dithiothreitol
EDC	1-Ethyl-3-(3-dimethylaminopropyl) carbodiimide
FISH	Fluorescence in situ hybridization
FO	Fluorescent oligonucleotide
HBTU	2-(1H-Benzotriazol-1-yl)-1,1,3,3-tetramethyluronium hexafluorophosphate
HNA	Hexitol nucleic acid
HOBT	1-Hydroxybenzotriazole

HPLC	High-performance liquid chromatography
LCAA	Long-chain alkyl amino
LNA	Locked nucleic acid
LOC	Lipid-oligonucleotide conjugate
MCOC	Metal complex–oligonucleotide conjugate
MMT	4-Monomethoxy trityl
NHS	*N*-Hydroxysuccinimidyl ester
NOC	Nanoparticle–oligonucleotide conjugate
ON	Oligonucleotide
PEG	Poly(ethylene glycol)
PNA	Peptide nucleic acid
POC	Peptide–oligonucleotide conjugate
QCMS	Quartz crystal microbalance
QD	Quantum dots
RES	Reticuloendothelial system
RNA	Ribonucleic acid
SAM	Self-assembled monolayer
SERS	Surface enhanced Raman scattering
shRNA	Small hairpin ribonucleic acid
siRNA	Small interfering ribonucleic acid
SMPT	*N*-Succinimidylocarbonyl-methyl-(2-pyridyldithio) toluene
SNALP	Stable nucleic acid lipid particle
SPDP	*N*-Succinimidyl 3-(2-pyridyldithio) propionate
SPR	Surface plasmon resonance
TCEP	Tris(2-carboxyethyl) phosphine
TOF-SIMS	Time-of-flight secondary ion mass spectrometry
Tr	Trityl
VEGF	Vascular endothelial growth factor

1 Introduction

Oligonucleotides cause selective inhibition of gene expression (Opalinska and Gewirtz 2002). It was Paterson, who first showed that exogenous nucleic acids modify gene expression (Paterson et al. 1977). Later, Zamecnik and Stephenson demonstrated that a short oligodeoxynucleotide (13-mer), antisense to Rous sarcoma virus, inhibits viral replication in cell culture (Stephenson and Zamecnik 1978). This pathbreaking discovery generated interest in the development of technologies employing synthetic oligonucleotides ($<$ 40 nucleotides) for manipulating gene expression in living cells. It was thought that these oligonucleotides would be immensely useful for treating viral diseases and cancer.

In antisense mechanism, reverse-complementary (antisense) oligonucleotides are introduced into the cells, where they hybridize to complementary mRNA of targeted gene (Opalinska and Gewirtz 2002; Juliano et al. 2008). After hybridization, these oligonucleotides block gene expression by following two

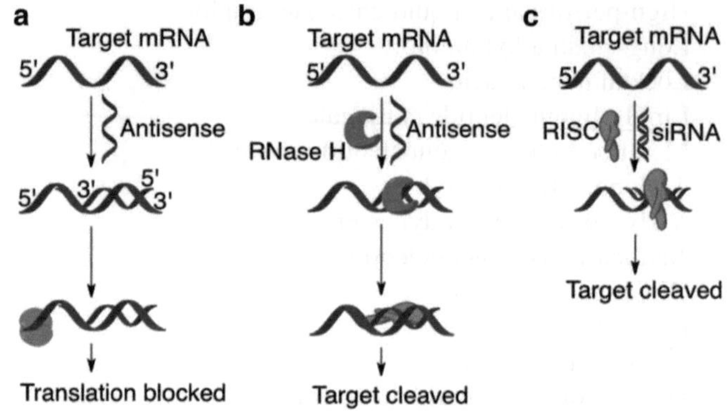

Fig. 1 Inhibition of gene expression by (**a**) and (**b**) antisense and (**c**) RNA interference (RNAi) mechanisms

mechanisms (Corey 2007): (1) sterically obstructing ribosomes, necessary for transcription and (2) forming a RNA–DNA duplex, which is a substrate for RNase H enzyme (Fig. 1). RNase H enzyme binds to RNA–DNA complex and cleaves the RNA strand of this duplex. Antisense oligonucleotides are being evaluated in clinical trials for the treatment of conditions like avian flu, muscular dystrophy, asthma, diabetes, cardiovascular diseases (high cholesterol level), rheumatoid arthritis, CMV retinitis, HIV, HCV, Crohn's disease, leukemia, cancer, etc. (Corey 2007; Rayburn and Zhang 2008; Jones 2011).

Antisense is not the only mechanism by which oligonucleotides inhibit gene expression. Another mechanism that has generated huge interest is RNA interference (RNAi) or posttranscriptional gene silencing, a natural process in cells by which double-stranded small interfering RNA (siRNA) targets mRNA for cleavage in a sequence-specific manner (Dorsett and Tuschl 2004; Whitehead et al. 2009; Paroo and Corey 2004). Fire et al. (1998) observed RNAi in a nematode worm, *Caenorhabditis elegans*. Tuschl and coworkers were the first to demonstrate sequence-specific gene knockdown using siRNA in a mammalian cell line (Elbashir et al. 2001). The therapeutic potential of siRNA lies in the fact that synthetic siRNA can be directly introduced into the cells (Fig. 1). Once inside the cytoplasm, it is incorporated into a large multicomponent complex called RNA-induced silencing complex (RISC) (Whitehead et al. 2009). A multifunctional protein (argonaute 2) in the complex unwinds the sense strand of siRNA, thus activating the complex. This activated complex, containing antisense siRNA strand, selectively degrades target (complementary) mRNA. It then seeks additional mRNA targets for cleavage. The capability of siRNA to knock down targets in diseases like hypercholesterolemia, liver cirrhosis, hepatitis B virus, human papillomavirus, ovarian cancer, and bone cancer has already been shown in vivo, and several siRNA candidates are undergoing clinical trials (Whitehead et al. 2009).

Aptamers are also oligonucleotides, but they are different from antisense and siRNA in that their mechanism of action does not involve sequence-specific base recognition (Watson–Crick or Hoogstein base pairing) (Bouchard et al. 2010; Dausse et al. 2009). These oligonucleotides fold into complex three-dimensional structure because of intramolecular interactions, and tightly and specifically bind to target molecules, ranging from small molecules to multimeric structures. Aptamers bound to target proteins are able to interfere with their biological activity. These are obtained by in vitro selection from combinatorial oligonucleotide libraries following a process called systematic evolution of ligands by exponential enrichment (SELEX). The Szostak and Gold groups first demonstrated the process of SELEX in 1990 (Ellington and Szostak 1990; Tuerk and Gold 1990). Several aptamers are being evaluated in clinical trial for indications like inflammatory diseases, tumor angiogenesis and growth, age-related macular degeneration (AMD), neoplasia, etc. (Dausse et al. 2009). Besides therapeutics, aptamers also find use in analytical applications (Tombelli and Mascini 2009).

The above three agents are the ones which have generated maximum interest. There are also others, which are not covered here, like triplex-forming oligonucleotides (TFO), micro-RNA (miRNA), ribozymes, and DNAzymes (Opalinska and Gewirtz 2002).

Oligonucleotides have achieved very limited clinical success because of their poor pharmacokinetic and pharmacodynamic properties (see Sect. 2) (Cunningham et al. 2011; Juliano et al. 2008; Whitehead et al. 2009). Despite the promise shown and tremendous amount of work done to this effect, only two products, fomivirsen (Crooke 1998) and pegaptanib (Ng et al. 2006), have been approved for clinical use (see Sect. 6) so far. Covalent attachment (conjugation) of molecules (see Sect. 2) with relevant biological or material properties to oligonucleotide is being explored for improving existing oligonucleotide properties or to impart new properties (Goodchild 1990; Da Ros et al. 2005; Singh et al. 2008, 2010b). Diverse oligonucleotide conjugates have been designed, developed, and evaluated for therapeutic, diagnostic, and material applications. In this chapter, we discuss the rationale for designing oligonucleotide conjugates, synthetic approaches available for their synthesis, and their representative examples in early development. In addition, oligonucleotide conjugates in clinical trial and those approved for clinical use are also reviewed.

2 Oligonucleotide Conjugates

Unmodified oligonucleotides show extremely poor plasma half-life because phosphodiester backbone is degraded by nucleases present in biological fluids (Cunningham et al. 2011; Juliano et al. 2008; Whitehead et al. 2009). They also exhibit poor penetration across cell membrane because of their hydrophilic character. In fact, poor delivery has been recognized as major impediment in their clinical success. There are other concerns also, like target selectivity, immunogenic response, and uptake by the cells of reticuloendothelial system (RES).

Chemical modifications have been used to improve oligonucleotide stability and inhibit immune stimulation (Kurreck 2003). Oligonucleotides are modified at sugar, base, or phosphodiester backbone. Most of the oligonucleotides being developed for clinical applications have phosphorothioate or morpholino backbone instead of natural phosphodiester. Other examples of structural modifications are 2′-*O*-methyl, peptide nucleic acid (PNA), locked nucleic acid (LNA), hexitol nucleic acid (HNA), etc. In fact, several advanced antisense oligonucleotides, called as gapmers, contain a central DNA portion to recruit RNase H and flanking 2′-*O*-methyl modifications to improve stability and affinity (Kurreck 2003).

Where feasible, oligonucleotides are delivered locally to overcome the challenge of poor delivery. Both approved oligonucleotide products, fomivirsen (Crooke 1998) and pegaptanib (Ng et al. 2006) (see Sect. 6), are delivered via intraocular injection. Direct administration to brain, respiratory system, and tumor has also been reported (Cunningham et al. 2011). Systemic delivery is achieved using viral vectors (e.g., retroviruses, adenoviruses, adeno-associated viruses, baculoviruses, vaccinia, *Salmonella, Clostridium*) (Hedley et al. 2007). Viral vectors are efficient, but there are concerns like reversion to wild type, generation of immune responses, lack of target cell specificity, and manufacturing costs. To overcome these concerns, several nonviral vectors have been developed for systemic oligonucleotide delivery. Most extensively used nonviral vectors are cationic lipid-based delivery systems (e.g., lipofectin, SNALP) (Semple et al. 2010). Polypetides (e.g., polylysine) and cationic polymers (e.g., polyethylenimine, polyamidoamine, imidazole-containing polymers, cyclodextin-containing polymers, membrane-disruptive polymers like polyethylacrylic acid, polymethyl acrylic acid, polyacrylic acid, polypropyl acrylic acid) are also being developed for nonviral oligonucleotide delivery (Pack et al. 2005).

Another major approach for improving oligonucleotide properties is conjugation (Goodchild 1990; Da Ros et al. 2005; Singh et al. 2008, 2010b). Conjugation is achieved by attaching molecules with relevant properties (promoiety) to oligonucleotide, using covalent bonds. Both chemical and enzymatic methods are used for conjugate synthesis. Promoieties like antibodies, peptides and proteins, hormones, enzymes, growth factors, carbohydrates, lipids, polymers, drugs, fluorophores, photoprobes, metal complexes, nanoparticles, metal surfaces, etc., are attached to oligonucleotide to obtain conjugates for various applications (Goodchild 1990). Conjugation aims to either improve existing oligonucleotide properties (e.g., stability, cell uptake, targeted delivery) or endow it with entirely new property (e.g., fluorescence). For instance, oligonucleotides conjugated to cell-penetrating peptides show improved cellular uptake (Gait 2003; Tung and Stein 2000). Oligonucleotides conjugated to carbohydrates exhibit improved cell/tissue-specific delivery by receptor-mediated endocytosis (Zatsepin and Oretskaya 2004; Lonnberg 2009), and those conjugated to lipophilic moieties exhibit enhanced cell permeation (Rosemeyer 2005; Manoharan 2002). PEGylated oligonucleotide conjugates demonstrate improved plasma half-life, lower renal clearance, reduced immunogenicity, and diminished uptake by cells of RES (Bonora et al. 2003). Covalent attachment of fluorophores, on the other hand, makes oligonucleotide fluorescent that aids in their detection (Asseline 2006).

Conjugates must be distinguished from prodrugs. Conjugates often contain stable (nondegradable) covalent bonds unlike prodrugs, which invariably contain unstable (degradable) bonds. However, in many cases, this distinction is blurred, particularly when conjugation is achieved using degradable bonds. Also, conjugates retain their activity (often slightly different), whereas prodrugs are inactive and converted to active form in the body. Conjugates must also be distinguished from oligonucleotide complexes (with proteins, cationic lipids, peptides, cationic polymers, etc.), where oligonucleotides are held by ionic bonds and not covalent bonds.

3 Syntheses of Oligonucleotide Conjugates

Oligonucleotides are conjugated at 5′- and 3′-termini and also at internal positions (sugar, base, and internucleosidic phosphodiester backbone). Because of their easy accessibility, 5′- and/or 3′-termini are preferred for conjugation. It has been reported that conjugation at 3′- enhances the exonuclease resistance. Similarly, 2′-OH modifications are particularly useful in antisense strategy because the ribose conformation of 2′-modified oligonucleotide is essentially RNA-like (sugar conformation is C3′-endo), which enhances the affinity toward the target RNA.

Oligonucleotide conjugation is not straightforward because subtle changes in structure influence biological properties. A particular care must be taken to ensure that oligonucleotide affinity for the target sequence is not compromised after conjugation. A number of other factors, like accessibility of conjugation sites, chemical compatibility of oligonucleotide and promoiety, linker length, and covalent linkage (nondegradable or degradable) used for conjugation, must also be considered in designing oligonucleotide conjugates. Several synthetic approaches have been developed to prepare oligonucleotide conjugates, which can be grouped into two main categories (Singh et al. 2008, 2010b): (1) on-support conjugation, where conjugation is achieved using support-bound oligonucleotides and (2) solution-phase conjugation, where oligonucleotides are cleaved from the support, deprotected, and purified, prior to solution-phase conjugation (Fig. 2). A brief overview of different chemical approaches developed for oligonucleotide conjugation is provided below.

3.1 Solid-Phase Conjugate Synthesis

The on-support conjugation is achieved by following methods: (1) incorporation of promoiety during the automated oligonucleotides synthesis, (2) in-line or stepwise solid-phase conjugate synthesis, and (3) on-support fragment conjugation (Fig. 2).

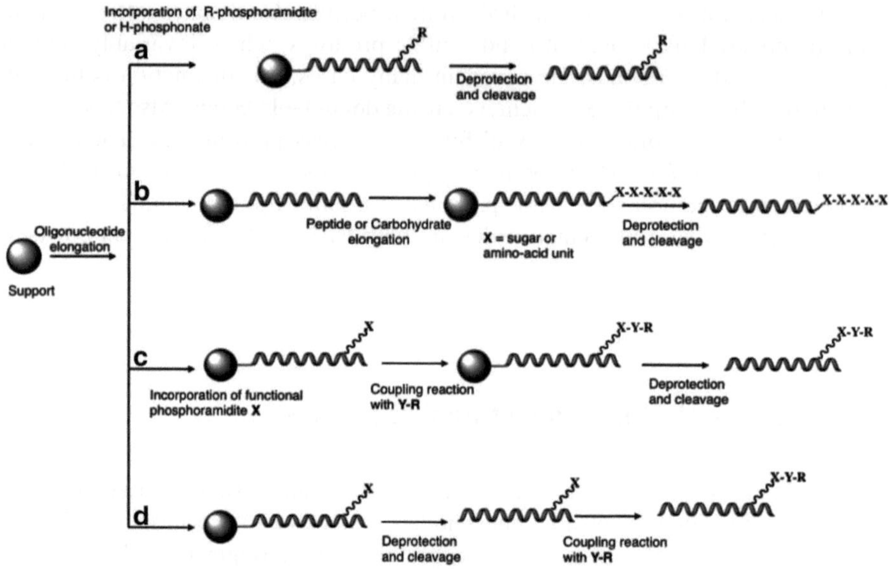

Fig. 2 Schematic representation of chemical strategies used for oligonucleotide conjugation with promoiety R: (**a**) incorporation of phosphoramidite or H-phosphonate derivative of R, (**b**) in-line or stepwise conjugate synthesis (R = peptide or saccharide), (**c**) on-support fragment conjugation, and (**d**) solution-phase coupling reaction (X and Y: mutually reactive functional groups)

3.1.1 Incorporation of Promoiety During the Solid-Phase Synthesis

In this approach, modified phosphoramidite or H-phosphonate derivatives of promoiety (nucleosidic or non-nucleosidic) to be conjugated are prepared and incorporated into the oligonucleotide chain using automated solid-phase synthesis protocols (Fig. 2a). This method is extensively used for conjugation with lipophilic molecules (i.e., cholesterol), fluorescent labels, dyes, and PEGs. The major advantage of this approach is that vast repertoire of modified "ready-to-use" building blocks and solid supports are commercially available (Fig. 3). However, modified building blocks for every conceivable application are not available and, therefore, must be prepared in laboratory. In addition, the incorporated promoiety must remain stable during the automated oligonucleotide synthesis and subsequent deprotection, the latter carried out in drastic basic conditions. Therefore, this method is limited only to molecules, which are chemically robust and could be obtained by simple synthetic procedure.

3.1.2 In-Line or Stepwise Solid-Phase Conjugate Synthesis

In this approach, elongation of oligonucleotide and promoiety is achieved on same solid support (Fig. 2b). It is mainly used to prepare peptide–oligonucleotide and carbohydrate–oligonucleotide conjugates (POCs and COCs) (Lonnberg 2009).

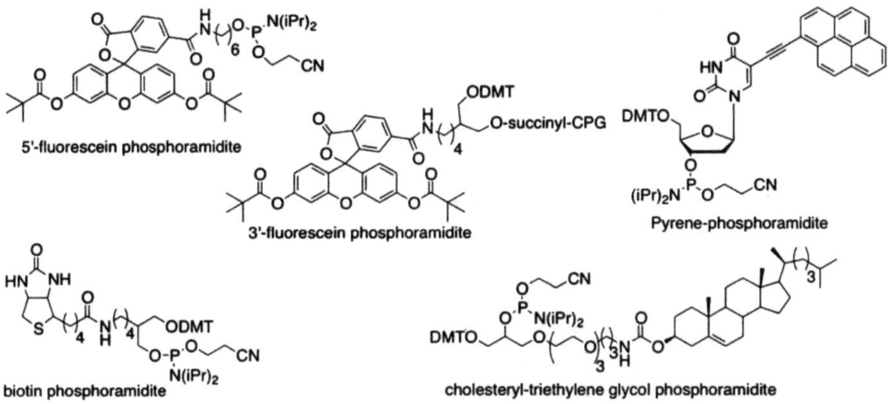

Fig. 3 Representative examples of commercially available modified phosphoramidite for oligonucleotide conjugation

Either oligonucleotide is synthesized first followed by elongation of promoiety or vice versa. Often, a branched linker is incorporated to achieve independent assembly of two fragments. The main advantage of this approach is certainly the ease of purification, and these approaches are considered indispensable for synthesizing combinatorial conjugate libraries. However, these approaches have stringent requirements. For instance, the protecting group used for reactive functions present on promoiety must be stable to conditions employed for oligonucleotide synthesis, and their removal should not affect oligonucleotide stability as well as protecting groups present on oligonucleotide. Furthermore, conjugate prepared must remain stable under the basic conditions employed for oligonucleotide deprotection. These approaches are difficult to use when the two fragments have incompatible chemistries.

Some representative examples of stepwise solid-phase oligonucleotide conjugate synthesis are described here. Truffert et al. used this approach to prepare POC, which can act as artificial nuclease (Truffert et al. 1996). A peptide was assembled first on CPG support. The Dde group protecting the amino side chain on lysine was deprotected, and a linker was introduced to assemble oligonucleotide using standard phosphoramidite chemistry. Final deprotection was achieved using sodium hydroxide. On the contrary, Ocampo et al. first assembled oligonucleotide using standard phosphoramidite protocols and introduced an amino group at 5′-terminus using N-6-MMT-aminohexyl phosphoramidite (Ocampo et al. 2005). After removing 5′-MMT in mild acidic conditions, a peptide was assembled on support-bound oligonucleotide, starting from the free amino group at 5′-terminus. The conjugate was obtained after standard deprotection with ammonia.

3.1.3 On-Support Fragment Conjugation

This approach is only slightly different from stepwise solid-phase conjugate synthesis. Here, oligonucleotide is assembled on solid support, and reactive functional group is incorporated during oligonucleotide elongation. The protective group masking the reactive group is removed, but oligonucleotide is retained protected and support bound. A promoiety modified with complementary reactive group is coupled to support-bound oligonucleotide. The conjugate is obtained after cleavage and deprotection of oligonucleotide (Fig. 2c). The reactive groups commonly used are primary amine and carboxylic acid, which are introduced using appropriate phosphoramidites. The on-support coupling reaction is carried out using activators like HBTU/HOBt or DCC (Kachalova et al. 2002).

Kachalova et al. prepared oligonucleotide modified with 5′-carboxylic group, which was activated by HBTU/HBT and conjugated to promoieties (biotin, dansylcadaverine, 1-pyrenylmethylamine, spermine, histamine, octadecylamine, and peptides) containing amino groups (Kachalova et al. 2002). The conjugates were obtained after deprotection with ammonia and purification. Recently, copper-catalyzed 1,3-dipolar cycloaddition (CuAAC) reaction was used to prepare COC by this approach, and microwave was used to fasten the process (Bouillon et al. 2006).

3.2 Solution-Phase Conjugate Synthesis

In this approach, oligonucleotide and promoiety containing mutually reactive groups are prepared and purified independently using most appropriate method. The two fragments are coupled in solution leading to the formation of a covalent bond between oligonucleotide and promoiety (Fig. 2d). The major benefit of this approach is that promoieties with incompatible chemistries can be conjugated to oligonucleotide. Besides, the individual fragments are prepared, purified, and characterized separately using the most efficient chemistry available. However, major limitation with this approach is that promoiety might not be highly soluble in aqueous solutions like oligonucleotide. This inconsistency in solubility behavior may lower coupling yield. Conjugates prepared by this approach usually need exhaustive purification that further lowers the yield.

Despite limitations, solution-phase approaches are most extensively used for oligonucleotide conjugation. A large number of coupling reactions have been developed for solution-phase oligonucleotide conjugation, as described below.

3.2.1 Amide and Thiourea Linkages

Amide and thiourea linkages are formed by the reaction of an amino group with an activated carboxylic and isothiocyanate group, respectively (Fig. 4). Usually,

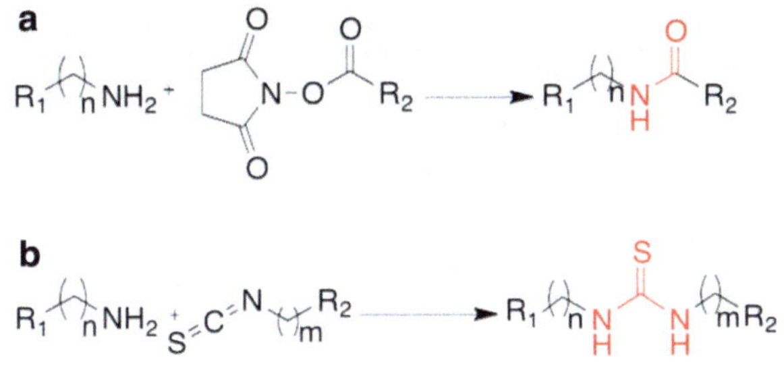

Fig. 4 Solution-phase oligonucleotide conjugation using (**a**) amide and (**b**) thiourea linkages

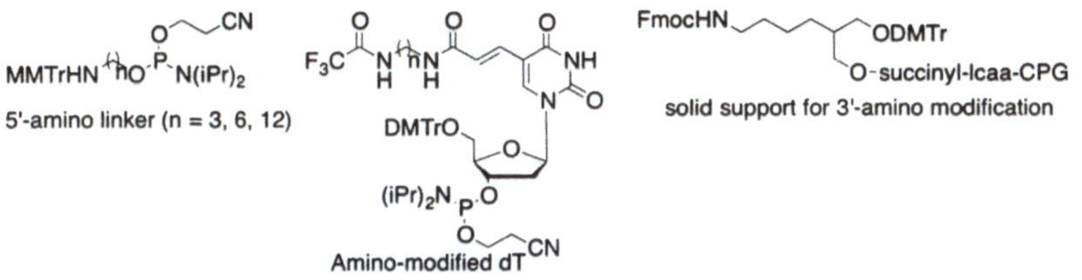

Fig. 5 Phosphoramidites and solid support to incorporate an amino group in oligonucleotides

amino group is incorporated into the oligonucleotide using several phosphoramidites and solid supports available (Fig. 5), whereas promoiety contains activated carboxylic and isothiocyanate groups. Reactions are usually carried out at alkaline pH. Amide linkages are preferred over thiourea because the latter are more prone to competing hydrolysis in aqueous alkaline medium.

Although extensively used for oligonucleotide conjugation with easy-to-access promoiety (several activated ester and isothiocyanate derivatives are commercially available), the use of amino groups for conjugation has certain limitations. The nucleophilicity of amino groups is pH dependent, which means that these groups are protonated at lower pH that reduces their reactivity toward electrophiles. The reaction of amino groups involve carbodiimide-activated intermediates, which are known to give side reactions, and lastly, the activated esters are unstable in aqueous medium, which lowers coupling yield (in most cases, activated ester is used in large excess, which makes purification arduous).

Zatsepin et al. used amide linkage to prepare POCs. An oligonucleotide containing 2′-amine was prepared by incorporating phosphoramidite, 2′-*O*-(2,3-aminopropyl) uridine, and coupled to a peptide containing carboxylic group (Zatsepin et al. 2003). A reverse strategy was also developed. Acharya et al. used thiourea linkage to prepare oligonucleotide conjugate of aminoglycosides. The oligonucleotide with amine group was prepared using 5′-amino-5′-deoxythymidine and conjugated to isothiocyanate derivatives of kanamycin and neomycin (Charles et al. 2002).

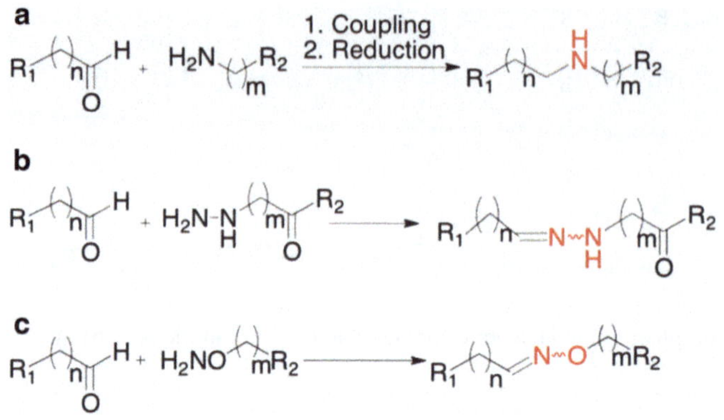

Fig. 6 Solution-phase oligonucleotide conjugation using (**a**) imine, (**b**) hydrazone, and (**c**) oxime linkages

3.2.2 Imine, Hydrazone, and Oxime Linkages

Nucleophilic addition on carbonyl group is an efficient way to conjugate oligonucleotides (Zatsepin et al. 2005; Singh et al. 2008, 2010b). The most common nucleophilic additions used in oligonucleotide conjugation involve formation of imine (Schiff base), hydrazone, and oxime linkages, which are formed by the reaction of an aldehyde group with an amino, hydrazide, and aminooxy group, respectively (Fig. 6). The imine linkages are reversible but can be stabilized by in situ reduction to secondary amines, using sodium borohydride. The hydrazone and oxime linkages are more stable, and therefore, further reduction to secondary amines, even though possible, is not required. The reactions between aldehyde or keto and aminooxy or hydrazide groups are chemoselective and therefore very efficient for oligonucleotide conjugation. Furthermore, the reactions are carried out in mild acidic conditions (pH ~4).

The aldehyde group is generally introduced in oligonucleotides as 1,2 diol precursor. This route is more convenient than the use of protected aldehyde because diols can be readily converted to aldehyde by mild periodate oxidation, whereas harsh conditions are often required to deprotect aldehyde groups. Modified phosphoramidites **1a,b** were developed using a two-step synthetic procedure to introduce protected 1,2-diol group at 5′-terminus (Fig. 7). The benzylidene group was found to be stable during oligonucleotide deprotection, and it also aided in HPLC purification of oligonucleotide on account of its hydrophobic character. The 5′-diol moiety was deprotected using acetic acid after oligonucleotide purification, and 5′-aldehyde was generated by periodate oxidation (Edupuganti et al. 2004). Uridine phosphoramidite derivatives **2a,b** were developed by Zatsepin et al. to introduce aldehyde precursor at 2′-position in oligonucleotide (Zubin et al. 2005). Benzoyl and acetyl groups, which are cleaved during ammonia treatment, were used to protect the hydroxyl groups on 1,2 diol. The 2′-aldehyde was generated by oxidation of diol.

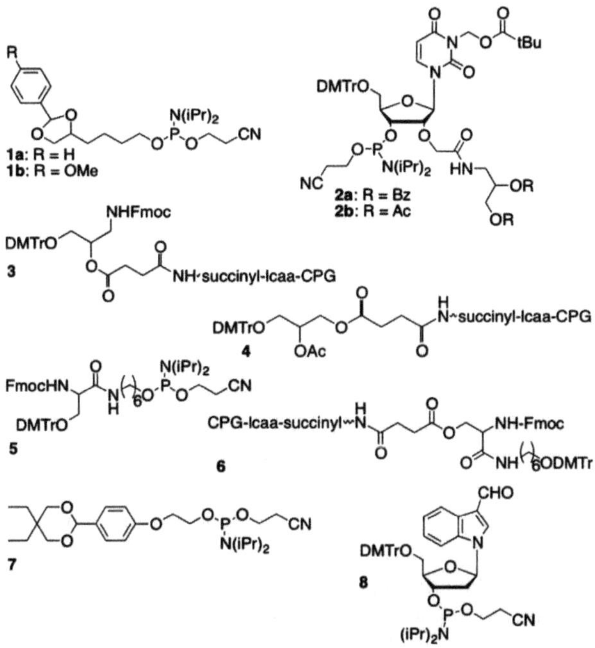

Fig. 7 Phosphoramidites and solid supports to incorporate an aldehyde group in oligonucleotides

The modification at 2′-position was chosen due to the fact that 2′-*O*-alkyloligonucleotides improve the binding affinity toward the target RNA and allow the possibility of multiple conjugations. Introduction of aldehyde group at the 3′- was achieved using 1,2-amino alcohol and glyceryl solid supports **3** and **4**, respectively, that allows the formation of aldehydic group by periodate oxidation, after cleavage from support **9** (Fig. 7) (Edupuganti et al. 2004).

Glyoxylic aldehyde was introduced at 5′- and 3′-termini of oligonucleotide using phosphoramidite **5** and solid support **6**, respectively (Fig. 7) (Singh et al. 2004; Spinelli et al. 2007). The advantage of using glyoxylic aldehyde over aldehyde is that it is more stable to air oxidation and does not show side reactions. Recently, phosphoramidites **7** and **8** were developed to introduce aromatic aldehyde, and these are now commercially available. The acetal-protecting group in **7** is readily removed after oligonucleotide synthesis using standard detritylation with 80% aqueous acetic acid. On the contrary, formylindole nucleoside analogue **8** has no protecting group on the aldehyde, and deprotection of oligonucleotide is achieved by ammonia treatment.

Ketone groups were also used instead of aldehyde for oligonucleotide conjugation. Dey and Sheppard employed ketone functionality for postsynthetic modification of DNA. Methylketone containing phosphoramidite was prepared by modifying C-5 position of uridine (Dey and Sheppard 2001).

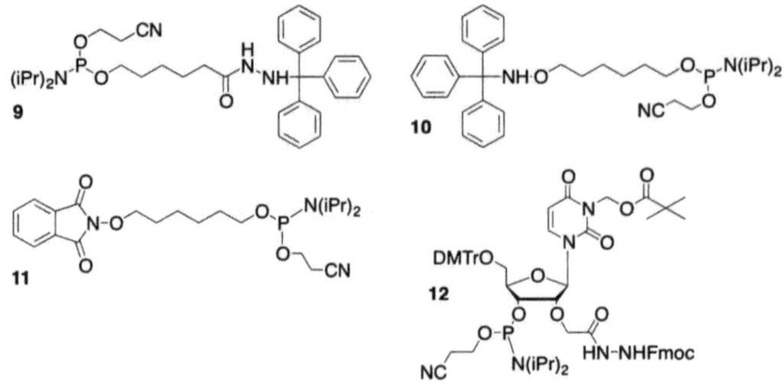

Fig. 8 Phosphoramidites for incorporating hydrazide and aminooxy groups in oligonucleotide

The oligonucleotides containing aldehyde, glyoxylic aldehyde, or ketone groups are reacted with promoiety containing an amino, hydrazide, or aminooxy group (Singh et al. 2008; Zatsepin et al. 2005). However, the reverse approach is also possible where promoiety contains the aldehyde group and oligonucleotide is modified with an amino, hydrazide, or aminooxy group (Singh et al. 2008; Zatsepin et al. 2005). Hydrazide and aminooxy moieties are introduced in oligonucleotide by incorporating modified phosphoramidite in which reactive function is protected. The protecting group is often a trityl or MMTr as they are cleaved using mild acidic conditions after oligonucleotide synthesis and purification. Common examples are phosphoramidites **9** and **10** (Fig. 8). Other protecting groups like phthaloyl (e.g., **11**) and Fmoc (e.g., **12**) are also used. Oligonucleotides modified with aminooxy and hydrazide groups are difficult to purify due to the high nucleophilicity associated with these groups and resulting formation of side products.

3.2.3 Disulfide, Thioether, Thiazolidine, Native Ligation, and Metal–Thiol Linkages

The disulfide, thioether, thiazolidine, native ligation, and metal–thiol linkages are formed by the reaction of thiol group with other reactive groups (Fig. 9) (Singh et al. 2008). The disulfide linkage is formed by oxidative coupling of two sulfhydryl (thiol) groups; usually one of them is present in activated form. The advantage of using disulfide linkage is that it is unstable in the cytosol due to the reducing environment present in intracellular compartments and could be employed for selective release of oligonucleotides into the intracellular media. However, there is also the possibility of forming homo-conjugates of oligonucleotide and promoiety during oligonucleotide conjugation using disulfide linkage. The thioether linkage, on the other hand, is formed by nucleophilic addition of thiol to an electron-deficient carbon, provided by functional groups halogeno-acetyl, acrylic, vinylic, or maleimide (most used). Unlike disulfide, thioether linkage is relatively stable. Both linkages have been used to prepare oligonucleotide

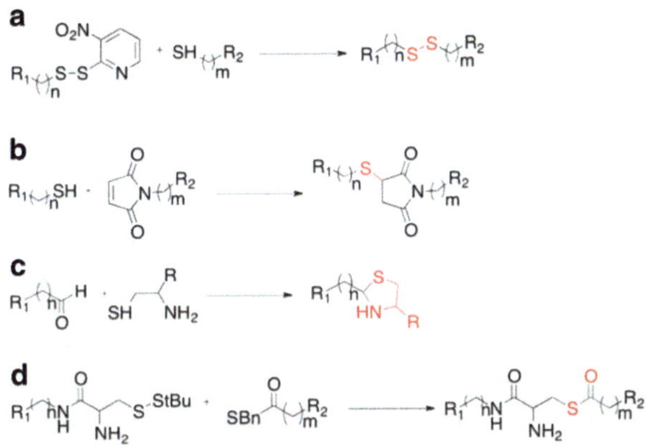

Fig. 9 Solution-phase oligonucleotide conjugation using (**a**) disulfide, (**b**) thioether, (**c**) thiazolidine, and (**d**) native ligation

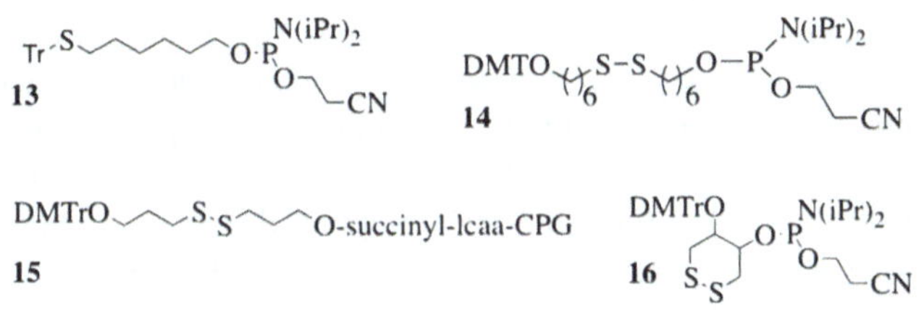

Fig. 10 Phosphoramidites and solid supports to incorporate thiol group in oligonucleotides

conjugate with peptides and proteins, carbohydrates, lipophilic moieties, enzymes, fluorescent groups, and metal surfaces (Singh et al. 2008).

The oligonucleotide modified with 5′-thiol group is usually prepared by incorporating *S*-trityl phosphoramidite derivative **13** (Fig. 10) during automated oligonucleotide synthesis, followed by deprotection with silver nitrate. The removal of *S*-trityl with silver nitrate could be troublesome, and therefore, the use of disulfide was proposed. Modified phosphoramidite **14** was also developed to incorporate thiol group at 5′-terminus of oligonucleotide using automated oligonucleotide synthesis. The free thiol is generated by DTT reduction. The oligonucleotide containing 3′-thiol group is prepared using modified solid supports (e.g., **15**). Recently developed dithiol phosphoramidite (DTPA, **16**) is particularly useful for incorporating multiple thiol groups at any position within the oligonucleotide. This was employed to tether oligonucleotides to gold surface. It is also possible to use promoieties modified with thiol group for coupling with oligonucleotide containing thiol-reactive groups (activated thiol, maleimide, etc.). Several linkers are available to incorporate activated thiol or maleimide groups in oligonucleotide (Singh et al. 2008).

Fig. 11 Solution-phase oligonucleotide conjugation using Diels–Alder cycloadditions

Besides disulfide and thioether, there are thiazolidine linkages, which are formed by reaction between aldehyde and 1,2-aminothiol groups. This linkage has been extensively used for chemical ligation of peptides because N-terminal cysteine-containing peptides are readily available by chemical synthesis. Thiazolidine linkage is stable at neutral to basic pH but prone to hydrolysis at acidic pH. Oligonucleotide conjugation using thiazolidine linkage was achieved by reacting cysteine-derived promoiety with oligonucleotide modified with aldehyde groups (Forget et al. 2001; Zatsepin et al. 2002).

Similarly, the native ligation involves joining of two largely unprotected peptides, one containing C-terminal thioester group and other containing N-terminal cysteine. The coupling is carried out in aqueous solution, and efficient coupling is achieved even in the presence of denaturing agents and organic solvent additives. Gait's group used this approach to prepare POCs (Stetsenko and Gait 2000). An oligonucleotide containing 5′-cysteine was prepared and coupled to a peptide containing thioester group in the presence of water-soluble reagent TCEP. Lastly, the affinity of alkyl thiols for metals like gold, silver, copper, palladium, and platinum is used to prepare self-assembled monolayers of oligonucleotide. Niemeyer's group used this approach to obtain oligonucleotide-loaded gold nanoparticles (Niemeyer et al. 2003).

3.2.4 Diels–Alder [4 + 2] Cycloadditions

The Diels–Alder cycloadditions involve reaction between a diene carrying an electron-donating group and a dienophile carrying an electron-withdrawing group. These reactions are particularly useful for oligonucleotide conjugation because they are fast and efficient in aqueous media, are chemoselective, and do not require catalysts (Fig. 11). The Diels–Alder reactions are performed using unprotected oligonucleotide and promoiety because diene and dienophile groups are not present in biomolecules.

In most cases, diene group is introduced in oligonucleotide because it is stable during solid-phase oligonucleotide synthesis (Graham and Enright 2006). This is achieved using phosphoramidites containing a furan or a hexadiene moiety. This approach was used for the synthesis of POCs and labeled oligonucleotides (fluorescence and SERS applications) (Graham and Enright 2006). Recently, this approach was used to perform template reactions in order to simultaneously ligate three oligonucleotides, and it was shown that reaction is very rapid (1 min) at room temperature (El-Sagheer et al. 2011). Inverse-electron-demand Diels–Alder

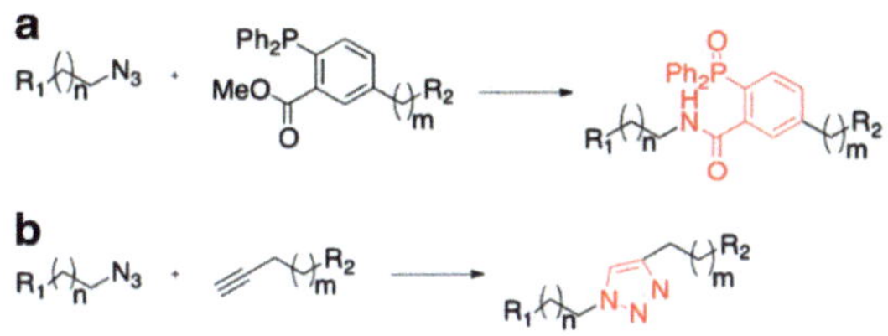

Fig. 12 Solution-phase oligonucleotide conjugation using (**a**) Staudinger ligation and (**b**) 1,3-dipolar cycloadditions

reaction was also investigated for oligonucleotide conjugation. Norbornene-containing phosphoramidite was developed to incorporate dienophile groups at both termini and internal positions within oligonucleotide, and conjugation reaction was performed with water-stable tetrazine dienes (Schoch et al. 2010).

3.2.5 Staudinger Ligation and 1,3-Dipolar Cycloadditions

The Staudinger ligation involves reaction of an azide group with phosphane to form an aza-ylide linkage (Fig. 12a) (Schilling et al. 2011). The reaction is carried out in aqueous environment at room temperature. These reactions demonstrate more stability against oxygen and tolerance toward superior functional groups. The main advantage of using azide group is its bio-orthogonal characteristics. Azide groups are poor electrophile and, therefore, do not react with amines or other nucleophiles commonly found in biological systems. Ju et al. used Staudinger ligation to obtain fluorescent oligonucleotides (FOs) (Seo et al. 2003). Oligonucleotides containing azide group were prepared and reacted with fluorescein phosphane under aqueous conditions. The azide group was incorporated by post-functionalization of 5′-amino oligonucleotide with azidovalerate in presence of EDC.

The Huisgen or 1,3-dipolar cycloadditions involve reaction between alkyne and azide groups to form 1,2,3-triazoles (Fig. 12b) (El-Sagheer et al. 2011). The Huisgen reactions are accomplished in water and are nearly quantitative. Both azide and alkyne groups are stable, and reaction is orthogonal to most typical organic transformations. However, achieving complete regioselectivity (1,4-disubstituted triazoles over 1,5-regioisomer) is difficult. Sharpless and Meldal showed that use of Cu(I) salt considerably enhances reaction yield and regioselectivity (1,4- over 1,5-) (Rostovtsev et al. 2002). Usually, the alkyne moiety is incorporated in oligonucleotide using appropriate alkyne-containing phosphoramidites, which is due to the fact that azide group react with P(III) derivatives (Staudinger ligation) during solid-phase oligonucleotide synthesis using phosphoramidite approach. A number of alkyne-modified phosphoramidites and solid supports (**17-19**) have been developed, and some of them are commercially available (Fig. 13). Carell et al. used Cu(I)-catalyzed Huisgen reaction for multiple postsynthetic labeling of

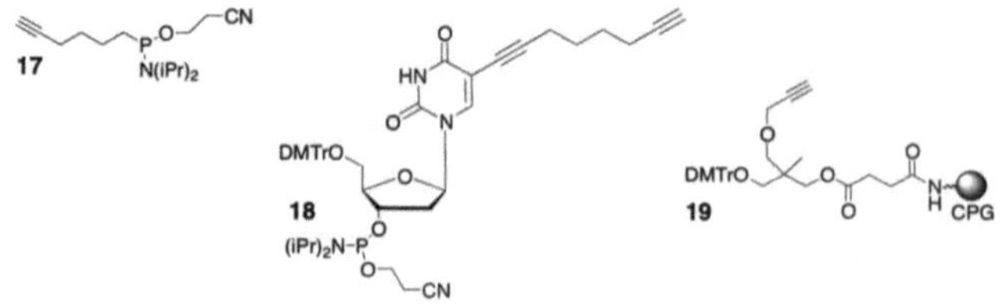

Fig. 13 Phosphoramidites and solid supports to incorporate alkyne group in oligonucleotides

oligonucleotides (Gramlich et al. 2008). This reaction was also used for oligonucleotide conjugation with carbohydrates, peptides, and other promoieties to obtain DNA analogues with modified nucleobases and backbones (El-Sagheer et al. 2011).

4 Oligonucleotide Conjugates in Early Development

Several types of oligonucleotide conjugates are being developed and evaluated for improving the therapeutic, diagnostic, and material properties of oligonucleotide. An exhaustive review of these conjugates is beyond the scope of this chapter. In this section, we provide a brief overview of different classes of oligonucleotide conjugate along with some representative examples.

4.1 Fluorescent Oligonucleotides

FOs play an important role in DNA sequencing as well as DNA-based clinical diagnostics such as Fluorescent In Situ Hybridization (FISH) and molecular beacons (Asseline 2006; Wang et al. 2009). Human Genome Mapping Project (HGMP) has tremendously raised the demand for fluorescently labeled oligonucleotides. The labeling is achieved at 5′- or 3′-termini, nucleobases (mainly at C-5 position of pyrimidines), or sugar. Mostly, labeling is accomplished by incorporating modified phosphoramidites or by solution-phase coupling. A large number of commercially available fluorophores and modified supports are available for preparing FOs covering very broad absorbance (494–683 nm) and emission (525–707 nm) wavelengths, ranging from UV–VIS to near infrared (NIR) (Goncalves 2009). For instance, fluorescein phosphoramidites can be used for labeling of 5′-terminus or internal positions, whereas TAMRA-functionalized succinyl resin can be used for 3′- labeling. The incorporation of fluorescent probes during oligonucleotide synthesis requires that the probe remains stable to reagent and conditions used for automated synthesis, cleavage, and deprotection. Appropriately derivatized fluorescence probes (e.g., *N*-hydroxysuccinimide ester, isothiocyanate, maleimide) are also available for

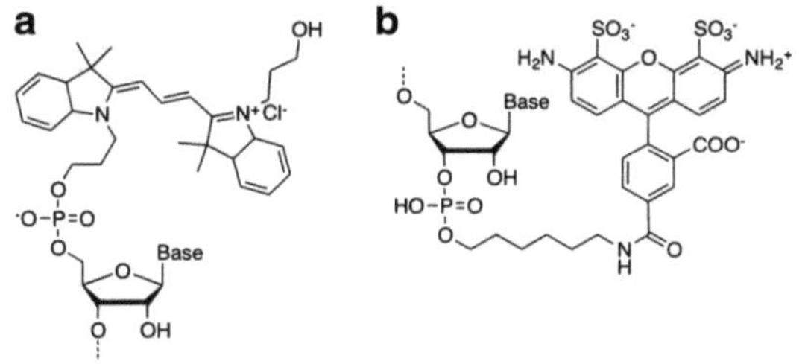

Fig. 14 (**a**) 5′-Cy3-modified siRNA and (**b**) 3′-Alexa488-modified siRNA

solution-phase labeling of oligonucleotides. The 1,3-dipolar cycloadditions have also been used to obtain FOs (Weisbrod and Marx 2008).

Fluorescently labeled RNAs have been used to unravel the mechanism of RNA silencing by RNA interference (RNAi). Berezhna et al. used RNA duplexes in different pair wise combination of labeled (Alexa Fluor 488 at 3′- of sense strand and Cy5 at 3′-end of antisense strand) and unlabeled strands to visualize the localization of siRNA in live human cells (Berezhna et al. 2006). They showed that siRNA either localizes in cytoplasm or translocates to nucleus, depending on where the silencing target RNA resides. The siRNAs localized in nucleus stayed in their duplex form, whereas the ones in cytoplasm were quickly converted to single strand (leading to the degradation of sense strand). Chiu et al. used 5′-Cy3-modified sense strands to determine the uptake efficiency of siRNA and showed that interactions with RISC dictate siRNA localization even when synthetic siRNAs (POCs) are used (Fig. 14) (Chiu et al. 2004). Ohrt et al. used small RNAs labeled at 5′- or 3′- with Alexa488-TFP or Cy5-NHS to investigate how siRNAs are processed within human cells (Fig. 14) (Ohrt et al. 2006). By using quantitative fluorescence microscopy, they showed that siRNAs are excluded from non-nucleolar areas of the nucleus in an exportin-5-dependent process that specifically recognizes key structural features shared by these and other small RNAs such as miRNAs.

4.2 Peptide–Oligonucleotide Conjugates

The interest in POCs stems from the fact that there are certain peptides, called cell-penetrating peptides (CPP), that possess the remarkable ability to translocate across cell membranes by a mechanism that does not involve receptor-mediated endocytosis, at least for a free peptide (Gait 2003). Some examples of this kind of peptide are penetratin (residues 43–58 of the third helix of *Antennapedia* homeodomain), the highly arginine/lysine-rich region of HIV-1 TAT protein, or transportan (a hybrid section of the neuropeptide galanin and the wasp venom peptide mastoparan). Consequently, there is a tremendous interest in the design and

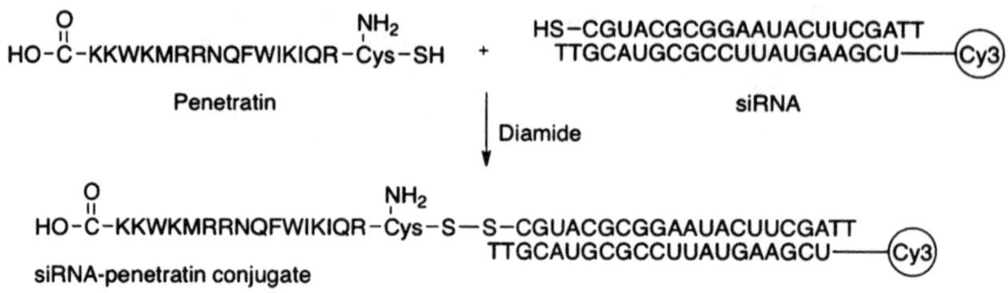

Fig. 15 Synthesis of disulfide-linked siRNA–penetratin conjugate, labeled with Cy3, for siRNA cellular delivery study

development of POCs for improving oligonucleotide delivery (Marlin et al. 2010). However, the synthesis of POCs is not straightforward because peptide and oligonucleotide chemistries are not fully compatible (Lonnberg 2009).

Moschos et al. investigated peptide–siRNA conjugates for siRNA delivery in lung (Moschos et al. 2007). They used disulfide-linked constructs synthesized from Npys-activated C-terminal Cys containing TAT (48-60 GRKKRRQRRRPPQC) and penetratin (RQIKIWFQNRRMKWKKC) peptides with N-terminal acetyl and C-terminal amide functions, and conjugated to 5′- of sense strand of a thiol-modified oligonucleotide using disulfide exchange under denaturing conditions in the presence of urea (Fig. 15). The administration of penetratin or TAT (48-60) peptides alone caused a significant reduction in p38 MAP kinase mRNA expression, while the penetratin–siRNA conjugate activated the immune response. Although the use of CPP to enhance the cellular uptake of siRNA seems to be efficient, the use of CPP may be limited due to yet uncharacterized effects on gene expression and potential for immune activation.

Muratovska et al. used CPP–siRNA conjugates to target the firefly *luciferase* or GFP transgenes in COS-7, C166-GFP, and EOMA-GFP cell lines (Muratovska and Eccles 2004). It was shown that CPP–siRNA conjugates efficiently silence *luciferase* and GFP genes in different cells. Moreover, the use of cationic polymers (Lipofectamine 2000) allowed a more efficient transfection in other cell types. The siRNA had a thiol group attached at the 5′- of one strand and a Cy3 label attached to the 5′- of complementary strand to enable the identification of the siRNA within cells by fluorescence microscopy. The reaction between the thiol groups on siRNA and CPP was catalyzed by oxidant diamide (Fig. 15).

4.3 Carbohydrate–Oligonucleotide Conjugates

Major pathways for cellular recognition are based on carbohydrate–protein interactions. Cell surfaces are covered by sugar-binding proteins (i.e., lectins), which are involved in essential biological processes like signal transduction and

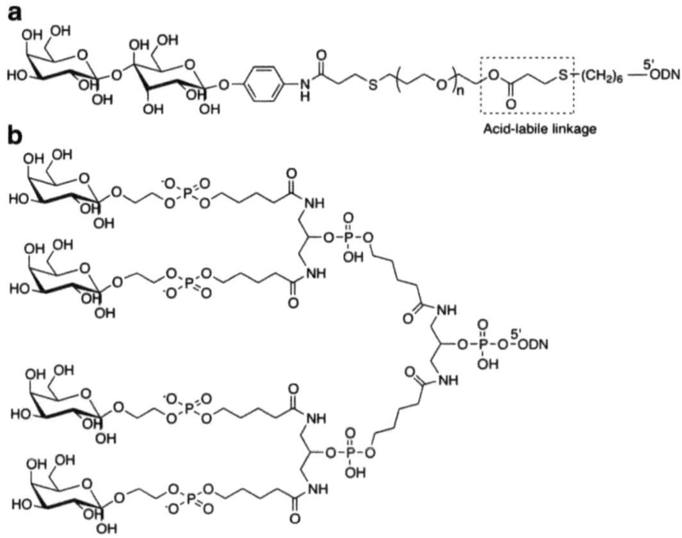

Fig. 16 (**a**) Carbohydrate–siRNA conjugate to prepare pH-sensitive and targetable PIC micelles and (**b**) multivalent carbohydrate–siRNA conjugate

cell adhesion. These lectins are able to recognize and internalize glycoproteins bearing specific carbohydrate moieties through endocytosis. Therefore, there is an interest in design and development of COCs for improving the poor cell or tissue-specific delivery of oligonucleotide through receptor-mediated endocytosis (Lonnberg 2009; Zatsepin and Oretskaya 2004).

Oishi et al. prepared siRNA conjugated to lactosylated PEG at 5′- of sense strand using a β-thiopropionate linker (Fig. 16), followed by complexation with poly(L-lysine), and used to develop pH-sensitive and targetable PIC micelles of siRNA (Oishi et al. 2005). The acid-labile linkage was cleaved at pH corresponding to that in intracellular endosomal compartment (pH = 5.5). This system caused efficient transfection of siRNA in HuH-7 cells (human hepatoma cells) possessing asialoglycoprotein (ASGP) receptors, which recognize compounds bearing terminal galactose moieties. It was observed that the presence of covalently attached carbohydrate resulted in improved uptake of RNAs and that the acid-labile group caused recovery of fully potent interfering RNA.

It should be noted that the individual protein–carbohydrate binding is weak, and therefore, natural processes employ multivalent interactions, where carbohydrates adopt an oligomeric form assembled on natural scaffolds (the glycocluster effect) (Lundquist and Toone 2002). This approach was used for improving the cellular uptake of siRNAs. Avino et al. developed carbohydrate–siRNA conjugates, containing one, two, or four glucose and galactose residues at the 5′- (synthesized using phosphoramidite chemistry), against tumor necrosis factor (Fig. 16) (Avino et al. 2011). It was observed that siRNA duplexes containing galactose were

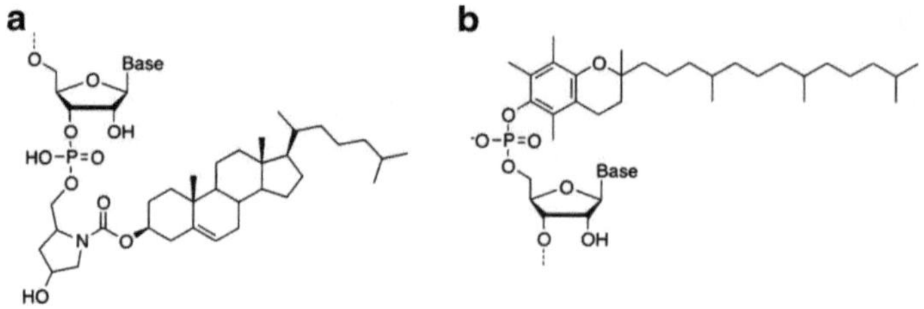

Fig. 17 (**a**) 3′-RNA–cholesterol conjugate for developing antagomirs and (**b**) 5′-α-tocopherol-siRNA conjugate

efficiently delivered to HeLa cells, and these conjugates entered the RNAi pathway to silence gene expression. The results demonstrated the receptor-mediated uptake of siRNA containing a galactose cluster, although the efficacy of the process was not enough for efficient RNA interference experiments.

4.4 *Lipophilic–Oligonucleotide Conjugates*

The cell membranes act as a selective barrier and regulate entry of molecules into the cells. It consists mainly of phospholipid bilayers, and these hydrophobic barriers prevent the entry of hydrophilic molecules into the cells. Due to its polyanionic nature, oligonucleotides are highly hydrophilic, and therefore, they show extremely poor penetration across cell membranes. The use of lipophilic carrier has been proposed to improve cellular uptake by (1) reducing the hydrophilic character of oligonucleotides and/or (2) using the lipoprotein-mediated endocytosis pathway (Manoharan 2002; Rosemeyer 2005).

The synthetic RNAs, known as antagomirs, were used to study the role of miRNA (Fig. 17). Krützfeldt et al. developed modified RNAs, conjugated to cholesterol at 3′- by hydroxypropinol linker, that are efficient and specific silencers of endogenous miRNAs in mice (Krutzfeldt et al. 2005). Efficient in vivo delivery of siRNA to the liver was successfully achieved using α-tocopherol-RNA conjugates. Nishina et al. used siRNA conjugated to α-tocopherol, which was covalently attached to 5′- of antisense strand, using a modified phosphoramidite, to target *apolipoprotein B* (Fig. 17) (Nishina et al. 2008). α-Tocopherol (a form of vitamin E) was chosen to utilize the physiological pathways of vitamin E transport to the liver that involves α-tocopherol-associated proteins (SeC14L2, SEC14L3, and SEC14L4) and albumin-related protein afamin.

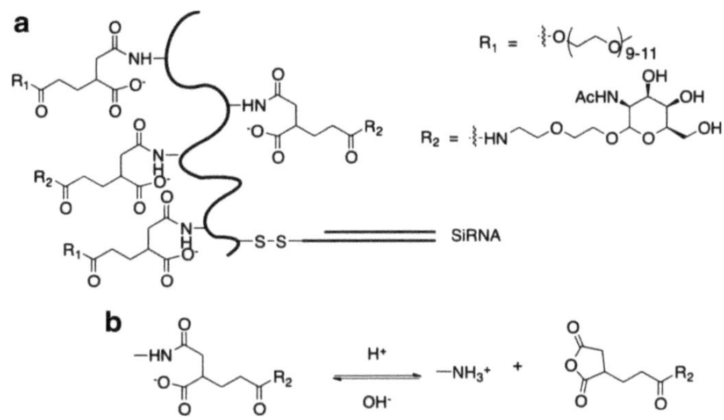

Fig. 18 (**a**) siRNA polyconjugate and (**b**) hydrolysis of maleic anhydride derivatives. In vivo studies demonstrated a maximal knockdown of gene (80–90%) with a dose of 2.5 mg/kg

4.5 Polymer–Oligonucleotide Conjugates

Polymers are used as carrier for oligonucleotide delivery (Pack et al. 2005; Smith et al. 2011). Polymers reduce the toxicity and immunogenicity and improve pharmacokinetic characteristics of oligonucleotides upon conjugation. Among various polymers available, PEG, in particular, is considered the most useful polymer for pharmaceutical applications (Bonora et al. 2003; Harris and Chess 2003). Another advantage of using polymers is the presence of multiple attachment sites.

A poly(vinyl ether) polymer was modified with siRNA, using disulfide linkage, PEG, and *N*-actetylgalactosamine (Fig. 18) (Rozema et al. 2007). PEG was used to provide stealth properties, and *N*-actetylgalactosamine was selected for hepatocyte targeting, and both were linked using maleic anhydride derivative because it is cleaved at lower pH (pH of mature endosome). This hydrolysis led to unmasking of positively charged amines, leading to the release of siRNA into the cytoplasm.

4.6 Metal Complex–Oligonucleotide Conjugates

Oligonucleotides conjugated to metal complexes are useful in electron transfer studies (O'Neill and Barton 2005), as artificial nucleases (Aiba et al. 2011), photoprobes, supramolecular scaffolds (Ghosh and Defrancq 2010; Megger et al. 2011; Bandy et al. 2011), and asymmetrical catalysts (Boersma et al. 2010; Park and Sugiyama 2010; Silverman 2010). Europium complexes have been used for developing a luminescence assay to detect oligonucleotide hybridization (Fig. 19) (Karhunen et al. 2010). Two conjugate probes were synthesized (thiourea linkage): one containing europium ion chelate (europium probe) and the other containing a

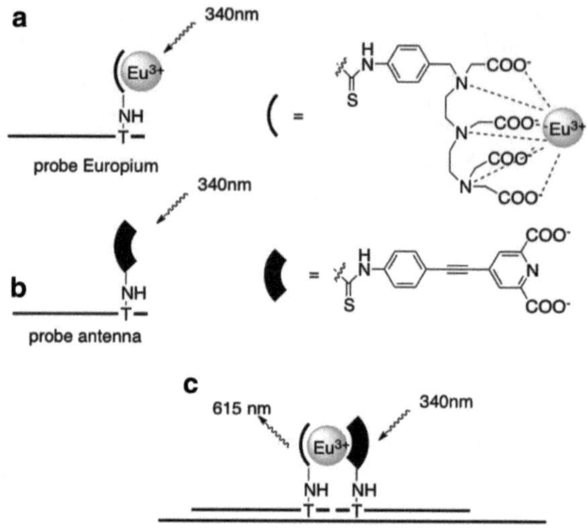

Fig. 19 Europium-based luminescence assay for detecting oligonucleotide hybridization. (**a**) Europium probe, (**b**) antenna probe, and (**c**) luminescent assay

ligand (antenna). Without target oligonucleotide, both probes were non-luminescent, but addition of the target oligonucleotide induced the templated formation of highly luminescent mixed chelate complex (excitation at 340 nm and emission at 615 nm), with detection limit of hybridization being 22 pM.

Ruthenium conjugates have been used as photoprobes (Le Gac et al. 2009). The $Ru(TAP)_2$phen-oligonucleotide conjugates were prepared by solution-phase conjugation between aminooxy-metal complex and oligonucleotide containing aldehyde at 3′- or 5′-. After hybridization, a photoadduct was formed with guanine on the target sequence upon illumination. In the absence of hybridization, the photoadduct was formed with guanine on its own sequence. Sequence-specific cleavage was achieved by exploiting the hydrolytic properties of lanthanides (Jiang et al. 2007). The Ce(III) ethylendiamine *N,N,N′,N′*-tetrakis(methylene-phosphonic acid) (EDTP)–oligonucleotide conjugates caused site-selective cleavage of oligonucleotide (Fig. 20) (Lonnberg et al. 2010). EDTP ligand was conjugated to oligonucleotides at 5′- or 3′- using on-support oxime bond formation (Lonnberg et al. 2008). Conjugates were hybridized to target and then Ce(III) was complexed to EDTP. During incubation, Ce(III) was air oxidized to Ce(IV), leading to the hydrolysis of phosphodiester bond on the target.

The metal complex–oligonucleotide conjugates (MCOCs) have been used for inhibiting zinc finger transcription factor (Harney et al. 2009). A cobalt (III) Schiff base complex [Co(III)-sb] was attached to oligonucleotide that recognizes a gene promoter region, which itself is recognized by zinc finger transition factor family (Snail) (Fig. 21). These transition factors are involved in development and epithelial tumor metastasis. The conjugate was hybridized with the promoter and it bound

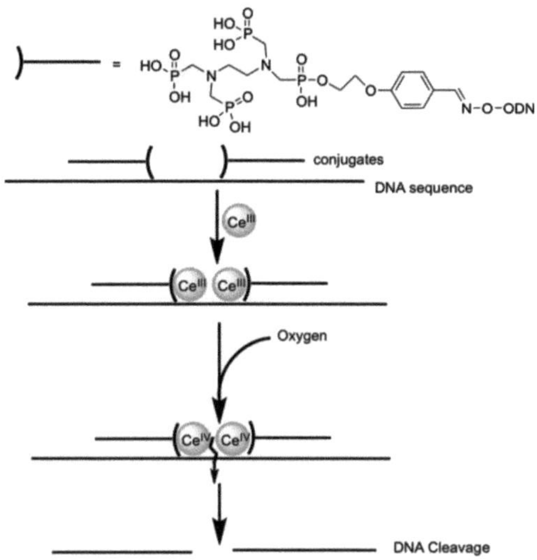

Fig. 20 Site-selective oligonucleotide cleavage using Ce^{III} EDTP conjugates

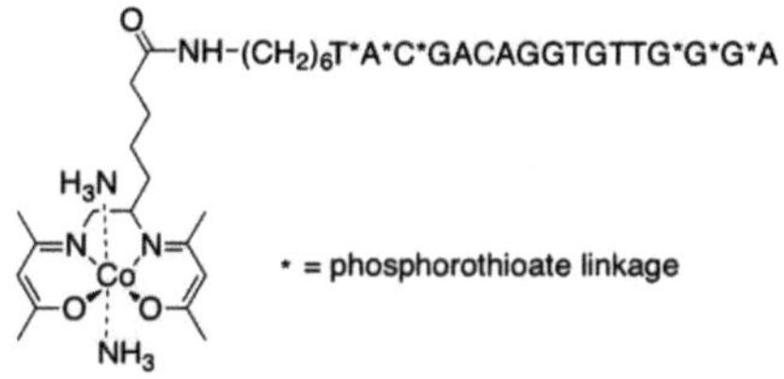

Fig. 21 Metal complex–oligonucleotide conjugate for inhibiting zinc finger transcription factor

irreversibly to zinc finger through a ligand exchange on Co(III). In vitro assay on lysates of frog embryos demonstrated that the conjugate complex was 150-fold more specific than the unconjugated complex.

4.7 Nanoparticle–Oligonucleotide Conjugates

Conjugation of nanoparticles with oligonucleotide is used to develop nanomaterials that combine recognition properties of oligonucleotide with electronic, photonic, and catalytic characteristics of materials. Nanoparticle formulations containing unmodified (unconjugated) oligonucleotide have been used for oligonucleotide delivery (Howard 2009; Yuan et al. 2011). For example, Davis et al. developed formulation composed of unmodified siRNA, cyclodextrin-based polymer, human transferrin protein, and PEG (clinical trial NCT00689065). This formulation was administered systemically to patients with solid cancer, where it reduced the

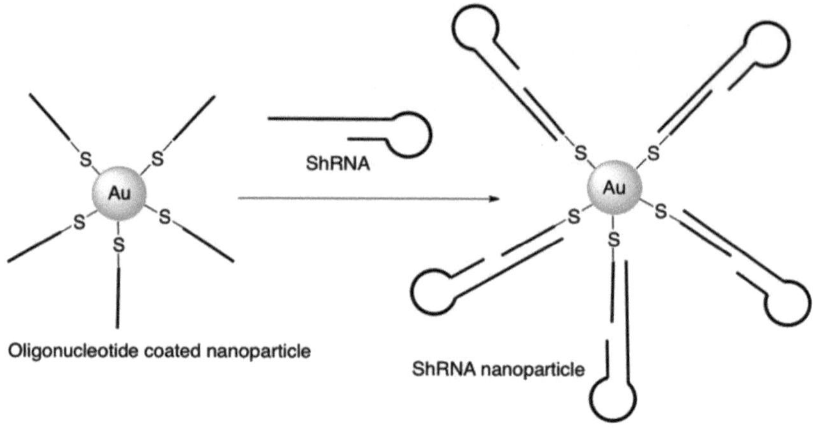

Fig. 22 Preparation of shRNA–gold nanoparticles

expression of target protein, and this represents the first mechanistic evidence of RNAi in human, using administered siRNA (Davis et al. 2010).

Covalent attachment of oligonucleotide to nanoparticles has also been used. This is achieved using following techniques: specific recognition (e.g., biotin–streptavidine), physisorption, and chemisorption. Biotin–oligonucleotide conjugates have been extensively used for immobilization on streptavidine-coated nanoparticles. Cholesterol-conjugated siRNA was used for preparation of phospholipid-based nanoparticles (Yang et al. 2011). Nanoparticles were made with phospholipids, a peptide mimicking apolipoprotein A–I, cholesteryl oleate, and cholesteryl siRNA. siRNAs were trapped into the phospholipid monolayer using its cholesteryl moiety. It was observed that this carrier delivered siRNA more efficiently in cytosol than classical lipofectamine. Cellular gene silencing assay demonstrated that these particles reduce the expression of targeted protein to 35%, whereas cholesteryl siRNA alone reduced it to 84%.

Chemisorption of oligonucleotide on nanoparticles depends on their nature (gold, quantum dots, magnetic nanoparticles, etc.). Gold nanoparticles have been extensively studied for delivery of drug, gene, and siRNA (Ghosh et al. 2008; Duncan et al. 2010). The simplest strategy for coating gold nanoparticles is by chemisorption of thiolated oligonucleotides on gold surface (Fritzsche and Taton 2003). Gold nanoparticles have proved to be efficient system for siRNA delivery. In a recent example, thiolated single-strand oligonucleotide-coated gold nanoparticles and shRNA with a tail at 5′- complementary to single-strand oligonucleotide were developed (Ryou et al. 2010) (Fig. 22). It delivered shRNA efficiently into the xenograft tumors implanted in mice (Ryou et al. 2011). These particles successfully halted the aggressive growth of tumor, as their volume decreased to 36%.

Quantum dots (QD) possess unique photoluminescent properties, which allow their detection by optical imaging (Akerman et al. 2002). These properties were used to study the influence of linker between particle and siRNA on gene silencing (Singh et al. 2010a) (Fig. 23).

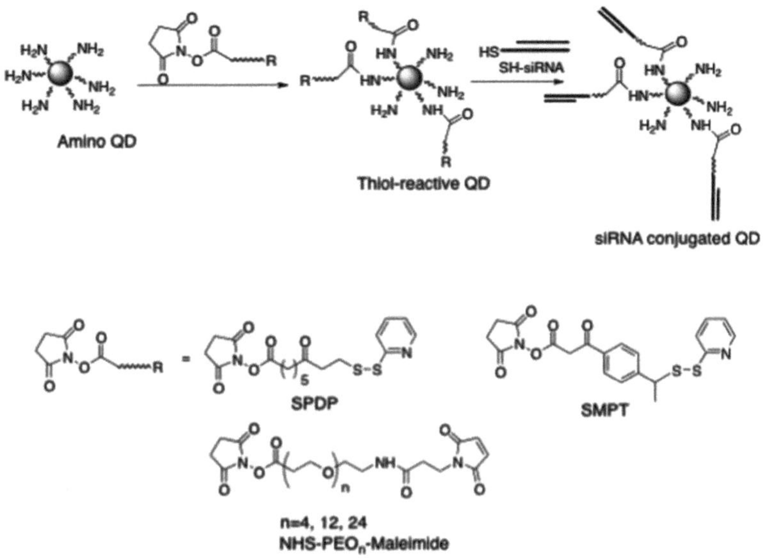

Fig. 23 Preparation of siRNA-quantum dots to investigate the influence of linker on siRNA release

Amino QDs were modified with five different hetero-bifunctional linkers containing NHS for anchoring QDs and thiol-reactive functional group for anchoring thiol-modified siRNA. Two linkers, SPDP and SMPT, contained pyridylsulfenyl group that leads to the formation of glutathione-labile disulfide bond upon addition of SH–siRNA. The three others contained a variable PEG linker and maleimide group that leads to the formation of non-labile thioether linkage. The particles were tested in vitro, and it was shown that the best gene silencing occurs with either labile linkers or with long ($n = 24$) non-labile linkers. In addition, the effect of thiol position in siRNA was studied, and it was demonstrated that thiol could be carried either at sense strand or antisense strand. Finally, it was also shown that if these conjugates were incubated for 8 h in serum (mimicking bloodstream) before delivery to cells, only those containing a long non-labile PEG ($n = 24$) thioether linker were still active, demonstrating that a labile linker can lead to the release of siRNA before they reach cells.

4.8 DNA Biochips

DNA chips have received considerable interest for diagnostic applications (Beaucage 2001). It contains oligonucleotide probes immobilized on surface (e.g., glass, gold) and is used for hybridization with complementary target sequences

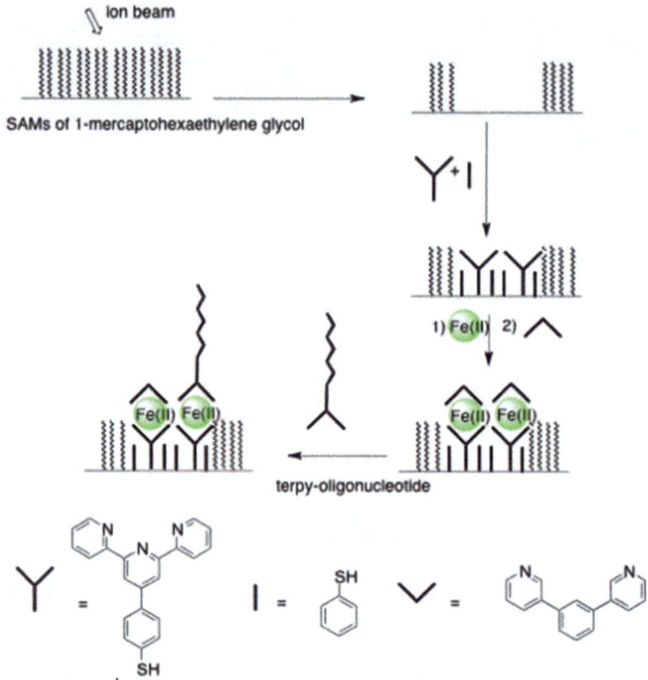

Fig. 24 Patterning of surface by formation of Fe(II)-terpyridine oligonucleotide conjugate

obtained from experimental or clinical samples. Hybridization is detected (transduction) using optical (e.g., fluorescence, SPR) or electrochemical methods.

The two major challenges associated with this technology are immobilization of probes on surface and transduction of hybridization. Oligonucleotide conjugation is used to address both challenges. For example, FOs complementary to a part of target (sandwich assay) can be used for transduction. Oligonucleotide conjugation protocols are employed for immobilizing probes on surface, as they are highly efficient in controlling spatial distribution without perturbing hybridization properties. These strategies have been reviewed earlier (Beaucage 2001).

MCOCs [Fe(II) terpyridine complexes] were used for immobilization or patterning of oligonucleotides (Tuccitto et al. 2011) (Fig. 24). SAMs of 1-mercapto hexaethylene glycol were formed on gold surface. Patterns of bare gold were etched with an ion beam, and a mixture of [4′-(4-mercaptophenyl)-2, 2′:6′2″-terpyridine] (MPTP) and mercaptobenzene (MB) was immobilized on these patterns. MPTP was then complexed with Fe(II) and terpyridine. Tolylterpyridine-tagged oligonucleotide probes were then immobilized on these patterns using exchange with terpyridine. Surfaces were characterized using TOF-SIMS and QCM-D, which demonstrated the controlled immobilization of oligonucleotide on patterns.

DNA chips are not restricted to detection of target oligonucleotide. A preformed DNA chip can be used for controlled immobilization of other molecules (peptides,

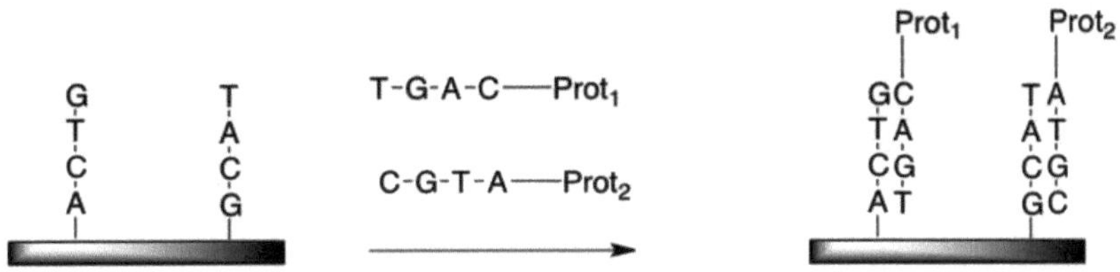

Fig. 25 DNA-directed immobilization of protein array

proteins, carbohydrates), and this approach is known as DNA-directed immobilization (DDI) (Fig. 25). Niemeyer developed this approach (Niemeyer et al. 1999) and used it to fabricate protein (Bano et al. 2009) and carbohydrate biochips (Zhang et al. 2009).

Recently, another type of DNA biochip, known as aptasensor, was reported (Lim et al. 2010; Sassolas et al. 2011; Song et al. 2008), which contains aptamers immobilized on surface. Aptasensors could be particularly useful for protein detection, as they have the advantage of small size, chemical synthesis and modification, and stability (compared to antibodies). Immobilization techniques used in aptasensors are similar to those used in "classical" DNA array.

5 Oligonucleotide Conjugates in Clinical Trial

About 52 antisense oligonucleotide, 24 siRNA, and 21 aptamer candidates were being evaluated in clinical trials in January 2012. As far as oligonucleotide conjugates are concerned, only four candidates are currently undergoing clinical trial (Table 1). Three candidates are PEGylated aptamers, ARC1779 (Gilbert et al. 2007), NOX-A12 (Sayyed et al. 2009), and NOX-E36 (Hoffmann et al. 2011), which is not surprising considering the previous success with PEGylated proteins (e.g., Adagen, Oncaspar, Pegasys, PEG-Intron, Krystexxa) (Harris and Chess 2003) and aptamer (pegaptanib) (Ng et al. 2006). The remaining conjugate in clinical trial is a conjugate of synthetic immunostimulatory DNA with protein (ragweed allergen) (Higgins et al. 2006). The limited number of conjugates in clinical trial points to the difficulties in synthesizing conjugates.

6 Oligonucleotide Conjugates in Market

The market approval of oligonucleotide-based drugs has been less than promising. So far, only two oligonucleotide drugs have been approved for clinical use. The first drug was a phosphorothioate antisense oligonucleotide (fomivirsen/Vitravene) for

Table 1 Oligonucleotide conjugates in clinical trials

Name (identi fier)	Oligonucleotide	Promoiety	Company	Indications	Status
ARC1779 (NCT00632242)	Aptamer	PEG (20 kDa)	Archemix Corp.	Purpura, thrombotic thrombocyto penic, von Willebrand dis ease type 2B	Phase II
NOX-A12 (NCT00976378)	45-mer L-RNA aptamer	PEG (40 kDa)	Noxxon Pharma AG	Autologous stem cell transplantation	Phase I
NOX-E36 (NCT0097629)	40-mer L-RNA aptamer	PEG (40 kDa)	Noxxon Pharma AG	Chronic in flammatory disease, type 2 diabetes melli tus, systemic lupus erythe matosus	Phase I
AIC/Amb a1 immunostimula tory conjugate (NCT00346086)	Synthetic im munostimula tory DNA	Protein (ragweed allergen)	Johns Hop kins Uni versity	Seasonal aller gic rhinitis	Phase II

Data were collected from the website http://clinicaltrials.gov in January 2012. The search terms used were "antisense oligonucleotide," "siRNA," "aptamer," and "oligonucleotide conjugate," respectively

treating cytomegalovirus (CMV)-induced retinitis (Crooke 1998), whereas the second drug was a PEG conjugate of an aptamer (pegaptanib/Macugen) for treating AMD (Ng et al. 2006).

Fomivirsen sodium injection (Vitravene) was the first oligonucleotide drug approved by the FDA in 1998 for the treatment of CMV-induced retinitis (Crooke 1998). CMV retinitis is a viral infection of the retina that occurs in persons with AIDS, and if untreated, it leads to blindness. Vitravene is a 21-mer phosphorothioate oligonucleotide (5′-GCG TTT GCT CTT CTT CTT GCG-3′) designed to inhibit human CMV replication by an antisense mechanism. It was administered intravitreally once weekly during induction and every 2 or 4 weeks during maintenance period. The drug showed safety and efficacy in newly diagnosed patients and patients who have failed prior therapy with available CMV drugs. The drug was manufactured by ISIS Pharmaceuticals, Carlsbad, CA, but was withdrawn from the market on account of poor demand.

Pegaptanib sodium injection (Macugen) is another oligonucleotide drug approved by the FDA in 2004 for the treatment of all types of neovascular (wet) AMD (Ng et al. 2006). AMD is characterized by the growth of normal and leaky vessels under macula (central part of retina), which break and leak blood and fluids, causing loss of central vision. Pegaptanib is a PEGylated RNA aptamer directed against vascular endothelial growth factor (VEGF)-165, an isoform of which is responsible for pathological ocular neovascularization and vascular permeability. The aptamer is 28 nucleotides in length (5′-CGG AAU CAG UGA AUG CUU AUA CAU CCG dT-3′) and contains 2′-F pyrimidines and 2′-*O*-Me purines. It terminates into pentylamino linker to which 40 kDa PEG is conjugated. This is achieved by covalently attaching two 20 kDa monomethoxy PEGs to two amino groups on lysine, and free carboxylic group on lysine is covalently attached to the free amino group on pentylamino linker present on aptamer. PEGs are used to prolong plasma half-life, reduce immunogenicity, lower renal clearance, and decrease uptake of oligonucleotides by cells of RES. The drug was discovered by Gilead and is marketed by Eyetech Pharmaceuticals/Pfizer. It is administered intravitreally once every 6 weeks. It shows higher efficacy when compared to usual care irrespective of angiographic subgroups.

Both approved oligonucleotide drugs were locally delivered, which again points to the fact that delivery remains a major concern for oligonucleotide therapeutics.

7 Conclusions

Diverse oligonucleotide conjugates have been prepared and evaluated for various diagnostic, therapeutic, and material applications. Promoieties like fluorescent dyes, peptides, carbohydrates, lipids, polymers, metal complexes, nanoparticles, metal surfaces, etc., have been conjugated to oligonucleotide to either improve their existing properties or to endow them with new properties for different applications. Use of oligonucleotide conjugates in nucleic acid sequencing, FISH, and microarray technology is now fairly established, but most of oligonucleotide conjugates being developed for clinical applications are still in their early phase of development. The data available are mostly from "proof of principle" in vitro studies and to a limited extent in vivo studies. Nevertheless, enough evidence supporting the rationale for oligonucleotide conjugate development has been collected. Approval of pegaptanib, a PEG conjugate of an aptamer, has provided a fresh impetus to the field. However, it must be said that progress made in this field is far from satisfactory. There are several reasons for this limited success. Tremendous progress has been made so far as development of synthesis protocols for oligonucleotide conjugates is concerned, but conjugate synthesis, particularly of combinatorial libraries, still remains tricky. Major breakthrough in stepwise solid-phase synthesis approaches is needed to address this issue. There is a serious lack of structure–activity relationship studies, and it is difficult to make general predictions about the utility of conjugates. Both approved oligonucleotide drugs are delivered locally. Despite expectations, conjugation has not been able to fully resolve poor cell uptake and delivery of oligonucleotides. Thus, systemic delivery of oligonucleotides still remains a major challenge.

Acknowledgments We gratefully acknowledge our students, postdoctoral fellows, and colleagues who contributed to this work. Financial support to Prof. Eric Defrancq came from Cluster Région Rhone-Alpes, the Centre National pour la Recherche Scientifique (CNRS), and Université Joseph Fourier Grenoble 1. Dr. Yashveer Singh is supported by grant from NIH HIT-IT program (R01AI084137-01) to Prof. Patrick J. Sinko (PI).

References

Aiba Y, Sumaoka J, Komiyama M (2011) Artificial DNA cutters for DNA manipulation and genome engineering. Chem Soc Rev 40:5657–5668

Akerman ME, Chan WC, Laakkonen P et al (2002) Nanocrystal targeting in vivo. Proc Natl Acad Sci USA 99:12617–12621

Asseline U (2006) Development and applications of fluorescent oligonucleotides. Curr Org Chem 10:491–518

Avino A, Ocampo SM, Lucas R et al (2011) Synthesis and in vitro inhibition properties of siRNA conjugates carrying glucose and galactose with different presentations. Mol Divers 15:751–757

Bandy TJ, Brewer A, Burns JR et al (2011) DNA as supramolecular scaffold for functional molecules: progress in DNA nanotechnology. Chem Soc Rev 40:138–148

Bano F, Fruk L, Sanavio B et al (2009) Toward multiprotein nanoarrays using nanografting and DNA directed immobilization of proteins. Nano Lett 9:2614–2618

Beaucage SL (2001) Strategies in the preparation of DNA oligonucleotide arrays for diagnostic applications. Curr Med Chem 8:1213–1244

Berezhna SY, Supekova L, Supek F et al (2006) siRNA in human cells selectively localizes to target RNA sites. Proc Natl Acad Sci USA 103:7682–7687

Boersma AJ, Megens RP, Feringa BL et al (2010) DNA-based asymmetric catalysis. Chem Soc Rev 39:2083–2092

Bonora GM, Ballico M, Campaner P et al (2003) Synthesis of oligonucleotide-peptide PEG-conjugated: the EGG (oligonucleotide)-chicken (peptide) dilemma? Nucleosides Nucleotides Nucleic Acids 22:1255–1257

Bouchard PR, Hutabarat RM, Thompson KM (2010) Discovery and development of therapeutic aptamers. Annu Rev Pharmacol Toxicol 50:237–257

Bouillon C, Meyer A, Vidal S et al (2006) Microwave assisted "click" chemistry for the synthesis of multiple labeled-carbohydrate oligonucleotides on solid support. J Org Chem 71:4700–4702

Charles I, Xue L, Arya DP (2002) Synthesis of aminoglycoside-DNA conjugates. Bioorg Med Chem Lett 12:1259–1262

Chiu YL, Ali A, Chu CY et al (2004) Visualizing a correlation between siRNA localization, cellular uptake, and RNAi in living cells. Chem Biol 11(8):1165–1175

Corey DR (2007) RNA learns from antisense. Nat Chem Biol 3:8–11

Crooke ST (1998) Vitravene—another piece in the mosaic. Antisense Nucleic Acid Drug Dev 8(4):vii–viii

Cunningham JJ, Crocker LS, Leone A (2011) Tackling the challenges of nucleic acid delivery: progress and new approaches. Drug Dev Deliv 11:36–40

Da Ros T, Spalluto G, Prato M et al (2005) Oligonucleotides and oligonucleotide conjugates: a new approach for cancer treatment. Curr Med Chem 12:71–88

Dausse E, Da Rocha GS, Toulme JJ (2009) Aptamers: a new class of oligonucleotides in the drug discovery pipeline? Curr Opin Pharmacol 9:602–607

Davis ME, Zuckerman JE, Choi CH et al (2010) Evidence of RNAi in humans from systemically administered siRNA via targeted nanoparticles. Nature 464:1067–1070

Dey S, Sheppard TL (2001) Ketone-DNA: a versatile postsynthetic DNA decoration platform. Org Lett 3:3983–3986

Dorsett Y, Tuschl T (2004) siRNAs: applications in functional genomics and potential as therapeutics. Nat Rev Drug Discov 3:318–329

Duncan B, Kim C, Rotello VM (2010) Gold nanoparticle platforms as drug and biomacromolecule delivery systems. J Control Release 148:122–127

Edupuganti OP, Singh Y, Defrancq E et al (2004) New strategy for the synthesis of 3′,5′-bifunctionalized oligonucleotide conjugates through sequential formation of chemoselective oxime bonds. Chem Eur J 10:5988–5995

El-Sagheer AH, Cheong VV, Brown T (2011) Rapid chemical ligation of oligonucleotides by the Diels-Alder reaction. Org Biomol Chem 9:232–235

Elbashir SM, Harborth J, Lendeckel W et al (2001) Duplexes of 21-nucleotide RNAs mediate RNA interference in cultured mammalian cells. Nature 411:494–498

Ellington AD, Szostak JW (1990) In vitro selection of RNA molecules that bind specific ligands. Nature 346:818–822

Fire A, Xu S, Montgomery MK et al (1998) Potent and specific genetic interference by double-stranded RNA in *Caenorhabditis elegans*. Nature 391:806–811

Forget D, Boturyn D, Defrancq E et al (2001) Highly efficient synthesis of peptide-oligonucleotide conjugates: chemoselective oxime and thiazolidine formation. Chem Eur J 7:3976–3984

Fritzsche W, Taton TA (2003) Metal nanoparticles as labels for heterogeneous, chip-based DNA detection. Nanotechnology 14:R63–R73

Gait MJ (2003) Peptide-mediated cellular delivery of antisense oligonucleotides and their analogues. Cell Mol Life Sci 60:844–853

Ghosh P, Han G, De M et al (2008) Gold nanoparticles in delivery applications. Adv Drug Deliv Rev 60:1307–1315

Ghosh S, Defrancq E (2010) Metal-complex/DNA conjugates: a versatile building block for DNA nanoarrays. Chem Eur J 16:12780–12787

Gilbert JC, DeFeo-Fraulini T, Hutabarat RM et al (2007) First-in-human evaluation of anti von Willebrand factor therapeutic aptamer ARC1779 in healthy volunteers. Circulation 116:2678–2686

Goncalves MS (2009) Fluorescent labeling of biomolecules with organic probes. Chem Rev 109:190–212

Goodchild J (1990) Conjugates of oligonucleotides and modified oligonucleotides: a review of their synthesis and properties. Bioconjug Chem 1:165–187

Graham D, Enright A (2006) Cycloadditions as a method for oligonucleotide conjugation. Curr Org Synth 3:175–213

Gramlich PM, Wirges CT, Manetto A et al (2008) Postsynthetic DNA modification through the copper-catalyzed azide-alkyne cycloaddition reaction. Angew Chem Int Ed 47:8350–8358

Harney AS, Lee J, Manus LM et al (2009) Targeted inhibition of Snail family zinc finger transcription factors by oligonucleotide-Co(III) Schiff base conjugate. Proc Natl Acad Sci USA 106:13667–13672

Harris JM, Chess RB (2003) Effect of pegylation on pharmaceuticals. Nat Rev Drug Discov 2:214–221

Hedley D, Ogilvie L, Springer C (2007) Carboxypeptidase-G2-based gene-directed enzyme-prodrug therapy: a new weapon in the GDEPT armoury. Nat Rev Cancer 7:870–879

Higgins D, Rodriguez R, Milley R et al (2006) Modulation of immunogenicity and allergenicity by controlling the number of immunostimulatory oligonucleotides linked to Amb a 1. J Allergy Clin Immunol 118:504–510

Hoffmann S, Hoos J, Klussmann S, Vonhoff S (2011) RNA aptamers and spiegelmers: synthesis, purification, and post-synthetic PEG conjugation. Curr Protoc Nucleic Acid Chem 46:1–30, Chapter 4:Unit 4

Howard KA (2009) Delivery of RNA interference therapeutics using polycation-based nanoparticles. Adv Drug Deliv Rev 61:710–720

Jiang Q, Xiao N, Shi P et al (2007) Design of artificial metallonucleases with oxidative mechanism. Coord Chem Rev 251:1951–1972
Jones D (2011) The long march of antisense. Nat Rev Drug Discov 10:401–402
Juliano R, Alam MR, Dixit V et al (2008) Mechanisms and strategies for effective delivery of antisense and siRNA oligonucleotides. Nucleic Acids Res 36:4158–4171
Kachalova AV, Stetsenko DA, Romanova EA et al (2002) A new and efficient method for synthesis of 5′-conjugates of oligonucleotides through amide bond formation on solid phase. Helv Chim Acta 85:2409–2416
Karhunen U, Jaakkola L, Wang Q et al (2010) Luminescence switching by hybridization-directed mixed lanthanide complex formation. Anal Chem 82:751–754
Krutzfeldt J, Rajewsky N, Braich R et al (2005) Silencing of microRNAs in vivo with 'antagomirs'. Nature 438:685–689
Kurreck J (2003) Antisense technologies. Improvement through novel chemical modifications. Eur J Biochem 270:1628–1644
Le Gac S, Rickling S, Gerbaux S et al (2009) A photoreactive ruthenium(II) complex tethered to a guanine-containing oligonucleotide: a biomolecular tool that behaves as a "Seppuku molecule". Angew Chem Int Ed 48:1122–1125
Lim YC, Kouzani AZ, Duan W (2010) Aptasensors: a review. J Biomed Nanotechnol 6:93–105
Lonnberg H (2009) Solid-phase synthesis of oligonucleotide conjugates useful for delivery and targeting of potential nucleic acid therapeutics. Bioconjug Chem 20:1065–1094
Lonnberg T, Aiba Y, Hamano Y et al (2010) Oxidation of an oligonucleotide-bound Ce(III)/multiphosphonate complex for site-selective DNA scission. Chem Eur J 16:855–859
Lonnberg T, Suzuki Y, Komiyama M (2008) Prompt site-selective DNA hydrolysis by Ce(IV)-EDTA using oligonucleotide multiphosphonate conjugates. Org Biomol Chem 6:3580–3587
Lundquist JJ, Toone EJ (2002) The cluster glycoside effect. Chem Rev 102:555–578
Manoharan M (2002) Oligonucleotide conjugates as potential antisense drugs with improved uptake, biodistribution, targeted delivery, and mechanism of action. Antisense Nucleic Acid Drug Dev 12:103–128
Marlin F, Simon P, Saison-Behmoaras T et al (2010) Delivery of oligonucleotides and analogues: the oligonucleotide conjugate-based approach. Chembiochem 11:1493–1500
Megger DA, Guerra CF, Hoffmann J et al (2011) Contiguous metal-mediated base pairs comprising two Ag(I) ions. Chem Eur J 17:6533–6544
Moschos SA, Jones SW, Perry MM et al (2007) Lung delivery studies using siRNA conjugated to TAT(48-60) and penetratin reveal peptide induced reduction in gene expression and induction of innate immunity. Bioconjug Chem 18:1450–1459
Muratovska A, Eccles MR (2004) Conjugate for efficient delivery of short interfering RNA (siRNA) into mammalian cells. FEBS Lett 558:63–68
Ng EW, Shima DT, Calias P et al (2006) Pegaptanib, a targeted anti-VEGF aptamer for ocular vascular disease. Nat Rev Drug Discov 5:123–132
Niemeyer CM, Boldt L, Ceyhan B et al (1999) DNA-directed immobilization: efficient, reversible, and site-selective surface binding of proteins by means of covalent DNA-streptavidin conjugates. Anal Biochem 268:54–63
Niemeyer CM, Ceyhan B, Hazarika P (2003) Oligofunctional DNA-gold nanoparticle conjugates. Angew Chem Int Ed 42:5766–5770
Nishina K, Unno T, Uno Y et al (2008) Efficient in vivo delivery of siRNA to the liver by conjugation of alpha-tocopherol. Mol Ther 16:734–740
O'Neill MA, Barton JK (2005) Sequence-dependent DNA dynamics: the regulator of DNA-mediated charge transport. In: Wagenknecht HA (ed) Charge transfer in DNA: from mechanism to application. Wiley-VCH, Germany, pp 27–75
Ocampo SM, Albericio F, Fernandez I et al (2005) A straightforward synthesis of 5′-peptide oligonucleotide conjugates using N(alpha)-fmoc-protected amino acids. Org Lett 7:4349–4352
Ohrt T, Merkle D, Birkenfeld K et al (2006) In situ fluorescence analysis demonstrates active siRNA exclusion from the nucleus by Exportin 5. Nucleic Acids Res 34:1369–1380

Oishi M, Nagasaki Y, Itaka K et al (2005) Lactosylated poly(ethylene glycol)-siRNA conjugate through acid-labile beta-thiopropionate linkage to construct pH-sensitive polyion complex micelles achieving enhanced gene silencing in hepatoma cells. J Am Chem Soc 127:1624–1625

Opalinska JB, Gewirtz AM (2002) Nucleic-acid therapeutics: basic principles and recent applications. Nat Rev Drug Discov 1:503–514

Pack DW, Hoffman AS, Pun S et al (2005) Design and development of polymers for gene delivery. Nat Rev Drug Discov 4:581–593

Park S, Sugiyama H (2010) DNA-based hybrid for asymmetric organic synthesis. Angew Chem Int Ed 49:3870–3878

Paroo Z, Corey DR (2004) Challenges for RNAi in vivo. Trends Biotechnol 22:390–394

Paterson BM, Roberts BE, Kuff EL (1977) Structural gene identification and mapping by DNA-mRNA hybrid-arrested cell-free translation. Proc Natl Acad Sci USA 74:4370–4374

Rayburn ER, Zhang R (2008) Antisense, RNAi, and gene silencing strategies for therapy: mission possible or impossible? Drug Discov Today 13:513–521

Rosemeyer H (2005) Nucleolipids: natural occurrence, synthesis, molecular recognition, and supramolecular assemblies as potential precursors of life and bioorganic materials. Chem Biodivers 2:977–1063

Rostovtsev VV, Green LG, Fokin VV et al (2002) A stepwise huisgen cycloaddition process: copper(I)-catalyzed regioselective "ligation" of azides and terminal alkynes. Angew Chem Int Ed 41:2596–2599

Rozema DB, Lewis DL, Wakefield DH et al (2007) Dynamic PolyConjugates for targeted in vivo delivery of siRNA to hepatocytes. Proc Natl Acad Sci USA 104:12982–12987

Ryou SM, Kim S, Jang HH et al (2010) Delivery of shRNA using gold nanoparticle-DNA oligonucleotide conjugates as a universal carrier. Biochem Biophys Res Commun 398:542–546

Ryou SM, Park M, Kim JM et al (2011) Inhibition of xenograft tumor growth in mice by gold nanoparticle-assisted delivery of short hairpin RNAs against Mcl-1 L. J Biotechnol 156:89–94

Sassolas A, Blum LJ, Leca-Bouvier BD (2011) Optical detection systems using immobilized aptamers. Biosens Bioelectron 26:3725–3736

Sayyed SG, Hagele H, Kulkarni OP et al (2009) Podocytes produce homeostatic chemokine stromal cell-derived factor-1/CXCL12, which contributes to glomerulosclerosis, podocyte loss and albuminuria in a mouse model of type 2 diabetes. Diabetologia 52:2445–2454

Schilling CI, Jung N, Biskup M et al (2011) Bioconjugation via azide–Staudinger ligation: an overview. Chem Soc Rev 40:4840–4871

Schoch J, Wiessler M, Jaschke A (2010) Post-synthetic modification of DNA by inverse-electron-demand Diels-Alder reaction. J Am Chem Soc 132:8846–8847

Semple SC, Akinc A, Chen J et al (2010) Rational design of cationic lipids for siRNA delivery. Nat Biotechnol 28:172–176

Seo TS, Li Z, Ruparel H et al (2003) Click chemistry to construct fluorescent oligonucleotides for DNA sequencing. J Org Chem 68:609–612

Silverman SK (2010) DNA as a versatile chemical component for catalysis, encoding and stereocontrol. Angew Chem Int Ed 49:7180–7201

Singh N, Agrawal A, Leung AK et al (2010a) Effect of nanoparticle conjugation on gene silencing by RNA interference. J Am Chem Soc 132:8241–8243

Singh Y, Defrancq E, Dumy P (2004) New method to prepare peptide-oligonucleotide conjugates through glyoxylic oxime formation. J Org Chem 69:8544–8546

Singh Y, Murat P, Defrancq E (2010b) Recent developments in oligonucleotide conjugation. Chem Soc Rev 39:2054–2070

Singh Y, Spinelli N, Defrancq E (2008) Chemical strategies for oligonucleotide conjugate synthesis. Curr Org Chem 12:263–290

Smith D, Holley AC, McCormick CL (2011) RAFT-synthesized copolymers and conjugates designed for therapeutic delivery of SiRNA. Polym Chem 2:1428–1441

Song S, Wang L, Li J et al (2008) Aptamer-based biosensors. Trends Anal Chem 27:108–117
Spinelli N, Edupuganti OP, Defrancq E et al (2007) New solid support for the synthesis of 3′-oligonucleotide conjugates through glyoxylic oxime bond formation. Org Lett 9:219–222
Stephenson ML, Zamecnik PC (1978) Inhibition of Rous sarcoma viral RNA translation by a specific oligodeoxyribonucleotide. Proc Natl Acad Sci USA 75:285–288
Stetsenko DA, Gait MJ (2000) Efficient conjugation of peptides to oligonucleotides by "native ligation". J Org Chem 65:4900–4908
Tombelli S, Mascini M (2009) Aptamers as molecular tools for bioanalytical methods. Curr Opin Mol Ther 11:179–188
Truffert JC, Asseline U, Brack A et al (1996) Synthesis, purification, and characterization of two peptide oligonucleotide conjugates as potential artificial nuclease. Tetrahedron 52:3005–3016
Tuccitto N, Giamblanco N, Ghosh S et al (2011) Controlled density patterning of tolylterpyridine-tagged oligonucleotides. Langmuir 27:8595–8599
Tuerk C, Gold L (1990) Systematic evolution of ligands by exponential enrichment: RNA ligands to bacteriophage T4 DNA polymerase. Science 249:505–510
Tung CH, Stein S (2000) Preparation and applications of peptide-oligonucleotide conjugates. Bioconjug Chem 11:605–618
Wang K, Tang Z, Yang CJ et al (2009) Molecular engineering of DNA: molecular beacons. Angew Chem Int Ed 48:856–870
Weisbrod SH, Marx A (2008) Novel strategies for the site-specific covalent labelling of nucleic acids. Chem Commun 44:5675–5685
Whitehead KA, Langer R, Anderson DG (2009) Knocking down barriers: advances in siRNA delivery. Nat Rev Drug Discov 8:129–138
Yang M, Jin H, Chen J et al (2011) Efficient cytosolic delivery of siRNA using HDL-mimicking nanoparticles. Small 7:568–573
Yuan X, Naguib S, Wu Z (2011) Recent advances of SiRNA delivery by nanoparticles. Exp Opin Drug Deliv 8:521–536
Zatsepin TS, Oretskaya TS (2004) Synthesis and applications of oligonucleotide-carbohydrate conjugates. Chem Biodivers 1:1401–1417
Zatsepin TS, Romanova EA, Stetsenko DA et al (2003) Synthesis of 2′-modified oligonucleotides containing aldehyde or ethylenediamine groups. Nucleosides Nucleotides Nucleic Acids 22:1383–1385
Zatsepin TS, Stetsenko DA, Arzumanov AA et al (2002) Synthesis of peptide-oligonucleotide conjugates with single and multiple peptides attached to 2′-aldehydes through thiazolidine, oxime, and hydrazine linkages. Bioconjug Chem 13:822–830
Zatsepin TS, Stetsenko DA, Gait MJ et al (2005) Use of carbonyl group addition—elimination reactions for synthesis of nucleic acid conjugates. Bioconjug Chem 16:471–489
Zhang J, Pourceau G, Meyer A et al (2009) DNA-directed immobilisation of glycomimetics for glycoarrays application: comparison with covalent immobilisation, and development of an on-chip ic50 measurement assay. Biosens Bioelectron 24:2515–2521
Zubin EM, Stetsenko DA, Zatsepin TS et al (2005) Oligonucleotides containing 2′-O-[2-(2,3-dihydroxypropyl)amino-2-oxoethyl]uridine as suitable precursors of 2′-aldehydo oligonucleotides for chemoselective ligation. Bioorg Med Chem 13:4912–4920

CRISPR: A Bacterial Immunity System Based on Small RNAs

Rolf Wagner and Ümit Pul

Contents

Abstract CRISPR is the acronym for a recently discovered defense system against the invasion of foreign genetic elements, such as transposons, phages, or plasmids. Originally observed during genome sequencing of bacteria and archaea, the CRISPR system has been proven to confer adaptive and inheritable immunity against foreign nucleic acid. Although quite different in the components and mechanisms involved, it resembles the RNA interference phenomenon of eukaryotes (RNAi), an important control and defense tool with widespread applications in research and medical developments. The CRISPR system consists of genomic clusters of short repeat sequences flanking unique spacer sequences of similar length (CRISPR = *C*lustered *R*egularly *I*nterspaced *S*hort *P*alindromic *R*epeats). The spacer sequences

R. Wagner (✉) • Ü. Pul
Heinrich-Heine-University Düsseldorf, Molecular Biology of Prokaryotes, Universitätsstr. 1, 40225 Düsseldorf, Germany
e-mail: r.wagner@rz.uni-duesseldorf.de; pul@uni-duesseldorf.de

V.A. Erdmann and J. Barciszewski (eds.), *From Nucleic Acids Sequences to Molecular Medicine*, RNA Technologies, DOI 10.1007/978-3-642-27426-8_6,

derive from the DNA of previous invaders and are taken up in the bacterial CRISPR array by mechanisms not yet solved. In case of a new infection, CRISPR arrays are transcribed and the RNA is processed and assembled into ribonucleoprotein complexes, which target the DNA of the invader through base pair complementarity. The targeted DNA is then inactivated by hydrolytic destruction.

Keywords Cas proteins • Horizontal Gene Transfer • Prokaryotic RNAi • Phage • Regulatory RNA

1 Introduction

Interactions between phages and microbes are of exceptional importance, not only for the understanding of basic mechanisms in molecular biology but also because of their direct implications for biotechnology and food industries, which often rely on resistant and robust microbial strains. Bacteria have evolved a variety of defense mechanisms to survive phage infections, which, due to the extremely high phage abundance on earth and their exceptional efficiency in killing microbes, represent a mortal threatening for all microbial life. There are about 10^{30} phage infections per day, and within 2 h a single T7 phage is able to destroy a culture of about ten billion bacterial cells, producing the incredible number of ten trillion phage particles (Young 2008). A number of bacterial defense mechanisms, such as blocking absorption, inhibiting DNA uptake, restriction of incorporated DNA, or proteins that control horizontal gene transfer, are known for quite some time and contribute to the limited spreading of phages (Chibani-Chennoufi et al. 2004). More recently, a new potential system in the combat between foreign genetic elements and microbes, termed CRISPR, has been discovered in almost all archaea and about half of the known eubacteria. The acronym CRISPR stands for *C*lustered *R*egularly *I*nterspaced *S*hort *P*alindromic *R*epeats. Organisms that contain CRISPR systems are characterized by the presence of one or several such CRISPR arrays on their genome. Operationally, the defense system resembles eukaryotic immune systems and confers resistance to those bacteria, which have survived an encounter with phages or other foreign genetic elements. During such encounters, short DNA sequences (protospacers) of the invader are integrated into the microbial genome as unique spacers between palindromic repeats of the CRISPR array, where they act as inheritable molecular memory conferring immunity against subsequent exposures to homolog nucleic acids. As a result of a complex reaction, the unique spacer sequences of the CRISPR arrays provide short RNAs, which target invading nucleic acids by sequence complementarity. Hence, immunity is based on small RNAs carrying a spacer. These RNAs guide the defense apparatus to foreign nucleic acids that match the spacers. The system is thus sequence specific, adaptive, and inheritable. Moreover, information on different challenging invaders may be stored simultaneously by the acquired spacer-repeat units of a growing CRISPR

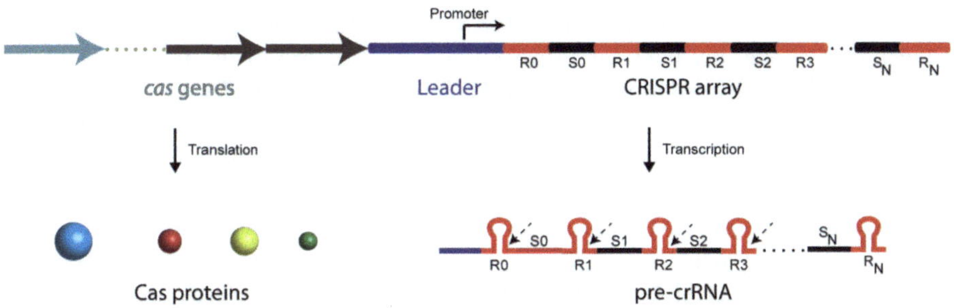

Fig. 1 Typical organization of a bacterial CRISPR defense system. A typical CRISPR system consists of an array of (palindromic) repeat units, shown in *red* (R0, R1, R2,. . .) interspaced by unique spacer sequences originating from invading DNA, depicted in black (S0, S1, S2,. . .). The CRISPR array is flanked by a leader region (*blue*), harboring the promoter sequence, which directs transcription of the complete precursor CRISPR RNA. A set of *cas* genes is found adjacent to the leader region, which encode Cas proteins involved in the specific stages of the immunization, maturation, and interference reactions

array. Hence, in a way, the CRISPR array reflects the history of encounters with invading genetic elements for the respective organism (He and Deem 2010).

The CRISPR arrays are preceded by a sequence element, termed leader, which contain the transcription start sites from which a primary transcript of the CRISPR spacer-repeat unit is initiated. The primary CRISPR RNA is subsequently processed to small crRNAs comprising spacers flanked by truncated repeat sequences. In addition to the CRISPR spacer-repeat sequences, the active defense system requires the action of specific proteins closely associated with the CRISPR array, termed Cas proteins (*C*RISPR-*as*sociated proteins) (Fig. 1). Cas proteins typically contain functional domains known from nucleases, helicases, polymerases, or nucleotide-binding proteins. Cas proteins and CRISPR repeats constitute highly diverse families, which have recently been grouped into three major types with further division into several subtypes characterizing different CRISPR/Cas systems (Makarova et al. 2011). The diversity of the Cas proteins among different species indicates that CRISPR systems are also highly diverse and a multitude of different mechanisms can be expected, resulting in a great versatility of CRISPR defense systems.

Cas proteins are involved in all essential steps of the CRISPR-directed defense, such as the integration of new DNA sequences into the CRISPR array (immunization) and the processing of precursor CRISPR RNA transcripts to short crRNAs, as well as the interference and hydrolysis reactions leading to the inactivation of the foreign nucleic acids.

Functionally, the CRISPR-mediated immunity of prokaryotes against foreign DNA can be divided into three stages (van der Oost et al. 2009; Al-Attar et al. 2011). The first stage represents the adaptation/immunization of the host with invading foreign DNA by integration of new spacer sequences into the CRISPR array. The second stage comprises the expression of the CRISPR elements (Cas proteins and CRISPR arrays) and the processing of the pre-crRNAs to the small crRNAs, which are complementary to the invader DNA. The third stage, designated interference stage, involves the targeting and inactivation of the foreign DNA by crRNA-mediated recognition and hydrolysis of the foreign DNA.

Within the last 2 years, CRISPR systems have been studied intensively in different organisms, and tremendous progress has been made with respect to the characterization of the components involved. Moreover, several specific steps underlying the expression and interference stages have been elucidated. Remarkable progress has also been made in solving the three-dimensional structures of many Cas proteins at high resolution, but also low-resolution structures of several composite structures of Cas proteins with bound crRNA, acting as functional units for CRISPR RNA processing and interference, are available. Yet, many open questions still remain to be answered. In the following chapter, we will summarize the known details of the components that constitute typical CRISPR systems and illuminate the known individual steps and complex mechanisms underlying this fascinating prokaryotic immunity system. Moreover, we will outline some open and unsolved questions and give a brief perspective on future scientific and technical prospects related to the CRISPR defense system.

2 The Components Involved

2.1 Organization and Structure of CRISPR Arrays

The main principle of the CRISPR defense system, the identification and destruction of foreign nucleic acid by crRNA-mediated interference, seems to be similar in CRISPR-containing organisms, although there are notable mechanistic differences between the three major CRISPR types (Fig. 2). The mechanistic variation is reflected by the modular organization of the CRISPR operons. In general, a CRISPR operon consists of a CRISPR array and a leader region. In addition, it is associated with a set of *cas* genes, encoding for the Cas proteins (Fig. 1).

CRISPR arrays do not contain open reading frames but are generally composed of identical, often palindromic, repeat sequences between about 20 and 45 bp in length, which are interspaced by highly variable sequences of similar length (between about 20 and 70 bp) (Karginov and Hannon 2010). The number of repeat-spacer units is highly variable and can basically differ between a few and several hundred, the average number being around 60 (Marraffini and Sontheimer 2010a). Frequently, single organisms exhibit more than one CRISPR array in their genome. In that case, similar repeat sequences are generally shared between the different arrays. Within a single array, the repeat sequences are highly conserved, with the exception of the last repeat, which is sometimes truncated (Jansen et al. 2002). Transcription of the palindromic repeat sequences gives rise to stem–loop secondary structures of the CRISPR RNA transcripts, which likely serve as recognition elements for processing and maturation reactions. Repeat sequences are not

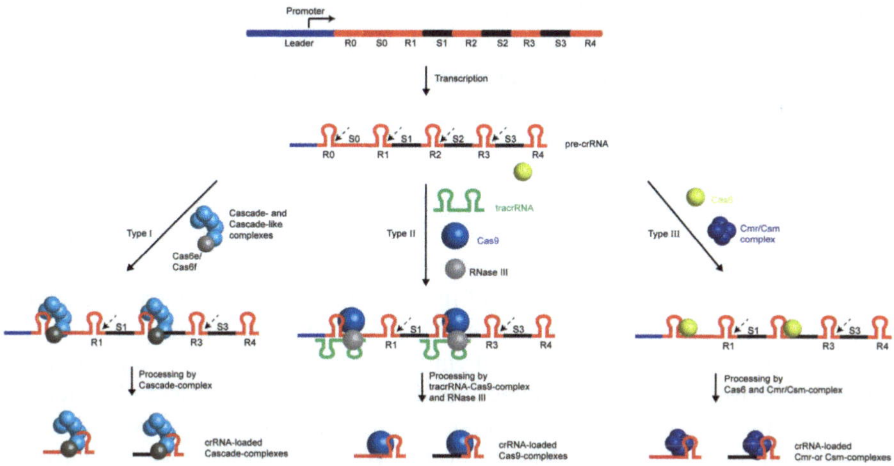

Fig. 2 Processing of CRISPR RNAs: three major types of the CRISPR system. Three types of CRISPR system have been defined to classify ten CRISPR subtypes (Makarova et al. 2011). The type I system comprises six subtypes (I-A to I-F), which use higher-order Cas protein complexes (Cascade) for processing of the pre-crRNA by cleavage at the stem–loop structures (indicated by *arrows*). The catalytic subunit of the complex is the endonuclease Cas6 (*gray*). The processed crRNAs remain bound to the Cascade complex. In type II systems (II-A, II-B), which lack the endonuclease Cas6, a trans-encoded noncoding RNA, called tracrRNA, is involved in processing of the pre-crRNA through base pairing with repeat sequences. Instead of a high-order protein complex, the processing of the pre-crRNA is catalyzed by the multidomain protein Cas9 and the non-Cas protein RNase III. In type III systems (III-A, III-B), the Cas6 protein processes the pre-crRNAs. The final crRNAs are transferred to multiprotein complexes Csm (III-A) or Cmr (III-B), respectively (see Sect. 2.2.5)

conserved between different organisms, however, and not all repeat sequences are palindromic. On the other hand, related species generally have similar repeat structures.

The analysis of spacer sequences revealed that they are nonidentical and often match foreign mobile genetic elements. Spacers are thus generally considered to derive from phages and plasmids, although only a few spacer sequences have been directly identified by sequencing as of phage or plasmid origin. The obvious reason for that is the small number of known phage sequences in comparison to the enormous number of existing phages in nature. Note, however, that frequently, spacer sequences have been identified, which match endogenous genomic sequences of the microbe itself, raising the question of autoimmunity or self-regulation (see Sect. 4.2). In a pioneering study of the *Streptococcus thermophilus* CRISPR system, it was experimentally demonstrated that spacer sequences matching sequences from a phage genome (protospacer) provide immunity against that particular phage. Moreover, direct evidence was obtained that new spacer-repeat sequences from the phage were acquired upon phage infection (Barrangou et al. 2007).

Interestingly, several studies consistently show that spacer sequences within CRISPR arrays correspond to both template and nontemplate strands of phage genomes without apparent preference for particular phage genes or intergenic regions (Makarova et al. 2006; Shah et al. 2009). This observation was taken as evidence that DNA, not RNA, is the target of the CRISPR defense, a fact that was experimentally verified for *Escherichia coli*, where crRNAs could be isolated from the coding or noncoding strand of phage lambda (Brouns et al. 2008). Note, however, that this is not the case for all CRISPR systems and RNA could be identified as target in the CRISPR system of *Pyrococcus furiosus* (see Sect. 3.3.1).

A leader region is found upstream of the first repeat-spacer element of CRISPR arrays. Leader sequences do not contain any protein-encoding reading frames and generally have a high adenine and thymine content. The leader sequences are conserved between similar prokaryotes but differ significantly between less-related species. Within the leader region, promoter elements are located, which direct the transcription of the entire CRISPR array. While in *Sulfolobus solfataricus*, *Pyrococcus furiosus*, and several other species transcription of the CRISPR RNA seems to be constitutive, transcription of the *E. coli* K12 CRISPR array is under control of the regulator protein H-NS (Pul et al. 2010). It is also known that the absence of a leader restricts the incorporation of new spacer-repeat elements into the CRISPR array, indicating that leader sequences are also important for the incorporation of new spacers (Lillestøl et al. 2006; Marraffini and Sontheimer 2008).

2.2 Higher-Order Cas Protein Complexes

Although Cas proteins are highly diverse, it has become clear that in almost all systems functional complexes exist, which are assembled from more than one individual Cas protein. Whether or not non-Cas proteins are also involved in constituting functional complexes for the CRISPR-mediated defense is presently studied intensively. The only documented case thus far is RNase III, which seems to interact with Cas9 (Csn1) during processing of *Streptococcus pyogenes* pre-crRNA (see Sect. 2.2.4).

According to a recent convention, the heterogeneous Cas complexes have been classified into three distinct types: type I, type II, and type III (Table 1) (Makarova et al. 2011). In type I CRISPR systems, the processing of the pre-crRNA is achieved by a higher-order Cas protein complex, termed Cascade (*C*RISPR-*as*sociated *c*omplex for *a*ntiviral *de*fense). Together with the type I-specific Cas protein Cas3, the crRNA-loaded Cascade complexes recognize and inactivate the target DNA. In contrast, type II CRISPR systems lack Cascade components and Cas3. In type II, the processing of the pre-crRNA is catalyzed by the multidomain protein Cas9 and the RNase III protein. Moreover, the processing requires a small noncoding RNA (tracrRNA) (see Sect. 2.2.4). The Cas9 protein has also been shown to mediate the targeting and inactivation of the target DNA by cleavage. Type III

Table 1 Cas protein composition and repeat sequences of different CRISPR/Cas system types

		Effector proteins[a] (1)	Adaptation proteins[b]	CRISPR consensus repeat sequence[c] (2)
Type I	**I-A**	Csa5, Cas8a1, Cas7, Cas5, Cas8a2, Cas6, Cas3′-Cas3″	Cas1, Cas2, Cas4	GCATA**TCCCT**AA**AGGGA**ATAGAAAG *Aeropyrum pernix* K1
	I-B	Cas6, Cas8b, Cas7, Cas5, Cas3′-Cas3″	Cas1, Cas2, Cas4	GTTTTT**AT**C**TTCCT**AAG**AGGAA**T**AT**GAAC *Thermotoga neapolitana* DSM 4359
	I-C	Cas5, Cas8c, Cas7, Cas3	Cas1, Cas2, Cas4	GT**CGCCCC**TCACGCG**GGGGCG**TGGATAGAAAC *Desulfovibrio vulgaris* str. Miyazaki F
	I-D	Cas10d, Csc2, Csc1, Cas6, Cas3	Cas1, Cas2, Cas4	CTTTCCTTCTACTAA**TCCCG**GCGAT**CGGGA**CTGAAAC *Synechocystis* sp. PCC 6803plasmid pSYSA
	I-E	Cse1, Cse2, Cas7, Cas5, Cas6e, Cas3	Cas1, Cas2	CGGTTTA**TCCCCGC**TGGC**GCGGGGA**ACAC *Escherichia coli* K12
	I-F	Csy1, Csy2, Csy3, Cas6f, Cas3	Cas1, Cas2	TTTCTAAG**CTGCC**TGTGC**GGCAG**TGAAC *Yersinia pestis* Z176003
Type II	**II-A**	Cas9	Csn2	**GTTTT**AGAGCTATGCTGTTTTGAATGGTCCC**AAAAC** *Streptococcus pyogenes* M1 GAS
	II-B	Cas9	Cas4	GCAACACTTTATAGCAAATCCGCTTAGCCTGTGAAAC *Wolinella succinogenes* DSM 1740
Type III	**III-A**	Cas6, Cas10, Csm2, Csm3, Csm4, Csm5 Csm6	Cas1, Cas2	AGTTTCCGT**CCCC**TCTC**GGGG**TTTTGGGTCTGACGAC *Mycobacterium tuberculosis* CDC1551
	III-B	Cmr1, Cas10, Cmr3, Cmr4, Cmr5, Cmr6, Cas6	Cas1,Cas2	CTTCAATTCTTTTGTAGTCTTATTGGAAC *Pyrococcus furiosus* DSM 3638

(1) Cascade proteins underlined

(2) Palindromic sequences are underlined; representative organisms are indicated

[a]The table is adapted from recent CRISPR and Cas protein classification by Makarova and colleagues (Makarova et al. 2011)

[b]Note that Cas1 has been proposed, in addition to its role in antiviral defense, to function in DNA repair (Babu et al. 2011)

[c]Consensus repeat sequences are taken from CRISPRdb (http://crispr.u-psud.fr/) or CRISPI (http://crispi.genouest.org/) databases

CRISPR systems also use a higher-order Cas protein complex for processing, termed Cmr/Csm complex. In contrast to type I and type II systems, which target DNA, the psiRNA–Cmr complexes type III-B bind and cleave complementary RNA.

The following sections review three functional Cascade complexes of type I, the CRISPR RNA maturation by Cas9, RNase III and tracrRNA in type II, and the type III crRNA processing by the Csm and Cmr complexes.

2.2.1 The *E. coli* Cascade Complex (Cascade of Type I-E)

In *E. coli* K12 (Cse subtype), a functional complex of the five Cas proteins Cse1 (CasA), Cse2 (CasB), Cas7 (CasC), Cas5 (CasD), and Cas6e (CasE), termed Cascade (see Table 1), is responsible for the maturation of the pre-crRNA (Brouns et al. 2008). The complex has a nonuniform subunit composition with one molecule each Cse1, Cas5, and Cas6e, six Cas7, and two Cse2 molecules ($CasA_1B_2C_6D_1E_1$). The catalytic activity for the processing reaction resides in Cas6e, which has endoribonuclease activity (Fig. 2). In fact, mutation of the invariant residue His20 of Cas6e abolishes the processing reaction without interfering Cascade complex assembly. The cleaved small crRNAs remain bound to the Cascade complex after processing has occurred, enabling the recognition of invaded nucleic acids, which match the crRNA sequence. Supported by Cas3, which has predicted nuclease/helicase activity, the crRNA-loaded Cascade complex binds and subsequently cleaves foreign DNA (Brouns et al. 2008). A low-resolution structure of the 404-kDa Cascade complex together with a 61-nucleotide-long crRNA has been solved, displaying an unusual seahorse-shaped architecture. The backbone of the complex is composed of six Cas7 subunits that are arranged in an arch (Jore et al. 2011). According to a recent refinement of the structure, the architecture of the complex also reflects the trajectory of the crRNA along the backbone of the complex formed by the Cas proteins (Wiedenheft et al. 2011a).

2.2.2 The Archaeal Cascade of *Sulfolobus solfataricus* (Cascade of Type I-A)

Sulfolobus solfataricus contains the type I-A (Apern-type) CRISPR system. Similar to the *E. coli* Cascade, the proteins Cas8a1/2, Csa5, Cas7, Cas5, and Cas6 assemble into a functional complex, with Cas6 as the catalytic subunit and a backbone formed by multimers of Cas7 (see Table 1) (Lintner et al. 2011). Moreover, the Cas7 multimers are arranged into a right-handed helical structure, which very likely spans the complete crRNA. The formation of the Cascade complex depends on the presence of crRNA, suggesting a cooperative Cascade assembly, mediated by crRNA (Lintner et al. 2011). The processed crRNA remains bound to the Cascade targeting the foreign DNA by sequence complementarity.

2.2.3 The Csy Complex in *Pseudomonas aeruginosa* (Cascade of Type I-F)

Pseudomonas aeruginosa encodes four Cas proteins (Csy1, Csy2, Csy3, and Cas6f), which have been shown to form a 350-kDa complex responsible for the sequence-specific targeting of foreign nucleic acids (see Table 1). The complex has a stoichiometry of one Csy1, one Csy2, six Csy3, and one Cas6f molecules each. A low-resolution structure of this complex has been solved, indicating a crescent-shaped protomer. The overall structure exhibits striking similarity to the *E. coli* Cascade complex, although the sequences of the Cas proteins are not conserved (Wiedenheft et al. 2011b). Apparently, conserved higher-order structures of Cas complexes exist between unrelated CRISPR/Cas systems of type I (Makarova et al. 2011). The responsible Cas protein for the endoribonucleolytic processing of the pre-crRNA is Cas6f for which a high-resolution structure is available (Haurwitz et al. 2010). Cleavage of the pre-crRNA occurs at the 3′-end of each stem–loop structure, resulting in crRNAs, 60 nucleotides in length, which include the 32-nucleotide spacer sequence. Attached to Cas6f of the Csy complex, the processed crRNA forms sequence-specific hybrids with complementary targets.

2.2.4 Alternative Processing by Noncoding RNAs (tracrRNA) in *Streptococcus pyogenes* in Effector Complexes Type II

Organisms, such as *Streptococcus pyogenes*, containing CRISPR/Cas system of type II lack the endoribonucleases Cas6, which is responsible for crRNA maturation, and have evolved a different processing strategy. A small (24 nucleotides) *trans*-encoded RNA, termed tracrRNA, has been identified, which interacts by base pair complementarity with the repeat structures of the pre-crRNA transcript. The resulting double-strand structure of the RNA–RNA complex is susceptible to hydrolysis by the non-Cas nuclease RNase III (Fig. 2). The reaction requires the action of the multifunctional Cas protein Cas9. In most cases, tracrRNA genes are located upstream of the first Cas protein gene. Processed crRNAs between 39 and 42 nucleotides in length can be isolated from *Streptococcus pyogenes* cells, which are likely resulting from two cleavage reactions: one within the repeat and the other within the spacer sequence. The mature crRNAs contain at their 5′-end a 20-nucleotide spacer-derived sequence that serves as guide element and a 19–22-nucleotide repeat-derived 3′ sequence element. Hence, the 3′-end is different compared to mature crRNAs processed by CRISPR/Cas effector complexes containing Cas6 subtype protein from *E. coli*, *Pyrococcus furiosus*, or *Staphylococcus epidermidis*, which have an eight-nucleotide tag sequence from the upstream repeat (see Sect. 3.2.1). The system represents an alternative processing mechanism and is the first reported case, where a non-Cas protein host factor (RNase III) is involved in the CRISPR immunity reaction (Deltcheva et al. 2011). The tracrRNA-mediated processing of pre-crRNA is found in other bacterial species with type II CRISPR systems, such as *Streptococcus mutans*, *Listeria innocua*, *Neisseria meningitidis*, or *Streptococcus thermophilus* as well.

2.2.5 Type III Effector Complexes with Different Target Recognition

Two different type III CRISPR/Cas systems have been described (type III-A and type III-B), which differ in the nature of the nucleic acid that is recognized and cleaved. Both type III systems have Cascade-like effector complexes but are constituted either by Csm or Cmr subtype Cas proteins (Fig. 2). The Csm complex of *Staphylococcus epidermidis* (type III-A) is composed of Cas6, Cas10, Csm2, Csm3, Csm4, and Csm5, and the Cmr complex of *Pyrococcus furiosus* (type III-B) consists of the RAMP module Cas proteins Cmr1-Cas10-Cmr3-Cmr4-Cmr5-Cmr6. Both complexes contain the Cas6-processed crRNA as guide molecules. In *Pyrococcus*, Cas6 has been identified as endoribonuclease, which cleaves the pre-crRNA transcript into crRNA units including a partial (eight-nucleotide-long) repeat sequence at the 5′-end as part of the Cas–crRNA complex (Hale et al. 2009). While the type III-A system of *Staphylococcus epidermidis* has elegantly been demonstrated to target DNAs (Marraffini and Sontheimer 2008), the *Pyrococcus furiosus* (type III-B) system was shown to target and degrade complementary RNA (Hale et al. 2009) (see Sect. 3.3.1).

3 The Three Stages of CRISPR Defense

3.1 First Stage: Immunization and Uptake of Spacer Sequences

Although the acquisition of new spacers from phages or plasmids can be demonstrated experimentally in several species, the mechanisms how such foreign nucleic acids are recognized and how protospacers are selected, cleaved, or otherwise extracted and finally integrated into the CRISPR array are still widely unknown, making the acquisition of new spacer-repeat units the most elusive part of the CRISPR-mediated defense (Fig. 3).

The occurrence of template and nontemplate phage sequences within CRISPR arrays makes dsDNA the most likely source of new spacers. If mRNAs were the source, the action of a reverse transcriptase must be postulated. Although putative reverse transcriptase genes are found associated with some *cas* genes, many CRISPR/Cas loci lack such genes (Makarova et al. 2006). Therefore, a central question not yet solved is related to the discrimination between self and nonself DNA. Unlike the situation for restriction–modification systems, where the determination is based on specific methylation pattern of the host DNA, no evidence for differences in methylation of self or foreign DNA is available for CRISPR immunization.

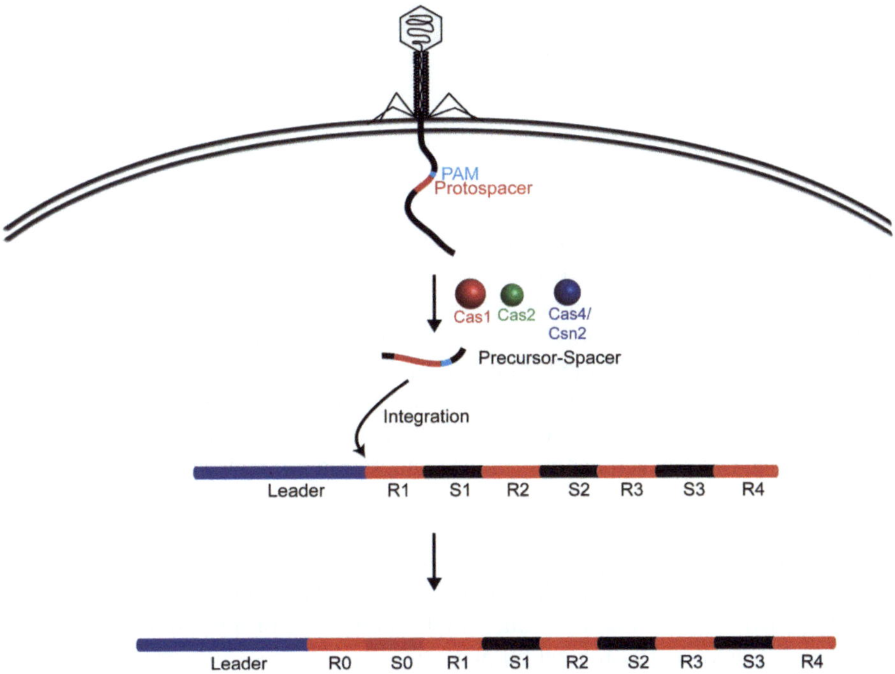

Fig. 3 Immunization of the host cell through uptake of new spacer DNA sequences. The immunization of the host is achieved by integration of short DNA segments (protospacer) from the genome of the invader into the CRISPR array. The PAM motif, located immediately or one position after the protospacer sequence, seems to direct the Cas proteins for cleavage of the invader DNA to yield spacer precursors. The two universal Cas proteins with known nuclease activity Cas1 (*red*) and Cas2 (*green*) are likely involved in selection and generation of the new spacer DNA fragments. In addition, the subtype-specific proteins Cas4 and Csn2 proteins are also considered to participate in spacer DNA uptake, but their exact roles are not yet known. The integration of a new spacer DNA element (S0) always occurs at the leader proximal site, suggesting the involvement of the leader in the integration procedure, for instance, by providing binding regions for Cas proteins. The acquisition of new spacer sequence is associated with duplication of the repeat sequence (R0) and shifting of the existing spacer sequences by one position

3.1.1 Importance of Protospacer-Adjacent Motifs for Spacer Acquisition

Systematic comparison of protospacer flanking sequences in phage genomes has led to the discovery of short conserved sequence motifs (2–3 nucleotides in length), termed Protospacer-Adjacent Motifs (PAM) (*p*rotospacer-*a*djacent *m*otifs). PAM sequences are located immediately or one position after the protospacer in different CRISPR types (Mojica et al. 2009; Shah et al. 2009). Studies have shown that PAM sequences are CRISPR/Cas type-specific and the orientation of the spacer-repeat array with respect to the PAM is conserved, even when the protospacers derive from opposite strands of the invader DNA (Mojica et al. 2009). This finding suggests that PAM sequences could participate in the acquisition and insertion of new spacers in CRISPR arrays (Fig. 3).

3.1.2 Involvement of Cas Proteins in Spacer-Repeat Acquisition

The question, which Cas proteins participate in the integration process, can partly be answered today. The universal and most highly conserved proteins Cas1 (putative nuclease/integrase) and Cas2 (predicted ssRNA-specific endonuclease) are prime candidates for the integration process. Both Cas proteins have been shown to be dispensable for the function of preexisting spacers in *E. coli* (Brouns et al. 2008). Studies with Cas protein mutants have indicated that the subtype-specific Csn2 is involved in the acquisition of new spacers (Barrangou et al. 2007). Moreover, Cas4 has been shown to impair the integration of new spacers, suggesting a role in acquisition and/or integration of new spacer-repeat units (Fig. 3).

Any mechanistic explanation how new spacers are integrated should rely on the key observation that the uptake of new spacer-repeat units into existing CRISPR arrays always occurs at the leader end upstream of the 5′-most spacer-repeat unit. This fact also suggests that the leader might recruit one or more of the above Cas proteins and serve as recombination platform.

3.2 Second Stage: Expression of CRISPR RNA and cas *Genes*

3.2.1 Transcription and Maturation of Precursor CRISPR RNA (Pre-CRISPR RNA)

With one exception, where transcription of both DNA strands of the CRISPR array had been reported in *Sulfolobus acidocaldarius* (Lillestøl et al. 2009), all cases investigated thus far showed that pre-crRNA transcription is unidirectional starting from a promoter structure within the leader region. CRISPR arrays are generally transcribed as long precursor transcripts (pre-CRISPR RNA) usually spanning the complete set of spacer-repeat units. The pre-CRISPR RNAs are subsequently processed endonucleolytically into smaller RNAs (crRNAs) of about 40–60 nucleotides in length, depending on the CRISPR system. Mature crRNAs from the Csy complex, for instance, consist of 32 nucleotides spacer sequence, flanked by eight repeat-derived nucleotides at the 5′-end (5′-handle) and 20 nucleotides forming a 3′-hairpin. The length of the 5′-handle is conserved. The structure of mature crRNAs derived from the various CRISPR effector complexes differs slightly not only in length but also in the chemical nature of the 5′- and 3′-ends. For instance, the crRNAs generated by Csy6f have a terminal 3′ phosphate. In contrast, crRNAs derived from the Cascade complex of type I-E (Cascade from *E. coli* K12) are 61 nucleotides in length, consisting of 32-nucleotide central spacer sequence, flanked by eight repeat-derived nucleotides at the 5′ -terminus (5′-handle) and 21 repeat-derived nucleotides forming a hairpin with a tetranucleotide loop at the 3′-terminus (3′-handle). In contrast to the free 5′-phosphate of the crRNAs generated by Csy6f, crRNAs of type I-E have a hydroxyl group at the 5′-end and a 2′,3′-cyclic phosphate at the 3′-end (Jore et al. 2011).

Irrespective of the different crRNA ends and the Cas protein composition of the effector complexes, in the CRISPR defense systems of type I-E and type II, a single-stranded RNA portion of the crRNA containing the spacer, often initiated by a "seed" sequence (see Sect. 3.3), acts as guide to target the invading nucleic acid by base pair complementarity.

3.2.2 Regulation of CRISPR RNA Transcription in *E. coli* K12

Expression of pre-crRNAs has been reported to be constitutive or slightly enhanced during stationary growth (Lillestøl et al. 2006) for many CRISPR systems, and although RNA polymerase binding sites have been identified (Oshima et al. 2006) or possible promoters predicted (Nonaka et al. 2006; Wade et al. 2006), no comprehensive information about transcription start sites, promoter specificities, and/or putative regulatory elements is available for CRISPR leader promoters. One notable exception is the *E. coli* K12 CRISPR I and II leader promoters, which have been characterized in detail (Pul et al. 2010). Based on footprinting, transcription start site determination, and protein binding studies, the mapped CRISPR promoter was shown to be sigma70 dependent and located about 50 bp upstream of the first nucleotide from the first CRISPR repeat sequence within the leader region (Pul et al. 2010). Furthermore, binding sites for the regulatory protein H-NS, which is also known to silence horizontally acquired foreign DNA elements (Navarre et al. 2007), have been characterized within the AT-rich sequence of the leader. Mutational studies supported the conclusion that transcription of the *E. coli* K12 CRISPR arrays, albeit not completely repressed, is under negative regulation of H-NS.

3.2.3 Regulation of *cas* Gene Expression

Presently, there are no common rules that allow predicting the general expression of *cas* genes, and one rather has to consider that individual *cas* gene expression varies from one species to the other, in line with the diverse classes identified. Often, however, *cas* genes are not transcribed as single genes but as operons. The expression and regulation of one such operon, encoding the *E. coli* Cascade genes (Cse1, Cse2, Cas7, Cas5, and Cas6e), has been studied in detail (Pul et al. 2010; Westra et al. 2010). A sigma70-dependent promoter (P*cas*), located upstream of the Cse1 gene, directs a polycistronic transcript including the downstream *cas* genes Cse2, Cas7, Cas5, and Cas6e. Independent evidence indicates that the two *cas* genes, *cas1* and *cas2*, encoded directly downstream of the Cascade genes, are also included in the polycistronic transcript (Shimada et al. 2009). Under normal growth conditions, the P*cas* promoter-directed transcript in *E. coli* K12 is cryptic, however, due to the efficient silencing by the transcriptional regulator H-NS, which binds to the upstream regulatory region of P*cas* (Pougach et al. 2010; Pul et al. 2010). Hence, the expression of functional Cas proteins requires an antisilencing mechanism releasing the H-NS-dependent repression in order to become effective. Although it was shown that the transcription activator LeuO could relieve H-NS-mediated repression of the Cascade proteins in *E. coli* K12, acting as an H-NS antagonist, the induction of the P*cas* promoter upon phage infection is not yet completely understood (Westra et al. 2010). A very similar antagonistic regulation of *cas* genes transcription has been reported for *Salmonella enterica* involving the transcriptional regulators LRP and H-NS and the antagonist LeuO (Medina-Aparicio et al. 2011).

The fact that the transcriptional regulator LeuO is functionally linked to membrane stress and the possible involvement of a second two-component antisilencing regulatory system BaeSR may indicate that envelope stress triggers a potential signal transduction mechanism (Perez-Rodriguez et al. 2011).

Genome-wide transcriptional analysis of *Thermus thermophilus* after phage infection revealed the upregulation of several cAMP receptor protein (CRP)-regulated genes, including two *cas* operons, while global bacterial metabolism decreased. Since the CRP level did not increase upon phage infection, the authors concluded that cAMP might be the signaling molecule that transmits information in response to phage infection to CRP, which then upregulates these genes (Shinkai et al. 2007; Agari et al. 2010). However, the involvement of cAMP in activating CRISPR leader or *cas* gene promoters has not been shown for other CRISPR systems so far.

If and how expression of *cas* genes is regulated on the translational level is not precisely known, although some posttranscriptional regulation has to occur in order to explain the different stoichiometry of Cas proteins in various effector complexes, which are encoded on the same transcription unit with no internal promoters known. Again, a notable example are the *E. coli* Cascade genes (Cse1, Cse2, Cas7, Cas5, and Cas6e), which have overlapping translation start segments by either three or five codons or extremely short intergenic distances (between 13 and 2 bp) highly suggestive for translational coupling.

3.3 *Third Stage: Interference/Destruction of Foreign Genetic Elements*

The identification of the target DNA relies on complementarity of the Cascade-bound crRNA to the protospacer region of the foreign DNA (Fig. 4). It has been shown for the *E. coli* CRISPR/Cas system (type I-E) in vitro that the crRNA-loaded Cascade complex binds to double-stranded DNA by formation of an R-loop through specific base pairing between crRNA and the targeted DNA strand with

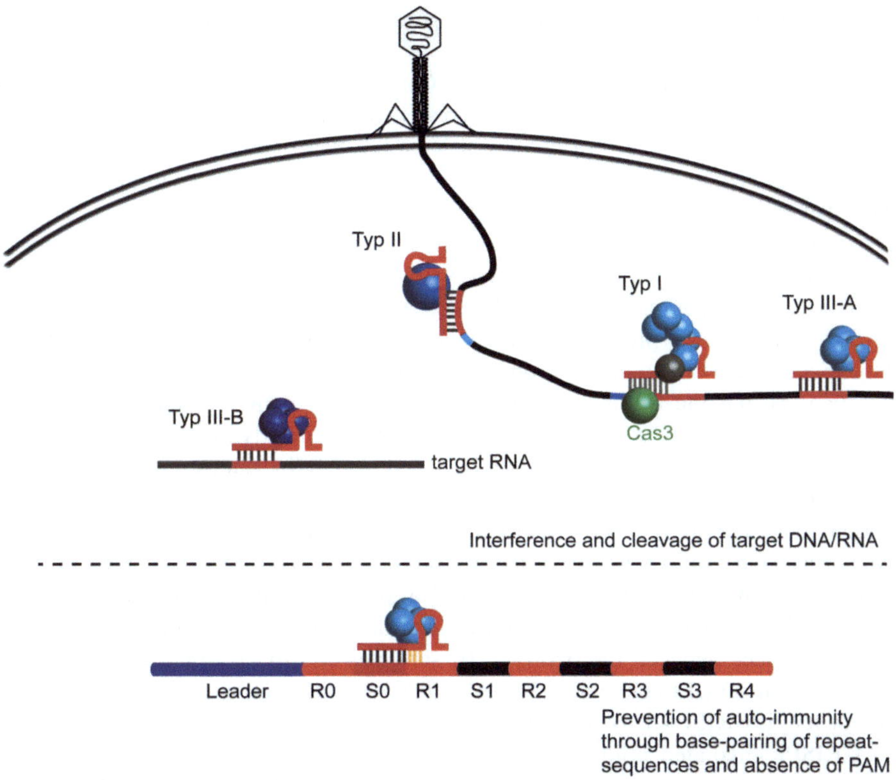

Fig. 4 CRISPR RNA-mediated DNA/RNA interference. The recognition and inactivation of the target DNA is directed by base pairing of crRNAs with the complementary protospacer (*colored red*) of the invading DNA. In type I systems, the rapid detection of the protospacer by crRNA-loaded Cascade is achieved by ~7-nucleotide-long seed sequences (Semenova et al. 2011; Wiedenheft et al. 2011b). After identification of the protospacer, the base pairing extends along the entire crRNA sequence, whereby the noncomplementary DNA strand becomes displaced, forming an R-loop structure (Jore et al. 2011). The single-stranded DNA is very likely cleaved by the HD-nuclease domain of the Cas3 protein. In type II systems, the crRNA–Cas9 ribonucleoprotein complex inactivates the target DNA by cutting at the PAM site in a protospacer-specific way (Garneau et al. 2010). Whether tracrRNA and/or RNase III is also involved in targeting is not known. In type III-A Csm system, the ribonucleoprotein complex targets and inactivates double-stranded DNA (Marraffini and Sontheimer 2008), whereas type III-B Cmr ribonucleoprotein complex binds and cuts complementary RNA molecules in vitro (Hale et al. 2008). As shown for the Csm system (lower part in the figure), the self-targeting by interference with the host CRISPR array is prevented by additional base pairing of the repeat residues of the crRNA with the repeat sequences on the array. The additional base pairing inhibits the nuclease activity and thus prevents autoimmunity of the CRISPR system (Marraffini and Sontheimer 2010b)

simultaneous displacement of the nontargeted DNA strand. The inactivation of the targeted DNA occurs very likely by cleavage of the single-stranded displaced DNA strand by the HD-nuclease domain of Cas3 (Jore et al. 2011). In fact, the Cas3 protein has been shown to exhibit a single-stranded DNase activity (Sinkunas et al. 2011).

The type II system does not contain the Cas3 protein, and the effector complex is a single Cas protein, Cas9 (Fig. 4). In vivo analysis has revealed that a crRNA–Cas9 complex is sufficient to inactivate the target DNA by cleavage within the protospacer region leading to production of linear DNA (Garneau et al. 2010). Whether the effector complex includes the tracrRNA and/or RNase III, which are required for processing of the pre-crRNA, and whether the targeting mechanism bases also on the formation of an R-loop structure, is not known.

The effector complex of type III-B is assembled by the crRNA (originally termed psiRNA for prokaryotic silencing RNA) and high-order Cmr complex (see Sect. 2.2.5). The most pronounced difference to the other CRISPR/Cas subtypes is that the ribonucleoprotein complex Cmr–psiRNA binds and cleaves complementary RNA in vitro (Fig. 4), suggesting the inactivation of invader RNA, e.g., viral mRNA, by a prokaryotic RNA-mediated RNA interference mechanism (Hale et al. 2009) (see also Sect. 3.3.1). Interestingly, the type III-A system has been shown to act exclusively on DNA (Marraffini and Sontheimer 2008). According to Makarova and colleagues, the effector complex has also been proposed to be a higher-order protein complex formed by Csm proteins (Table 1, Fig. 4) (Makarova et al. 2011). Further analyses are required to unravel the molecular details that cause the differences between the two type III subtypes and to explore the mechanistic variations of target nucleic acid (DNA or RNA) identification and inactivation by the different CRISPR/Cas subtypes.

CRISPR-mediated immunity requires that the spacer sequences of the crRNA match the protospacer sequence in the phage genome. Whether or not this match has to be perfect, in order to cause interference, was initially unclear due to some controversial findings. It had been shown, for instance, that mismatches in the PAM sequence abolish CRISPR-mediated immunity, even if the spacer sequence exhibits a perfect match with the protospacer. Moreover, analyses in *Streptococcus thermophilus* demonstrated that single point mutations in the protospacers of phage genomes causing mismatches within the spacer sequence outside of the PAM sequence abolished immunity (Deveau et al. 2008). In contrast to these findings, it was reported that single mismatches between CRISPR spacers and plasmid protospacers did not abolish immunity of a *Sulfolobus solfataricus* CRISPR system (Gudbergsdottir et al. 2011; Manica et al. 2011). A recent study with the *E. coli* CRISPR system could explain the apparent discrepancy (Semenova et al. 2011). It was shown that in order to confer immunity, a perfect match between protospacer DNA and the crRNA was only necessary proximal to the PAM sequence. Single or even multiple mismatches at distal positions within the spacer sequence were tolerated, however. The authors concluded that for efficient interaction between the protospacer and the spacer, a nucleation site close to the PAM region is required. Similar nucleation sequences are known as seed regions in Argonaute complexes during eukaryotic RNA interference (RNAi) (Parker et al. 2009). Hence, target recognition between CRISPR/Cas systems and eukaryotic RNAi may not be too different.

Likewise, experiments performed with the Csy complex from *Pseudomonas aeruginosa* support the existence of seed sequences, which also facilitate efficient target recognition in this CRISPR defense system. Hybridization experiments with different crRNA targets revealed a high-affinity binding site at the 5′-end of the crRNA spacer sequence. This region, spanning a single helical turn, functions as seed sequence. While base pairing between target DNA and the crRNA from the Csy complex is also observed downstream of this sequence, the seed appears to serve as a nucleation site for efficient recognition of matching foreign sequences (Wiedenheft et al. 2011b). Hence, the Csy complex too shows functional similarity to Argonaute, which recognizes mRNA sequences during RNAi in eukaryotes (Lambert et al. 2011). The Csy complex binds ssDNA with significant higher affinity than dsDNA, suggesting that an additional unwinding activity is required for interference or that the complex interacts with intermediates of the phage life cycle that transiently are single-stranded. Whether the importance of seed sequences for rapid scanning of the complementarity of invading DNA is restricted to the type I CRISPR/Cas subtypes (I-E and I-F) or whether this nucleation is also valid for the other subtypes of the three major CRISPR types remains to be shown.

3.3.1 Target Nucleic Acids: DNA and/or RNA?

There is compelling evidence from studies of the CRISPR systems in *Staphylococcus* (type III-A), *E. coli* (type I-E), and *S. thermophilus* (type II-A) that target for the interference reaction is DNA (Brouns et al. 2008; Marraffini and Sontheimer 2008). This finding is consistent with the observation that in the absence of dsRNA intermediates, spacer sequences are found from both the coding and noncoding strands of phage genes. Experimental proof for DNA as target was obtained by artificially interrupting the protospacer of a plasmid gene by a self-splicing intron, which would reconstitute the target sequence in the spliced mRNA. However, insertion blocked interference, indicating that the interference machinery targets DNA (Marraffini and Sontheimer 2008).

In contrast, results from studies of the archaeon *Pyrococcus furiosus* CRISPR system revealed that the CRISPR/Cas effector complex cleaved complementary target RNAs rather than DNA (Hale et al. 2009). The *Pyrococcus furiosus* CRISPR system belongs to the Cmr subtype characterized by RAMP module proteins (repeat-associated mysterious proteins) (Haft et al. 2005). Processing of pre-crRNA results in invader-targeting small RNAs, which differ from crRNAs and are termed psiRNAs (*p*rokaryotic *si*lencing RNAs). Mature psiRNAs are either 36 or 45 nucleotides long and, in addition to 31- or 37-nucleotide invader-targeting guide sequences, contain a conserved 5′ sequence element derived from the CRISPR repeat. These RNAs assemble with six RAMP module Cas proteins forming a Cmr1-6~psiRNA complex (see Table 1 and Sect. 2.2.5). This effector complex binds and cleaves invader RNA in vitro, not DNA. The system may therefore be effective for viral mRNA or RNA phages. Interestingly, the target

RNA is cleavage at a fixed distance from 3′-end of crRNA (14 nucleotides), suggesting a ruler-like mechanism (Hale et al. 2009). It is also interesting to note that PAM sequences may not be necessary in Cmr subtype CRISPR systems targeting RNA.

4 Anomalies and Prospects

4.1 Phage Coevolution

It is quite clear that the CRISPR/Cas system has a strong impact on phage–bacteria coevolution and/or horizontal gene transfer in general. Phage-challenge experiments have shown that mutations at PAM or protospacer sequences of the phage genome allow overcoming the bacterial CRISPR immunity (Garneau et al. 2010). Recent studies on the seed sequences (see Sect. 3.3) have revealed that not all positions of the spacer sequence contribute to the same extent on CRISPR effectiveness (Gudbergsdottir et al. 2011). The CRISPR system tolerates single or multiple mismatches downstream of the PAM and seed sequences, which further limits the mutational escape response of phages (Semenova et al. 2011). Another option of phages to overcome the CRISPR system is the inhibition of CRISPR expression. The bacterial H-NS protein controls horizontal gene transfer by binding to AT-rich foreign DNA and inhibits the expression of foreign genes (Navarre et al. 2006). In addition to its role as silencer of foreign DNA, the H-NS protein inhibits in *E. coli* and *Salmonella typhimurium* the expression of the CRISPR system (Pul et al. 2010; Medina-Aparicio et al. 2011). The control of CRISPR expression by H-NS fits its role as a genome "sentinel" (Dorman 2007). An uncontrolled expression of the CRISPR system as an effective defense weapon constitutes a potential danger for the host itself, e.g., by misintegration of spacers targeting the own genome (see Sect. 4.2). On the other hand, regulation of CRISPR expression by H-NS protein provides phages another feasibility to escape the bacterial CRISPR defense. In fact, recently a phage-encoded *hns* has been reported and proposed to use H-NS protein for repression of bacterial antiviral systems, like CRISPR or restriction–-modification system (Skennerton et al. 2011). The bacterial CRISPR system on the one hand and the anti-CRISPR system of phages on the other hand contribute to the balanced "arms race" of microbes and phages (Sorek et al. 2008).

4.2 Autoimmunity

Acquired immunity to combat foreign invaders is always linked to two important problems, and the CRISPR/Cas system is no exception to this. The two questions are as follows: (1) What is the mechanism to distinguish foreign from own genetic material during immunization? (2) How does the cell avoid destruction of the foreign spacer sequences, which have been taken up into the CRISPR array of the host genome? There is no satisfying answer to the first point yet (see Sect. 3.1.1). With respect to point 2, however, some answers can be provided. A molecular basis for the discrimination of spacer sequences in CRISPR arrays and protospacer sequences of invaders (self-targeting) has been proposed, which relies on mismatches outside of the spacer sequence (Marraffini and Sontheimer 2010b). The protospacer sequences and spacer sequences within CRISPR arrays differ in principle by flanking PAM sequences found only adjacent to protospacers and flanking repeat sequences linked to spacers within CRISPR arrays. Self and nonself DNA therefore differs in the presence or absence of repeat sequences. Hence, sequence complementarity of the transcribed and processed crRNA guides is only perfect with the DNA of CRISPR arrays but exhibits mismatches outside of the protospacer sequence of the invader DNA (Fig. 4). This proposal was experimentally verified for the CRISPR system of *Staphylococcus epidermidis* and explains at least in part that CRISPR arrays are not targeted by autoimmunity mechanisms (Marraffini and Sontheimer 2010b).

There is still the possibility that CRISPR systems can self-target genes or even regulate self-gene expression by (accidental?) uptake of endogenous gene sequences. In fact, the analysis of more than 300 organisms revealed that self-targeting occurs in 18% of all organisms that do have CRISPRs and one in every 250 spacers fully match endogenous sequences of the host genome (Stern et al. 2010).

Since there is no conservation among the identified self-targeting spacers, the authors excluded that self-targeting represents a regulatory phenomenon. Instead, it was observed that integration of self-targeting spacers seems to be linked to inactivation of the respective CRISPR/Cas system. Either *cas* genes were lost or mutated, leader sequences mutated, or the CRISPR system could be otherwise inactivated, like base changes in flanking repeats of self-targeting spacers, resulting in aberrant processing or change in PAM sequences, which are prone to mutations. It appears that self-targeting causing autoimmunity is a "flaw" in the CRISPR mechanism (Stern et al. 2010), explaining the relative abundance of defective CRISPR systems found throughout all organisms.

4.3 *Importance and Application in Biotechnology/Medicine and Food Industry*

The CRISPR/Cas system offers many potential applications in research and development. Besides the possibilities to create robust starter strains with resistance against phages for industrial biotechnology, the system may also be used for directed gene knockout and/or specific gene regulation (Sorek et al. 2008).

The destruction of bacterial population by phages has led very early to the idea to using phages for treatment of infections with pathogenic bacteria (Merril et al. 2003). The prevalent use of antibiotics put this vision into the background, but the increasing number of drug-resistant bacteria reinforces the search of alternative strategies apart from the finding of new antibiotics. The fact that multiresistant pathogenic bacteria lack the CRISPR system underlines the potential of this system to limit the spread of resistance genes by horizontal gene transfer (Palmer and Gilmore 2010). Thus, the construction of strains targeting multidrug resistance genes could be considered as a potential application to limit the spread of antibiotic resistance. Another possibility relies on construction of engineered phages to apply them as adjuvants for antibiotic therapy. In fact, engineered phage targeting bacterial DNA repair system has been shown to enhance the antibiotic therapy and survival of infected mice (Lu and Collins 2009). A similar strategy could be used with engineered phages encoding for CRISPR/Cas proteins and spacer sequences matching bacterial genes. The type II system could be the most appropriate CRISPR system for an approach like this because the Nmeni type requires only one Cas protein (Cas9) in addition to the tracrRNA and the array for interference with phage proliferation (Garneau et al. 2010; Deltcheva et al. 2011). Furthermore, the Nmeni-type CRISPR/Cas system seems to be restricted to pathogenic bacteria and commensals (Haft et al. 2005), which indicates that this subtype may be the most appropriate CRISPR system to act in pathogens.

4.4 Similarity Between Eukaryotic RNA Interference and CRISPR/Cas Systems

Similarities between the CRISPR/Cas system and eukaryotic RNA interference to silence foreign nucleic acids seem obvious. However, apart from the fact that both systems use small RNAs to guide a ribonucleoprotein complex to the target nucleic acid and that functionally analogous proteins can be found in Dicer, Slicer, and the RNA-induced silencing complex (RISC), several key differences exist. For instance, the short RNA duplexes resulting as processing products from dsRNA during RNA interference are typically 21–28 nucleotide in length (Meister and Tuschl 2004), whereas crRNAs, which are flanked by partial repeats, are longer (23–47 nucleotide). Moreover, evidence for RNA-dependent transcription resulting in dsRNA and use of cleaved target RNA during RNA interference is lacking in CRISPR/Cas systems. In contrast to the RNAi, bacterial CRISPR systems make use of DNA-encoded spacers and CRISPR arrays are inheritable to the progeny microbes.

References

Agari Y, Sakamoto K, Tamakoshi M et al (2010) Transcription profile of *Thermus thermophilus* CRISPR systems after phage infection. J Mol Biol 395:270–281

Al-Attar S, Westra ER, van der Oost J et al (2011) Clustered regularly interspaced short palindromic repeats (CRISPRs): the hallmark of an ingenious antiviral defense mechanism in prokaryotes. Biol Chem 392:277–289

Babu M, Beloglazova N, Flick R et al (2011) A dual function of the CRISPR-Cas system in bacterial antivirus immunity and DNA repair. Mol Microbiol 79:484–502

Barrangou R, Fremaux C, Deveau H et al (2007) CRISPR provides acquired resistance against viruses in prokaryotes. Science 315:1709–1712

Brouns SJ, Jore MM, Lundgren M et al (2008) Small CRISPR RNAs guide antiviral defense in prokaryotes. Science 321:960–964

Chibani-Chennoufi S, Bruttin A, Dillmann ML et al (2004) Phage-host interaction: an ecological perspective. J Bacteriol 186:3677–3686

Deltcheva E, Chylinski K, Sharma CM et al (2011) CRISPR RNA maturation by trans-encoded small RNA and host factor RNase III. Nature 471:602–607

Deveau H, Barrangou R, Garneau JE et al (2008) Phage response to CRISPR-encoded resistance in *Streptococcus thermophilus.* J Bacteriol 190:1390–1400

Dorman CJ (2007) H-NS, the genome sentinel. Nat Rev Microbiol 5:157–161

Garneau JE, Dupuis ME, Villion M et al (2010) The CRISPR/Cas bacterial immune system cleaves bacteriophage and plasmid DNA. Nature 468:67–71

Gudbergsdottir S, Deng L, Chen Z et al (2011) Dynamic properties of the Sulfolobus CRISPR/Cas and CRISPR/Cmr systems when challenged with vector-borne viral and plasmid genes and protospacers. Mol Microbiol 79:35–49

Haft DH, Selengut J, Mongodin EF et al (2005) A guild of 45 CRISPR-associated (Cas) protein families and multiple CRISPR/Cas subtypes exist in prokaryotic genomes. PLoS Comput Biol 1:e60

Hale C, Kleppe K, Terns RM et al (2008) Prokaryotic silencing (psi)RNAs in *Pyrococcus furiosus.* RNA 14:2572–2579

Hale CR, Zhao P, Olson S et al (2009) RNA-Guided RNA Cleavage by a CRISPR RNA-Cas Protein Complex. Cell 139:945–956

Haurwitz RE, Jinek M, Wiedenheft B et al (2010) Sequence- and structure-specific RNA processing by a CRISPR endonuclease. Science 329:1355–1358

He J, Deem MW (2010) Heterogeneous diversity of spacers within CRISPR (clustered regularly interspaced short palindromic repeats). Phys Rev Lett 105:128102

Jansen R, Embden JD, Gaastra W et al (2002) Identification of genes that are associated with DNA repeats in prokaryotes. Mol Microbiol 43:1565–1575

Jore MM, Lundgren M, van Duin E et al (2011) Structural basis for CRISPR RNA-guided DNA recognition by Cascade. Nature Struct Mol Biol 18:529–536

Karginov FV, Hannon GJ (2010) The CRISPR system: small RNA-guided defense in bacteria and archaea. Mol Cell 37:7–19

Lambert NJ, Gu SG, Zahler AM (2011) The conformation of microRNA seed regions in native microRNPs is prearranged for presentation to mRNA targets. Nucleic Acids Res 39:4827–4835

Lillestøl RK, Redder P, Garrett RA et al (2006) A putative viral defence mechanism in archaeal cells. Archaea 2:59–72

Lillestøl RK, Shah SA, Brugger K et al (2009) CRISPR families of the crenarchaeal genus *Sulfolobus*: bidirectional transcription and dynamic properties. Mol Microbiol 72:259–272

Lintner NG, Frankel KA, Tsutakawa SE et al (2011) The structure of the CRISPR-associated protein Csa3 provides insight into the regulation of the CRISPR/Cas system. J Mol Biol 405:939–955

Lu TK, Collins JJ (2009) Engineered bacteriophage targeting gene networks as adjuvants for antibiotic therapy. Proc Natl Acad Sci USA 106:4629–4634

Makarova KS, Grishin NV, Shabalina SA et al (2006) A putative RNA-interference-based immune system in prokaryotes: computational analysis of the predicted enzymatic machinery, functional analogies with eukaryotic RNAi, and hypothetical mechanisms of action. Biol Direct 1:7

Makarova KS, Haft DH, Barrangou R et al (2011) Evolution and classification of the CRISPR-Cas systems. Nat Rev Microbiol 9:467–477

Manica A, Zebec Z, Teichmann D et al (2011) In vivo activity of CRISPR-mediated virus defence in a hyperthermophilic archaeon. Mol Microbiol 80:481–491

Marraffini LA, Sontheimer EJ (2008) CRISPR interference limits horizontal gene transfer in *staphylococci* by targeting DNA. Science 322:1843–1845

Marraffini LA, Sontheimer EJ (2010a) CRISPR interference: RNA-directed adaptive immunity in bacteria and archaea. Nat Rev Genet 11:181–190

Marraffini LA, Sontheimer EJ (2010b) Self versus non-self discrimination during CRISPR RNA-directed immunity. Nature 463:568–571

Medina-Aparicio L, Rebollar-Flores JE, Gallego-Hernandez AL et al (2011) The CRISPR/Cas immune system is an operon regulated by LeuO, H-NS and LRP in *Salmonella enterica* serovar Typhi. J Bacteriol 193:2396–2407

Meister G, Tuschl T (2004) Mechanisms of gene silencing by double-stranded RNA. Nature 431:343–349

Merril CR, Scholl D, Adhya SL (2003) The prospect for bacteriophage therapy in Western medicine. Nat Rev Drug Discov 2:489–497

Mojica FJ, Diez-Villasenor C, Garcia-Martinez J et al (2009) Short motif sequences determine the targets of the prokaryotic CRISPR defence system. Microbiology 155:733–740

Navarre WW, Porwollik S, Wang Y et al (2006) Selective silencing of foreign DNA with low GC content by the H-NS protein in *Salmonella*. Science 313:236–238

Navarre WW, McClelland M, Libby SJ et al (2007) Silencing of xenogeneic DNA by H-NS-facilitation of lateral gene transfer in bacteria by a defense system that recognizes foreign DNA. Genes Dev 21:1456–1471

Nonaka G, Blankschien M, Herman C et al (2006) Regulon and promoter analysis of the *E. coli* heat-shock factor, sigma32, reveals a multifaceted cellular response to heat stress. Genes Dev 20:1776–1789

Oshima T, Ishikawa S, Kurokawa K et al (2006) *Escherichia coli* histone-like protein H-NS preferentially binds to horizontally acquired DNA in association with RNA polymerase. DNA Res 13:141–153

Palmer KL, Gilmore MS (2010) Multidrug-resistant enterococci lack CRISPR-cas. MBio 1:e00227–10

Parker JS, Parizotto EA, Wang M et al (2009) Enhancement of the seed-target recognition step in RNA silencing by a PIWI/MID domain protein. Mol Cell 33:204–214

Perez-Rodriguez R, Haitjema C, Huang Q et al (2011) Envelope stress is a trigger of CRISPR RNA-mediated DNA silencing in *Escherichia coli*. Mol Microbiol 79:584–599

Pougach K, Semenova E, Bogdanova E et al (2010) Transcription, processing and function of CRISPR cassettes in *Escherichia coli*. Mol Microbiol 77:1367–1379

Pul U, Wurm R, Arslan Z et al (2010) Identification and characterization of *E. coli* CRISPR-*cas* promoters and their silencing by H-NS. Mol Microbiol 75:1495–1512

Semenova E, Jore MM, Datsenko KA et al (2011) Interference by clustered regularly interspaced short palindromic repeat (CRISPR) RNA is governed by a seed sequence. Proc Natl Acad Sci USA 108:10098–10103

Shah SA, Hansen NR, Garrett RA (2009) Distribution of CRISPR spacer matches in viruses and plasmids of crenarchaeal acidothermophiles and implications for their inhibitory mechanism. Biochem Soc Trans 37:23–28

Shimada T, Yamamoto K, Ishihama A (2009) Involvement of the leucine response transcription factor LeuO in regulation of the genes for sulfa drug efflux. J Bacteriol 191:4562–4571

Shinkai A, Kira S, Nakagawa N et al (2007) Transcription activation mediated by a cyclic AMP receptor protein from *Thermus thermophilus* HB8. J Bacteriol 189:3891–3901

Sinkunas T, Gasiunas G, Fremaux C et al (2011) Cas3 is a single-stranded DNA nuclease and ATP-dependent helicase in the CRISPR/Cas immune system. EMBO J 30:1335–1342
Skennerton CT, Angly FE, Breitbart M et al (2011) Phage encoded H-NS: a potential achilles heel in the bacterial defence system. PLoS One 6:e20095
Sorek R, Kunin V, Hugenholtz P (2008) CRISPR–a widespread system that provides acquired resistance against phages in bacteria and archaea. Nat Rev Microbiol 6:181–186
Stern A, Keren L, Wurtzel O et al (2010) Self-targeting by CRISPR: gene regulation or autoimmunity? Trends Genet 26:335–340
van der Oost J, Jore MM, Westra ER et al (2009) CRISPR-based adaptive and heritable immunity in prokaryotes. Trends Biochem Sci 34:401–407
Wade JT, Roa DC, Grainger DC et al (2006) Extensive functional overlap between sigma factors in *Escherichia coli*. Nat Struct Mol Biol 13:806–814
Westra ER, Pul U, Heidrich N et al (2010) H-NS-mediated repression of CRISPR-based immunity in *Escherichia coli* K12 can be relieved by the transcription activator LeuO. Mol Microbiol 77:1380–1393
Wiedenheft B, Lander GC, Zhou K et al (2011a) Structures of the RNA-guided surveillance complex from a bacterial immune system. Nature 477:486–489
Wiedenheft B, van Duijn E, Bultema J et al (2011b) RNA-guided complex from a bacterial immune system enhances target recognition through seed sequence interactions. Proc Natl Acad Sci USA 108:10092–10097
Young RF 3rd (2008) Molecular biology. Secret weapon. Science 321:922–923

Antisense Technology: From Unique Laboratory Tool to Novel Anticancer Treatments

Christine Di Cresce, Colin Way, Mateusz Rytelewski, Saman Maleki Vareki, Supritha Nilam, Mark D. Vincent, James Koropatnick, and Peter J. Ferguson

Contents

C. Di Cresce (✉) • C. Way • M. Rytelewski • S. Maleki Vareki
Department of Microbiology and Immunology, University of Western Ontario, London, ON, Canada N6A 4L6

London Regional Cancer Program and Lawson Health Research Institute, London, ON, Canada
e-mail: cdicresc@gmail.com

S. Nilam
Faculty of Science, University of Western Ontario, London, ON, Canada

M.D. Vincent • P.J. Ferguson (✉)
Department of Oncology, University of Western Ontario, London, ON, Canada

London Regional Cancer Program and Lawson Health Research Institute, London, ON, Canada
e-mail: peter.ferguson@uwo.ca

J. Koropatnick
Department of Microbiology and Immunology, University of Western Ontario, London, ON, Canada N6A 4L6

Department of Oncology, University of Western Ontario, London, ON, Canada

London Regional Cancer Program and Lawson Health Research Institute, London, ON, Canada
e-mail: jkoropat@uwo.ca

V.A. Erdmann and J. Barciszewski (eds.), *From Nucleic Acids Sequences to Molecular Medicine*, RNA Technologies, DOI 10.1007/978-3-642-27426-8_7,

Abstract Antisense reagents and technology have developed as extraordinarily useful tools for the analysis of gene function. The capacity of antisense to reduce expression of RNA (including protein-encoding mRNA and noncoding RNA) important in a multitude of diseases has led to the concept of using antisense molecules as drugs to treat those diseases. Antisense oligonucleotides (ASOs) are being developed for this purpose, with single-stranded DNA ASOs currently the most advanced in clinical testing. Phase I to III clinical trials of ASOs are either completed or in progress for a number of diseases, including cancer. In this review, we focus on progress in developing antisense drugs to downregulate genes mediating malignant characteristics in tumors originating in multiple tissues. In addition, we review progress in (1) ASO targeting of microRNAs (miRNAs) to repress malignant characteristics in multiple tumor types including use of MT-AMOs (multiple-target anti-miRNA oligonucleotides); (2) combining ASOs with each other to generate "synthetic lethality" that enhances chemotherapeutic drug activity; (3) the use of RASONs (radiolabeled antisense ODNs) to image tumors for diagnostic purposes and to monitor therapeutic activity; (4) "on-target" and "off-target" effects of ASOs that lead to both decreases and increases in therapeutic benefit; (5) the chemistries of ASOs that enhance ASO stability, specificity, and activity and reduce undesirable toxicity; and (6) the relative advantages and disadvantages of RNAi-dependent and other ASOs for future application as therapeutic agents to diagnose, monitor, and treat cancer.

Keywords Anticancer drug/therapeutic • Antisense • Oligodeoxynucleotide(s) • RNA interference • Synthetic lethality

Abbreviation List

ACRBP	Acrosin-binding protein
AD–PEG	Adamantane–polyethylene glycol
AD–PEG–Tf	Adamantane–polyethylene glycol–transferrin

Ago 2	Argonaute 2
AKT	Serine–threonine protein kinase B
AMO	Anti-miRNA antisense oligodeoxynucleotide inhibitors
APC/C	Anaphase-promoting complex/cyclosome
APE1	Apurinic/apyrimidinic endonuclease-1
ATM	Ataxia telangiectasia mutated
ATR	Ataxia telangiectasia and Rad3 related
AuNPs	Gold nanoparticles
Bcl-2	B-cell lymphoma-2
Bcl-xL	B-cell lymphoma extra long
BRCA1	Breast cancer susceptibility gene 1
BRCA2	Breast cancer susceptibility gene 2
CDK	Cyclin-dependent kinase
Chk1	Checkpoint kinase 1, S- and G2-checkpoint protein
Chk2	Checkpoint kinase 2, a serine–threonine kinase
c-Raf	Proto-oncogene c-Raf
DDB1	Damage-specific DNA-binding protein 1
DHFR	Dihydrofolate reductase
DNA-PKcs	DNA-dependent protein kinase catalytic subunit
DR5	Death receptor 5, officially TNFRSF10B, or tumor necrosis factor receptor superfamily, member 10b
ECT2	Epithelial cell transforming sequence 2 (exchange factor for GTPases)
EGFR	Epidermal growth factor receptor; *Her-1*
FANCD2	Fanconi anemia complementation group D2
FEN	Flap 5′-endonuclease
GEF-H1	Rho/Rac guanine nucleotide exchange factor
GSK-3B	Glycogen synthase kinase 3 beta
HIV	Human immunodeficiency virus
HR	Homologous recombination
H-Ras	GTPase H-Ras (a.k.a. transforming protein p21)
HSP90	90-kDa heat shock protein
IPMK	Inositol polyphosphate multikinase
KSP	Kinesin spindle protein
Ki-RAS	RAS-family member K
LNA	Locked nucleic acid
MDR	Multidrug resistance
MGMT	O6-methylguanine-DNA methyltransferase
MK2	Officially MAPKAPK2, mitogen-activated protein kinase-activated protein kinase 2
MLH1	mutL homolog 1, colon cancer, nonpolyposis type 2
MMR	Mismatch repair
MPO or PMO	Morpholino phosphoroamide
MRI	Magnetic resonance imaging

MSH2	mutS homolog 2, nonpolyposis type 1
MSH6	mutS homolog 6
mTOR	Mammalian target of rapamycin
NGR peptide	Asparagine–glycine–arginine peptide
NOTCH-1	Notch homolog 1
NSCLC	Non-small cell lung carcinoma
PAK3	p21 protein (Cdc42/Rac)-activated kinase 3
PARP	Poly(ADP-ribose) polymerase
PEI-C	Catechol-conjugated polyethylenimine
Pgp	P-glycoprotein
PINK1	PTEN-induced putative kinase 1
PKCα	Protein kinase C alpha
PLK1	Polo-like kinase 1
pol	Polymerase
PSMA	Prostate-specific membrane antigen
PSMA6	Proteasome (prosome, macropain) subunit, alpha type, 6, a subunit of proteasome 20S
PTEN	Phosphatase–tensin homolog
PTP-1B	Protein tyrosine phosphotase-1B
RAD51	DNA repair protein, homolog of *Saccharomyces cerevisiae* rad51 protein that is a determinant of radiation sensitivity
RAD52	DNA repair protein, homolog of *Saccharomyces cerevisiae* rad52 protein that is a determinant of radiation sensitivity
RB	Retinoblastoma protein
ROS	Reactive oxygen species
RRM2	Ribonucleotide reductase M2
SGK2	Serum/glucocorticoid regulated kinase 2
SKP2	S-phase kinase-associated protein 2 (p45)
SNAIL2	Also known as SNAI2, homolog of Drosophila sna, a zinc-finger, DNA-binding transcriptional factor
SNALP	Stable nucleic acid-lipid particle
STK33	Serine–threonine kinase
tb-FGF	Truncated human basic fibroblast growth factor
TBK1	TANK-binding kinase
TGF-β2	Transforming growth factor-beta 2
TNF-α	Tumor necrosis factor-alpha
topo	Topoisomerase
TS	Thymidylate synthase
TSC2	Tuberous sclerosis 2
TUBGCP2	Tubulin, gamma complex associated protein 2
VEGF	Vascular endothelial growth factor
VHL	von Hippel–Landau protein
WEE1	G2-checkpoint protein (mitotic inhibitor) that is a homolog of the wee1 yeast kinase

XAB2	XPA-binding protein 2a, also known as pre-mRNA-splicing factor SYF1; a transcription-coupled DNA repair protein
XIAP	X-linked inhibitor of apoptosis
XRCC1	X-ray repair cross-complementing protein 1

1 Introduction

Single-stranded, covalently linked nucleic acid bases (antisense oligonucleotides and antisense oligodeoxynucleotides) administered as exogenous agents in the process of *antisense targeting* are extraordinarily valuable tools to illuminate the function of specific mRNAs and proteins. Progress is being made toward using them, not only for discovery of the function of specific messenger and other RNAs but also as targeted therapeutic agents for a number of human diseases. Antisense molecules affect gene expression by specific, complementary base pairing with target RNA. This was initially demonstrated by the use of antisense molecules in a cell-free system to inhibit translation of mRNA (Paterson et al. 1977), followed by in vitro demonstration in chick embryo fibroblast cells (antisense complementary to a subregion of Rous sarcoma virus 35S RNA inhibited production of infectious virus particles) (Stephenson and Zamecnik 1978; Zamecnik and Stephenson 1978). These experiments were among the first to show the promise of antisense oligonucleotides (ASOs) as a tool for the discovery of gene function; a promise that has been fulfilled by the widespread use of ASOs in research laboratories to specifically decrease targeted mRNAs and the proteins they encode (Izant and Weintraub 1984). The usefulness of ASOs in laboratory investigations of basic biological and physiological gene function strongly and immediately suggested that antisense molecules have potential for use as therapeutics. This has led to the exploration of antisense molecules as drugs to ameliorate or cure a broad range of maladies including infectious diseases, cancer, cardiovascular disease, neurodegenerative disease, dyslipidemia, atherosclerosis, and retinitis.

ASOs are broadly defined as oligonucleotides between 8 and 50 nucleotides in length, designed to bind to target RNA by Watson–Crick base pairing in order to mediate the destruction of that target RNA (by endogenous nucleases such as RNase H or Argonaute 2, or cleavage mechanisms inherent in the ASO such as ribozyme activity) and/or reduce the function of target RNA (Usman and Blatt 2000; Bennett and Swayze 2010). Oligonucleotides that modulate gene expression by nonphysiological binding to protein (aptamer oligonucleotides), physiological binding to transcription factors to reduce their capacity to mediate gene transcription (squelching), or other events are excluded from the ASO family of molecules. ASOs mediating RNA interference (RNAi) are a subset of the larger ASO family. RNAi ASOs recruit Argonaute 2 within the *R*NA-*i*nducible *s*ilencing *c*omplex (RISC) to cleave target RNA. Other ASOs that recruit RNase H to mediate RNA destruction are antisense molecules but are not considered to be RNAi ASOs.

RNAi includes small interfering RNA (siRNA) and other forms of RNA-based antisense molecules [i.e., microRNA (miRNA) and short-hairpin RNA (shRNA)] to specifically and potently downregulate target mRNA and other RNA molecules important in regulating gene expression. First discovered in nematodes (Fire et al. 1998), RNAi is employed for both transient and stable downregulation of gene expression. It can be used both to study the function of genes associated with human disease and as a therapeutic treatment (Cerutti 2003; Dykxhoorn et al. 2003; Di Cresce and Koropatnick 2010). Both ASOs utilizing RNase H and other non-RISC mechanisms to cleave RNA and/or reduce RNA function, and RNAi ASOs, are extensively employed for discovery of gene function and are in development for therapeutic application in the treatment of multiple human diseases. In this chapter, we will review multiple RNAi and non-RNAi ASO strategies to target the expression of genes important in development, progression, and malignancy of human tumors, and the potential advantages and disadvantages of RNAi and non-RNAi ASO approaches to eventual treatment of cancer in the clinic.

2 Mechanism of Action of Antisense Molecules

ASOs bind to specific RNA targets by complementary Watson–Crick base pairing, both to inhibit events essential for protein production (translation, splicing events, mRNA maturation mediated by 5′-cap formation and polyadenylation, and ribosomal read-through) and to induce RNA cleavage. RNase H-dependent cleavage is the mechanism by which the majority of ASOs currently in clinical trials act. RNase H is present in all mammalian cells and in two forms in human cells: RNase H1 and RNase H2. Both cleave RNA strands involved in RNA–DNA heteroduplexes (Cerritelli and Crouch 2009), but RNase H1 is responsible for cleavage of RNA bound to ASOs with five or more consecutive complementary nucleotides (Wu et al. 2004). RNAi-dependent ASOs also induce target RNA destruction, but do so by recruiting Argonaute 2 (Ago 2) to cleave RNA by a mechanism similar to that of RNase H (Meister et al. 2004a). Ago 2 functions in the context of the RISC complex by mechanisms that have been reasonably well characterized and reviewed elsewhere (Kim et al. 2007; Bennett and Swayze 2010).

Ribozymes and DNAzymes have also been developed to selectively cleave RNA targets. These oligonucleotides have inherent catalytic activity that does not depend on endogenous cellular enzymes (Breaker and Joyce 1994). Although effective in cultured cell experiments, their early promise has not been borne out in vivo as well as other approaches that recruit endogenous RNA-cleaving enzymes. To our knowledge, none has proceeded to development as therapeutic agents to target RNA.

ASOs can also reduce activity of mRNA by blocking translation (translation or hybridization arrest) without requiring RNA degradation. Although translation arrest has been shown in cell-free assays, evidence that this is the primary process by which ASOs reduce RNA activity in cultured cells or in whole organisms is limited, and arrest of translation has not generally been invoked as a mechanism of action for ASOs under therapeutic development (Bennett and Swayze 2010). In contrast, noncoding miRNAs present in eukaryotes are a form of ASOs that occur naturally and are transcribed from chromosomes. miRNAs promote degradation of target RNA but can also block mRNA translation to reduce the level of protein produced without reducing the amount of mRNA template (Carthew and Sontheimer 2009). Application of ASOs that bind to miRNAs to antagonize their activity has been explored as a therapeutic strategy.

Most mRNAs require multiple processing steps (splicing, polyadenylation, 5′G-methylation) prior to translation, and alternative processing pathways exist (particularly with respect to splicing and polyadenylation) that lead to multiple possible proteins with different biological activities (Sharp 2009). ASOs have been employed to bind to pre-mRNA and modulate RNA splicing (Roberts et al. 2006) and polyadenylation (Vickers et al. 2001). In strategies of this kind, ASOs may be particularly suited to therapy of human diseases involving aberrant RNA metabolism (Cooper et al. 2009). Furthermore, ASOs have been demonstrated to bind to target RNA and disrupt secondary or tertiary structure important for function in mediating pathological conditions (Cooper et al. 2009; O'Rourke and Swanson 2009). This suggests that ASOs have potential as therapeutic agents in the treatment of disease (including diseases such as HIV) where RNA structure is important (Ivanova et al. 2007) and where RNA is capable of adopting multiple conformations, each of which has different consequences for disease progression.

3 RNAi-Dependent and RNAi-Independent ASOs: Relative Merits

ASOs mediating RNAi-dependent inactivation of target RNA and ASOs inducing RNA inactivation by other mechanisms (chiefly those dependent on RNase H) are both effective at decreasing target mRNA levels. RNAi-dependent ASOs that involve Ago 2 and the RISC complex are often more potent than RNase H ASOs when used to target RNAs in cultured cells. It has been suggested that, in cell culture, RNAi-dependent ASOs (primarily double-stranded siRNAs) may be more effective in many instances due to RISC-mediated facilitation of ASO binding to target RNA, in spite of the fact that Ago 2 does not appear to cleave RNA as efficiently as RNase H (Bennett and Swayze 2010). In vivo, however, charged, single-stranded RNAi-independent first-, second-, and third-generation ASOs (see below) distribute to distant tissues and target organs and tumors and are taken up into cells without the help of formulations to protect the ASOs *en route*, unlike uncharged ASOs and double-stranded siRNAs. This confers a significant pharmacokinetic advantage to RNase H-dependent ASOs over RNAi-dependent ASOs. Furthermore, although duplex siRNA is more stable within cells than single-stranded antisense RNA molecules (Bertrand et al. 2002), RNase H-recruiting

DNA ASOs are more stable in the intracellular milieu than siRNA ASOs, and modifications to the chemical structure of DNA ASOs enhance their systemic and intracellular stability even further (see below). Moreover, many recent studies describe the pharmacokinetics and toxicology of RNase H-dependent ASOs in animal models and the experience of dozens of clinical trials (Mansoor and Melendez 2008), reports of preclinical and clinical studies of pharmacokinetics, toxicity, and effectiveness of RNAi-dependent antisense therapeutics are much less extensive (Oh and Park 2009). Other relative advantages of RNase H-dependent DNA ASOs over RNAi-dependent ASOs include better specificity for individual mRNA target molecules, flexibility in target choice and reagent design, and relative ease of chemical synthesis in vivo that facilitates large-scale production of pharmaceutical-grade material (Grunweller et al. 2003). Single-stranded DNA ASOs are half the molecular weight of double-stranded siRNAs and do not require an additional and costly hybridization step in synthesis. Although further analyses will undoubtedly reveal different conditions where one or the other strategy (or a combination of both) will be better suited for therapeutic application, use of RNase H-dependent AS ODNs is presently more advanced under in vivo (including human) conditions than RNAi-dependent ASOs at the present time.

4 Chemistry of Antisense Oligodeoxynucleotide Molecules

Depending on the chemistry, antisense molecules act by inhibiting translation or inducing degradation of complementary target mRNA. RNase H-dependent ASOs are single-stranded oligodeoxynucleotides, 15 to 25 nucleotide bases in length in order to maximize specificity and binding affinity. Synthetic RNase H-dependent ASOs mediate mRNA degradation by activation of endogenous RNase H and subsequent exonuclease cleavage of the target mRNA (Stein and Cohen 1988; Stein and Cheng 1993). Chemical modifications to the DNA structure such as phosphorothioation of the backbone and 2′-methoxyethylation (2′MOE) of ribose sugar groups enhance uptake, stability, and ultimate effectiveness of ASOs (Agrawal 1999; Geary et al. 2001).

RNase H-dependent ASOs have undergone progressive stages of chemical development beginning with the so-called "first-generation" ASOs, in which a non-bridging oxygen atom in the backbone phosphate group is replaced with (1) a sulfur (*phosphorothioate*), or (2) a methyl (*methyl phosphonate*) group, or (3) an amine (*phosphoamidates*). Phosphorothioation does not impede RNase H cleavage, and phosphorothioated ASOs were the most widely used first-generation single-stranded DNA antisense molecules. The only antisense drug currently approved by the US Food and Drug Administration (FDA) in the USA (Vitravene, for treatment of viral retinitis) is of this type (Marwick 1998). However, the emergence of undesirable nonspecific toxicities and side effects (immune stimulation, complement activation) (Jason et al. 2004) associated with phosphorothioates has led to the development of "second-generation" RNase H-dependent ASOs with alkyl modifications (2′-O-methyl [2′-OME] or 2′-O-methoxyethyl [2′-MOE]) on ribose groups of selected bases. However, alkyl substitutions on bases reduce recruitment of RNase H and inhibit the ability of ASOs to mediate RNA cleavage. To circumvent this limitation and, indeed, take advantage of it, chimeric ASOs have been generated that leave a sequence of consecutive 8–12 unmodified bases, or "gap," in the center of the ASO. The resulting "gapmers" include alkyl modification at the 2′ position of 5 or 6 ribose molecules at each end of the ODN. The unmodified DNA bases in the central region provide a suitable target for RNase H with resulting cleavage of bound RNA targets, with the desirable effect of localizing cleavage the center of the ASO-bound RNA. This promotes release of RNA after cleavage to allow rebinding of the ASO to additional targets and degradation of multiple RNA molecules by the same ASO (Crooke et al. 2007). Gapmers retain the advantages of first-generation ASOs in that they minimize off-target effects and have the high degree of stability associated with alkyl substitutions on riboses. Gapmer ODNs have progressed through preclinical development stages including pharmacokinetic and toxicological studies. At least 13 RNase H-dependent ASOs are currently in human clinical trials (Table 1). Of particular importance and interest, both first- and second-generation ASOs of this type do not require delivery systems to facilitate their uptake into tumor cells and other tissues in vivo (Hossini et al. 2003), although such systems are required for in vitro delivery. This characteristic has contributed to preferential use of second-generation ASOs in clinical trials.

"Third-generation" ASOs are currently under development and include the addition of furanose rings to improve nuclease stability, target affinity, and pharmacokinetic characteristics. The most commonly used ASOs of this type are *locked nucleic acids* (LNAs), *morpholino phosphoroamidates* (MPOs or PMOs), and *peptide nucleic acids* (PNAs). Unlike second-generation RNase H-dependent ASOs, third-generation ASOs do not recruit RNAse H and are thought to exert their biological effects by mediating translational arrest of mRNAs (Summerton 1999). Chimeric ASOs, containing bases with chemical modifications characteristic of both second- and third-generation ASOs, are under development to take maximum advantage of the desirable characteristics of both.

5 Delivery Methods of Antisense Oligodeoxynucleotides and Small Interfering RNAs

Although agents that facilitate transfer of antisense molecules from extracellular medium into the interior of cells are usually required to introduce antisense molecules into cells in culture, single-stranded DNA ASOs are able to accumulate in cells without the assistance of a delivery agent if used at sufficiently high concentrations (Yu et al. 2001; Yin et al. 2010; Carroll et al. 2011). In vivo, DNA ASOs accumulate in tumor cells and some normal tissues without the

Table 1 ASOs approved or in development for treatment of human disease. ** is used to indicate the use of www.clinicaltrials.gov

Compound	Company	Target	Clinical trials	Phase	References
Vitravene (Fomivirsen)	Isis Pharmaceuticals	CMV 1E2 protein	Retinitis caused by CMV in HIV-infected patients	Approved *N.B. the only FDA-approved antisense drug to date*	Crooke (1998), De Clercq (2004)**
			For the treatment of cancer		
G3139 (Genasense; oblimersen, augmerosen)	Genta	Bcl-2	Malignant melanoma, CLL, multiple myeloma, hepatocellular carcinoma	Phase III	O'Brien et al. (2005), Moreira et al. (2006), O'Brien et al. (2007), Knox et al. (2008), Chanan-Khan et al. (2009), O'Brien et al. (2009))**
OGX-011 (Custirsen)	OncoGenex Technologies	Clusterin	Prostate, breast, and lung cancers	Phase III	Chi et al. 2008b, Chia et al. (2009), Chi et al. (2010), Saad et al. (2011)**
AP12009 (Trabedersen)	Antisense Pharma	TGF-β2	Malignant glioma, metastatic melanoma, pancreatic cancer, and colorectal cancer	Phase III	Bogdahn et al. (2011), Jaschinski et al. (2011)**
Aprinocarsen (Affinitak, LY900003, formerly ISIS 3521)	Eli Lilly Pharmaceuticals (formerly with Isis Pharmaceuticals)	PKCα	NSCLC, colorectal cancer, ovarian carcinoma	Phase II/III *N.B. discontinued in 2005 for lack of efficacy* (Kelland 2006)	Advani et al. (2004), Marshall et al. (2004), Villalona-Calero et al. (2004), Paz-Ares et al. (2006)**
ISIS-5132 (CGP 69846A)	Isis Pharmaceuticals	c-raf-kinase	Lung cancers (small cell and NSCLC), ovarian, colorectal, brain and pancreatic cancer, melanoma	Phase II *N.B. discontinued in 2005 for lack of efficacy* (Kelland 2006)	Coudert et al. (2001), Cripps et al. (2002), Oza et al. (2003)

ISIS 2503	Isis Pharmaceuticals	H-Ras	Pancreatic adenocarcinoma, colorectal cancer	Phase II	Saleh et al. (2000), Adjei et al. (2003), Alberts et al. (2004)**
AEG35156	Aegera Therapeutics	XIAP	Advanced solid tumors and AML	Phase I/II	LaCasse et al. (2006), Dean et al. (2009), Carter et al. (2011); Schimmer et al. (2011)**
EGFR AS [EGFR antisense (AS) gene sequence under U6 promoter control]	University of Pittsburgh	EGFR	Head and neck cancer (squamous cell carcinoma)	Phase I/II	Lai et al. (2009)**
ALN-VSP02	Alnylam Pharmaceuticals, Tekmira	VEGF, KSP (kinesin spindle protein)	Primary and secondary liver cancer, solid tumors	Phase I	Gollob et al. (2010)**
TAG (TGF Antisense + rhGMCSF) *Note: TAG is a plasmid-transfected, autologous tumor cell vaccine*	Mary Crowley Medical Research Center	TGF-β2	Currently recruiting patients with histologically confirmed advanced or metastatic non-curable solid tumor (not candidates for surgery or radiation)	Phase I	Olivares et al. (2011)**
For the treatment of other diseases					
AVI-4658 (splice-switching morpholino ODN)	Imperial College London	Dystrophin (exon 51)	Duchenne's muscular dystrophy	Phase II	Kinali et al. (2009); Cirak et al. (2011, 2012)**
ISIS 113715 (in combination with metformin, glipizide, rosiglitazone)	Isis Pharmaceuticals	Protein tyrosine phosphotase-1B (PTP-1B)	Type 2 diabetes	Phase II	Geary et al. (2006)**
ISIS 104838	Isis Pharmaceuticals	TNF-α	Rheumatoid arthritis	Phase II	Sewell et al. (2002), Geary et al. (2003)**

assistance of a delivery system (Geary 2009; Koller et al. 2011). siRNAs, in contrast, require a delivery agent for in vivo delivery to target cells and tissues. Use of naked siRNA in vivo is hindered by a number of factors, including (1) inefficient cellular uptake, (2) nonspecific uptake into non-tumor tissues, (3) enzymatic degradation, (4) renal elimination, and (5) difficulty crossing the capillary endothelium to gain proximity to cells other than those lining the vascular system (Oh and Park 2009; Higuchi et al. 2010). To increase the efficiency of RNAi-dependent ASOs as in vivo therapeutic agents, effective delivery systems must maximize both ASO stability and specific ASO delivery to target cells (Aigner 2006; Sanguino et al. 2008). Mechanisms to achieve these goals can be categorized into (1) physical delivery methods, (2) conjugation to targeting molecules, and (3) carrier-mediated delivery methods. Some of these mechanisms can also be adapted for use with RNAi-independent ASOs to improve their delivery (Kurreck 2003; Zaghloul et al. 2010).

5.1 Physical Delivery Methods

5.1.1 Hydrodynamic Injection

Hydrodynamic injection involves the intravascular injection of a high volume of diluting agent (buffer) in which siRNA is dissolved (McCaffrey et al. 2002). In mouse models, this requires an injection volume (in ml) equivalent to 10% of body weight (in gm), administered over a 5- to 7-s period of time (Lewis and Wolff 2005). Most studies have been conducted in rodents and involve tail vein injection of ASOs, resulting in preferential accumulation in certain organs such as kidney and liver in a fashion that does not depend on ASO sequence or the level of expression of the targeted RNA (Liu et al. 1999; Lewis et al. 2002). However, tissue-specific delivery of ASOs to desired target organs (other than reticuloendothelial tissues such as liver and kidney) can be improved through use of more localized injection in the vicinity of target cells, as demonstrated in both rats and larger animals such as rhesus monkey. This method is tissue-specific in the sense that specificity is achieved by physically concentrating ASOs in the tissue at the point of entry and not by targeting ASOs to specific cell markers (i.e. receptor-mediated) following injection at distant sites (see below) (Satkauskas et al. 2001). In studies involving delivery of plasmid DNA rather than ASOs, intra-arterial delivery of plasmid DNA into the femoral artery was well tolerated and increased expression of plasmid-encoded genes in skeletal muscle (Acsadi et al. 1991; Zhang et al. 2001). In rats, the injected volume required for successful gene delivery to the liver could be reduced from 10% to less than 2% of their body weight when administered locally (portal vein) over tail-vein injection (Zhang et al. 2004a). The ability of hydrodynamic injection to enhance ASO delivery is reflected by its use in effective delivery of siRNAs in rodent models (including capacity of the method to partially overcome limitations to in vivo siRNA delivery attributed to siRNA instability) (Rychahou and Evers 2010). However, information on local delivery is limited, and there has been limited clinical application of the hydrodynamic approach to date due to its invasive nature. However, the use of computer-controlled injection devices may facilitate this therapeutic approach (Suda et al. 2008; Kamimura et al. 2009).

5.1.2 Mechanical Massage

Mechanical massage refers to the direct application of light, physical pressure to an organ to enhance accumulation of exogenous ASOs injected intravenously into that organ. This was initially demonstrated using expression vectors (plasmids) directing expression of luciferase in mouse kidney, in which a single application of pressure enhanced uptake of siRNA or plasmid DNA following intravenous injection (Mukai et al. 2008). This was later repeated in the spleen, with little production of pro-inflammatory cytokines or tissue-damaging enzymes (Mukai et al. 2009). It has been suggested that physical pressure activates transcriptional factors that may increase plasmid accumulation by the cells of the massaged organ (Mukai et al. 2010).

5.1.3 Electroporation

Electroporation involves the use of electrical pulses to aid in cellular uptake of ASOs into cells (Golzio et al. 2009). The possible clinical utility of electroporation is based on its ability to limit delivery within an electric field, making it selective to a specific site or organ, and its minimal side effects (Paganin-Gioanni et al. 2011). In mice bearing tumor xenografts, siRNA targeting VEGF was injected directly into tumor tissue using a "fork and plate" electrode apparatus that sent a current through the tumor to increase transfection and subsequent accumulation of ASOs (Takei et al. 2008). In that study, siRNA delivery to tumors, enhanced by in vivo electroporation, suppressed growth of high VEGF-expressing tumors by more than 90% compared to control tumors and, supported further exploration of electroporation as a method to enhance in vivo delivery of ASOs.

5.2 Conjugation to Targeted Molecules

An alternative method to enhance selective delivery of ASOs to target cells involves covalent conjugation of the ASO to specific ligands for receptors expressed at a sufficiently high level on target cells and to moieties that increase cellular uptake and endosomal release by other mechanisms (Juliano et al. 2008; Alam et al. 2011). For example, to increase tissue-specific delivery of RNAi molecules, an aptamer-siRNA chimeric RNA was designed to bind to cells expressing cell-surface receptor prostate-specific membrane antigen (PSMA), a tumor-associated antigen overexpressed in prostate cancer (McNamara et al. 2006). These RNAs selectively bound to, and were internalized by, cells expressing PSMA and functioned as antisense molecules (McNamara et al. 2006). An additional strategy linking cell-penetrating peptides (CPP) with siRNA to increase their transport across the cellular plasma membrane has also been reported. In this technique, molecules are conjugated through active thiol groups, and upon uptake into the cytoplasm, these groups are reduced and release active siRNA (Muratovska and Eccles 2004). This system delivered siRNA with the same efficiency as cationic liposomes (see 5.3.1 below) but was simpler to synthesize (Muratovska and Eccles 2004). Since CPPs enter cells by endocytosis, new derivatives are currently under development to limit the fraction of peptides (and ASOs associated with them) that remain trapped in the endosomes and are thus not released to carry out their antisense function. Stearylation of CPPs with transportan 10 increases transfection efficiency by enhancing endosomal escape and may be of future value in enhancing the activity of siRNAs delivered by CPPs (Lehto et al. 2011).

The combined use of tumor-targeting ligands, CPPs, and nanoparticles is expected to enhance siRNA targeting, delivery, and uptake, as the combinations may act complementary to increase transfection efficiency (Cheng and Saltzman 2011).

5.3 Carrier-Mediated Delivery Methods

5.3.1 Cationic Liposomes

Cationic liposomes are positively charged lipid spheres that complex with and encapsulate molecules of a wide range of sizes, including antisense molecules. Cationic carriers are popular tools in both the laboratory and the clinic due to the ease of complex formation via electrostatic interactions (positive liposome + negative DNA or RNA) (Oh and Park 2009; Higuchi et al. 2010). Targeting or increased internalization can be achieved through the conjugation of different molecules to the surface of the cationic liposome. For example, the addition of polyethylene glycol (PEG) residues into cationic liposomes has been shown in vivo to increase half-life and transfection efficiency into prostate tumor xenografts by preventing both systemic clearance by phagocytes and opsonization (Sonoke et al. 2008).

Stable nucleic acid-lipid particles (SNALPs) combine a mixture of cationic and fusogenic lipids that permits intracellular uptake and endosomal release of siRNA (Morrissey et al. 2005). SNALPs are initially coated with PEG during siRNA incorporation to provide a neutral, hydrophilic environment during formation, and the PEG molecules are then detached postinjection, usually in a pH-mediated reaction, leaving a functional siRNA vehicle (Ambegia et al. 2005). In in vivo, studies with siRNA targeting hepatitis B virus have shown that SNALP particles increased the serum half-life of siRNA from 49 min to 6.5 h and decreased the hepatitis B serum antigen levels by 86% compared to the control groups (Morrissey et al. 2005).

Targeting of specific cells can be achieved by complexing ASOs to cationic liposome decorated with ligands for specific, overexpressed receptors on the cancer cell surface. By this method, significant tumor suppression was achieved by coupling a truncated human basic fibroblast growth factor peptide (tb-FGF) to a cationic liposome and targeting cancer cells that overexpressed the fibroblast growth factor receptor (Chen et al. 2010a). An interesting prospect in liposome targeting involves conjugating various ligands via a reversible, pH-dependent cleavable bond that, after directing the liposome to the specific cell type, can be cleaved, revealing secondary ligands for intracellular targeting (Biswas et al. 2011). This could potentially lead to effective multifunctional liposomal systems.

Further studies, such as screening of lipidoid libraries for transfection efficiency and in vivo delivery (Akinc et al. 2008), are promising approaches likely to aid in developing optimal lipid carrier molecules.

5.3.2 Cationic Polymers

Cationic polymer delivery systems act in a similar fashion as cationic liposomes by utilizing their positive charge to form a strong electrostatic interaction with anionic ASOs, termed polyplexes, which can be taken up by cells (Varkouhi et al. 2012). There are two main types of cationic polymers: natural and synthetic. Natural polymers have the advantage of being biocompatible and, biodegradable, thus achieving low toxicity. The benefit to synthetic polymers is that they can be designed with specifications that result in high transfection efficiency and reduced enzymatic degradation. However, synthetic polymers may not ultimately be useful for in vivo administration of ASOs due to the increased toxicity associated with their low biodegradability (Higuchi et al. 2010). When left intact, they have the tendency to precipitate in large clusters (2–6 μm in diameter), adhere to cell surfaces, and promote non-specific necrosis as soon as 30 min after administration (Fischer et al. 1999).

Atelocollagen is one of the most widely studied natural cationic polymers. Atelocollagen–siRNA complexes are resistant to endonuclease degradation for up to 3 days and can deliver siRNA to target tissues in a sustained fashion for as long as 24-h postinjection in an in vivo mouse model (Takeshita et al. 2005). In tumor-bearing mice, intravenous or intratumoral administration of atelocollagen–siRNA complexes targeting Bcl-xL displayed selective uptake into tumor tissues without inciting immunogenic activation or renal or hepatic toxicity in the animals (Mu et al. 2009).

Polyethylenimine (PEI) is a synthetic polymer that is positively charged under physiological conditions, facilitating effective binding to negatively charged antisense molecules. The PEI-siRNA produced is protected from degradation in serum and subsequently during cellular uptake (Intra and Salem 2008). PEI–siRNA complexes reduce tumor growth *in vivo* against a model in which naked siRNA had failed (Urban-Klein et al. 2005). PEI was used to condense DNA ASOs to form nanosized complexes that were later modified with an asparagine–glycine–arginine (NGR) peptide (Zhou et al. 2011). These NGR–PEI–DNA ASO complexes were able to efficiently target tumor cells and act as a tumor-targeted nano-delivery complex (Zhou et al. 2011). One limitation of PEI-based vector delivery methods is the inclusion of free or unbound PEI in the mixture that can contribute to nonspecific toxicity. Therefore, strategies for purifying or limiting unbound PEI have been implemented (Boeckle et al. 2004). Alternatively, certain combinations of preformed PEI–siRNA complexes with liposomes (lipopolyplexes) not only reduce toxicity of the PEI–RNA molecules but also enhance their transfection efficiency (Schafer et al. 2010).

The potential toxicity associated with the use of synthetic polymers has led to the development of "star polymers" with PEG arms and a degradable cationic core to deliver ASOs safely. In mouse models, star polymers are only mildly toxic: over 80% of cells in mouse organs exposed to them remained viable 48 h after administration (Cho et al. 2011).

Future research will focus on optimizing synthetic cationic polymer systems for maximal delivery of ASOs under conditions that minimize potential totoxic effects. One cutting edge example is the synthesis and use of catechol-conjugated PEI (PEI-C) polymers coated with gold nanoparticles (AuNPs) for the delivery of siRNA. The catechol group allows the PEI-coated AuNPs to form micelles in aqueous solutions as well as stable complexes with siRNAs. They have shown silencing effects in cancer cells and display low cytotoxicity (Lee et al. 2011).

5.3.3 Carbon Nanotubes

Carbon nanotubes (CNTs) have a nano-needle structure that has been suggested to enter cells through nonspecific physical penetration. However, recent studies indicate they enter through an energy-dependent endocytic pathway (Yaron et al. 2011). The development of functionalized carbon [containing chemical functional groups (e.g. ammonium)] may lead to advancements in targeting, diagnosis, and drug transport (Prato et al. 2008). Although CNTs have the inherent problem of not being soluble in aqueous solutions, the adsorption of phospholipid molecules with PEG chains confers solubility (Kam et al. 2005). Terminal *amine* or *maleimide* groups may also be added, upon which to conjugate a wide variety of molecules including nucleic acids. These functionalized CNTs conferred higher silencing efficiencies of siRNA than a comparable Lipofectamine transfection of siRNA (Kam et al. 2005). Improved tumor-cell targeting of CNTs via ligand conjugates promises to make these molecules valuable for therapeutic use in delivery of ASOs (Herrero et al. 2009).

5.4 *Future Perspectives on Antisense Delivery: From the Lab to the Clinic*

The first clinical trials of ASOs involved direct injection of naked siRNA into the tissue of interest, such as the eye (Michels et al. 2006). In 2008, a phase 1 trial was undertaken for experimental drug CALAA-01 (Calando Pharmaceuticals-01), which was the first time for the targeted delivery of siRNA in humans (Davis 2009). The delivery system was composed of a cyclodextrin-containing polymer (CDP), adamantane–PEG (AD–PEG), and adamantane–PEG–transferrin (AD–PEG–Tf). The CDP molecule confers many important characteristics to the overall functioning of the delivery system including electrostatic interaction with the siRNA molecule, while the cyclodextrin constituents act as scaffolding for incorporating other agents. The AD–PEG molecule conveys overall steric stabilization, and the transferrin in AD–PEG–Tf acts as the targeting ligand (Davis 2009). The siRNA chosen targets the M2 subunit of ribonucleotide reductase (RRM2), a proven anticancer target, and initial testing on patient tumor tissues indicates a reduction in target mRNA as well as protein levels (Davis et al. 2010).

6 Clinical Use of Single-Stranded DNA ASOs

Treatment of cancer with small molecule drugs continues to be a mainstay in cancer therapy. However, limitations to the effectiveness of such drugs include (1) variation among individual patients with respect to drug metabolism and pharmacokinetics, that are mediated by genetic factors (pharmacogenomic considerations); (2) variations between tumors, and heterogeneity within tumors, with respect to sensitivity to anticancer drug treatment; and (3) acquired resistance to chemotherapeutics arising from spontaneous mutations in tumor cell populations, which decreases the effectiveness of repeated treatment with the same drugs (or different drugs acting on the same tumor cell target molecules) even with combination therapies (O'Connor 2009). For example, defects in cell death pathways in human tumor cells (such as overexpression of normal proteins that suppress apoptosis) confer resistance to apoptosis that would normally be induced by chemotherapeutic drugs (Hanahan and Weinberg 2000). The potential advantages of using RNAi-dependent and other ASOs in the clinical treatment of cancer (and other diseases) include (1) increased target selectivity and lower association with nontarget molecules, since target selection is sequence-based; (2) the ability to reduce what are currently considered "non-druggable" protein targets (i.e., proteins against which small molecule drugs are problematic due to lack of unique targetable protein sequences or structure and/or immune-targetable epitopes); (3) the possibility that when mutations arise in the sequence of RNAs targeted by ASOs, new ASOs can be designed to target other sequences in the same RNA, or even the mutated sequence if it contributes to increased pathology; and (4) the ability to use antisense molecules in multi-targeted cancer treatments in combination with current chemotherapeutic drugs to: target the same protein or different proteins in a related pathway(s), increase sensitivity to coadministered drugs, and potentially decrease the dose of drug needed for efficacy (Coppelli and Grandis 2005; Aagaard and Rossi 2007; Grimm and Kay 2007; Bennett and Swayze 2010).

The use of antisense technology has the potential to expand the repertoire of anticancer drugs to include ASOs that directly and specifically suppress both normal and mutant genes contributing to the tumorigenic phenotype, and possibly reverse mechanisms of resistance. One promising strategy for antisense therapy in cancer treatment is to combine antisense reagents with drugs that do not target RNA to achieve synergistic toxicity to tumor cells (Jason et al. 2004; Zhang et al. 2004b; Rayburn and Zhang 2008; Geary 2009).

7 Antisense in Cancer Clinical Trials

By combined downregulation of targeted genes with both ASOs that target RNAs and drugs that bind and inhibit the protein encoded by that target RNA (or proteins in connected biological pathways), synergistic inhibition of tumor cell growth may be achieved, possibly with improved selectivity against cancer cells. These potential combinations are discussed below. With respect to anticancer trials, ASOs have been used most frequently to target mRNAs encoded by oncogenes, mutant tumor suppressor genes (the expression of which interferes with normal tumor suppressor function), and genes that encode signaling molecules that, when upregulated, mediate tumor cell resistance to apoptosis, proliferation, metastasis, and other malignant characteristics (Rayburn et al. 2006; Tamm and Wagner 2006; Bruserud 2007; Leonetti and Zupi 2007; Alvarez-Salas 2008). ASOs that decrease expression of oncoproteins, including Bcl-2, c-Raf, H-Ras, and XIAP, have been evaluated (Table 1). ASOs that target mRNAs encoding cell signaling molecules important in initiating cancers and promoting cancer progression (VEGF, TGF-β2, PKA, and PKCα) and ASOs targeting cancer-related mRNAs such as survivin, RRM2, DNA methyltransferase, and clusterin have also been assessed (Rayburn et al. 2006; Leonetti and Zupi 2007). Many of the ASOs targeting a wide variety of mRNAs have undergone preclinical development and early stage clinical trials, and a high proportion have shown significant biological activity against their mRNA target and against the disease throughout this process (Rayburn et al. 2006; Leonetti and Zupi 2007). Regardless of that promise, only one ASO molecule has been approved in the USA, Canada, and Europe for clinical use to date (Table 1).

Possibly, the most effective use of antisense molecules will be as a part of a multi-targeted approach to cancer treatment (i.e., in combination with other antisense molecules and chemotherapeutic drugs). Both RNAi-dependent and other ASOs can be used in combination with each other and/or small molecule drugs to target RNA and protein in the same, different, or complementary biological pathways. However, unknown and known off-target interactions (based on sequence) among antisense reagents could limit the use of antisense molecules and combinations of antisense molecules with each other and/or current chemotherapeutic drugs (Pandyra et al. 2007). For example, in in vitro studies combining siRNAs targeting Bcl-2 and thymidylate synthase in conjunction with drugs that target TS enzyme; Bcl-2 siRNA nonreciprocally antagonized the ability of TS-siRNA to downregulate TS mRNA and protein. This consequently reduced the sensitization effects seen when antisense and small molecule drugs targeting TS mRNA and protein (respectively) are combined (Pandyra et al. 2007). Although simultaneous antisense targeting of multiple mRNAs is therapeutically promising (Di Cresce et al. 2011), the unanticipated consequences of those combinations (including antagonistic effects) highlight the need for thorough preclinical assessment of the biological consequences before clinical studies.

ASOs listed in Table 1 that have progressed through to phase III of clinical trials and have not been discontinued are discussed below in more detail. Most of these show promise in combination with small-molecule chemotherapeutic drugs. However, some have shown benefit as single agents (e.g., Trabedersen for treatment of cancers and AVI-4658 for treatment of muscular dystrophy).

7.1 Genasense (G3139)

G3139 is an 18-mer, first-generation phosphorothioated ASO that targets the mRNA of the anti-apoptotic protein Bcl-2. Overexpression of BCL2 correlates with resistance to chemotherapies, increased tumor aggressiveness, and poor prognosis and patient survival (Hockenbery et al. 1990; Fesik 2005; Yip and Reed 2008). Bcl-2 is overexpressed in many cancers including melanoma, CLL, small cell lung carcinoma, multiple myeloma, and cancers of the prostate, ovary, bladder, cervix, pancreas, breast, and colon (Fesik 2005).

In a phase III clinical trial of G3139 in combination with fludarabine and cyclophosphamide (FC) in patients with relapsed or refractory CLL, G3139 significantly increased the number of complete responses (CR) and nodular partial responses (nPR) compared to using FC alone (O'Brien et al. 2007). Out of 120 patients randomly assigned to receive G3139, 17% showed CR/nPR compared to 8% in the chemotherapy-only group ($P = 0.025$). There was no significant difference in 5-year survival between treatment arms. However, G3139 conferred a 40% improvement in 5-year survival among patients who had achieved CR/nPR. Patients who had previously responded to FC alone (but had subsequently relapsed) were four times more likely to respond to the G3139–FC combination (CR/nPR) than patients who had initially failed FC. Those patients (FC responders and relapsers) also achieved a 50% reduction in the risk of death compared to those who initially failed FC (O'Brien et al. 2007, 2009).

7.2 *Custirsen (OGX-011)*

There are wide variations in the cellular/tissue expression and function of clusterin throughout the body, which indicates that its regulation is tissue specific and many pathways may exert regulatory control in healthy and diseased cells (Rizzi et al. 2009). Consequently, while its functions are diverse and controversial, we will focus on clusterin in the context of cancer and its relation to being targeted with OGX-011 in the treatment of prostate cancer. With this said, clusterin functions as both (1) a chaperone protein that helps to preserve protein function during cellular stress by preventing protein aggregation and precipitation (Shannan et al. 2006) and (2) an anti-apoptotic protein as it can inhibit activated BAX to reduce caspase activation and thus prevent apoptosis (Trougakos et al. 2009). Clusterin is overexpressed in many human cancers including lung, breast, prostate, and kidney (Chi et al. 2008a).

In prostate cancer patients, clusterin suppresses apoptosis in prostate cancer cells after radiation or chemotherapy and/or androgen withdrawal (Miyake et al. 2000, 2003; Zellweger et al. 2003), and antitumor activity of cytotoxic agents is enhanced when combined with ASOs targeting clusterin to treat prostate cancer (Di Cresce and Koropatnick 2010). OGX-011 is a second-generation gapmer ASO with a high affinity for exon 2 of human clusterin mRNA (Zellweger et al. 2001). In a randomized phase II clinical trial of docetaxel and prednisone (DPr) with or without OGX-011 in patients with metastatic, castration-resistant prostate cancer, the addition of OGX-011 to DPr treatment was well tolerated in patients, and OGX-011 conferred a 25% reduction in serum clusterin levels (Chi et al. 2010). More importantly, patients who received OGX-011 had significantly higher mean progression-free survival times (7.3 months vs. 6.1 months) and overall survival times (23.8 months vs. 16.9 months). With this in mind, two phase III studies involving OGX-011 are ongoing, and outcomes have not yet been reported.

7.3 *Trabedersen (AP12009)*

AP12009 is a phosphorothioated ASO targeted against transforming growth factor-beta 2 (TGF-β2) mRNA. TGF-β2 is an immunosuppressive cytokine overexpressed by many tumors and mediates tumor cell evasion of the immune system (Hanahan and Weinberg 2011). Preclinical analysis revealed that AP12009 increased tumor recognition by the immune system (lymphocyte proliferation and cytotoxicity) and inhibited tumor growth and metastasis (Jachimczak et al. 1993). Following extensive preclinical *in vivo* evaluation (Schlingensiepen et al. 2005), AP12009 is now in clinical trials as a single-agent treatment for malignant glioma (phase III), malignant melanoma (phase I/II), and pancreatic cancer (phase I/II).

The greatest success achieved by AP12009 has been in the treatment of malignant glioma (Schlingensiepen et al. 2006; Bogdahn et al. 2011; Jaschinski et al.

2011). In phase I/II trials, AP12009 was better tolerated than standard chemotherapy (temozolomide or procarbazine/lomustine/vincristine), and 2.41 mg/m^2 AP12009 showed superior safety and efficacy compared to standard chemotherapy or 19.81 mg/m^2 AP12009. A 63-center, multinational phase III clinical trial is now under way comparing AP12009 infusion to temozolomide or carmustine infusion (Mulholland 2011). Phase I/II trials in metastatic melanoma and colorectal and pancreatic cancers have begun.

7.4 AVI-4658

Duchenne's muscular dystrophy is a debilitating neuromuscular disease affecting boys. The disease is caused by point mutations, deletions, and/or rearrangements in the open reading frame of the X-linked dystrophin gene (Bushby et al. 2010). In muscles, the dystrophin protein is responsible for connecting the subsarcolemmal cytoskeleton to the sarcolemma: lack of functional dystrophin causes muscular degeneration, inflammation, and replacement of the muscle tissue with fibroadipose tissue (Hoffman et al. 1987). Splice-switching DNA ODNs such as AVI-4658 have been used to target regions of the pre-mRNA transcript and induce targeted exon skipping to restore the open reading frame and produce a more viable dystrophin protein (length of the protein produced will vary depending on the region that is skipped) (Muntoni and Wood 2011). RNA-based ASO therapies that can induce exon skipping and restore a viable open reading frame have advanced through in vitro and in vivo testing, and clinical trials are under way (Muntoni and Wood 2011).

AVI-4658 is a phosphorodiamidate, splice-switching (morpholino) ODN that targets exon 51 of dystrophin and results in production of truncated, but still partially functional, dystrophin protein molecules similar to those found in milder (and Duchenne's-related) Becker muscular dystrophy (Hoffman et al. 1987; Monaco et al. 1988; Arechavala-Gomeza et al. 2007). Because this DNA ASO targets exon 51, it is appropriate for use in patients with deletion mutations of dystrophin that are amenable to exon 51 skipping. In a phase II trial involving 19 patients with ambulatory Duchenne's muscular dystrophy, intravenous administration of AVI-4658 was well tolerated, and no serious adverse events were reported (Cirak et al. 2011, 2012). In this dose escalation study, the ASO induced exon skipping and variable protein production in those cohorts given $\geq$2.0 mg/kg, compared to protein in pretreatment muscle biopsy. As AVI-4658 was only given for 12 weeks, there was no clinical improvement of symptoms. Further improvements in methodology and additional trials are therefore necessary. Indeed, it should be noted that all therapies for Duchenne's muscular dystrophy are anticipated to require lifelong application. There are positive indications that the ASO-mediated treatment described above will induce clinical response if given for a sufficiently long period of time, because a concomitant increase in dystrophin-associated glycoprotein complex proteins (necessary for sarcolemma connections) was observed along with a decrease in inflammatory infiltrate in patients with the greatest response to treatment at 12 weeks (Cirak et al. 2011, 2012).

8 General Strategies for Antisense Treatment of Cancer

The lack of selectivity of anticancer chemotherapy against cancer cells versus normal cells remains the greatest obstacle to successful optimization of treatment. However, major gains over the last few decades in understanding the biology of cancer have yielded potential, selective targets. Tumor cells are partially characterized by genetic instability and oncogene addiction, that is, dependency on an upregulated oncogene or a downregulated tumor suppressor. If the expression level of an oncogene or suppressor could be reversed, it should result in growth inhibition and/or cell death (Nijman 2011). However, putting this concept into practice using single agents has been difficult, if not impossible. An alternative approach with increased potential for success may involve endogenous over- or underexpression of an important gene product to render tumor cells exquisitely sensitive to inhibition of a second cellular function. This strategy is the basis underlying *synthetic lethal* approaches to cancer therapy.

Synthetic lethality, from the time that the phenomenon was originally described in *Drosophila* (Shaheen et al. 2011), yeast, nematodes, and bacteria (Brough et al. 2011; Chan and Giaccia 2011), has been defined in general terms to describe a relationship between two genes, the loss of either of which is not lethal, but concomitant loss of both leads to the death of a cell or organism (Evers et al. 2010; Brough et al. 2011; Nijman 2011). This definition has been modified in the context of developing anticancer treatments to mean that a tumor that is deficient in a particular gene product, and thereby dependent on a second gene product to maintain viability, is vulnerable to inhibition of that second gene product by a drug. In many cases, the loss of the first function contributes directly to tumor progression (e.g., loss of pro-apoptotic proteins, increased genetic instability that leads to tumorigenic mutations, and loss of negative regulators of proliferation). Increased tumorigenic behavior (proliferation, survival, invasiveness) due to such a loss is often dependent on compensatory or, at least, concomitant activity of a second pro-survival or protective function. For the purpose of this discussion, the definition of synthetic lethality has been broadened slightly to accommodate numerous scenarios that involve eliminating a gene product using antisense technology, and exploitation of the resultant dependency on a second gene product as a potential tumor-selective target (Berg et al. 2003; Fong et al. 2009; Tutt et al. 2010; Brough et al. 2011; Chan and Giaccia 2011; Dedes et al. 2011; Shaheen et al. 2011). Note that this is in contrast to the concept of using antisense to knock down a gene product that is, itself, the target of the drug used for the second treatment (Ferguson et al. 2001; Berg et al. 2003; Flynn et al. 2006; Jason et al. 2008).

The ability to create synthetic lethality in cancer cells requires knowledge of potentially exploitable targets. For example, a cell that has lost p53 function is often dependent on increased pro-survival mechanisms or loss of pro-apoptotic mechanisms. Identification of such survival mechanisms could reveal drug targets to which the cancer cell may be selectively sensitive, a so-called Achilles heel, compared to normal somatic cells. Similarly, a cell devoid of active BRCA1 or BRCA2 (enzymes involved in DNA repair mediated by homologous recombination [HR]) is heavily dependent on DNA repair that does not involve HR, making the cell and its progeny selectively sensitive to inhibitors of poly(ADP-ribose) polymerase (PARP) (Farmer et al. 2005). In these cases, spontaneous loss of a single gene function makes the cancer cell dependent on a second function that is a potentially selective drug target. If synthetic lethality could be induced by ASO-mediated knockdown of an appropriate gene product, then cancer cells could be rendered selectively sensitive to a drug (Chan and Giaccia 2011). The first step to exploiting this phenomenon for therapeutic benefit is to identify pairs of gene products and/or metabolic functions that are interdependent for cell survival. For the purpose of designing a chemotherapy regimen, it is expected that synthetic lethality should involve at least one drug against a specified target. However, combinations of antisense molecules may decrease the amounts, and thereby the activities, of a pair of proteins below a level required for cell viability.

Tables 2 and 3 list examples, determined or theorized, where ASO or small molecule targeting to exploit preexisting loss of one molecule (*preexisting synthetic lethality*, Table 2) or simultaneous ASO/small molecule knockdown of two endogenous cancer cell molecules (*induced synthetic lethality*, Table 3) may be of therapeutic value in treatment of cancer. It should be noted that this field is rapidly progressing and new examples of preexisting and induced synthetic lethality are continuously emerging, and these lists should be considered partial.

RNAi has been widely used as a laboratory tool to identify potential synthetic lethal combinations. Cells treated with libraries of shRNAs to downregulate both known and unknown targets were subsequently screened for sensitivity to antineoplastic agents or a second antisense molecule (Neshat et al. 2001; Gleave and Jansen 2003; Morgan-Lappe et al. 2006; Dong et al. 2010; Whitehurst et al. 2010; Brough et al. 2011; Chan and Giaccia 2011). The purpose of such shRNA screens was to uncover potential vulnerabilities that might present as synthetic lethal drug targets if detected in tumors. We propose that ASOs themselves may serve as excellent inducers of synthetic lethality, exploitable by simultaneous treatment with ASOs/small molecule drugs to knock down a second cancer cell target for synergistic inhibition and/or killing of cancer cells, particularly as methodologies for targeting delivery to cancer cells continue to improve.

Table 3 also lists abnormal expressions of various protein activities (some of which are increases) that render cells hypersensitive to inhibition of a second target with a drug. Except in instances in which the first anomaly is gene overexpression rather than reduced or lost expression, the synthetic lethal situation can be recapitulated by downregulating both the primary and secondary gene products using ASOs.

Many of the targets presented in Tables 2 and 3 are involved in DNA repair. A good understanding of the interactions of repair pathways will help to predict

Table 2 Gene pairs with potential for *induced synthetic lethality* (opportunity for simultaneous therapeutic targeting of primary and secondary gene products)

Primary target	Secondary target	References	Comments
ATM (ataxia telangiectasia mutated)	DNA-PKcs (DNA-dependent protein kinase catalytic subunit)	Plummer (2010)	DNA-PKcs is a DNA repair enzyme
	PARP (poly(ADP-ribose) polymerase)	Hartwell et al. (1997), McCabe et al. (2006)	
BRCA1 (breast cancer susceptibility gene 1)	PARP	Farmer et al. (2005)	
	Tankyrase	Plummer (2010)	PARP-family member involved in telomere maintenance
BRCA2 (breast cancer susceptibility gene 2)	PARP	Bryant et al. (2005), Farmer et al. (2005)	
Checkpoint kinase 1 (Chk1)	DNA (in general)	Plummer (2010)	Checkpoint kinase 1, S- and G2-checkpoint protein
	WEE1 (G2-checkpoint protein)	Wang et al. (2004)	Synergistic with DNA-damaging agents; in clinical trials
Checkpoint kinase 2 (Chk2)	PARP	McCabe et al. (2006), Anderson et al. (2011)	Serine–threonine kinase
CSNK1G3	AKT	Morgan-Lappe et al. (2006)	AKT is a serine–threonine protein kinase with 3 isoforms; part of PI3K signaling cascade
HR repair	Topo I and II	Hartwell et al. (1997)	
HSP90 (90-kDa heat shock protein)	Androgen receptor	Mizuarai et al. (2008)	
IPMK (inositol polyphosphate multikinase)	AKT	Morgan-Lappe et al. (2006)	
MLH1 (mutL homolog 1, colon cancer, nonpolyposis type 2)	Pol-gamma	Martin et al. (2010)	MLH1 is frequently mutated in hereditary nonpolyposis colon cancer
	PINK-1 (PTEN-induced putative kinase 1)	Martin et al. (2011)	PTEN = phosphatase–tensin homolog, *inhibits phosphorylation of Akt*
MSH2 (mutS homolog 2, nonpolyposis type 1)	Polβ (DNA polymerase β)	Martin et al. (2010)	MSH2 is frequently mutated in hereditary nonpolyposis colon cancer; combines with MSH6 to form DNA mismatch recognition complex
	PINK-1	Martin et al. (2011)	

MSH6 (mutS homolog 6)	PINK-1	Martin et al. (2011)	MSH6 is frequently mutated in hereditary nonpolyposis colon cancer
Notch-1 (Notch homolog 1)	EGFR (epidermal growth factor receptor; *Her-1*)	Dong et al. (2010)	Notch-1 is a membrane signaling protein and the receptor for Jagged ligand
p53 (53-kDa checkpoint protein and transcriptional regulator)	GEF-H1(ARHGEF2) (Rho/Rac guanine nucleotide exchange factor)	Ren et al. (1998), Krendel et al. (2002)	Activator of Rho-family proteins
	Chk1 or Chk2	Reinhardt et al. (2007), Luo et al. (2009)	Hypersensitive to doxorubicin and cisplatin
	SGK2	Baldwin et al. (2010)	
	PAK3	Baldwin et al. (2010))	
	ATR	Ruzankina et al. (2009)	
	MK2	Reinhardt et al. (2007)	
	ATR	Nghiem et al. (2001), Mizuarai et al. (2008)	
	ATM	Reinhardt et al. (2007), Jiang et al. (2009)	
	PLK1 (STPK13) (polo-like kinase 1)	Sur et al. (2009)	PLK1 is a mitosis checkpoint protein
	WEE1	Mizuarai et al. (2009)	
PALB2	PARP	Rehman et al. (2010)	Found in pancreatic cancers
PTEN	PARP	Mendes-Pereira et al. (2009)	PTEN inhibits phosphorylation of Akt
	mTOR (mammalian target of rapamycin)	Neshat et al. (2001), Shi et al. (2002)	mTOR is part of PI3K signaling cascade
RAD51 (DNA repair protein)	PARP	McCabe et al. (2006)	
	BRCA1	Chan and Giaccia (2011)	
	FANCD2 (Fanconi's anemia complementation group D2)	Chan and Giaccia (2011)	
RAD52 (DNA repair protein)	BRCA2	Feng et al. (2011)	
RAD54B	FEN1 (Flap 5′-endonuclease, removes single-stranded RNA and DNA loose ends [flaps])	Rehman et al. (2010)	FEN1 is involved in DNA repair; demonstrated by antisense because there are no drugs available

(continued)

Table 2 (continued)

Primary target	Secondary target	References	Comments
Ki-RAS (RAS-family member, *overactivity of which is most often associated with cancer, due to a constitutively activating mutation*)	APC/C (anaphase-promoting complex/cyclosome)	Luo et al. (2009)	Ki-RAS overactivity is most often associated with cancer, due to a constitutively activating mutation
	STK33 (serine–threonine kinase)	Scholl et al. (2009)	There is controversy over whether STK33 is essential in RAS-mutant cells
	TBK1 (TANK-binding kinase)	Barbie et al. (2009)	TBK1 is an NF-kB activator
	SNAIL2 (also known as SNAI2)	Wang et al. (2010)	SNAIL2 is one of a family of transcription factors involved in epithelial to mesenchymal transition and associated with poor prognosis of breast cancer
	Survivin	Sarthy et al. (2007)	
	PLK1	Luo et al. (2009)	
	CDK (cyclin-dependent kinase)	Puyol et al. (2010)	
Rb (retinoblastoma protein)	TSC2 (tuberous sclerosis 2)	Weidle et al. (2011)	Rb mutation is commonly associated with a variety of cancers and is important for cell cycle progression checkpoint control and chromosome condensation. TSC2 is associated with renal carcinoma and is a tumor suppressor and negative regulator of mTORC1 and involved in accumulation of ROS
	SKP2	Bauzon and Zhu (2010)	SKP2 = S-phase kinase-associated protein 2 (p45)
	ECT2 (exchange factor for GTPases)	Eguchi et al. (2007)	
VHL (von Hippel–Landau) protein	mTOR	Martin et al. (2009)	VHL is involved in DNA repair and is commonly associated with renal cell carcinoma
Interactive protein kinase networks		Weidle et al. (2011)	Summarized in a table in a review Weidle et al. (2011)
myc (overexpression)	Aurora kinase	Yang et al. (2010)	
	GSK-3β (glycogen synthase kinase 3β)	Rottmann et al. (2005)	

Table 3 Gene pairs with potential for *preexisting synthetic lethality* (preexisting loss of the first target creates an opportunity for therapeutic ASO/small molecule targeting of the second)

Primary genetic anomaly	Drug against second target	Drug target	References	Comments
Loss of gene product				
APE1 (apurinic/apyrimidinic endonuclease-1)	Temozolomide		Wilson and Simeonov (2010)	APE1 involved in base-excision repair; inhibitor currently in clinical trials
BRCA1	Cisplatin	DNA	Ganesan et al. (2002)	In p53-negative cells
BRCA2	Cisplatin, mitoxantrone, amsacrine, etoposide	DNA	Abbott et al. (1998), Lee et al. (1999)	
Clusterin	Various chemotherapy drugs		Gleave and Jansen (2003)	
DNA ligases	Alkylating agents	DNA	Chen et al. (2008)	
DNA-repair proteins	Bortezomib	Proteasome	Chen et al. (2010b)	
Gamma-tubulin ring complex TUBGCP2 (a subunit of the γ tubulin complex)	Paclitaxel	Tubulin	Whitehurst et al. (2007)	
CT-antigen ACRBP (OY-TES-1) (acrosin-binding protein)	Paclitaxel	Tubulin	Whitehurst et al. (2007)	ACRBP binds with mitotic complex; contributes to resistance to paclitaxel in ovarian cancer
PSMA6 (a subunit of proteasome 20S)	Paclitaxel	Tubulin	Whitehurst et al. (2007)	
Ki-RAS (activated)	Paclitaxel	Tubulin	Luo et al. (2009)	
	Bortezomib	Proteasome	Luo et al. (2009)	
VHL	Chromomycin A3	RNA synthesis	Sutphin et al. (2007)	
	STF-62247	Autophagy inducer	Turcotte et al. (2008)	
MGMT (O6-methylguanine-DNA methyltransferase)	Alkylating agents	DNA	Kaina et al. (2010)	Toxicity to bone marrow

(continued)

Table 3 (continued)

Primary genetic anomaly	Drug against second target	Drug target	References	Comments
Mismatch repair (in general) (includes MLH1, MSH2, and MSH6)	Oxaliplatin	DNA cross-linking	Sargent et al. (2010)	Resistance to some alkylators (e.g., oxaliplatin [V25])
MSH2	Methotrexate	DHFR; ROS	Martin et al. (2009)	
XAB2 (a transcription-coupled DNA repair protein)	Olaparib	PARP	Lord et al. (2008)	mRNA splicing
XRCC1 (X-ray repair complementing defective repair)	Olaparib	PARP	Lord et al. (2008)	XRCC1 is involved in repair of single-strand DNA breaks caused by ionizing radiation and alkylating agents as well as base excision repair
DDB1 (damage-specific DNA-binding protein 1)	Olaparib	PARP	Lord et al. (2008)	DDB1 is involved in nucleotide excision repair
Overexpression of gene product				
myc	Chloroquine	Autophagy	Maclean et al. (2008)	
	Activator of DR5	DR5	Wang et al. (2004)	
NOTCH-1	Gamma-secretase inhibitor		Nickoloff et al. (2005)	In some melanomas

appropriate antisense therapeutic approaches. For example, cells in which spontaneous, preexisting DNA repair defects create synthetic lethal situations exploitable by ASO knockdown of a second target could become resistant to therapy if alternative DNA repair mechanisms are upregulated (Shaheen et al. 2011). Some repair pathways are more prone to error than others, and diverting repair to a more error-prone pathway in transformed cells could eventually lead to cancer cell death, a desirable outcome that could be considered a useful treatment.

Impaired repair in untransformed cells, in contrast, contributes to enhanced tumorigenesis and subsequent tumor aggressiveness due to a high rate of spontaneous mutations (the so-called mutator phenotype). Cells with impaired repair may rely on cell cycle checkpoint mechanisms to prevent induction of apoptosis, and RNA encoding proteins mediating cell cycle checkpoint arrest are potential targets to induce and exploit synthetic lethality (Evers et al. 2010).

There are multiple advantages to using antisense to exploit synthetic lethality in cancer cells, including:

1. Exploitation of tumors that harbor endogenous mutations leading to synthetic lethality requires screening methodologies to detect those mutations. However, if deficiencies in these proteins are induced by antisense, then the necessity to screen is eliminated. Indeed, far from restricting the opportunity offered by synthetic lethality to a possibly small subset of tumors that serendipitously display the necessary mutations, tumors may be "forced" into synthetic lethality by purposeful ASO-mediated reduction of gene products.
2. Many of the targets for which synthetic lethality would be an optimal treatment have no pharmacological inhibitors. ASO-mediated knockdown of the target gene product eliminates most, if not all, targets from being "non-druggable."
3. It is possible that spontaneous reacquisition of an endogenously lost or mutated gene product could render a cell resistant to a synthetic lethal treatment. For example, cell lines with defective BRCA1 have developed resistance to PARP inhibitors by acquiring a deletion mutation in BRCA1 that restores its function (Aly and Ganesan 2011; Leung et al. 2011). This phenomenon could be avoided in situations in which synthetic lethality was artificially induced with ASOs by targeting sequences of BRCA1 mRNA in regions where the mutation does not occur.

It is acknowledged that there are examples of naturally occurring synthetic lethality for which the artificial situation may be challenging to replicate. Creating synthetic lethality selectively in a cancer cell depends on knocking down the drug target only in cancer cells and not normal cells (perhaps by targeted, selective delivery only to tumor cells). In addition, some instances of synthetic lethality arise due to an activating mutation. In these situations, such gene products would obviously not be targeted by antisense, but its lethal partner could be targeted using an antisense molecule. This is not to say that it would not be possible to take advantage of potential synthetic lethal situations as they arise spontaneously in patient tumors. The development of a tumor often depends on mutations in two mutually dependent pathways (Stavropoulou et al. 2010). Although they are potential antisense targets for synthetic lethality, the appearance of mutations in tumors will vary considerably among patients with different tumors and even within a given tumor type. However, with the rapid advancements in genetic screening techniques, such targets could be identified for individual patients and could potentially contribute to the design of personalized, cancer-selective antisense molecules to synergistically induce death of cancer cells (personalized medicine).

Antisense technology is, therefore, not only a powerful laboratory tool with which to identify potential tumor targets for established anticancer drugs, but is poised to soon become part of the clinical armament to treat tumor cells while sparing normal cells by application in conjunction with nonselective drugs in current clinical use. ASOs are progressing through clinical trials for treatment of conditions other than cancer (de Fougerolles 2008) and have been used in anticancer trials with the intention of enhancing antitumor activity of cytotoxic drugs (Chi et al. 2010) (discussed above). Therefore, the routine use of antisense technology to create and exploit synthetic lethality in anticancer treatment is imminent and potentially powerful.

9 Novel Uses for Antisense Technology

9.1 Targeting MicroRNAs in Cancer

miRNAs are small, endogenously expressed RNA molecules (approximately 19–25 nucleotides long) that act to negatively regulate gene expression posttranscriptionally by complementary base pairing with specific mRNA molecules to inhibit translation and/or catalyze cleavage of the target mRNA in the RISC complex (Bartel 2004; Pasquinelli et al. 2005). miRNAs help to regulate many cellular processes ranging from cellular differentiation and proliferation to senescence and apoptosis, partly due to their role in regulation of both oncogenes and tumor suppressor genes (Bartel 2004; Feliciano et al. 2011). miRNAs have different degrees of complementarity with their target mRNA strands, allowing a single miRNA to regulate the level of many different gene products. miRNA dysregulation has been determined as an underlying mechanism in oncogenesis, as miRNAs can function as tumor suppressors [e.g., the *miR-15a–miR-16-1* cluster in CLL and prostate cancer (Calin et al. 2005; Bonci et al. 2008)] or as oncogenes [e.g., the *miR-17-92* cluster in lung cancer and CLL (Hayashita et al. 2005; Venturini et al. 2007)].

Although there are some instances of miRNA upregulation associated with malignancy and tumor aggressiveness (e.g., *miR-103/107*) (Calin and Croce 2006; Stenvang and Kauppinen 2008; Croce 2009; Martello et al. 2010; Lennox and Behlke 2011) there is a general trend toward miRNA downregulation (e.g., *let-7*) in the majority of cancer types tested (Takamizawa et al. 2004; Lu et al. 2005; Kumar et al. 2007). Therefore, a potential mode of cancer treatment could be to replace the missing miRNA with a synthetic therapeutic miRNA polynucleotide (a miRNA agonist). It may also be possible to specifically decrease the level of an oncogenic miRNA by treating with an ASO that is complementary to that miRNA (an miRNA antagonist) (Wang 2011). The use of antisense molecules to modulate the expression of miRNAs involved in cancer and other diseases is being investigated both in vitro and in vivo (Meister et al. 2004b; Matsubara et al. 2007; Duchaine and Slack 2009; Kota et al. 2009; Segura et al. 2009; Lin et al. 2010). Other therapeutic applications based on miRNA antagonists and agonists are addressed elsewhere in this volume.

9.2 *Multiple-Target Anti-miRNA Antisense Oligodeoxynucleotide Inhibitor Technology*

A potential limitation to the use of single-target anti-miRNA antisense oligodeoxynucleotide inhibitors (AMOs) to silence an oncogenic miRNA is that multiple miRNAs may be simultaneously overexpressed in the same cancer cell (Volinia et al. 2006). To overcome this, design of single AMOs with multiple RNA targets, referred to as multiple-target AMO technology (MT-AMO), has been proposed (Lu et al. 2009).

To design an MT-AMO against a particular cancer type, it is first necessary to determine which miRNA(s) is overexpressed, based on historical evidence or an *ad hoc* screen of miRNAs with aberrantly high levels in specific tumor classes (Wang 2011). AMOs specific for the selected miRNAs would then be designed and linked together in a 5′-end to 3′-end orientation, forming an MT-AMO. Each AMO can be separated from the next AMO by a linker designed to inhibit formation of secondary structure and avoid nonspecific binding (e.g., linker sequences such as CTTAAATG). A modified MT-AMO containing approximately five base mismatches at the 5′-end must be used as a negative control to verify the specificity and effectiveness of the candidate MT-AMO (Wang 2011). It is also possible to target multiple miRNAs with three covalently linked antisense molecules and still maintain inhibitory activity. For example, an MT-AMO that simultaneously targets miR-21, miR-155, and miR-17-5p showed greater inhibitory effects on the growth of MCF-7 breast cancer cell line than each of the single-target AMOs administered alone or in combination of all three single-target AMOs (Lu et al. 2009).

Although MT-AMO technology has great potential as an effective strategy for antisense cancer therapy, there are aspects of MT-AMO technology that require further development:

1. Is there a maximum number of target miRNAs against which an MT-AMO can be successfully targeted? This may be limited by formation of intra-sequence secondary structures within MT-AMOs and effects of MT-AMO molecular weight on cellular uptake and target-binding efficiency. At present, this is not known.

2. Are certain delivery methods more efficient for transfecting target cells with MT-AMOs? It is not known whether the higher molecular weight of MT-AMOs (compared to traditional RNAi-dependent or other ASOs) poses a challenge to their entry into tumor or other target cells, particularly in vivo.
3. Would MT-AMOs activate the intracellular immune sensors when introduced into the target cells? As MT-AMOs are large molecules, they could potentially activate innate immune mechanisms that recognize foreign nucleic acids in a host cell.

9.3 *Radiolabeled Antisense Oligonucleotides*

The use of radiolabeled antisense oligonucleotides (RASONS) for imaging and therapy is still a relatively new and undeveloped application of antisense-based technology. However, this approach displays several favorable aspects that render it worthy of further investigation. From a therapeutic perspective, RASONS not only degrade specific mRNA sequences through RNase H or RNAi, but they are also a vector for the targeted delivery of micro-radiotherapy (e.g., high LET radionuclides that induce toxic ionization events in target tumor cells) (Kairemo et al. 2000). Thus, they constitute a potential method of delivering a "two-hit therapy" that not only knocks down specific protein expression but also induces DNA damage in a tumor cell. RASON technology can also be used for functional tumor imaging by detecting specific mRNA transcripts which produce known, tumor associated proteins (Bai et al. 2004). Therefore, imaging with RASONs allows for the specific identification of cancer markers to aid patient diagnosis and prognosis, in contrast to traditional imaging techniques such as MRI, since RASONS identify specific gene products instead of gross anatomical features (Fu et al. 2010).

9.3.1 Therapeutic Applications

An important consideration in designing a RASON as a therapeutic intervention is determining which radionuclides will be most suitable for micro-radiotherapy. An appropriate radionuclide must exhibit limited dose variation in regard to the amount of radiation that a cell absorbs (Kairemo et al. 1996a, b, 2000). Beta-particle-emitting radionuclides such as ^{35}S or ^{33}P display the lowest nuclear dose variation among the isotopes tested, including α- and γ-emitters (Kairemo et al. 1996a, b, 2000). Beta-particle emitters, especially ^{35}S, were more effective than Auger-emitting radionuclides for nuclear targeting and thus are the most suitable candidates for further investigation (Kairemo et al. 1998).

Although the therapeutic arm of RASON technology is relatively underdeveloped, one study attempted to incorporate RASONs into a two-step radioimmunotherapy for cancer. This strategy involved the administration of an antibody-DNA conjugate, followed by the addition of a complementary radiolabelled ODN. In principle, an antibody against a known tumor antigen would have to be generated and this antibody would be bound to a DNA sequence. The radiolabelled ODN would recognize the DNA sequence attached to the antibody, thus targeting the cell with micro-radiotherapy (Kuijpers et al. 1993).

9.3.2 Imaging Applications

The primary objective of RASON-mediated imaging involves the identification of diagnostic and prognostic gene markers of cancer. Several studies have looked at the exciting potential for using RASONs to detect changes in gene expression to evaluate the severity of disease and predict the response of both the patient and their cancer to treatment (Bai et al. 2004; Fu et al. 2010). For example, multidrug resistance of cancer cells, although often mediated by ATP-dependent cassette transporters such as Pgp and MRP, may also be a more complex, multifactorial phenomenon. Therefore, conventional screening approaches using radiolabeled cationic compounds such as ^{99}mTc-MIBI may not be able to adequately distinguish between different factors leading to unforeseen MDR. RASONS can potentially be used for more effective screening of tumors to detect the presence and quantity of mRNA(s) for various MDR-associated proteins (Bai et al. 2004).

A second potential target for RASON imaging is MDM2 (mouse double minute 2), considered to be an oncoprotein because it promotes p53 ubiquitination and degradation (Fu et al. 2010). Overexpression of MDM2 is associated with invasive, metastatic, and recurrent cancers; thus, it is an important prognostic marker (Fu et al. 2010). In an experimental human breast cancer xenograft model, liposome-coated ^{99}mTc-MDM2 ASOs accumulated specifically in breast cancer tissue and were eliminated mainly through the urinary tract over time (Fu et al. 2010). Tumors were clearly imaged at 1-, 4-, and 10-h postinjection using this antisense strategy.

Furthermore, since p21 is expressed in response to DNA damage, it was postulated that the level of intracellular accumulation of ^{18}F-labeled ODNs targeting p21 mRNA could be indicative of the health of tumor cells following treatment with chemotherapy or radiation (Koslowsky et al. 2011). Also, ^{111}In-labeled ODNs targeting p21 mRNA accumulated preferentially in MDA-MB-468 human breast cancer cells treated with epidermal growth factor (EGF) to induce p21 (Wang et al. 2003). Successful imaging of EGF-induced, p21-expressing tumors was achieved at 48-h postinjection of labeled DNA ASO. Therefore, p21 mRNA expression can be imaged as a surrogate marker of DNA damage induced in patient tumors, using imaging techniques that are currently available.

10 Conclusions

Antisense molecules have progressed from the laboratory toward the clinic at a rapid pace. As better antisense chemistries and delivery methods are developed, we expect to see the extraordinary specificity of antisense molecules, and their ability to circumvent the barriers of "undruggability" (inherent when proteins rather than RNAs are targets), to result in the clinical approval of antisense as drugs to treat human diseases. Antisense molecules have the potential to be invaluable agents to exploit and induce synthetic lethality in combination with other drugs, modulate miRNAs for therapeutic benefit, and expand the repertoire of disease-ameliorating therapeutic agents into a new realm.

References

Aagaard L, Rossi JJ (2007) RNAi therapeutics: principles, prospects and challenges. Adv Drug Deliv Rev 59:75–86

Abbott DW, Freeman ML, Holt JT (1998) Double-strand break repair deficiency and radiation sensitivity in BRCA2 mutant cancer cells. J Natl Cancer Inst 90:978–985

Acsadi G, Dickson G, Love DR et al (1991) Human dystrophin expression in mdx mice after intramuscular injection of DNA constructs. Nature 352:815–818

Adjei AA, Dy GK, Erlichman C et al (2003) A phase I trial of ISIS 2503, an antisense inhibitor of H-ras, in combination with gemcitabine in patients with advanced cancer. Clin Cancer Res 9:115–123

Advani R, Peethambaram P, Lum BL et al (2004) A Phase II trial of aprinocarsen, an antisense oligonucleotide inhibitor of protein kinase C alpha, administered as a 21-day infusion to patients with advanced ovarian carcinoma. Cancer 100:321–326

Agrawal S (1999) Importance of nucleotide sequence and chemical modifications of antisense oligonucleotides. Biochim Biophys Acta 1489:53–68

Aigner A (2006) Gene silencing through RNA interference (RNAi) in vivo: strategies based on the direct application of siRNAs. J Biotechnol 124:12–25

Akinc A, Zumbuehl A, Goldberg M et al (2008) A combinatorial library of lipid-like materials for delivery of RNAi therapeutics. Nat Biotechnol 26:561–569

Alam MR, Ming X, Fisher M et al (2011) Multivalent cyclic RGD conjugates for targeted delivery of small interfering RNA. Bioconjug Chem 22:1673–1681

Alberts SR, Schroeder M, Erlichman C et al (2004) Gemcitabine and ISIS-2503 for patients with locally advanced or metastatic pancreatic adenocarcinoma: a North Central Cancer Treatment Group phase II trial. J Clin Oncol 22:4944–4950

Alvarez-Salas LM (2008) Nucleic acids as therapeutic agents. Curr Top Med Chem 8:1379–1404

Aly A, Ganesan S (2011) BRCA1, PARP, and 53BP1: conditional synthetic lethality and synthetic viability. J Mol Cell Biol 3:66–74

Ambegia E, Ansell S, Cullis P et al (2005) Stabilized plasmid-lipid particles containing PEG-diacylglycerols exhibit extended circulation lifetimes and tumor selective gene expression. Biochim Biophys Acta 1669:155–163

Anderson VE, Walton MI, Eve PD et al (2011) CCT241533 is a potent and selective inhibitor of CHK2 that potentiates the cytotoxicity of PARP inhibitors. Cancer Res 71:463–472

Arechavala-Gomeza V, Graham IR, Popplewell LJ et al (2007) Comparative analysis of antisense oligonucleotide sequences for targeted skipping of exon 51 during dystrophin pre-mRNA splicing in human muscle. Hum Gene Ther 18:798–810

Bai J, Yokoyama K, Kinuya S et al (2004) In vitro detection of mdr1 mRNA in murine leukemia cells with 111In-labeled oligonucleotide. Eur J Nucl Med Mol Imaging 31:1523–1529

Baldwin A, Grueneberg DA, Hellner K et al (2010) Kinase requirements in human cells: V. Synthetic lethal interactions between p53 and the protein kinases SGK2 and PAK3. Proc Natl Acad Sci USA 107:12463–12468

Barbie DA, Tamayo P, Boehm JS et al (2009) Systematic RNA interference reveals that oncogenic KRAS-driven cancers require TBK1. Nature 462:108–112

Bartel DP (2004) MicroRNAs: genomics, biogenesis, mechanism, and function. Cell 116:281–297

Bauzon F, Zhu L (2010) Racing to block tumorigenesis after pRb loss: an innocuous point mutation wins with synthetic lethality. Cell Cycle 9:2118–2123

Bennett CF, Swayze EE (2010) RNA targeting therapeutics: molecular mechanisms of antisense oligonucleotides as a therapeutic platform. Annu Rev Pharmacol Toxicol 50:259–293

Berg RW, Ferguson PJ, Vincent MD et al (2003) A "combination oligonucleotide" antisense strategy to downregulate thymidylate synthase and decrease tumor cell growth and drug resistance. Cancer Gene Ther 10:278–286

Bertrand JR, Pottier M, Vekris A et al (2002) Comparison of antisense oligonucleotides and siRNAs in cell culture and in vivo. Biochem Biophys Res Commun 296:1000–1004

Biswas S, Dodwadkar NS, Sawant RR et al (2011) Development of the novel PEG-PE-based polymer for the reversible attachment of specific ligands to liposomes: synthesis and in vitro characterization. Bioconjug Chem 22:2005–2013

Boeckle S, von Gersdorff K, van der Piepen S et al (2004) Purification of polyethylenimine polyplexes highlights the role of free polycations in gene transfer. J Gene Med 6:1102–1111

Bogdahn U, Hau P, Stockhammer G et al (2011) Targeted therapy for high-grade glioma with the TGF-beta2 inhibitor trabedersen: results of a randomized and controlled phase IIb study. Neuro Oncol 13:132–142

Bonci D, Coppola V, Musumeci M et al (2008) The miR-15a-miR-16-1 cluster controls prostate cancer by targeting multiple oncogenic activities. Nat Med 14:1271–1277

Breaker RR, Joyce GF (1994) A DNA enzyme that cleaves RNA. Chem Biol 1:223–229

Brough R, Frankum JR, Costa-Cabral S et al (2011) Searching for synthetic lethality in cancer. Curr Opin Genet Dev 21:34–41

Bruserud O (2007) Introduction: RNA and the treatment of human cancer. Curr Pharm Biotechnol 8:318–319

Bryant HE, Schultz N, Thomas HD et al (2005) Specific killing of BRCA2-deficient tumours with inhibitors of poly(ADP-ribose) polymerase. Nature 434:913–917

Bushby K, Finkel R, Birnkrant DJ et al (2010) Diagnosis and management of Duchenne muscular dystrophy, part 1: diagnosis, and pharmacological and psychosocial management. Lancet Neurol 9:77–93

Calin GA, Croce CM (2006) MicroRNA signatures in human cancers. Nat Rev Cancer 6:857–866

Calin GA, Ferracin M, Cimmino A et al (2005) A MicroRNA signature associated with prognosis and progression in chronic lymphocytic leukemia. N Engl J Med 353:1793–1801

Carroll JB, Warby SC, Southwell AL et al (2011) Potent and selective antisense oligonucleotides targeting single-nucleotide polymorphisms in the huntington disease gene/allele-specific silencing of mutant huntingtin. Mol Ther 19:2178–2185

Carter BZ, Mak DH, Morris SJ et al (2011) XIAP antisense oligonucleotide (AEG35156) achieves target knockdown and induces apoptosis preferentially in CD34 + 38- cells in a phase 1/2 study of patients with relapsed/refractory AML. Apoptosis 16:67–74

Carthew RW, Sontheimer EJ (2009) Origins and mechanisms of miRNAs and siRNAs. Cell 136:642–655

Cerritelli SM, Crouch RJ (2009) Ribonuclease H: the enzymes in eukaryotes. FEBS J 276:1494–1505

Cerutti H (2003) RNA interference: traveling in the cell and gaining functions? Trends Genet 19:39–46

Chan DA, Giaccia AJ (2011) Harnessing synthetic lethal interactions in anticancer drug discovery. Nat Rev Drug Discov 10:351–364

Chanan-Khan AA, Niesvizky R, Hohl RJ et al (2009) Phase III randomised study of dexamethasone with or without oblimersen sodium for patients with advanced multiple myeloma. Leuk Lymphoma 50:559–565

Chen X, Zhong S, Zhu X et al (2008) Rational design of human DNA ligase inhibitors that target cellular DNA replication and repair. Cancer Res 68:3169–3177

Chen X, Wang X, Wang Y et al (2010a) Improved tumor-targeting drug delivery and therapeutic efficacy by cationic liposome modified with truncated bFGF peptide. J Control Release 145:17–25
Chen S, Blank JL, Peters T et al (2010b) Genome-wide siRNA screen for modulators of cell death induced by proteasome inhibitor bortezomib. Cancer Res 70:4318–4326
Cheng CJ, Saltzman WM (2011) Enhanced siRNA delivery into cells by exploiting the synergy between targeting ligands and cell-penetrating peptides. Biomaterials 32:6194–6203
Chi KN, Zoubeidi A, Gleave ME (2008a) Custirsen (OGX-011): a second-generation antisense inhibitor of clusterin for the treatment of cancer. Expert Opin Investig Drugs 17:1955–1962
Chi KN, Siu LL, Hirte H et al (2008b) A phase I study of OGX-011, a 2′-methoxyethyl phosphorothioate antisense to clusterin, in combination with docetaxel in patients with advanced cancer. Clin Cancer Res 14:833–839
Chi KN, Hotte SJ, Yu EY et al (2010) Randomized phase II study of docetaxel and prednisone with or without OGX-011 in patients with metastatic castration-resistant prostate cancer. J Clin Oncol 28:4247–4254
Chia S, Dent S, Ellard S et al (2009) Phase II trial of OGX-011 in combination with docetaxel in metastatic breast cancer. Clin Cancer Res 15:708–713
Cho HY, Srinivasan A, Hong J et al (2011) Synthesis of biocompatible PEG-based star polymers with cationic and degradable core for siRNA delivery. Biomacromolecules 12:3478–3486
Cirak S, Arechavala-Gomeza V, Guglieri M et al (2011) Exon skipping and dystrophin restoration in patients with Duchenne muscular dystrophy after systemic phosphorodiamidate morpholino oligomer treatment: an open-label, phase 2, dose-escalation study. Lancet 378:595–605
Cirak S, Feng L, Anthony K et al (2012) Restoration of the dystrophin-associated glycoprotein complex after exon skipping therapy in Duchenne muscular dystrophy. Mol Ther 20:462–467
Cooper TA, Wan L, Dreyfuss G (2009) RNA and disease. Cell 136:777–793
Coppelli FM, Grandis JR (2005) Oligonucleotides as anticancer agents: from the benchside to the clinic and beyond. Curr Pharm Des 11:2825–2840
Coudert B, Anthoney A, Fiedler W et al (2001) Phase II trial with ISIS 5132 in patients with small-cell (SCLC) and non-small cell (NSCLC) lung cancer. A European Organization for Research and Treatment of Cancer (EORTC) early clinical studies group report. Eur J Cancer 37:2194–2198
Cripps MC, Figueredo AT, Oza AM et al (2002) Phase II randomized study of ISIS 3521 and ISIS 5132 in patients with locally advanced or metastatic colorectal cancer: a National Cancer Institute of Canada clinical trials group study. Clin Cancer Res 8:2188–2192
Croce CM (2009) Causes and consequences of microRNA dysregulation in cancer. Nat Rev Genet 10:704–714
Crooke ST (1998) Vitravene—another piece in the mosaic. Antisense Nucleic Acid Drug Dev 8: vii–viii
Crooke ST, Vickers T, Lima W et al (2007) In: Crooke ST (ed) Antisense drug technology: principles, strategies and applications. Taylor & Francis Group, Boca Raton, FL, pp 5–46
Davis ME (2009) The first targeted delivery of siRNA in humans via a self-assembling, cyclodextrin polymer-based nanoparticle: from concept to clinic. Mol Pharm 6:659–668
Davis ME, Zuckerman JE, Choi CH et al (2010) Evidence of RNAi in humans from systemically administered siRNA via targeted nanoparticles. Nature 464:1067–1070
De Clercq E (2004) Antiviral drugs in current clinical use. J Clin Virol 30:115–133
de Fougerolles AR (2008) Delivery vehicles for small interfering RNA in vivo. Hum Gene Ther 19:125–132
Dean E, Jodrell D, Connolly K et al (2009) Phase I trial of AEG35156 administered as a 7-day and 3-day continuous intravenous infusion in patients with advanced refractory cancer. J Clin Oncol 27:1660–1666
Dedes KJ, Wetterskog D, Ashworth A et al (2011) Emerging therapeutic targets in endometrial cancer. Nat Rev Clin Oncol 8:261–271
Di Cresce C, Koropatnick J (2010) Antisense treatment in human prostate cancer and melanoma. Curr Cancer Drug Targets 10:555–565

Di Cresce C, Figueredo R, Ferguson PJ et al (2011) Combining small interfering RNAs targeting thymidylate synthase and thymidine kinase 1 or 2 sensitizes human tumor cells to 5-fluorodeoxyuridine and pemetrexed. J Pharmacol Exp Ther 338:952–963
Dong Y, Li A, Wang J et al (2010) Synthetic lethality through combined Notch-epidermal growth factor receptor pathway inhibition in basal-like breast cancer. Cancer Res 70:5465–5474
Duchaine TF, Slack FJ (2009) RNA interference and micro RNA-oriented therapy in cancer: rationales, promises, and challenges. Curr Oncol 16:61–66
Dykxhoorn DM, Novina CD, Sharp PA (2003) Killing the messenger: short RNAs that silence gene expression. Nat Rev Mol Cell Biol 4:457–467
Eguchi T, Takaki T, Itadani H et al (2007) RB silencing compromises the DNA damage-induced G2/M checkpoint and causes deregulated expression of the ECT2 oncogene. Oncogene 26:509–520
Evers B, Helleday T, Jonkers J (2010) Targeting homologous recombination repair defects in cancer. Trends Pharmacol Sci 31:372–380
Farmer H, McCabe N, Lord CJ et al (2005) Targeting the DNA repair defect in BRCA mutant cells as a therapeutic strategy. Nature 434:917–921
Feliciano A, Sanchez-Sendra B, Kondoh H et al (2011) MicroRNAs regulate key effector pathways of senescence. J Aging Res 2011:205378
Feng Z, Scott SP, Bussen W et al (2011) Rad52 inactivation is synthetically lethal with BRCA2 deficiency. Proc Natl Acad Sci USA 108:686–691
Ferguson PJ, DeMoor JM, Vincent MD et al (2001) Antisense-induced down-regulation of thymidylate synthase and enhanced cytotoxicity of 5-FUdR in 5-FUdR-resistant HeLa cells. Br J Pharmacol 134:1437–1446
Fesik SW (2005) Promoting apoptosis as a strategy for cancer drug discovery. Nat Rev Cancer 5:876–885
Fire A, Xu S, Montgomery MK et al (1998) Potent and specific genetic interference by double-stranded RNA in *Caenorhabditis elegans*. Nature 391:806–811
Fischer D, Bieber T, Li Y et al (1999) A novel non-viral vector for DNA delivery based on low molecular weight, branched polyethylenimine: effect of molecular weight on transfection efficiency and cytotoxicity. Pharm Res 16:1273–1279
Flynn J, Berg RW, Wong T et al (2006) Therapeutic potential of antisense oligodeoxynucleotides to down-regulate thymidylate synthase in mesothelioma. Mol Cancer Ther 5:1423–1433
Fong PC, Boss DS, Yap TA et al (2009) Inhibition of poly(ADP-ribose) polymerase in tumors from BRCA mutation carriers. N Engl J Med 361:123–134
Fu P, Shen B, Zhao C et al (2010) Molecular imaging of MDM2 messenger RNA with 99mTc-labeled antisense oligonucleotides in experimental human breast cancer xenografts. J Nucl Med 51:1805–1812
Ganesan S, Silver DP, Greenberg RA et al (2002) BRCA1 supports XIST RNA concentration on the inactive X chromosome. Cell 111:393–405
Geary RS (2009) Antisense oligonucleotide pharmacokinetics and metabolism. Expert Opin Drug Metab Toxicol 5:381–391
Geary RS, Watanabe TA, Truong L et al (2001) Pharmacokinetic properties of 2′-O-(2-methoxyethyl)-modified oligonucleotide analogs in rats. J Pharmacol Exp Ther 296:890–897
Geary RS, Yu RZ, Watanabe T et al (2003) Pharmacokinetics of a tumor necrosis factor-alpha phosphorothioate 2′-O-(2-methoxyethyl) modified antisense oligonucleotide: comparison across species. Drug Metab Dispos 31:1419–1428
Geary RS, Bradley JD, Watanabe T et al (2006) Lack of pharmacokinetic interaction for ISIS 113715, a 2′-0-methoxyethyl modified antisense oligonucleotide targeting protein tyrosine phosphatase 1B messenger RNA, with oral antidiabetic compounds metformin, glipizide or rosiglitazone. Clin Pharmacokinet 45:789–801
Gleave M, Jansen B (2003) Clusterin and IGFBPs as antisense targets in prostate cancer. Ann NY Acad Sci 1002:95–104

Gollob J, Infante, JR, Shapiro G et al (2010) Interim safety and pharmacodynamic results for ALN-VSP02, a novel RNAi therapeutic for solid tumors with liver involvement. J Clin Oncol 28 (suppl): abstr 3042
Golzio M, Mazzolini L, Paganin-Gioanni A et al (2009) Targeted gene silencing into solid tumors with electrically mediated siRNA delivery. Methods Mol Biol 555:15–27
Grimm D, Kay MA (2007) Combinatorial RNAi: a winning strategy for the race against evolving targets? Mol Ther 15:878–888
Grunweller A, Wyszko E, Bieber B et al (2003) Comparison of different antisense strategies in mammalian cells using locked nucleic acids, 2′-O-methyl RNA, phosphorothioates and small interfering RNA. Nucleic Acids Res 31:3185–3193
Hanahan D, Weinberg RA (2000) The hallmarks of cancer. Cell 100:57–70
Hanahan D, Weinberg RA (2011) Hallmarks of cancer: the next generation. Cell 144:646–674
Hartwell LH, Szankasi P, Roberts CJ et al (1997) Integrating genetic approaches into the discovery of anticancer drugs. Science 278:1064–1068
Hayashita Y, Osada H, Tatematsu Y et al (2005) A polycistronic microRNA cluster, miR-17-92, is overexpressed in human lung cancers and enhances cell proliferation. Cancer Res 65:9628–9632
Herrero MA, Toma FM, Al-Jamal KT et al (2009) Synthesis and characterization of a carbon nanotube-dendron series for efficient siRNA delivery. J Am Chem Soc 131:9843–9848
Higuchi Y, Kawakami S, Hashida M (2010) Strategies for in vivo delivery of siRNAs: recent progress. BioDrugs 24:195–205
Hockenbery D, Nunez G, Milliman C et al (1990) Bcl-2 is an inner mitochondrial membrane protein that blocks programmed cell death. Nature 348:334–336
Hoffman EP, Brown RH Jr, Kunkel LM (1987) Dystrophin: the protein product of the Duchenne muscular dystrophy locus. Cell 51:919–928
Hossini AM, Eberle J, Fecker LF et al (2003) Conditional expression of exogenous Bcl-X(S) triggers apoptosis in human melanoma cells in vitro and delays growth of melanoma xenografts. FEBS Lett 553:250–256
Intra J, Salem AK (2008) Characterization of the transgene expression generated by branched and linear polyethylenimine-plasmid DNA nanoparticles in vitro and after intraperitoneal injection in vivo. J Control Release 130:129–138
Ivanova G, Arzumanov A, Gait MJ et al (2007) Comparative studies of tricyclo-DNA- and LNA-containing oligonucleotides as inhibitors of HIV-1 gene expression. Nucleosides Nucleotides Nucleic Acids 26:747–750
Izant JG, Weintraub H (1984) Inhibition of thymidine kinase gene expression by anti-sense RNA: a molecular approach to genetic analysis. Cell 36:007–1015
Jachimczak P, Bogdahn U, Schneider J et al (1993) The effect of transforming growth factor-beta 2-specific phosphorothioate-anti-sense oligodeoxynucleotides in reversing cellular immunosuppression in malignant glioma. J Neurosurg 78:944–951
Jaschinski F, Rothhammer T, Jachimczak P et al (2011) The antisense oligonucleotide trabedersen (AP 12009) for the targeted inhibition of TGF-beta2. Curr Pharm Biotechnol 12:2203–2213
Jason TL, Koropatnick J, Berg RW (2004) Toxicology of antisense therapeutics. Toxicol Appl Pharmacol 201:66–83
Jason TL, Figueredo R, Ferguson PJ et al (2008) ODN 491, a novel antisense oligodeoxynucleotide that targets thymidylate synthase, exerts cell-specific effects in human tumor cell lines. DNA Cell Biol 27:229–240
Jiang H, Reinhardt HC, Bartkova J et al (2009) The combined status of ATM and p53 link tumor development with therapeutic response. Genes Dev 23:1895–1909
Juliano R, Alam MR, Dixit V et al (2008) Mechanisms and strategies for effective delivery of antisense and siRNA oligonucleotides. Nucleic Acids Res 36:4158–4171
Kaina B, Margison GP, Christmann M (2010) Targeting O-methylguanine-DNA methyltransferase with specific inhibitors as a strategy in cancer therapy. Cell Mol Life Sci 67:3663–3681

Kairemo KJ, Tenhunen M, Jekunen AP (1996a) Dosimetry of radionuclide therapy using radiophosphonated antisense oligodeoxynucleotide phosphorothioates based on animal pharmacokinetic and tissue distribution data. Antisense Nucleic Acid Drug Dev 6:215–220

Kairemo KJ, Tenhunen M, Jekunen AP (1996b) Oligoradionuclidetherapy using radiolabelled antisense oligodeoxynucleotide phosphorothioates. Anticancer Drug Des 11:439–449

Kairemo KJ, Tenhunen M, Jekunen AP (1998) Gene therapy using antisense oligodeoxynucleotides labeled with Auger-emitting radionuclides. Cancer Gene Ther 5:408–412

Kairemo KJ, Jekunen AP, Tenhunen M (2000) Dosimetry and optimization of in vivo targeting with radiolabeled antisense oligodeoxynucleotides: oligonucleotide radiotherapy. Methods Enzymol 314:506–524

Kam NW, Liu Z, Dai H (2005) Functionalization of carbon nanotubes via cleavable disulfide bonds for efficient intracellular delivery of siRNA and potent gene silencing. J Am Chem Soc 127:12492–12493

Kamimura K, Suda T, Xu W et al (2009) Image-guided, lobe-specific hydrodynamic gene delivery to swine liver. Mol Ther 17:491–499

Kelland L (2006) Discontinued drugs in 2005: oncology drugs. Expert Opin Investig Drugs 15:1309–1318

Kim K, Lee YS, Carthew RW (2007) Conversion of pre-RISC to holo-RISC by Ago2 during assembly of RNAi complexes. RNA 13:22–29

Kinali M, Arechavala-Gomeza V, Feng L et al (2009) Local restoration of dystrophin expression with the morpholino oligomer AVI-4658 in Duchenne muscular dystrophy: a single-blind, placebo-controlled, dose-escalation, proof-of-concept study. Lancet Neurol 8:918–928

Knox JJ, Chen XE, Feld R et al (2008) A phase I-II study of oblimersen sodium (G3139, Genasense) in combination with doxorubicin in advanced hepatocellular carcinoma (NCI # 5798). Invest New Drugs 26:193–194

Koller E, Vincent TM, Chappell A et al (2011) Mechanisms of single-stranded phosphorothioate modified antisense oligonucleotide accumulation in hepatocytes. Nucleic Acids Res 39:4795–4807

Koslowsky I, Shahhosseini S, Mirzayans R et al (2011) Evaluation of an 18F-labeled oligonucleotide probe targeting p21(WAF1) transcriptional changes in human tumor cells. Oncol Res 19:265–274

Kota J, Chivukula RR, O'Donnell KA et al (2009) Therapeutic microRNA delivery suppresses tumorigenesis in a murine liver cancer model. Cell 137:1005–1017

Krendel M, Zenke FT, Bokoch GM (2002) Nucleotide exchange factor GEF-H1 mediates cross-talk between microtubules and the actin cytoskeleton. Nat Cell Biol 4:294–301

Kuijpers WH, Bos ES, Kaspersen FM et al (1993) Specific recognition of antibody-oligonucleotide conjugates by radiolabeled antisense nucleotides: a novel approach for two-step radioimmunotherapy of cancer. Bioconjug Chem 4:94–102

Kumar MS, Lu J, Mercer KL et al (2007) Impaired microRNA processing enhances cellular transformation and tumorigenesis. Nat Genet 39:673–677

Kurreck J (2003) Antisense technologies. Improvement through novel chemical modifications. Eur J Biochem 270:1628–1644

LaCasse EC, Cherton-Horvat GG, Hewitt KE et al (2006) Preclinical characterization of AEG35156/GEM 640, a second-generation antisense oligonucleotide targeting X-linked inhibitor of apoptosis. Clin Cancer Res 12:5231–5241

Lai SY, Koppikar P, Thomas SM et al (2009) Intratumoral epidermal growth factor receptor antisense DNA therapy in head and neck cancer: first human application and potential antitumor mechanisms. J Clin Oncol 27:1235–1242

Lee H, Trainer AH, Friedman LS et al (1999) Mitotic checkpoint inactivation fosters transformation in cells lacking the breast cancer susceptibility gene, Brca2. Mol Cell 4:1–10

Lee Y, Lee SH, Kim JS et al (2011) Controlled synthesis of PEI-coated gold nanoparticles using reductive catechol chemistry for siRNA delivery. J Control Release 155:3–10

Lehto T, Simonson OE, Mager I et al (2011) A peptide-based vector for efficient gene transfer in vitro and in vivo. Mol Ther 19:1457–1467

Lennox KA, Behlke MA (2011) Chemical modification and design of anti-miRNA oligonucleotides. Gene Ther 18:1111–1120
Leonetti C, Zupi G (2007) Targeting different signaling pathways with antisense oligonucleotides combination for cancer therapy. Curr Pharm Des 13:463–470
Leung M, Rosen D, Fields S et al (2011) Poly(ADP-ribose) polymerase-1 inhibition: preclinical and clinical development of synthetic lethality. Mol Med 17:854–862
Lewis DL, Wolff JA (2005) Delivery of siRNA and siRNA expression constructs to adult mammals by hydrodynamic intravascular injection. Methods Enzymol 392:336–350
Lewis DL, Hagstrom JE, Loomis AG et al (2002) Efficient delivery of siRNA for inhibition of gene expression in postnatal mice. Nat Genet 32:107–108
Lin H, Dai T, Xiong H et al (2010) Unregulated miR-96 induces cell proliferation in human breast cancer by downregulating transcriptional factor FOXO3a. PLoS One 5:e15797
Liu F, Song Y, Liu D (1999) Hydrodynamics-based transfection in animals by systemic administration of plasmid DNA. Gene Ther 6:1258–1266
Lord CJ, McDonald S, Swift S et al (2008) A high-throughput RNA interference screen for DNA repair determinants of PARP inhibitor sensitivity. DNA Repair (Amst) 7:2010–2019
Lu J, Getz G, Miska EA et al (2005) MicroRNA expression profiles classify human cancers. Nature 435:834–838
Lu Y, Xiao J, Lin H et al (2009) A single anti-microRNA antisense oligodeoxyribonucleotide (AMO) targeting multiple microRNAs offers an improved approach for microRNA interference. Nucleic Acids Res 37:e24
Luo J, Emanuele MJ, Li D et al (2009) A genome-wide RNAi screen identifies multiple synthetic lethal interactions with the Ras oncogene. Cell 137:835–848
Maclean KH, Dorsey FC, Cleveland JL et al (2008) Targeting lysosomal degradation induces p53-dependent cell death and prevents cancer in mouse models of lymphomagenesis. J Clin Invest 118:79–88
Mansoor M, Melendez AJ (2008) Advances in antisense oligonucleotide development for target identification, validation, and as novel therapeutics. Gene Regul Syst Biol 2:275–295
Marshall JL, Eisenberg SG, Johnson MD et al (2004) A phase II trial of ISIS 3521 in patients with metastatic colorectal cancer. Clin Colorectal Cancer 4:268–274
Martello G, Rosato A, Ferrari F et al (2010) A microRNA targeting dicer for metastasis control. Cell 141:1195–1207
Martin SA, McCarthy A, Barber LJ et al (2009) Methotrexate induces oxidative DNA damage and is selectively lethal to tumour cells with defects in the DNA mismatch repair gene MSH2. EMBO Mol Med 1:323–337
Martin SA, McCabe N, Mullarkey M et al (2010) DNA polymerases as potential therapeutic targets for cancers deficient in the DNA mismatch repair proteins MSH2 or MLH1. Cancer Cell 17:235–248
Martin SA, Hewish M, Sims D et al (2011) Parallel high-throughput RNA interference screens identify PINK1 as a potential therapeutic target for the treatment of DNA mismatch repair-deficient cancers. Cancer Res 71:1836–1848
Marwick C (1998) First "antisense" drug will treat CMV retinitis. JAMA 280:871
Matsubara H, Takeuchi T, Nishikawa E et al (2007) Apoptosis induction by antisense oligonucleotides against miR-17-5p and miR-20a in lung cancers overexpressing miR-17-92. Oncogene 26:6099–6105
McCabe N, Turner NC, Lord CJ et al (2006) Deficiency in the repair of DNA damage by homologous recombination and sensitivity to poly(ADP-ribose) polymerase inhibition. Cancer Res 66:8109–8115
McCaffrey AP, Meuse L, Pham TT et al (2002) RNA interference in adult mice. Nature 418:38–39
McNamara JO 2nd, Andrechek ER, Wang Y et al (2006) Cell type-specific delivery of siRNAs with aptamer-siRNA chimeras. Nat Biotechnol 24:1005–1015
Meister G, Landthaler M, Patkaniowska A et al (2004a) Human Argonaute2 mediates RNA cleavage targeted by miRNAs and siRNAs. Mol Cell 15:185–197
Meister G, Landthaler M, Dorsett Y et al (2004b) Sequence-specific inhibition of microRNA- and siRNA-induced RNA silencing. RNA 10:544–550

Mendes-Pereira AM, Martin SA, Brough R et al (2009) Synthetic lethal targeting of PTEN mutant cells with PARP inhibitors. EMBO Mol Med 1:315–322

Michels S, Schmidt-Erfurth U, Rosenfeld PJ (2006) Promising new treatments for neovascular age-related macular degeneration. Expert Opin Investig Drugs 15:779–793

Miyake H, Nelson C, Rennie PS et al (2000) Testosterone-repressed prostate message-2 is an antiapoptotic gene involved in progression to androgen independence in prostate cancer. Cancer Res 60:170–176

Miyake H, Hara I, Kamidono S et al (2003) Resistance to cytotoxic chemotherapy-induced apoptosis in human prostate cancer cells is associated with intracellular clusterin expression. Oncol Rep 10:469–473

Mizuarai S, Irie H, Schmatz DM et al (2008) Integrated genomic and pharmacological approaches to identify synthetic lethal genes as cancer therapeutic targets. Curr Mol Med 8:774–783

Mizuarai S, Yamanaka K, Itadani H et al (2009) Discovery of gene expression-based pharmacodynamic biomarker for a p53 context-specific anti-tumor drug Wee1 inhibitor. Mol Cancer 8:34

Monaco AP, Bertelson CJ, Liechti-Gallati S et al (1988) An explanation for the phenotypic differences between patients bearing partial deletions of the DMD locus. Genomics 2:90–95

Moreira JN, Santos A, Simoes S (2006) Bcl-2-targeted antisense therapy (Oblimersen sodium): towards clinical reality. Rev Recent Clin Trials 1:217–235

Morgan-Lappe S, Woods KW, Li Q et al (2006) RNAi-based screening of the human kinome identifies Akt-cooperating kinases: a new approach to designing efficacious multitargeted kinase inhibitors. Oncogene 25:1340–1348

Morrissey DV, Lockridge JA, Shaw L et al (2005) Potent and persistent in vivo anti-HBV activity of chemically modified siRNAs. Nat Biotechnol 23:1002–1007

Mu P, Nagahara S, Makita N et al (2009) Systemic delivery of siRNA specific to tumor mediated by atelocollagen: combined therapy using siRNA targeting Bcl-xL and cisplatin against prostate cancer. Int J Cancer 125:2978–2990

Mukai H, Kawakami S, Hashida M (2008) Renal press-mediated transfection method for plasmid DNA and siRNA to the kidney. Biochem Biophys Res Commun 372:383–387

Mukai H, Kawakami S, Kamiya Y et al (2009) Pressure-mediated transfection of murine spleen and liver. Hum Gene Ther 20:1157–1167

Mukai H, Kawakami S, Takahashi H et al (2010) Key physiological phenomena governing transgene expression based on tissue pressure-mediated transfection in mice. Biol Pharm Bull 33:1627–1632

Mulholland P (2011) "A trial looking at Trabedersen (AP12009) for anaplastic astrocytoma that has come back after treatment or continued to grow during treatment (SAPPHIRE)." from http://cancerhelp.cancerresearchuk.org/trials/a-trial-Trabedersen-AP12009-anaplastic-astrocytoma-come-back-after-treatment-or-continued-grow-during-treatment

Muntoni F, Wood MJ (2011) Targeting RNA to treat neuromuscular disease. Nat Rev Drug Discov 10:621–637

Muratovska A, Eccles MR (2004) Conjugate for efficient delivery of short interfering RNA (siRNA) into mammalian cells. FEBS Lett 558:63–68

Neshat MS, Mellinghoff IK, Tran C et al (2001) Enhanced sensitivity of PTEN-deficient tumors to inhibition of FRAP/mTOR. Proc Natl Acad Sci USA 98:10314–10319

Nghiem P, Park PK, Kim Y et al (2001) ATR inhibition selectively sensitizes G1 checkpoint-deficient cells to lethal premature chromatin condensation. Proc Natl Acad Sci USA 98:9092–9097

Nickoloff BJ, Hendrix MJ, Pollock PM et al (2005) Notch and NOXA-related pathways in melanoma cells. J Investig Dermatol Symp Proc 10:95–104

Nijman SM (2011) Synthetic lethality: general principles, utility and detection using genetic screens in human cells. FEBS Lett 585:1–6

O'Brien SM, Cunningham CC, Golenkov AK et al (2005) Phase I to II multicenter study of oblimersen sodium, a Bcl-2 antisense oligonucleotide, in patients with advanced chronic lymphocytic leukemia. J Clin Oncol 23:7697–7702

O'Brien S, Moore JO, Boyd TE et al (2007) Randomized phase III trial of fludarabine plus cyclophosphamide with or without oblimersen sodium (Bcl-2 antisense) in patients with relapsed or refractory chronic lymphocytic leukemia. J Clin Oncol 25:1114–1120

O'Brien S, Moore JO, Boyd TE et al (2009) 5-year survival in patients with relapsed or refractory chronic lymphocytic leukemia in a randomized, phase III trial of fludarabine plus cyclophosphamide with or without oblimersen. J Clin Oncol 27:5208–5212

O'Connor R (2009) A review of mechanisms of circumvention and modulation of chemotherapeutic drug resistance. Curr Cancer Drug Targets 9:273–280

O'Rourke JR, Swanson MS (2009) Mechanisms of RNA-mediated disease. J Biol Chem 284:7419–7423

Oh YK, Park TG (2009) siRNA delivery systems for cancer treatment. Adv Drug Deliv Rev 61:850–862

Olivares J, Kumar P, Yu Y et al (2011) Phase I trial of TGF-{beta}2 antisense GM-CSF gene-modified autologous tumor cell (TAG) vaccine. Clin Cancer Res 17:183–192

Oza AM, Elit L, Swenerton K et al (2003) Phase II study of CGP 69846A (ISIS 5132) in recurrent epithelial ovarian cancer: an NCIC clinical trials group study (NCIC IND.116). Gynecol Oncol 89:129–133

Paganin-Gioanni A, Bellard E, Escoffre JM et al (2011) Direct visualization at the single-cell level of siRNA electrotransfer into cancer cells. Proc Natl Acad Sci USA 108:10443–10447

Pandyra AA, Berg R, Vincent M et al (2007) Combination silencer RNA (siRNA) targeting Bcl-2 antagonizes siRNA against thymidylate synthase in human tumor cell lines. J Pharmacol Exp Ther 322:123–132

Pasquinelli AE, Hunter S, Bracht J (2005) MicroRNAs: a developing story. Curr Opin Genet Dev 15:200–205

Paterson BM, Roberts BE, Kuff EL (1977) Structural gene identification and mapping by DNA-mRNA hybrid-arrested cell-free translation. Proc Natl Acad Sci USA 74:4370–4374

Paz-Ares L, Douillard JY, Koralewski P et al (2006) Phase III study of gemcitabine and cisplatin with or without aprinocarsen, a protein kinase C-alpha antisense oligonucleotide, in patients with advanced-stage non-small-cell lung cancer. J Clin Oncol 24:1428–1434

Plummer R (2010) Perspective on the pipeline of drugs being developed with modulation of DNA damage as a target. Clin Cancer Res 16:4527–4531

Prato M, Kostarelos K, Bianco A (2008) Functionalized carbon nanotubes in drug design and discovery. Acc Chem Res 41:60–68

Puyol M, Martin A, Dubus P et al (2010) A synthetic lethal interaction between K-Ras oncogenes and Cdk4 unveils a therapeutic strategy for non-small cell lung carcinoma. Cancer Cell 18:63–73

Rayburn ER, Zhang R (2008) Antisense, RNAi, and gene silencing strategies for therapy: mission possible or impossible? Drug Discov Today 13:513–521

Rayburn ER, Wang H, Zhang R (2006) Antisense-based cancer therapeutics: are we there yet? Expert Opin Emerg Drugs 11:337–352

Rehman FL, Lord CJ, Ashworth A (2010) Synthetic lethal approaches to breast cancer therapy. Nat Rev Clin Oncol 7:718–724

Reinhardt HC, Aslanian AS, Lees JA et al (2007) p53-deficient cells rely on ATM- and ATR-mediated checkpoint signaling through the p38MAPK/MK2 pathway for survival after DNA damage. Cancer Cell 11:175–189

Ren Y, Li R, Zheng Y et al (1998) Cloning and characterization of GEF-H1, a microtubule-associated guanine nucleotide exchange factor for Rac and Rho GTPases. J Biol Chem 273:34954–34960

Rizzi F, Coletta M, Bettuzzi S (2009) Chapter 2: Clusterin (CLU): from one gene and two transcripts to many proteins. Adv Cancer Res 104:9–23

Roberts J, Palma E, Sazani P et al (2006) Efficient and persistent splice switching by systemically delivered LNA oligonucleotides in mice. Mol Ther 14:471–475

Rottmann S, Wang Y, Nasoff M et al (2005) A TRAIL receptor-dependent synthetic lethal relationship between MYC activation and GSK3beta/FBW7 loss of function. Proc Natl Acad Sci USA 102:15195–15200

Ruzankina Y, Schoppy DW, Asare A et al (2009) Tissue regenerative delays and synthetic lethality in adult mice after combined deletion of Atr and Trp53. Nat Genet 41:1144–1149

Rychahou PG, Evers BM (2010) Hydrodynamic delivery protocols. Methods Mol Biol 623:189–195

Saad F, Hotte S, North S et al (2011) Randomized phase II trial of Custirsen (OGX-011) in combination with docetaxel or mitoxantrone as second-line therapy in patients with metastatic castrate-resistant prostate cancer progressing after first-line docetaxel: CUOG trial P-06c. Clin Cancer Res 17:5765–5773

Saleh M, Posey J, Pleasant L et al (2000) A phase II trial of ISIS 2503, an antisense inhibitor of H-ras, as first line therapy for advanced colorectal carcinoma. Proc Am Soc Clin Oncol 19:abstr 1258

Sanguino A, Lopez-Berestein G, Sood AK (2008) Strategies for in vivo siRNA delivery in cancer. Mini Rev Med Chem 8:248–255

Sargent DJ, Marsoni S, Monges G et al (2010) Defective mismatch repair as a predictive marker for lack of efficacy of fluorouracil-based adjuvant therapy in colon cancer. J Clin Oncol 28:3219–3226

Sarthy AV, Morgan-Lappe SE, Zakula D et al (2007) Survivin depletion preferentially reduces the survival of activated K-Ras-transformed cells. Mol Cancer Ther 6:269–276

Satkauskas S, Bureau MF, Mahfoudi A et al (2001) Slow accumulation of plasmid in muscle cells: supporting evidence for a mechanism of DNA uptake by receptor-mediated endocytosis. Mol Ther 4:317–323

Schafer J, Hobel S, Bakowsky U et al (2010) Liposome-polyethylenimine complexes for enhanced DNA and siRNA delivery. Biomaterials 31:6892–6900

Schimmer AD, Herr W, Hanel M et al (2011) Addition of AEG35156 XIAP antisense oligonucleotide in reinduction chemotherapy does not improve remission rates in patients with primary refractory acute myeloid leukemia in a randomized phase II study. Clin Lymphoma Myeloma Leuk 11:433–438

Schlingensiepen R, Goldbrunner M, Szyrach MN et al (2005) Intracerebral and intrathecal infusion of the TGF-beta 2-specific antisense phosphorothioate oligonucleotide AP 12009 in rabbits and primates: toxicology and safety. Oligonucleotides 15:94–104

Schlingensiepen KH, Schlingensiepen R, Steinbrecher A et al (2006) Targeted tumor therapy with the TGF-beta 2 antisense compound AP 12009. Cytokine Growth Factor Rev 17:129–139

Scholl C, Frohling S, Dunn IF et al (2009) Synthetic lethal interaction between oncogenic KRAS dependency and STK33 suppression in human cancer cells. Cell 137:821–834

Segura MF, Hanniford D, Menendez S et al (2009) Aberrant miR-182 expression promotes melanoma metastasis by repressing FOXO3 and microphthalmia-associated transcription factor. Proc Natl Acad Sci USA 106:1814–1819

Sewell KL, Geary RS, Baker BF et al (2002) Phase I trial of ISIS 104838, a 2'-methoxyethyl modified antisense oligonucleotide targeting tumor necrosis factor-alpha. J Pharmacol Exp Ther 303:1334–1343

Shaheen M, Allen C, Nickoloff JA et al (2011) Synthetic lethality: exploiting the addiction of cancer to DNA repair. Blood 117:6074–6082

Shannan B, Seifert M, Leskov K et al (2006) Challenge and promise: roles for clusterin in pathogenesis, progression and therapy of cancer. Cell Death Differ 13:12–19

Sharp PA (2009) The centrality of RNA. Cell 136:577–580

Shi Y, Gera J, Hu L et al (2002) Enhanced sensitivity of multiple myeloma cells containing PTEN mutations to CCI-779. Cancer Res 62:5027–5034

Sonoke S, Ueda T, Fujiwara K et al (2008) Tumor regression in mice by delivery of Bcl-2 small interfering RNA with pegylated cationic liposomes. Cancer Res 68:8843–8851

Stavropoulou V, Brault L, Schwaller J (2010) Insights into molecular pathways for targeted therapeutics in acute leukemia. Swiss Med Wkly 140:w13068

Stein CA, Cheng YC (1993) Antisense oligonucleotides as therapeutic agents—is the bullet really magical? Science 261:1004–1012

Stein CA, Cohen JS (1988) Oligodeoxynucleotides as inhibitors of gene expression: a review. Cancer Res 48:2659–2668

Stenvang J, Kauppinen S (2008) MicroRNAs as targets for antisense-based therapeutics. Expert Opin Biol Ther 8:59–81

Stephenson ML, Zamecnik PC (1978) Inhibition of Rous sarcoma viral RNA translation by a specific oligodeoxyribonucleotide. Proc Natl Acad Sci USA 75:285–288

Suda T, Suda K, Liu D (2008) Computer-assisted hydrodynamic gene delivery. Mol Ther 16:1098–1104

Summerton J (1999) Morpholino antisense oligomers: the case for an RNase H-independent structural type. Biochim Biophys Acta 1489:141–158

Sur S, Pagliarini R, Bunz F et al (2009) A panel of isogenic human cancer cells suggests a therapeutic approach for cancers with inactivated p53. Proc Natl Acad Sci USA 106:3964–3969

Sutphin PD, Chan DA, Li JM et al (2007) Targeting the loss of the von Hippel-Lindau tumor suppressor gene in renal cell carcinoma cells. Cancer Res 67:5896–5905

Takamizawa J, Konishi H, Yanagisawa K et al (2004) Reduced expression of the let-7 microRNAs in human lung cancers in association with shortened postoperative survival. Cancer Res 64:3753–3756

Takei Y, Nemoto T, Mu P et al (2008) In vivo silencing of a molecular target by short interfering RNA electroporation: tumor vascularization correlates to delivery efficiency. Mol Cancer Ther 7:211–221

Takeshita F, Minakuchi Y, Nagahara S et al (2005) Efficient delivery of small interfering RNA to bone-metastatic tumors by using atelocollagen in vivo. Proc Natl Acad Sci USA 102:12177–12182

Tamm I, Wagner M (2006) Antisense therapy in clinical oncology: preclinical and clinical experiences. Mol Biotechnol 33:221–238

Trougakos IP, Lourda M, Antonelou MH et al (2009) Intracellular clusterin inhibits mitochondrial apoptosis by suppressing p53-activating stress signals and stabilizing the cytosolic Ku70-Bax protein complex. Clin Cancer Res 15:48–59

Turcotte S, Chan DA, Sutphin PD et al (2008) A molecule targeting VHL-deficient renal cell carcinoma that induces autophagy. Cancer Cell 14:90–102

Tutt A, Robson M, Garber JE et al (2010) Oral poly(ADP-ribose) polymerase inhibitor olaparib in patients with BRCA1 or BRCA2 mutations and advanced breast cancer: a proof-of-concept trial. Lancet 376:235–244

Urban-Klein B, Werth S, Abuharbeid S et al (2005) RNAi-mediated gene-targeting through systemic application of polyethylenimine (PEI)-complexed siRNA in vivo. Gene Ther 12:461–466

Usman N, Blatt LM (2000) Nuclease-resistant synthetic ribozymes: developing a new class of therapeutics. J Clin Invest 106:1197–1202

Varkouhi AK, Mountrichas G, Schiffelers RM et al (2012) Polyplexes based on cationic polymers with strong nucleic acid binding properties. Eur J Pharm Sci 45:459–466

Venturini L, Battmer K, Castoldi M et al (2007) Expression of the miR-17-92 polycistron in chronic myeloid leukemia (CML) CD34+ cells. Blood 109:4399–4405

Vickers TA, Wyatt JR, Burckin T et al (2001) Fully modified 2′ MOE oligonucleotides redirect polyadenylation. Nucleic Acids Res 29:1293–1299

Villalona-Calero MA, Ritch P, Figueroa JA et al (2004) A phase I/II study of LY900003, an antisense inhibitor of protein kinase C-alpha, in combination with cisplatin and gemcitabine in patients with advanced non-small cell lung cancer. Clin Cancer Res 10:6086–6093

Volinia S, Calin GA, Liu CG et al (2006) A microRNA expression signature of human solid tumors defines cancer gene targets. Proc Natl Acad Sci USA 103:2257–2261

Wang Z (2011) The concept of multiple-target anti-miRNA antisense oligonucleotide technology. Methods Mol Biol 676:51–57

Wang J, Chen P, Mrkobrada M et al (2003) Antisense imaging of epidermal growth factor-induced p21(WAF-1/CIP-1) gene expression in MDA-MB-468 human breast cancer xenografts. Eur J Nucl Med Mol Imaging 30:1273–1280

Wang Y, Decker SJ, Sebolt-Leopold J (2004) Knockdown of Chk1, Wee1 and Myt1 by RNA interference abrogates G2 checkpoint and induces apoptosis. Cancer Biol Ther 3:305–313
Wang Y, Ngo VN, Marani M et al (2010) Critical role for transcriptional repressor Snail2 in transformation by oncogenic RAS in colorectal carcinoma cells. Oncogene 29:4658–4670
Weidle UH, Maisel D, Eick D (2011) Synthetic lethality-based targets for discovery of new cancer therapeutics. Cancer Genomics Proteomics 8:159–171
Whitehurst AW, Bodemann BO, Cardenas J et al (2007) Synthetic lethal screen identification of chemosensitizer loci in cancer cells. Nature 446:815–819
Whitehurst AW, Xie Y, Purinton SC et al (2010) Tumor antigen acrosin binding protein normalizes mitotic spindle function to promote cancer cell proliferation. Cancer Res 70:7652–7661
Wilson DM 3rd, Simeonov A (2010) Small molecule inhibitors of DNA repair nuclease activities of APE1. Cell Mol Life Sci 67:3621–3631
Wu H, Lima WF, Zhang H et al (2004) Determination of the role of the human RNase H1 in the pharmacology of DNA-like antisense drugs. J Biol Chem 279:17181–17189
Yang D, Liu H, Goga A et al (2010) Therapeutic potential of a synthetic lethal interaction between the MYC proto-oncogene and inhibition of aurora-B kinase. Proc Natl Acad Sci USA 107:13836–13841
Yaron PN, Holt BD, Short PA et al (2011) Single wall carbon nanotubes enter cells by endocytosis and not membrane penetration. J Nanobiotechnology 9:45
Yin H, Betts C, Saleh AF et al (2010) Optimization of peptide nucleic acid antisense oligonucleotides for local and systemic dystrophin splice correction in the mdx mouse. Mol Ther 18:819–827
Yip KW, Reed JC (2008) Bcl-2 family proteins and cancer. Oncogene 27:6398–6406
Yu RZ, Geary RS, Leeds JM et al (2001) Comparison of pharmacokinetics and tissue disposition of an antisense phosphorothioate oligonucleotide targeting human Ha-ras mRNA in mouse and monkey. J Pharm Sci 90:182–193
Zaghloul EM, Viola JR, Zuber G et al (2010) Formulation and delivery of splice-correction antisense oligonucleotides by amino acid modified polyethylenimine. Mol Pharm 7:652–663
Zamecnik PC, Stephenson ML (1978) Inhibition of Rous sarcoma virus replication and cell transformation by a specific oligodeoxynucleotide. Proc Natl Acad Sci USA 75:280–284
Zellweger T, Miyake H, Cooper S et al (2001) Antitumor activity of antisense clusterin oligonucleotides is improved in vitro and in vivo by incorporation of 2′-O-(2-methoxy)ethyl chemistry. J Pharmacol Exp Ther 298:934–940
Zellweger T, Kiyama S, Chi K et al (2003) Overexpression of the cytoprotective protein clusterin decreases radiosensitivity in the human LNCaP prostate tumour model. BJU Int 92:463–469
Zhang G, Budker V, Williams P et al (2001) Efficient expression of naked dna delivered intraarterially to limb muscles of nonhuman primates. Hum Gene Ther 12:427–438
Zhang X, Dong X, Sawyer GJ et al (2004a) Regional hydrodynamic gene delivery to the rat liver with physiological volumes of DNA solution. J Gene Med 6:693–703
Zhang Z, Wang H, Prasad G et al (2004b) Radiosensitization by antisense anti-MDM2 mixed-backbone oligonucleotide in in vitro and in vivo human cancer models. Clin Cancer Res 10:1263–1273
Zhou T, Jia X, Li H et al (2011) New tumor-targeted nanosized delivery carrier for oligonucleotides: characteristics in vitro and in vivo. Int J Nanomedicine 6:1527–1534

Antisense-Mediated Reduction of Eukaryotic Noncoding RNAs

Xue-hai Liang, Timothy A. Vickers, and Stanley T. Crooke

Contents

Abstract The human genome has been demonstrated to be transcribed largely as noncoding RNAs. Accumulating evidence suggests important roles for many noncoding RNAs in different biological processes as well as in human diseases. Thus, it is important to be able to manipulate the expression of the noncoding RNAs, both for biological functionalization of these RNAs and for treatment of diseases related to noncoding RNAs. In this review, we summarize the approaches for reducing expression of noncoding RNAs in different eukaryotic organisms and the insights provided into the functions of these noncoding RNAs.

X.-h. Liang (✉) • T.A. Vickers • S.T. Crooke
Department of Core Antisense Research, ISIS Pharmaceuticals, Inc., 2855 Gazelle Court, Carlsbad, CA 92010, USA
e-mail: lliang@isisph.com

V.A. Erdmann and J. Barciszewski (eds.), *From Nucleic Acids Sequences to Molecular Medicine*, RNA Technologies, DOI 10.1007/978-3-642-27426-8_8,

Keywords Antisense • ASO • Noncoding RNAs • Ribozyme • RNase H • siRNA

1 Introduction

Recent discoveries of numerous non-protein-coding RNAs (ncRNAs) in eukaryotic cells have dramatically changed the view of the eukaryotic transcriptome. Tiling-array analyses and genome-wide cDNA sequencing have demonstrated that most of the mammalian genome is transcribed and the majority of the mammalian transcriptome consists of ncRNAs, whereas only ~2% encodes for proteins (Jacquier 2009). NcRNAs have been sorted into different groups including small nuclear RNAs (snRNAs), small nucleolar RNAs (snoRNAs), small Cajal body RNAs (scaRNAs), microRNAs (miRNAs), piwi-interacting RNAs (piRNAs), and long noncoding RNAs (lncRNAs), based on cellular localization, size, function, or protein interactions. It has been demonstrated that ncRNAs are involved in various biological processes, including chromosome stabilization, transcriptional and post-transcriptional regulation, translation, protein translocation, and subcellular organelle formation (Chen and Carmichael 2010). In addition, some ncRNAs have been implicated in human diseases including cancer (Galasso et al. 2010). These findings imply the importance of ncRNAs in gene expression regulation and strong potential of these RNAs as therapeutic targets for treatment of different diseases.

Despite the importance of the ncRNAs, defining biological functions and understanding the roles of the vast majority of ncRNAs in human diseases remain elusive. One obstacle to functionalizing ncRNAs has been the lack of efficient and specific approaches to deplete ncRNAs in eukaryotic cells, especially small ncRNAs in mammalian cells (Ploner et al. 2009). However, several approaches, developed in recent years and summarized in this review, have made it possible to efficiently deplete ncRNAs in multiple organisms.

2 Noncoding RNAs and Their Biological Functions

2.1 Small Nuclear RNAs

Small nuclear RNAs are a class of small noncoding RNAs that localize in the nucleus of eukaryotic cells where they form small nuclear RNA–protein complexes (snRNPs). The spliceosomal snRNAs associate with sm core proteins (for U1, U2, U4, and U5) or sm-like proteins (for U6) and snRNP-specific proteins (Will and Luhrmann 2001). These snRNAs are required for nuclear pre-mRNA splicing (Sharp 1994). In addition to the spliceosomal snRNAs, another U snRNA, U7, was found to be required for processing of replication-dependent histone mRNAs in metazoan, with the assistance of U2 snRNP (Dominski and Marzluff 2007; Friend et al. 2007).

In addition to the role in pre-mRNA splicing, the U1 snRNP has been proposed to play a role in transcription (Kwek et al. 2002). A recent study showed that U1 could be recruited to the transcription unit independent of splicing in human cells (Spiluttini et al. 2010), further supporting a role of U1 snRNA in the regulation of transcription. The involvement of small nuclear RNAs in transcription has also been established for several non-spliceosomal RNAs. For example, a stem loop of a yeast small RNA (FC RNA) can be inserted into the cleft of the RNA polymerase, a position that otherwise should be occupied by template DNA, leading to inhibition of transcription initiation (Kettenberger et al. 2006). Similarly to FC RNA, mouse B2 RNA also directly interacts with core RNA polymerase II and can inhibit transcription in vitro (Allen et al. 2004). A recent study showed that B2 RNA suppresses phosphorylation of Ser5 residue in the Pol II carboxy-terminal domain (CTD) performed by TFIIH, contributing to transcription suppression (Yakovchuk et al. 2011).

NcRNAs can also target transcription factors to regulate transcription. For example, a short transcript that accumulates upon serum starvation can repress expression of the human dihydrofolate reductase (DHFR) gene, by binding to TFIIB. As a result, the pre-initiation complex formed on the DHFR promoter can be dissociated (Martianov et al. 2007). In another case, the 5′ region of 7SK RNA, an abundant ncRNA, interacts with HEXIM1, a protein present in an inactive form of positive transcription elongation factor b (P-TEFb). Binding of 7SK RNA with HEXIM1 converts the protein to a P-TEFb inhibitor, leading to inhibition of transcription (Michels et al. 2004). In addition to 7SK RNA, some recently identified nuclear long ncRNAs have also been shown to regulate transcription.

2.2 *Small Nucleolar RNAs*

Hundreds of snoRNAs have been identified in eukaryotic cells. Based on the conserved sequence motifs, snoRNAs are sorted into two groups: the C/D box and H/ACA box snoRNAs (Fig. 1). C/D snoRNAs typically exhibit a stem-loop structure and contain the conserved C (RUGAUGA, where R is a purine) and D (CUGA) boxes in the 5′ and 3′ regions of the RNAs. H/ACA RNAs typically have two hairpin structures, linked by the H box (ANANNA) in the hinge region, and an ACA box near the 3′-end of the RNAs (Bertrand and Fournier 2004). In protists such as trypanosomes and euglena, most H/ACA-like RNAs have only a single hairpin and an AGA box at the 3′-end (Russell et al. 2004; Uliel et al. 2004). Both C/D and H/ACA snoRNAs are present in snoRNP complexes. C/D snoRNAs associate with Nop56, Nop58, snu13/15.5K, and Nop1/fibrillarin, with the latter being a methyltransferase. H/ACA snoRNAs also associate with four common core proteins specific to this family: Gar1, Nop10, NHP2, and Cbf5/dyskerin, a pseudouridylase (Bachellerie et al. 2002).

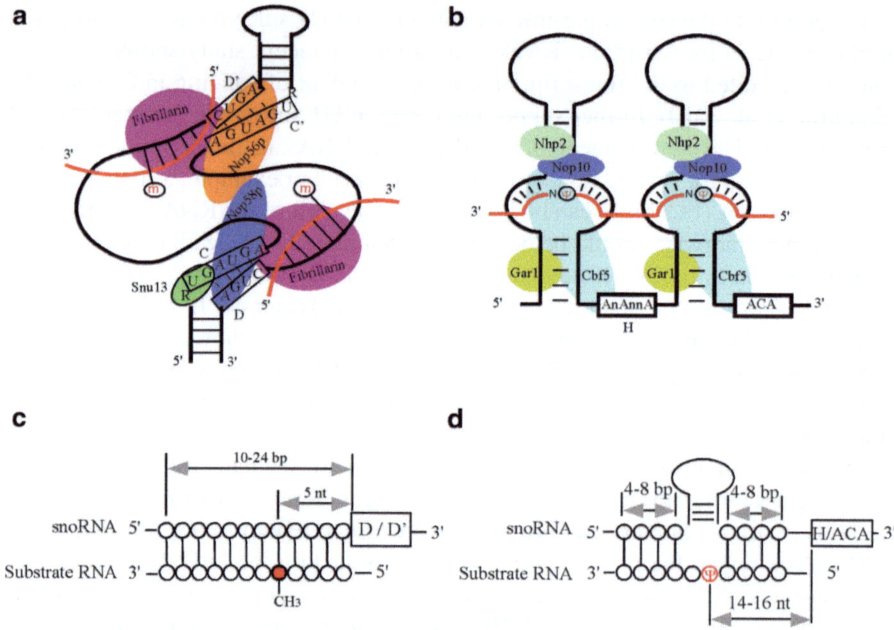

Fig. 1 Schematic representation of snoRNPs and the rules for guiding modification. (**a**) C/D snoRNPs. The snoRNA conserved box sequences and snoRNP proteins are shown. The substrate RNA is indicated in *red*. The methylation is marked with a *circled m*. (**b**) H/ACA snoRNPs. (**c**) The 2′-O-methylation guide rule for C/D snoRNAs. D/D′, conserved D or D′ box in a snoRNA. The nucleotide to be methylated is marked in *red*. (**d**) The guide rule for H/ACA RNAs. The pseudouridylation site is indicated with a *circled Ψ*

2.2.1 SnoRNAs Direct RNA Nucleotide Modification and Pre-rRNA Processing

Most snoRNAs are required for posttranscriptional nucleotide modifications in rRNAs, with C/D snoRNAs directing 2′-O-methylation and H/ACA box RNAs guiding pseudouridylation (Bertrand and Fournier 2004). In both cases, the nucleotide to be modified is site-specifically selected by snoRNA through base pairing with the substrate RNA, while the modification reaction is catalyzed by a snoRNP protein (Bachellerie et al. 2002). Although many rRNA modifications are conserved, a significant number of modifications are species-specific. While most single nt modifications have no significant effect on cell growth, depletion of multiple nt modifications in functionally important domains of the rRNAs can have broad effects on pre-rRNA processing, ribosome assembly, translation rate and fidelity, and rRNA stability (King et al. 2003; Liang et al. 2007).

Spliceosomal snRNAs also undergo nt modifications. 2′-O-methylation and pseudouridylation reactions of U snRNAs are guided by small Cajal body RNAs (scaRNAs), a group of RNAs that are similar to snoRNAs both in function and in RNP composition, but are accumulated in the Cajal bodies. Interestingly, snoRNA-guided modifications were also found for pre-mRNA-like RNAs in trypanosomes and tRNAs in archaea (Liang et al. 2002; Singh et al. 2004).

A few snoRNAs are required for pre-rRNA processing. These include the C/D box U3 and U14 snoRNAs and a H/ACA box U17/snR30 RNA. These snoRNAs are conserved in eukaryotes and are essential for cell growth (Gerbi and Borovjagin 2004). Several processing snoRNAs are less conserved, including U8 and U22 in mammals and snR10 in yeast. These snoRNAs mediate processing by base pairing with pre-rRNA and most likely serve as RNA chaperones to facilitate pre-rRNP folding (Gerbi and Borovjagin 2004).

2.2.2 Other Functions of snoRNPs and Implications in Human Diseases

Many snoRNAs (so-called orphan snoRNAs) do not have the potential to guide modifications in rRNAs and snRNA. This suggests potential new biological roles of snoRNAs. Indeed, snoRNAs have recently been found to be sources of miRNA production in different organisms (e.g., Ender et al. 2008; Saraiya and Wang 2008). A recent study also proposed that several C/D box snoRNAs in the RPL13a locus (U32a, U33, and U35) may serve as regulators of metabolic stress response pathways in mammalian cells (Michel et al. 2011). Interestingly, snoRNAs can also regulate expression of protein-coding genes. For example, HBII-52, an imprinted, brain-specific snoRNA, has been shown to regulate alternative splicing and/or mRNA editing of serotonin receptor 2C [5-HT(2C)R] (Vitali et al. 2005; Kishore and Stamm 2006). This snoRNA, together with several other snoRNAs, is encoded in a genomic region related to Prader–Willi syndrome and Angelman syndrome, disorders that are characterized by developmental, behavioral, and intellectual problems. Patients with Prader–Willi syndrome do not express HBII-52, and they have different 5-HT(2C)R mRNA isoforms than healthy individuals (Kishore and Stamm 2006). Though the exact mechanism of this regulation is unknown, snoRNA/pre-mRNA base pairing may play a role in guiding pre-mRNA modification, facilitating folding or regulating the recruitment of transacting factor(s) required for normal splicing or editing. Interestingly, a recent study showed that MBII-52 (mouse homologue of HBII-52) is further processed into smaller RNAs in mouse brain and can regulate alternative splicing of several other pre-mRNAs as well (Kishore et al. 2010). These smaller RNAs, associated with hnRNP proteins but not with canonical C/D snoRNP proteins, were proposed to be the actual regulators for alternative splicing (Kishore et al. 2010).

Point mutations in DKC1, a gene encoding the H/ACA snoRNP protein dyskerin, have been implicated in dyskeratosis congenita (Gupta and Kumar 2010). Mutations of DKC1 cause a selective defect in translation of a subgroup of IRES (internal ribosome entry site) containing mRNAs, many of which encode tumor suppressors, including P27 and P53 (Montanaro et al. 2010). The defect in P53 mRNA translation appears to be associated with impairment in translation initiation mediated by the IRES element, which is most likely related to loss of pseudouridylation in rRNAs, but seems to be independent of another role of dyskerin in telomerase function, which requires the telomerase RNA, a special nucleolar RNA bound to H/ACA proteins (Montanaro et al. 2010). Reduction of the telomerase activity or expression of a low level of mutant telomerase RNA that produces mutant telomere repeats has been shown to control tumor cell growth or to cause robust apoptotic response, respectively (Xu et al. 2010).

2.3 The MRP RNA and RNase P RNA

Another nucleolar RNA that is distinct from C/D and H/ACA snoRNAs is the MRP RNA, which is only present in eukaryotic cells and is essential for cell growth. Without C/D or H/ACA boxes, MRP RNA associates with a set of proteins (ten in yeast *Saccharomyces cerevisiae* and nine in human) that are different from the snoRNP proteins (Esakova and Krasilnikov 2010). MRP, an endonuclease, is localized primarily in the nucleolus but also found in mitochondria. In mitochondria, MRP cleaves RNA transcripts to produce RNA primers that are required for DNA synthesis of the heavy strand (Lee and Clayton 1998). In the nucleolus, this endonuclease cleaves pre-rRNA at a site upstream to 5.8S rRNA. Interestingly, MRP was demonstrated to specifically cleave the 5′-UTR of CLB2 mRNA in yeast, allowing rapid degradation of the mRNA by the 5′–3′ exonuclease Xrn1 (Gill et al. 2004) and suggesting that additional MRP substrates may exist. The MRP RNA is the first nuclear ncRNA found to cause a genetic disease (McKusick et al. 1965). Mutations in the MRP RNA have been shown to be the cause of an inherited disease, cartilage–hair hypoplasia (CHH), and have been implicated in several other inherited diseases, such as anauxetic dysplasia and Omenn syndrome (Martin and Li 2007).

MRP RNA is closely related to another nucleolar RNA, the RNase P RNA, which is also present in a RNP complex. The majority of protein components of RNase P are shared with MRP complex, but complex-specific proteins do exist (Esakova and Krasilnikov 2010). In addition to the proteins, the RNA components of the two complexes bear structural similarities, suggesting a close evolutionary relationship (Esakova and Krasilnikov 2010). Like MRP, the RNase P complex is also an endonuclease with ribozyme activity. RNase P cleaves Pol III-transcribed tRNA precursors to produce mature 5′-ends. In addition, Pol II-transcribed RNAs can also be RNase P substrates. These include yeast intronic C/D snoRNAs that are processed in a splicing-independent manner (Coughlin et al. 2008) and human Malat1 RNA, a nuclear long ncRNA (Wilusz et al. 2008). Additionally, RNase P appears to play a role in transcription by Pol I and Pol III in mammalian cells. Depletion of RNase P significantly reduced the level of transcription of some Pol III transcripts in HeLa cell extracts, and protein components of RNase P associate with chromatin of actively transcribed tRNA and 5S rRNA genes (Jarrous and Reiner 2007). The ribozyme activity of the RNase P RNA has been used to reduce other genes by introducing external guide sequence (EGS) containing oligonucleotides into cells.

2.4 *The Signal Recognition Particle RNA (SRP RNA) and Alu RNA*

The cytoplasmic SRP RNA (7SL RNA) forms the core of the signal recognition particle and serves as the platform for binding of the SRP proteins (Nagai et al. 2003). SRP interacts with the signal sequence of nascent peptides of membrane or secretory proteins emerging from the ribosome. This interaction leads to the formation of a ribosome–nascent chain–SRP complex, resulting in a translation arrest (Nagai et al. 2003). The complex then associates with the SRP receptor anchored to the ER membrane, and the signal peptide enters the ER lumen and translation resumes.

The 5′- and 3′-ends of the SRP/7SL RNA form an Alu domain. The Alu sequence is the most abundant repetitive element in the human genome. It emerged from a 5′ and 3′ fusion of the 7SL RNA gene and has been amplified throughout the human genome by retrotransposition (Hasler and Strub 2006a). In addition to their effects on genome stability, Alu elements can play other roles, i.e., in pre-mRNA alternative splicing, RNA editing, and translation regulation (Hasler and Strub 2006a). Alu RNAs, transcribed from Alu elements, have been found to be present at low levels in the cytoplasm of normal primate cells and are expressed at increased levels under various stresses (Liu et al. 1995). Two forms of Alu RNA exist in cells, a full-length Alu RNA and a shorter cytoplasmic Alu RNA (Maraia et al. 1993). Alu RNAs in the cytoplasm associate with SPR14/SRP9 to form a RNP complex and can regulate translation initiation (Hasler and Strub 2006b). Interestingly, a recent study showed that upon heat shock, full-length human Alu RNA in the nucleus can inhibit transcription of some mRNA genes by binding to Pol II (Mariner et al. 2008).

2.5 *Endogenous siRNAs and miRNAs*

RNA interference (RNAi) is a process by which dsRNA triggers specific mRNA degradation (Fire et al. 1998). RNAi is mediated by small interfering RNAs (siRNAs), which are 21–24 nt double-stranded RNAs cleaved from long dsRNAs by an RNase III-type protein, Dicer (Zamore et al. 2000). The double-stranded siRNAs are incorporated into the RNA-induced silencing complex (RISC) containing Ago2, where the passenger strand siRNA is cleaved by Ago2 and is subsequently released from RISC (Rand et al. 2005). The guide strand siRNA remains associated with Ago2 and directs the RISC to substrate mRNA, which is cleaved by Ago2, resulting in sequence-specific mRNA degradation. Early studies focused on siRNAs derived from exogenous dsRNAs; however, endogenous siRNAs were also identified in plants, flies, and animals (Ghildiyal and Zamore 2009).

In addition to repression of gene expression, siRNAs or dsRNAs can also activate gene expression. dsRNAs targeting gene promoters can induce potent transcriptional activation of associated genes (Li et al. 2006; Janowski et al. 2007), likely by reducing antisense transcripts which overlap the promoter regions

(Schwartz et al. 2008). In human cells, this was demonstrated using synthetic dsRNAs (Janowski et al. 2007) but has also been shown for endogenous miRNAs (Place et al. 2008).

In plants, the *cis*-acting siRNAs (casiRNAs) are produced from transposons, repetitive elements, and tandem repeats. The majority of casiRNAs are 24 nt in size and are methylated at 3′-ends of both strands. casiRNAs promote heterochromatin formation by directing DNA methylation and histone modification of the region where the siRNAs are encoded. Trans-acting siRNAs (tasiRNAs) in plants are generated by the convergence of the miRNA and siRNA pathways (Ghildiyal and Zamore 2009). miRNA-directed cleavage of certain transcripts can recruit the RNA-dependent RNA polymerase, which produces a complementary strand of the cleaved transcript to form dsRNA. The dsRNA is then cleaved by Dicer to generate 21-nt siRNAs (Allen et al. 2005). Endogenous siRNAs have also been detected in Drosophila (e.g., Czech et al. 2008). Fly siRNAs can be generated in a Dcr-2-dependent manner, from transposons, heterochromatic sequences, intergenetic regions, as well as mRNAs. Endogenous siRNAs have also been found in mouse oocytes where siRNAs can be derived from different sources including protein-coding genes (Tam et al. 2008; Watanabe et al. 2008). The exact roles of most endogenous siRNAs are largely unknown; however, they may contribute to the regulation of mRNA levels either by specific cleavage or serve as miRNAs to regulate expression of multiple genes.

Closely related to siRNAs are miRNAs. More than 17,000 distinct mature miRNA sequences have been registered thus far in the miRNA database (http://www.mirbase.org). Both siRNA and miRNAs are incorporated into RISC containing Argonaute proteins. In contrast to siRNAs, miRNAs modulate gene expression either by downregulating mRNA levels or by modulating mRNA translation, depending on which Ago protein a miRNA binds and on the degree of complementarity of miRNAs to the target mRNAs (Chekulaeva and Filipowicz 2009). Perfect or near-perfect base pairing promotes cleavage, whereas partial base pairing causes translation repression (Fabian et al. 2010). Although in most cases miRNAs were found to suppress protein expression, miRNAs mediated translation activation have also been demonstrated (e.g., Peng et al. 2011).

In plants, miRNAs normally have extensive complementarity with their targets and result in target degradation. Most miRNAs have target sites within open reading frames of mRNAs, with one site per mRNA. In a few cases, miRNA target sites can be predicted in the 3′- or 5′-UTR regions (Sunkar and Zhu 2004). However, partial base pairing of miRNA/mRNA that causes translation repression has also been found in plants (Moxon et al. 2008; Lanet et al. 2009). In contrast, only a few fly and mammalian miRNAs have been identified as having nearly perfect base pairing with their mRNA targets (e.g., Mansfield et al. 2004). Instead, most fly and mammalian miRNAs base pair with the 3′-UTR regions of the substrate mRNAs, via a 7-nt “seed region” at positions 2–8 of the miRNA (Williams 2008). Since the target specificity is primarily determined by the relatively small size of the seed region of miRNAs, a single miRNA can regulate many different genes, or target multiple sites of the same mRNA. miRNAs with partial complementarity to the target can also promote mRNA deadenylation and degradation (Eulalio et al. 2009). In addition, miRNAs have been reported to cause histone modification and DNA methylation of promoter sites and therefore can regulate expression of targeted genes through transcriptional regulation (Hawkins and Morris 2008; Tan et al. 2009). Not surprisingly, miRNAs have been implicated in a number of human diseases, including cancer and metabolic, neurological, and developmental disorders (Taft et al. 2010).

2.6 piRNAs

The piRNAs, or piwi-interacting RNAs, are a recently identified group of small RNAs that associate with the PIWI proteins (Piwi, Aub, and Ago3 in flies; MILI, MIWI, and MIWI2 in mice; and HILI, HIWI, HIWI2, and HIWI3 in humans). Initially identified in flies, piRNAs were later also discovered in a variety of organisms including mammals (Seto et al. 2007; Ghildiyal and Zamore 2009). In contrast to most miRNAs and siRNAs, piRNAs are localized in the nucleus, are 26–32 nt in length, and do not require Dicer activity for biogenesis (Samji 2009; Siomi et al. 2010). In addition, piRNAs are not conserved, can be either sense or antisense, and can be derived from transposons and other intergenic repetitive elements, whereas miRNAs are conserved and are produced from their own genes (Siomi et al. 2010).

In flies, PIWI–piRNA complexes have been found mainly to target and silence transposon transcripts and to protect the genomes of gametes from their invasion. However, piRNAs can also target protein-coding genes (Siomi et al. 2010). Antisense piRNAs, associated with Aub or PIWI, can lead to cleavage of target RNA, resulting in the formation of sense piRNAs that associate with Ago3. Interaction of sense piRNA with an antisense strand also leads to cleavage of the antisense strand by Ago3, creating antisense piRNAs that associate with Aub or PIWI. This amplification loop results in transposon silencing and piRNA accumulation (Ghildiyal and Zamore 2009; Siomi et al. 2010). In mammals, piRNAs can be divided into prepachytene and pachytene piRNAs, based on the stage of meiosis at which they are expressed in developing spermatocytes (Ghildiyal and Zamore 2009). Prepachytene piRNAs predominantly correspond to repetitive sequences and are implicated in silencing transposons, similar to fly piRNAs. However, different from the cleavage-mediated silencing mechanism for drosophila piRNAs, mouse piRNAs may be involved in transcriptional silencing of target genes through DNA methylation (Kim et al. 2009). In contrast, pachytene piRNAs are derived from unannotated regions of the genome, and their functions remain to be determined.

2.7 Long Noncoding RNAs

In addition to the small ncRNAs and classical non-protein-coding genes, i.e., rRNAs and tRNAs, ncRNAs greater than 200 nt in size are predominant in the transcriptome (Wilusz et al. 2009). To date, ~35,000 noncoding transcripts have been identified from ~10,000 distinct loci that bear many signatures of mRNAs, including 5′ capping, splicing, and polyadenylation, but have little or no open reading frame (Carninci et al. 2005). However, a significant portion (~40%) of lncRNAs is not polyadenylated (Cheng et al. 2005). Many lncRNAs have been found to be transcribed from the antisense strand of genes encoding proteins or sense ncRNAs. While the majority of lncRNAs are localized and may function in the nucleus, at least some are transported to the cytoplasm (Tycowski et al. 1996; Pelczar and Filipowicz 1998). Although the functions of most lncRNAs are unknown, accumulating evidence indicates that lncRNAs might play important roles in a variety of biological processes, including chromatin remodeling, transcriptional regulation, integrity of subcellular structures, and mRNA stability (Chen and Carmichael 2010). lncRNAs have also been implicated in human diseases and recognized as cancer biomarkers (Lin et al. 2007; Mallardo et al. 2008).

The expression of a lncRNA itself could be sufficient to affect expression of nearby genes. For example, transcription of a ncRNA across the promoter region of a downstream protein-coding gene can directly interfere with binding of transcription factors, thus preventing expression of the protein-coding gene (Martens et al. 2004; Wilusz et al. 2009). Even if not directly interfering with a nearby promoter, transcription of ncRNAs can induce histone modifications that repress transcription of overlapping protein-coding genes (e.g., Wilusz et al. 2009). On the other hand, transcription of lncRNA can also induce chromatin remodeling that is critical for transcriptional activation of downstream protein-coding gene (Hirota et al. 2008).

In addition to reducing transcription, lncRNAs have also been determined to enhance neighboring gene expression (Feng et al. 2006). Depletion of a number of ncRNAs has been demonstrated to decrease expression of their neighboring protein-coding genes. The enhancer-like activity has been further supported by heterologous transcription assays that showed the requirement of the ncRNAs for activation of gene expression (Orom et al. 2010). lncRNAs can also act posttranscriptionally. A recent study showed that a cytoplasmic lncRNA containing Alu element could base pair with the 3′-UTR of ADP-ribosylation factor 1 (ARF1) mRNA to form a Staufen1 (STAU1) binding site, thus leading to STAU1-mediated mRNA decay (SMD) (Gong and Maquat 2011). In addition, lncRNAs also promote alternative splicing, polyadenylation, and serve as structural RNAs (Beltran et al. 2008; Sasaki et al. 2009; Sunwoo et al. 2009).

3 Antisense-Mediated Reduction of ncRNAs

Functionalization of a gene requires approaches to efficiently disrupt its expression. For ncRNAs, several approaches have been explored, including ribozyme, siRNA-mediated target degradation, and oligonucleotide-directed RNase H-based mechanisms. These approaches utilize different mechanisms, yet all require antisense hybridization with the target RNA. Thus, binding specificity depends on the nucleotide sequences of the antisense molecules that should be designed to minimize off-target effects. In all cases, the efficiency of target reduction is determined by the delivery efficiency and stability of the antisense molecules, accessibility to the target RNA, as well as the binding affinity of the antisense molecule to the target RNA (Crooke et al. 2006).

3.1 Ribozyme-Directed Cleavage of ncRNAs

Catalytic RNAs (ribozymes) that exhibit certain structures can specifically cleave RNA either *in cis* or *in trans*. Natural ribozymes perform various enzymatic reactions and are classified by their relative size. Large ribozymes (>300 nt) include group I and group II introns and RNase P. Small ribozymes include hammerhead, hairpin, hepatitis delta virus (HDV), Varkud satellite (VS), and glmS ribozymes (Mulhbacher et al. 2010). The hammerhead ribozyme is the smallest and best known RNA catalytic motif with endonucleolytic activity. This motif exhibits a secondary structure containing three stems that are formed between a 17 nt highly conserved catalytic domain and the substrate RNA (Citti and Rainaldi 2005). The cleavage reaction, which does not require other cellular components, occurs downstream from a NUH site within the RNA substrate (Fig. 2a). The hammerhead ribozyme requires not only accessibility to the target RNA but also the formation of a proper structure for catalytic activity, thus ensuring a high specificity. Indeed, hammerhead ribozymes are able to discriminate between targets differing by a single nucleotide when placed inside or in close proximity to the cleavage site (Phylactou et al. 1998).

Hammerhead ribozymes (~30 nt) have been designed to target cellular RNAs in *trans*. Selection of active sites in substrate RNAs involves identification of NUH triple nucleotides and selection of flanking sequence that ensures hybridization specificity and accessibility. Since experimental verification of ribozyme activity is difficult, expensive, and time-consuming, computing methods have been developed to predict the most possible sites on substrate RNAs for ribozymes (Mercatanti et al. 2002; Cruz and Westhof 2011). However, catalytic activity (target reduction) must be experimentally verified either in cells or in cell-free system, due to the strict structural requirements of a ribozyme. Engineered ribozymes can be introduced into cells by transient transfection. Unmodified ribozymes are subject to degradation; however, the stability of the exogenous ribozyme can be increased by chemical modification, such as 2′-*O*-methyl, 2′-amino, 2′-fluoro, and 2′-methylene, that can confer some nuclease resistance without affecting the catalytic activity (Sproat 1996). Since the phosphate groups contribute to the three-dimensional structure of ribozyme, phosphorothioate substitutions (PS) need to be avoided in the catalytic core but can be introduced to the target binding domain (Shimayama et al. 1993). In addition, ribozymes can also be expressed in cells. However, limitations may include low transcription efficiency, incorrect localization, and frequent silencing of the target gene (Citti and Rainaldi 2005; Val et al. 2011). Only a few ncRNAs have been targeted with hammerhead ribozymes. For example, telomerase RNA, which is overexpressed in cancer cells, has been efficiently depleted using

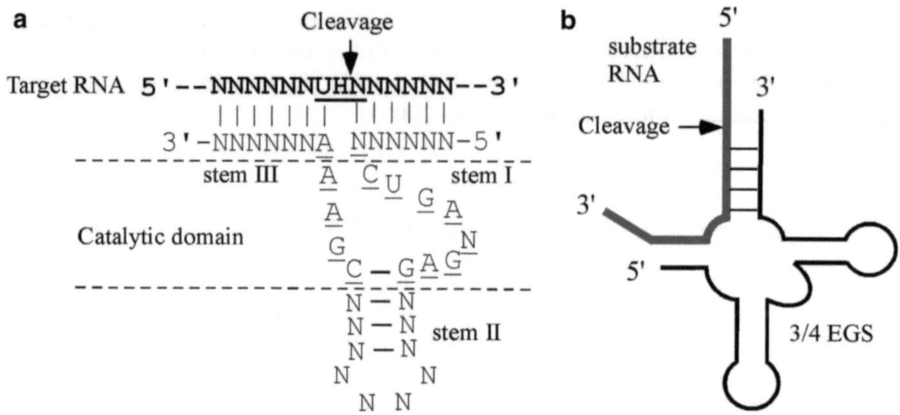

Fig. 2 Schematic prediction of ribozymes. (**a**) A hammerhead ribozyme is depicted based on Citti and Rainaldi (2005). The substrate RNA is shown in *bold letters*. The nucleotides forming the catalytic core of the ribozyme are *underlined*. The conserved nucleotides are shown. N, any of the four nucleotides. H, any nucleotide except G. (**b**) A RNase P RNA-based ribozyme is depicted according to Lundblad and Altman (2010). The substrate RNA is shown in *thick gray*, whereas the ribozyme is in *black*. EGS, external guide sequence

hammerhead ribozymes in human and mouse cancer cells (Li et al. 2007). In addition, a target snoRNA was dramatically depleted by expression of a ribozyme within a snoRNA vector (Samarsky et al. 1999).

Another better studied ribozyme is RNase P RNA. Since RNase P recognizes the acceptor stem and possibly the T-stem loop regions in precursor tRNAs, it is possible to introduce an external guide sequence (EGS) that is complementary to the target RNA to guide RNase P cleavage. Indeed, it was found that cleavage of a target mRNA can be induced when three fourths of a tRNA is used as an EGS in human cells (Fig. 2b) (Yuan et al. 1992), and exogenous administration of 2′-*O*-methylated EGS was shown to efficiently inhibit expression of PKC-α and Bcl-xl proteins in human bladder carcinoma cells (Ma et al. 2000). Directed cleavage can also be achieved by expression of a fused RNA containing the catalytic M1 RNA (derived from *Escherichia coli* RNase P RNA) and a guide sequence that is complementary to the target. Success with this approach has been reported in bacteria, cell culture, and in mice (Lundblad and Altman 2010). Although mainly mRNAs have been targeted, in principle, this approach can also be used to target ncRNAs, since the natural targets of RNase P are often these types of RNAs. Similar to hammerhead ribozymes, the design of EGS to induce RNase P cleavage of a range of RNAs also requires maintenance of a proper secondary structure.

3.2 *siRNA/shRNA-Mediated Reduction of ncRNAs*

Synthetic siRNAs are usually introduced into cells by transient transfection, while expressed small hairpin RNAs (shRNAs) are processed into mature siRNAs using the cellular machinery. siRNAs are widely used to target mRNAs in the cytoplasm, and it is believed that RISC complex is not active in the nucleus, especially in mammals (Zeng and Cullen 2002; Chiu et al. 2004). However, in a few cases, nuclear ncRNAs have been shown to be reduced by siRNAs, though a nuclear mechanism supporting the reduction of RNAs in the nucleus has not been demonstrated.

It was found that expression of dsRNA, which is further processed into siRNAs, could trigger the degradation of some targeted snoRNAs in *Trypanosoma brucei* (Liang et al. 2003). However, not all tested snoRNAs could be similarly reduced in this organism. snoRNAs are synthesized in the nucleus and accumulate in the nucleolus without trafficking to the cytoplasm, inferring that siRNAs can operate in the nucleus in this organism. This is further supported by a recent study showing that siRNA-mediated snoRNA degradation requires Argonaute 1 (the single Ago protein in this organism) and TBDCL2, the nuclear Dicer protein, but not the cytoplasmic Dicer protein TBDLC1 that is required for RNAi in the cytoplasm (Gupta et al. 2010). However, transfection of siRNAs or expression of shRNAs failed to deplete snoRNAs in mammalian cells (Ploner et al. 2009; Liang et al. 2011).

There have been a few reports of reduction of nuclear ncRNAs by transfection of siRNAs in mammalian cells. The nucleoplasmic 7SK RNA was reduced by more than 90% in human cells transfected with siRNA (Robb et al. 2005). In the same study, another tested snRNA, the U6 snRNA, was only reduced by ~50% with similar siRNA treatment, suggesting that not all nuclear ncRNAs can be efficiently targeted by siRNAs. In addition, nuclear lncRNAs have also been targeted with siRNAs. For example, NEAT1 was reduced by ~60–80% in HeLa cells transfected with siRNAs (Clemson et al. 2009), and ~70% reduction was observed for an antisense lncRNA, HOTAIRM1, in human cells expressing shRNAs (Zhang et al. 2009). Xist, a 17-kb lncRNA, was reduced by ~90% in mouse embryonic stem cells expressing shRNAs (Zhao et al. 2008). Interestingly, a shorter nuclear ncRNA (~1.6 kb), RepA, containing overlapping sequence with Xist and being targeted by the same shRNA, appeared to be only moderately reduced (~50%). In addition, ~55–80% reduction was observed for six different lncRNAs in several different types of human cells transfected with siRNAs (Orom et al. 2010).

In many cases, the reduction efficiency for ncRNAs by siRNA/shRNAs is low compared with reduction for mRNAs. For example, gadd7 was reduced by <50% using shRNA (Brookheart et al. 2009). Although the above-mentioned examples imply that certain siRNAs may be functional in the nucleus, it has also been shown that siRNAs targeting intron regions of multiple nuclear pre-mRNAs are not active in mammalian cells (Vickers et al. 2003). This failure is not likely due to the inaccessibility of the intron regions, since RNase H-based antisense oligonucleotides targeting the same regions can cause significant pre-mRNA reduction. These observations suggest that siRNAs are not active in targeting nuclear pre-mRNAs and that siRNA-mediated ncRNA reduction may occur through a different mechanism. It is also possible that some nuclear ncRNAs may have a cytoplasmic phase during biogenesis, such as some spliceosomal snRNAs (Will and Luhrmann 2001). Thus, nuclear ncRNAs may be targeted by siRNAs in the cytoplasm.

3.3 Single-Stranded Antisense RNA-Mediated Target Reduction

Antisense RNA had been used to inactivate target RNAs in different organisms, for example, to block mRNA translation in frog oocytes, or to reduce gene expression in mouse cells. In addition, in *Leptomonas collosoma* and *Leishmania major*, overexpression of antisense RNA complementary to snoRNAs resulted in ~90% reduction of some snoRNAs (Liang et al. 2003). Notably, no siRNA-mediated RNAi activity exists in these two trypanosomatid species. However, snoRNA degradation is accompanied by the appearance of both sense and antisense small RNAs with sizes similar to siRNAs, suggesting that the snoRNAs are cleaved by an endonuclease activity via a mechanism different from RNAi but yielding cleavage products similar in size to siRNAs.

Antisense RNA has also been employed to downregulate miRNAs (anti-miRs). Different chemical modifications have been incorporated into anti-miRs to inhibit the function of miRNAs, including 2′-*O*-methyl, 2′-*O*-methyl mixed with PS backbone, 2′-*O*-methoxyethyl in combination with PS backbone, 2′-fluoro, locked nucleotide acid (LNA), and morpholino. It appears that anti-miRs act on mature miRNAs, but not on precursors (Krutzfeldt et al. 2005). Interestingly, inhibition of miRNA function by antisense oligonucleotides utilizes different mechanisms, for example, some modified anti-miRs inhibit miRNA activity without reducing the level of miRNAs, suggesting that anti-miRs sequester the miRNAs (Esau 2008). It has been proposed that high-affinity anti-miRs (K-PNA-K3, LNA/DNA, and LNA/OMe) sequester the target miRNA without causing degradation, while a lower-affinity anti-miR (2′-*O*-Me and 2′-F) promotes degradation through unknown pathway(s) (Su et al. 2011; Torres et al. 2011).

3.4 RNase H-Mediated Reduction of NcRNAs

RNase H hydrolyzes the RNA component within a RNA/DNA duplex. All RNase H enzymes function as endonucleases exhibiting limited sequence specificity, require divalent cations, and generate products with 5′-phosphate and 3′ hydroxyl termini (Lima et al. 2008). Several thousand different mRNAs have been targeted in mammalian cells using synthesized oligonucleotides that induce RNase H activity (Crooke et al. 2006). Since RNase H is present in cytoplasm, nucleus, and mitochondria, it is capable of cleaving target RNAs present in different cellular compartments in the presence of complementary DNA oligonucleotides. RNase H-mediated cleavage has been employed to characterize the roles of different spliceosomal snRNAs in splicing in a cell-free system (e.g., Rinke et al. 1984; Black et al. 1985). Different delivery approaches have been used to reduce ncRNA in cells. Antisense oligonucleotides (ASOs) have been microinjected into *Xenopus* oocytes to degrade mRNAs, snRNAs, and snoRNAs (e.g., Tycowski et al. 1994; Hulstrand et al. 2010). A semi-in vivo system has also been established for

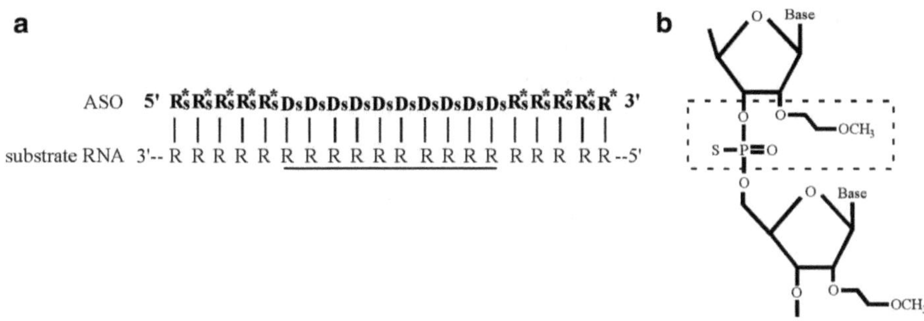

Fig. 3 Second-generation RNase H ASOs. (**a**) Depiction of a typical 5–10–5 RNA/DNA chimeric ASO. The ASO is shown in *bold*. *R* ribonucleotide, *D* deoxyribonucleotide, *S* phosphorothioate backbone, *asterisk* 2′-*O*-MOE modification. The base-paired substrate is shown, and the region to be cleaved is *underlined*. (**b**) Depiction of the chemical modifications. PS linkage and 2′-*O*-MOE modifications are *boxed*

trypanosomes to introduce oligonucleotides targeting several small ncRNAs into permeabilized cells (Ullu and Tschudi 1990). In mammalian cells, ASOs are often introduced using lipid-mediated chemical transfection or electroporation.

Different chemical modifications have been evaluated to improve ASO stability and binding affinity. Phosphorothioate DNA (PS) represents by far the most widely used backbone modification and is the major modification for first generation of ASOs. A PS backbone-containing deoxynucleotide is able to efficiently elicit RNase H cleavage of the target RNA and possesses increased stability in plasma, tissues, and cells (Swayze and Bhat 2006). Although the PS backbone modification reduces the binding affinity of the oligonucleotides to the target RNAs, higher affinity can be achieved by introducing ribose modifications, such as 2′-*O*-methyl (Me), LNA, 2′-F, and 2′-*O*-methoxyethyl (MOE), to the RNA portion of an ASO. A 5–10–5 or 3–12–3 RNA/DNA/RNA composition of chimeric oligonucleotides is commonly used, with the RNA portion increasing binding activity and the cleavage sites restricted to the DNA binding region (Bennett 2006; Hulstrand et al. 2010). The 2′-*O*-MOE modification is currently used in the second-generation ASOs, together with PS backbone modification (Fig. 3). These modifications significantly increase the ASO binding potential to substrate RNA and the stability of the ASO (Bennett 2006; Swayze and Bhat 2006), leading to high efficiency in reduction of the target RNAs.

Recently, mammalian ncRNAs, including long ncRNAs and highly structured, protein-coated small ncRNAs, have been successfully targeted with RNase H ASOs. For example, using 5–10–5 2′-MOE gapmer ASOs, Malat1, a lncRNA, was reduced by 85–90% in human and mouse cells (Tripathi et al. 2010; Koller et al. 2011). Similarly, NEAT1 was reduced by ~70% in human cells (Sunwoo et al. 2009). Highly structured small ncRNAs have also been successfully depleted using RNase H-based ASOs. These include snoRNAs, which are small in size, present in stable RNP complexes, reside in the nucleolus, and are thus difficult to target. Other technologies have not successfully reduced this type of ncRNAs. For example, it was demonstrated that expression of shRNAs or treatment by transfection of siRNAs failed to reduce tested snoRNAs in human cells (Ideue et al. 2009; Ploner et al. 2009; Liang et al. 2011). Further, transfection of 200-nM LNA gapmer oligonucleotides that contain DNA core flanked by LNA nucleotides, or expression of ribozymes, only partially (~60%) reduced a tested snoRNA (U81) in HEK-293T cells, without causing loss of function (Ploner et al. 2009). In contrast, modified, RNase H-based ASOs have proven to efficiently deplete snoRNAs and snRNAs in *Xenopus* oocytes and in mammalian cells.

Microinjection of antisense oligodeoxynucleotides in *Xenopus* oocytes has been employed to deplete some snoRNAs and snRNAs (e.g., Tycowski et al. 1994; Borovjagin and Gerbi 1999). In mammalian cells, RNase H-based ASOs targeting ncRNAs were mainly delivered by chemical transfection or electroporation. Using 2′-MOE gapmer ASOs, snoRNAs could be efficiently depleted in HeLa cells using nucleofection, an electroporation-based approach, but not by lipid-mediated chemical transfection, as demonstrated in a recent study (Ideue et al. 2009). This observation suggests that MOE gapmer ASOs are capable of degrading snoRNAs in the nucleus as long as they can be delivered into cells and have sufficient accessibility to the target RNA sequence. Indeed, we have found that all tested snRNAs and snoRNAs, including C/D box and H/ACA box snoRNAs as well as scaRNAs, could be efficiently reduced by treatment with active ASOs that were preselected through whole RNA sequence screening (Liang et al. 2011). More than 80% reduction was achieved in different human cells and in mouse cells when transfected with different chemical transfection reagents (Liang et al. 2011). The action of ASOs in targeting snoRNA degradation is prompt, as reduction was observed as early as 2 h after transfection of the ASOs. For snoRNAs that are linked to enzymatic activity of snoRNPs, sufficient reduction needs to be reached in order to observe functional loss, since a ~60% reduction of a snoRNA had no obvious effect on the level of modification guided by the snoRNA, whereas ~80% or more reduction of a snoRNA resulted in significant decrease in the directed modification signal (Ploner et al. 2009; Liang et al. 2011).

Most mammalian snoRNAs are encoded in introns of host genes; therefore, RNase H-based ASOs, which are active in the nucleus, may also target precursors of snoRNAs, such as pre-mRNAs. However, ASOs targeting intronic snoRNAs do not act on pre-mRNAs. For example, in cells transfected with ASOs, the levels of targeted intronic snoRNAs were dramatically reduced; however, the levels of mature mRNAs and proteins encoded by the host genes, and neighboring snoRNAs encoded in the same pre-mRNAs, were not affected (Liang et al. 2011), indicating that these ASOs do not act on pre-mRNA before splicing. Although ASOs for snoRNAs do not act on pre-mRNA, they appear to be able to act on pre-snoRNAs after splicing. Indeed, we found that depletion of Rat1/Xrn2, a $5' \rightarrow 3'$ exonuclease that is required for snoRNA maturation, resulted in accumulation of pre-snoRNAs containing snoRNA-coding region and 5′ intron, but not 3′ intron. This pre-snoRNA, along with mature snoRNA, could be efficiently degraded by ASOs (our unpublished data), indicating that this form of pre-snoRNA is accessible to the RNase H ASOs. Difference in ASO activities for targeting pre-mRNA and pre-snoRNA may stem from different protein composition in different RNPs, and/ or different RNP structures at different maturation stages.

The high specificity of RNase H ASOs has also been documented by mutational analysis. ASOs bearing 2–3 mismatches in the DNA core region exhibit little or no activity in directing RNase H cleavage of the target ncRNAs (Ideue et al. 2009; Liang et al. 2011). In certain cases, this high specificity may limit the utilization when several ncRNAs that have similar sequences need to be depleted at the same time, since many ncRNAs, especially snRNAs, have multiple functional isoforms in mammals. However, at least five ncRNAs were depleted simultaneously in human cells by transfection of ASO pools containing different ASOs, with more than 85% reduction for all the five targeted snoRNAs (Liang et al. 2011). Combined depletion of snoRNAs is important for functional analysis of RNA modifications, since in most cases a significant effect could be observed only when a subset of modifications in a particular domain of the substrate RNA was depleted (King et al. 2003; Liang et al. 2007).

In addition to the numerous cytoplasmic mRNAs targeted, ncRNAs present in the nucleolus, nucleoplasm, and Cajal bodies have also been efficiently depleted in mammalian cells using second-generation RNase H ASOs (Ideue et al. 2009; Liang et al. 2011). These findings indicate a high capability of ASOs in targeting various ncRNAs localized in different cellular compartments. Importantly, ASOs were shown to be highly effective in depletion of nuclear and nucleolar RNAs in vivo, with systematic administration of ASOs. Subcutaneous injection of a single dose of ASOs resulted in dramatic reduction of targeted ncRNAs in mouse liver, with more than 90% reduction for two snoRNAs and a long nuclear ncRNA, Malat1 (Koller et al. 2011; Liang et al. 2011). The IC_{50} for these RNAs was less than 3 mg/kg [(Koller et al. 2011) and our unpublished data]. Moderate reduction was also observed in other tissues, such as white adipose tissue and kidney, suggesting that ASOs can also reach other tissues and act in the nucleus (Liang et al. 2011).

4 Concluding Remarks

Several approaches to target various types of ncRNAs for functionalization and therapeutic treatments have been developed. siRNAs are generally more potent than other approaches when targeting mRNAs in the cytoplasm. siRNA design is relatively simple, and chemical modifications are not required in most cases. However, the RISC pathway does not exist in certain organisms, and only in a few cases have siRNAs been shown to be able to reduce nuclear ncRNAs. In addition, off-target and secondary effects of siRNAs need to be considered seriously (Svoboda 2007). This property also makes it difficult to find a truly inactive siRNA for control experiments. In addition, due to the competition of siRNA with miRNAs for binding to Ago proteins, transfection of siRNAs may also globally disturb gene regulation by endogenous miRNAs. Thus, caution needs to be taken when using siRNAs, and the concentration of siRNAs used to deplete RNAs should be as low as possible.

As compared with siRNAs, ribozymes and RNase H ASO can be used in most, if not all, organisms, and they do not utilize a cellular mechanism that involves large amount of endogenous regulatory RNAs and thus have less potential to induce off-target and secondary effects. In addition, ribozymes and RNase H ASOs are more specific, since longer and perfect or near-perfect match is required for their activities in directing target reduction. In the case of ribozymes, the requirement of proper structure between ribozyme/substrate RNA for cleavage activity further increases target specificity. However, this property also increases the difficulty in finding an active ribozyme. Design and large-scale screening for active ribozymes are thus time-consuming and laborious. Ribozymes can be used to target ncRNAs in the nucleus (and nucleolus) when delivered by a vector RNA; however, this may limit the application in developing drugs since constant expression of a "drug-RNA" might be avoided. RNase H ASOs have potency similar to ribozymes and somewhat less than siRNAs but can act on almost any type of RNAs localized in different cellular compartments and can be easily delivered and screened in a large scale.

The different approaches described above all have their unique advantages; thus, the best approach for reducing a ncRNA may vary for different RNAs and organisms. In our laboratory, we typically confirm conclusions by using numerous control oligonucleotides and more than one mechanism, for example, RNase H ASOs and siRNAs. Further methodological development and optimization is still needed to meet the needs for functionalization of huge numbers of ncRNAs and for targeting ncRNAs for therapeutic treatments.

Acknowledgments We apologize to those whose paper could not be cited due to space limitations. We thank Donna Parrett for excellent editing.

References

Allen TA, Von Kaenel S, Goodrich JA et al (2004) The SINE-encoded mouse B2 RNA represses mRNA transcription in response to heat shock. Nat Struct Mol Biol 11:816–821

Allen E, Xie Z, Gustafson AM, Carrington JC (2005) microRNA-directed phasing during trans-acting siRNA biogenesis in plants. Cell 121:207–221

Bachellerie JP, Cavaille J, Huttenhofer A (2002) The expanding snoRNA world. Biochimie 84:775–790

Beltran M, Puig I, Pena C et al (2008) A natural antisense transcript regulates Zeb2/Sip1 gene expression during Snail1-induced epithelial-mesenchymal transition. Genes Dev 22:756–769

Bennett CF (2006) Pharmacological properties of 2′-O-methoxyethyl-modified oligonucleotides. In: Crooke ST (ed) Antisense drug technology—principles, strategies, and applications. CRC, Boca Raton, FL, pp 273–304

Bertrand E, Fournier MJ (2004) The snoRNPs and related machines: ancient devices that mediate maturation of rRNA and other RNAs. In: Olson MOJ (ed) The nucleolus. Landes Bioscience, Georgetown, TX, pp 225–261

Black DL, Chabot B, Steitz JA (1985) U2 as well as U1 small nuclear ribonucleoproteins are involved in premessenger RNA splicing. Cell 42:737–750

Borovjagin AV, Gerbi SA (1999) U3 small nucleolar RNA is essential for cleavage at sites 1, 2 and 3 in pre-rRNA and determines which rRNA processing pathway is taken in Xenopus oocytes. J Mol Biol 286:1347–1363

Brookheart RT, Michel CI, Listenberger LL et al (2009) The non-coding RNA gadd7 is a regulator of lipid-induced oxidative and endoplasmic reticulum stress. J Biol Chem 284:7446–7454

Carninci P, Kasukawa T, Katayama S et al (2005) The transcriptional landscape of the mammalian genome. Science 309:1559–1563

Chekulaeva M, Filipowicz W (2009) Mechanisms of miRNA-mediated post-transcriptional regulation in animal cells. Curr Opin Cell Biol 21:452–460

Chen LL, Carmichael GG (2010) Decoding the function of nuclear long non-coding RNAs. Curr Opin Cell Biol 22:357–364

Cheng J, Kapranov P, Drenkow J et al (2005) Transcriptional maps of 10 human chromosomes at 5-nucleotide resolution. Science 308:1149–1154

Chiu YL, Ali A, Chu CY et al (2004) Visualizing a correlation between siRNA localization, cellular uptake, and RNAi in living cells. Chem Biol 11:1165–1175

Citti L, Rainaldi G (2005) Synthetic hammerhead ribozymes as therapeutic tools to control disease genes. Curr Gene Ther 5:11–24

Clemson CM, Hutchinson JN, Sara SA et al (2009) An architectural role for a nuclear noncoding RNA: NEAT1 RNA is essential for the structure of paraspeckles. Mol Cell 33:717–726

Coughlin DJ, Pleiss JA, Walker SC et al (2008) Genome-wide search for yeast RNase P substrates reveals role in maturation of intron-encoded box C/D small nucleolar RNAs. Proc Natl Acad Sci USA 105:12218–12223

Crooke ST, Vickers T, Lima W et al (2006) Mechanisms of antisense drug action, an introduction. In: Crooke ST (ed) Antisense drug technology—principles, strategies, and application. CRC, Roca Raton, FL

Cruz JA, Westhof E (2011) Sequence-based identification of 3D structural modules in RNA with RMDetect. Nat Methods 8:513–21

Czech B, Malone CD, Zhou R et al (2008) An endogenous small interfering RNA pathway in Drosophila. Nature 453:798–802

Dominski Z, Marzluff WF (2007) Formation of the 3′ end of histone mRNA: getting closer to the end. Gene 396:373–390

Ender C, Krek A, Friedlander MR et al (2008) A human snoRNA with microRNA-like functions. Mol Cell 32:519–528

Esakova O, Krasilnikov AS (2010) Of proteins and RNA: the RNase P/MRP family. RNA 16:1725–1747

Esau CC (2008) Inhibition of microRNA with antisense oligonucleotides. Methods 44:55–60

Eulalio A, Huntzinger E, Nishihara T et al (2009) Deadenylation is a widespread effect of miRNA regulation. RNA 15:21–32

Fabian MR, Sonenberg N, Filipowicz W (2010) Regulation of mRNA translation and stability by microRNAs. Annu Rev Biochem 79:351–379

Feng J, Bi C, Clark BS et al (2006) The Evf-2 noncoding RNA is transcribed from the Dlx-5/6 ultraconserved region and functions as a Dlx-2 transcriptional coactivator. Genes Dev 20:1470–1484

Fire A, Xu S, Montgomery MK et al (1998) Potent and specific genetic interference by double-stranded RNA in *Caenorhabditis elegans*. Nature 391:806–811

Friend K, Lovejoy AF, Steitz JA (2007) U2 snRNP binds intronless histone pre-mRNAs to facilitate U7-snRNP-dependent 3′ end formation. Mol Cell 28:240–252

Galasso M, Elena Sana M, Volinia S (2010) Non-coding RNAs: a key to future personalized molecular therapy? Genome Med 2:12

Gerbi SA, Borovjagin AV (2004) Pre-ribosomal RNA processing in multicellular organisms. In: Olson MOJ (ed) The nucleolus. Landes Bioscience, Georgetown, TX, pp 170–198

Ghildiyal M, Zamore PD (2009) Small silencing RNAs: an expanding universe. Nat Rev Genet 10:94–108

Gill T, Cai T, Aulds J et al (2004) RNase MRP cleaves the CLB2 mRNA to promote cell cycle progression: novel method of mRNA degradation. Mol Cell Biol 24:945–953

Gong C, Maquat LE (2011) lncRNAs transactivate STAU1-mediated mRNA decay by duplexing with 3′ UTRs via Alu elements. Nature 470:284–288

Gupta V, Kumar A (2010) Dyskeratosis congenita. Adv Exp Med Biol 685:215–219

Gupta SK, Hury A, Ziporen Y et al (2010) Small nucleolar RNA interference in *Trypanosoma brucei*: mechanism and utilization for elucidating the function of snoRNAs. Nucleic Acids Res 38:7236–7247

Hasler J, Strub K (2006a) Alu elements as regulators of gene expression. Nucleic Acids Res 34:5491–5497

Hasler J, Strub K (2006b) Alu RNP and Alu RNA regulate translation initiation in vitro. Nucleic Acids Res 34:2374–2385

Hawkins PG, Morris KV (2008) RNA and transcriptional modulation of gene expression. Cell Cycle 7:602–607

Hirota K, Miyoshi T, Kugou K et al (2008) Stepwise chromatin remodelling by a cascade of transcription initiation of non-coding RNAs. Nature 456:130–134

Hulstrand AM, Schneider PN, Houston DW (2010) The use of antisense oligonucleotides in Xenopus oocytes. Methods 51:75–81

Ideue T, Hino K, Kitao S et al (2009) Efficient oligonucleotide-mediated degradation of nuclear noncoding RNAs in mammalian cultured cells. RNA 15:1578–1587

Jacquier A (2009) The complex eukaryotic transcriptome: unexpected pervasive transcription and novel small RNAs. Nat Rev Genet 10:833–844

Janowski BA, Younger ST, Hardy DB et al (2007) Activating gene expression in mammalian cells with promoter-targeted duplex RNAs. Nat Chem Biol 3:166–173

Jarrous N, Reiner R (2007) Human RNase P: a tRNA-processing enzyme and transcription factor. Nucleic Acids Res 35:3519–3524

Kettenberger H, Eisenfuhr A, Brueckner F et al (2006) Structure of an RNA polymerase II-RNA inhibitor complex elucidates transcription regulation by noncoding RNAs. Nat Struct Mol Biol 13:44–48

Kim VN, Han J, Siomi MC (2009) Biogenesis of small RNAs in animals. Nat Rev Mol Cell Biol 10:126–139

King TH, Liu B, McCully RR et al (2003) Ribosome structure and activity are altered in cells lacking snoRNPs that form pseudouridines in the peptidyl transferase center. Mol Cell 11:425–435

Kishore S, Stamm S (2006) The snoRNA HBII-52 regulates alternative splicing of the serotonin receptor 2C. Science 311:230–232

Kishore S, Khanna A, Zhang Z et al (2010) The snoRNA MBII-52 (SNORD 115) is processed into smaller RNAs and regulates alternative splicing. Hum Mol Genet 19:1153–1164

Koller E, Vincent TM, Chappell A et al (2011) Mechanisms of single-stranded phosphorothioate modified antisense oligonucleotide accumulation in hepatocytes. Nucleic Acids Res 39:4795–4807

Krutzfeldt J, Rajewsky N, Braich R et al (2005) Silencing of microRNAs in vivo with 'antagomirs'. Nature 438:685–689

Kwek KY, Murphy S, Furger A et al (2002) U1 snRNA associates with TFIIH and regulates transcriptional initiation. Nat Struct Biol 9:800–805

Lanet E, Delannoy E, Sormani R et al (2009) Biochemical evidence for translational repression by Arabidopsis microRNAs. Plant Cell 21:1762–1768

Lee DY, Clayton DA (1998) Initiation of mitochondrial DNA replication by transcription and R-loop processing. J Biol Chem 273:30614–30621

Li LC, Okino ST, Zhao H et al (2006) Small dsRNAs induce transcriptional activation in human cells. Proc Natl Acad Sci USA 103:17337–17342

Li S, Nosrati M, Kashani-Sabet M (2007) Knockdown of telomerase RNA using hammerhead ribozymes and RNA interference. Methods Mol Biol 405:113–131

Liang XH, Xu YX, Michaeli S (2002) The spliced leader-associated RNA is a trypanosome-specific sn(o) RNA that has the potential to guide pseudouridine formation on the SL RNA. RNA 8:237–246

Liang XH, Liu Q, Michaeli S (2003) Small nucleolar RNA interference induced by antisense or double-stranded RNA in trypanosomatids. Proc Natl Acad Sci USA 100:7521–7526

Liang XH, Liu Q, Fournier MJ (2007) rRNA modifications in an intersubunit bridge of the ribosome strongly affect both ribosome biogenesis and activity. Mol Cell 28:965–977

Liang XH, Vickers TA, Guo S et al (2011) Efficient and specific knockdown of small non-coding RNAs in mammalian cells and in mice. Nucleic Acids Res 39:e13

Lima W, Wu H, Crooke ST (2008) The RNase H mechanism. In: Crooke ST (ed) Antisense drug technology—principles, strategies, and applications. CRC, Boca Raton, FL, pp 47–74

Lin R, Maeda S, Liu C et al (2007) A large noncoding RNA is a marker for murine hepatocellular carcinomas and a spectrum of human carcinomas. Oncogene 26:851–858

Liu WM, Chu WM, Choudary PV et al (1995) Cell stress and translational inhibitors transiently increase the abundance of mammalian SINE transcripts. Nucleic Acids Res 23:1758–1765

Lundblad EW, Altman S (2010) Inhibition of gene expression by RNase P. Nat Biotechnol 27:212–221

Ma M, Benimetskaya L, Lebedeva I et al (2000) Intracellular mRNA cleavage induced through activation of RNase P by nuclease-resistant external guide sequences. Nat Biotechnol 18:58–61

Mallardo M, Poltronieri P, D'Urso OF (2008) Non-protein coding RNA biomarkers and differential expression in cancers: a review. J Exp Clin Cancer Res 27:19

Mansfield JH, Harfe BD, Nissen R et al (2004) MicroRNA-responsive 'sensor' transgenes uncover Hox-like and other developmentally regulated patterns of vertebrate microRNA expression. Nat Genet 36:1079–1083

Maraia RJ, Driscoll CT, Bilyeu T et al (1993) Multiple dispersed loci produce small cytoplasmic Alu RNA. Mol Cell Biol 13:4233–4241

Mariner PD, Walters RD, Espinoza CA et al (2008) Human Alu RNA is a modular transacting repressor of mRNA transcription during heat shock. Mol Cell 29:499–509

Martens JA, Laprade L, Winston F (2004) Intergenic transcription is required to repress the *Saccharomyces cerevisiae* SER3 gene. Nature 429:571–574

Martianov I, Ramadass A, Serra Barros A et al (2007) Repression of the human dihydrofolate reductase gene by a non-coding interfering transcript. Nature 445:666–670

Martin AN, Li Y (2007) RNase MRP RNA and human genetic diseases. Cell Res 17:219–226

McKusick VA, Eldridge R, Hostetler JA et al (1965) Dwarfism in the Amish II. Cartilage-hair hypoplasia. Bull Johns Hopkins Hosp 116:285–326

Mercatanti A, Rainaldi G, Mariani L et al (2002) A method for prediction of accessible sites on an mRNA sequence for target selection of hammerhead ribozymes. J Comput Biol 9:641–653

Michel CI, Holley CL, Scruggs BS et al (2011) Small nucleolar RNAs U32a, U33, and U35a are critical mediators of metabolic stress. Cell Metab 14:33–44

Michels AA, Fraldi A, Li Q et al (2004) Binding of the 7SK snRNA turns the HEXIM1 protein into a P-TEFb (CDK9/cyclin T) inhibitor. EMBO J 23:2608–2619

Montanaro L, Calienni M, Bertoni S et al (2010) Novel dyskerin-mediated mechanism of p53 inactivation through defective mRNA translation. Cancer Res 70:4767–4777

Moxon S, Jing R, Szittya G et al (2008) Deep sequencing of tomato short RNAs identifies microRNAs targeting genes involved in fruit ripening. Genome Res 18:1602–1609

Mulhbacher J, St-Pierre P, Lafontaine DA (2010) Therapeutic applications of ribozymes and riboswitches. Curr Opin Pharmacol 10:551–556

Nagai K, Oubridge C, Kuglstatter A et al (2003) Structure, function and evolution of the signal recognition particle. EMBO J 22:3479–3485

Orom UA, Derrien T, Beringer M et al (2010) Long noncoding RNAs with enhancer-like function in human cells. Cell 143:46–58

Pelczar P, Filipowicz W (1998) The host gene for intronic U17 small nucleolar RNAs in mammals has no protein-coding potential and is a member of the 5′-terminal oligopyrimidine gene family. Mol Cell Biol 18:4509–4518

Peng S, Chen LL, Lei XX et al (2011) Genome-wide studies reveal that Lin28 enhances the translation of genes important for growth and survival of human embryonic stem cells. Stem Cells 29:496–504

Phylactou LA, Tsipouras P, Kilpatrick MW (1998) Hammerhead ribozymes targeted to the FBN1 mRNA can discriminate a single base mismatch between ribozyme and target. Biochem Biophys Res Commun 249:804–810

Place RF, Li LC, Pookot D et al (2008) MicroRNA-373 induces expression of genes with complementary promoter sequences. Proc Natl Acad Sci USA 105:1608–1613

Ploner A, Ploner C, Lukasser M et al (2009) Methodological obstacles in knocking down small noncoding RNAs. RNA 15:1797–1804

Rand TA, Petersen S, Du F et al (2005) Argonaute2 cleaves the anti-guide strand of siRNA during RISC activation. Cell 123:621–629

Rinke J, Appel B, Blocker H et al (1984) The 5′-terminal sequence of U1 RNA complementary to the consensus 5′ splice site of hnRNA is single-stranded in intact U1 snRNP particles. Nucleic Acids Res 12:4111–4126

Robb GB, Brown KM, Khurana J et al (2005) Specific and potent RNAi in the nucleus of human cells. Nat Struct Mol Biol 12:133–137

Russell AG, Schnare MN, Gray MW (2004) Pseudouridine-guide RNAs and other Cbf5p-associated RNAs in *Euglena gracilis*. RNA 10:1034–1046

Samarsky DA, Ferbeyre G, Bertrand E et al (1999) A small nucleolar RNA:ribozyme hybrid cleaves a nucleolar RNA target in vivo with near-perfect efficiency. Proc Natl Acad Sci USA 96:6609–6614

Samji T (2009) PIWI, piRNAs, and germline stem cells: what's the link? Yale J Biol Med 82:121–124

Saraiya AA, Wang CC (2008) snoRNA, a novel precursor of microRNA in Giardia lamblia. PLoS Pathog 4:e1000224

Sasaki YT, Ideue T, Sano M et al (2009) MENepsilon/beta noncoding RNAs are essential for structural integrity of nuclear paraspeckles. Proc Natl Acad Sci USA 106:2525–2530

Schwartz JC, Younger ST, Nguyen NB et al (2008) Antisense transcripts are targets for activating small RNAs. Nat Struct Mol Biol 15:842–848

Seto AG, Kingston RE, Lau NC (2007) The coming of age for Piwi proteins. Mol Cell 26:603–609

Sharp PA (1994) Split genes and RNA splicing. Cell 77:805–815

Shimayama T, Nishikawa F, Nishikawa S et al (1993) Nuclease-resistant chimeric ribozymes containing deoxyribonucleotides and phosphorothioate linkages. Nucleic Acids Res 21:2605–2611

Singh SK, Gurha P, Tran EJ et al (2004) Sequential 2′-O-methylation of archaeal pre-tRNATrp nucleotides is guided by the intron-encoded but trans-acting box C/D ribonucleoprotein of pre-tRNA. J Biol Chem 279:47661–47671

Siomi MC, Miyoshi T, Siomi H (2010) piRNA-mediated silencing in Drosophila germlines. Semin Cell Dev Biol 21:754–759

Spiluttini B, Gu B, Belagal P et al (2010) Splicing-independent recruitment of U1 snRNP to a transcription unit in living cells. J Cell Sci 123:2085–2093

Sproat BS (1996) Synthetic catalytic oligonucleotides based on the hammerhead ribozyme. In: Eckstein F, Lilley DMJ (eds) Catalytic RNA, vol 10. Springer, Berlin, pp 265–281

Su J, Baigude H, McCarroll J et al (2011) Silencing microRNA by interfering nanoparticles in mice. Nucleic Acids Res 39:e38

Sunkar R, Zhu JK (2004) Novel and stress-regulated microRNAs and other small RNAs from Arabidopsis. Plant Cell 16:2001–2019

Sunwoo H, Dinger ME, Wilusz JE et al (2009) MEN epsilon/beta nuclear-retained non-coding RNAs are up-regulated upon muscle differentiation and are essential components of paraspeckles. Genome Res 19:347–359

Svoboda P (2007) Off-targeting and other non-specific effects of RNAi experiments in mammalian cells. Curr Opin Mol Ther 9:248–257

Swayze EE, Bhat B (2006) The medicinal chemistry of oligonucleotides. In: Crooke ST (ed) Antisense drug technology—principles, strategies, and applications. CRC, Boca Raton, FL, pp 143–182

Taft RJ, Pang KC, Mercer TR et al (2010) Non-coding RNAs: regulators of disease. J Pathol 220:126–139

Tam OH, Aravin AA, Stein P et al (2008) Pseudogene-derived small interfering RNAs regulate gene expression in mouse oocytes. Nature 453:534–538

Tan Y, Zhang B, Wu T et al (2009) Transcriptional inhibition of Hoxd4 expression by miRNA-10a in human breast cancer cells. BMC Mol Biol 10:12

Torres AG, Fabani MM, Vigorito E et al (2011) MicroRNA fate upon targeting with anti-miRNA oligonucleotides as revealed by an improved Northern-blot-based method for miRNA detection. RNA 17:933–943

Tripathi V, Ellis JD, Shen Z et al (2010) The nuclear-retained noncoding RNA MALAT1 regulates alternative splicing by modulating SR splicing factor phosphorylation. Mol Cell 39:925–938

Tycowski KT, Shu MD, Steitz JA (1994) Requirement for intron-encoded U22 small nucleolar RNA in 18S ribosomal RNA maturation. Science 266:1558–1561

Tycowski KT, Shu MD, Steitz JA (1996) A mammalian gene with introns instead of exons generating stable RNA products. Nature 379:464–466

Uliel S, Liang XH, Unger R et al (2004) Small nucleolar RNAs that guide modification in trypanosomatids: repertoire, targets, genome organisation, and unique functions. Int J Parasitol 34:445–454

Ullu E, Tschudi C (1990) Permeable trypanosome cells as a model system for transcription and trans-splicing. Nucleic Acids Res 18:3319–3326

Val R, Wyszko E, Valentin C et al (2011) Organelle trafficking of chimeric ribozymes and genetic manipulation of mitochondria. Nucleic Acids Res 39:9262–9274

Vickers TA, Koo S, Bennett CF et al (2003) Efficient reduction of target RNAs by small interfering RNA and RNase H-dependent antisense agents. A comparative analysis. J Biol Chem 278:7108–7118

Vitali P, Basyuk E, Le Meur E et al (2005) ADAR2-mediated editing of RNA substrates in the nucleolus is inhibited by C/D small nucleolar RNAs. J Cell Biol 169:745–753

Watanabe T, Totoki Y, Toyoda A et al (2008) Endogenous siRNAs from naturally formed dsRNAs regulate transcripts in mouse oocytes. Nature 453:539–543

Will CL, Luhrmann R (2001) Spliceosomal UsnRNP biogenesis, structure and function. Curr Opin Cell Biol 13:290–301

Williams AE (2008) Functional aspects of animal microRNAs. Cell Mol Life Sci 65:545–562

Wilusz JE, Freier SM, Spector DL (2008) 3′ end processing of a long nuclear-retained noncoding RNA yields a tRNA-like cytoplasmic RNA. Cell 135:919–932

Wilusz JE, Sunwoo H, Spector DL (2009) Long noncoding RNAs: functional surprises from the RNA world. Genes Dev 23:1494–1504

Xu T, Xu Y, Liao CP et al (2010) Reprogramming murine telomerase rapidly inhibits the growth of mouse cancer cells in vitro and in vivo. Mol Cancer Ther 9:438–449

Yakovchuk P, Goodrich JA, Kugel JF (2011) B2 RNA represses TFIIH phosphorylation of RNA polymerase II. Transcription 2:45–49

Yuan Y, Hwang ES, Altman S (1992) Targeted cleavage of mRNA by human RNase P. Proc Natl Acad Sci USA 89:8006–8010

Zamore PD, Tuschl T, Sharp PA et al (2000) RNAi: double-stranded RNA directs the ATP-dependent cleavage of mRNA at 21 to 23 nucleotide intervals. Cell 101:25–33

Zeng Y, Cullen BR (2002) RNA interference in human cells is restricted to the cytoplasm. RNA 8:855–860
Zhang X, Lian Z, Padden C, Gerstein MB et al (2009) A myelopoiesis-associated regulatory intergenic noncoding RNA transcript within the human HOXA cluster. Blood 113:2526–2534
Zhao J, Sun BK, Erwin JA et al (2008) Polycomb proteins targeted by a short repeat RNA to the mouse X chromosome. Science 322:750–756

Antisense Oligonucleotides in the Treatment of Malignant Gliomas

Gerardo Caruso, Mariella Caffo, Giuseppe Raudino, Federica Raudino, Mario Venza, and Francesco Tomasello

Contents

G. Caruso (✉) • M. Caffo • G. Raudino • F. Raudino • M. Venza • F. Tomasello
Neurosurgical Clinic, Department of Neurosciences, Psychiatry and Anesthesiology, A.O.U. Policlinico "G. Martino", University of Messina School of Medicine, Via Consolare Valeria, 1, 98125 Messina, Italy
e-mail: gcaruso@unime.it

V.A. Erdmann and J. Barciszewski (eds.), *From Nucleic Acids Sequences to Molecular Medicine*, RNA Technologies, DOI 10.1007/978-3-642-27426-8_9,

Abstract Brain tumor therapy, despite the recent progress in neuro-oncology, in neurosurgical and neuroradiological techniques, represents a hard challenge for molecular biomedicine. Gene expression in neoplastic cells is a complex multistep process. During this event, single or multiple gene mutation can developed, resulting in clonal neoplastic cells selection. Studies on epigenetic regulation of gene expression during normal cell, tissue, and organism differentiation and evolution have, in the last decade, induced to the possibility to modulate therapeutically genome expression in cancer cells to revert or modify these toward normal cell phenotype. This kind of approach points the attention to the development of molecular therapeutic strategies based on the interaction with specific nucleotide sequences, at level of DNA or RNA, by specular nucleotide-based short molecules capable of hybridizing specific genomic DNA sequences or mRNA transcripted and translated in proteins.

Antisense oligonucleotides (AONs) are synthetic stretches of DNA which hybridize with specific mRNA strands. The specificity of hybridization makes antisense method an interesting strategy to selectively modulate the expression of genes involved in tumorigenesis. The efficacy of classical anticancer strategies in high-grade gliomas is seriously limited by the lack of specific therapies against malignant cells, and the prognosis in patients affected by high-grade gliomas is still very unfavorable. Glial tumors seem to be able to create a favorable environment for the invasion of glioma cells in cerebral parenchyma when they combine with the extracellular matrix via cell surface receptors. In this chapter, we will focus on the mechanisms of action of AONs and other short nucleotide-based molecules, such as siRNA (small interference RNA), miRNA (microRNA), and long noncoding RNA, and their impact in cerebral gliomas treatment.

Keywords Angiogenesis • Antisense oligonucleotides • Cerebral gliomas • Gene therapy • Ribozyme • RNA interference

Abbreviations

AONs	Antisense oligonucleotides
BBB	Blood–brain barrier
BM	Basement membrane
CNS	Central nervous system
ECM	Extracellular matrix
EGFR	Epidermal growth factor receptor
FAK	Focal adhesion kinase
FGFR	Fibroblast growth factor receptor
GBM	Glioblastoma multiforme
HGF	Hepatocyte growth factor
HIF	Hypoxia-inducible factor
hTR	Human telomerase

IGF	Insulin-like growth factor
miRNA	MicroRNA
MMP	Matrix metalloproteinase
NP	Nanoparticle
RNAi	RNA interference
shRNA	Short hairpin RNA
siRNA	Short interfering RNA
TGF	Transforming growth factor
VEGF	Vascular endothelial growth factor

1 Introduction

The treatment of gliomas represents one of the most challenging areas in neurosurgery and oncology. Malignant gliomas involve multiple aberrant signaling pathways, and the blood–brain barrier restricts the delivery of many chemotherapeutic agents. The effectiveness of the actual chemotherapeutic approach and multimodal targeted therapies remains very modest in malignant gliomas. Considering the multitude of molecular entities and signaling pathways regulating the proliferation and cellular survival/cell death, the inhibition of a singular target gene or transcriptional factor could not be sufficient to suppress neoplastic progression. In this intricate and complex field, it seems to be very important to improve specific selective drug delivery systems to lead to diffusion of drugs, antisense oligonucleotides (AONs), small interference RNAs, engineered monoclonal antibodies, and other therapeutic molecules into CNS overcoming the blood–brain barrier.

Completing the sequencing of the human genome further enhances our ability to identify those genes that are important for development and maintenance of the malignant phenotype. These studies have created a wealth of potential molecular targets for drug discovery efforts. Unfortunately, many of the genes of interest are not approachable by conventional means. The use of AONs as therapeutic agents in oncology has been proposed for longer than 10 years. AONs are short DNA or RNA fragments with a complementary nucleotide sequence to DNA sequence sense or mRNA (Jansen and Zangemeister-Wittke 2002; Tamm 2006; Wacheck and Zangemeister-Wittke 2006). A DNA single strand or positive sense is normally called sequence sense, when an RNA version of the same sequence exists and is then translated into protein. Antisense molecules interact with complementary strands of nucleic acids, modifying expression of genes. The AONs approach to high-grade gliomas could represent an interesting and promising strategy.

Malignant gliomas, including GBM and anaplastic astrocytomas (AA), are the most common primary brain tumors in adults up to an extent of 78% of all primary malignant brain tumors (Kleihues et al. 2002). Mortality is still close to 100%, and the average survival of patients with GBM is less than 1 year when classical treatment is used. Despite progresses in radiotherapy and in chemotherapy

(Cheng et al. 2005b; Stupp et al. 2006; Wen et al. 2006) as well as in experimental immunotherapy (Kjaergaard et al. 2005), an effective curative treatment of gliomas does not yet exist. Indeed, progress in treatment of this disease has led only to a slight increase in average survival up to 15–18 months (Sanson et al. 2006; Stegh et al. 2007).

In this chapter, we will first introduce a summary about glioma biology and then review of DNA- and RNA-based antisense therapies. A brief summary about the most important preclinical data and AONs clinical application in cerebral gliomas is also reported.

2 Gliomas Biology

2.1 Genetic Alterations

Features of the glioma cells are the loss of tumor-suppressor genes, which are critical for the cell growth, differentiation, and function (Table 1). In last year different mutated genes regarding malignant gliomas have been identified. Classically, GBM has been described as result endpoint from two main gliomagenesis pathways (Lee et al. 2008; Liang et al. 2005). Primary GBM shows amplification of the epidermal growth factor receptor (EGFR), deletion or mutation of homozygous cyclin-dependent kinase (CDK) inhibitor $p16^{INK4A}$/(CDKN2A), alterations in tumor-suppressor PTEN on chromosome 10, and deletion in the INK4a gene with loss of p14 and p16 (Ihanahan and Weinberg 2010; Holland 2001). Progression from low-grade to high-grade astrocytomas involves inactivating mutations of tumor-suppressor gene TP53 and elevated expression of platelet-derived growth factor (PDGF) ligands and receptors, accumulation of genetic alteration of retinoblastoma-associated cell cycle regulatory pathways, including deletion or mutations of cyclin-dependent kinase inhibitor $p16^{INK4A}$/(CDKN2A) or the retinoblastoma susceptibility locus 1 (pRB1), as well as amplification or overexpression of cyclin-dependent kinase 4 (CDK4) and human double minute 2 (HDM2). Evolution to secondary GBM is associated with deletion of chromosome 10, which includes tumor-suppressor phosphatase and tensin homologue (PTEN). Additionally, BcI2-like 12 (BcI2L12) increased expression inhibits apoptosis (Stegh et al. 2007). The astrocyte-elevated gene-1 (AEG-1) is overexpressed in the majority of gliomas and stimulates cellular transformation and invasion together with the Ha-ras family of retrovirus-associated DNA sequences (RAS) (Ohgaki and Kleihues 2007). Furthermore, oncogenic Ha-ras induces AEG-1 expression by modulating the phosphatidylinositol 3-kinase (PI3K)-Akt signaling pathway and contributes to the growth of gliomas. Mitogenetic signals activate a molecular cascade known as ras–mitogen-activated protein kinase (Ras/MAPK). MAPK inhibits the Rb gene and activates the transcriptional factor E2F, and cells enter the S phase. The INK4a

Table 1 Principal genes involved in gliomagenesis

Molecular target	Function	Mutation	Effect of mutation
Bcl-2	Inhibitor of apoptosis Bcl-2 protein prevents the release of cytochrome C from the mitochondria	Overexpression	Higher inhibition of apoptosis; increase of resistance to chemo- and radiotherapies in cancer cells
PKC-α	Member of family of cytoplasmic serine–threonine protein kinases, involved in growth-factor-mediated signal transduction pathways	Overexpression	Upregulation of cancer cell, proliferation, and tumor growth
TGF-β2 (glioblastoma-derived T-cell suppressor factor, G-TSF, BSC-1 cell growth inhibitor, polyergin)	Extracellular glycosylated protein that performs many cellular functions and has a vital role during embryonic development. It suppresses the effects of interleukin-dependent T-cell tumors	Overexpression	Downregulation of apoptosis and tumor growth. During tumorigenesis, TGF-β increases extracellular matrix invasion and induces escape of tumor from immunosurveillance
Telomerase (h-TR)	Reverse transcriptase that carries its own RNA molecule, which is used as a template when it elongates telomeres	Overexpression	Cancer cell immortalization and tumor growth
Focal adhesion kinase (FAK) and insulin-like growth factor I receptor (IGF-IR)	FAK is a protein tyrosine kinase which is recruited at an early stage to focal adhesion, leading the cytoskeleton to interact with proteins of the extracellular matrix and initiate events such as cell migration and anchorage-dependent differentiation IGF-1R is a transmembrane receptor that is activated by IGF-1 and by the related growth factor IGF-2. This receptor mediates the effects of IGF-1 (growth and anabolic effects in adults)	Overexpression, gene amplification	Cancer cell proliferation, downregulation of apoptosis, and tumor growth

(continued)

Table 1 (continued)

Molecular target	Function	Mutation	Effect of mutation
PTEN (MMAC 1)	Tumor-suppressor gene located on human chromosome 10q23.3, regulating cell growth, apoptosis, and interaction with extracellular matrix and inhibiting cell migration	Nonsense mutation, LOH	Cancer cell proliferation, downregulation of apoptosis, tumor growth, invasion
PI3K/AKT/PKB (pathway of bFGF, PDGF, IGF, VEGF, EGFR)	Crucial pathway, involved in cell survival. AKT has an antiapoptotic effect by inactivating the proapoptotic BAD, procaspase-9, and forkhead transcription factors (activation of Fas ligand). AKT also induces NF-kB-dependent pro-survival gene expression (BCL-xL, C-Myb), modifies cell cycle progression by inhibiting p27 and p21, upregulates transcription of cyclin D, and promotes nuclear entry of mdm2	Overexpression, gene amplification	Upregulation of cancer cell proliferation, loss of cell cycle control, increased telomerase activity, cell invasiveness and angiogenesis by stimulating secretion of MMPs and by activating endothelial nitric oxide synthase (eNOS)

gene influences the Rb pathway by activating three cyclin kinase inhibitors: p15, p16, and p19.

The final result is the blockage of cyclin-dependent kinase 2, 4, and 6, triggering cell cycle progression by inhibiting pRb. The discovery of $CD133^+$ glioma stem cells inside the tumor gross could represent an important key point in gliomagenesis knowledge and in GBM selective and targeted strategies directed mainly against this particular stem cell cluster (Yan et al. 2011).

2.2 *Glioma Progression and Extracellular Matrix*

The cellular and molecular events that initiate and promote malignant glioma development are not completely understood. The treatment modalities designed to promote its demise are all ultimately ineffective, leading to disease progression. Vasculogenesis and angiogenesis potentially play distinct roles in the etiology of primary and recurrent malignant gliomas, suggesting that patient therapy should perhaps be tailored specifically against the predominant vasculature pathway at a given specific stage of gliomagenesis. Cerebral gliomas show a unique pattern of invasion and, with rare exceptions, do not metastasize outside of the brain. How invasive glioma cells survive in the setting of invasion, evading immune detection, thwarting cytotoxic therapies, and deferring commitment to proliferation remains unknown. Invading glioma cells normally migrate to distinct anatomical structures. These structures include the basement membrane (BM) of blood vessels, the subependymal space, the glial limitans externa, and parallel and intersecting nerve fiber tracts in the white matter. Glioma invasion is well depicted by three passages such as adhesion of glioma cells to proteins of the surrounding ECM mediated by cell adhesion receptors, proteases secretion by glioma cells which locally degrade the ECM components, and migration of glioma cells into the newly created space through the ECM. The proteolytic degradation of the BM and of ECM is mediated by proteases, such as the MMPs, secreted by tumor and stromal cells (Uhm et al. 1997; Idbaih et al. 2008). MMPs play an important role in human brain tumor invasion, probably due to an imbalance between the production of MMPs and tissue inhibitor of metalloproteinases-1 (TIMP-1) by the tumor cells. MMP-1 is the crucial enzyme able to initiate breakdown of the interstitial collagen types; in this way, it activates the other MMPs, which allows the glioma cells to infiltrate normal brain tissue.

ECM is a complex network of different collagens, proteoglycans, hyaluronic acid, laminin, fibronectin, and many other glycoproteins, including proteolytic enzymes involved in degradation and remodeling of the ECM (Chintala et al. 1996). Extracellular matrix exists in two forms: interstitial matrix that fills in the intercellular space and the more specialized BM, which is a thin sheet of extracellular matrix underlying the epithelium. ECM provides the microenvironment for the cells and serves as a tissue scaffold, guiding cell migration during embryonic development and wound repair. Beyond that, it also functions as the repository and modulator of growth factors and cytokines and therefore is responsible for transmitting environmental signals to the cells. Critical factors, in glioma invasion, include the synthesis and deposition of ECM components by glioma and mesenchymal cells, the release of ECM-degrading activities for remodeling interstitial spaces, and the presence of adhesion molecules. ECM modification aids the loss of contact inhibition, allowing tumor cells to freely migrate and invade the surrounding tissues. Various ECM components, such us hyaluronan, vitronectin, tenascin-C, and osteopontin, are regulated within the tumor stroma and at the advancing edge of the tumor within brain parenchyma. These proteins may be considered crucial targets of molecular nanoplatform-based therapies able to downregulate invasion pathways during brain tumor infiltration. Changes in these ECM components are felt to modulate brain tumor growth, proliferation, and invasion, although specific interactions and exact mechanisms are unknown. Cell adhesion is the binding of the cells to each other and to the ECM through cell adhesion molecules such as integrins, selectins, cadherins, the Ig (immunoglobulin) superfamily, and lymphocyte homing receptors. Cell adhesion mediates cell attachment, migration, and signaling to and from the extracellular matrix. Adhesion complexes include focal adhesions, adherens junctions, tight junctions, desmosomes, hemidesmosomes, and gap junctions. A focal adhesion is a large dynamic cluster of integrins and other adhesion molecules that connect the ECM to the cytoskeleton. It transmits both mechanical force and regulatory signals. Adherens junctions are cell-to-cell junctions between epithelial cells and are composed of cadherins and alpha-, beta-, and delta-catenins. Tight junctions are the connections between two adjacent cells whose membranes are joined together. Desmosomes are spot-like adhesions between epithelial cells and make epithelium to resist shearing forces. The cell adhesion proteins of the desmosome are members of the cadherin family. Hemidesmosomes connect the basal part of the cells to the BM. Gap junctions directly connect the cytoplasm of two cells, allowing small molecules to pass between the two adjacent cells. The best characterized adhesion molecules are the integrins, and the best characterized adhesion complexes are focal adhesions. The extracellular ligands that anchor these adhesions include laminin, fibronectin, vitronectin, and various collagens. Focal adhesions can be considered both as sensors of force and as sites that originate cytoskeletal forces through anchored actin-microfilament bundles. Focal adhesions have many SH2-containing components (Src kinases, PI3K, SHP-2) as well as many tyrosine-phosphorylated molecules (focal adhesion kinase paxillin, tensin, CAS, caveolin). Therefore, tyrosine phosphorylation of these sites, which is induced by integrin clustering, growth factor stimulation, or applied force, could potentially stimulate focal adhesion assembly. Inhibitors of tyrosine phosphorylation block adhesion-complex formation and recruitment of a large subset of focal adhesion components. Integrins are heterodimers of α- and β-subunits that bind to the ECM through large extracellular domains and connect intracellularly to the actin cytoskeleton filaments. Integrins regulate many aspects of the cell behavior including survival, proliferation, migration, and differentiation. Integrins of the $\beta 1$- and αv-classes are expressed on different cell types, including neurons, glial cells, meningeal cells, and endothelial cells. $\beta 2$-Integrins are specifically expressed by leukocytes, and they are found on microglia and on infiltrating leukocytes within the CNS; $\alpha v \beta 3$-integrins are associated with elevated levels of their ligands and with vitronectin in human gliomas (Gladson 1996). Antibody block against the $\beta 1$-subunit inhibits adhesion, motility, and invasion of glioma cells in vitro (Rooprai et al. 1999). It has also been demonstrated that an antisense strategy in a C6 glioma model has produced a decrease of the $\beta 1$ subunit (Bello et al. 2001). Downregulated $\beta 1$-integrin protein levels in vivo probably affect interactions of glioma cells with ECM components, leading to reduced migration along vascular basement membranes. These data can be interpreted as contributing to the locally invasive behavior of astrocytic tumors favoring the regulation of proteases activation.

2.3 *Angiogenesis*

Angiogenesis is the formation of new capillaries from already existing vessels. During angiogenesis, the capillary plexus is remodeled by sprouting, microvascular growth, and fusion into a mature and functional vascular bed (Yancopoulos et al.

2000). Angiogenesis also includes vessel penetration into avascular regions of the tissue and is critically dependent on the correct interactions among endothelial cells, pericytes, and surrounding cells and their association with the ECM and the vascular BM. Angiogenesis is primarily driven by tissue hypoxia and occurs under physiological conditions, such as during wound healing, endometrial growth in a menstrual cycle, or following tissue grafting and ischemia. Under pathological conditions, angiogenesis can be induced by cancer growth; this is referred to as tumor angiogenesis. Cerebral gliomas are characterized by extensive microvascular proliferation and a higher degree of vasculature. The formation of new blood vessels from existing microvessels is a histological indicator of the degree of malignancy and prognosis of patients. The presence of endothelial glomeruloid-like proliferation and of positive immunoreaction at level of BM of tumor vascular channel is predictive of active tumor invasiveness (Caffo et al. 2004). Glioma vasculature is structurally and functionally abnormal, and it correlates and leads to vasogenic edema, increased interstitial pressure, and heterogeneous delivery of oxygen and drugs (Jain et al. 2007). Endothelial cells are guided into avascular areas via macromolecules such as VEGF-A, a potent proangiogenic factor and endothelial cell mitogen (Lee et al. 2005; Ito et al. 2007). VEGF-A activation causes endothelial cell differentiation into a "tip" cell, and a VEGF-A gradient induces "stalk" cell proliferation along an opening in the BM in forming of a new vessel sprout (Gerhardt et al. 2003). VEGF also induces expression of the delta-like ligand, DLL-4, in tip cells that binds to its receptors, Notch 1 and Notch 4, on adjacent "stalk" endothelial cells. DLL-4-Notch signaling functions as a dampening mechanism in preventing excess angiogenesis and promoting orderly development of new vessels. Membrane type 1 matrix metalloproteinase (MT1-MMP) expressed by the tip cell of an endothelial sprout opens up the surrounding matrix and is only later downregulated when stalk cells come into contact with pericytes. Reestablished contact between endothelial cells and pericytes induces expression of tissue inhibitor of metalloproteinases-2 (TIMP-2) in endothelial cells and TIMP-3 in pericytes in switching off the proteolytic phenotype in endothelial cells (Saunders et al. 2008). The VEGF-A is regulated in cerebral gliomas, and it is secreted by tumor cells as well as by stromal and inflammatory cells. VEGF A can be linked in the ECM through the interaction with proteoglycans or glycosaminoglycans. The ligands for VEGF3 (VEGF-C and VEGF-D) are expressed by multiple cell types that surround the angiogenic vessels, suggesting the existence of a novel proangiogenic paracrine signaling pathways in these neoplasms. VEGF expression is stimulated by hypoxia and acidosis, and probably correlates with many other growth factors and their specific receptors (EGFR, HGFR, PDGFR, C-Kit, IGFR) and their downstream signaling pathway (PI3K-Akt, Ras–MAPK) upregulation and activation in gliomas. MT1-MMPs on the endothelial cell surface are also required for the subsequent step in the angiogenesis cascade of tube formation by playing a role in endothelial intracellular vacuole and lumen formation. The vessel lumen is tightly sealed by adjoining endothelial cells held together by tight and adherens junctions. A basement membrane is then produced by endothelial cells in cooperation with surrounding cells in providing structural support and maintaining endothelial cell quiescence. The BM is built up of scaffolding laminins and essential components such as collagen IV, perlecan, nidogens, and collagen XVIII (Davis and Senger 2005; Miner 2006). During the maturation of new capillaries, pruning of excess or unneeded vessels for optimal perfusion is proposed to represent an adjustment to oxygen surplus while the surrounding ECM exerts mechanical strain providing traction and orientation for angiogenic microvessels (Krishnan et al. 2008). Part of the final steps of angiogenesis is the recruitment of pericytes as their association with endothelial and vascular smooth muscle cells is essential for the maturation of endothelial tubes into blood vessels. Pericytes and vascular smooth muscle cells expressing angiopoietin-1 (Ang-1) on their surface are bound tightly to their corresponding receptor Tie-2 expressed on the surface of endothelial cells. This tight interaction between pericytes and endothelial cells mediates blood vessel stabilization but can be antagonized by Ang-2 expression (Mao and Schwarzbauer 2005). Maintenance of an optimal number of pericytes is necessary for successful angiogenesis in stabilizing newly formed vessels from further sprouting and to support adequate blood flow. In brain tumor growth, it is possible to observe two vascular phases. In the first, the vessels are native cerebral vessels, which are co-opted by tumor cells, while in the second phase, there is true neovascularization arising from existing vessels. During the transition period between these two phases, hypoxia-driven HIF-1 expression occurs, which results in VEGF secretion and in the induction of neovascularization. In stage IV, angiogenesis adjacent to the necrotic area is triggered in response to increased expression of HIF-1α and VEGF.

Basic fibroblast growth factor (bFGF) is expressed by vascular cells and, focally, by the tumor cells. The receptors for bFGF include FGFR1, expressed by the tumor cells and the tumor endothelial cells, and FGFR2, expressed only by the tumor cells. The binding of VEGF on endothelial cells activates the phosphatidylinositol-3-hydroxyl kinase (PI3K)/protein kinase B (Akt) pathway, whereas the bFGF receptors are predominantly shown through the protein kinase C-α (PKC-α) pathway. The activation of endothelial cells results in increased expression of cell adhesion receptors, such as αvβ3- and α5β1-integrins, and in increased cell survival, proliferation, and migration responses (Idbaih et al. 2008). Another group of endothelial growth factors is represented by angiopoietins. Their signal transduction pathway passes via the Tie-2 receptor tyrosine kinase expressed on endothelial cells. Ang-1-mediated activation of Tie-2 is required for stabilization, remodeling, and maturation of blood vessels, promotes angiogenesis and tumor growth, and is associated with an increased number of highly branched vessels covered by pericytes. While VEGF and Ang-1 may act in concert (proliferation and maturation), Ang-2 has been implicated in further remodeling of the initial microvasculature (Holash et al. 1999). Increased expression of Ang-2 on GBM microvasculature appears early during glioma angiogenesis. Ang-2 and Tie-2 expressions are absent in the normal brain vasculature but are induced in tumor endothelium of co-opted tumor vessels prior to their regression. The discovery of hypoxia-inducible factor-1 (HIF-1) and the observation that hypoxia-induced HIF-1α expression in pseudopalisading cells, in tumoral necrotic areas, was concomitant with the expression of one of its target genes, VEGF, established a biological link between hypoxia and angiogenesis (Semenza 2003). The process of blood vessel formation involves an initial VEGF-mediated increase of vascular permeability leading to extravasation of plasma proteins with dilatation of native vessels and reduction in their pericyte coverage. For this cascade, deposition of proangiogenic matrix for the newly sprouting vessel is essential. This involves breakdown of the vascular basement membrane and ECM through the action of cathepsin B, MMPs, and other enzymes (Ljubimova et al. 2006). The matrix metalloproteinases and a disintegrin and metalloproteinase with thrombospondin motifs (ADAMTS) family are often associated with ECM degradation and remodeling. The inhibitors of MMPs are called tissue inhibitors of metalloproteinases (TIMPs), which are comprised of TIMP-1, TIMP-2, TIMP-3, and TIMP-4. The interactions between these proteases and their inhibitors play important roles in cell morphogenesis, angiogenesis, tissue remodeling, tissue repair, tumor metastasis, cirrhosis, and arthritis. The features of the ECM are determined both by the cells that produce the matrix and by the cells growing in it.

3 The Antisense Strategy

The antisense strategy was proposed in 1970s by Zamecnik and Stephenson (1978). They have utilized short AONs to block the expression of targeted genes within intact cells and used a tridecamer oligonucleotide as a hybridization competitor to inhibit Rous sarcoma virus replication. This work provided the first evidence that oligonucleotides, although highly negatively charged, could be transported into cells (Zamecnik and Stephenson 1978).

AONs are short DNA or RNA fragments, 13–25 nucleotides (nt) long, with a complementary nucleotidic sequence to sense DNA or mRNA. A DNA single strand or positive sense is denominated "sequence sense" when an RNA version of the same sequence exists, and it is then translated into protein. Antisense molecules interact with complementary strands of nucleic acids modifying expression of genes. At cellular level, a strand of the DNA serves as a template for the synthesis of a complementary strand of RNA. Because the DNA is double-stranded, the strand complementary to the antisense sequence is called non-transcribed strand and has the same sense sequence as the mRNA transcript. These antisense molecules may be introduced into a cell to silence genic functions through inhibiting translation of a complementary mRNA. This approach has been applied in various kinds of tumors and showed a very high specificity and efficacy against cancer cells (Pawlak et al. 2000). Many AONs have shown sharp in vitro reduction in target gene expression and promising activity against a variety of human malignancies. Several experimental in vitro and in vivo studies in cell line cultures and animal models showed inhibition of genes involved in cell proliferation, apoptosis, and angiogenesis. Since AONs inhibit gene expression in a sequence-specific manner, it is possible to selectively alter the expression of genes with closely related sequences. The antisense strategy allows the detailed analysis of signal transduction pathways, which often comprise families of highly homologous proteins. Furthermore, it may lead to the identification of new therapeutic targets and provide the corresponding drug at the same time. Since most tumor cells have a different pattern of gene expression compared with normal cells, AONs can theoretically be used to specifically target tumor-associated genes or mutated genes without altering gene expression of normal cells. In addition, the selective inhibition of a single member of a protein family that plays a role in cancer progression allows other members of the family to perform their normal cellular function. Also, most of the proteins involved in the pathogenesis of cancer operate inside the cell and thus are not accessible to protein-based drugs. Targeting those genes by AONs might be more effective in vivo. The goal of antisense molecule-based approach is to selectively suppress the expression of a protein by exploiting the genetic sequence in which it is encoded acting at translational level (Tamm 2006; Wacheck and Zangemeister-Wittke 2006). RNA- and DNA-based oligonucleotides are the most prevalent and most practical antisense drugs. Interactions of RNA-based AONs with target mRNA inhibit gene expression by interfering with protein translation without necessarily altering mRNA stability. AONs inhibit mRNA function by several mechanisms including modulation of splicing and inhibiting protein translation by disrupting ribosome assembly. Double-stranded DNA/RNA in mammalian cells activates RNase H-mediated degradation of the target mRNA. RNase H recognizes the mRNA–DNA duplex and cleaves the mRNA strand leaving the AONs intact (Jason et al. 2004; Passioura and Symonds 2004). However, the most important mechanism appears to be the utilization of endogenous RNase H enzymes (Wacheck and Zangemeister-Wittke 2006). In some cases, binding of antisense sequences to the 5′ end of mRNAs was found to inhibit protein synthesis directly, presumably by interfering with the initiation of translation.

Two different and adjunctive antisense-based approaches to modulate gene expression into cancer cells are RNA interference (RNAi) and ribozymes. RNAi can be intracellular synthesized or exogenously administered and cleaved by the enzyme DICER into 21–25 nucleotides. siRNAs become associated with the RNA-induced silencing complex (RISC), which uses the antisense strand of the siRNA to bind to and cleave the target mRNA. The siRNAs can also be used as primers for the generation of new dsRNA by RNA-dependent RNA polymerase (RdRp). This newly formed dsRNA can then also serve as a target for the Dicer enzyme. RNAi has emerged as an efficient and selective technique to provide stable inhibition of gene expression (Jackson et al. 2003). This strategy is normally characterized by an endogenous gene silencing mechanism physiologically used by eukaryotes to regulate gene expression. RNAi plays a fundamental role in diverse eukaryotic functions including viral defense, chromatin remodeling, genome rearrangement, developmental timing, brain morphogenesis, and stem cell maintenance. In mammalian cells, RNAi can exist under two distinct forms: short interfering RNAs (siRNAs) and short hairpin RNAs (shRNAs). These last molecules then are processed into siRNAs by the multidomain enzyme DICER in an ATP-dependent process (Vidal et al. 2005). This enzyme is an endoribonuclease that cleaves double-stranded RNA and pre-microRNA (miRNA) into short double-stranded RNA fragments called small interfering RNA. DICER then helps load its small RNA products into large multiprotein complexes denominated RISCs. RISC and RISC-like complexes use the small RNAs as guides for the sequence-specific silencing of cognate genes through mRNA degradation, translational inhibition, and heterochromatin formation (MacRae et al. 2006). dsRNAs activate endogenous dsRNA-dependent protein kinase (PKR), leading to phosphorylation of the translation initiation factor EIF2α and finally to a suppression of protein synthesis and to an upregulation of apoptosis. This is not always a side effect. The objectives of the use of RNAi in cancer therapeutics include finding good targets, delivery, and minimizing toxicity. Recent studies include adenoviral or retroviral delivery of shRNAs targeting Hec1 that promoted efficacy on adenocarcinoma-induced tumor growth (Hu-Lieskovan et al. 2005), aptamer-siRNA chimeric RNAs that resulted in tumor regression by specific delivery to prostate cancer cells by aptamer binding to surface-expressed tumor cell marker (PSMA) (McNamara et al. 2006), atelocollagen-complexed siRNA (Takei et al. 2004) that effectively silenced vascular endothelial growth factor and limited angiogenesis and tumor growth in vivo, and transferrin receptor-targeted cyclodextrin nanoparticles that delivered anti-Ews–Fli1 siRNAs to human Ewing's tumors, effectively blocking metastasis (Gurzov and Izquierdo 2006). Interestingly, RNAi may also be exploited to silence pathways that facilitate the effects of traditional cancer drugs. This includes targeting of the multidrug resistance gene (MDR1) for resensitization to chemotherapy (Nieth et al. 2003) and silencing of double-strand break repair enzymes for enhanced effects of radio- and chemotherapy (Collis et al. 2003). The human EGFR plays an oncogenic role in solid cancer, including cerebral gliomas. Transvascular nonviral gene therapy in combination with EGFR-RNA interference represents a new therapeutic approach to silencing oncogenic genes in solid cancers. This is achieved with pegylated immunoliposomes (PILs) carrying short hairpin RNA expression plasmids driven by the U6 RNA polymerase promoter and directed to target EGFR expression by RNAi. The tissue target specificity of PILs is given by conjugation of 1% of the PEG residues to monoclonal antibodies (mAbs) that bind to specific endogenous receptors located in the brain vascular endothelium, which forms the blood–brain barrier, and brain cellular membranes. Treatment of an experimental human brain tumor model in SCID mice is possible with weekly intravenous RNAi gene therapy causing reduced tumor expression of EGFR and 88% increase in survival time of these mice with advanced intracranial brain cancer (Boado 2005).

The limit in therapeutic use of siRNA to modify strongly gene expression is represented from the transient effect of siRNA. These antisense molecules, for the lack of the RdRp, should require the continuous transfection into target cells by plasmid vectors. A peculiar distinctive characteristic between RNA interference and the other antisense approaches (AONs and ribozymes) is the extreme selectivity for siRNA. In this way, siRNA may inhibit selectively the expression of oncogene containing a single point mutation, without suppressing the expression of the wild-type transcript (Jason et al. 2004; Passioura and Symonds 2004).

Ribozymes are RNA molecules that act as enzymes, molecular scissors that cut RNA, and the molecular messages given by genes in order to produce proteins. Some ribozymes have a self-cleavage catalytic action while other ones are true catalysts and can carry out RNA slicing by transesterification (splicesome) and peptidyl transfer (in ribosomes) (Doudna and Lorsch 2005). This was seen as a major discovery since until then proteins were thought to be the only entity capable of behaving as enzymes. Ribozymes are now utilized to study gene function in HIV disease and in cancer research.

The use of triple-helix forming oligonucleotides or siRNA molecules is promising alternatives to AONs. The discovery of some RNA molecules represent an important input in biomedical applications. These molecules, transcripted from DNA sequences different, should play a crucial role in the epigenetic mechanisms of gene expression and in cell function. In particular, the discovery of microRNA and their role in the complex control of gene expression may offer new interesting and efficacious strategies to change or modulate gene expression profiles (Wurdinger and Tannous 2009).

3.1 Antisense Oligonucleotides

The first generation of antisense agents contains backbone modifications such as replacement of the oxygen atoms of the phosphate linkage by sulfur (phosphorothioates), methyl group (methylphosphonates), or amines (phosphoroamidates). The phosphorothioates have been most widely used for gene silencing because of their sufficient resistance to nucleases and ability to induce RNase H functions (Baker and Monia 1999; Dias and Stein 2002; Dean and Bennett 2003). On the contrary, binding to the target sequence is less satisfactory. Most AONs are currently in clinical testing and are first-generation antisense.

The second generation of antisense modifications was aimed at improving these properties; substitutions of position 2′ of ribose with an alkoxyl group were most successful. 2′-*O*-Methyl and 2′MOE (2′-*O*-methoxyethyl) derivates can be further combined with a phosphorothioate linkage. 2′MOE AONs show an improved resistance against nuclease-mediated metabolism as well as tissue half-life in vivo, which produces a longer duration of action (Baker and Monia 1999; Dias and Stein 2002; Dean and Bennett 2003).

The third generation of AONs contains structural elements, such as zwitterionic oligonucleotides, locked nucleic acids/bridged (LNAs), morpholino, PNA (peptic nucleic acid), and hexitol nucleic acids. All of the modifications enhanced AONs in terms of nuclease resistance, specific binding, and cellular uptake with agents such as PNA and morpholino. Morpholino (PMO: phosphorodiamidate morpholino oligo) oligomers are usually 25 bases in length. Structurally, the difference between the morpholino and DNA is that bases in morpholino are bound to morpholino rings instead of deoxyribose rings and linked through phosphorodiamidate groups instead of phosphates. These modifications eliminate ionization in the usual physiological pH range, so morpholinos in organisms or cells are uncharged molecules. Morpholino can interfere with progression of the ribosomal initiation complex from the 5′ cap to the start codon, with pre-mRNA processing steps by preventing splice-directing small nuclear ribonucleoprotein (snRNP) complexes (Summerton 1999).

PNA is composed of repeating *N*-(2-aminoethyl)-glycine units linked by peptide bonds. The various purine and pyrimidine bases are linked to the backbone by methylene carbonyl bonds (Vidal et al. 2005). Since the backbone of PNA contains no charged phosphate groups, the binding between PNA/DNA strands is stronger than between DNA/DNA strands due to the lack of electrostatic repulsion. Because of their high binding strength, it is not necessary to design long PNA oligomers, but it is necessary to design usually oligonucleotide probes of 20–25 bases (Nielsen 2003). LNA is a modified RNA nucleotide, characterized by a ribose modification with an extra bridge connecting the 2′ and 4′ carbons (Orum and Wengel 2001).

AONs are taken up primarily by cells via endocytosis. Because of the hydrophilic and macromolecular nature of the cell membrane, permeation of AONs across is relatively difficult. Different techniques such as scrape loading, electroporation, microinjection, or the binding to particular peptides with membrane translocation properties have been developed to overcome this permeation problem (Pardridge et al. 1996; Pooga et al. 1998). More recently, nanoparticles and AONs conjugates have shown improved cellular uptake, biodistribution, and targeted delivery, especially in cancer treatment. Inhalation and topical applications of AONs in patients have shown satisfactory profiles of uptake and distribution. Generally speaking, systemic AON treatment is well tolerated, and side effects are dose dependent. Dose-limiting toxicities include thrombocytopenia, hypotension, fever, and asthenia (Cotter 1999; Waters et al. 1999). Furthermore, elevation of the liver enzymes aspartate aminotransferase and alanine aminotransferase, as well as complement activation and a prolonged partial thromboplastin time, has been reported (Levin 1999).

4 Molecular Targets

The choice of target gene is crucial to the potential success of the therapeutic approach. Shown in Table 1 are the functions of the most studied molecules and signaling pathways that regulate the biological cellular cycle and that could represent interesting targets for molecular approach. Malignant gliomas exhibit marked and aberrant blood vessel formation, indicating angiogenic endothelial cells as a potential target for brain tumor treatment. Other target molecules of different pathways, such as cell immortalization, apoptosis, and invasion have been studied as possible novel targets of antisense strategy.

4.1 *Bcl-2*

Bcl-2 gene is a proto-oncogene located at the breakpoints of t(14;18) chromosomal translocations in low-grade B-cell non-Hodgkin's lymphomas. Bcl-2 encodes an oncogenic mitochondrial protein with transport functions, and its expression is very increased in follicular non-Hodgkin's lymphoma, leukemias, lung, breast, colorectal, gastric, prostate, and renal cancer, and neuroblastoma. Bcl-2 protein prevents the release of cytochrome C from the mitochondria (Gross 99) leading to inhibition of apoptosis. In response to DNA damage, Bcl-2 induces the release of cytochrome C from mitochondrial matrix to cytosol, where it activates caspase-9 and caspase-3 through the activation of Apaf-1. The antisense modulation of Bcl-2 expression could increase the effectiveness of conventional chemotherapeutic agent (Pirollo et al. 2003). Phase I and II studies are also being done to test G3139 in combination with docetaxel in patients with advanced breast cancer, hormone-refractory prostate cancer, and other solid tumors. By using AONs against the first six codons of the human Bcl-2 gene transfected into malignant glioma cells (Jon52 and Roc glioblastoma cell lines), a decrease in cell growth and an increase in apoptotic death have been demonstrated (Julien et al. 2000). There is a link between resistance to chemotherapy in glioblastoma and expression of antiapoptotic Bcl-2 family members, including Bcl-XL. Anti-Bcl-XL AONs (ISIS 16009, ISIS 16967) in M059K GBM cell lines have been used and demonstrated a valid correlation between reduction of Bcl-XL protein expression, induction of intrinsic apoptotic pathway, and enhancement of cytotoxic responses to paclitaxel treatment, resulting in a chemosensitizing effect of anti-Bcl-XL therapy (Guensberg et al. 2002).

4.2 *Protein Kinase C-α*

The protein kinase C-α (PKC-α), translated from the mRNA of PKC-α gene, located on chromosome 16p11.2–q12.1, is involved in growth-factor-mediated signal transduction pathways. Its overexpression has been implicated in various kinds of tumors such as breast, colon, lung, ovarian cancer, melanoma, and brain tumors (Pirollo et al. 2003). ISIS 3521, an AON directed against PKC-α, is the most experimented, and the effects of the PKC-α downregulation are evident in ovarian cancer, NSCLCs, kidney clear cell carcinoma, and low-grade lymphoma (Nemunaitis et al. 1999). The transcription factor c-Jun is one of the major targets of JNK phosphorylation. JNK activation and c-Jun phosphorylation are involved in the induction of apoptosis (Potapova et al. 2000). Inhibition of PKC-α expression by a synthetic AON inhibits proliferation of C6 glioma cells. In addition, inhibition of PKC-α expression, has also showed growth inhibition of transformed U-87 cells transfected with antisense anti-PKC-α oligonucleotide (Baltuch et al. 1995; Pollack et al. 1996). Grossman et al. studied the therapeutic efficacy and toxicity of a phosphorothioate AON (Aprinocarsen, Eli Lilly LY9000003) directed against

PKC-α in patients with recurrent malignant gliomas. In this clinical phase II study, no therapeutic benefit was seen, probably because of tumor growth or the effect of Aprinocarsen on BBB depends on mechanisms unknown (Grossman et al. 2004). Other potential therapeutic targets are human C-raf kinase and c-Ki-RAS proteins and signal transduction kinase proteins, involved in the control of proliferation, cellular migration and differentiation, and cytoskeletal rearrangements.

4.3 *TGF-β2*

The transforming growth factor-β (TGF-β) normally induces growth arrest and therefore is considered as a typical tumor-suppressor protein. In human malignant gliomas, overexpression and secretion of the TGF-β2 isoforms have been described as hallmarks of TGF-β2 pathobiology (Platten et al. 2001). Jachimczak et al. demonstrated the efficacy of specific antisense therapy in downregulation of TGF-β2 through stimulation of antitumoral immunosurveillance. In HTZ-153, HTZ-209, and HTZ-243 glioblastoma and malignant astrocytomas cell lines, TGF-β2-specific phosphorothioate AONs enhanced lymphocyte proliferation up to 2.5-fold and autologous tumor cytotoxicity up to 60% (Jachimczak et al. 1996). TGF-β2 binds to TGF-β receptors (TBR) and gives the start to a signaling cascade via cytoplasmic signaling mediators (SMAD) into the nucleus, inducing regulation of target gene expression (Nickl-Jockschat et al. 2007). During tumorigenesis, TGF-β increases extracellular matrix invasion and induces escape of tumor from immunosurveillance. The antisense compound AP 12009 (AON anti-TGF-β 2 mRNA) enhanced the immune-cell-mediated cytotoxic antitumor response (Schlingensiepen et al. 2006; Hau et al. 2007). Schneider et al. have examined a "double-punched" approach to overcome the escape of glioblastoma cells to the immune surveillance, through an active specific immunization (ASI) with Newcastle-Disease-Virus-infected tumor cells, and blocked the TGF-β production by delivery of TGF-β AONs using polybutyl cyanoacrylate NPs. This approach induced a significant decrease in plasma TGF-β2 level and, at the same time, an increase in rate of high-affinity IL-2 receptor (CD25) on lymphocytes and consequently of antitumoral cytotoxicity (Schneider et al. 2008).

4.4 *Human Telomerase*

The role of human telomerase (hTR) is to maintain telomeric length and promote cell proliferation potential. The activation of telomerase is considered a critical factor in cellular immortality. It has been demonstrated that anti-hTR antisense treatment suppressed tumor cell growth and survival by inducing apoptosis by treating malignant glioma cells (U251-MG cells) in culture with 2-5A AON against telomerase RNA. In vitro study showed that human telomerase antisense inhibition induced two different pathways: apoptosis and differentiation in subpopulations of malignant glioma cells that escape from apoptosis. Probably, malignant glioma cells are cells in G2/M phase in contrast with the major percentage of apoptotic cells that are in S phase of the cell cycle. These results indicate that anti-hTR antisense strategy reduced cell cycling and caused an accumulation of cells in G2/M. In the same study, treatment of surviving nonapoptotic cells with AONs against p27 (a cyclin-dependent kinase inhibitors—CDKIs) induced apoptotic cell death; this suggests that p27 may have a protective role for the survival of differentiating glioma cells (Kondo et al. 1998).

Antisense hTR cDNA inhibited TJ905 human glioma cell proliferation and tumor growth in vitro and in vivo (You et al. 2006). The efficacy of 2-5A anti-hTR and recombinant adenovirus p53 (Ad5CMV-p53) was demonstrated in malignant glioma cell lines with mutant p53 and in vivo GBM murine model (Komata et al. 2000). By using an AON against hTER (RNA component of hTR), many telomerase-positive tumor cells in the vast majority of malignant gliomas were detected (Mukai et al. 2000). In a recent in vitro (U87-MG and U373-MG human malignant glioma cells cultures, cell cultures of human astrocytes expressing telomerase with or without oncogenic Ras) and in vivo study, it was demonstrated that treatment with 2-5A-anti-hTR in the presence of N,N′-bis (2-chloroethyl)-N-nitrosourea (BCNU), cisplatin (CDDP), and temozolomide (TMZ) better enhanced the cell-killing effect if done sequentially. Treatment with 2-5A-anti-hTR in the presence of paclitaxel (PTX) and γ irradiation (IR) always enhanced this effect. The enhanced cell killing resulted by upregulation of apoptosis. In an animal xenograft study model, TMZ enhanced the antitumor effect of 2-5A-anti-hTR via induction of apoptosis (Iwado et al. 2007). In a murine model, created by implantation of U-87 MG malignant glioma cells in mouse, it has been demonstrated the suppression of the ability of glioma cells to form tumors in mice. This result was obtained after transfection of antisense VEGF cDNA in an antisense orientation through the recombinant adenoviral vector Ad5CMV-αVEGF. Infection of U-87 MG malignant glioma cells caused the reduction of the endogenous VEGF mRNA level and of the targeted VEGF secretory form production (Im et al. 1999).

4.5 FAK and IGF

Neuroglial cell growth and CNS development are normally strongly regulated by IGF-1 and IGF-2 via IGF-1R. IGF-1R AON-treated autologous glioma cells, collected during surgical treatment, were induced toward apoptosis and that, in an in vivo murine model, they gave rise to a host response (Andrews et al. 2001). The crucial involvement of two signal transduction pathways, activated by focal adhesion kinase (FAK) and insulin-like growth factor I receptor (IGF-IR) kinase into gliomagenesis, has been evidenced in an in vitro and in vivo study (U87 and LN229 glioma cell lines and mice bearing intracranial glioma xenografts). The authors demonstrated that TAE226, an ATP-competitive inhibitor of several tyrosine kinases, downregulated proliferation and invasion of glioma cells and inhibited cell cycle progression at the G2-M checkpoint. TAE226 also gave rise to a concomitant reduction of the expression of p-cdc2 (Tyr15) and cyclin B1, increase in apoptosis, and reduction in glioma invasion in an in vitro Matrigel (Liu et al. 2007). Anti-FAK phosphorothioate AONs packaging into liposomes in U251 MG cells downregulate the expression levels of FAK and activation of apoptosis, through increase in caspase-3 activity, the key mediator of apoptosis in mammalian cells. In fact, FAK activates the PI3K survival pathway with the concomitant activation of nuclear factor Kappa B (NF-kB) and induction of inhibitor of apoptosis (Wu et al. 2006).

4.6 PI3K/AKT/PKB Pathway

PI3K/AKT/protein kinase B (PKB) pathway plays an important role in cell survival. AKT is a subfamily of the serine/threonine kinases that includes AKT1, AKT2, and AKT3. Growth factors and their receptors led to the activation of PI3K, which in turn phosphorylates phosphatidylinositol-3,4-biphosphate (PIP2) to phosphatidylinositol-3,4,5-triphosphate (PIP3). AKT has an antiapoptotic effect by phosphorylating and inactivating the proapoptotic BAD, procaspase-9, and forkhead transcription factors, leading to the activation of Fas ligand. Moreover, AKT also induces NF-kB-dependent transcription of pro-survival genes including BCL-xL, C-Myb, modifies cell cycle progression by inhibiting p27 and p21, upregulates transcription of cyclin D, and promotes nuclear entry of mdm2. AKT enhances telomerase activity and promotes cell invasiveness and angiogenesis by stimulating secretion of MMPs and by activating endothelial nitric oxide synthase.

In a murine model, the downregulation of proliferation rate of C6 glioma cells transfected with antisense AKT2 cDNA construct was shown. Parental C6 cells and C6 cells, transfected with antisense construct, were implanted through lipofectamine complexes. Antisense inhibition of AKT2 resulted in the reduction of growth rate and proliferation of C6 cells, as well as in the upregulation of GFAP expression and in the induction of apoptosis (Pu et al. 2006).

4.7 Extracellular Matrix Proteins

Effects of antisense inhibition of laminin-8 expression in glioma therapy through an in vitro model using human GBM cell lines M059K and U-87MG co-cultured with normal human brain microvascular endothelial cells (HBMVEC) have been observed. During progression of human gliomas, the expression of capillary BM laminins containing $\alpha 4$-chain switches from the predominant laminin-9 into laminin-8. Laminin-8 and its receptors, $\alpha 3\beta 1$- and $\alpha 6\beta 1$-integrins, are important for the functioning of endothelial cell BMs, which play a role in the maintenance of the BBB. Laminin-8 plays an important role in glioma cell invasiveness, in combination with other proteins associated with glioma progression, such as tenascin-C, MMP-2, and MMP-9. In fact, upregulation of laminin-8 expression in both glioma cells and glioma-adjacent capillary endothelial cells may reduce glial cell adhesion and enhance migration. The result of the study was a significant reduction of invasion of co-cultures through Matrigel and through the use of morpholino oligos against α4- and β1-chains of laminin-8 (Khazenzon et al. 2003). α4-Chain is a component of laminin-8 and laminin-9, both of which are overexpressed in approximately 75% of GBM cases and in low-grade gliomas, respectively. The downregulation of laminin α4-chain using AONs inhibits the motility of human glioma cells. Laminin α4-chain appears to be an important factor in glioma migration and invasion, both in vitro and in vivo (Nagato et al. 2005).

4.8 Other Possible Targets

PTEN or MMAC 1 is a tumor-suppressor gene located on human chromosome 10q23.3, regulating cell growth, apoptosis, interaction with extracellular matrix, and inhibiting cell migration, as well as spreading and focal adhesion. PTEN has two phosphatase activities: a lipid phosphatase and a protein phosphatase. Substrates of these phosphatase activities are phosphatidylinositol 3,4,5-triphosphate and phosphatidylinositol 3,4-biphosphate; they are both involved in the PI3K FAK pathways. It was demonstrated that PTEN has an important role in the inhibition of hyaluronic acid-induced MMP-9 secretion, in the downregulation of basal levels of MMP-2, and in the increase in expression levels of the tissue inhibitor of MMPs-1 and-2 (Park et al. 2002).

HER1/EGFR is a tyrosine kinase receptor that can show various kinds of alterations such as gene amplification or overproduction of HER1/EGFR ligands. EGFRvIII, a mutated isoform, is the most studied mutant form which lacks a ligand-binding domain and is constitutively activated. Studies in human GBM cells suggest that EGFRvIII promotes tumor growth and progression via constitutive activation of the PI3-K/Akt pathway; it also induces upregulation of cell proliferation via the MAPK/ERK1/2 signal transduction pathway. The use of antisense RNA into murine models of human glioma facilitates tumor targeting and induces reduction of HER1/EGFR expression increasing survival (Halatsch et al. 2006).

c-Met is a receptor tyrosine kinase whose ligand is hepatocyte growth factor (HGF). This pathway is very critical in cellular proliferation, motility, and invasion, and it is overexpressed in gliomas. It was demonstrated in vitro and in vivo that AONs against c-Met (FAM-labeled c-Met nonsense ODNs–LIPOFECTAMINE PLUS™) markedly suppressed the expression of c-Met mRNA in human glioma cells and cell growth and enhanced significantly the cytotoxic effect of radiation on human U251 glioma cells in culture (Chu et al. 2007).

5 MicroRNAs

5.1 Clinical Applications

Although knowledge of the specific mRNA targets for the 250–300 known mammalian miRNAs is limited, one emerging function of this extensive regulatory network is its control on cell proliferation and differentiation during normal development. A potential role for miRNAs in malignancy has been suggested by the location of the genes for several miRNAs at sites of translocation breakpoints or deletions linked to human leukemias (Calin et al. 2004a, b). Indeed, altered miRNA expression has now been reported in leukemia, lung cancer, and colon cancer (Michael et al. 2003; Calin et al. 2004a, b). Strong overexpression of a specific miRNA, miR-21, in human GBM tumor tissues, in early-passage glioblastoma cultures, and in six established GBM cell lines (A172, U87, U373, LN229, LN428, and LN308) compared with nonneoplastic fetal and adult brain tissues and compared with cultured nonneoplastic glial cells was observed. Knockdown of miR-21 in cultured glioblastoma cells leads to the activation of caspases and leads to increased apoptotic cell death. These data suggest that aberrant expressed miR-21 may contribute to the malignant phenotype by blocking expression of critical apoptosis-related genes, and the high expression of miR-21 is a common feature of GBM and that miR-21 acts as an antiapoptotic factor in cultured GBM cells (Chan et al. 2005). Aberrant increased expression of miR-21 may downregulate the translation or stability of mRNA(s) coding for apoptosis-related genes. Besides, low miR-21 positivity is present in normal white matter, in the mature brain, and in other primary brain tumors. Moreover, miR-21 promotes glioma invasive phenotype, by regulating also genes associated not only with glioma cell apoptosis but also with migration and invasiveness, including the RECK and TIMP3 genes, inhibitors of matrix metalloproteinases (MMPs) (Gabriely et al. 2008). Interestingly, miR-21 and its precursor also have been reported to be strongly expressed in cell lines established from colorectal carcinoma (HCT-116) and cervical adenocarcinoma but are relatively weakly expressed in cell lines from promyelocytic leukemia (HL-60), chronic myelogenous leukemia (K562), and prostatic adenocarcinoma (LNCAP). A modest upregulation of miR-21 expression in colon, kidney, and breast tumor, although limited for few samples, was observed if compared with GBM (Chan et al. 2005; Cheng et al. 2005a). However, in contrast to the effect in glioblastoma cells, miR-21 suppression increased growth of HeLa cells without affecting their apoptosis. The different biological effects of any particular miRNA, including miR-21, in different cells are likely to be highly dependent on the cell-specific repertoire of target genes, and overexpressed miR-21 may function as a micro-oncogene in glioblastomas (Chan et al. 2005). Suppression of miR-21 might sensitize gliomas for cytotoxic tumor therapy. The combined suppressive effect of miR-21 and NPC-S-TRAIL leads to a synergistic increase in caspase activity and significantly decreased cell viability in human glioma cells (Corsten et al. 2007). miR-21 targets multiple important components of the p53, transforming growth factor-β (TGF-β), and mitochondrial apoptosis tumor-suppressive pathways. Downregulation of miR-21 in GBM cells leads to downregulation of these pathways, causing repression of growth, increased apoptosis, and cell cycle arrest. Moreover, this study shows that miR-21 is in the nexus of direct and indirect negative regulation of a plethora of genes involved in p53, TGF-β, and apoptotic signaling, which establishes its function as an oncogene (Papagiannakopoulos et al. 2008). Several genes within this TGF-β pathway were predicted targets of miR-21, including TGFBR2, TGFBR3, and DAXX. The TGF-β pathway induces apoptosis in cancer cells and during development in response to TGF-β ligand binding to its receptors, TGFBR2 and TGFBR3, which in turn can inhibit growth and activate apoptosis either through activation of SMAD transcription or DAXX (Siegel and Massague 2003). p53 pathway predicted targets included the tumor-suppressive homologue of p53, TP73L (TAp63), and activating cofactors of p53, such as JMY, TOPORS, HNRPK, and TP53BP2 (Coutts and La Thangue 2005; Stiewe 2007). All the above genes are known to associate with p53 and assist in the transcriptional activation of antiproliferative and proapoptotic genes in response to DNA damage. The mitochondrial-apoptotic pathway genes included important apoptotic inducers, such as APAF1, caspase-8, and proteins that control mitochondria permeability and cytochrome c release, such as VDAC1 and PPIF (Newmeyer and Ferguson-Miller 2003). Cell proliferation and cell cycle was significantly repressed by down-regulation of miR-21 in U251 and U87 cell lines. These data are a consequence of p21 up-regulation, key protein of p53 pathway. Both the apoptotic and cell cycle phenotypes, which were observed in the absence of miR-21, were partly dependent on two miR-21 targets: HNRPK and TAp63. HNRPK is a known coactivator of p53 in response to DNA damage, assisting in the transactivation of p21 and GADD45A. TAp63 is a homologue of p53 and is known to mediate both cell cycle arrest and apoptosis. However, the inability of their knockdown to completely rescue the proapoptotic and cell cycle arrest phenotypes, which are observed in the absence of miR-21, suggests that there are other critical miR-21 targets involved in mediating miR-21 oncogenic phenotypes. These data suggest that miR-21 upregulation may be a key step leading to oncogenesis in glioblastoma and other types of cancer, by acting, functionally, as classic oncogenes, through a loss of function of apoptotic and cell-control cycle pathways (Papagiannakopoulos et al. 2008). Analysis of profile expression of 192 miRNAs in human nonneoplastic brain tissues and malignant gliomas using quantitative reverse transcription polymerase chain reaction (RT-PCR) identified a number of miRNAs that have been described previously, such as miR-10b and the apoptosis regulator miR-21, and also identified a number of miRNAs, including miR-124 and miR-137, which have not been described in prior GBM profiling studies. In this study, the authors revealed that expression levels of microRNA-124 and microRNA-137 were significantly decreased in anaplastic astrocytomas and GBM relative to nonneoplastic brain tissue and were increased 8- to 20-fold during differentiation of cultured mouse neural stem cells following growth factor withdrawal. Expression of microRNA-137 was increased 3- to 12-fold in GBM cell lines U87 and U251 following inhibition of DNA methylation with 5-aza-2′-deoxycytidine (5-aza-dC).

Transfection of microRNA-124 or microRNA-137 induced morphological changes and marker expressions consistent with neuronal differentiation in mouse neural stem cells, mouse oligodendroglioma-derived stem cells derived from S100β-v-*erbB* tumors, and cluster of differentiation 133^+ human GBM-derived stem cells (SF6969). Transfection of microRNA-124 or microRNA-137 also induced G1 cell cycle arrest in U251 and SF6969 GBM cells, which was associated with decreased expression of cyclin-dependent kinase 6 and phosphorylated retinoblastoma (pSer 807/811) proteins. MicroRNA-124 and microRNA-137 induce differentiation of adult mouse neural stem cells, mouse oligodendroglioma-derived stem cells, and human GBM-derived stem cells and induce GBM cell cycle arrest. These results suggest that targeted delivery of microRNA-124 and/or microRNA-137 to GBM cells may be therapeutically efficacious for the treatment of this disease, leading to a cell cycle regulation and control and inducing toward to a normal differentiation of very high undifferentiated pathological cell phenotypes. The discovery of a rare, highly tumorigenic, self-renewing subpopulation of GBM cells that express the cell surface marker cluster of differentiation (CD) 133 (Galli et al. 2004; Singh et al. 2004) suggests that therapeutic approaches that effectively inhibit or kill $CD133^+$ GBM stem cells may lead to marked improvements in patient outcome. It has been demonstrated that induction of differentiation of $CD133^+$ GBM cells by bone morphogenetic protein 4 can effectively inhibit intracerebral GBM tumor growth in mice (Piccirillo et al. 2006). Those results show that miR-124 and miR-137 can induce neuronal differentiation of OSCs and GBM stem cells and inhibit proliferation of GBM cell lines. In particular, the enhancement of neurogenesis, the differentiation of stem cells, and G1 cell cycle arrest in standard GBM cell lines are higher in growth-factor-deprived human cells. It seems that miR-124 and miR-137 may be useful therapeutic agents for the treatment of GBMs (Silber et al. 2008).

MicroRNA-7 (miR-7) is a potential tumor suppressor in GBM, targeting critical cancer pathways, and inhibits EGFR expression and the Akt pathway via targeting upstream regulators. Transfection with miR-7 decreased viability and invasiveness of primary glioblastoma lines. Additionally, miR-7 transfection impaired viability and invasiveness of glioma lines. These results show miR-7 as a regulator of EGFR expression and Akt pathway with potential therapeutic implications (Kefas et al. 2008).

6 Ribozymes

6.1 *Clinical Applications*

An important role of the PKC-α ribozyme on cell proliferation and on PKC-α gene expression was observed by showing downregulation of cell proliferation and increase of apoptosis (Leirdal and Sioud 1999). Protein kinase C is a multigene family of serine/threonine kinases that are implicated in many cell growth and differentiation pathways. The increased PKC activity in glioma cells has been used as a target for inhibition by PKC inhibitors. Ribozymes hybridize specifically to complementary mRNA and block the encoded protein by cleaving its mRNA. By introducing 2-amino pyrimidine residues into ribozymes, the authors developed a nuclease-resistant PKC-α ribozyme that blocked glioma tumor growth in inbred syngeneic rats (Sioud and Sørensen 1998). The inhibition of PKC-α leads to a decrease in Bcl-xL gene expression and consequent induction of apoptosis. Morphological examination of glioma cell lines treated with the PKC-α ribozymes indicated alteration in cellular morphology. Cells became rounded and displayed condensation of the nuclear chromatin. These morphological changes are reminiscent of apoptosis. In cells treated with ribozyme for 48 h, the amount of apoptotic nuclei was 87% as compared to only 5% in control cells (Leirdal and Sioud 1999).

Abounader et al. studied the expression of scatter factor (SF), also known as HGF, and its receptor, c-Met, often associated with malignant progression of human tumors, including gliomas. The only known receptor for SF/HGF is the c-met proto-oncogene product, a transmembrane tyrosine kinase receptor. SF/HGF and c-met are found in a wide variety of normal human tissues. SF/HGF is mainly expressed and secreted by a multitude of mesenchymally derived cells, whereas c-met messenger RNA (mRNA) and protein are detected in the epithelium of almost all tissues. These expression patterns are consistent with the paracrine role of SF/HGF and c-met in developmental mesenchymal–epithelial interactions, such as in branching morphogenesis of liver and breast. It was shown that human GBM can be SF/HGF–c-met dependent and that a reduction in endogenous SF/HGF or c-met expression can lead to inhibition of tumor growth and tumorigenicity by transfecting glioblastoma cells with chimeric transgenes consisting of U1 small nuclear RNA, a hammerhead ribozyme, and antisense sequences. The authors have used a novel chimeric U1 small nuclear RNA (U1snRNA), ribozyme, and antisense construct to inhibit SF/HGF and c-met gene expression in human GBM cells possessing an autocrine SF/HGF–c-met loop (U-87 MG) and in human GBM cells expressing c-met but not SF/HGF (U-373 MG). In tumor cells that express both SF/HGF and c-met, SF/HGF can activate c-met receptors on the tumor cell itself and act on c-met receptors present on surrounding tissues, such as vascular endothelial cells. SF/HGF is a potent angiogenic factor. In this way, SF/HGF can act stimulating various downstream events such as the induction of genes that affect tumorigenicity, angiogenic factor secretion, tumor growth, and tumor cell migration and invasion. These downstream responses vary, depending on the target cell type, and include the expression of vascular endothelial growth factor that is a prominent angiogenic/vascular permeability factor in malignant gliomas, matrix-degrading proteases such as plasminogen activators and matrix metalloproteinases, and apoptosis-inhibiting factors (Abounader et al. 1999).

The epidermal growth factor receptor (HER1/EGFR) is known to be dysregulated in a large subgroup of GBM cases. Dysregulation of HER1/EGFR is related to malignant transformation and tumor growth in various human cancers, including malignant glioma. The most common mutant form of HER1/EGFR, named, variant III (EGFRvIII), results from an 801-bp in-frame deletion in the DNA sequence encoding the extracellular ligand-binding domain. Independent of ligand binding, EGFRvIII is constitutively activated and beyond external control. Since its cellular expression was shown to relate enhanced tumorigenicity, various therapeutic strategies were developed to target EGFRvIII, including monoclonal antibodies, vaccination therapies, and small-molecule tyrosine kinase inhibitors. Unfortunately actually there is again no strong evidence of efficacy (Karpel-Massler et al. 2010). Recently, ribozymes' role in human gliomas treatment has been applied on hypoxia-induced angiogenesis during gliomagenesis, (Chen et al. 2011), by focusing a strong attention on HIF-1. To investigate the inhibitory effect of hypoxia-induced factor-1α (HIF-1α) ribozyme on the growth of glioma in nude mice, the glioma cells (C6 cells) were inoculated into nude mice and the eukaryotic expression plasmid carrying HIF-1α ribozyme gene was encapsulated by liposomes. These plasmids were injected into the tumor for consecutive 20 days, and the tumor size and activity of HIF-1α were determined. Reverse transcription - polymerase chain reaction (RT-PCR) was used to detect the expression of HIF-1α. After injection of HIF-1α ribozyme plasmid, the activity of HIF-1α in the cancers was significantly decreased, and the angiogenesis was inhibited, which led to inhibition of glioma growth (Jaakkola et al. 2001). HIF-1α ribozyme has the potential to become an inhibitor of HIF-1α and may play an important role in the treatment of cancers. HIF-1α gene was used as a target in the gene therapy, and previously prepared plasmids carrying HIF-1α hammerhead ribozyme gene were encapsulated by liposomes which were then employed to inoculate C6 cell-induced glioma in nude mice. The results revealed the plasmids could enter the cancer cells after being encapsulated by liposomes, and the ribozyme was efficiently expressed in the glioma, which decreased the HIF-1α expression to 20% of initial level. In addition, the number of microvessels was markedly decreased and the cancer size reduced resulting in suppression of cancer growth. Abounader et al. (2002) designed SF/HGF-specific ribozyme which could specifically incise target genes in vitro and in vivo and then inhibit the proliferation and migration of glioma cells.

7 Conclusion and Perspectives

Malignant glioma treatment, and in particular GBM, remains one of the most challenging areas of neurosurgery and also of malignant tumor therapy. Multimodal therapeutic approaches and molecular-based targeted therapies are modern and complex strategies potentially very efficacious and applicable to upregulate the selectivity of therapeutic effects and downregulate systemic toxicity and side effect toward peritumoral and safe brain tissue. The studies reported are few and, however, are limited to an experimental phase. Evidently, there exists an objective lack of information about the real efficacy and the long-term safety of these novel therapeutic approaches. Considering the multitude of actually used chemotherapeutic approaches and the results in terms of recurrences and survival, the field of glioma therapy is again subject to high rate of investigation and development with the hope to postulate new possible molecular targets and therapeutic schemes to attack glioma cells. Understanding the genetic bases of gliomas and of the invasive behavior, in particular the differentiated gene expression in distinct areas of the same tumor during glioma progression, may suggest new molecular targets to overcome the mechanisms of multidrug resistance of the actual therapeutic approaches to gliomas and to attack at the same time different crucial biological events of gliomagenesis. Glioma gene expression and its development during gliomagenesis may help to better understand the role of important molecules involved in tumor-safe brain parenchyma relationships. These molecules, such as ECM proteases, cell adhesion molecules, and their related signaling pathways, show an important role in glioma cell migration and invasion and could selectively attack to inhibit the glioma invasive rim. It is clear that the complexity and cross talk between signal transduction pathways limit the potential efficacy of targeting a single receptor or molecule. Besides, we think it could be very interesting and effective to detect more molecular targets of the same pathway due to multiple phenotypes inside glioma. The recent discovery of many RNA molecules noncoding for protein, transcripted from DNA sequences different than II class genes (gene coding for protein), has modified strongly the classic idea of human genome, genome-phenotype correlation, and gene expression regulation. Particularly interesting is the role of microRNA, siRNA, ribozymes, and antisense RNA and DNA oligonucleotides in cell-specific gene expression regulation and phenotype determination. These important aspects are very crucial during gliomagenesis and tumorigenesis in general and can be used as a new multimodal targeted molecular-based approach in glioma therapy. Recently long (more than 21 nucleotides), RNA regulator molecules named long non-coding RNA, have been discovered. All these nucleotidic sequences could easily move into the cell and among different cells to regulate gene expression and cellular phenotype.

We are investigating the role of HIF (hypoxic inducing factor) isoforms and IL-8 (Interleukin-8) in glioma progression and the possibility to block their pathways through a only antisense molecule loaded into a nanoparticle-based system and directed against both these molecular targets. HIF-1α actively regulates downstream processes, and it is also influenced by the tumor microenviroment in many different ways. Antisense inhibition of HIF may be a strong target for antiangiogenic therapy. In fact, HIF is the crucial molecule and transcriptional factor produced in response to hypoxic conditions within HIF-1α/VEGF-regulation expression-dependent angiogenic pathway. In this molecular cascade, HIF-1α upregulated levels finally induce upregulation of VEGF expression and, at the same time, may stimulate gene expression of other genes involved in gliomagenesis. High expression levels of PGES-1 (Prostaglandin E 1 Synthase) and IL-8 in malignant glioma cells and microglial cells, strongly correlated with grading tumor, are demonstrated (Caruso et al. 2010a, b). IL-8 expression is first observed in low-grade astrocytoma in perivascular tumor areas expressing inflammatory cytokines. In malignant gliomas, IL-8 further localizes in oxygen-deprived cells surrounding necrosis. Macrophages are known to produce high levels of IL-8, which has a tumorigenic activity, by inducing tumor growth and angiogenesis; IL-8 is an inflammatory chemoattractant responding to the tumor microenvironment. Tumor pseudopalisading cells secrete HIF which induces IL-8 secretion. On the basis of our preliminary results, we hypothesize an important role of IL-8 as crucial angiogenesis mediator within HIF-1α pathway and cross talk between hypoxia-induced high levels of HIF-1α and VEGF expression. A simultaneous selective antisense strategy against two molecular targets involved at different levels in the same pathological pathway could be an attractive therapeutic mechanism aimed at enhancing an antiangiogenic and antitumoral effect.

A very interesting new field of research and therapeutic application in glioma treatment is represented by studying and knowledge of differential phenotypic transcriptional profiles expressed into gliomas and GBM in particular. Hoelzinger et al. in their study have shown two different gene expression profiles into two distinct subpopulations of glioma cells, respectively, the glioma cells located inside the stationary and proliferative tumor core and the glioma cells located inside the motile and invading tumor rim. The authors report two different patterns of gene expression, characterized respectively by high levels of IGFBP2 and vimentin into tumor core glioma cells and upregulated expression of genes involved in adhesion, extracellular signal transduction, cytoskeletal rearrangement, intracellular signal transduction, and apoptosis. In particular, genes upregulated in invasive cells are autotoxin (ATX), the autocrine motility factor, protein tyrosine kinase 2 beta (PYK2), ephrin B3, antiapoptotic factor B-cell lymphoma-w (BCLW), and death-associated protein 3 (DAP3). The existence of two different gene expression profiles into distinct areas of the same tumor suggests the possibility to attack different molecular pathways and steps of glioma progression at the same time, targeting selectively different subsets of glioma cells. This introduces the possibility of multiagent treatment modalities, specifically targeting invasive cells, through new molecular therapeutic approaches in conjunction with classic treatments (Hoelzinger et al. 2005). The use of antisense strategies to modulate gene expression in glioma cells, to change glioma cells phenotype toward the wild-type and functional aspect, and to silence angiogenic factors and neovascularization inside the tumor should be reconsidered. The therapeutic efficacy of much kind of RNA and antisense molecules remains to be determined and requires the development of efficient targeting methods to specific cells. Currently, studies are focusing on elucidating tumor-induced changes in miRNA expression in endothelial blood vessel cells. On the basis of these data, the future of neuro-oncology will be probably centered on an improvement of knowledge of gene expression regulation in normal cells to understand better pathways and molecular approaches to modify pathological phenotype of cancer cells in normal ones.

References

Abounader R, Lal B, Luddy C et al (2002) In vivo targeting of SF/HGF and c-met expression via U1sn RNA/Rybozymes inhibits glioma growth and angiogenesis and promotes tumor apoptosis. Faseb J 16:108–110

Andrews DW, Resnixoff M, Flanders AE et al (2001) Results of a pilot study involving the use of an antisense oligodeoxynucleotide directed against the insulin-like growth factor type I receptor in malignant astrocytomas. J Clin Oncol 19:2189–2200

Baker BF, Monia BP (1999) Novel mechanisms for antisense-mediated regulation of gene expression. Biochim Biophys Acta 1489:3–18

Baltuch GH, Dooley NP, Rostworowski KM (1995) Protein kinase C isoform alpha overexpression in C6 glioma cells and its role in cell proliferation. J Neurooncol 24:241–250

Bello L, Francolini M, Marthyn P et al (2001) Alpha(v)beta3 and alpha(v)beta5 integrin expression in glioma periphery. Neurosurgery 49:380–389

Boado RJ (2005) RNA interference and nonviral targeted gene therapy of experimental brain cancer. NeuroRx 2:139–150

Caffo M, Germanò A, Caruso G et al (2004) An immunohistochemical study of extracellular matrix proteins laminin, fibronectin and type IV collagen in paediatric glioblastoma multiforme. Acta Neurochir (Wien) 146:1113–1118

Calin GA, Liu CG, Sevignani C et al (2004a) MicroRNA profiling reveals distinct signatures in B cell chronic lymphocytic leukemias. Proc Natl Acad Sci USA 101:11755–11760

Calin GA, Sevignani C, Dumitru CD et al (2004b) Human microRNA genes are frequently located at fragile sites and genomic regions involved in cancers. Proc Natl Acad Sci USA 101:2999–3004

Caruso G, Caffo M, Raudino G et al (2010a) Antisense oligonucleotides as an innovative therapeutic strategy in the treatment of high-grade gliomas. Recent Pat CNS Drug Discov 5:53–69

Caruso G, Raudino G, Caffo M et al (2010b) Nanotechnology platforms in diagnosis and treatments of primary brain tumors. Recent Pat Nanotechnol 4:119–122

Chan JA, Krichevsky AM, Kosik KS (2005) MicroRNA-21 is an antiapoptotic factor in human glioblastoma cells. Cancer Res 65:6029–6033

Chen M, Sun J, Lu C et al (2011) Inhibitory effect of hypoxia-induced factor-1α ribozyme on the growth of glioma in nude mice. Sci Res Essays 6:3045–3048

Cheng AM, Byrom MW, Shelton J et al (2005a) Antisense inhibition of human miRNAs and indications for an involvement of miRNA in cell growth and apoptosis. Nucleic Acids Res 33:1290–1297

Cheng CL, Johnson SP, Keir ST et al (2005b) Poly(ADP-ribose) polymerase-1 inhibition reverses temozolomide resistance in a DNA mismatch repair-deficient malignant glioma xenograft. Mol Cancer Ther 4:1364–1368

Chintala SK, Sawaya R, Gokaslan ZL et al (1996) Immunohistochemical localization of extracellular matrix proteins in human glioma, both in vivo and in vitro. Cancer Lett 101:107–114

Chu S, Zang H, Ma Y et al (2007) c-Met antisense oligonucleotides as a novel therapeutic agent for glioma: *in vitro* and *in vivo* studies of uptake, effects, and toxicity. J Surg Res 141:284–288

Collis SJ, Swartz MJ, Nelson WG et al (2003) Enhanced radiation and chemotherapy-mediated cell killing of human cancer cells by small inhibitory RNA silencing of DNA repair factors. Cancer Res 63:1550–1554

Corsten MF, Miranda R, Kasmieh R et al (2007) MicroRNA-21 knockdown disrupts glioma growth *in vivo* and displays synergistic cytotoxicity with neural precursor cell-derived S-TRAIL in human gliomas. Cancer Res 67:8994–9000

Cotter FE (1999) Antisense therapy of hematologic malignancies. Semin Hematol 36:9–14

Coutts AS, La Thangue NB (2005) The p53 response: emerging levels of co-factor complexity. Biochem Biophys Res Commun 331:778–785

Davis GE, Senger DR (2005) Endothelial extracellular matrix: biosynthesis, remodeling, and functions during vascular morphogenesis and neovessel stabilization. Circ Res 97:1093–1107
Dean NM, Bennett FC (2003) Antisense oligonucleotide-based therapeutics for cancer. Oncogene 22:9087–9096
Dias N, Stein CA (2002) Antisense oligonucleotides: basic concept and mechanisms. Mol Cancer Ther 1:347–355
Doudna J, Lorsch J (2005) Ribozyme catalysis: not different, just worse. Nat Struct Mol Biol 12:95
Gabriely G, Wurdinger T, Kesari S et al (2008) MicroRNA 21 promotes glioma invasion by targeting matrix metalloproteinase regulators. Mol Cell Biol 28:5369–5380
Galli R, Binda E, Orfanelli U et al (2004) Isolation and characterization of tumorigenic, stem-like neural precursors from human glioblastoma. Cancer Res 64:7011–7021
Gerhardt H, Golding M, Fruttiger M et al (2003) VEGF guides angiogenic sprouting utilizing endothelial tip cell filopodia. J Cell Biol 161:1163–1177
Gladson CL (1996) Expression of integrin alpha v beta 3 in small blood vessels of glioblastoma tumors. J Neuropathol Exp Neurol 55:1143–1149
Grossman SA, Alavi JB, Supko JG et al (2004) Efficacy and toxicity of the antisense oligonucleotide aprinocarsen directed against protein kinase C-α delivered as a 21-day continuous intravenous infusion in patients with recurrent high-grade astrocytomas. Neuro Oncol 6:32–40
Guensberg P, Wacheck V, Lucas T et al (2002) Bcl-xL antisense oligonucleotides chemosensitize human glioblastoma cells. Chemotherapy 48:189–195
Gurzov EN, Izquierdo M (2006) RNA interference against Hec1 inhibits tumor growth in vivo. Gene Ther 13:1–7
Halatsch ME, Schmidt U, Behnke-Mursch J et al (2006) Epidermal growth factor receptor inhibition for the treatment of glioblastoma multiforme and other malignant brain tumors. Cancer Treat Rev 32:74–89
Hau P, Jachimczak P, Schlingensiepen R et al (2007) Inhibition of TGF-β2 with AP 12009 in recurrent malignant gliomas: from preclinical to phase I/II studies. Oligonucleotides 17:201–212
Holash J, Maisonpierre PC, Compton D et al (1999) Vessel cooption, regression, and growth in tumors mediated by angiopoietins and VEGF. Science 284:1994–1998
Holland EC (2001) Gliomagenesis: genetic alterations and mouse models. Nat Rev Genet 2:120–129
Hoelzinger DB, Mariani L, Weis J et al (2005) Gene expression profile of glioblastoma multiforme invasive phenotype points to new therapeutic targets. Neoplasia 7:7–16
Hu-Lieskovan S, Heidel JD, Bartlett DW et al (2005) Sequence-specific knockdown of EWS-FLI1 by targeted, nonviral delivery of small interfering RNA inhibits tumor growth in a murine model of metastatic Ewing's sarcoma. Cancer Res 65:8984–8992
Idbaih A, Ducray F, Sierra del Rio M et al (2008) Therapeutic application of noncytotoxic molecular targeted therapy in gliomas: growth factor receptors and angiogenesis inhibitors. Oncologist 13:978–992
Ihanahan D, Weinberg RA (2010) The hallmarks of cancer. Cell 100:57–70
Im SA, Gomez-Manzano C, Fueyo J et al (1999) Antiangiogenesis treatment for gliomas: transfer of antisense-vascular endothelial growth factor inhibits tumor growth *in vivo*. Cancer Res 59:895–900
Ito TK, Ishii G, Chiba H et al (2007) The VEGF angiogenic switch of fibroblasts is regulated by MMP-7 from cancer cells. Oncogene 26:7194–7203
Iwado E, Daido S, Kondo Y et al (2007) Combined effect of 2-5A-linked antisense against telomerase RNA and conventional therapies on human malignant glioma cells *in vitro* and *in vivo*. Int J Oncol 31:1087–1095
Jaakkola P, Mole DR, Tian YM et al (2001) Targeting of HIF-alpha to the von Hippel–Lindau ubiquitylation complex by O_2 -regulated prolyl hydroxylation. Science 292:468–472

Jachimczak P, Hessdorfer B, Fabel-Schulte K et al (1996) Transforming growth factor-beta mediated autocrine growth regulation of gliomas with phosphorothioate antisense oligonucleotides. Int J Cancer 65:332–337

Jackson AL, Bartz SR, Schelter J et al (2003) Expression profiling reveals off-target gene regulation by RNAi. Nat Biotechnol 21:635–637

Jain RK, di Tomaso E, Duda DG et al (2007) Angiogenesis in brain tumours. Nat Rev Neurosci 8:610–622

Jansen B, Zangemeister-Wittke U (2002) Antisense therapy for cancer—the tome of truth. Lancet Oncol 3:672–683

Jason TL, Koropatnick J, Berg RW (2004) Toxicology of antisense therapeutics. Toxicol Appl Pharmacol 201:66–83

Julien T, Frankel B, Longo S et al (2000) Antisense-mediated inhibition of the bcl-2 gene induces apoptosis in human malignant glioma. Surg Neurol 53:360–369

Karpel-Massler G, Wirtz CR, Halatsch ME (2010) Ribozyme-mediated inhibition of 801-bp deletion-mutant epidermal growth factor receptor mRNA expression in glioblastoma multiforme. Molecules 15:4670–4678

Kefas B, Godlewski J, Comeau L et al (2008) MicroRNA-7 inhibits the epidermal growth factor receptor and the Akt pathway and is down-regulated in glioblastoma. Cancer Res 68:3566–3572

Khazenzon NM, Ljubimov AV, Lakhter AJ et al (2003) Antisense inhibition of laminin-8-expression reduces invasion of human gliomas *in vitro*. Mol Cancer Ther 2:985–994

Kjaergaard J, Wang LX, Kuriyama H et al (2005) Active immunotherapy for advanced intracranial murine tumors by using dendritic cell-tumor cell fusion vaccines. J Neurosurg 103:156–164

Kleihues P, Louis DN, Scheithauer BW et al (2002) The WHO classification of tumors of the nervous system. J Neuropathol Exp Neurol 61:215–229

Komata T, Kondo Y, Koga S et al (2000) Combination therapy of malignant glioma cells with 2-5A-antisense telomerase RNA and recombinant adenovirus p53. Gene Ther 7:2071–2079

Kondo S, Tanaka Y, Kondo Y et al (1998) Antisense telomerase treatment: induction of two distinct pathways, apoptosis and differentiation. FASEB J 12:801–811

Krishnan L, Underwood CJ, Maas S et al (2008) Effect of mechanical boundary conditions on orientation of angiogenic microvessels. Cardiovasc Res 78:324–332

Lee S, Jilani SM, Nikolova GV et al (2005) Processing of VEGF-a by matrix metalloproteinases regulates bioavailability and vascular patterning in tumors. J Cell Biol 169:681–691

Leirdal M, Sioud M (1999) Ribozyme inhibition of the protein kinase Ca triggers apoptosis in glioma cells. Br J Cancer 80:1558–1564

Levin AA (1999) A review of issue in the pharmacokinetics and toxicology of phosphorothioate antisense oligonucleotides. Biochim Biophys Acta 1489:69–84

Lee Y, Scheck AC, Cloughesy TF et al (2008) Gene expression analysis of glioblastomas identifies the major molecular basis for the prognostic benefit of younger age. BMC Med Genomics 1:52. doi:10.1186/1755-8794-1-52

Liang Y, Diehn M, Watson N et al (2005) Gene expression profiling reveals molecularly and clinically distinct subtypes of glioblastoma multiforme. Proc Natl Acad Sci USA 102:5814–5819

Liu TJ, La Fortune T, Honda T et al (2007) Inhibition of both adhesion kinase and insulin-like growth factor-I receptor kinase suppresses glioma proliferation *in vitro* and *in vivo*. Mol Cancer Ther 6:1357–1367

Ljubimova JY, Fujita M, Khazenzon NM et al (2006) Changes in laminin isoforms associated with brain tumor invasion and angiogenesis. Front Biosci 11:81–88

Mao Y, Schwarzbauer JE (2005) Fibronectin fibrillogenesis, a cell-mediated matrix assembly process. Matrix Biol 24:389–399

MacRae IJ, Zhou K, Li F et al (2006) Structural basis for double-stranded RNA processing by Dicer. Science 311:195–198

McNamara JO II, Andrechek ER, Wang Y et al (2006) Cell type-specific delivery of siRNAs with aptamer-siRNA chimeras. Nat Biotechnol 24:1005–1015
Michael MZ, O'Connor SM, van Holst Pellekaan NG et al (2003) Reduced accumulation of specific micro-RNAs in colorectal neoplasia. Mol Cancer Res 1:882–891
Miner JH (2006) Laminins and their roles in mammals. J Cell Biol 175:179–191
Mukai S, Kondo Y, Koga S et al (2000) 2-5A antisense telomerase RNA therapy for intracranial malignant gliomas. Cancer Res 60:4461–4467
Nagato S, Nakagawa K, Harada H et al (2005) Downregulation of laminin α4 chain expression inhibits glioma invasion *in vitro* and *in vivo*. Int J Cancer 117:41–50
Nemunaitis J, Jt H, Kraynak M et al (1999) Phase I evolution of ISIS 3521, an antisense oligodeoxynucleotide to protein kinase-c-α, in patients with advanced cancer. J Clin Oncol 17:3586–3595
Newmeyer DD, Ferguson-Miller S (2003) Mitochondria: releasing power for life and unleashing the machineries of death. Cell 112:481–490
Nickl-Jockschat T, Arslan F, Doerfelt A et al (2007) An imbalance between SMAD and MAPK pathways is responsible for TGF-β tumor promoting effects in high-grade gliomas. Int J Oncol 30:499–507
Nielsen PE (2003) Antisense properties of peptide nucleic acid. Methods Enzymol 313:156–164
Nieth C, Priebsch A, Stege A et al (2003) Modulation of the classical multidrug resistance (MDR) phenotype by RNA interference (RNAi). FEBS Lett 545:144–150
Ohgaki K, Kleihues P (2007) Genetic pathways to primary and secondary glioblastoma. Am J Pathol 170:1445–1453
Orum H, Wengel J (2001) Locked nucleic acids: a promising molecular family for gene-function analysis and antisense drug development. Curr Opin Mol Ther 3:329–343
Papagiannakopoulos T, Shapiro A, Kosik A (2008) MicroRNA-21 targets a network of key tumor suppressive pathways in glioblastoma cells. Cancer Res 68:8164–8172
Pardridge M, Vincent A, Matthews P (1996) A simple method for delivering morpholino antisense oligos into the cytoplasm of cells. Antisense Nucleic Acid Drug Dev 6:169–175
Park MJ, Kim MS, Park IC et al (2002) PTEN suppresses hyaluronic acid-induced matrix metalloproteinases-9 expression in U87MG glioblastoma cells through focal adhesion kinase dephosphorylation. Cancer Res 62:6318–6322
Passioura T, Symonds G (2004) Cancer suppression strategies: issues and potential. Curr Issues Mol Biol 6:89–102
Pawlak W, Zolnierek J, Sarosiek T et al (2000) Antisense therapy in cancer. Cancer Treat Rev 26:333–350
Piccirillo SG, Reynolds BA, Zanetti N et al (2006) Bone morphogenetic proteins inhibit the tumorigenic potential of human brain tumour-initiating cells. Nature 444:761–765
Pirollo KF, Rait A, Sleer SL et al (2003) Antisense therapeutics: from theory to clinical practice. Pharmacol Ther 99:55–77
Platten M, Wick W, Weller M (2001) Malignant glioma biology: role for TGF-β in growth, motility, angiogenesis and immune escape. Microsc Res Tech 52:401–410
Pollack IF, Kawecki S, Lazo JS (1996) Blocking of glioma proliferation in vitro and in vivo and potentiating the effects of BCNU and cisplatin: UCN-01, a selective protein kinase C inhibitor. J Neurosurg 84:1024–1032
Pooga M, Soomets U, Hallbrink M et al (1998) Cell penetrating PNA constructs regulate galanin receptor levels and modify pain transmission in vivo. Nat Biotechnol 16:857–861
Potapova O, Gorospe M, Dougherty RH et al (2000) Inhibition of c-Jun N-terminal kinase 2 expression suppresses growth and induces apoptosis of human tumor cells in a p53-dependent manner. Mol Cell Biol 20:1713–1722
Pu P, Kang C, Li J et al (2006) The effects of antisense AKT2 RNA on the inhibition of malignant glioma cell growth *in vitro* and *in vivo*. J Neurooncol 76:1–11
Rooprai HK, Vanmeter T, Panou C et al (1999) The role of integrin receptors in aspects of glioma invasion in vitro. Int J Dev Neurosci 17:613–623

Sanson M, Laigle-Donadey F, Benouaich-Amiel A (2006) Molecular changes in brain tumors: prognostic and therapeutic impact. Curr Opin Oncol 18:623–630
Saunders WB, Bohnsack BL, Faske SB et al (2008) Coregulation of vascular tube stabilization by endothelial cell TIMP-2 and pericyte TIMP-3. Microsc Res Tech 71:349–356
Schlingensiepen KH, Schlingensiepen R, Steinbrecher A et al (2006) Targeted tumor therapy with the TGF-β2 antisense compound AP 12009. Cytokine Growth Factor Rev 17:129–139
Schneider T, Becker A, Ringe K et al (2008) Brain tumor therapy by combined vaccination and antisense oligonucleotide delivery with nanoparticles. J Neuroimmunol 195:21–27
Semenza GL (2003) Targeting HIF-1 for cancer therapy. Nat Rev Cancer 3:721–732
Siegel PM, Massague J (2003) Cytostatic and apoptotic actions of TGF-beta in homeostasis and cancer. Nat Rev Cancer 3:807–820
Silber J, Lim DA, Petritsch C et al (2008) miR-124 and miR-137 inhibit proliferation of glioblastoma multiforme cells and induce differentiation of brain tumor stem cells. BMC Med 24:6–14
Singh SK, Hawkins C, Clarke ID et al (2004) Identification of human brain tumour initiating cells. Nature 432:396–401
Sioud M, Sørensen DR (1998) A nuclease -resistant protein kinase C alpha ribozyme blocks glioma cell growth. Nat Biotechnol 16:556–561
Stegh AH, Kim H, Bachoo RM et al (2007) BcI2L12 inhibits post-mitochondrial apoptosis signaling in glioblastoma. Genes Dev 21:98–111
Stiewe T (2007) The p53 family in differentiation and tumorigenesis. Nat Rev Cancer 7:165–167
Stupp R, Hegi ME, van den Bent MJ et al (2006) Changing paradigms: an update on the multidisciplinary management of malignant glioma. Oncologist 11:165–180
Summerton J (1999) Morpholino antisense oligomers: the case for an RNase H-independent structural type. Biochim Biophys Acta 1489:141–158
Takei Y, Kadomatsu K, Yuzawa Y et al (2004) A small interfering RNA targeting vascular endothelial growth factor as cancer therapeutics. Cancer Res 64:3365–3370
Tamm I (2006) Antisense therapy in malignant disease: status quo and quo vadis? Clin Sci 110:427–442
Uhm JH, Dooley NP, Villemure J-G et al (1997) Mechanisms of glioma invasion: role of matrix metalloproteinases. Can J Neurol Sci 24:3–15
Vidal L, Blagden S, Attard G et al (2005) Making sense of antisense. Eur J Cancer 41:2812–2818
Wacheck V, Zangemeister-Wittke U (2006) Antisense molecules for targeted cancer therapy. Crit Rev Oncol Hematol 59:65–73
Waters JS, Webb A, Cunningham D et al (1999) Phase I clinical and pharmacokinetic study of bcl-2 antisense oligonucleotide therapy in patients with non-Hodgkin's lymphoma. J Clin Oncol 18:1812–1823
Wen PY, Yung WK, Lamborn KR et al (2006) Phase I/II study of imatinib mesylate for recurrent malignant gliomas: North American Brain Tumor Consortium Study 99–08. Clin Cancer Res 12:4899–4907
Wu ZM, Yuan XH, Jiang P et al (2006) Antisense oligonucleotides targeting the focal adhesion kinase inhibit proliferation, induce apoptosis and cooperate with cytotoxic drugs in human glioma cells. J Neurooncol 77:117–123
Wurdinger T, Tannous BA (2009) Special focus: angiogenesis in the central nervous system. Glioma angiogenesis. Toward a novel RNA therapeutics. Cell Adh Migr 3:230–235
Yan X, Ma L et al (2011) A CD133-related gene expression signature identifies an aggressive glioblastoma subtype with excessive mutations. Proc Natl Acad Sci USA 108(4):1591–1596. doi:doi/10.1073/pnas.1018696108
Yancopoulos GD, Davis S, Gale NW et al (2000) Vascular-specific growth factors and blood vessel formation. Nature 407:242–248
You Y, Pu P, Huang Q et al (2006) Antisense telomerase RNA inhibits the growth of human glioma cells *in vitro* and *in vivo*. Int J Oncol 28:1225–1232
Zamecnik PC, Stephenson ML (1978) Inhibition of Rous sarcoma virus replication and cell transformation by a specific oligodeoxynucleotide. Proc Natl Acad Sci USA 75:280–284

Natural Antisense Transcripts Mediate Regulation of Gene Expression

Marco Magistri and Mohammad Ali Faghihi

Contents

Abstract Mammalian genomes encode numerous natural antisense transcripts, but the function of these transcripts is not well understood. Functional validation studies indicate that antisense transcripts are not a uniform group of regulatory RNAs but instead belong to multiple categories with some common features. Recent evidence indicates that antisense transcripts are frequently functional and

Karolinska Institutet previously published part of this work as Mohammad Ali Faghihi's thesis.

M. Magistri • M.A. Faghihi (✉)
Department of Psychiatry & Behavioral Sciences, Hussman Institute for Human Genomics, Leonard M. Miller School of Medicine, University of Miami, Biomedical Research Building (BRB) Room#429, 1501 NW 10th Avenue, Miami, FL 33136, USA
e-mail: mmagistri@med.miami.edu

V.A. Erdmann and J. Barciszewski (eds.), *From Nucleic Acids Sequences to Molecular Medicine*, RNA Technologies, DOI 10.1007/978-3-642-27426-8_10,

use diverse transcriptional and posttranscriptional gene regulatory mechanisms to carry out a wide variety of biological roles.

Keywords lncRNAs • Natural Antisense Transcripts • NATs • noncoding RNAs • ncRNAs

1 Introduction

Natural antisense transcripts (also known as antisense RNAs or antisense transcripts) are RNA molecules that are transcribed from the opposite DNA strand relative to conventional mRNA.

Recent research on natural antisense transcripts, including several large-scale expression-profiling studies, has conclusively established the existence of antisense transcripts in eukaryotic genomes. Transcription of antisense RNA, similar to sense mRNA, is tissue and cell line specific, and a large fraction of antisense transcripts is expressed in specific regions of the brain, supporting involvement of these regulatory RNAs in sophisticated brain functions as well as in complex neurological disorders.

There is still no extensive characterization of the large number of non-protein-coding RNAs found in mammalian genomes, but there is a growing body of evidence indicating that natural antisense transcripts can be involved in a myriad of processes in cells having important regulatory functions. Natural antisense transcripts appear to be utilizing various cellular pathways, but it is still not clear which intrinsic properties of antisense RNA molecules, or extrinsic features such as protein interactions and cellular and developmental context, are decisive for the selection of any given pathway. Considering tissue- and cell type-specific expression patterns of antisense RNAs and their heterogeneous proposed functions, natural antisense transcripts appear to be a heterogeneous group of regulatory RNAs with a wide variety of biological roles.

2 Pervasive Transcription of Mammalian Genomes

Since the human and other mammalian genomes sequencing were completed, different large-scale functional genomic approaches were developed in order to identify the complete network of transcribed sequences. An unexpected level of complexity in the transcriptional output of the mammalian genomes has recently emerged, thanks to the development and application of new powerful techniques like large-scale sequencing of cDNA clones (Okazaki et al. 2002; Kiyosawa et al. 2003; Carninci et al. 2005; Katayama et al. 2005), high resolution tiling array

(Rinn et al. 2003; Bertone et al. 2004; Cawley et al. 2004; Cheng et al. 2005), bioinformatics analysis of RefSeq and EST databases (Fahey et al. 2002; Lehner et al. 2002; Shendure and Church 2002; Yelin et al. 2003; Chen et al. 2004; Klimov et al. 2006; Li et al. 2006; Zhang et al. 2006), hybridization techniques (Rosok and Sioud 2004; Yu et al. 2008), SAGE libraries (Quere et al. 2004; Ge et al. 2006), strand-specific microarrays (Gyorffy et al. 2007; Vallon-Christersson et al. 2007; Ge et al. 2008), and most recently, asymmetric strand-specific analysis of gene expression (ASSAGE) (He et al. 2008) and global run-on sequencing (GRO-seq) (Core et al. 2008).

Up to 93% of the human genome is transcribed with a high proportion of transcripts originating from intergenic and intragenic regions that were previously thought to be silent and meaningless (Birney et al. 2007; Mattick 2009a). The FANTOM-3 mouse transcriptome-sequencing consortium identified natural antisense transcripts for more than 70% of transcriptional units (TUs), the majority of which represent non-protein-coding RNAs (Katayama et al. 2005). Moreover, antisense transcripts have been documented for many active promoters (Core et al. 2008; Preker et al. 2008; Seila et al. 2008). Therefore, the presence of antisense transcripts is no longer a curiosity but rather a pervasive feature of mammalian genomes.

Natural antisense transcripts can overlap in part with sense RNA, sense promoter, or other sense regulatory elements. Both sense and antisense RNAs can encode proteins or be non-protein-coding transcripts; however, the most prominent form of antisense transcription in the mammalian genome is a non-protein-coding antisense RNA partner of a protein-coding mRNA transcript.

The widespread transcription occurring in mammalian genome challenges the canonical "gene" definition and how genes and their regulatory elements are arranged. Mammalian transcriptomes appear to contain an abundance of nested and overlapping gene structures, giving rise to both coding and noncoding transcripts. In addition, it seems that the RNA regulatory elements that control the expression of a gene are frequently distributed within or beyond other genes in both directions.

2.1 Transcription at Gene Boundaries

Both ends of protein-coding genes have a propensity for natural antisense transcription (Finocchiaro et al. 2007; He et al. 2008); specifically, antisense transcription is enriched 250 nucleotides upstream of the transcription start site (TSS) (Core et al. 2008; Seila et al. 2008) and 1.5 kb downstream of sense genes (Sun et al. 2005; Core et al. 2008).

Non-protein-coding RNAs (ncRNAs) overlapping gene promoters may represent a distinct class of ncRNAs with specific functions in transcriptional control. Promoter-associated RNAs (paRNAs) are broadly defined as ncRNA transcripts originating within a few hundred bases of protein-coding transcription start sites

(TSSs) (Xu et al. 2009). Relatively small RNAs enriched at gene boundaries were initially identified by high-density tiling arrays (Kapranov et al. 2007). In this pioneering study, two different categories of new transcripts were indentified: small RNA molecules of 20–200 nucleotides in length and long RNA molecules longer than 1 kb (Kapranov et al. 2007). Small and long ncRNAs associated within gene promoters were called "promoter-associated short RNAs" (PASRs) and "promoter-associated long RNAs" (PALRs), respectively. Small RNAs clustering in the proximity of the 3′ ends were called "termini-associated short RNAs" (TASRs) (Kapranov et al. 2007). RNA polymerase II (RNAPII) that stalls at promoters produces these ncRNAs, which can be antisense RNAs or short sense RNAs.

Subsequently, other groups have identified other classes of short ncRNAs, overlapping the promoter regions of known protein-coding genes. These short ncRNAs have been called TSSa-RNAs, NRO-RNAs, and tiRNAs (Core et al. 2008; Seila et al. 2008; Taft et al. 2009). TSSa-RNAs are low abundant 18–90 nucleotides long RNAs and 5′ phosphorylated RNA molecules that are transcribed in the opposite direction to the transcripts of known protein-coding genes (Seila et al. 2008). NRO-RNAs were identified by global run-on sequencing and are bidirectionally transcribed small RNAs elongating from the TSSs of many active protein-coding genes (Core et al. 2008). tiRNAs were isolated in human, mouse, chicken, and drosophila, are 18 nucleotides in length, and have their highest density at 20 nucleotide downstream of TSSs. tiRNAs are associated with highly expressed genes, regions of RNAPII and Sp1 transcriptions factor binding, are weakly expressed, and show bidirectional distributions that mirror RNAPII (Taft et al. 2009). Another class of paRNAs was called "promoter upstream transcripts" (PROMPTs) that are very unstable and were identified by blocking the exosome (Preker et al. 2008). They are expressed from a broad region upstream of TSS of the majority of active genes, and they are generally bidirectionally transcribed. PROMPTs are longer than 500 nucleotides and do not have upstream promoters, suggesting the existence of alternative mechanism of PROMTs generation (Preker et al. 2008).

The different classes of promoter-associated RNAs could represent separate entities involved in similar functions, or longer transcripts could be processed and generate short RNAs that act locally. Whether the widespread transcription that occurs at eukaryotic gene boundaries has a function remains unclear, but there are concrete possibilities that ncRNAs could be involved in the epigenetic regulation of transcription of protein-coding genes. Another possibility is that bidirectional transcription may serve to maintain an open chromatin structure at the promoters, thereby facilitating the access of transcription regulatory factors.

3 Characteristics of Natural Antisense Transcripts

Natural antisense transcripts have been suggested to regulate gene expression by causing transcriptional and posttranscriptional changes in sense mRNA levels (Kiyosawa et al. 2003; Rosok and Sioud 2005; Chan et al. 2006; Li et al. 2006).

Evolutionary studies and numerous well-documented experiments have ruled out the possibility that this genome-wide natural antisense transcription represents a leakage of the RNA transcription machinery (Dahary et al. 2005). In fact, the high abundance of antisense transcripts reported in gene expression studies has been established as real overlap which has considerably affected vertebrate genome evolution (Dahary et al. 2005), and antisense transcripts are suggested to display pivotal role in complexity of the higher organism nervous system (St Laurent and Wahlestedt 2007).

The basal expression levels of sense and antisense transcripts in different tissues and cell lines may be either positively or negatively correlated (Katayama et al. 2005; Okada et al. 2008), suggesting the use of alternative transcriptional regulatory elements and promoters (He et al. 2008). Antisense RNAs have a tendency to undergo fewer splicing events and typically show lower abundance compared with sense transcripts (He et al. 2008). Interestingly, knockdown or blockade of endogenous antisense transcripts can have multiple outcomes, with the corresponding sense transcript concentration showing either an increase (discordant regulation) or a decrease (concordant regulation) (Wahlestedt 2006). It has been proposed that discordant derepression of sense transcript expression, resulting in upregulation of sense RNA expression, can be achieved by removal or steric blockade of many but not all antisense transcripts (Katayama et al. 2005; Wahlestedt 2006).

These and other variable intrinsic properties imply that antisense-mediated regulation of gene expression must operate through a variety of mechanisms and further suggest that antisense transcripts are a heterogeneous group of regulatory RNAs.

4 Classification of Natural Antisense Transcripts

In the absence of a univocal classification of all the non-protein-coding transcripts in mammalian genomes, natural antisense transcripts can be classified based on their size, genomic localization, expression pattern, alignment with regard to the sense mRNA, coding potentials, and regulatory function or mechanism of action. Classifications of antisense transcripts are useful for defining various transcripts and more importantly for predicting their regulatory function. Here, there are examples of different possible classifications.

4.1 Cis- *vs.* Trans-*antisense Transcripts*

Most natural antisense transcripts are *cis*-encoded antisense RNA (Kumar and Carmichael 1998; Vanhee-Brossollet and Vaquero 1998). By definition, *cis*-natural antisense transcripts are complementary RNA with an overlapping transcriptional unit (TU) at the same chromosomal locus. *Trans*natural antisense transcripts are complementary RNA transcribed from different chromosomal locations (Rosok and Sioud 2004; Makalowska et al. 2005; Li et al. 2008).

4.2 Sense–Antisense Format

There is enrichment for the presence of natural antisense transcripts in both 5′ and 3′ ends of the sense genes. Some reports indicate that over 70% of *cis*-encoded antisense transcripts have a tail-to-tail 3′ overlap, while 15% have a 5′ head-to-head format with the 3′ UTR or 5′ UTR of their target mRNA, respectively. The remaining antisense transcripts have full, intronic, or coding overlap. These figures vary widely, and some reports (Katayama et al. 2005) revealed more prevalent head-to-head orientation among natural antisense transcripts.

4.3 Concordant vs. Discordant

There are broadly two types of regulation between sense and antisense transcripts, namely, concordant and discordant regulation.

In concordant regulation, like in the case of *BACE1*-AS (Faghihi et al. 2008), aHIF1 (Neckers 1999), asEPO-R (Zhang et al. 2008), and Zeb2-AS (Beltran et al. 2008), antisense transcripts augment the level of the sense mRNA or corresponding protein levels. In discordant regulation, e.g., naPINK1 (Scheele et al. 2007), RevΔ5-desaturase (Dreesen et al. 2006), zfh-5AS (Komine et al. 2006), or antisense-*frq* (Kramer et al. 2003), the antisense transcripts have negative (opposing) effects on sense transcripts. Interestingly, in most cases, both types of regulation (concordant and discordant) are primarily unidirectional in that the antisense transcript regulates the sense RNA (Wahlestedt 2006).

In most reported cases of concordant regulation, the natural antisense transcripts show a low degree of RNA processing. In these cases, antisense RNAs are not spliced or display short introns (Chen et al. 2005) and have a much shorter half-life time than their coding sense partner (Faghihi et al. 2008). Therefore, it might take much shorter time for antisense transcription and processing compared to sense mRNA transcripts. The rapid transcription and processing of this antisense transcripts class, similar to what suggested for ("nimble" genes) (Chen et al. 2005), implies that they are implicated in acute stress responses. In contrast, discordant natural antisense transcripts are mainly fully processed RNA transcripts with multiple exons and are more prominent in the cytoplasm. High cytoplasmic abundance of this group of antisense transcripts suggests a long-lasting and housekeeping regulatory role through cytoplasmic RNA duplex formation (see proposed cytoplasmic mechanisms).

5 Regulatory Roles of Antisense Transcripts

The natural antisense transcripts have been suggested to regulate gene expression at different levels, including chromatin architecture/epigenetic memory, transcription, transcript localization, translation, and turnover (Kiyosawa et al. 2003; Rosok and Sioud 2005; Chan et al. 2006; Li et al. 2006).

There are a number of proposed mechanisms for antisense-mediated regulation of sense mRNA expression (Faghihi and Wahlestedt 2009). The mechanisms can be categorized into four main groups: mechanisms related to transcription, RNA–DNA interactions, RNA–RNA interactions in the nucleus, and RNA–RNA interactions in the cytoplasm. Each type of mechanism will be discussed in this chapter.

5.1 Transcription-Related Mechanism

According to this mechanism, the act of transcription in the antisense direction, but not the antisense RNA molecule itself, modulates transcription of sense RNA. Transcription in the antisense direction is suggested to cause alterations in sense RNA and is also proposed to be involved in genomic rearrangements.

The transcriptional collision model is based on the assumption that during *cis*-antisense RNA transcription, RNA polymerases bind to the promoters of convergent genes on opposing strands of DNA and then proceed toward the 3′ end of sense and antisense genes. RNA polymerase complexes collide in the overlapping region, blocking further transcription (Shearwin et al. 2005; Osato et al. 2007). Collision of RNA polymerases has been observed in *Escherichia coli* and in *Saccharomyces cerevisiae* (Prescott and Proudfoot 2002), where transcription from the antisense direction halts transcription of the sense mRNA. A bioinformatics study found that, in humans and mice, the length of overlapping regions between sense and antisense transcripts is inversely correlated to the expression levels of antisense transcript. This finding may suggest a possible clash of RNA polymerases, pertaining to the point that the likelihood of collision increases as the length of overlapping region increases (Osato et al. 2007).

Immunoglobulin production in B-lymphocytes and receptor selection by T-lymphocytes depend on transcription from hypervariable regions. To generate variability, T- and B-lymphocytes need a recombination process that occurs through hypermutation in the variable regions of immunoglobulin and T-cell receptor genes. Activation-induced cytidine deaminase (AID), which deaminates deoxycytidine to deoxyuridine in single-stranded DNA (ssDNA), is required for both somatic hypermutation (SHM) and class switch recombination (CSR), two processes that are involved in secondary immune response. DNA must locally uncoil or melt and become single stranded for DNA recombination, replication, and transcription to occur. The transcription complex consists of an RNA polymerase molecule enclosing approximately 17 $\pm$ 1 melted bases (the transcription bubble) of the template DNA (Sastry and Hoffman 1995). The AID enzyme optimally deaminates a small ssDNA bubble,

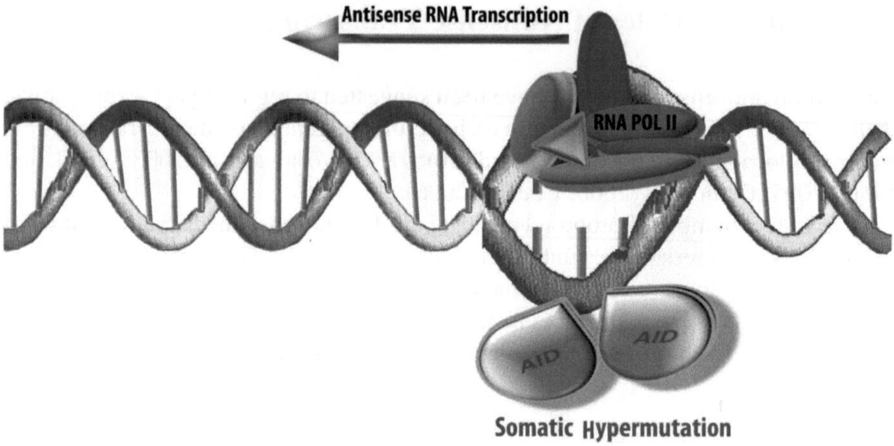

Fig. 1 Genomic DNA rearrangement in hypervariable regions of B- and T-lymphocytes. Natural antisense transcription from the opposite strand will open the transcription bubble and expose the single-stranded DNA (ssDNA), which can be recognized by the activation-induced cytidine deaminase (AID). This enzyme converts deoxycytidine to deoxyuridine, which underlies hypermutation processes

in vitro, consisting of ~5 nucleotides with a ssDNA motif, called WRC (Larijani and Martin 2007). It has been hypothesized that antisense transcription in the variable region could make the ssDNA accessible for AID (Ronai et al. 2007; Perlot et al. 2008) (Fig. 1).

Antisense transcription is frequently observed in the variable (Julius et al. 1988; Apel et al. 1992; Bolland et al. 2004, 2007; Roa et al. 2008), but not in the constant, regions, and it may be initiated in preparation for SHM and CSR [reviewed in Roa et al. (2008)]. According to this model, antisense transcription is proposed to open the transcription bubble and expose the ssDNA, which is a substrate for AID (Bolland et al. 2004, 2007), or antisense transcription may be involved in remodeling of the chromatin structure to make the DNA sequence accessible for recombination (Bolland et al. 2004, 2007; Ronai et al. 2007). In both of these scenarios, transcription from the antisense direction, but not the antisense RNA molecule, is important for genomic rearrangement.

5.2 *RNA–DNA Interactions*

Evidence from different model systems demonstrates that non-protein-coding RNAs play a significant role in the control of epigenetic regulation and chromosomal dynamics, and have an impact on normal cellular differentiation and organism development (Mattick et al. 2010). A common theme that is emerging is that natural antisense transcripts may function as scaffolds to recruit and retain epigenetic effectors and transcriptional regulatory complexes at specific genomic loci,

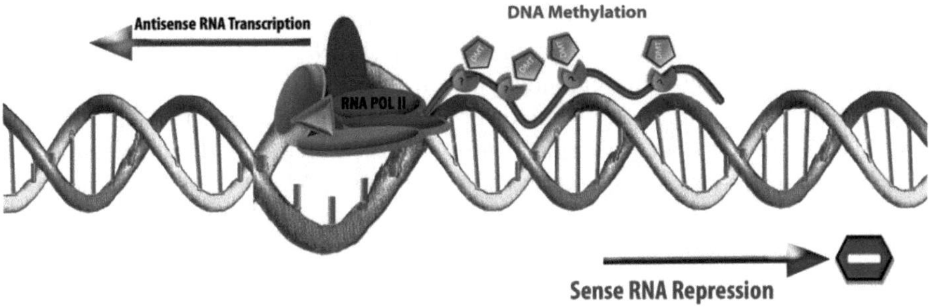

Fig. 2 Induction of DNA methylation by natural antisense transcripts: RNA–DNA interaction model in which newly formed antisense RNA transcript can interact, directly or indirectly, with DNA methyltransferases (DMT) and guide methylation of DNA, resulting in repression of sense RNA transcription

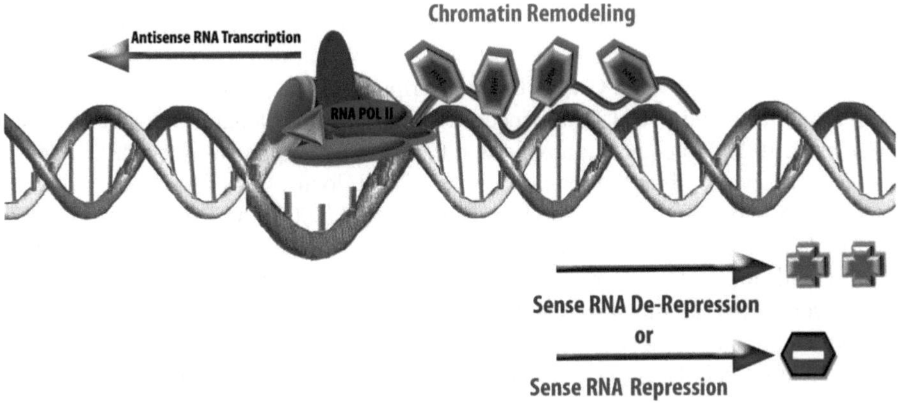

Fig. 3 Induction of chromatin modifications by natural antisense transcripts: Alternatively, newly formed antisense RNA transcript can recruit histone-modifying enzymes (HMEs)

establishing differentiation and developmental specific patterns of transcriptional activation and repression. This includes very well-studied mechanisms such as dosage compensation in flies and mammals, imprinting, and gene silencing by RNA interference (RNAi) phenomena in a variety of organisms (Bernstein and Allis 2005).

The catalog of functional natural antisense transcripts encoded in the human genome is rapidly expanding as well as the understanding of their roles in mediating the function of chromatin-modifying complexes. Differentially expressed natural antisense transcripts could bind diverse chromatin-modifying complexes that are ubiquitously expressed and help in establishing specific epigenetic states among different cell types and conditions.

Natural antisense transcripts have been proposed to cause DNA methylation (Tufarelli et al. 2003), DNA demethylation (Imamura et al. 2004), and chromatin modifications of non-imprinted autosomal loci (Morris et al. 2008; Yu et al. 2008) (Figs. 2 and 3). The involvement of noncoding RNAs in the epigenetic regulation of chromatin structure is suggested by the observation that many histone-modifying complexes lack DNA-binding domains but possess RNA-binding motifs (Bernstein and Allis 2005). RNA immunoprecipitation (RIP) experiments with antibodies directed against proteins involved in chromatin modification (PRC2, CoREST, and SMCX) revealed that a large fraction of newly discovered natural antisense transcripts are associated with one of these complexes. Moreover, siRNA-mediated depletion of natural antisense transcripts associated with the PRC2 complex induced the activation of several genes known to be targeted by PRC2 (Huarte et al. 2010).

5.2.1 ANRIL

The large antisense noncoding RNA in the *INK4b/ARF/INK4a* tumor suppressor locus ANRIL is a natural antisense noncoding RNA overlapping the INK4b/ARF/INK4a locus (Mattick 2009b). INK4b/ARF/INK4a locus encodes for two cyclin-dependent kinase inhibitors, p15INK4b and p16INK4a, and a regulator of the p53 pathway, ARF (Popov and Gil 2010). Transcription of this locus is controlled by the action of PRC2 and PRC1 that respectively initiate and maintain the silenced state (Cao et al. 2008). The chromobox protein CBX7 is a component of the PRC1, and it has been shown to directly repress the INK4a/ARF locus. CBX7 has been shown to possess RNA-binding activity in vitro (Bernard et al. 2005). *ANRIL* ncRNA, CBX7, and EZH2 were found to be upregulated in several tissue specimens from prostate cancer with an inverse correlation to the expression of INK4a (Yap et al. 2010). Moreover, *ANRIL* co-localized with EZH2 and CBX7 to the promoter region of INK4a in prostate cancer cells, where it is specifically associated with the PRC1 component CBX7. Thus, the natural antisense transcript *ANRIL* is involved in *cis* in the recruitment of PRC complexes to induce INK4a gene silencing (Yap et al. 2010).

In the case of p15, an inverse relationship (discordant regulation) was observed between the level of the p15 mRNA and a natural antisense transcript in the p15 locus (p15AS) in normal and leukemic cells (Yu et al. 2008). Experimentally increasing the level of p15AS in mouse ES cells reduced mRNA transcription due to the induction of repressive histone modifications in the p15 promoter (Yu et al. 2008). The silencing of the tumor suppressor gene *p15*, via DNA methylation and heterochromatin formation, was mediated by the natural antisense transcript *p15*AS. Epigenetic silencing of *p15* was not caused by transcriptional interference and was not Dicer dependent, excluding intermediate small RNAs originating from *p15*AS as mediating the silencing effect (Yu et al. 2008).

5.2.2 P21-AS transcript

Bidirectional transcription in the p21 locus generated an antisense transcript along with sense p21 mRNA (Morris et al. 2008). Targeting the antisense transcript with small RNAs led to increased p21 transcription in a breast cancer cell line that normally expresses the gene at a low level. Transcriptional activation in this case was due to the AGO2-dependent knockdown of the p21 antisense transcript, rather than binding of the small RNA to p21 promoter as previously proposed by Li et al. The p21 antisense ncRNA was shown to maintain the gene repressed in through an AGO1-dependent mechanism involving histone methylation. Consistently, transcriptional activation consequent to the siRNA-mediated knockdown of the p21 antisense ncRNA was associated with the loss of the suppressive H3K27me3 mark in the p21 promoter, although the mechanism by which the H3K27me3 mark was removed from the promoter was not determined (Morris et al. 2008). Since bidirectional transcription at gene promoters is widely present in the human genome, this mechanism of transcriptional regulation might be common and have an important pathogenetic role in human diseases, like cancer.

5.2.3 HOTAIR Transcript

The human genome encodes a number of sense and antisense large intergenic noncoding RNAs (lincRNAs) that are highly conserved and most likely exert important regulatory functions (Mattick 2009b). Several lincRNAs were identified using a method based on their association with a specific signature for actively transcribed genes. This signature, called the K4-K36 domain, allowed the identification of about 4,000 lincRNAs with clear evolution conservation (Guttman et al. 2009).

A large intergenic natural antisense transcript, encoded in the HOX-C cluster, termed *HOTAIR* is required for the correct expression of HOX-D cluster (Rinn et al. 2007). *HOTAIR* associates with the PRC2 complex, which comprises EZH2, SUZ12, and EDD, to silence and maintain a large domain of heterochromatin in the HOX-D cluster and on other chromosomes (Beuchle et al. 2001). In addition to the repressive polycomb-mediated regulation, HOX genes are also epigenetically regulated by trithorax proteins (Fanti et al. 2008). This protein complex enzymatically modifies histone H3 lysine 4 trimethylation, establishing an epigenetic mark of transcriptional active chromatin (Fanti et al. 2008). Different studies in *Drosophila melanogaster* and in mouse embryonic stem cells pointed out the ability of large intergenic noncoding RNAs to bind proteins of the TRX complex and to target them to specific loci (Sanchez-Elsner et al. 2006; Breiling et al. 2007).

It was recently shown in primary foreskin fibroblast and HeLa cells that *HOTAIR* acts as a modular scaffold ncRNA that interacts with both PRC2 and LSD1 complexes (Tsai et al. 2010). In this way, *HOTAIR* coordinates the binding of PRC2 and LSD1 to target genes on chromatin where they respectively catalyze methylation of H3K27 and demethylation of H3K4me2 (Tsai et al. 2010). Interestingly, altered *HOTAIR* expression has been found to be involved in human cancer-promoting genomic re-localization of the PRC2 complex and of the H3K27me3 mark (Gupta et al. 2010).

5.2.4 Genomic Imprinting

Imprinted genes are genes for which only one allele, maternal or paternal, is actively transcribed. Natural antisense transcripts are often associated with imprinted genes, with a frequency up to 81% in one study (Katayama et al. 2005) (Text Box-3). Fifteen percent of imprinted genes were previously reported to have an associated antisense transcript (Reik and Walter 2001). In a more recent reports, this fraction significantly increased to 24–47% of human and mouse imprinted loci (Zhang et al. 2006) and up to 81% when antisense transcripts to introns were included (Katayama et al. 2005). More than 160 imprinted genes have been identified so far in humans and mice, most of which are organized into clusters (http://igc.otago.ac.nz/home.html). Some imprinted genes, including insulin-like growth factor type 2 receptor (*IGFER*) (Sleutels et al. 2002) and potassium voltage-gated channel, KQT-like (*Kcnq1*) imprinting control region (Pandey et al. 2008), exhibit guided chromatin and DNA modification by antisense RNA that expands to include neighboring genes. These effects are not mediated through the RNAi pathway (Tomari and Zamore 2005). Instead, antisense RNA appears to recruit repressor complexes that modify chromatin into an inactive state. Suppressive chromatin modifications spread in both directions to neighboring genes, similar to X chromosome inactivation but with limited penetrance (Pandey et al. 2008).

There are several studies suggesting a pivotal role for antisense transcripts in imprinted genes, such as IGF2R (Wutz et al. 1997), KCNQ1 (Smilinich et al. 1999), UBE3A (Rougeulle et al. 1998), ATP10C, MKRN3, MAGEL2, NDN (Runte et al. 2001), Slc22a2, Slc22a3 (Sleutels et al. 2002), GNAS (Hayward and Bonthron 2000), and Gnas (Wroe et al. 2000). Although some controversy still exists (Rougeulle and Heard 2002; Sleutels et al. 2003), guided chromatin and DNA modification by antisense RNA and spreading to the neighboring genes was clearly shown for some imprinted genes, such as IGFER (Sleutels et al. 2002) and potassium voltage-gated channel, KQT-like (Kcnq1) imprinting control region (Thakur et al. 2004).

5.2.5 Kcnq1ot1 Transcript

Antisense transcription and its functional role in bidirectional silencing of Kcnq1 imprinted locus is a well-studied case that provides a model for involvement of antisense transcripts in allelic exclusion (Kanduri 2008). The antisense RNA, *Kcnq1ot1*, is essential for the silencing activity of *Kcnq1* imprinting control region, which in turn controls the imprinting of a cluster of neighboring genes on chromosome 11 (Thakur et al. 2004).

The effect of antisense RNA is not mediated through RNAi. Natural antisense transcripts rather appear to recruit repressor complexes, turning chromatin into an inactive state. Suppressive chromatin modifications are spreading in both directions to the neighboring genes, similar to the X chromosome inactivation but with a limited penetrance. *Kcnq1ot1* antisense RNA was shown to be involved in both establishment of transcriptional silencing as well as in maintenance of silencing through subsequent cell divisions (Pandey et al. 2008). Furthermore, hypomethylation of this region was documented in 40% of sporadic cases of Beckwith–Wiedemann syndrome (BWS) patients (Lee et al. 1999; Smilinich et al. 1999).

5.2.6 Air Transcript

A region that contains three imprinted, maternally expressed protein-coding genes (*Igf2r*/*Slc22a2*/*Slc22a3*) on chromosome 6 has been shown to be controlled by a paternally expressed noncoding *Air* RNA (Sleutels et al. 2002). Expression of *Air* (named Air, for antisense Igf2r RNA) correlates with repression of all three genes (Lyle et al. 2000). *Air* is a 108-kb unspliced and repeat-rich transcript, which overlaps with just one of these genes in an antisense orientation (Zwart et al. 2001). However, *Air* is required for silencing of all three genes, which is likely through a similar mechanism shown for *Kcnq1* imprinted locus.

5.2.7 UBE3A-ATS Transcript

In the case of ubiquitin ligase E3A (*UBE3A*), natural antisense transcript, *UBE3A-ATS*, is a very long (460-kb), spliced noncoding RNA located on chromosome 15. Paternal expression of *UBE3A-ATS* is responsible for monoallelic (maternal) expression of the *UBE3A* gene in the brain (Rougeulle et al. 1998; Chamberlain and Brannan 2001; Yamasaki et al. 2003). *UBE3A-ATS* lies within a highly complex locus containing several other imprinted genes including *ATP10C* and *UBE3A*, both show genomic imprinting (Herzing et al. 2001; Rougeulle and Heard 2002). Association studies have revealed an important link to *ATP10C* in autistic patients (Nurmi et al. 2003; Kato et al. 2008). Prader–Willi syndrome and Angelman syndrome also result from the disturbance of *UB3EA* loci-imprinted gene expression (Johnstone et al. 2006; Lalande and Calciano 2007). These human disorders emphasize the importance of studying the role of antisense transcripts in genomic imprinting.

5.2.8 X Chromosome Inactivation

Dosage compensation is a process that equalizes the output of transcription from the unequal number of X chromosomes found in male and females. This process takes place through non-protein-coding RNAs that orchestrate heterochromatin formation along the X chromosome to be inactivated. Two long non-protein-coding genes are transcribed from the X chromosome inactivation center, *XIST* (X-inactive specific transcript) and *TSIX* [X (inactive)-specific transcript, antisense]. These two sense and antisense transcripts control the silencing of the X chromosome (Ohhata et al. 2008). The X chromosome inactivation center is necessary and sufficient for X chromosome inactivation.

The *Tsix* gene encodes a large non-protein-coding RNA, transcribed antisense to *Xist* (Chow et al. 2005). *Tsix* is a negative regulator of *Xist* through a mechanism requiring overlapping transcription and blocks inactivation on the future active X chromosome. It has been shown that *Tsix* silences *Xist* through modification of the chromatin structure in the *Xist* promoter region. Premature termination of *Tsix* transcription, by introduction of polyadenylation signal, abolishes the repressive chromatin configuration at the *Xist* promoter on the mutated X chromosome (Ohhata et al. 2008).

The *XIST* gene is the only gene expressed from the inactive X chromosome but not from the active X chromosome (Brown et al. 1992). X chromosomes that lack the *Xist* gene cannot be inactivated. Prior to inactivation, both X chromosomes weakly express *Xist* RNA. During the inactivation process, the future active X chromosome ceases to express *Xist*, whereas the future inactive X chromosome dramatically increases *Xist* RNA production. On the future inactive X chromosome, *Xist* RNA progressively coats the chromosome, spreading out from the X inactivation center. The silencing of genes along inactive X chromosome occurs soon after coating by *Xist* RNA. Therefore, dosage compensation occurs through heterochromatin formation along the inactive X chromosome, which is coated by *Xist*.

Like *Xist*, prior to inactivation, both X chromosomes weakly express *Tsix* RNA. At the onset of X inactivation, the future inactive X chromosome ceases to express *Tsix* RNA (thereby increasing *Xist* expression), whereas active X chromosome continues to express *Tsix* for several days. Alterations in *Tsix* expression lead to distorted inactivation patterns, with deletions of the *Tsix* promoter resulting in preferential silencing of the mutant chromosome (Lee and Lu 1999; Lee 2000; Sado et al. 2001) and upregulation of *Tsix* expression, inhibiting X chromosome inactivation (Luikenhuis et al. 2001; Stavropoulos et al. 2001).

Overall, natural antisense transcripts may have a common role for interacting with chromatin-modifying complexes (PRC2 and TRX), or DNA methyltransferases, in order to coordinate the epigenetic regulation of certain genomic domains. This proposed mechanism might also explain functionality even if the abundance of antisense RNA molecule is low (Morris et al. 2008; Preker et al. 2008; Seila et al. 2008). Unlike RNAs that can be present in cells in many copies, there are two copies of DNA for any given gene in cells; therefore, only two molecules of antisense RNA per cell are sufficient to bind to the corresponding DNA strand and to exert a regulatory function. These proposed local effects of antisense transcripts might sometimes extend to the neighboring regions, such as in the case of the imprinted locus (Kanduri 2008). These examples suggest the possibility of local and extended antisense RNA-induced DNA and chromatin modifications.

5.3 *Nuclear RNA Duplex Formation*

The third type of mechanism for antisense transcript-mediated gene regulation is based on the predicted nuclear sense–antisense RNA duplex formation (Fig. 4). The

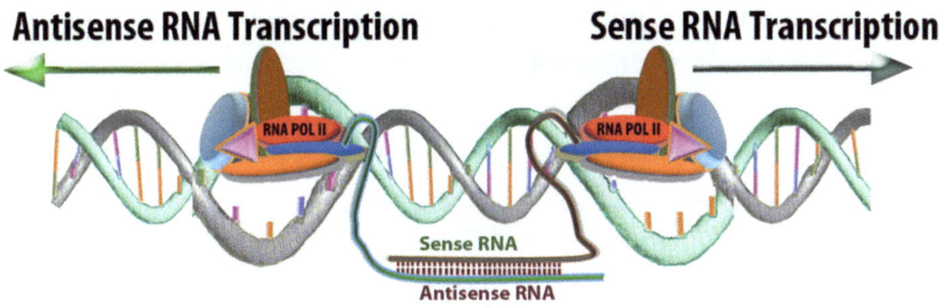

Fig. 4 Nuclear sense–antisense RNA pairing

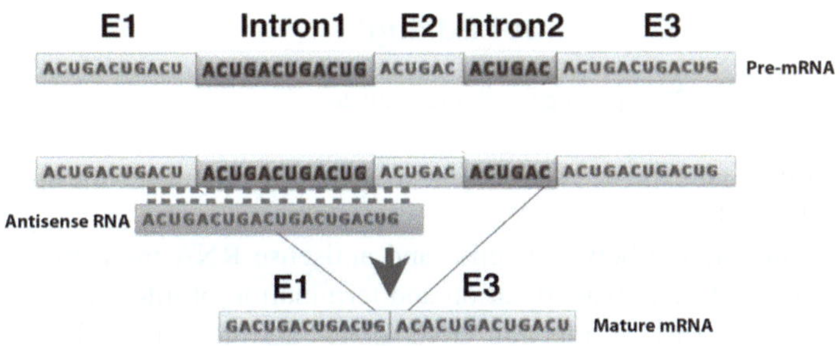

Fig. 5 Sense–antisense RNA pairing as splicing regulator

duplex RNA between sense and antisense transcripts may result in several outcomes, all of which modulate sense mRNA expression. Nuclear RNA duplex may produce mRNA transcripts with alternative splicing. Additionally, hybrid nuclear RNA is a substrate for editing enzymes that can alter the localization, transport, and stability of the sense mRNA transcript. Antisense transcript can potentially form a triple helix with DNA strands and cause alternative initiation and termination of the sense RNA (Enerly et al. 2005).

Nuclear RNA duplex formation might occur locally after transcription, thereby inhibiting sense RNA processing. The RNA processing includes capping, polyadenylation, nuclear localization, and transport, which all might be affected by nuclear sense–antisense RNA duplex formation.

5.3.1 Alternative Splicing and Transcription Termination

Antisense RNA may also bind to the sense RNA, masking the splice sites and thereby changing the balances between splice variants. This is the case of the thyroid hormone receptor alpha gene (*TRα*), where the antisense transcript *RevErbAα* influences splicing of *TRα1* and *TRα2* mRNAs (Hastings et al. 1997) (Fig. 5). Also in the case of N-Myc antisense, N-cym, the antisense transcript forms an RNA duplex with the

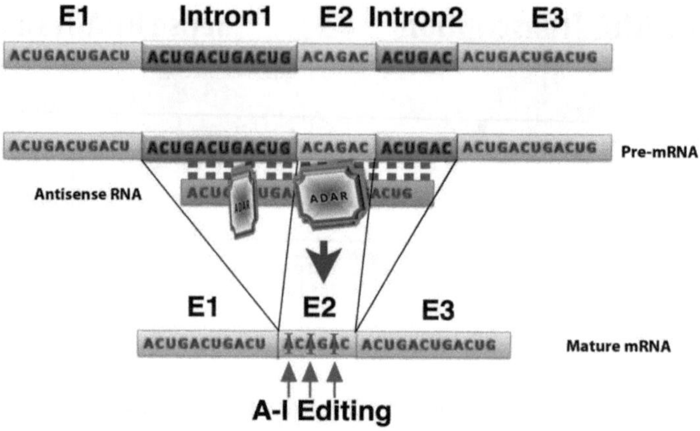

Fig. 6 Sense–antisense RNA pairing as editing regulator

donor site of the first exon of N-Myc gene and thereby modulates the splice variants (Krystal et al. 1990).

A similar interaction between sense and antisense RNA molecules can potentially cause alternative polyadenylation and termination of mRNA transcription.

Natural antisense transcripts can cover donor and acceptor splice sites in the sense pre-mRNA transcript to change alternative splicing patterns.

5.3.2 RNA Transport, Nuclear Retention, and Editing

Natural antisense transcripts may modulate mRNA nuclear transport by a mechanism that involves duplex formation between sense and antisense RNAs. For instance, the antisense RNA for the non-spliced p53 RNA binds to intron-1, preventing transport of p53 RNA to the cytosol (Khochbin et al. 1992).

Some cellular stressors, such as hypoxia, serum starvation, and hydrogen peroxide, can change the nuclear retention pattern of antisense transcripts, thereby altering the levels of their sense partners (Faghihi et al. 2008). Direct interactions with nuclear proteins or other nuclear RNAs likely cause nuclear retention of antisense transcripts, although the precise mechanism for each case has not yet been discovered.

Natural antisense transcripts have also been linked to mRNA editing (Hatzoglou et al. 2002; Peters et al. 2003) (Fig. 6). Interaction between the *D. melanogaster 4f-rnp* gene and its antisense transcript, *sas-10*, is reported to induce adenosine-to-inosine (A-to-I) editing in the overlapping region of *4f-rnp* mRNA (Peters et al. 2003).

RNA editing is another possible consequence of nuclear RNA duplex formation, which causes A-to-I editing by the ADAR (adenosine deaminase that acts on RNA) enzyme. RNA editing can induce nuclear retention and degradation of hyper-edited transcript by inosine-specific nucleases, or alternatively, editing can result in alteration of the amino acid sequence.

A-to-I RNA editing is induced by double-stranded RNA formation, resulting in the recruitment of the enzyme ADAR (adenosine deaminase that act on RNA), which deaminates the targeted adenosine to inosine (Ohman 2007).

There are few documented reports on involvement of natural antisense transcripts in A-to-I RNA editing (Kimelman and Kirschner 1989; Dolnick 1993; Peters et al. 2003), and the overlapping regions of many sense–antisense pairs contain A–G mismatches (Hatzoglou et al. 2002; Faghihi et al. 2008), suggesting possible RNA editing. However, there is a bioinformatics report arguing that the overlapping regions of the sense–antisense transcripts are not extensively edited, which may still not exclude the possibility of alternative forms or degrees of editing of sense–antisense transcripts (Neeman et al. 2005). Nevertheless, editing is sometimes followed by the degradation of sense RNA via inosine-specific RNAse activity (Kumar and Carmichael 1997; Scadden and Smith 1997).

5.4 Cytoplasmic RNA Duplex Formation

In the fourth proposed mechanism of sense–antisense interference, a duplex forms between sense and antisense RNA in the cytoplasm. Cytoplasmic RNA hairpins may affect sense mRNA stability or translation. The sense–antisense RNA duplex may also, in theory, cover microRNA (miRNA)-binding sites or serve as a hairpin template for generating endogenous small interfering RNA (siRNA).

5.4.1 Changes in mRNA Stability

Cytoplasmic sense–antisense duplex formation can alter sense mRNA stability. We previously showed that cytoplasmic sense–antisense duplexes are not normally subject to Dicer cleavage in mammalian cells and thus do not generally form a basis for endogenous RNA interference (Faghihi and Wahlestedt 2006). However, duplex formation is still biologically relevant should antisense transcripts modulate the stability of their sense partners.

5.4.2 BACE1-AS Is an Example of Alteration in mRNA Stability

The overlapping region may affect mRNA stability by reducing mRNA decay, whereby mRNA undergoes endo- or exonucleolytic degradation by various RNases. Indeed, we have recently demonstrated that an antisense *BACE1* transcript (*BACE1*-AS) increases the stability of *BACE1* mRNA through a mechanism involving RNA duplex formation. We hypothesize that transient RNA duplex formation may alter the secondary or tertiary structure of *BACE1* and thereby increase its stability (Faghihi et al. 2008).

Antisense transcripts for inducible nitric oxide synthase (*iNOS*), an important gene in inflammatory diseases, increase the stability of *iNOS* mRNA (Matsui et al. 2008). Enhancement of *iNOS* mRNA stability is mediated through interactions of antisense RNA molecule with the AU-rich element (ARE)-binding HuR protein. The HuR protein, in turn, may suppress RNA degradation by inhibiting deadenylase or exonuclease enzymes (Matsui et al. 2008). Alterations in the secondary structure of HIF-1α sense transcript by an antisense RNA may expose the ARE and reduce the stability of the transcript by making this RNA prone to degradation (Uchida et al. 2004).

Natural antisense transcript for basic fibroblast growth factor (bFGF) has been also reported to alter the stability of bFGF mRNA (Volk et al. 1989; Zuniga Mejia Borja et al. 1993), possibly through similar mechanism. The antisense transcript for hypoxia-inducible factor, HIF-1α (aHIF), is yet another example of decrease in RNA stability induced by antisense RNA. The aHIF non-protein-coding RNA destabilizes one isoform of HIF-α mRNA and shifts the balance in favor of the other variant (Cayre et al. 2003; Uchida et al. 2004). Two isoforms of HIF, HIF-1α and HIF-2α, are both responsive to hypoxia, but HIF-1α has a more ubiquitous expression pattern than HIF-2α. Prolonged hypoxia causes induction of antisense transcript, aHIF, which in turn binds to sense mRNA, leading to reduction of HIF-1α transcript. Importantly, destabilization takes place by exposing the AU-rich elements (ARE) in HIF-1α mRNA following antisense binding to its 3′ UTR (Rossignol et al. 2002, 2004; Uchida et al. 2004).

An AU-rich element (ARE) is a region in RNA transcript with frequent "A" and "U" nucleotides, such as "AUUUA," that targets the RNA for degradation. Alterations in HIF-1α secondary structure expose the ARE and make this RNA prone for degradation. Stabilization of mRNA by an opposite mechanism, covering the AU-rich element, has been suggested for antisense transcript of the Bcl-2/IgH hybrid gene (Capaccioli et al. 1996).

5.4.3 Cytoplasmic RNA Duplex Formation and "Masking" miRNA-Binding Sites

We propose that many natural antisense transcripts may have the ability to cover miRNA-binding sites following cytosolic RNA duplex formation. This appears to be the case for an antisense *BACE1* transcript (*BACE1*-AS) in addition to its ability to increase the stability of *BACE1* mRNA (Faghihi et al. 2008). Thus, one of the regulatory functions of antisense transcripts may be to "mask" the miRNA-binding site on the sense mRNA. The 3′ UTR mRNAs are frequently shown to contain target sites for miRNAs, and at least 34% of natural antisense transcripts in the FANTOM3 dataset showed tail-to-tail format with a 3′ UTR overlapping region.

In contrast with plant miRNA, most animal miRNAs are predicted to have their binding site in the 3′ UTR of target mRNA (Arteaga-Vazquez et al. 2006). Although most web tools for miRNA-binding sites are designed to screen 3′ UTR regions of transcripts, there is no evidence that miRNA does not bind to the coding region. Binding of miRNA to the coding region of the mRNA, or even 5′ UTR, has been shown in plants and recently in animals (Rhoades et al. 2002; Lytle et al. 2007). We propose that one of the regulatory functions of antisense transcripts could occur by an ability of the antisense transcript to "mask" the miRNA-binding site on the sense mRNA.

5.4.4 Generation of Endogenous siRNA

RNAi is a naturally occurring, evolutionary conserved, highly efficient, and specific pathway by which short double-stranded RNAs (dsRNA) trigger the inhibition of gene expression. In 1998, Fire and Mello discovered that the injection of long double-stranded RNAs into *Caenhorabditis elegans* triggered potent and specific gene silencing. The specificity was achieved by small RNA molecule (20–30 nucleotides long), which was termed small interfering RNAs (siRNAs) (Fire et al. 1998). siRNA-mediated silencing has now been observed in the vast majority of the eukaryotes, is widely used to analyze the function of proteins by knocking down their associated mRNAs, and offers the promise of new therapeutic strategies.

Short RNA-dependent pathways can act in the nucleus and the cytoplasm by mediating, respectively, transcriptional gene silencing (TGS) and posttranscriptional gene silencing (PTGS). In both situations, a similar set of steps is engaged from the activation of the pathway to the silencing of target gene. siRNA-mediated silencing pathways are activated by long double-stranded RNAs that are processed into siRNAs by Dicer, a type III ribonuclease. siRNAs are then incorporated onto RNAi effector complexes that use the siRNA sequence as a specificity determinant to associate with the complementary RNA target. The RNAi-induced silencing complex (RISC) and the RNAi-induced transcriptional silencing (RITS) complex mediate, respectively, posttranscriptional and transcriptional silencing (Grimm 2009). Both types of complexes contain an Argonaute protein, but unlike RISC, RITS localizes exclusively in the nucleus.

Once a perfect match occurs between the siRNA and a targeted RNA, the latter is cleaved (Cullen 2002). In fission yeast, *Neurospora crassa*, *C. elegans*, and plants, RNA-dependent RNA polymerases (RDRP) amplify the silencing response by using the target RNA as a template to synthesize more dsRNAs that fuel the siRNA pathway and amplify the siRNA-mediated silencing. Fly and vertebrate genomes do not possess genes encoding proteins homologous to RDRPs, indicating that these organisms might do without this amplification step (Duan et al. 2010). Another major distinction between nuclear and cytoplasmic siRNA-mediated silencing pathways is that in the first instance siRNAs can trigger a long-term and heritable gene silencing which is maintained in an epigenetic fashion (Hawkins and Morris 2008). To date, the ability of siRNA to modulate TGS has been shown in plants (Arabidopsis), yeast (*Schizosaccharomyces pombe*), worms (*C. elegans*), and more recently in mammalian cells (Pastori et al. 2010).

In mammals, where no RNA-dependent RNA polymerase (RdRP) activity has yet been found, biogenesis and function of endogenous siRNAs remain largely unknown

(Okamura and Lai 2008). Endogenous siRNAs derived from transposable elements and pseudogenes have also been identified in mouse oocytes and cultured human cells (Watanabe et al. 2006; Yang and Kazazian 2006; Kawaji et al. 2008; Tam et al. 2008; Watanabe et al. 2008). Endogenous siRNAs originating from mRNAs and their corresponding antisense were also recently identified in mouse oocytes (Watanabe et al. 2008) and human HepG2 liver carcinoma cells (Kawaji et al. 2008). Both Piwi-interacting RNAs (piRNAs, Dicer independent) (Aravin et al. 2007; Klattenhoff and Theurkauf 2008) and approximately 21-nucleotide siRNAs originating from mRNAs were found in mouse oocytes (Watanabe et al. 2008).

Transposable elements, inverted repeat structures, sense–antisense genes (*cis*-natural antisense transcript), and antisense transcripts from remote loci (*trans*-natural antisense transcripts) have also been recognized as sources of dsRNAs and subsequent, Dicer-dependent, endogenous siRNA production (Kawaji et al. 2008; Watanabe et al. 2008). Therefore, formation of endogenous siRNA from naturally occurring antisense transcripts appears to occur in mammalian cells.

6 Four Models for Antisense Transcripts Regulatory Functions

We presented examples of functional natural antisense transcripts in order to show the multilayered involvement of these molecules in regulating gene expression. It can be concluded that antisense transcripts are not a uniform group of regulatory elements, but that they display some common features. In summary, the proposed mechanisms of antisense transcript actions can be categorized into four models, each corresponding to a putative regulatory mechanism for certain antisense transcripts.

Although there are examples for each model, it is important to note that not all four models have equally strong experimental support. For instance, the frequently observed convergent promoter activity and engagement of multiple RNA polymerase II complexes in the opposite orientation of active promoters would argue against the transcriptional collision model as a widely used mode of action. On the other hand, nuclear and cytoplasmic RNA–RNA interactions with various outcomes, such as RNA editing, splicing, and dicing, have been observed for several functionally validated antisense transcripts, but it is not a predominant basis of natural antisense regulation. Furthermore, antisense RNA-induced alteration in sense promoter DNA methylation has only a few documented examples, and considering the static nature of DNA methylation, it might be operational only in certain early developmental stages. Conversely, antisense RNA-induced chromatin remodeling seems to be a more feasible and dynamic mode of action for many low-copy natural antisense transcripts. In the latter case, antisense RNA might predominantly act locally to maintain or modify chromatin structure and ultimately to activate or suppress sense gene expression. Natural antisense transcripts are involved in different gene regulatory pathways, but it is still not clear which intrinsic properties of natural antisense

RNA molecules, or extrinsic features such as protein interactions and cellular and developmental context, are decisive for any given pathway.

It is becoming clear that the expression of natural antisense transcripts and other non-protein-coding RNAs is pervasive in the human genome, although the mechanisms of their production and preferential sites of action are poorly understood. Particularly, the exact molecular machineries behind the antisense RNA-induced chromatin modifications are not currently known. It will be important to learn which histone-modifying enzymes might be involved, how they interact with antisense RNAs, and which histone modifications they induce over time. It will also be of interest to define the exact contexts in which perturbation of the same antisense transcript (Schwartz et al. 2008) results in either activation or repression of sense RNA expression. It is remarkable that these fundamental and powerful regulatory mechanisms relating to natural antisense transcription remain to be elucidated in large part. It can be safely assumed that examination of natural antisense phenomena will remain a rich field of investigation for the foreseeable future.

7 Conclusions

Natural antisense transcripts are functional RNA molecules with the potential to control sense mRNA expression at different levels.

Most natural antisense transcripts are noncoding RNAs that, in response to a variety of environmental, contribute to keep tight regulation of sense protein expression (allowing proteins to perform their physiological functions while avoiding the serious consequences of over or under expression).

There are several examples of functional natural antisense transcripts showing a multilayer involvement of these molecules in gene expression regulation. Non-protein-coding RNAs are shown to be involved in many complex human disorders, such as Alzheimer's disease and cancers. Furthermore, the level of antisense transcripts has been found to correlate both positively and negatively with the expression of the associated protein-coding gene. Thus, antisense ncRNAs are a very heterogeneous class of regulatory ncRNAs probably exerting distinct functions related to the transcriptional state of the gene (Katayama et al. 2005; Faghihi and Wahlestedt 2009). In a recent study, many genes expressed in normal and breast tumor tissues were found to have overlapping antisense transcripts (Maruyama et al. 2010). A large set of differentially expressed genes had altered ratios of sense and antisense transcripts between normal and breast tumor samples. Antisense transcripts were often more abundant in normal cells, suggesting a role for these transcripts in gene regulation and tumorigenesis (Maruyama et al. 2010).

Although protein synthesis was the first assigned function for RNA molecules, it has become increasingly evident that the more pervasive function of RNA molecules is a regulatory one. This hypothesis is considerably supported with the content of this current book in which we summarize proposed regulatory functions of naturally occurring antisense transcripts. Considering other reported functional long non-protein-coding RNAs (macroRNAs), and small regulatory RNA (such as miRNA, piRNA, rasiRNA) and their enormous physiological impact, it is feasible to claim that the more frequent function of RNA molecules is a regulatory role, which is far greater than their assigned functions in protein synthesis as messenger, transfer, and ribosomal RNA.

References

Apel TW, Mautner J, Polack A et al (1992) Two antisense promoters in the immunoglobulin mu-switch region drive expression of c-myc in the Burkitt's lymphoma cell line BL67. Oncogene 7:1267–1271

Aravin AA, Hannon GJ, Brennecke J (2007) The Piwi-piRNA pathway provides an adaptive defense in the transposon arms race. Science 318:761–764

Arteaga-Vazquez M, Caballero-Perez J, Vielle-Calzada JP (2006) A family of microRNAs present in plants and animals. Plant Cell 18:3355–3369

Beltran M, Puig I, Pena C et al (2008) A natural antisense transcript regulates Zeb2/Sip1 gene expression during Snail1-induced epithelial-mesenchymal transition. Genes Dev 22:756–769

Bernard D, Martinez-Leal JF, Rizzo S et al (2005) CBX7 controls the growth of normal and tumor-derived prostate cells by repressing the Ink4a/Arf locus. Oncogene 24:5543–5551

Bernstein E, Allis CD (2005) RNA meets chromatin. Genes Dev 19:1635–1655

Bertone P, Stolc V, Royce TE et al (2004) Global identification of human transcribed sequences with genome tiling arrays. Science 306:2242–2246

Beuchle D, Struhl G, Muller J (2001) Polycomb group proteins and heritable silencing of Drosophila Hox genes. Development 128:993–1004

Birney E, Stamatoyannopoulos JA, Dutta A et al (2007) Identification and analysis of functional elements in 1% of the human genome by the ENCODE pilot project. Nature 447:799–816

Bolland DJ, Wood AL, Johnston CM et al (2004) Antisense intergenic transcription in V(D)J recombination. Nat Immunol 5:630–637

Bolland DJ, Wood AL, Afshar R et al (2007) Antisense intergenic transcription precedes Igh D-to-J recombination and is controlled by the intronic enhancer Emu. Mol Cell Biol 27: 5523–5533

Breiling A, Sessa L, Orlando V (2007) Biology of polycomb and trithorax group proteins. Int Rev Cytol 258:83–136

Brown CJ, Hendrich BD, Rupert JL et al (1992) The human XIST gene: analysis of a 17 kb inactive X-specific RNA that contains conserved repeats and is highly localized within the nucleus. Cell 71:527–542

Cao Q, Yu J, Dhanasekaran SM et al (2008) Repression of E-cadherin by the polycomb group protein EZH2 in cancer. Oncogene 27:7274–7284

Capaccioli S, Quattrone A, Schiavone N et al (1996) A bcl-2/IgH antisense transcript deregulates bcl-2 gene expression in human follicular lymphoma t(14;18) cell lines. Oncogene 13:105–115

Carninci P, Kasukawa T, Katayama S et al (2005) The transcriptional landscape of the mammalian genome. Science 309:1559–1563

Cawley S, Bekiranov S, Ng HH et al (2004) Unbiased mapping of transcription factor binding sites along human chromosomes 21 and 22 points to widespread regulation of noncoding RNAs. Cell 116:499–509

Cayre A, Rossignol F, Clottes E et al (2003) aHIF but not HIF-1alpha transcript is a poor prognostic marker in human breast cancer. Breast Cancer Res 5:R223–R230

Chamberlain SJ, Brannan CI (2001) The Prader-Willi syndrome imprinting center activates the paternally expressed murine Ube3a antisense transcript but represses paternal Ube3a. Genomics 73:316–322

Chan WY, Wu SM, Ruszczyk L et al (2006) The complexity of antisense transcription revealed by the study of developing male germ cells. Genomics 87:681–692

Chen J, Sun M, Kent WJ et al (2004) Over 20% of human transcripts might form sense-antisense pairs. Nucleic Acids Res 32:4812–4820

Chen J, Sun M, Hurst LD et al (2005) Human antisense genes have unusually short introns: evidence for selection for rapid transcription. Trends Genet 21:203–207

Cheng J, Kapranov P, Drenkow J et al (2005) Transcriptional maps of 10 human chromosomes at 5-nucleotide resolution. Science 308:1149–1154

Chow JC, Yen Z, Ziesche SM et al (2005) Silencing of the mammalian X chromosome. Annu Rev Genomics Hum Genet 6:69–92

Core LJ, Waterfall JJ, Lis JT (2008) Nascent RNA sequencing reveals widespread pausing and divergent initiation at human promoters. Science 322:1845–1848

Cullen BR (2002) RNA interference: antiviral defense and genetic tool. Nat Immunol 3:597–599

Dahary D, Elroy-Stein O, Sorek R (2005) Naturally occurring antisense: transcriptional leakage or real overlap? Genome Res 15:364–368

Dolnick BJ (1993) Cloning and characterization of a naturally occurring antisense RNA to human thymidylate synthase mRNA. Nucleic Acids Res 21:1747–1752

Dreesen TD, Adamson AW, Tekle M et al (2006) A newly discovered member of the fatty acid desaturase gene family: a non-coding, antisense RNA gene to delta5-desaturase. Prostaglandins Leukot Essent Fatty Acids 75:97–106

Duan G, Saint RB, Helliwell CA et al (2010) Expression of *Caenorhabditis elegans* RNA-directed RNA polymerase in transgenic *Drosophila melanogaster* does not affect morphological development. Transgenic Res 19:1121–1128

Enerly E, Sheng Z, Li KB (2005) Natural antisense as potential regulator of alternative initiation, splicing and termination. In Silico Biol 5:367–377

Faghihi MA, Wahlestedt C (2006) RNA interference is not involved in natural antisense mediated regulation of gene expression in mammals. Genome Biol 7:R38

Faghihi MA, Wahlestedt C (2009) Regulatory roles of natural antisense transcripts. Nat Rev Mol Cell Biol 10:637–643

Faghihi MA, Modarresi F, Khalil AM et al (2008) Expression of a noncoding RNA is elevated in Alzheimer's disease and drives rapid feed-forward regulation of beta-secretase. Nat Med 14:723–730

Fahey ME, Moore TF, Higgins DG (2002) Overlapping antisense transcription in the human genome. Comp Funct Genomics 3:244–253

Fanti L, Perrini B, Piacentini L et al (2008) The trithorax group and Pc group proteins are differentially involved in heterochromatin formation in Drosophila. Chromosoma 117:25–39

Finocchiaro G, Carro MS, Francois S et al (2007) Localizing hotspots of antisense transcription. Nucleic Acids Res 35:1488–1500

Fire A, Xu S, Montgomery MK et al (1998) Potent and specific genetic interference by double-stranded RNA in *Caenorhabditis elegans*. Nature 391:806–811

Ge X, Wu Q, Jung YC et al (2006) A large quantity of novel human antisense transcripts detected by LongSAGE. Bioinformatics 22:2475–2479

Ge X, Rubinstein WS, Jung YC et al (2008) Genome-wide analysis of antisense transcription with Affymetrix exon array. BMC Genomics 9:27

Grimm D (2009) Small silencing RNAs: state-of-the-art. Adv Drug Deliv Rev 61:672–703

Gupta RA, Shah N, Wang KC et al (2010) Long non-coding RNA HOTAIR reprograms chromatin state to promote cancer metastasis. Nature 464:1071–1076

Guttman M, Amit I, Garber M et al (2009) Chromatin signature reveals over a thousand highly conserved large non-coding RNAs in mammals. Nature 458:223–227

Gyorffy A, Surowiak P, Tulassay Z et al (2007) Highly expressed genes are associated with inverse antisense transcription in mouse. J Genet 86:103–109

Hastings ML, Milcarek C, Martincic K et al (1997) Expression of the thyroid hormone receptor gene, erbAalpha, in B lymphocytes: alternative mRNA processing is independent of differentiation but correlates with antisense RNA levels. Nucleic Acids Res 25:4296–4300

Hatzoglou A, Deshayes F, Madry C et al (2002) Natural antisense RNA inhibits the expression of BCMA, a tumour necrosis factor receptor homologue. BMC Mol Biol 3:4

Hawkins PG, Morris KV (2008) RNA and transcriptional modulation of gene expression. Cell Cycle 7:602–607

Hayward BE, Bonthron DT (2000) An imprinted antisense transcript at the human GNAS1 locus. Hum Mol Genet 9:835–841

He Y, Vogelstein B, Velculescu VE et al (2008) The antisense transcriptomes of human cells. Science 322:1855–1857

Herzing LB, Kim SJ, Cook EH et al (2001) The human aminophospholipid-transporting ATPase gene ATP10C maps adjacent to UBE3A and exhibits similar imprinted expression. Am J Hum Genet 68:1501–1505

Huarte M, Guttman M, Feldser D et al (2010) A large intergenic noncoding RNA induced by p53 mediates global gene repression in the p53 response. Cell 142:409–419

Imamura T, Yamamoto S, Ohgane J et al (2004) Non-coding RNA directed DNA demethylation of Sphk1 CpG island. Biochem Biophys Res Commun 322:593–600

Johnstone KA, DuBose AJ, Futtner CR et al (2006) A human imprinting centre demonstrates conserved acquisition but diverged maintenance of imprinting in a mouse model for Angelman syndrome imprinting defects. Hum Mol Genet 15:393–404

Julius MA, Street AJ, Fahrlander PD et al (1988) Translocated c-myc genes produce chimeric transcripts containing antisense sequences of the immunoglobulin heavy chain locus in mouse plasmacytomas. Oncogene 2:469–476

Kanduri C (2008) Functional insights into long antisense noncoding RNA Kcnq1ot1 mediated bidirectional silencing. RNA Biol 5:208–211

Kapranov P, Cheng J, Dike S et al (2007) RNA maps reveal new RNA classes and a possible function for pervasive transcription. Science 316:1484–1488

Katayama S, Tomaru Y, Kasukawa T et al (2005) Antisense transcription in the mammalian transcriptome. Science 309:1564–1566

Kato C, Tochigi M, Ohashi J et al (2008) Association study of the 15q11-q13 maternal expression domain in Japanese autistic patients. Am J Med Genet B Neuropsychiatr Genet 147B:1008–1012

Kawaji H, Nakamura M, Takahashi Y et al (2008) Hidden layers of human small RNAs. BMC Genomics 9:157

Khochbin S, Brocard MP, Grunwald D et al (1992) Antisense RNA and p53 regulation in induced murine cell differentiation. Ann NY Acad Sci 660:77–87

Kimelman D, Kirschner MW (1989) An antisense mRNA directs the covalent modification of the transcript encoding fibroblast growth factor in Xenopus oocytes. Cell 59:687–696

Kiyosawa H, Yamanaka I, Osato N et al (2003) Antisense transcripts with FANTOM2 clone set and their implications for gene regulation. Genome Res 13:1324–1334

Klattenhoff C, Theurkauf W (2008) Biogenesis and germline functions of piRNAs. Development 135:3–9

Klimov D, Skoblov M, Ryazantzev A et al (2006) In silico search for natural antisense transcripts reveals their differential expression in human tumors. J Bioinform Comput Biol 4:515–521

Komine Y, Nakamura K, Katsuki M et al (2006) Novel transcription factor zfh-5 is negatively regulated by its own antisense RNA in mouse brain. Mol Cell Neurosci 31:273–283

Kramer C, Loros JJ, Dunlap JC et al (2003) Role for antisense RNA in regulating circadian clock function in *Neurospora crassa*. Nature 421:948–952

Krystal GW, Armstrong BC, Battey JF (1990) N-myc mRNA forms an RNA-RNA duplex with endogenous antisense transcripts. Mol Cell Biol 10:4180–4191

Kumar M, Carmichael GG (1997) Nuclear antisense RNA induces extensive adenosine modifications and nuclear retention of target transcripts. Proc Natl Acad Sci USA 94:3542–3547

Kumar M, Carmichael GG (1998) Antisense RNA: function and fate of duplex RNA in cells of higher eukaryotes. Microbiol Mol Biol Rev 62:1415–1434

Lalande M, Calciano MA (2007) Molecular epigenetics of Angelman syndrome. Cell Mol Life Sci 64:947–960

Larijani M, Martin A (2007) Single-stranded DNA structure and positional context of the target cytidine determine the enzymatic efficiency of AID. Mol Cell Biol 27:8038–8048

Lee JT (2000) Disruption of imprinted X inactivation by parent-of-origin effects at Tsix. Cell 103:17–27

Lee JT, Lu N (1999) Targeted mutagenesis of Tsix leads to nonrandom X inactivation. Cell 99:47–57

Lee MP, DeBaun MR, Mitsuya K et al (1999) Loss of imprinting of a paternally expressed transcript, with antisense orientation to KVLQT1, occurs frequently in Beckwith-Wiedemann syndrome and is independent of insulin-like growth factor II imprinting. Proc Natl Acad Sci USA 96:5203–5208

Lehner B, Williams G, Campbell RD et al (2002) Antisense transcripts in the human genome. Trends Genet 18:63–65

Li YY, Qin L, Guo ZM et al (2006) In silico discovery of human natural antisense transcripts. BMC Bioinformatics 7:18

Li JT, Zhang Y, Kong L et al (2008) Trans-natural antisense transcripts including noncoding RNAs in 10 species: implications for expression regulation. Nucleic Acids Res 36:4833–4844

Luikenhuis S, Wutz A, Jaenisch R (2001) Antisense transcription through the Xist locus mediates Tsix function in embryonic stem cells. Mol Cell Biol 21:8512–8520

Lyle R, Watanabe D, te Vruchte D et al (2000) The imprinted antisense RNA at the Igf2r locus overlaps but does not imprint Mas1. Nat Genet 25:19–21

Lytle JR, Yario TA, Steitz JA (2007) Target mRNAs are repressed as efficiently by microRNA-binding sites in the 5' UTR as in the 3' UTR. Proc Natl Acad Sci USA 104:9667–9672

Makalowska I, Lin CF, Makalowski W (2005) Overlapping genes in vertebrate genomes. Comput Biol Chem 29:1–12

Maruyama R, Shipitsin M, Choudhury S et al (2010) Breast cancer special feature: altered antisense-to-sense transcript ratios in breast cancer. Proc Natl Acad Sci USA [Epub ahead of print]

Matsui K, Nishizawa M, Ozaki T et al (2008) Natural antisense transcript stabilizes inducible nitric oxide synthase messenger RNA in rat hepatocytes. Hepatology 47:686–697

Mattick JS (2009a) Deconstructing the dogma: a new view of the evolution and genetic programming of complex organisms. Ann NY Acad Sci 1178:29–46

Mattick JS (2009b) The genetic signatures of noncoding RNAs. PLoS Genet 5:e1000459

Mattick JS, Taft RJ, Faulkner GJ (2010) A global view of genomic information–moving beyond the gene and the master regulator. Trends Genet 26:21–28

Morris KV, Santoso S, Turner AM et al (2008) Bidirectional transcription directs both transcriptional gene activation and suppression in human cells. PLoS Genet 4:e1000258

Neckers LM (1999) aHIF: the missing link between HIF-1 and VHL? J Natl Cancer Inst 91:106–107

Neeman Y, Dahary D, Levanon EY et al (2005) Is there any sense in antisense editing? Trends Genet 21:544–547

Nurmi EL, Amin T, Olson LM et al (2003) Dense linkage disequilibrium mapping in the 15q11-q13 maternal expression domain yields evidence for association in autism. Mol Psychiatry 8(624–634):570

Ohhata T, Hoki Y, Sasaki H et al (2008) Crucial role of antisense transcription across the Xist promoter in Tsix-mediated Xist chromatin modification. Development 135:227–235

Ohman M (2007) A-to-I editing challenger or ally to the microRNA process. Biochimie 89:1171–1176

Okada Y, Tashiro C, Numata K et al (2008) Comparative expression analysis uncovers novel features of endogenous antisense transcription. Hum Mol Genet 17:1631–1640

Okamura K, Lai EC (2008) Endogenous small interfering RNAs in animals. Nat Rev Mol Cell Biol 9:673–678

Okazaki Y, Furuno M, Kasukawa T et al (2002) Analysis of the mouse transcriptome based on functional annotation of 60,770 full-length cDNAs. Nature 420:563–573
Osato N, Suzuki Y, Ikeo K et al (2007) Transcriptional interferences in cis natural antisense transcripts of humans and mice. Genetics 176:1299–1306
Pandey RR, Mondal T, Mohammad F et al (2008) Kcnq1ot1 antisense noncoding RNA mediates lineage-specific transcriptional silencing through chromatin-level regulation. Mol Cell 32:232–246
Pastori C, Magistri M, Napoli S et al (2010) Small RNA-directed transcriptional control: new insights into mechanisms and therapeutic applications. Cell Cycle 9:2353–2362
Perlot T, Li G, Alt FW (2008) Antisense transcripts from immunoglobulin heavy-chain locus V(D)J and switch regions. Proc Natl Acad Sci USA 105:3843–3848
Peters NT, Rohrbach JA, Zalewski BA et al (2003) RNA editing and regulation of Drosophila 4f-rnp expression by sas-10 antisense readthrough mRNA transcripts. RNA 9:698–710
Popov N, Gil J (2010) Epigenetic regulation of the INK4b-ARF-INK4a locus: in sickness and in health. Epigenetics 5:685–690
Preker P, Nielsen J, Kammler S et al (2008) RNA exosome depletion reveals transcription upstream of active human promoters. Science 322:1851–1854
Prescott EM, Proudfoot NJ (2002) Transcriptional collision between convergent genes in budding yeast. Proc Natl Acad Sci USA 99:8796–8801
Quere R, Manchon L, Lejeune M et al (2004) Mining SAGE data allows large-scale, sensitive screening of antisense transcript expression. Nucleic Acids Res 32:e163
Reik W, Walter J (2001) Genomic imprinting: parental influence on the genome. Nat Rev Genet 2:21–32
Rhoades MW, Reinhart BJ, Lim LP et al (2002) Prediction of plant microRNA targets. Cell 110:513–520
Rinn JL, Euskirchen G, Bertone P et al (2003) The transcriptional activity of human Chromosome 22. Genes Dev 17:529–540
Rinn JL, Kertesz M, Wang JK et al (2007) Functional demarcation of active and silent chromatin domains in human HOX loci by noncoding RNAs. Cell 129:1311–1323
Roa S, Kuang FL, Scharff MD (2008) Does antisense make sense of AID targeting? Proc Natl Acad Sci USA 105:3661–3662
Ronai D, Iglesias-Ussel MD, Fan M et al (2007) Detection of chromatin-associated single-stranded DNA in regions targeted for somatic hypermutation. J Exp Med 204:181–190
Rosok O, Sioud M (2004) Systematic identification of sense-antisense transcripts in mammalian cells. Nat Biotechnol 22:104–108
Rosok O, Sioud M (2005) Systematic search for natural antisense transcripts in eukaryotes (review). Int J Mol Med 15:197–203
Rossignol F, Vache C, Clottes E (2002) Natural antisense transcripts of hypoxia-inducible factor 1alpha are detected in different normal and tumour human tissues. Gene 299:135–140
Rossignol F, de Laplanche E, Mounier R et al (2004) Natural antisense transcripts of HIF-1alpha are conserved in rodents. Gene 339:121–130
Rougeulle C, Heard E (2002) Antisense RNA in imprinting: spreading silence through Air. Trends Genet 18:434–437
Rougeulle C, Cardoso C, Fontes M et al (1998) An imprinted antisense RNA overlaps UBE3A and a second maternally expressed transcript. Nat Genet 19:15–16
Runte M, Huttenhofer A, Gross S et al (2001) The IC-SNURF-SNRPN transcript serves as a host for multiple small nucleolar RNA species and as an antisense RNA for UBE3A. Hum Mol Genet 10:2687–2700
Sado T, Wang Z, Sasaki H et al (2001) Regulation of imprinted X-chromosome inactivation in mice by Tsix. Development 128:1275–1286
Sanchez-Elsner T, Gou D, Kremmer E et al (2006) Noncoding RNAs of trithorax response elements recruit Drosophila Ash1 to Ultrabithorax. Science 311:1118–1123
Sastry SS, Hoffman PL (1995) The influence of RNA and DNA template structures during transcript elongation by RNA polymerases. Biochem Biophys Res Commun 211:106–114

Scadden AD, Smith CW (1997) A ribonuclease specific for inosine-containing RNA: a potential role in antiviral defence? EMBO J 16:2140–2149

Scheele C, Petrovic N, Faghihi MA et al (2007) The human PINK1 locus is regulated in vivo by a non-coding natural antisense RNA during modulation of mitochondrial function. BMC Genomics 8:74

Schwartz JC, Younger ST, Nguyen NB et al (2008) Antisense transcripts are targets for activating small RNAs. Nat Struct Mol Biol 15:842–848

Seila AC, Calabrese JM, Levine SS et al (2008) Divergent transcription from active promoters. Science 322:1849–1851

Shearwin KE, Callen BP, Egan JB (2005) Transcriptional interference–a crash course. Trends Genet 21(6):339–345

Shendure J, Church GM (2002) Computational discovery of sense-antisense transcription in the human and mouse genomes. Genome Biol 3:RESEARCH0044

Sleutels F, Zwart R, Barlow DP (2002) The non-coding air RNA is required for silencing autosomal imprinted genes. Nature 415:810–813

Sleutels F, Tjon G, Ludwig T et al (2003) Imprinted silencing of Slc22a2 and Slc22a3 does not need transcriptional overlap between Igf2r and Air. EMBO J 22:3696–3704

Smilinich NJ, Day CD, Fitzpatrick GV et al (1999) A maternally methylated CpG island in KvLQT1 is associated with an antisense paternal transcript and loss of imprinting in Beckwith-Wiedemann syndrome. Proc Natl Acad Sci USA 96:8064–8069

St Laurent G 3rd, Wahlestedt C (2007) Noncoding RNAs: couplers of analog and digital information in nervous system function? Trends Neurosci 30:612–621

Stavropoulos N, Lu N, Lee JT (2001) A functional role for Tsix transcription in blocking Xist RNA accumulation but not in X-chromosome choice. Proc Natl Acad Sci USA 98:10232–10237

Sun M, Hurst LD, Carmichael GG et al (2005) Evidence for a preferential targeting of 3'-UTRs by cis-encoded natural antisense transcripts. Nucleic Acids Res 33:5533–5543

Taft RJ, Glazov EA, Cloonan N et al (2009) Tiny RNAs associated with transcription start sites in animals. Nat Genet 41:572–578

Tam OH, Aravin AA, Stein P et al (2008) Pseudogene-derived small interfering RNAs regulate gene expression in mouse oocytes. Nature 453:534–538

Thakur N, Tiwari VK, Thomassin H et al (2004) An antisense RNA regulates the bidirectional silencing property of the Kcnq1 imprinting control region. Mol Cell Biol 24:7855–7862

Tomari Y, Zamore PD (2005) Perspective: machines for RNAi. Genes Dev 19:517–529

Tsai MC, Manor O, Wan Y et al (2010) Long noncoding RNA as modular scaffold of histone modification complexes. Science 329:689–693

Tufarelli C, Stanley JA, Garrick D et al (2003) Transcription of antisense RNA leading to gene silencing and methylation as a novel cause of human genetic disease. Nat Genet 34:157–165

Uchida T, Rossignol F, Matthay MA et al (2004) Prolonged hypoxia differentially regulates hypoxia-inducible factor (HIF)-1alpha and HIF-2alpha expression in lung epithelial cells: implication of natural antisense HIF-1alpha. J Biol Chem 279:14871–14878

Vallon-Christersson J, Staaf J, Kvist A et al (2007) Non-coding antisense transcription detected by conventional and single-stranded cDNA microarray. BMC Genomics 8:295

Vanhee-Brossollet C, Vaquero C (1998) Do natural antisense transcripts make sense in eukaryotes? Gene 211:1–9

Volk R, Koster M, Poting A et al (1989) An antisense transcript from the Xenopus laevis bFGF gene coding for an evolutionarily conserved 24 kd protein. EMBO J 8:2983–2988

Wahlestedt C (2006) Natural antisense and noncoding RNA transcripts as potential drug targets. Drug Discov Today 11:503–508

Watanabe T, Takeda A, Tsukiyama T et al (2006) Identification and characterization of two novel classes of small RNAs in the mouse germline: retrotransposon-derived siRNAs in oocytes and germline small RNAs in testes. Genes Dev 20:1732–1743

Watanabe T, Totoki Y, Toyoda A et al (2008) Endogenous siRNAs from naturally formed dsRNAs regulate transcripts in mouse oocytes. Nature 453:539–543

Wroe SF, Kelsey G, Skinner JA et al (2000) An imprinted transcript, antisense to Nesp, adds complexity to the cluster of imprinted genes at the mouse Gnas locus. Proc Natl Acad Sci USA 97:3342–3346

Wutz A, Smrzka OW, Schweifer N et al (1997) Imprinted expression of the Igf2r gene depends on an intronic CpG island. Nature 389:745–749

Xu Z, Wei W, Gagneur J et al (2009) Bidirectional promoters generate pervasive transcription in yeast. Nature 457:1033–1037

Yamasaki K, Joh K, Ohta T et al (2003) Neurons but not glial cells show reciprocal imprinting of sense and antisense transcripts of Ube3a. Hum Mol Genet 12:837–847

Yang N, Kazazian HH Jr (2006) L1 retrotransposition is suppressed by endogenously encoded small interfering RNAs in human cultured cells. Nat Struct Mol Biol 13:763–771

Yap KL, Li S, Munoz-Cabello AM et al (2010) Molecular interplay of the noncoding RNA ANRIL and methylated histone H3 lysine 27 by polycomb CBX7 in transcriptional silencing of INK4a. Mol Cell 38:662–674

Yelin R, Dahary D, Sorek R et al (2003) Widespread occurrence of antisense transcription in the human genome. Nat Biotechnol 21:379–386

Yu W, Gius D, Onyango P et al (2008) Epigenetic silencing of tumour suppressor gene p15 by its antisense RNA. Nature 451:202–206

Zhang Y, Liu XS, Liu QR et al (2006) Genome-wide in silico identification and analysis of cis natural antisense transcripts (cis-NATs) in ten species. Nucleic Acids Res 34:3465–3475

Zhang Q, Zhang J, Moe OW et al (2008) Synergistic upregulation of erythropoietin receptor (EPO-R) expression by sense and antisense EPO-R transcripts in the canine lung. Proc Natl Acad Sci USA 105:7612–7617

Zuniga Mejia Borja A, Meijers C, Zeller R (1993) Expression of alternatively spliced bFGF first coding exons and antisense mRNAs during chicken embryogenesis. Dev Biol 157:110–118

Zwart R, Sleutels F, Wutz A et al (2001) Bidirectional action of the Igf2r imprint control element on upstream and downstream imprinted genes. Genes Dev 15:2361–2366

Activation and Deactivation of Antisense and RNA Interference Function with Light

Jeane M. Govan and Alexander Deiters

Contents

Abstract Oligonucleotides and oligonucleotide analogs have shown to be efficient tools for the silencing of gene expression in a wide range of cell lines and various model organisms. Such oligonucleotides include hairpin DNA, phosphorothioate DNA, morpholino oligonucleotides, peptide nucleic acids, and others. The common mode of action for all antisense agents is sequence-specific duplex formation with messenger RNA (mRNA), leading to the inhibition of translation and/or mRNA degradation and thus gene silencing. RNA interference (RNAi) is another tool to regulate gene expression through the site-specific degradation of mRNA. Several methods for the light regulation of oligonucleotide duplex formation and RNAi function have been developed, including the site-specific installation of light-removable protecting groups (caging groups) on nucleobases and photocleaveable inhibitor sequences. Light is an ideal external regulatory element as light irradiation can be easily and precisely controlled in timing, location, and amplitude. Through the engineering of light-activated oligonucleotides, their function can be regulated with high spatial and temporal resolution, allowing photochemical control of gene expression in biological systems with unprecedented precision.

J.M. Govan • A. Deiters (✉)
Department of Chemistry, North Carolina State University, Raleigh, NC 27695, USA
e-mail: alex_deiters@ncsu.edu

V.A. Erdmann and J. Barciszewski (eds.), *From Nucleic Acids Sequences to Molecular Medicine*, RNA Technologies, DOI 10.1007/978-3-642-27426-8_11,

Keywords Antisense agents • Gene expression • Light regulation • siRNA

1 Introduction

At the molecular level, all biological processes are tightly regulated in space and time, in both single cells and multicellular organisms. The activity of genes, proteins, and biological molecules is controlled with high precision by a variety of endogenous processes. Arguably, this is most evident in early embryogenesis, where the misregulation of a gene can cause severe congenital defects and/or embryonal lethality. Therefore, to study complex biological processes involved in embryogenesis, it is essential to have the ability to control gene function in high spatial and temporal control that is found in nature. Light represents an ideal external control element, as it is easily regulated in timing, location, and amplitude; thus, light can enable the precise activation and deactivation of gene function (Chen et al. 2005; Deiters 2009, 2010; Dmochowski and Tang 2007; Lee et al. 2009; Mayer and Heckel 2006; Meng et al. 2008; Priestman and Lawrence 2010; Riggsbee and Deiters 2010; Young and Deiters 2007).

Established research tools used to control gene expression include antisense oligonucleotides (ASOs). In eukaryotic organisms, DNA is transcribed into mRNA and is transported out of the nucleus into the cytoplasm, where the mRNA is translated into protein. ASOs are short oligonucleotides that bind to a targeted mRNA sequence of the gene of interest by Watson–Crick base pairing (Dean and Bennett 2003; Wacheck and Zangemeister-Wittke 2006). Once bound, the DNA: RNA hybrid either recruits RNaseH to bind and cleave the mRNA (Fig. 1a) or sterically blocks the ribosome from translating the mRNA (Fig. 1b).

Various modifications have been developed to improve the stability of ASOs, including phosphorothioate linkages (PS), locked nucleic acids (LNA), peptide nucleic acids (PNAs), and morpholinos (MO). Several methods have also been developed to facilitate oligonucleotide uptake and delivery in mammalian cells and model organisms. For example, it has been shown that oligonucleotides can be conjugated to various biomolecules, including cholesterol (Cheng et al. 2006), polyethylene glycol (Kim et al. 2006), antibodies (Palma and Cho 2007), nanoparticles (Jin et al. 2010), and dendrimers (Kumar et al. 2010) in order to overcome delivery and localization limitations in mammalian cells. One of the first-generation ASOs was PS DNA in which an oxygen atom is replaced with a sulfur atom on the phosphate backbone (Fig. 2). This modification provides a fast and cost-effective way to stabilize DNA but at high concentrations can be toxic to cells and cause off-target effects (Aboul-Fadl 2005; Summerton 2007). The next generation of ASOs includes LNAs, PNAs, and MOs. PNAs contain a peptide backbone instead of the sugar phosphate backbone and thus are not substrates for nuclease degradation but are able to still bind to their complementary DNA sequence through Watson–Crick base pairing (Summerton 2007). *MO* antisense agents contain a morpholino ring and a phosphorodiamidate backbone (Fig. 2). Due to

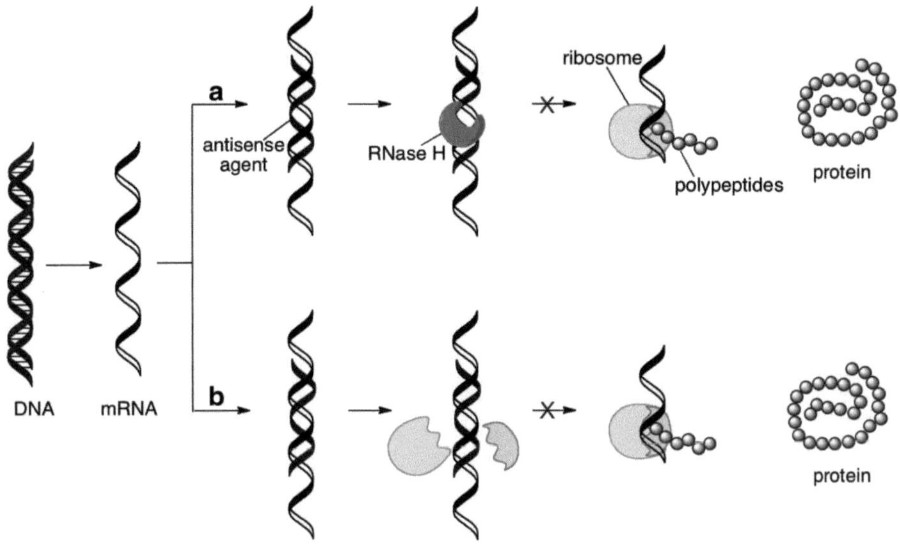

Fig. 1 Mechanism of gene regulation by antisense agents. DNA is transcribed into mRNA and then translated into protein. Antisense agents bind to the mRNA and either (**a**) recruit RNaseH, which binds to the DNA:RNA duplex and cleaves the mRNA, or (**b**) sterically block the ribosome from binding to the mRNA, thereby inhibiting gene expression

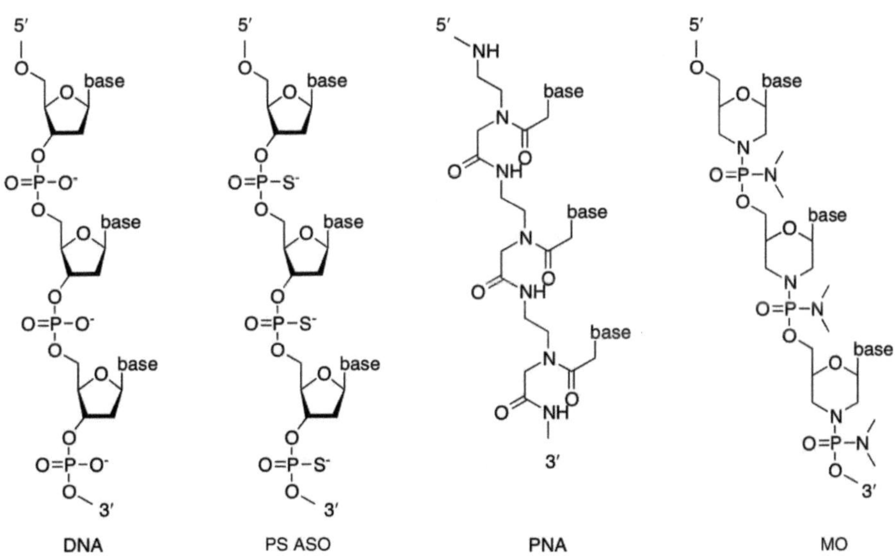

Fig. 2 Structures of antisense agents including DNA, phosphorothioate ASO (PS ASO), peptide nucleic acid (PNA), and morpholino (MO). Base = thymine, adenine, cytosine, or guanine

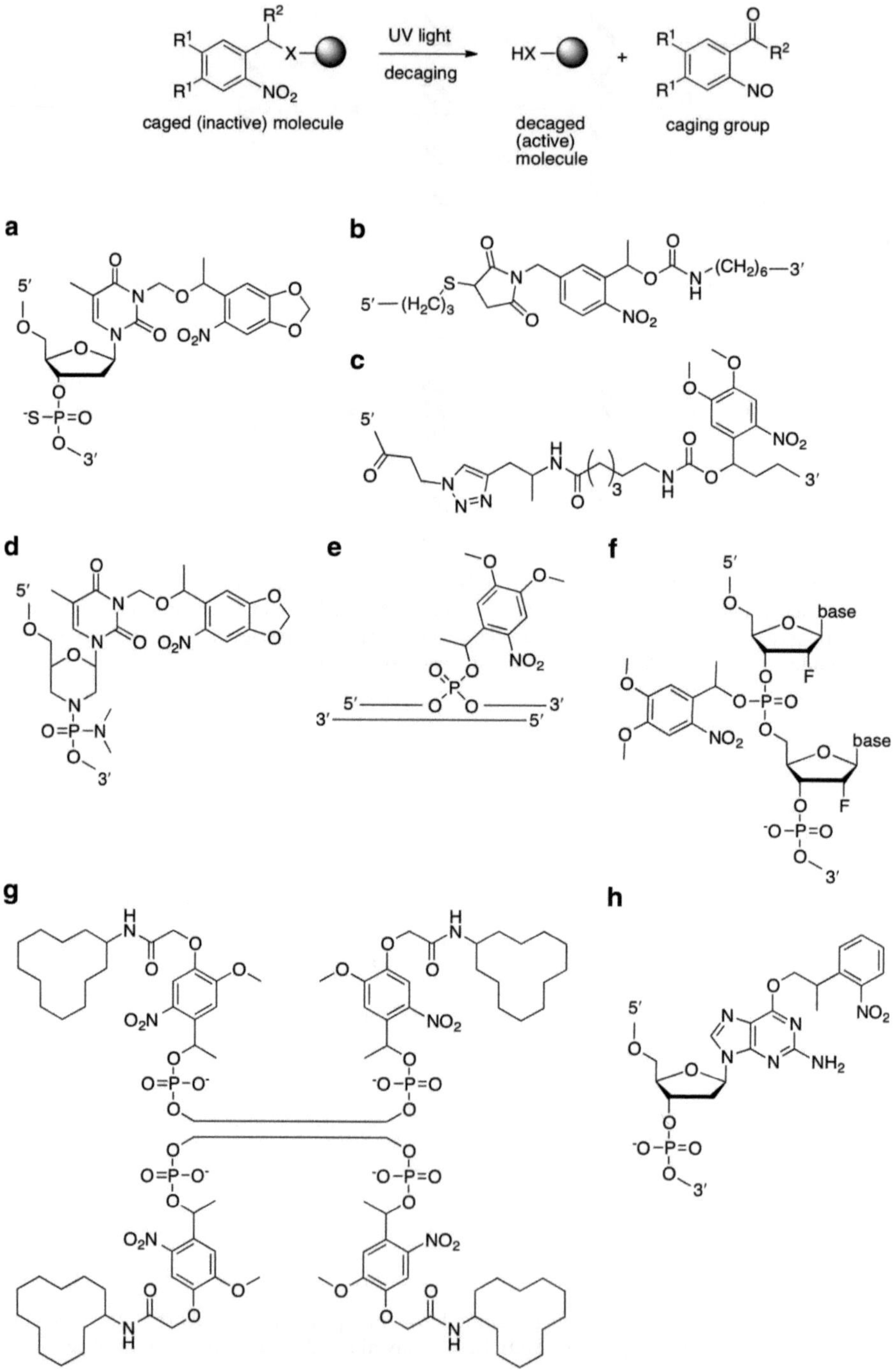

Fig. 3 General decaging reaction of a molecule (*sphere*), caged with a classical *ortho*-nitrobenzyl group. Chemical and photochemical properties of the caging group can be tuned with suitable

the phosphorodiamidate linkage, the oligonucleotide is stable in the presence of nucleases and exhibits an enhanced binding affinity to RNA. Due to the lack of off-targeting effects and an improved pharmacokinetic profile, MOs have been extensively used in animal studies (Moulton and Moulton 2010).

A general approach to achieve spatiotemporal regulation of biological molecules with light is through the attachment of a light-removable protecting group, commonly referred to as "caging" (Ellis-Davies 2007; Lee et al. 2009; Meng et al. 2008). This process typically renders the molecule biologically inactive. After irradiation with UV light of a long, nondamaging wavelength (>360 nm), the caging group is removed and the biological activity is restored (Fig. 3).

Several caging groups have been developed for this purpose, and the majority of these groups are based on an *ortho*-nitrobenzyl scaffold (Banerjee et al. 2003; Han et al. 2008; Meng et al. 2008), which can easily be installed on oxygen, sulfur, and nitrogen atoms. These caging groups show high light sensitivity, ease of synthesis, and compatibility with a wide range of functional groups, such as phosphates, carboxylates, hydroxyls, and amines (Adams and Tsien 1993). Additionally, photoactivation with UVA light (315–400 nm) does not induce deleterious effects in mammalian cells, although duration and strength of the light source should be taken into consideration when designing experiments (Forman et al. 2007). The caged molecules discussed in this chapter are shown in Fig. 3.

2 Light-Activated DNA Antisense Agents

Several approaches have been developed to photochemically regulate ASO function, including direct caging of nucleobases, circularizing the ASO with a photocleavable linker, and indirect caging using an inhibitor sequence with a photocleavable linker. One such approach is through the direct incorporation of caging groups onto specific nucleobases of the ASO sequence. Nucleobase-caged RNA has previously been synthesized and used to photoregulate RNA folding (Höbartner and Silverman 2005). More recently, it has been applied to the regulation of antisense activity, as the caging groups interfere with nucleic acid hybridization, specifically DNA:RNA hybridization, rendering the ASO inactive. Upon UV irradiation, the caging groups are removed, activating the ASO, resulting in the inhibition of gene expression (Fig. 4). This methodology has been used to photoregulate antisense activity in mammalian cells by introducing 6-nitropiperonyloxymethyl-caged thymidine (Fig. 3a) nucleotides throughout a *Renilla* luciferase-targeting PS ASO (Young et al. 2008). In the presence of 3–4 caging groups, DNA:RNA hybridization is disrupted and gene expression is

Fig. 3 (continued) substituents, including $R^1 = H, OCH_3$; $R^2 = H, CH_3, CO_2H$; $X = O, CO_2, S, NH, NHCO_2, OPO_3H$. (**a**) NPOM-caged thymidine, (**b, c**) photocleavable linkers, (**d**) NPOM-caged thymidine morpholino, (**e**) DMNPE-caged siRNA, (**f**) DMNPE-caged 2′-fluoro siRNA, (**g**) DMNPE-caged terminal phosphate on siRNAs, and (**h**) NPP-caged deoxyguanosine. Base = uracil, adenine, cytosine, or guanine

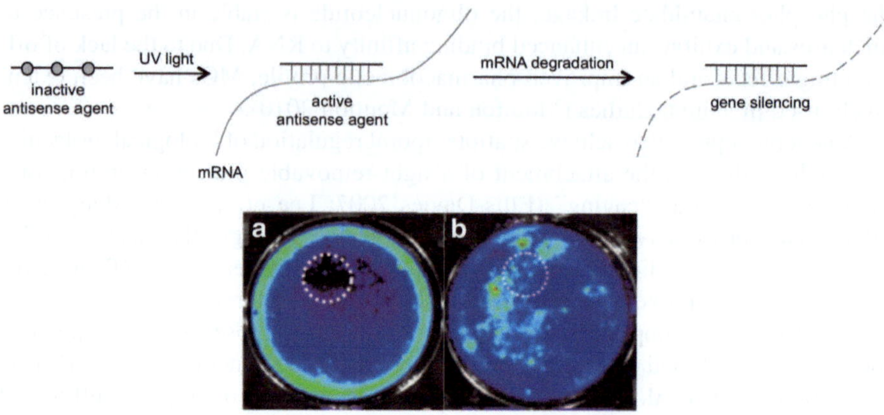

Fig. 4 Light activation of phosphorothioate antisense agents by the site-specific incorporation of caged thymidine nucleotides. Caged thymidine residues were incorporated into an antisense PS ASO. The caging groups are removed through UV irradiation, allowing the hybridization of the antisense agent to the mRNA, leading to the inhibition of gene expression. (**a**) Spatial control of *Renilla* luciferase expression using caged PS ASO. The cell monolayer within the *white dotted circle* was irradiated (365 nm, 5 min, 23 W), enabling the localized silencing of luciferase expression, compared to (**b**) negative control cells that were not transfected with the antisense agent but irradiated. Adapted from Young et al. (2008)

observed. After a brief UV irradiation at 365 nm, the caging groups are removed and a 70% inhibition of luciferase expression was achieved, the same level as in case of a noncaged ASO. Furthermore, spatial control over gene expression was demonstrated when a monolayer of mammalian cells was irradiated in a locally restricted area (Fig. 4a, b).

A structural modification to improve antisense stability are hairpin formation at both termini. Small hairpin loop structures flanking the antisense sequence have shown to stabilize antisense agents against nucleases (Abdelgany et al. 2007; Kwok et al. 2009) and are an inexpensive modification to antisense agents to study gene function. Three caged thymidine nucleotides were site-specifically incorporated into a phosphodiester DNA hairpin antisense agent targeting *DsRed* mRNA (Young et al. 2010). These caging groups prevent the hybridization of the antisense agent to the target mRNA, resulting in the inactivation of the antisense agent and observation of a DsRed reporter gene expression. Photocaged hairpin antisense agents have also been used to light-activate gene expression. Extending the antisense agent sequence and incorporating caged nucleotides into the complementary sequence, the photoactivation of gene expression can be achieved. In the absence of light irradiation, the caging groups prevent self-hybridization, thereby allowing the antisense agent to bind to its target mRNA and suppressing gene expression. After UV irradiation, the caging groups are removed and the antisense agent forms a hairpin structure, rendering the antisense agent inactive and activating gene expression (Fig. 5), as shown in the first example of a light-deactivated antisense agent (Young et al. 2010).

A different approach to photochemically regulate antisense function has been developed by employing a photocleavable linker between the ASO and an inhibitory

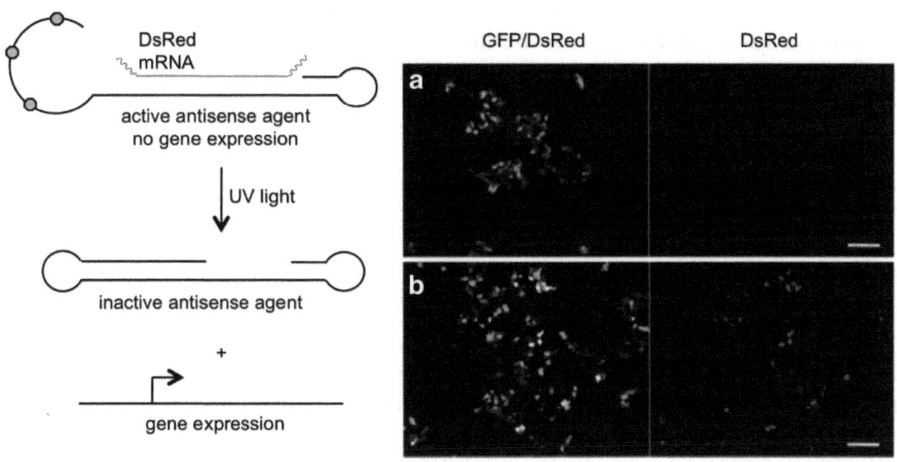

Fig. 5 Photochemical deactivation of an ASO in mammalian cells. (**a**) The caged hairpin antisense agent is active in the absence of light irradiation, silencing the DsRed gene. (**b**) After a brief UV irradiation (2 min, 23 W, 365 nm), the caging groups are removed, leading to the formation of a hairpin, which blocks the antisense agent for recognizing the mRNA target sequence. This inactivation of the antisense agent allows DsRed expression. Scale bar = 200 μm. Adapted from Young et al. (2010)

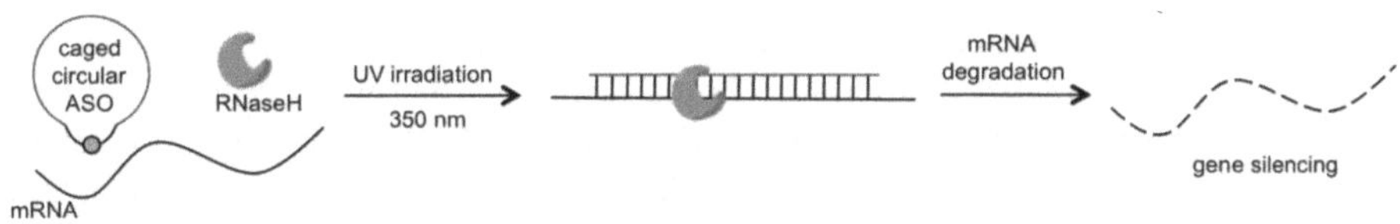

Fig. 6 Light regulation of antisense activity through circular antisense agents. In the circular form, the antisense agent does not bind to the target RNA; upon UV irradiation, the linker is cleaved, resulting in an active antisense agent. Adapted from Tang et al. (2010)

oligonucleotide. A photocleavable linker (Fig. 3b) was designed between the antisense DNA and a complementary sense sequence and sequentially tested in cell culture (Tang et al. 2008). This allows for the inhibition of the antisense agent until UV irradiation in which the linker is cleaved. The complementary sense sequence diffuses from the antisense sequence, releasing the antisense sequence to bind to the target mRNA and inhibiting gene expression. This technique resulted in an approximately 50% inhibition of *c-myc* expression after light irradiation. The advantage of this approach is that only one photolysis needs to occur to activate the antisense agent, whereas the caged nucleobase method requires multiple decaging reactions to restore native function. Nevertheless, the photocleavable inhibitor sequence is not trivial to design and to link to the antisense agent (Richards et al. 2008).

Another ASO photoregulation approach is based on a photocleavable linker joining the 5′ and 3′ termini of the ASO. In the circular form, the ASO cannot bind to its target mRNA. After UV irradiation, the circular ASO is linearized, and its activity is restored (Fig. 6). Although only in vitro experiments were performed (Richards et al. 2010;

Tang et al. 2010), this methodology has potential to photomodulate gene expression in vivo and has the potential to resolve the inhibitor design issues inherent to the light-cleavable complement approach (Richards et al. 2008).

3 Light-Activated PNA Antisense Agents

PNAs are pseudopeptide DNA mimics based on an aminoethyl glycine backbone developed in the late 1980s by Nielsen et al. (1991). PNAs bind to complementary DNA or RNA in a highly sequence-specific fashion through Watson–Crick or Hoogsteen base pairing and induce gene silencing by sterically blocking mRNA translation (Aboul-Fadl 2005; Karkare and Bhatnagar 2006).

A caged PNA was reported in 2007 (Tang et al. 2007). Due to the low water solubility of PNAs, a negatively charged PNA was synthesized and bound to a complementary 2′OMe RNA sequence through a photocleavable linker. In this caged form, the PNA antisense agent is inactive, resulting in gene expression. Upon UV irradiation, the inhibitor RNA is cleaved, releasing the PNA to bind to its complementary mRNA and inhibit gene expression. This approach is complementary to the previously described hairpin deactivating antisense agent (Young et al. 2010) in that an inhibitor sequence is used to quench the antisense activity. Based on this principle, zebrafish embryos were injected with caged PNA:2′OMe RNA targeting the *chordin* mRNA. Zebrafish embryos that were kept in the dark developed normally (Fig. 7). Upon irradiation, the 2′OMe RNA inhibitor is cleaved from the PNA leading to the inhibition of chordin expression and resulting in a "no chordin" phenotype.

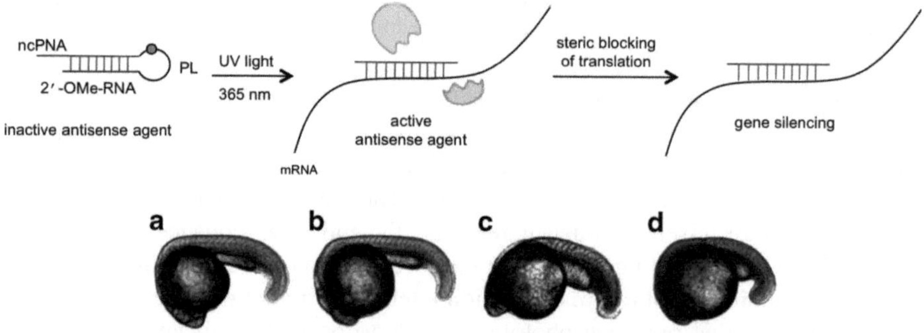

Fig. 7 Photoregulation of gene expression through caged PNAs. Negatively charged peptide nucleic acids (ncPNA) were covalently attached to complementary 2′OMe RNA through a photocleavable linker (PL). UV light irradiation induced linker cleavage and release of the 2′OMe RNA inhibitor and promoted ncPNA binding to the mRNA, thereby blocking gene expression in zebrafish embryos. (**a**) Uninjected embryos UV irradiated at 3 hours post fertilization (hpf) for 8 min developed normally. (**b**) Embryos injected with caged chordin ncPNA developed normally. (**c**) Embryos injected with chordin PNA exhibited a classical "no chordin" phenotype. (**d**) Embryos injected with caged chordin PNA followed by UV irradiation at 3 hpf also resulted in the "no chordin" phenotype. Adapted from Tang et al. (2007)

4 Light-Activated Morpholino Antisense Agents

Morpholino oligomers (MO) are established and widely used antisense agents in model organisms such as *Caenorhabditis elegans*, zebrafish, *Xenopus*, and mice (Heasman 2002; Karkare and Bhatnagar 2006). Due to the unnatural ring structure, morpholino oligonucleotides are highly resistant to nucleases. In addition, the binding affinity of MOs to RNA is greater than the hybridization affinity of DNA:RNA or RNA:RNA duplexes (Summerton 2007). Similar to ASOs, MOs bind to the target mRNA and inhibits gene expression by sterically blocking the ribosome from translating the mRNA.

The ability to spatially and temporally regulate morpholino activity is highly desired due to the versatility and powerful inhibitory activity of MOs in functional genomic studies. A similar design concept to the caged PNA (Fig. 7) was applied in which the MO is hybridized to another morpholino inhibitor sequence through a photocleavable linker (Fig. 3c) (Ouyang et al. 2009; Shestopalov et al. 2007). In the absence of light, the MO is inactive; however, upon UV irradiation, the linker is cleaved and the inhibitor is released from the MO. This renders the MO active and, therefore, free to bind to the target mRNA suppressing gene function (Fig. 8). Targeting the zebrafish homolog of the mouse *Brachyury* (*ntl*) gene, *ntl*-targeting caged MOs were injected into embryos. Upon injection with a noncaged, active *ntl* MO, the formation of the tail and notochord is severely impaired due to loss of

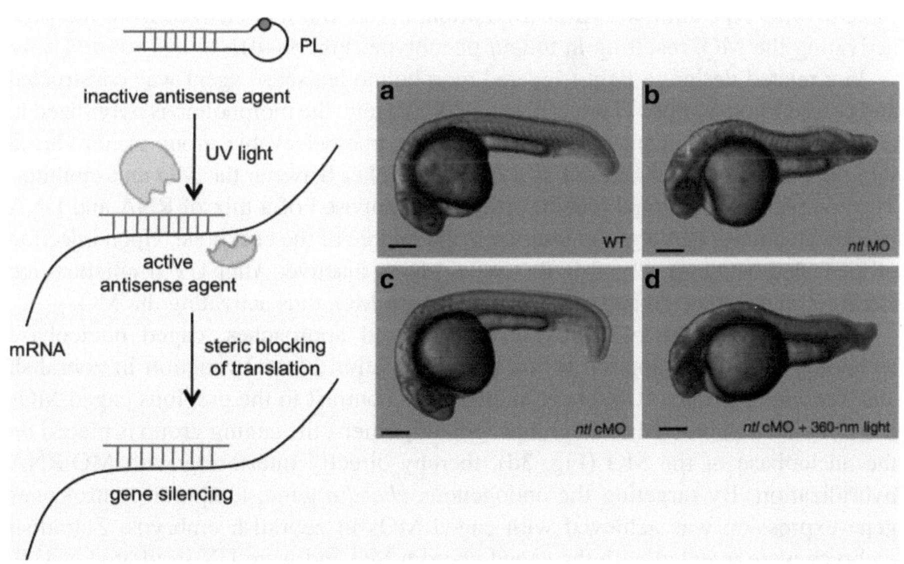

Fig. 8 A caged morpholino that targets *ntl* in zebrafish. (**a**) Wild-type zebrafish embryos show no phenotypic changes. (**b**) Injection of the *ntl* MO elicits a classical *ntl* phenotype. (**c**) Zebrafish embryos injected with the caged MO develop normally in the absence of UV irradiation. (**d**) Irradiation at 360 nm induces a classical *ntl* phenotype in embryos. All images were taken 24 hours post fertilization (hpf). Scale bar = 200 μm. Adapted from Ouyang et al. (2009)

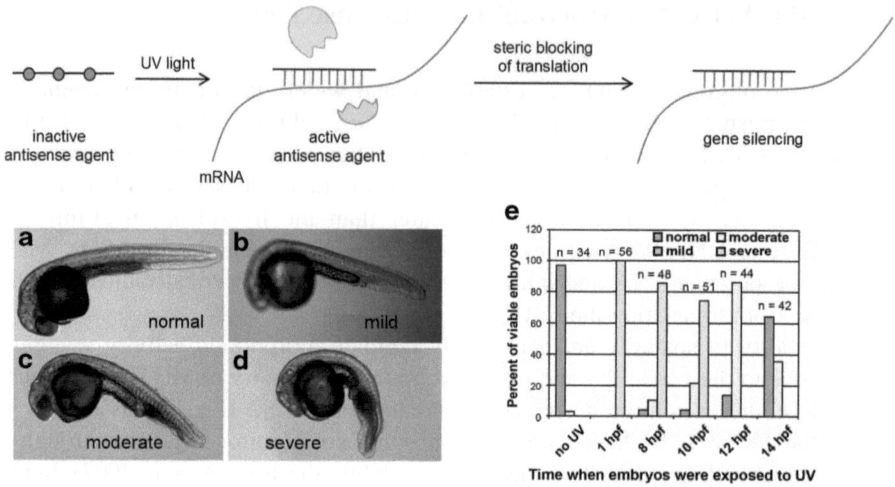

Fig. 9 Light activation of a nucleobase-caged chordin MO in zebrafish embryos. (**a**) Normal, (**b**) mild, (**c**) moderate, and (**d**) severe chordin *null* phenotypes. (**e**) Zebrafish embryos were microinjected with caged MOs during the 1–4 cell stage and were irradiated with UV light of 365 nm at various time points after fertilization. The phenotype was assessed at 24–28 hpf. Adapted from Deiters et al. (2010)

notochord cells and posterior structures. Prior to UV irradiation, the caged MO-injected zebrafish embryos show no *ntl* phenotype and are morphologically comparable to wild-type animals. After irradiation, the inhibitor is cleaved from the MO, activating the MO, resulting in the *ntl* phenotype (Fig. 8a–d).

In a related design, a light-triggered morpholino antisense agent was constructed and coined PhotoMorphs (Tomasini et al. 2009). Here, the morpholino is hybridized to an inhibitory complement strand; however, the photocleavable group is introduced into the inhibitor strand and not as a chemical linker between the MO and inhibitor. The complementary strand (caging strand) is composed of a mix of RNA and DNA bases with a nitrophenyl group inserted in the middle of the backbone. Upon injection of the caged MO into zebrafish, the caged MO is inactive. After UV irradiation, the caging strand is cleaved and dissociates from the MO, thus activating the MO.

In addition to the two previously discussed approaches, caged nucleobase technology has been applied to the photoregulation of MO function in zebrafish and *Xenopus* embryos (Deiters et al. 2010). In contrast to the previous caged MOs that contained photocleavable groups with oligomers, the caging group is placed on the nucleobase of the MO (Fig. 3d), thereby directly interfering with MO:RNA hybridization. By targeting the endogenous *chordin* gene, temporal control over gene expression was achieved with caged MOs in zebrafish embryos. Zebrafish embryos were injected with the caged chordin MO and were UV irradiated at 1, 8, 10, 12, and 14 hours post fertilization (hpf). Without exposure to UV light, the embryos developed normally. UV irradiation at 1 hpf resulted in the most severe chordin phenotype while embryos irradiated at 12 and 14 hpf only showed moderate to normal chordin phenotypes (Fig. 9). These observations support previous findings that chordin is only required in the early stages of development since by the end of the gastrulation stage the chordin mRNA is largely degraded (Schulte-Merker 1997). These experiments demonstrate the advantages of light-activated antisense agents in the temporal investigation of gene function.

5 Light-Activated siRNA Agents

Another important tool in the study of gene function is RNA interference (RNAi). Similar to antisense agents, RNAi targets transcribed mRNAs to silence gene expression. Antisense agents and small interfering RNAs (siRNAs) act through fundamentally different gene silencing mechanisms. As discussed previously, antisense agents inhibit translation either through steric hindrance or through an RNaseH-mediated pathway. RNAi proceeds through a dedicated RNAi pathway with dsRNA being cleaved by the enzyme Dicer into 21–23 nucleotide oligos, which are referred to as siRNAs. The siRNA is then loaded into the RNA-Induced Silencing Complex (RISC). RISC is composed of many different protein subunits; however, arguably, the most important subunit is argonaute, which cleaves the mRNA. Once bound, the siRNA unwinds and the sense strand is cleaved and leaves the complex. While bound to RISC, the antisense strand can base-pair with its complementary mRNA sequence forming an RNA:RNA duplex. The formation of this duplex signals argonaute to cleave the mRNA, effectively inhibiting gene expression (Fig. 10) (Aravin and Tuschl 2005; Harborth et al. 2003; Meister and Tuschl 2004).

It was found that synthetic siRNA can be transfected into mammalian cells and avoid an immune response (Elbashir et al. 2001; Harborth et al. 2003). Synthetic siRNAs are synthesized with 2′-deoxythymidine nucleotide overhangs to mimic the endogenous siRNAs that are cleaved by Dicer and to enhance the overall efficiency of the siRNA (Bolcato-Bellemin et al. 2007; Elbashir et al. 2001). With the addition of caging technology, RNAi holds the promise to tightly regulate gene function with high spatial and temporal resolution.

Based on a phosphate caging approach (Ando et al. 2001), the first caged siRNAs were developed through random covalent attachment of a 4,5-dimethyloxy-2-nitrophenylethyl (DMNPE) group to the phosphate backbone of an siRNA (Fig. 3e) (Ando et al. 2001; Shah et al. 2005). With an average of 1.4 caging groups (3% caging efficiency) per siRNA duplex, the authors speculated that the caging groups could prevent interaction between the siRNA and RISC; however, no experimental evidence was provided. The caged siRNA displayed decreased silencing activity but still showed a measurable amount of background activity. Upon UV irradiation, siRNA activity was comparable to that of noncaged siRNA. The authors reasoned that the caged siRNA background activity was due to the low number of modified caged phosphates on the siRNA. Indeed, with an average of seven caging groups (15% caging efficiency), the siRNA activity was completely inhibited until UV irradiation. Following these initial publications, it was reported that terminal

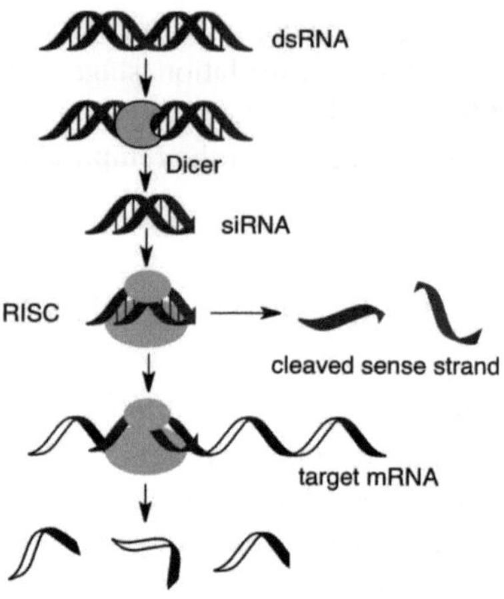

Fig. 10 The siRNA pathway in mammalian cells. DsRNAs are cleaved by Dicer into 21–23 nucleotides that are referred to as siRNAs. The siRNAs are loaded into the RISC where the sense strand is unwound and further degraded. In RISC, the antisense strand binds to its complementary mRNA target sequence and the mRNA is cleaved by argonaute and is subsequently degraded

phosphates are more readily alkylated than the internal phosphates (Shah et al. 2009). In fact, upon reinvestigation, the reported caged siRNA only had one caging group on the 5′ terminal phosphate of the antisense strand. The additional caging groups on the siRNA were still inactive after irradiation due to a slower decaging reaction of the internal caged phosphates.

Additionally, the nonspecific caging of internal phosphate groups using DMNPE was also described (Blidner et al. 2008) and performed in the presence of chemical modifications, such as 2′-fluoro siRNAs (siFNA). siFNAs maintain the A-type helical structure of RNA by exhibiting a C3′ *endo* sugar pucker (Chiu and Rana 2003). Furthermore, siFNAs have increased stability in serum and plasma by resisting sugar-specific enzymatic degradation (Cekaite et al. 2007; Layzer et al. 2004). Combining chemical modifications to improve pharmacological properties and a phosphate caging approach, caged siFNAs were designed to achieve spatial control over gene expression (Blidner et al. 2008). Through the nonspecific caging of siFNAs using DMNPE (Fig. 3f), with an average of eight caging groups per siRNA, siFNA function was completely inhibited. However, UV irradiation restored only 56% siFNA activity as measured in a cell-based reporter assay. Despite this small recovery in mammalian cells, caged siFNAs were also co-injected into zebrafish embryos and employed in the silencing of GFP. After UV irradiation, the siFNA inhibited GFP expression at a comparable level to noncaged siFNA.

Recently, the DMNPE group was improved by synthesizing a new, more sterically demanding cyclododecyl DMNPE (CD-DMNPE) caging group (Jain et al. 2010). The CD-DMNPE caging groups were covalently attached to both the 5′ and 3′ phosphates on the sense and the antisense siRNA strand through a diazo-DMNPE intermediate (Fig. 3g). This was based on previous findings, where a single caging group on the 5′ phosphate could reduce siRNA activity, but complete suppression of siRNA function was not achieved (Nguyen et al. 2006; Shah and Friedman 2007). The authors hypothesized that the addition of four bulky caging groups would fully inhibit the interaction between the siRNA precursor RNA and Dicer, thereby

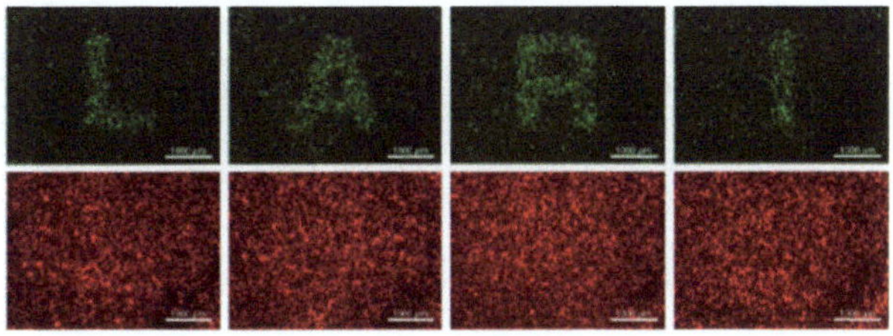

Fig. 11 Light-activated RNA interference (LARI) in HeLa cells. The caged siRNA is inactive in the absence of UV irradiation, enabling the patterning of the letters LARI. Upon irradiation, the siRNA is active, inhibiting gene expression. Micrographs taken on the GFP (target) and the RFP (control) channel. Scale bar = 1 mm. Adapted from Jain et al. (2010)

preventing the formation of active siRNAs, leading to an inhibition of the RNAi pathway (Fig. 10). Additionally, the caging groups are predicted to stabilize the dsRNA against nucleases. The CD-DMNPE-caged dsRNA showed no RNAi activity until UV irradiation, which removed the caging groups, and the dsRNA was able to be processed by Dicer, ultimately leading to the inhibition of gene expression. In addition, it was shown that the caged siRNA could be used to spatially regulate RNAi by creating a distinct patterning of reporter gene expression in a cell monolayer using a caged GFP siRNA (Fig. 11).

In a chemical analysis study of siRNAs, it was found that a single N^3-methyl uridine mutation at the 11 nucleotide from the 5′ end on the antisense strand completely inactivates the RNAi pathway (Chiu and Rana 2003). It was reasoned that this modification creates a disruption in the A-form helix of the RNA:RNA, which in turn affects the interaction between the RNA and RISC, thereby inhibiting gene silencing. Based on this observation, the nucleobase-caging approach was applied to the light regulation of siRNA activity through the site-specific introduction of a photocleavable 2-(2-nitrophenyl)propyl (NPP) group onto a deoxynucleobase of the antisense strand (Fig. 3h) (Mikat and Heckel 2007). The presence of the caging group rendered the siRNA agent completely inactive until it was removed through irradiation with UV light, thus restoring siRNA activity, which was comparable to the activity of noncaged RNA.

6 Future Directions

All discussed caging methodologies provide precise spatial and temporal control over gene function; however, such light-activation processes are irreversible. In contrast, most biological processes are reversible, and thus, reversible photoresponsive molecules are needed to turn gene expression on and off in order to closely mimic complex biological mechanisms such as embryogenesis. Azobenzene is a unique molecule that photochemically isomerizes from its *trans* to *cis* form in a reversible fashion. Through the direct incorporation of azobenzene residues within a strand of DNA, the hybridization of dsDNA can be photochemically controlled (Fig. 12).

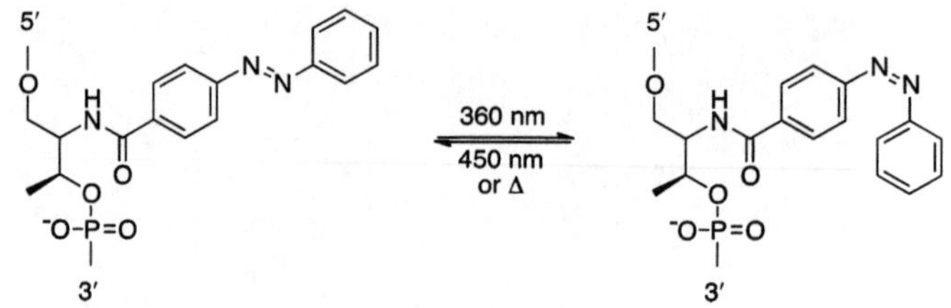

Fig. 12 D-Threonine azobenzene residues are incorporated into DNA. Azobenzene can photoisomerize between its *trans* and *cis* forms to control DNA hybridization

Based on this discovery, it was demonstrated that this technology can be applied to disrupt RNA:RNA and DNA:RNA hybridization and therefore can reversibly regulate RNaseH degradation (Ito et al. 2010; Matsunaga et al. 2004), setting the stage for the photochemical regulation of antisense agents and RNA interference. However, to date, azobenzene-modified DNA or RNA has not been demonstrated in cell culture or in a model organism to reversibly regulate gene function.

7 Summary

In summary, the photochemical regulation of gene expression can be achieved through a variety of different oligonucleotide caging methodologies enabling the precise study of biological processes. Caging groups have been applied to many different antisense agents including phosphorothioate DNA, hairpin DNA, PNAs, MOs, and siRNAs. All of these caged gene silencing agents have been tested in proof-of-principle experiments and in traditional reporter assays. The future of these caged antisense agents lies in their application to actual biological problems, most importantly in embryogenesis and developmental biology. As chemists and chemical biologists devise new ways to build and expand the light-regulation technology for oligonucleotides (e.g., through the application of two-photon caging groups, reversible photoswitching, new photocleavable linkers), the applicability of light-activated antisense agents and RNA interference greatly expands. These developments are driven by the need to precisely control biological processes with high spatial and temporal resolution in live cells and multicellular organisms.

References

Abdelgany A, Wood M, Beeson D (2007) Hairpin DNAzymes: a new tool for efficient cellular gene silencing. J Gene Med 9:727–738

Aboul-Fadl T (2005) Antisense oligonucleotides: the state of the art. Curr Med Chem 12: 2193–2214

Adams SR, Tsien RY (1993) Controlling cell chemistry with caged compounds. Annu Rev Physiol 55:755–784

Ando H, Furuta T, Tsien RY et al (2001) Photo-mediated gene activation using caged RNA/DNA in zebrafish embryos. Nat Genet 28:317–325

Aravin A, Tuschl T (2005) Identification and characterization of small RNAs involved in RNA silencing. FEBS Lett 579:5830–5840

Banerjee A, Grewer C, Ramakrishnan L et al (2003) Toward the development of new photolabile protecting groups that can rapidly release bioactive compounds upon photolysis with visible light. J Org Chem 68:8361–8367

Blidner RA, Svoboda KR, Hammer RP et al (2008) Photoinduced RNA interference using DMNPE-caged 2′-deoxy-2′-fluoro substituted nucleic acids in vitro and in vivo. Mol Biosyst 4:431–440

Bolcato-Bellemin AL, Bonnet ME, Creusat G et al (2007) Sticky overhangs enhance siRNA-mediated gene silencing. Proc Natl Acad Sci USA 104:16050–16065

Cekaite L, Furset G, Hovig E et al (2007) Gene expression analysis in blood cells in response to unmodified and 2′-modified siRNAs reveals TLR-dependent and independent effects. J Mol Biol 365:90–108

Chen X, Dudgeon N, Shen L et al (2005) Chemical modification of gene silencing oligonucleotides for drug discovery and development. Drug Discov Today 10:587–593

Cheng K, Ye ZY, Guntaka RV et al (2006) Enhanced hepatic uptake and bioactivity of type alpha 1(I) collagen gene promoter-specific triplex-forming oligonucleotides after conjugation with cholesterol. J Pharmacol Exp Ther 317:797–805

Chiu YL, Rana TM (2003) SiRNA function in RNAi: a chemical modification analysis. RNA 9:1034–1048

Dean NM, Bennett CF (2003) Antisense oligonucleotide-based therapeutics for cancer. Oncogene 22:9087–9096

Deiters A (2009) Light activation as a method of regulating and studying gene expression. Curr Opin Chem Biol 13:678–686

Deiters A (2010) Principles and applications of the photochemical control of cellular processes. Chembiochem 11:47–53

Deiters A, Garner RA, Lusic H et al (2010) Photocaged morpholino oligomers for the light-regulation of gene function in zebrafish and Xenopus embryos. J Am Chem Soc 132: 15644–15650

Dmochowski IJ, Tang XJ (2007) Taking control of gene expression with light-activated oligonucleotides. Biotechniques 43:161–171

Elbashir S, Harborth J, Lendeckel W et al (2001) Duplexes of 21-nucleotide RNAs mediate RNA interference in cultured mammalian cells. Nature 411:494–498

Ellis-Davies GC (2007) Caged compounds: photorelease technology for control of cellular chemistry and physiology. Nat Methods 4:619–628

Forman J, Dietrich M, Monroe WT (2007) Photobiological and thermal effects of photoactivating UVA light doses on cell cultures. Photochem Photobiol Sci 6:649–658

Han G, Mokari T, Ajo-Franklin C et al (2008) Caged quantum dots. J Am Chem Soc 130: 15811–15813

Harborth J, Elbashir SM, Vandenburgh K et al (2003) Sequence, chemical, and structural variation of small interfering RNAs and short hairpin RNAs and the effect on mammalian gene silencing. Antisense Nucleic Acid Drug Dev 13:83–105

Heasman J (2002) Morpholino oligos: making sense of antisense? Dev Biol 243:209–214

Höbartner C, Silverman SK (2005) Modulation of RNA tertiary folding by incorporation of caged nucleotides. Angew Chem Int Ed 44:7305–7309

Ito H, Liang X, Nishioka H et al (2010) Construction of photoresponsive RNA for photoswitching RNA hybridization. Org Biomol Chem 8:5519–5524

Jain PK, Shah S, Friedman SH (2010) Patterning of gene expression using new photolabile groups applied to light activated RNAi. J Am Chem Soc 133:440–446

Jin Y, Liu S, Yu B et al (2010) Targeted delivery of antisense oligodeoxynucleotide by transferrin conjugated pH-sensitive lipopolyplex nanoparticles: a novel oligonucleotide-based therapeutic strategy in acute myeloid leukemia. Mol Pharm 7:196–206

Karkare S, Bhatnagar D (2006) Promising nucleic acid analogs and mimics: characteristic features and applications of PNA, LNA, and morpholino. Appl Microbiol Biotechnol 71:575–586

Kim SH, Jeong JH, Lee SH et al (2006) PEG conjugated VEGF siRNA for anti-angiogenic gene therapy. J Control Release 116:123–129

Kumar A, Yellepeddi VK, Davies GE et al (2010) Enhanced gene transfection efficiency by polyamidoamine (PAMAM) dendrimers modified with ornithine residues. Int J Pharm 392:294–303

Kwok T, Heinrich J, Jung-Shiu J et al (2009) Reduction of gene expression by a hairpin-loop structured oligodeoxynucleotide: alternative to siRNA and antisense. Biochim Biophys Acta 1790:1170–1178

Layzer JM, McCaffrey AP, Tanner AK et al (2004) In vivo activity of nuclease-resistant siRNAs. RNA 10:766–771

Lee HM, Larson DR, Lawrence DS (2009) Illuminating the chemistry of life: design, synthesis, and applications of "caged" and related photoresponsive compounds. ACS Chem Biol 4:409–427

Matsunaga D, Asanuma H, Komiyama M (2004) Photoregulation of RNA digestion by RNase H with azobenzene-tethered DNA. J Am Chem Soc 126:11452–11453

Mayer G, Heckel A (2006) Biologically active molecules with a "light switch". Angew Chem Int Ed 45:4900–4921

Meister G, Tuschl T (2004) Mechanisms of gene silencing by double-stranded RNA. Nature 431:343–349

Meng XM, Chen XY, Fu Y et al (2008) Photolysis of caged compounds and its applications to chemical biology. Prog Chem 20:2034–2044

Mikat V, Heckel A (2007) Light-dependent RNA interference with nucleobase-caged siRNAs. RNA 13:2341–2347

Moulton HM, Moulton JD (2010) Morpholinos and their peptide conjugates: therapeutic promise and challenge for Duchenne muscular dystrophy. Biochim Biophys Acta 1798:2296–2303

Nguyen QN, Chavli RV, Marques JT et al (2006) Light controllable siRNAs regulate gene suppression and phenotypes in cells. Biochim Biophys Acta 1758:394–403

Nielsen PE, Egholm M, Berg RH et al (1991) Sequence-selective recognition of DNA by strand displacement with a thymine-substituted polyamide. Science 254:1497–1500

Ouyang X, Shestopalov IA, Sinha S et al (2009) Versatile synthesis and rational design of caged morpholinos. J Am Chem Soc 131:13255–13269

Palma E, Cho MJ (2007) Improved systemic pharmacokinetics, biodistribution, and antitumor activity of CpG oligodeoxynucleotides complexed to endogenous antibodies in vivo. J Control Release 120:95–103

Priestman MA, Lawrence DS (2010) Light-mediated remote control of signaling pathways. Biochim Biophys Acta 1804:547–558

Richards JL, Tang X, Turetsky A et al (2008) RNA bandages for photoregulating in vitro protein synthesis. Bioorg Med Chem Lett 18:6255–6258

Richards JL, Seward GK, Wang YH et al (2010) Turning the 10–23 DNAzyme on and off with light. Chembiochem 11:320–324

Riggsbee CW, Deiters A (2010) Recent advances in the photochemical control of protein function. Trends Biotechnol 28:468–475

Shah S, Friedman SH (2007) Tolerance of RNA interference toward modifications of the 5′ antisense phosphate of small interfering RNA. Oligonucleotides 17:35–43

Shah S, Rangarajan S, Friedman SH (2005) Light activated RNA interference. Angew Chem Int Ed 44:1328–1332

Shah S, Jain PK, Kala A et al (2009) Light-activated RNA interference using double-stranded siRNA precursors modified using a remarkable regiospecificity of diazo-based photolabile groups. Nucleic Acids Res 37:4508–4517

Shestopalov IA, Sinha S, Chen JK (2007) Light-controlled gene silencing in zebrafish embryos. Nat Chem Biol 3:650–651

Schulte-Merker S, Lee KJ, McMahon AP et al (1997) The zebrafish organizer requires Chordin. Nature 387:862–863

Summerton JE (2007) Morpholino, siRNA, and S-DNA compared: impact of structure and mechanism of action on off-target effects and sequence specificity. Curr Top Med Chem 7:651–660

Tang X, Maegawa S, Weinberg ES et al (2007) Regulating gene expression in zebrafish embryos using light-activated, negatively charged peptide nucleic acids. J Am Chem Soc 129:11000–11001

Tang X, Swaminathan J, Gewirtz AM et al (2008) Regulating gene expression in human leukemia cells using light-activated oligodeoxynucleotides. Nucleic Acids Res 36:559–569

Tang XJ, Su M, Yu LL et al (2010) Photomodulating RNA cleavage using photolabile circular antisense oligodeoxynucleotides. Nucleic Acids Res 38:3848–3855

Tomasini AJ, Schuler AD, Zebala JA et al (2009) PhotoMorphs: a novel light-activated reagent for controlling gene expression in zebrafish. Genesis 47:736–743

Wacheck V, Zangemeister-Wittke U (2006) Antisense molecules for targeted cancer therapy. Crit Rev Oncol Hematol 59:65–73

Young DD, Deiters A (2007) Photochemical control of biological processes. Org Biomol Chem 5:999–1005

Young DD, Lusic H, Lively MO et al (2008) Gene silencing in mammalian cells with light-activated antisense agents. Chembiochem 9:2937–2940

Young D, Lively M, Deiters A (2010) Activation and deactivation of DNAzyme and antisense function with light for the photochemical regulation of gene expression in mammalian cells. J Am Chem Soc 132:6183–6193

Ribozymes as Molecular Biology Reagents

Sanchita Bhadra, Arti Pothukuchy, and Andrew Ellington

Contents

Abstract Catalytic RNA molecules (ribozymes) can catalyze a number of biochemical processes, including tRNA processing, mRNA splicing and regulation, and of course peptide bond formation. While many of these reactions involve phosphodiester bond rearrangement, catalysts selected by directed evolution further expand the range of reactions available to ribozymes to include the formation of carbon–carbon bonds and redox reactions. This chapter reviews the adaptation of ribozymes into unique as well as alternative tools for (1) developing biosensors and reporters, (2) manipulation of target RNA, and (3) biocatalysis of non-phosphoryltransfer reactions.

S. Bhadra • A. Pothukuchy • A. Ellington (✉)
Institute for Cellular and Molecular Biology, University of Texas at Austin, Austin, TX 78712, USA
e-mail: andy.ellington@mail.utexas.edu

V.A. Erdmann and J. Barciszewski (eds.), *From Nucleic Acids Sequences to Molecular Medicine*, RNA Technologies, DOI 10.1007/978-3-642-27426-8_12,

Keywords Aptazyme • Biosensor • Diagnostic • Molecular biology • Reagent • Ribozyme

1 Introduction

Natural ribozymes include self-cleaving RNAs such as the hammerhead, hairpin, hepatitis delta virus (HDV), varkud satellite (VS), the glucosamine-6-phosphate-dependent *glm*S, *trans*-cleaving ribozymes such as RNase P, and self-splicing RNAs such as the group I and group II introns and group-I-like cleavage ribozymes (Serganov and Patel 2007; Wachter 2010; Topp and Gallivan 2010). Rational design, engineering, and in vitro selection have also yielded synthetic ribozymes that catalyze a wide spectrum of chemical reactions (Table 1). Several natural and synthetic ribozymes including the hammerhead, *Tetrahymena* group I intron, HDV, hairpin, and Diels–Alderase ribozymes have been converted to allosteric ribozymes (aptazymes) by inserting oligonucleotide sequences (aptamers) that regulate ribozyme activity upon ligand binding (Table 2). Aptazymes can also be temporally, spatially, and reversibly photoregulated using photocaged ribozyme or input molecules (Lee et al. 2007; Nierth et al. 2010; Young et al. 2009). Computational approaches and formalization of design principles have assisted with the precision design of aptazymes for biotechnological and clinical applications (Chen and Ellington 2009; Famulok et al. 2007; Hall et al. 2007; Penchovsky and Breaker 2005).

2 Ribozyme-Based Biosensors and Reporters

Effector-mediated catalysis opens the door to the application of ribozymes as reporters and biosensors. Ribozyme catalysis can be readily quantitated by monitoring cleavage or ligation of reporter oligonucleotides using analytically diverse schemes, from fluorescent substrates to label-free mass-sensitive detection (Furchak et al. 2008; Knudsen et al. 2006; Seetharaman et al. 2001).

2.1 Ribozyme-Based Biosensors for Analyte Detection and Quantitation

Among the biosensor applications of allosteric ribozymes, several stand out. Aptazymes are particularly useful for the detection of small analytes, such as metabolites, that are not amenable to antibody-based sandwich assays (Ferguson et al. 2004; Frauendorf and Jäschke 2001; Furchak et al. 2008). Both allosteric hammerhead and allosteric ligase ribozymes have been adapted to aptazyme arrays for the simultaneous detection of multiple small analytes (Hesselberth et al. 2003;

Table 1 Catalytic diversity of synthetic ribozymes

Ribozyme-catalyzed chemical reaction	References
RNA-templated RNA ligation and polymerization	Bagby et al. (2009), Fujita et al. (2009), Paul and Joyce (2002), Wochner et al. (2011), Zaher and Unrau (2007)
5′- and 2′-OH self-phosphorylation and 2′-OH *trans*-phosphorylation	Biondi et al. (2010), Cho and Burke (2006), Saran et al. (2005)
ATPγS hydrolysis	Saran et al. (2006)
5′-Phosphoryl coupling of RNA and nucleophiles with unblocked phosphates (RNA capping and cap exchange)	Huang and Yarus (1997a, b), Zaher et al. (2006)
Peptide bond formation	Cui et al. (2004), Sun et al. (2007)
Nucleotide synthesis	Unrau and Bartel (1998), Lau et al. (2004), Lau and Unrau (2009)
C–C bond formation in Diels–Alder, aldol, and Michael reactions	Fusz et al. (2005), Seelig and Jäschke (1999), Sengle et al. (2001)
Self and *trans*-RNA and tRNA acylation	Jenne and Famulok (1998), Lee et al. (2000), Murakami et al. (2003), Ohuchi et al. (2007), Turk et al. (2010)
RNA–protein phosphoamide bonding	Baskerville and Bartel (2002)
Synthesis of CoA, NAD, and FAD	Huang et al. (2000)
Thioester synthesis	Coleman and Huang (2002)
5′-Nucleotidyl transfer reaction forming 2′–5′ phosphodiester bond	Kang and Suga (2007)
NAD^{+}-dependent alcohol dehydrogenase	Tsukiji et al. (2003)

Table 2 Allosteric regulators of ribozymes

Allosteric regulators	References
Small molecules such as theophylline, tobramycin, and caffeine	Amontov and Jäschke (2006), Ausländer et al. (2010), Chen et al. (2010), Ferguson et al. (2004)
Metal ions such as Mn^{2+}, Co^{2+}, Ni^{2+}, Zn^{2+}, and Cd^{2+}	Zivarts et al. (2005)
Amino acids like L-histidine	Kawakami et al. (2006)
Guanine-quadruplex-inducing compounds	Wieland and Hartig (2006)
Proteins and peptides such as λN, HIV Rev, and α-thrombin	Atsumi et al. (2001), Najafi-Shoushtari and Famulok (2007)
Oligonucleotides	Penchovsky and Breaker (2005)
Organic cofactors such as redox-responsive FMN	Strohbach et al. (2006)
Feedback regulation by ribozyme substrate RNA as in maxizymes and specific on/off adaptor (SOFA) modules	Amontov and Jäschke (2006), Bergeron and Perreault (2005), Fiola et al. (2006), Kuwabara et al. (2001)
Dual competitive (oligonucleotide and FMN) or cooperative (FMN and theophylline) allosteric regulators	Jose et al. (2001), Najafi-Shoushtari and Famulok (2005)

Robertson and Ellington 2001; Seetharaman et al. 2001). Indeed, analytes as small as a proton can be detected. The pH cleavage rate curve of VS ribozyme is influenced by the identity and concentration of mono- or divalent cations. To evaluate the effect of monovalent salts on real changes in solution pH, the catalytic rate of this pH-sensitive ribozyme was used as an accurate pH sensor as opposed to glass electrode-based erroneous pH measurements in high salt concentrations (Duane Smith and Collins 2011). The ability to directly detect small analytes has also proven useful for the identification of posttranslational modifications, such as the phosphorylation status of extracellular regulated kinase 2 (ERK2) (Vaish et al. 2002).

Because of the many technologies surrounding RNA quantitation, RNA ligase aptazymes have been configured to create amplicons in response to a variety of targets, including proteins and small molecules. In the presence of allosteric regulators (including oligonucleotides and small-molecule analytes such as ATP), RNA ligase aptazymes ligate substrate RNA to generate nucleic acid templates that can be subsequently amplified by conventional amplification technologies like RT-PCR (Robertson and Ellington 1999). As little as 1 nM (30 fM) of oligonucleotide effectors and 1 μM of ATP could be specifically detected using allosteric L1 ligase ribozyme. Similarly, ribozymes can generally be adapted to sense hybridization. For example, sensitive (zeptomole-range detection of HCV sequences) virus detection assays have also been devised using RNA ligase-based half-ribozymes whose assembly and catalysis are dependent on binding target oligonucleotides found in the virus (Vaish et al. 2003). The relative ease of creating and optimizing allosteric RNA ligases should allow a variety of biomedically important effectors such as hormones or neurotransmitters to be transmogrified into amplicons that can then be quantitated using off-the-shelf amplification technologies such as PCR, LCR, 3SR, and NASBA.

Even without protein amplification, ribozyme self-amplification may prove useful. Joyce and coworkers constructed a system in which ligase half-molecules could be acted upon in *trans*, leading to a cascade in which each ligase could catalytically form additional copies of itself (Kim and Joyce 2004; Lincoln and Joyce 2009). This cascade was adapted to detection by generating ligase aptazymes that could kick-start the self- or cross-replication cascade through activation by theophylline and/or FMN (Lam and Joyce 2009) (Fig. 1).

2.2 *Ribozyme Reporters of Molecular Biology and Metabolism*

Ribozymes can serve as reporters of RNA metabolism such as transcription, co- and posttranscriptional processing, editing and splicing, structural assembly and chaperone-mediated folding, and RNA stability and decay (Wong and Pan 2009), including in vivo (Watson and Fedor 2009). As was suggested above (Sect. 2.1), since ribozymes are very good at detection of base-pairing interactions, they can be used for the quantitation of miRNA, gene expression, and viral RNAs (Hartig et al. 2004). For example, mammalian gene expression has been quantitated in vivo using a split *Tetrahymena* group I ribozyme-based RNA sensor that is composed of two ribozyme segments, each appended with one-half of a reporter transcript (Hasegawa et al. 2006). Upon target RNA binding, functional ribozymes assemble and the reporter is expressed (Fig. 2).

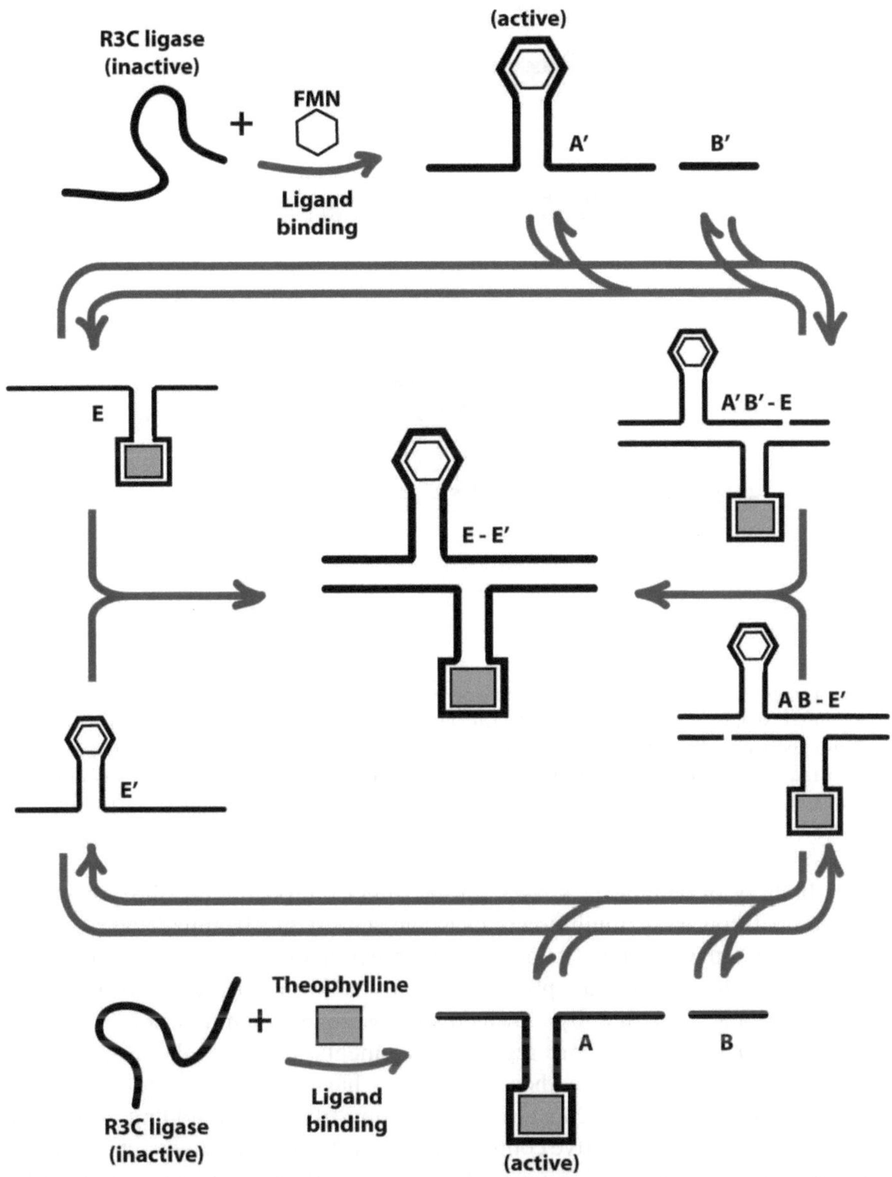

Fig. 1 Analyte detection by cross-catalytic replication of RNA ligase aptazymes (Lam and Joyce 2009). FMN-regulated aptazyme E′ catalyzes the ligation of substrate RNA A and B to form the theophylline-regulated aptazyme E. In the presence of theophylline, the E ligase in turn catalyzes the ligation of substrate RNA A′ and B′ to form the aptazyme E′. The newly formed E and E′ aptazymes can catalyze further cycles of ligand-induced cross-catalytic replication

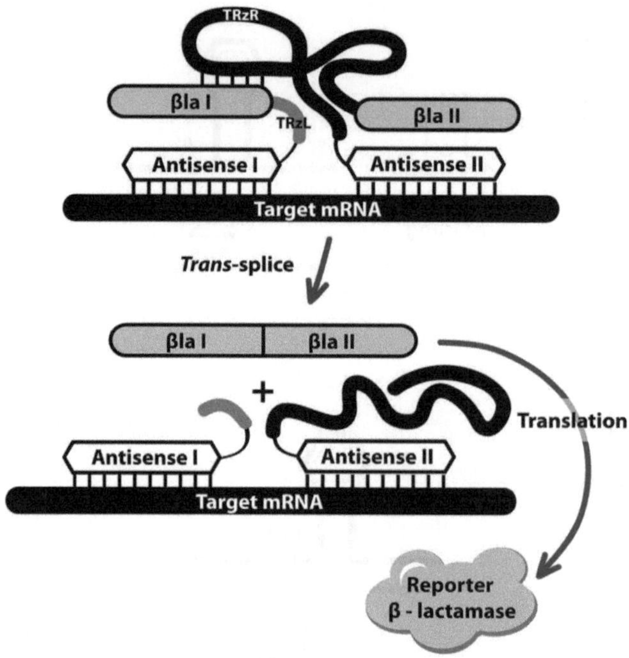

Fig. 2 *Tetrahymena* Group I intron ribozyme-based split ribozyme reporter for target RNA detection (Hasegawa et al. 2006). The *Tetrahymena* Group I ribozyme is appended both to an antisense sequence that allows targeting of an mRNA and a segment of the β-lactamase (bla) reporter transcript. Upon binding of the antisense to its target, the *left* (TRzL) and *right* (TRzR) segments of the ribozyme are assembled into an active complex that *trans*-splices the two reporter segments into a translatable transcript

The biology of cellular processes can be directly probed with ribozymes. In a novel competition assay, inhibition of cleavage activity of a hammerhead ribozyme placed between the 5′-splice site and intron branch point was used as a reporter to demonstrate cotranscriptional spliceosome activity in vivo (Lacadie et al. 2006). Cotranscriptional cleavage by downstream-positioned hammerhead ribozymes has been used to map intronic regulatory elements affecting alternate splicing (Gromak et al. 2008). Hammerhead ribozymes have also been used to monitor trypanosomatid editosomes (Moshiri and Salavati 2010). The refolding of inactive ribozyme conformers into active conformations is often used to study putative RNA chaperones both in vitro and in vivo (Prenninger et al. 2006; Tijerina et al. 2006; Zúñiga et al. 2007).

The same strategies used to generate sensors can also be adapted to actuation. A hammerhead ribozyme-based aptazyme that is inactivated upon xanthine binding was placed in the 3′-UTR of a GFP reporter gene. Xanthine could then be used to turn on gene expression, and the accumulation of GFP followed the conversion of xanthosine to xanthine in real time in xanthosine-fed yeast cultures (Win and Smolke 2007).

2.3 *Ribozyme Reporters for Screening*

The ability to detect small analytes has proven especially useful for high-throughput screening assays. Small molecules or proteins that bind to the target protein and prevent its interaction with the aptamer are readily selected due to modulation of reporter ribozyme activity (Hartig et al. 2002; Hartig and Famulok 2008). For instance, the HIV-1 reverse transcriptase (RT)-binding aptamer was joined to a hammerhead ribozyme, and the resultant negatively regulated aptazyme was used to screen for small molecule that could displace the RT from the aptamer, leading to activation and cleavage of a fluorogenic substrate (Yamazaki et al. 2007). Similarly, an ADP-dependent aptazyme was uniquely used to monitor the products of kinases or (or any enzyme that generates or destroys ADP; Srinivasan et al. 2004). Natural ribozymes can also be the targets of screens. For example, the *glm*S riboswitch is a promising target for development of antibacterial and fungal drugs via high-throughput screening of compound libraries for molecules that inhibit or stimulate ribozyme activity (Blount et al. 2006; Sudarsan et al. 2005). Such screens might even distinguish stereoisomers, as aptazymes have been described, whose self-cleaving activity is specifically activated by L-histidine and not D-histidine (Kawakami et al. 2006).

3 Ribozyme Tools for the Manipulation of Target RNAs

While RNA processing is generally dependent upon protein enzymes, these processes can be altered to be reliant instead upon ribozymes. Ribozyme processing of nascent RNA transcripts has several advantages (Fiola et al. 2006). First, ribozymes can be site-specifically targeted to almost any RNA using simple base-pairing rules. Second, ribozymes can be engineered to be allosterically regulated and thus can provide reversible, conditional, and partial modulation of gene expression (Chen et al. 2010). Third, ribozymes can be readily optimized by selection (Chen et al. 2009).

3.1 *Ribozyme Mechanisms Used to Manipulate Target RNAs*

Ribozyme-based RNA processing can be achieved by various mechanisms, including:

3.1.1 RNA Destabilization

1. Destabilization of transcripts via *trans*-acting ribozyme cleavages. This includes target RNA cleavage by endogenous ribonuclease P directed by its RNA component, a so-called external guide sequence that folds into a tRNA-like structure and can bind to complementary target RNA (Pei et al. 2008).

2. Destabilization via *cis*-acting ribozyme cleavages that target, for example, the 5′- or 3′-UTRs of an mRNA. If the ribozymes are aptazymes, cleavage can also be regulated by small molecules like toyocamycin or antisense oligonucleotides (Kim et al. 2005; Yen et al. 2004, 2006).

3.1.2 Template Reprogramming

Reprogramming the coding region of an mRNA by sequence-specific *trans*-splicing, for example, by engineered RNA exons tagged to the 3′-end of *trans*-splicing group I ribozymes (Fiskaa et al. 2006). Similarly, RNA transcripts can be edited and repaired by site-specific internal deletion or insertion of RNA pieces (including synthetic, modified RNAs) using the *Pneumocystis carinii* group I ribozyme-catalyzed *trans*-excision-splicing and *trans*-insertion-splicing reactions (Dotson et al. 2008; Johnson et al. 2005). A third approach involves the clever use of twin ribozymes (by tandem duplication of the Tobacco ringspot virus satellite RNA-derived hairpin ribozyme) to perform two *trans* chain-cleavage reactions and two ligations that ultimately lead to site-directed RNA recombination between the substrate and the repair strands provided in *trans*. Such twin ribozymes have been used to create site-specific insertion and deletion of four nucleotides from within a target RNA (Drude et al. 2007; Welz et al. 2003). But theoretically, this could be used to manipulate any accessible region of around ten residues or less of any given RNA sequence (Fig. 3).

A unique “lasso” reaction based on the hairpin ribozyme has been devised to enhance the efficiency of base-pairing and triplex antisense interactions (Dallas et al. 2008). In short, following target recognition, the lasso construct circularizes, locking in place the base-pairing and triplex interactions. Lassos can specifically and efficiently block translation in vitro (80–90%) and have been shown to alter splicing patterns in cell lines.

3.1.3 Targetron-Mediated DNA Insertion

DNA as well as RNA can be edited using bacterial group II intron ribozyme-based Targetron technologies (Jones et al. 2005; Yao et al. 2005). Targetrons have been used to knock genes out, insert introns, and as vehicles for the insertion of other DNA or RNA elements, such as Lox sites for recombination.

3.2 *Ribozymes as Research Tools*

These various mechanisms have been adapted to diverse applications.

3.2.1 Modification and Purification of RNAs

Ribozymes have been used for chemical modification of RNAs, including isotopic labeling and the incorporation of heavy atoms or modified nucleotides at multiple

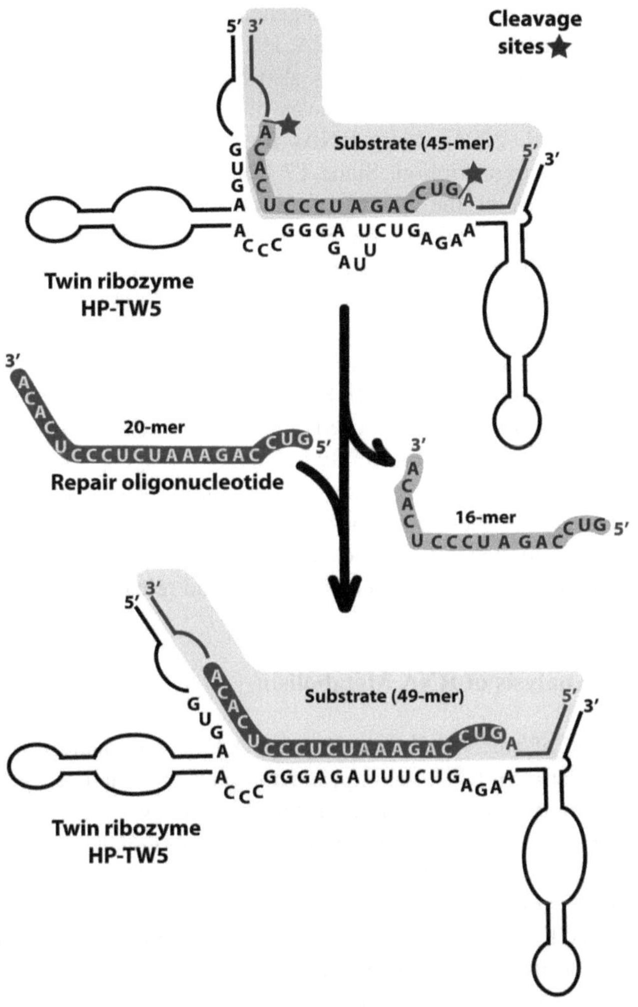

Fig. 3 Twin hairpin ribozyme-guided RNA reprogramming (Drude et al. 2007; Welz et al. 2003). Two ribozyme-catalyzed cleavage reactions remove a segment (shown in *light gray*) of the target RNA. Subsequent base-pairing of a new repair oligonucleotide leads to ribozyme-catalyzed ligation reaction and thus insertion (replacement) of nucleotides from the target RNA. This strategy can be used in a versatile way to either add or remove multiple residues from a transcript

internal sites (Duss et al. 2010). Substrate RNAs can be engineered with a 5′-end hammerhead ribozyme and a 3′-end VS ribozyme minimal cleavage sequence. Isotopically labeled and unlabeled RNAs are then co-transcriptionally cleaved by the ribozymes, generating 5′-OH and 2′/3′-cyclic phosphate ends. Acceptor (5′) and donor (3′) fragments are generated by 2′-*O*-methyl RNA/DNA oligonucleotide-guided RNase H cleavage. Religation using T4 RNA or DNA ligase results in two segmentally labeled RNAs.

Ribozymes can also be used to define the ends of transcripts. Gene expression vectors often incorporate hammerhead, hairpin, or HDV ribozymes at the 5′- and/or 3′-ends of transcripts (de Wit et al. 2007; Nandasoma et al. 2011). Posttranscriptional self-cleavage by these *cis*-acting ribozymes generates authentic and/or desired terminal nucleotides in the transcribed viral RNA, leading to optimized viral and reporter gene expression. The HDV ribozyme has also been used to generate tRNAs with homogeneous 3′-ends for the bulk synthesis of nonnatural amino acid-containing proteins (Röhrig et al. 2004). Long RNAs for crystallographic studies are often generated by in vitro transcription. Since T7 RNA polymerase can add additional, non-templated residues at the 3′-end of a transcript or can include unwanted nucleotides required for initiation at the 5′-end, self-cleaving hammerhead and HDV ribozymes flanking the coding sequence have been routinely used for cleaving transcripts and generating desired ends (Ke and Doudna 2004).

Ribozymes have also enabled the rapid native affinity purification of co-transcriptionally folded and conformationally pure RNAs (Batey and Kieft 2007; Di Tomasso et al. 2011; Pereira et al. 2010) (Fig. 4). Target RNAs are expressed as fusion transcripts containing the ligand-induced *glm*S riboswitch followed by an RNA sequence with complementarity or affinity to a capture reagent. Transcripts can then be captured on a solid support. Addition of glucosamine-6-phosphate induces site-specific self-cleavage of the *glm*S ribozyme, releasing co-transcriptionally folded target RNA with homogeneous 3′-ends from the solid support, similar to many schemes for protein purification by immobilization and release.

3.2.2 In Vivo Analysis of RNA Metabolism

To study the role of mRNA 5′-cap, 3′-poly(A) tail, and nuclear mRNA 3′-end processing in translation, mRNA stability, and nonsense-mediated decay, hammerhead or *Tetrahymena* group I ribozymes inserted at the 5′-end or within the coding region or the 3′-UTR have been used to generate cleaved transcripts that lack either the 5′-cap or the 3′-poly(A) tail (Baker and Parker 2006; Meaux and Van Hoof 2006). Similarly, the role of 5′-phosphorylation in regulating bacterial RNA decay has been studied using a self-cleaving 5′-end hammerhead ribozyme to generate reporter RNAs terminated by a 5′-OH group (Celesnik et al. 2007). HDV and hammerhead ribozyme-mediated cotranscriptional cleavage within introns and exons has been used to determine whether transcript continuity is required during pre-mRNA processing (Fong et al. 2009). HDV ribozyme-mediated cleavage of primary miRNAs (pri-miRNAs) from transcription sites demonstrated the cotranscriptional nature of pri-miRNA processing and the importance of transcription site retention (Pawlicki and Steitz 2009).

3.3 Ribozymes in Biomedical Applications

3.3.1 Ribozymes for Functional Genomics

Targeted inactivation of individual RNA species by ribozymes has been used to implicate genes such as human telomerase reverse transcriptase (hTERT)

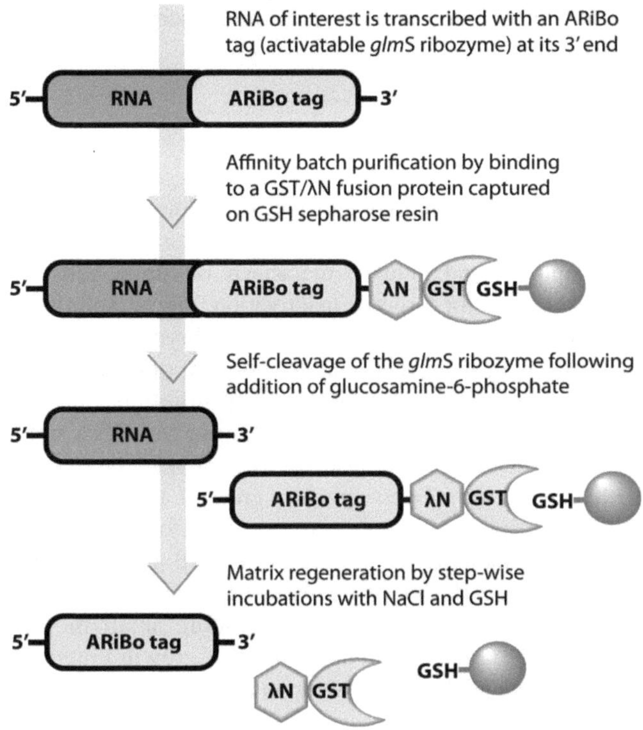

Fig. 4 Ribozyme-mediated rapid native affinity purification of RNA (Di Tomasso et al. 2011). This scheme is similar to affinity purification methods for proteins in which a tag is removed by protease cleavage

(Song et al. 2009) and the microRNA miR-21 (Suryawanshi et al. 2010) in carcinogenesis, tumor survival, invasiveness, and metastasis. RNA helicase-associated hammerhead ribozyme libraries with randomized target recognition sequences have been used to carry out genome-wide loss-of-function screens and have identified genes involved in metabolic diseases, drug activity and cellular apoptosis, migration, invasion, and differentiation (Kawasaki and Taira 2002; Suyama et al. 2003).

3.3.2 Ribozymes for Inhibition and Attenuation of Infectious Agents

Targeted inactivation of pathogen-derived RNA has been used to inhibit the growth of infectious agents such as HIV (Li et al. 2006; Mitsuyasu et al. 2009), Dengue virus (Carter et al. 2010), and mouse and human cytomegalovirus (Bai et al. 2011; Zou et al. 2003). Additionally, the inactivation of virulence genes can be used to create attenuated pathogens for therapeutic applications. Hammerhead ribozyme-mediated downregulation of the ICP27 gene was used to develop an attenuated, tumor cell-selective tetracycline-responsive oncolytic herpes simplex virus (HSV-1) (Yao et al. 2010).

3.3.3 Ribozymes for Gene Therapy

As with other gene-specific pharmaceuticals, ribozyme-mediated inactivation of genes implicated in carcinogenesis could provide therapeutic benefits including growth arrest, apoptosis, and reduced metastasis of tumor cells. *Trans*-splicing ribozymes have been used for targeted inactivation, reprogramming, and repair of cancer cell-specific transcripts such as the carcinoembryonic antigen (Jung and Lee 2006), hTERT (Song et al. 2009), and mutated versions of the tumor suppressor p53 (Watanabe and Sullenger 2000). Furthermore, specific replacement of a portion of these transcripts with the HSV thymidine kinase (tk) RNA not only reduces full-length target gene expression but sensitizes cancer cells to cytotoxic treatments with ganciclovir. *Trans*-splicing group I ribozymes have similarly been used for transcript repair and functional restoration of mutated genes involved in inherited disorders such as the mutant β-globin gene in sickle cell anemia (Lan et al. 1998). Allosteric ribozyme constructs have seen increasing use in regulation of other therapeutic RNAs, including small molecule-regulated conditional RNA interference and manipulation of growth kinetics of cells such as T lymphocytes. (Chen et al. 2010; Kumar et al. 2009; Wieland et al. 2010; Win and Smolke 2007).

3.3.4 Ribozymes for Imaging

Engineered *Tetrahymena* group I ribozyme has been used to *trans*-splice reporter mRNAs such as luciferase onto target mRNAs. The subsequent target mRNA-dependent translation of reporter enzymes can be imaged noninvasively and thereby used for detecting gene expression and silencing in cells and live animals (Hong et al. 2007; So et al. 2008).

3.3.5 Synthetic Biology and Metabolic Engineering

Sequence-specific RNA cleavage by the hammerhead, HDV, hairpin, RNase P, and the group I intron *trans*-splicing ribozymes has been used for gene regulation and metabolic engineering in bacteria, zebra fish, fungi, apicomplexan parasites, and mammalian and plant cells (Agop-Nersesian et al. 2008; Cheng et al. 2011; Fiola et al. 2006; Mueller et al. 2006). Allosteric ribozymes are favorites of synthetic biologists for engineering regulation of gene circuits, in part because of their size, modularity, and ease of manipulation. For example, nominal on/off switches have seen proof-of-principle in the regulation of *Escherichia coli* translation. The orthogonal ribosome made by Rackham and Chin (2005) was further modified with a thiamine-dependent aptazyme cleavase. In the presence of TPP, cleavage of the rRNA led to a reduction in the translation of messages specific for the orthogonal ribosome.

4 Synthetic Reactions

Ribozymes, like their protein counterparts, should be of use for biosynthesis. We examined how ribozymes can be used to manipulate mRNAs, but they can also act on smaller substrates. As an example, an interesting but underutilized reaction has been developed in the Yarus lab. The calcium-dependent ribozyme Iso6 catalyzes nucleophilic attack of phosphate on pyrophosphate linkages, resulting in pyrophosphoryl transfer (Huang and Yarus 1997a). This ribozyme can react with free GDP and a 5′-triphosphorylated RNA to form an unmethylated capped G(5′) pppRNA with the same structure as is formed by protein GTP:RNA guanylyltransferase. Iso6 has surprisingly little substrate specificity and also catalyzes decapping, cap exchange, and formation of 5′–5′-polyphosphate-linked oligonucleotides in *trans* (Huang et al. 1998). In fact, this ribozyme is an apparently universal 5′-phosphate-linker, reacting with any nucleophile containing an unblocked phosphate whose non-bridging oxygen attacks the 5′-α-phosphate of the ribozyme pppRNA with concurrent release of pyrophosphate (Huang and Yarus 1997b). In consequence, it can create 5′-phosphate–phosphate linkages with nucleotides; coenzymes including CoA, nicotinamide adenine dinucleotide phosphate, thiamine pyrophosphate, and flavin mononucleotide; macromolecules such as RNAs with 5′-diphosphates; and non-nucleotide molecules like Nε-phosphate arginine and 6-phosphate gluconic acid. This ribozyme could potentially be adapted into a unique catalyst for conjugating phosphate-containing molecules such as glucose phosphate and phosphorylated proteins.

Although most natural ribozymes catalyze phosphoryl transfer reactions, several synthetic ribozymes have been derived that accelerate a diversity of chemical reactions including C–C, C–N, C–O, and C–S bond formation or cleavage, and redox chemistry (Table 1). In vitro selection of such ribozymes has not only lent support to the evolutionary existence of RNA-based biosynthetic pathways but has also provided a veritable toolbox of catalysts with novel synthetic potential. One of the best examples of this is the flexizyme (Fx) that can perform de novo tRNA acylation using amino acid substrates with a standard, nonnatural leaving group. While chemical aminoacylation is laborious and technically difficult, the flexizyme has been shown to act on a variety of tRNAs and amino/hydroxyl acids, including N^{α}-methylated or octanoyl amino acids and peptides containing non-proteinogenic amino acids (Murakami et al. 2006; Goto et al. 2008; Goto and Suga 2009; Kawakami et al. 2008; Ohta et al. 2007). The remarkable Fx system has so far been used for (1) mRNA-encoded synthesis of nonnatural polypeptides, cyclic and bicyclic peptide libraries, and backbone-altered biopolymers such as polyesters, (2) drug discovery, (3) studying translation, and (4) translation initiation with exotic amino acids and peptides (Goto et al. 2008; Goto and Suga 2009; Kawakami et al. 2009; Ohuchi et al. 2007). Coupling RNase P and catalytic precursor tRNAs with *cis*-acting flexizymes has also allowed in situ generation of novel aminoacyl-tRNAs in reconstituted cell-free translation systems (Ohuchi et al. 2007). The flexizyme is now a major player in the development of synthetic, cyclic peptide libraries for screening (Goto et al. 2008; Kawakami et al. 2008).

Other ribozymes may also prove useful for generating libraries for screening. An in vitro-selected RNA urea synthase stereoselectively recognizes tripeptide substrates and forms a urea linkage between the RNA terminal 3′-cytidine amino group and the N terminus of the peptide (Nieuwlandt et al. 2003). Reactant-coupled ribozyme pools might conceivably be incubated with small-molecule reactant libraries for synthesis of pharmaceutical lead compound libraries (Eaton et al. 2007). It is amusing to note that it may now be possible to have wholly ribozyme-based high-throughput screening operations in which nucleic acid catalysts perform both synthesis and detection, as described above (Sect. 2.3).

5 Concluding Remarks

Synthetic and natural ribozymes are increasingly being adapted as tools for modern molecular biology, especially in situations where there are no comparable bio- or chemical catalysts. The increasing catalytic repertoire of natural and synthetic ribozymes presents further opportunities for developing unique applications, tools, and devices. For instance, the ATP-powered catalytic cycle of ATPγS-hydrolyzing ribozymes could potentially transduce chemical energy to mechanical work in nucleic acid nanodevices (Saran et al. 2006). Synthetic ribozymes that catalyze 5′-nucleotidyl transfer reactions might be used for 5′-modification of RNAs with unique chemical groups (Kang and Suga 2007). Ribozyme-catalyzed RNA–protein phosphoamide bonding may be used for affinity tagging, detection, and encoding of polypeptides (Baskerville and Bartel 2002). Large and chimeric RNA polymers could potentially be constructed using synthetic and natural RNA ligases and *Tetrahymena* and *Azoarcus* group I ribozyme-directed RNA recombination (Hayden et al. 2005; Vicens and Cech 2009). As the synthetic biology revolution continues, ribozymes should also increasingly serve as modular input/output devices for novel nucleic acid circuits.

Acknowledgments This work was sponsored by the Environmental Protection Agency (D5774-G1), the National Security Science and Engineering Faculty Fellowship (FA9550-10-1-0169), and the Welch Foundation (F-1654). Its contents are solely the responsibility of the authors and do not necessarily represent the official views of the sponsors. The authors would also like to acknowledge Dr. Angel Syrett for editing the figures.

References

Agop-Nersesian C, Pfahler J, Lanzer M et al (2008) Functional expression of ribozymes in Apicomplexa: towards exogenous control of gene expression by inducible RNA-cleavage. Int J Parasitol 38:673–681

Amontov S, Jäschke A (2006) Controlling the rate of organic reactions: rational design of allosteric Diels-Alderase ribozymes. Nucleic Acids Res 34:5032–5038

Atsumi S, Ikawa Y, Shiraishi H et al (2001) Design and development of a catalytic ribonucleoprotein. EMBO J 20:5453–5460

Ausländer S, Ketzer P, Hartig JS (2010) A ligand-dependent hammerhead ribozyme switch for controlling mammalian gene expression. Mol Biosyst 6:807–814

Bagby SC, Bergman NH, Shechner DM et al (2009) A class I ligase ribozyme with reduced Mg^{2+} dependence: selection, sequence analysis, and identification of functional tertiary interactions. RNA 15:2129–2146

Bai Y, Gong H, Li H et al (2011) Oral delivery of RNase P ribozymes by Salmonella inhibits viral infection in mice. Proc Natl Acad Sci USA 108:3222–3227

Baker KE, Parker R (2006) Conventional 3′ end formation is not required for NMD substrate recognition in Saccharomyces cerevisiae. RNA 12:1441–1445

Baskerville S, Bartel DP (2002) A ribozyme that ligates RNA to protein. Proc Natl Acad Sci USA 99:9154–9159

Batey RT, Kieft JS (2007) Improved native affinity purification of RNA. RNA 13:1384–1389

Bergeron LJ, Perreault J-P (2005) Target-dependent on/off switch increases ribozyme fidelity. Nucleic Acids Res 33:41240–41248

Biondi E, Nickens DG, Warren S et al (2010) Convergent donor and acceptor substrate utilization among kinase ribozymes. Nucleic Acids Res 38:6785–6795

Blount K, Puskarz I, Penchovsky R et al (2006) Development and application of a high-throughput assay for glmS riboswitch activators. RNA Biol 3:77–81

Carter JR, Keith JH, Barde PV et al (2010) Targeting of highly conserved Dengue virus sequences with anti-Dengue virus trans-splicing group I introns. BMC Mol Biol 11:84

Celesnik H, Deana A, Belasco JG (2007) Initiation of RNA decay in Escherichia coli by 5′ pyrophosphate removal. Mol Cell 27:79–90

Chen X, Ellington AD (2009) Design principles for ligand-sensing, conformation-switching ribozymes. PLoS Comput Biol 5:e1000620

Chen X, Denison L, Levy M et al (2009) Direct selection for ribozyme cleavage activity in cells. RNA 15:2035–2045

Chen YY, Jensen MC, Smolke CD (2010) Genetic control of mammalian T-cell proliferation with synthetic RNA regulatory systems. Proc Natl Acad Sci USA 107:8531–8536

Cheng X, Ko J-H, Altman S (2011) Inactivation of expression of two genes in Saccharomyces cerevisiae with the external guide sequence methodology. RNA 17:544–549

Cho B, Burke DH (2006) Topological rearrangement yields structural stabilization and interhelical distance constraints in the Kin.46 self-phosphorylating ribozyme. RNA 12:2118–2125

Coleman TM, Huang F (2002) RNA-catalyzed thioester synthesis. Chem Biol 9:1227–1236

Cui Z, Sun L, Zhang B (2004) A peptidyl transferase ribozyme capable of combinatorial peptide synthesis. Bioorg Med Chem 12:927–933

Dallas A, Balatskaya SV, Kuo T-C et al (2008) Hairpin ribozyme-antisense RNA constructs can act as molecular Lassos. Nucleic Acids Res 36:6752–6766

de Wit E, Spronken MIJ, Vervaet G et al (2007) A reverse-genetics system for Influenza A virus using T7 RNA polymerase. J Gen Virol 88:1281–1287

Di Tomasso G, Lampron P, Dagenais P et al (2011) The ARiBo tag: a reliable tool for affinity purification of RNAs under native conditions. Nucleic Acids Res 39:e18

Dotson PP 2nd, Frommeyer KN, Testa SM (2008) Ribozyme mediated trans insertion-splicing of modified oligonucleotides into RNA. Arch Biochem Biophys 478:81–84

Drude I, Vauléon S, Müller S (2007) Twin ribozyme mediated removal of nucleotides from an internal RNA site. Biochem Biophys Res Commun 363:24–29

Duane Smith M, Collins RA (2011) Use of ribozyme cleavage kinetics to measure salt-induced changes in solution pH. Anal Biochem 415:12–20

Duss O, Maris C, von Schroetter C et al (2010) A fast, efficient and sequence-independent method for flexible multiple segmental isotope labeling of RNA using ribozyme and RNase H cleavage. Nucleic Acids Res 38:e188

Eaton B, Tarasow TM, Nieuwlandt D, Dewey T (2007) Antibiotic compounds. US 7179907

Famulok M, Hartig JS, Mayer G (2007) Functional aptamers and aptazymes in biotechnology, diagnostics, and therapy. Chem Rev 107:3715–3743

Ferguson A, Boomer RM, Kurz M et al (2004) A novel strategy for selection of allosteric ribozymes yields RiboReporterTM sensors for caffeine and aspartame. Nucleic Acids Res 32:1756–1766

Fiola K, Perreault J-P, Cousineau B (2006) Gene targeting in the Gram-Positive bacterium Lactococcus lactis, using various delta ribozymes. Appl Environ Microbiol 72:869–879

Fiskaa T, Lundblad EW, Henriksen JR et al (2006) RNA reprogramming of alpha-mannosidase mRNA sequences in vitro by myxomycete group IC1 and IE ribozymes. FEBS J 273:2789–2800

Fong N, Ohman M, Bentley DL (2009) Fast ribozyme cleavage releases transcripts from RNA polymerase II and aborts co-transcriptional pre-mRNA processing. Nat Struct Mol Biol 16:916–922

Frauendorf C, Jäschke A (2001) Detection of small organic analytes by fluorescing molecular switches. Bioorg Med Chem 9:2521–2524

Fujita Y, Furuta H, Ikawa Y (2009) Tailoring RNA modular units on a common scaffold: a modular ribozyme with a catalytic unit for beta-nicotinamide mononucleotide-activated RNA ligation. RNA 15:877–888

Furchak JRW, Yang P, Jennings C et al (2008) Assay for glucosamine 6-phosphate using a ligand-activated ribozyme with fluorescence resonance energy transfer or CE-laser-induced fluorescence detection. Anal Chem 80:8195–8201

Fusz S, Eisenführ A, Srivatsan SG et al (2005) A ribozyme for the aldol reaction. Chem Biol 12:941–950

Goto Y, Suga H (2009) Translation initiation with initiator tRNA charged with exotic peptides. J Am Chem Soc 131:5040–5041

Goto Y, Ohta A, Sako Y et al (2008) Reprogramming the translation initiation for the synthesis of physiologically stable cyclic peptides. ACS Chem Biol 3:120–129

Gromak N, Talotti G, Proudfoot NJ et al (2008) Modulating alternative splicing by cotranscriptional cleavage of nascent intronic RNA. RNA 14:359–366

Hall B, Hesselberth JR, Ellington AD (2007) Computational selection of nucleic acid biosensors via a slip structure model. Biosens Bioelectron 22:1939–1947

Hartig JS, Famulok M (2008) Screening of molecular interactions using reporter hammerhead ribozymes. Methods Mol Biol 429:251–263

Hartig JS, Najafi-Shoushtari SH, Grune I et al (2002) Protein-dependent ribozymes report molecular interactions in real time. Nature 20:717–722

Hartig JS, Grüne I, Najafi-Shoushtari SH et al (2004) Sequence-specific detection of MicroRNAs by signal-amplifying ribozymes. J Am Chem Soc 126:722–723

Hasegawa S, Gowrishankar G, Rao J (2006) Detection of mRNA in mammalian cells with a split ribozyme reporter. Chembiochem 7:925–928

Hayden EJ, Riley CA, Burton AS et al (2005) RNA-directed construction of structurally complex and active ligase ribozymes through recombination. RNA 11:1678–1687

Hesselberth JR, Robertson MP, Knudsen SM et al (2003) Simultaneous detection of diverse analytes with an aptazyme ligase array. Anal Biochem 312:106–112

Hong S-H, Jeong J-S, Lee Y-J et al (2007) Molecular imaging of endogenous mRNA expression in a mouse tumor model by adenovirus harboring trans-splicing ribozyme. FEBS Lett 581:5396–5400

Huang F, Yarus M (1997a) 5′-RNA self-capping from guanosine diphosphate. Biochemistry 36:6557–6563

Huang F, Yarus M (1997b) Versatile 5′ phosphoryl coupling of small and large molecules to an RNA. Proc Natl Acad Sci USA 94:8965–8969

Huang F, Yang Z, Yarus M (1998) RNA enzymes with two small-molecule substrates. Chem Biol 5:669–678

Huang F, Bugg CW, Yarus M (2000) RNA-catalyzed CoA, NAD, and FAD synthesis from phosphopantetheine, NMN, and FMN. Biochemistry 39:15548–15555

Jenne A, Famulok M (1998) A novel ribozyme with ester transferase activity. Chem Biol 5:23–34

Johnson AK, Sinha J, Testa SM (2005) Trans insertion-splicing: ribozyme-catalyzed insertion of targeted sequences into RNAs. Biochemistry 44:10702–10710

Jones JP 3rd, Kierlin MN, Coon RG et al (2005) Retargeting mobile group II introns to repair mutant genes. Mol Ther 11:687–694

Jose AM, Soukup GA, Breaker RR (2001) Cooperative binding of effectors by an allosteric ribozyme. Nucleic Acids Res 29:1631–1637

Jung H-S, Lee S-W (2006) Ribozyme-mediated selective killing of cancer cells expressing carcinoembryonic antigen RNA by targeted trans-splicing. Biochem Biophys Res Commun 349:556–563

Kang TJ, Suga H (2007) In vitro selection of a 5′-purine nucleotide transferase ribozyme. Nucleic Acids Symp Ser 51:379–380

Kawakami J, Maiya C, Sugimoto N (2006) Functional role of the cofactor on activation process of L-histidine dependent ribozyme. Nucleic Acids Symp Ser 50:241–242

Kawakami T, Murakami H, Suga H (2008) Messenger RNA-programmed incorporation of multiple N-methyl-amino acids into linear and cyclic peptides. Chem Biol 15:32–42

Kawakami T, Ohta A, Ohuchi M et al (2009) Diverse backbone-cyclized peptides via codon reprogramming. Nat Chem Biol 5:888–890

Kawasaki H, Taira K (2002) A functional gene discovery in the Fas-mediated pathway to apoptosis by analysis of transiently expressed randomized hybrid-ribozyme libraries. Nucleic Acids Res 30:3609–3614

Ke A, Doudna JA (2004) Crystallization of RNA and RNA-protein complexes. Methods 34:408–414

Kim DE, Joyce GF (2004) Cross-Catalytic Replication of an RNA Ligase Ribozyme. Chem Biol 11:1505–1512

Kim DS, Gusti V, Pillai SG et al (2005) An artificial riboswitch for controlling pre-mRNA splicing. RNA 11:1667–1677

Knudsen SM, Lee J, Ellington AD et al (2006) Ribozyme-mediated signal augmentation on a mass-sensitive biosensor. J Am Chem Soc 128:15936–15937

Kumar D, An C-I, Yokobayashi Y (2009) Conditional RNA interference mediated by allosteric ribozyme. J Am Chem Soc 131:13906–13907

Kuwabara T, Tanabe T, Warashina M et al (2001) Allosterically controllable maxizyme-mediated suppression of progression of leukemia in mice. Biomacromolecules 2:1220–1228

Lacadie SA, Tardiff DF, Kadener S et al (2006) In vivo commitment to yeast cotranscriptional splicing is sensitive to transcription elongation mutants. Genes Dev 20:2055–2066

Lam BJ, Joyce GF (2009) Autocatalytic aptazymes enable ligand-dependent exponential amplification of RNA. Nat Biotechnol 27:288–292

Lan N, Howrey RP, Lee S et al (1998) Ribozyme-mediated repair of sickle β-globin mRNAs in erythrocyte precursors. Science 280:1593–1596

Lau MWL, Unrau PJ (2009) A promiscuous ribozyme promotes nucleotide synthesis in addition to ribose chemistry. Chem Biol 16:815–825

Lau MW, Cadieux KE, Unrau PJ (2004) Isolation of fast purine nucleotide synthase ribozymes. J Am Chem Soc 126:15686–15693

Lee N, Bessho Y, Wei K et al (2000) Ribozyme-catalyzed tRNA aminoacylation. Nat Struct Biol 7:28–33

Lee H-W, Robinson SG, Bandyopadhyay S et al (2007) Reversible photo-regulation of a hammerhead ribozyme using a diffusible effector. J Mol Biol 371:1163–1173
Li M, Li H, Rossi JJ (2006) RNAi in combination with a ribozyme and TAR decoy for treatment of HIV infection in hematopoietic cell gene therapy. Ann NY Acad Sci 1082:172–179
Lincoln TA, Joyce GF (2009) Self-sustained replication of an RNA enzyme. Science 323:1229–1232
Meaux S, Van Hoof A (2006) Yeast transcripts cleaved by an internal ribozyme provide new insight into the role of the cap and poly(A) tail in translation and mRNA decay. RNA 12:1323–1337
Mitsuyasu RT, Merigan TC, Carr A et al (2009) Phase 2 gene therapy trial of an anti-HIV ribozyme in autologous CD34+ cells. Nat Med 15:285–292
Moshiri H, Salavati R (2010) A fluorescence-based reporter substrate for monitoring RNA editing in trypanosomatid pathogens. Nucleic Acids Res 38:e138
Mueller D, Stahl U, Meyer V (2006) Application of hammerhead ribozymes in filamentous fungi. J Microbiol Methods 65:585–595
Murakami H, Saito H, Suga H (2003) A versatile tRNA aminoacylation catalyst based on RNA. Chem Biol 10:655–662
Murakami H, Ohta A, Goto Y et al (2006) Flexizyme as a versatile tRNA acylation catalyst and the application for translation. Nucleic Acids Symp Ser 50:35–36
Najafi-Shoushtari SH, Famulok M (2005) Competitive regulation of modular allosteric aptazymes by a small molecule and oligonucleotide effector. RNA 11:1514–1520
Najafi-Shoushtari SH, Famulok M (2007) DNA aptamer-mediated regulation of the hairpin ribozyme by human alpha-thrombin. Blood Cells Mol Dis 38:19–24
Nandasoma U, McCormick C, Griffin S et al (2011) Nucleotide requirements at positions +1 to +4 for the initiation of hepatitis C virus positive-strand RNA synthesis. J Gen Virol 92:1082–1086
Nierth A, Singer M, Jäschke A (2010) Efficient photoactivation of a Diels-Alderase ribozyme. Chem Commun 46:7975–7977
Nieuwlandt D, West M, Cheng X et al (2003) The first example of an RNA urea synthase: selection through the enzyme active site of human neutrophile elastase. Chembiochem 4:651–654
Ohta A, Murakami H, Higashimura E et al (2007) Synthesis of polyester by means of genetic code reprogramming. Chem Biol 14:1315–1322
Ohuchi M, Murakami H, Suga H (2007) In situ generation of aminoacyl-tRNAs assisted by ribozymes in translation apparatus. Nucleic Acids Symp Ser 51:115–116
Paul N, Joyce GF (2002) A self-replicating ligase ribozyme. Proc Natl Acad Sci USA 99:12733–12740
Pawlicki JM, Steitz JA (2009) Subnuclear compartmentalization of transiently expressed polyadenylated pri-microRNAs: processing at transcription sites or accumulation in SC35 foci. Cell Cycle 8:345–356
Pei D-S, Sun Y-H, Long Y et al (2008) Inhibition of no tail (ntl) gene expression in zebrafish by external guide sequence (EGS) technique. Mol Biol Rep 35:139–143
Penchovsky R, Breaker RR (2005) Computational design and experimental validation of oligonucleotide-sensing allosteric ribozymes. Nat Biotechnol 23:1424–1433
Pereira MJB, Behera V, Walter NG (2010) Nondenaturing purification of co-transcriptionally folded RNA avoids common folding heterogeneity. PLoS One 5:e12953
Prenninger S, Schroeder R, Semrad K (2006) Assaying RNA chaperone activity in vivo in bacteria using a ribozyme folding trap. Nat Protoc 1:1273–1277
Rackham O, Chin JW (2005) A network of orthogonal ribosome x mRNA pairs. Nat Chem Biol 1:159–166
Robertson MP, Ellington AD (1999) In vitro selection of an allosteric ribozyme that transduces analytes to amplicons. Nat Biotechnol 17:62–66
Robertson MP, Ellington AD (2001) In vitro selection of nucleoprotein enzymes. Nat Biotechnol 19:650–655

Röhrig CH, Retz OA, Meergans T et al (2004) In vitro non-natural amino acid mutagenesis using a suppressor tRNA generated by the cis-acting hepatitis delta virus ribozyme. Biochem Biophys Res Commun 325:731–738
Saran D, Nickens DG, Burke DH (2005) A trans acting ribozyme that phosphorylates exogenous RNA. Biochemistry 44:15007–15016
Saran D, Held DM, Burke DH (2006) Multiple-turnover thio-ATP hydrolase and phospho-enzyme intermediate formation activities catalyzed by an RNA enzyme. Nucleic Acids Res 34:3201–3208
Seelig B, Jäschke A (1999) A small catalytic RNA motif with Diels-Alderase activity. Chem Biol 6:167–176
Seetharaman S, Zivarts M, Sudarsan N et al (2001) Immobilized RNA switches for the analysis of complex chemical and biological mixtures. Nat Biotechnol 19:336–341
Sengle G, Eisenführ A, Arora PS et al (2001) Novel RNA catalysts for the Michael reaction. Chem Biol 8:459–473
Serganov A, Patel DJ (2007) Ribozymes, riboswitches and beyond: regulation of gene expression without proteins. Nat Rev Genet 8:776–790
So M-K, Gowrishankar G, Hasegawa S et al (2008) Imaging target mRNA and siRNA-mediated gene silencing in vivo with ribozyme-based reporters. Chembiochem 9:2682–2691
Song M-S, Jeong J-S, Ban G et al (2009) Validation of tissue-specific promoter-driven tumor-targeting trans-splicing ribozyme system as a multifunctional cancer gene therapy device in vivo. Cancer Gene Ther 16:113–125
Srinivasan J, Cload ST, Hamaguchi N et al (2004) ADP-Specific Sensors Enable Universal Assay of Protein Kinase Activity. Chem Biol 11:499–508
Strohbach D, Novak N, Müller S (2006) Redox-active riboswitching: allosteric regulation of ribozyme activity by ligand-shape control. Angew Chem Int Ed Engl 45:2127–2129
Sudarsan N, Cohen-Chalamish S, Nakamura S et al (2005) Thiamine pyrophosphate riboswitches are targets for the antimicrobial compound pyrithiamine. Chem Biol 12:1325–1335
Sun L, Cui Z, Li C et al (2007) Ribozyme-catalyzed dipeptide synthesis in monovalent metal ions alone. Biochemistry 46:3714–3723
Suryawanshi H, Scaria V, Maiti S (2010) Modulation of microRNA function by synthetic ribozymes. Mol Biosyst 6:1807–1809
Suyama E, Kawasaki H, Nakajima M et al (2003) Identification of genes involved in cell invasion by using a library of randomized hybrid ribozymes. Proc Natl Acad Sci USA 100:5616–5621
Tijerina P, Bhaskaran H, Russell R (2006) Nonspecific binding to structured RNA and preferential unwinding of an exposed helix by the CYT-19 protein, a DEAD-box RNA chaperone. Proc Natl Acad Sci USA 103:16698–16703
Topp S, Gallivan JP (2010) Emerging applications of riboswitches in chemical biology. ACS Chem Biol 5:139–148
Tsukiji S, Pattnaik SB, Suga H (2003) An alcohol dehydrogenase ribozyme. Nat Struct Biol 10:713–717
Turk RM, Chumachenko NV, Yarus M (2010) Multiple translational products from a five-nucleotide ribozyme. Proc Natl Acad Sci USA 107:4585–4589
Unrau PJ, Bartel DP (1998) RNA-catalysed nucleotide synthesis. Nature 395:260–263
Vaish NK, Dong F, Andrews L et al (2002) Monitoring post-translational modification of proteins with allosteric ribozymes. Nat Biotechnol 20:810–815
Vaish NK, Jadhav VR, Kossen K et al (2003) Zeptomole detection of a viral nucleic acid using a target-activated ribozyme. RNA 9:1058–1072
Vicens Q, Cech TR (2009) A natural ribozyme with 3′,5′ RNA ligase activity. Nat Chem Biol 5:97–99
Wachter A (2010) Riboswitch-mediated control of gene expression in eukaryotes. RNA Biol 7:67–76
Watanabe T, Sullenger BA (2000) Induction of wild-type p53 activity in human cancer cells by ribozymes that repair mutant p53 transcripts. Proc Natl Acad Sci USA 97:8490–8494
Watson PY, Fedor MJ (2009) Determination of intracellular RNA folding rates using self-cleaving RNAs. Meth Enzymol 468:259–286

Welz R, Bossmann K, Klug C et al (2003) Site-directed alteration of RNA sequence mediated by an engineered twin ribozyme. Angew Chem Int Ed Engl 42:2424–2427

Wieland M, Hartig JS (2006) Turning inhibitors into activators: a hammerhead ribozyme controlled by a guanine quadruplex. Angew Chem Int Ed Engl 45:5875–5878

Wieland M, Berschneider B, Erlacher MD et al (2010) Aptazyme-mediated regulation of 16S ribosomal RNA. Chem Biol 17:236–242

Win MN, Smolke CD (2007) A modular and extensible RNA-based gene-regulatory platform for engineering cellular function. Proc Natl Acad Sci USA 104:14283–14288

Wochner A, Attwater J, Coulson A et al (2011) Ribozyme-catalyzed transcription of an active ribozyme. Science 332:209–212

Wong TN, Pan T (2009) RNA folding during transcription: protocols and studies. Meth Enzymol 468:167–193

Yamazaki S, Tan L, Mayer G et al (2007) Aptamer displacement identifies alternative small-molecule target sites that escape viral resistance. Chem Biol 14:804–812

Yao J, Zhong J, Lambowitz AM (2005) Gene targeting using randomly inserted group II introns (targetrons) recovered from an Escherichia coli gene disruption library. Nucleic Acids Res 33:3351–3362

Yao F, Murakami N, Bleiziffer O et al (2010) Development of a regulatable oncolytic herpes simplex virus type 1 recombinant virus for tumor therapy. J Virol 84:8163–8171

Yen L, Svendsen J, Lee J-S et al (2004) Exogenous control of mammalian gene expression through modulation of RNA self-cleavage. Nature 431:471–476

Yen L, Magnier M, Weissleder R et al (2006) Identification of inhibitors of ribozyme self-cleavage in mammalian cells via high-throughput screening of chemical libraries. RNA 12:797–806

Young DD, Garner RA, Yoder JA et al (2009) Light-activation of gene function in mammalian cells via ribozymes. Chem Commun 5:568–570

Zaher HS, Unrau PJ (2007) Selection of an improved RNA polymerase ribozyme with superior extension and fidelity. RNA 13:1017–1026

Zaher HS, Watkins RA, Unrau PJ (2006) Two independently selected capping ribozymes share similar substrate requirements. RNA 12:1949–1958

Zivarts M, Liu Y, Breaker RR (2005) Engineered allosteric ribozymes that respond to specific divalent metal ions. Nucleic Acids Res 33:622–631

Zou H, Lee J, Umamoto S et al (2003) Engineered RNase P ribozymes are efficient in cleaving a human cytomegalovirus mRNA in vitro and are effective in inhibiting viral gene expression and growth in human cells. J Biol Chem 278:37265–37274

Zúñiga S, Sola I, Moreno JL et al (2007) Coronavirus nucleocapsid protein is an RNA chaperone. Virology 357:215–227

RNA Technologies for Mitochondrial Genetics

André Dietrich, Katarzyna Rolle, Marta Gabryelska, Eliza Wyszko, Romain Val, Maciej Szymanski, Clarisse Valentin, Anne Cosset, and Jan Barciszewski

Contents

A. Dietrich (✉) • R. Val • C. Valentin • A. Cosset
Institut de Biologie Moléculaire des Plantes, CNRS and Université de Strasbourg, 12 rue du Général Zimmer, 67084 Strasbourg, France
e-mail: andre.dietrich@ibmp-cnrs.unistra.fr

K. Rolle • M. Gabryelska • E. Wyszko • M. Szymanski • J. Barciszewski
Institute of Bioorganic Chemistry, Polish Academy of Science, Noskowskiego 12/14, 61-704 Poznan, Poland

V.A. Erdmann and J. Barciszewski (eds.), *From Nucleic Acids Sequences to Molecular Medicine*, RNA Technologies, DOI 10.1007/978-3-642-27426-8_13,

Abstract Mitochondria ensure fundamental functions in eukaryotic cells. They possess their own genetic system that provides a number of essential polypeptides of the oxidative phosphorylation chain. As a consequence, the respiratory complexes are built from both nuclear-encoded and organellar-encoded subunits. Mitochondrial biogenesis and response to the energetic or metabolic demands of the cell thus relies on elaborate regulation networks and continuous cross talk with the nucleus, as well as with plastids in plants. Main questions remain open regarding these control mechanisms and their relation with the complex transcriptional and posttranscriptional processes that take place in mitochondria. In particular, the field of organelle noncoding RNAs is growing and points to an involvement of mitochondria in cellular RNA interference. To gain further knowledge, manipulating the organelle genetic system is the obvious way to go, but mitochondrial transformation remains restricted to a couple of unicellular organisms. The scientific challenge is of wide relevance, as mutations or rearrangements in the mitochondrial genome cause incurable neurodegenerative diseases in human or lead to cytoplasmic male sterility in plants. RNA technologies may provide another path, as mitochondria naturally import several types of RNA, depending on the organism. Exploiting the natural mechanisms to target customized RNAs to or into mitochondria has begun. Considering their high specificity versus interfering RNAs, catalytic RNAs are of particular interest for such strategies. Major breakthrough has already been obtained in silencing mitochondrial RNAs through the import of *trans*-cleaving hammerhead ribozymes.

Keywords Cytoplasmic male sterility • Mitochondria • Mitochondrial disease • Noncoding RNA • Regulation • Ribozyme • RNA editing • RNA import

1 Introduction

Mitochondria have fundamental roles in cell homeostasis and response to the environment. Their efficiency, balance, and cross talk with the nucleus are essential and rely on proper functioning of their own genetic system. "Omics" approaches and functional analyses generate increasing new information in the field of mitochondria. However, filling the big gaps in the understanding of the organelle genetic processes is restricted by the current inability to transform mitochondria in animal or plant cells. Given the importance of the organelles for cell survival, manipulating mitochondrial genetics has been a long-standing but still poorly reached goal. The characterization of fungal and plant mitochondrial plasmids in the 1980s (Brown and Zhang 1995; Griffiths 1995) raised the idea that these autonomous small-size genetic units could become convenient transformation vectors. The development of biological ballistics (Lorence and Verpoorte 2004; Olmedo-Monfil et al. 2004) was subsequently the basis for many mitochondrial transformation trials using different selection markers. All these attempts were unsuccessful for higher eukaryotes, and to date, only two unicellular organisms (yeast and the alga *Chlamydomonas reinhardtii*) are amenable to mitochondrial genetic transformation with this conventional methodology (Bonnefoy et al. 2007; Zhou et al. 2010b). The challenge brought research groups to explore alternative strategies to transfer DNA into mitochondria in vitro or in intact cells (Mileshina et al. 2011), but mitochondrial transformation per se still has not been accessed (Lightowlers 2011). Thus, mitochondrial transfection and genetic investigation remain the subject of cutting-edge basic research. This context is appealing for alternative strategies based on RNA technology.

Accumulating genomic data in a variety of organisms revealed numerous transcripts that do not function as messenger RNAs (mRNAs), transfer RNAs (tRNAs), or ribosomal RNAs (rRNAs). Commonly called noncoding or non-protein-coding RNAs (ncRNAs), they play roles in epigenetic processes and chromatin structure, in transcriptional and posttranscriptional regulation of gene expression, in RNA processing and modification, in translation control, as well as in modulation of protein activity (Costa 2010; Zhou et al. 2010a). Hundreds of ncRNAs have been characterized from bacteria to higher organisms. In eukaryotes, apart from the particular case of guide RNAs in trypanosomatid mitochondrial RNA editing, investigations essentially focused on nuclear genetic systems, whereas the existence and function of ncRNAs in cytoplasmic organelles remained mostly speculative. Although their genomes are small, mitochondria and chloroplasts possess elaborate genetic systems that need to cross talk with other compartments. Organelle gene expression is controlled at both the transcriptional and posttranscriptional level (Barbrook et al. 2010; Binder and Brennicke 2003; Lipinski et al. 2010; Millar et al. 2011; Valkov et al. 2009). Intercompartment coordination proceeds through anterograde regulation (nucleus to organelles) and retrograde signaling (organelles to nucleus) (Giraud et al. 2009; Liu and Butow 2006; Pfannschmidt 2010; Woodson and Chory 2008). All these complex pathways are essential for ensuring proper energy supply and cell adaptation to the environment. Organelle gene control mechanisms and signaling cascades have been tentatively addressed, but the field suffers from a lack of directed and customized approaches. In particular, the question of RNA-mediated regulation in organelles can be asked. Mutations in the human mitochondrial genome cause severe neurodegenerative diseases through the production of nonfunctional tRNAs and mRNAs (Greaves et al. 2012; Tuppen et al. 2010). Understanding the pathogenic mechanisms and rescuing the defects constitute a further challenge. Finally, a number of traits in plants, some of which are highly relevant for agronomical purposes, depend on organelle/nucleus interactions (Chase 2007; Frei et al. 2004). Cytoplasmic male sterility (CMS) is the most renowned of these traits and is usually associated with the production of abnormal mitochondrial transcripts. Novel RNA technologies have a high potential to address all these mechanisms and pathways. The present chapter highlights a number of open questions in mitochondrial genetics that would particularly benefit from such approaches and details recent progress made toward that end, in particular through the use of catalytic RNAs.

2 The Noncoding RNA World in Mitochondria

2.1 Mitochondrial Guide RNAs in Kinetoplastid Protozoa

The best established group of organelle ncRNAs is the set of guide RNAs in the mitochondria of *Kinetoplastida*. Kinetoplastids constitute an order of flagellated protozoans in the class *Zoomastigophorea* and include the *Trypanosoma* and *Leishmania* pathogens. Kinetoplastid species possess a single mitochondrion that contains a two-component genome consisting of maxicircle (about 20 kb in size) and minicircle (about 1 kb in size) catenated DNA molecules (e.g., Lukes et al. 2002, 2005; Shlomai 2004). While the number of maxicircles is in the range of tens, thousands of minicircles populate the *Trypanosoma brucei* mitochondrion. Maxicircles encode respiratory chain subunits and rRNAs. However, the majority of the maxicircle genes are cryptogenes, and the transcripts they generate require RNA editing in order to become functional mRNAs. Editing is a posttranscriptional mechanism that occurs mainly in mitochondria and chloroplasts [for reviews, see for instance Aphasizhev and Aphasizheva (2011), Barbrook et al. (2010), Chateigner-Boutin and Small (2010, 2011), and Hajduk and Ochsenreiter (2010)]. It changes the RNA sequence versus that of the corresponding gene. At a lower frequency, such processes also affect nuclear transcripts, including protein-coding sequences, interfering RNAs, and microRNAs. Whereas editing of organelle transcripts is needed to provide functional RNAs, posttranscriptional modification of nuclear RNAs is believed to contribute to regulation mechanisms (Farajollahi and Maas 2010; Nishikura 2010). Various forms of sequence changes occur in different organisms and genetic systems (Horton and Landweber 2002). In kinetoplastid mitochondria, mRNA editing consists of the insertion or deletion of uridines, a process that can be massive for some transcripts (Hajduk and Ochsenreiter 2010; Madison-Antenucci et al. 2002; Stuart et al. 2005).

The information for mRNA editing in kinetoplastid organelles is carried by ncRNAs called guide RNAs (gRNAs). The genes for these RNAs are essentially distributed within the minicircles (Corell et al. 1993; Hong and Simpson 2003), but still a number of gRNAs are encoded by the maxicircles within intergenic regions or mRNA-coding sequences themselves (Blum et al. 1990; Clement et al. 2004; van der Spek et al. 1991). Whereas about 150 gRNAs would be sufficient to account for the editing of all mitochondrial mRNAs (Corell et al. 1993), more than 250 minicircle sequence classes have been characterized, which potentially provides up to a thousand different gRNAs (Hajduk and Ochsenreiter 2010). Redundant gRNAs, differing in sequence but targeting the same site, have been identified, and alternative editing appears to be extensive (Hong and Simpson 2003; Ochsenreiter et al. 2008). Although the precise structure of the primary transcripts remains to be characterized, gRNAs appear to be synthesized as polycistronic precursors that need 5′- and 3′-end processing (Grams et al. 2000). A number of protein factors involved in gRNA biogenesis and metabolism have been characterized, including the mitochondrial RNA polymerase, as well as members of the RNase III endoribonuclease and RNase D exoribonuclease families (Hashimi et al. 2009; Madina et al. 2011; Zimmer et al. 2011).

Kinetoplastid gRNAs have three domains. The 5′-region constitutes the "anchor" domain and is complementary to a sequence in the relevant mRNA just downstream of the site to be edited. The central region is the guiding domain. It specifies the edited sequence of the mRNA (Seiwert and Stuart 1994). Finally, the 3′-region is a short U-tail that is added and trimmed after 3′ processing of the precursor (Aphasizheva and Aphasizhev 2010; Blum and Simpson 1990). Beyond sequence heterogeneity, gRNAs have a common, imperfect double stem-loop secondary structure (Schmid et al. 1995). Hybridization of the gRNA anchor domain to its target region in the mRNA initiates editing (Seiwert et al. 1996). The MRP1/MRP2 gRNA-binding complex is responsible for gRNA/mRNA matchmaking (Schumacher et al. 2006; Zikova et al. 2008), whereas the GRBC complex ensures gRNA stabilization (Weng et al. 2008). Several further gRNA/pre-mRNA-interacting factors have been characterized (Aphasizhev and Aphasizheva 2011; Hajduk and Ochsenreiter 2010), including RNA helicases and chaperones (Ammerman et al. 2008; Hernandez et al. 2010). It was proposed that formation of the gRNA/preedited mRNA complex undergoes a proofreading step (Golden and Hajduk 2006). Structural analyses revealed that the mRNA/gRNA hybrid forms a conserved three-helical conformation (Reifur and Koslowsky 2008). On the other hand, the structure of the mRNA around the anchor binding site strongly affects the affinity for the cognate gRNA, and the set of accessory factors required for efficient editing might differ between the various gRNA/mRNA pairs (Reifur et al. 2010). Some gRNAs encoded by the maxicircles appear to function in *cis* (Golden and Hajduk 2005). The guiding domain has partial complementarity to the preedited mRNA, yielding mismatched bulges upstream of the anchor region. Bulges in the mRNA lead to U deletions, whereas bulges in the gRNA account for U insertions (Aphasizhev and Aphasizheva 2011). In mRNA regions covered by overlapping gRNAs, the first editing event often generates the binding site for the next gRNA (Maslov and Simpson 1992), while a helicase helps to displace the previous one (Li et al. 2011a). The dynamics of the gRNA structure during the early steps of the editing process has been probed, suggesting that a stem-loop structure is maintained through the first events (Yu and Koslowsky 2006). The editing reaction itself is carried out by a 20S multiprotein core complex, the "editosome," that is likely to occur in several forms differing in their endonuclease content and editing substrates (Aphasizhev and Aphasizheva 2011; Hajduk and Ochsenreiter 2010). Association with the substrate mRNA, gRNAs, and cofactors generates a larger, 35–40S complex (Göringer et al. 2011). The reaction starts with an endonucleolytic cleavage in the mRNA immediately upstream of the sequence annealed to the gRNA anchor domain (Seiwert et al. 1996). Determinants for RNA association with the editing complexes and cleavage have been characterized (Hernandez et al. 2008). The turnover of gRNAs seems to follow a specific pathway with a biphasic degradation (Ryan et al. 2006). The first step rapidly eliminates the 3′-oligo(U) tail, whereas subsequent degradation of the encoded part of the gRNA is slower.

Although a wealth of data have been collected on gRNAs and RNA editing in kinetoplastid mitochondria, a number of issues remain open and may benefit from novel approaches to manipulate organelle RNAs. In particular, regulation aspects have been little explored. No correlation has been observed between gRNA expression and developmentally regulated accumulation of the corresponding edited mRNA (Koslowsky et al. 1992; Riley et al. 1994). Conversely, it has been suggested that the level of a given gRNA may in some instances control the occurrence of the corresponding editing events (Penschow et al. 2004). Regulation of gRNA expression can potentially occur during transcription, processing, or turnover. Positively or negatively manipulating the level of specific gRNAs through mitochondrial targeting of customized RNAs would be a potent approach to further decipher these processes. Also, the role of the oligo(U)-tail is not clear (Aphasizhev and Aphasizheva 2011). The hypothesis has been proposed that the U-tail plays important roles in both stabilization of the gRNA/mRNA complex and tethering of the 5′ mRNA cleavage product during the editing reaction (Koslowsky et al. 2004). It indeed interacts with upstream purine-rich regions in the preedited mRNA (Blum and Simpson 1990; Leung and Koslowsky 2001a), which seems to stabilize the gRNA/mRNA complex in vitro (Leung and Koslowsky 1999, 2001a, b). It can recruit high-affinity binding factors (Kala and Salavati 2010; Vanhamme et al. 1998). However, gRNA/mRNA complex stabilization is variable, depending on the substrate (Koslowsky et al. 2004), and the oligo(U)-tail is dispensable for in vitro editing (Cruz-Reyes et al. 2001; Seiwert et al. 1996) as well as for in vivo gRNA stability (Aphasizheva and Aphasizhev 2010). Mitochondrial targeting of tagged gRNAs with various 3′-tails would be of great help on this issue. Last but not least, systematic analysis of the small ncRNA transcriptome of kinetoplastid mitochondria not only identified a further series of gRNAs, but also revealed a significant number of gRNA-like, oligo-uridylated RNAs without a predictable target (Madej et al. 2007, 2008; Ochsenreiter et al. 2007). The latter are encoded by maxicircles or minicircles, on either DNA strand, and in regions so far considered as untranscribed. A number of these oligo-uridylated gRNA-like transcripts map in antisense orientation to known gRNA genes or to nonedited protein-coding genes. While their expression has been confirmed, the biological role of such orphan transcripts remains elusive (Madej et al. 2007, 2008). Although some of them may support alternative editing events, it is tempting to speculate on a putative function in regulation mechanisms. RNA-based mitochondrial reverse genetics would be a valuable strategy to explore this hypothesis.

2.2 *Noncoding RNAs in Mammalian Mitochondria*

Contrasting with the complex organization occurring in kinetoplastids, mammalian mitochondria contain a compact, single-component genome of 16.5 kb encoding 13 polypeptides of the oxidative phosphorylation complexes, 22 tRNAs, and 2 rRNAs. It has no intergenic regions or introns and possesses a single noncoding region (the "displacement loop" or "D-loop") carrying the transcription promoters and one of the replication origins (Montoya et al. 1983; Smeitink et al. 2001). Gene expression in mammalian organelles involves a number of specific transcriptional and post-transcriptional steps (Fernandez-Silva et al. 2003), but RNA editing is extremely rare. In marsupials, the (GCC) anticodon of a mitochondrial tRNA was shown to undergo partial C to U editing, providing a functional $tRNA^{Asp}$ and allowing a single gene to encode two tRNAs with a different identity (Börner et al. 1996; Janke and Pääbo 1993). In human mitochondria, the nucleotide at the discriminator position in $tRNA^{Tyr}$ is added through an editing reaction that occurs after processing of a precursor in which this tRNA overlaps by one nucleotide with $tRNA^{Cys}$ (Reichert et al. 1998). Searching the mammalian mitochondrial genome failed to identify sequences potentially encoding gRNAs for such editing events (Janke and Pääbo 1993). Conversely, evidence has been raised for the presence of other types of ncRNAs in mammalian organelles.

Both strands of the mtDNA are entirely transcribed in mammalian mitochondria. The resulting transcripts are processed by "tRNA punctuation," that is, according to the recognition of tRNA secondary structures (Rossmanith et al. 1995). As there are no intergenic regions, processing of the tRNAs releases the individual mRNAs from the primary transcript. Based on the fact that often the sense and the antisense sequence of a tRNA can both form a cloverleaf structure (Paquette et al. 1990) and on structure/activity predictions, it has been hypothesized that functional antisense tRNAs are produced in mammalian organelles (Seligmann 2010).

A puzzling ncRNA made of the mitochondrial 16S rRNA and a 121 nucleotide 5′-trailor fragment deriving from the complementary strand to the gene encoding this rRNA has been characterized in mouse testis and shown to be present in sperm, liver, kidney, brain, and spleen (Villegas et al. 2000). Complementarity between the two components generates a long inverted repeat. Strikingly, this 1,685-nucleotide chimeric mitochondrial transcript seems to localize to the nucleus. It was proposed to be generated through a *trans*-splicing reaction (Villegas et al. 2000) and to undergo a putative editing event (Villegas et al. 2002). The function of this RNA remains unknown, although there has been some speculation on a role in the development of the mouse germ line (Villegas et al. 2002). A similar but longer ncRNA is overexpressed in proliferating human tumor cells (Villegas et al. 2007). In this case, the mitochondrial 16S rRNA is associated with an 815-nucleotide 5′-trailor fragment deriving from the complementary strand. The resulting 2,374 nucleotide chimeric RNA is polyadenylated and forms a long double-stranded stem with a 40-nucleotide loop. The data imply that this transcript is synthesized in mitochondria. Aphidicolin-mediated cell cycle arrest both blocks cell proliferation and downregulates the chimera (Villegas et al. 2007). Even more striking, normal proliferating cells express two transcripts containing inverted repeats linked to the 5′-region of the antisense 16S rRNA (Burzio et al. 2009). Whereas the above chimeric sense RNA is expressed in both normal and cancer cells, the expression of the antisense transcripts is downregulated in tumor cells. The sense and antisense RNAs are both present in mitochondria but also outside of the organelles and especially in the nucleus, suggesting a role in mitochondrial–nuclear communication and retrograde signaling (Landerer et al. 2011). Three further long mitochondrial ncRNAs, complementary to the *ND5*, *ND6*, and *cytb* mRNAs, were detected in human cells upon deep sequencing analysis (Rackham et al. 2011). Remarkably, their expression seems to be regulated by nuclear-encoded mitochondrial proteins involved in RNA processing, whereas their abundance is cell- and tissue-specific.

Besides these long transcripts, a number of short-size ncRNAs have been characterized in mammalian mitochondria. Initially, six candidates, ranging from 68 to 23 nucleotides, were identified from a *Mus musculus* cDNA library representative for small RNAs (10–500 nucleotides) (Lung et al. 2006). Supposedly encoded by the mitochondrial genome, these map to the D-loop (heavy or light strand) or are antisense to the *ND4* and *ND6* mRNAs. Molecular regulation of mitochondrial metabolism, structure, and function is controlled by microRNAs (miRNAs) in the nucleo-cytosolic compartment (Li et al. 2011b). Nevertheless, further studies by different groups implied the outsourcing of specific nuclear-encoded miRNAs into the mitochondria in rat, mouse, and human, suggesting that the organelles may constitute a reservoir of selected miRNAs for specific regulation pathways (Bandiera et al. 2011b; Barrey et al. 2011; Bian et al. 2010; Kren et al. 2009). Computational analyses did not predict nuclear genes encoding mitochondrial proteins to be preferential targets for these sets of miRNAs. Instead, bioinformatic data pointed to the expression of genes associated with apoptosis, cell proliferation and differentiation (Kren et al. 2009), or the expression of the mitochondrial genome itself, with target sites in most of the organelle mRNAs (Bandiera et al. 2011b; Barrey et al. 2011). A series of further data supports the concept that mitochondria are indeed involved in cellular RNA interference (RNAi) (Bandiera et al. 2011a). P-bodies, that is, mRNA-processing cytoplasmic granules linked to RNAi, interact with mitochondria (Huang et al. 2011). Inactivation of mitochondria leads to a strong decrease in RNAi efficiency associated with a delocalization of the Argonaute 2 (Ago2) component of the interference machinery from the P-bodies (Huang et al. 2011). On the other hand, immunoblotting data and microscopy imaging support the presence of Ago2 in mitochondria of human cells (Bandiera et al. 2011b). Also, Ago2 binds to mitochondrial transcripts (Beitzinger et al. 2007; Maniataki and Mourelatos 2005).

These data altogether open new prospects for the understanding of regulation pathways, cell homeostasis control, and intercompartment communication. They also raise fundamental questions, especially about subcellular trafficking of RNAs and pathway integration. The long ncRNAs appear to be generated in mitochondria, but they operate in the nucleus, implying RNA export from the organelles. Conversely, most of the mitochondrial miRNAs are assumed to be nuclear-encoded and to operate on nuclear genes. If the organelles play a role of miRNA reservoir, this implies both import into and export from mitochondria, including an appropriate control mechanism. A number of miRNAs enriched in the organelles potentially have mitochondrial targets, but it has not been established whether their transcription occurs in the nucleus, in mitochondria, or in both compartments (Bandiera et al. 2011b; Barrey et al. 2011), which leaves open further questions on trafficking.

From the data, mitochondria appear as novel players in the overall tuning of RNAi, and the crucial issue will be to define whether they are instrumental to RNAi or a target per se (Bandiera et al. 2011a). Finally, a better understanding of these mechanisms may eventually lead to miRNA-based therapy for the treatment of mitochondria-related diseases (see below) (Bandiera et al. 2011b; Barrey et al. 2011; Li et al. 2011b). Modulating the levels of long ncRNAs, miRNAs, and/or putative targets in mitochondria through specific RNA technologies would be a potent approach to take the challenge and decipher these complex processes.

2.3 Plant Mitochondrial Noncoding RNA Candidates

Contrasting in turn with the minimal mtDNA present in mammals, higher plant mitochondria possess large-size genomes (up to 1,556 kb so far in cucumber) with long intergenic regions, multiple promoters, and introns (Alverson et al. 2011; Clifton et al. 2004; Ogihara et al. 2005; Sugiyama et al. 2005; Unseld et al. 1997). These large genomes still contribute less than 60 identified genes (Kubo and Newton 2008). In many plant species, mitochondria also possess one or several extrachromosomal plasmids of essentially undetermined function and phylogenetic origin (Brown and Zhang 1995; Handa 2008). Plant mitochondrial posttranscriptional processes are more complex, involving both *cis*- and *trans*-splicing of introns (Bonen 2008), RNA processing and surveillance (Holec et al. 2006, 2008), as well as extensive RNA editing (Chateigner-Boutin and Small 2011). However, editing evolved independently in different lineages. Instead of the massive insertion or deletion of uridines taking place in kinetoplastid protozoa, RNA editing in mitochondria of flowering plants is essentially occurring through substitution of single cytidines for uridines. The reaction is currently considered to be mediated by cytidine deamination, but the editing enzyme has not been identified. Editing specificity in plant mitochondria is not guided by RNAs. No equivalent of trypanosomatid gRNAs has ever been detected in plants. Both in vitro experiments and genetics point to a protein-mediated guiding. Most editing specificity factors in plants are likely to be members of the PPR (pentatricopeptide repeat) protein family, a large group of RNA-binding proteins involved in different steps of organelle gene expression in land plants (Chateigner-Boutin and Small 2010, 2011; Schmitz-Linneweber and Small 2008).

Remarkably, about half of the sequences in plant mitochondrial genomes, representing hundreds of kb, are of unknown origin and have no homologs in databases (Alverson et al. 2011; Clifton et al. 2004; Handa 2003; Ogihara et al. 2005; Sugiyama et al. 2005; Unseld et al. 1997). Most of these sequences that lack known genes are nevertheless actively transcribed (Holec et al. 2006, 2008), and the question may be asked as to whether some of the resulting transcripts have a function. Considering the complexity of the genetic processes in plant mitochondria and the recent observations made in mammalian organelles (see above), it seems legitimate in particular to envisage the existence of plant mitochondrial regulatory

RNAs. Analysis of an *Arabidopsis thaliana* cDNA library representative for small RNAs (50–500 nucleotides) indeed identified five putative mitochondrial ncRNAs (Marker et al. 2002). One of these showed a short overlap with an exon of a known gene, whereas the other four matched to intergenic regions in the mtDNA. From the latter, three were present in two copies. In the absence of appropriate genetic tools, it remains a big challenge to determine whether any of these transcripts have a significance and a function in the organelles. Here again, RNA-based mitochondrial direct or reverse genetics would be a valuable strategy to explore the hypothesis.

3 Mitochondrial Regulation Processes

Adjustment of mitochondrial functions to the energetic and metabolic demands of the cell requires coordinated expression of the nuclear and organellar genomes. In particular, the oxidative phosphorylation complexes are built from both nuclear-encoded and mitochondrial-encoded subunits and hence the need for regulation mechanisms ensuring the availability of proper amounts of proteins for assembly. The control of genes encoding organelle proteins occurs through anterograde and retrograde regulatory pathways. Anterograde mechanisms control gene expression in organelles in response to endogenous or environmental signals that are perceived by the nucleus (Woodson and Chory 2008). Retrograde regulation refers to signals sent by organelles to communicate their functional and developmental state to the nucleus, which can then modulate anterograde control and cellular metabolism accordingly (Liu and Butow 2006; Woodson and Chory 2008). Retrograde signaling is usually mediated by metabolic features (energy production, oxygen availability, levels of reactive oxygen species, haem status, cytosolic calcium level) and activates transcription factors. A further, puzzling mechanism of nuclear-mitochondrial cross talk has been characterized in yeast. Called intergenomic signaling, it is responsive to the mtDNA (Woo et al. 2009). Through which kind of signals the mitochondrial genome can affect nuclear gene expression independently of the oxidative phosphorylation system remains to be elucidated. As a whole, a highly interconnected and responsive network of transcription factors regulates the nuclear genes encoding mitochondrial proteins (Giraud et al. 2009; Hock and Kralli 2009; Ljubicic et al. 2010; Millar et al. 2011; Scarpulla 2008, 2011). These include the PGC-1 family coactivators, which have also a role in neuroprotection, maintenance of cardiac function, and cancer cell fate control (Jones et al. 2011). The response to different models of plant mitochondrial dysfunction and environmental conditions has been analyzed at the general transcriptome level, revealing both common functional regulation and unique pathways (Leister et al. 2011; Schwarzländer et al. 2011; Van Aken et al. 2009). Remarkable features of the network of transcription factors regulating mitochondrial biogenesis and function include an extensive use of feed-forward and feedback loops, as well as parallel activation of multiple signaling pathways (Hock and Kralli 2009).

Contrasting with the growing information generated on nuclear gene expression networks, how mitochondrial transcription is regulated to meet the metabolic needs of the cell or respond to exogenous constraints is largely unknown (Asin-Cayuela and Gustafsson 2007; Barbrook et al. 2010; Millar et al. 2011). The main features of mitochondrial transcription have been characterized in different organisms, with all enzymes and cofactors being nuclear-encoded. In mammalian organelles, both strands of the compact mtDNA are entirely transcribed from the promoters located in the D-loop region (Asin-Cayuela and Gustafsson 2007; Gaspari et al. 2004). Besides promoter efficiency and transcription factor synergy (Litonin et al. 2010), transcription termination might be an important control mechanism (Guja and Garcia-Diaz 2011; Peralta et al. 2011). Evidence also suggests that the tumor suppressor p53 can be imported into mitochondria and may be involved in modulating the expression of mtDNA genes (Heyne et al. 2004; Ljubicic et al. 2010). Finally, hormone receptors have been identified in the organelles, but their mode of action remains obscure (Asin-Cayuela and Gustafsson 2007). Thirteen main primary transcripts are synthesized in *Saccharomyces cerevisiae* mitochondria, mostly as polycistronic units (Lipinski et al. 2010). The RNA polymerase binds to nonanucleotide promoter sequences that determine the transcription rate (Mangus et al. 1994). Regulation seems to occur through the organellar ATP concentration, as adenosine is the first nucleotide of all RNAs to be synthesized (Amiott and Jaehning 2006). Attenuation might play a role, but the yeast mitochondrial termination mechanism is not known (Lipinski et al. 2010). The transcription machineries of plant organelles have been described, and a number of consensus promoters have originally been identified in plant mitochondrial genomes (Fey and Maréchal-Drouard 1999; Kuhn et al. 2005; Liere et al. 2011). Nevertheless, transcription of the large plant mtDNA finally turned out to be surprisingly relaxed, with little control or modulation (Holec et al. 2006, 2008). Most of the sequences are actually expressed, including large regions without any known gene or long intergenic regions (Fujii et al. 2011).

As mentioned above, mitochondrial transcripts undergo multiple posttranscriptional processes, including precursor processing, *cis*- and *trans*-splicing, editing, and steady-state control, all of which are carried out by nuclear-encoded protein factors (Bonen 2008; Chateigner-Boutin and Small 2011; Fernandez-Silva et al. 2003; Holec et al. 2008; Lipinski et al. 2010; Rossmanith et al. 1995; Sanchez et al. 2011). Transcript synthesis in mitochondria generally does not correlate with the levels of the corresponding proteins and rather is in excess, so that the data favor the concept that the posttranscriptional and posttranslational mechanisms mostly determine the organelle protein levels (Barbrook et al. 2010; Giegé et al. 2005; Holec et al. 2008; Liere et al. 2011; Lipinski et al. 2010; Millar et al. 2011; Woodson and Chory 2008). However, analysis of the whole plant or human organelle transcriptome further illustrated the complexity of the regulation, expression, and processing of mitochondrial RNAs (Fujii et al. 2011; Mercer et al. 2011). Organelle RNA levels do vary in different conditions, suggesting that some transcriptional regulation occurs (Millar et al. 2011; Woodson and Chory 2008). Also, decreased levels of specific mitochondrial RNAs were correlated with reduced abundance of the corresponding protein complexes in an organelle RNA polymerase mutant (Kuhn et al. 2009). Enhanced transcription rate has been observed in mammalian organelles in response to mtDNA depletion, emphasizing the importance of transcriptional regulation in relation to the onset and severity of mitochondrial diseases (Peralta et al. 2011). The existence of further, yet to be identified, factors that influence organelle gene transcription has been proposed (Asin-Cayuela and Gustafsson 2007). High-depth and high-resolution DNaseI protection analysis indeed revealed many binding motifs in the human mtDNA, for which the majority of interacting proteins remain to be discovered (Mercer et al. 2011). Finally, as detailed above, a number of ncRNAs have been detected in mitochondria, including miRNAs that potentially target the organellar DNA or transcripts in mammals. The possibility of RNA-mediated regulation mechanisms in mitochondria has thus to be considered. Positively or negatively manipulating the levels of specific organellar RNAs, or introducing customized RNAs, would be of particular efficiency to characterize the control and coordination of gene expression in mitochondria.

4 Human Mitochondrial Diseases

Deciphering the complex mitochondrial genetic processes is of fundamental interest but has also major applications in areas of primary importance. In humans, mutations in the organelle genome cause severe neurodegenerative diseases that are currently incurable and await the development of gene therapy strategies (Greaves et al. 2012; Schon and DiMauro 2007; Tuppen et al. 2010). Sequence analyses identified over 300 pathogenic mutations in the human mtDNA (compiled on http://www.mitomap.org). They fall into three groups. The first group corresponds to missense mutations in protein-coding genes that are in particular associated with two common ophthalmologic manifestations, LHON (Leber's hereditary optic neuropathy) and NARP (neuropathy, ataxia, and *retinitis pigmentosa*). Three primary mtDNA mutations, that is, m.3460G > A in the *ND1* gene, m.11778G > A in the *ND4* gene, and m.14484T > C in the *ND6* gene, are present in at least 95% of LHON cases (Man et al. 2002). The second group of mtDNA pathogenic alterations brings together point mutations in tRNA or rRNA genes that impair mitochondrial protein synthesis. The example diseases for tRNA point mutations are MERRF (myoclonic epilepsy and ragged-red fibers) and MELAS (mitochondrial encephalopathy, lactic acidosis, and stroke-like episodes). MERRF is caused most commonly by the m.8344A > G point mutation in the *TK* ($tRNA^{Lys}$) gene (Silvestri et al. 1993), whereas over 80% of the patients with MELAS have the m.3243A > G mutation in the *TL1* ($tRNA^{Leu}$) gene (Wallace 1992). The third group of mutations refers to duplications or deletions. Single or multiple large-scale mtDNA deletions can produce the Pearson syndrome (refractory sideroblastic anemia, pancytopenia, defective oxidative phosphorylation, exocrine pancreatic insufficiency, and variable hepatic, renal, and endocrine failure) or the CPEO syndrome (chronic progressive external ophthalmoplegia) (Tuppen et al. 2010). CPEO may be part of KSS

(Kearns–Sayre syndrome), a more severe and fatal multisystem disorder (muscle weakness, cerebellar damage, and heart failure). Deletions in the mtDNA are a primary cause of mitochondrial disease and are likely to have a central role in the aging of postmitotic tissues. Mitochondrial dysfunction associated with mtDNA deletions has also been characterized in neurons of patients with Parkinson's disease (Bender et al. 2006) or multiple sclerosis (Campbell et al. 2011). Finally, mtDNA mutations can promote tumor cell proliferation, support tumor adaptation to new environments, or regulate metastasis (Brandon et al. 2006; Imanishi et al. 2011).

Mutations can affect all mtDNA copies in the cell or only some copies, resulting in a homoplasmic or a heteroplasmic state, respectively. Heteroplasmy is the most general situation, and the onset of clinical symptoms is determined by the ratio of wild-type to mutant mtDNA, with a typical threshold effect (Wong 2007). This means that within a tissue there may be a mixture of respiratory-competent and deficient cells.

RNA technologies may contribute major advances in these domains, especially through the replacement of mutant transcripts with functional sequences. Also, it can be hypothesized that, in an heteroplasmic state, mitochondrial targeting of short antigenomic RNAs directed against mutant sequences would suppress the replication of mutated mtDNA copies and favor wild-type mtDNA propagation (Smirnov et al. 2008). Finally, whichever the complementation approach chosen, a crucial aspect that could not be addressed so far is that the nonfunctional mRNAs and tRNAs regularly synthesized from the mutated genes in the mtDNA poison the mitochondrial genetic system. Developing an RNA-mediated strategy to eliminate these aberrant RNAs in the organelles would thus be of primary relevance.

5 Mitochondrial Genetic Traits: Plant Cytoplasmic Male Sterility

Alterations in the plant mtDNA that are not lethal commonly lead to cytoplasmic male sterility (CMS), that is, to the inability to generate functional pollen (Chase 2007; McCauley and Olson 2008; Pelletier and Budar 2007). The CMS trait is of great interest in crop breeding to avoid self-pollination or more generally self-pollenizing, so as to create and maintain hybrid lines. Such a trait is commonly associated with the appearance of chimeric genetic units in the mitochondrial genome, which are thought to result from aberrant recombination events. These genetic units generate additional transcripts carrying unusual open reading frames that often associate regions from authentic mitochondrial genes with sequences of unknown origin. The protein products of such open reading frames have been identified in CMS lines from a number of species (Schnable and Wise 1998). These polypeptides are expressed throughout the plant, but remarkably only male gametophytic development is impaired, whereas the phenotype is otherwise normal. For instance, the highly hydrophobic protein ORF138 thought to be responsible for the Ogura CMS in *Brassicaceae* strongly associates with the mitochondrial inner membrane in all organs of male-sterile plants, but no morphological or respiratory defect is detected in the vegetative tissues of these plants (Duroc et al. 2005). As a whole, the CMS mechanism appears to be complex and cannot be explained solely through the expression of chimeric polypeptides (Pelletier and Budar 2007). This is especially true when CMS cannot be associated with the appearance of a specific chimeric genetic unit in the mtDNA, as it is the case for maize (*Zea mays*) CMS-C (Allen et al. 2007). Conversely, maize CMS-S mitochondria express a specific transcript carrying two open reading frames, *orf355* and *orf77* (Gabay-Laughnan et al. 2009; Zabala et al. 1997). Interestingly, *orf77* contains several regions from the mitochondrial *atp9* gene interspersed with sequences of unknown origin (Gallagher et al. 2002) and was originally thought to be the main factor of CMS-S. However, synthesis of the corresponding polypeptide could not be detected (Zhang et al. 2004). On the other hand, the transcript carrying *orf355* and *orf77* is present in both sense and antisense orientation in CMS-S organelles (Zhang et al. 2004), raising the possibility of RNA-based CMS mechanisms. Developing specific RNA technologies to manipulate the level and/or the content of CMS-associated transcripts, or to introduce CMS transcripts into the mitochondria of fertile plants, would in turn be a powerful approach to understand these complex mechanisms.

6 Targeting RNAs to the Mitochondrial Surface

Subcellular localization of mRNAs currently appears as a general process mediating a further step in the control of gene expression (Corral-Debrinski 2007; Kloc et al. 2002; St Johnston 2005). In particular, it has been established in yeast, mammals, and plants that a large number of nuclearly synthesized mRNAs encoding mitochondrial proteins are actually targeted to the organelle surface for translation (Marc et al. 2002; Michaud et al. 2010; Sylvestre et al. 2003). Localization to the mitochondria involves *cis*-acting signals mostly carried by the 3′-UTR of the mRNA (Sylvestre et al. 2003) and a number of *trans*-acting protein factors, including the Puf3 RNA-binding protein (Quenault et al. 2011), the Ssa1 chaperone (Eliyahu et al. 2012), and proteins of the outer membrane translocase (Eliyahu et al. 2010; Gadir et al. 2011). A number of mRNAs might be translated while on the way to the organelles (Eliyahu et al. 2012). The mitochondrial presequence and further elements in the synthesized polypeptide may contribute to organellar anchoring of the mRNA (Eliyahu et al. 2010; Garcia et al. 2010), but translation is not always essential (Gadir et al. 2011). These processes as a whole support a spatio-temporal regulation of mitochondrial protein import (Ahmed and Fisher 2009; Devaux et al. 2010).

Targeting mRNAs to the organelle surface to improve translation and import has been one of the first RNA technologies developed to manipulate mitochondrial genetics in mammalian cells. Circumventing the absence of an organelle transformation methodology, indirect approaches were designed to complement pathogenic mtDNA mutations through nuclear transgenesis. The strategy was to transcribe in the nucleus appropriately recoded mitochondrial genes completed with a sequence encoding an organelle-targeting N-terminal transit peptide. The transcript was subsequently expected to be translated in the cytosol, whereas the resulting polypeptide would be transported into the mitochondria through the regular protein import pathway. Such allotopic expression approaches were developed in human cells with the *ATP6* and *ND4* mitochondrial genes, leading to conflicting results. In both cases, initial positive reports were contradicted by subsequent studies (Bokori-Brown and Holt 2006; Guy et al. 2002; Manfredi et al. 2002; Oca-Cossio et al. 2003; Ojaimi et al. 2002). A further attempt based on nuclear expression of the mitochondrial *ND6* gene in mouse cells was clearly negative (Perales-Clemente et al. 2010).

A major difficulty in these approaches might be the high hydrophobicity of the proteins normally encoded by the mtDNA, which may hinder trafficking and organelle import. This hypothesis led to the idea of adding appropriate 3′-UTR *cis*-acting signals to the sequence, so as to target the nuclear transgene-derived transcript to the mitochondrial surface, facilitate cotranslational translocation of the polypeptide, and prevent the formation of cytosolic aggregates. Expressing the *ATP6*-coding sequence combined with the mitochondrial targeting sequence and the 3′-UTR *cis*-acting signals from the nuclear *SOD2* gene allowed localization of the hybrid mRNA to the mitochondrial surface and efficient organelle translocation of the resulting protein in human cells (Kaltimbacher et al. 2006). The strategy was thus extended to human cells carrying pathogenic mutations in the mtDNA. Nuclear expression of the *ATP6*-coding sequence combined with the mitochondrial targeting sequence and the 3′-UTR *cis*-acting signals from the nuclear *SOD2* gene led to long-lasting rescue of organelle dysfunction in human fibroblasts harboring the m.8993T > G *ATP6* mutation (Bonnet et al. 2007). Similarly, nuclear expression of *ND1* and *ND4* combined with the mitochondrial targeting sequence and the 3′-UTR *cis*-acting signals from the nuclear *COX10* gene allowed to rescue the m.3460G > A *ND1* mutation and the m.11778G > A *ND4* mutation, respectively (Bonnet et al. 2007, 2008). Finally, compensation of the m.11778G > A mutation upon nuclear expression of *ND4* combined with the *COX10* signals was validated in a rat model (Ellouze et al. 2008). However, the issue remains somehow controversial, as a number of reports do not unequivocally show that the protein expressed from the nuclear transgene is indeed imported into the organelles and assembled into the relevant mitochondrial complex (Perales-Clemente et al. 2010).

7 Targeting RNAs into Mitochondria: RNA Import Pathways

In many organisms, the mitochondrial genome does not code for a complete set of tRNAs, so that nuclear-encoded tRNAs are taken up from the cytosol [reviewed for instance in Salinas et al. (2008) and Schneider (2011)]. These tRNAs are usually shared between the two compartments, that is, they are not exclusively destined to the mitochondria. The number of imported tRNAs ranges between a single tRNA species in marsupials (Dörner et al. 2001) and all mitochondrial tRNA species in trypanosomatids (Schneider and Maréchal-Drouard 2000). In higher plants, mitochondria selectively import one third to one half of their tRNAs from the cytosol (Kumar et al. 1996; Maréchal-Drouard et al. 1990). The targeting and transport mechanisms that underlie these different situations have been investigated in different organisms. Although many aspects remain unknown, it seems likely that there is no unique pathway.

In the yeast *S. cerevisiae*, mitochondria take up one of the cytosolic $tRNAs^{Lys}$ (Tarassov and Entelis 1992) and likely the cytosolic $tRNAs^{Gln}$ (Rinehart et al. 2005). A recently developed model proposes that the cytosolic $tRNA^{Lys}$(CUU) would be recognized and targeted to the mitochondrial surface by the glycolytic enzyme enolase-2, handed over to the precursor of the mitochondrial lysyl-tRNA synthetase and translocated into the mitochondrial matrix, possibly through the protein uptake channel (Entelis et al. 2006). The import of $tRNA^{Lys}$ is regulated by the ubiquitin proteasome system (Brandina et al. 2007). Conversely, a direct uptake mechanism, independent from the protein import pathway, was put forward for the $tRNAs^{Gln}$ (Rinehart et al. 2005). Whereas it has long been considered that human mitochondria encode all their tRNAs, functional evidence was recently reported supporting organelle import of cytosolic $tRNAs^{Gln}$ in human cells as well (Rubio et al. 2008). As in yeast, the uptake would proceed through a mechanism distinct from protein import. Transport of tRNAs into trypanosomatid mitochondria also relies on recognition of specific sequence elements (Crausaz Esseiva et al. 2004; Mahapatra et al. 1998). The putative role of thiomodification of the wobble nucleotide in the mechanism of tRNA sorting has been debated (Bruske et al. 2009; Kaneko et al. 2003; Paris et al. 2009). Recognition of the relevant determinants by elongation factor 1a mediates the specificity of mitochondrial tRNA import in *T. brucei* (Bouzaidi-Tiali et al. 2007). Analyzing the mechanism of tRNA translocation into trypanosomatid organelles led to controversial results. Transport across the inner membrane in *Leishmania tropica* was proposed to occur through a large RNA-specific import complex involving respiratory components, including the Rieske Fe–S protein (Chatterjee et al. 2006; Mukherjee et al. 2007). Such a complex could never be characterized in other *Leishmania* species or in *Trypanosoma*, and the Rieske protein was shown to be nonessential for the mitochondrial tRNA import machinery in *T. brucei* (Paris et al. 2009). In plants, mitochondrial uptake of tRNAs requires both aminoacylation determinants and import-specific determinants (Delage et al. 2003), whereas translocation through the outer membrane is potentially mediated by the mitochondrial voltage-dependent anion channel (VDAC) together with subunits of the protein import TOM complex (Salinas et al. 2006).

Beyond tRNAs, further organelle RNA trafficking has been characterized in mammals. Mitochondrial import of nuclear-encoded 5S rRNA was established in several mammalian species (Entelis et al. 2001; Magalhaes et al. 1998; Yoshionari et al. 1994). The biological significance of this process is still intriguing, but import of 5S rRNA appeared to be critical for mitochondrial translation, and evidence was obtained for association of this RNA with the mammalian organelle ribosomes (Smirnov et al. 2010, 2011). Experimental conditions for 5S rRNA import into mammalian mitochondria are reminiscent of $tRNA^{Lys}$(CUU) uptake in yeast (Entelis et al. 2001). The protein factors recruited to transport the RNA to the organelles include the mitochondrial enzyme rhodanese (Smirnov et al. 2010) and the ribosomal protein L18 (Smirnov et al. 2011).

Early studies supported mitochondrial import of the nuclear-encoded catalytic RNA components of RNase P, the ribonucleoprotein that processes the 5′-end of tRNAs (Doersen et al. 1985), and RNase MRP, the endoribonuclease that generates the RNA primers for the replication of the mtDNA leading strand (Chang and Clayton 1989). These data became a subject of controversy (Kiss and Filipowicz 1992), but were subsequently supported by further observations (Li et al. 1994; Lu et al. 2010; Puranam and Attardi 2001). The question became again puzzling when it was shown that human mitochondrial RNase P is actually a protein enzyme that does not require a *trans*-acting RNA component for catalysis (Holzmann et al. 2008). Whether the RNase P RNA has another, yet unidentified, role in the organelles or a protein-only and an RNA-containing RNase P coexist in mammalian mitochondria remains to be established. Nevertheless, mitochondrial import of the RNase P/RNase MRP RNA was again confirmed by functional analyses of the uptake mechanism (Wang et al. 2010, 2011). Unexpectedly, the latter showed that polynucleotide phosphorylase (PNPASE), originally an exoribonuclease and poly (A) polymerase, is located in the mitochondrial intermembrane space and mediates organelle import of small RNAs, including the RNase P/RNase MRP RNAs and the 5S rRNA (Wang et al. 2010, 2011).

The existence of multiple natural pathways dedicated to mitochondrial transport of specific RNAs provides a strong potential for the development of organelle RNA technologies. Targeting recombinant tRNAs of interest to mitochondria represents the most direct exploitation of these pathways and has already been used in human cellular models for the rescue of organelle deficiencies (see Sect. 8). A further step is to design customized importable RNAs. In this respect, combining a tRNA structure with a cargo RNA has recently been successful for the manipulation of mitochondrial gene expression in plants (see Sects. 9 and 10).

8 Targeting tRNAs into Mitochondria for Functional Rescue

Many mitochondrial disorders in humans are due to mutations in organelle tRNA-encoding genes (http://www.mitomap.org). There has thus been a special interest in rescuing the resulting dysfunctions by importing relevant functional tRNAs from the cytosol. As mentioned, whereas the human mtDNA was initially considered to encode all tRNAs needed for the organellar translation, evidence was reported for the import of cytosolic $tRNAs^{Gln}$ (Rubio et al. 2008). Furthermore, isolated human mitochondria proved able to import yeast cytosolic $tRNA^{Lys}$(CUU), synthetic variants of this tRNA, as well as human organelle $tRNA^{Lys}$, in the presence of appropriate yeast or human protein factors (Entelis et al. 2001; Kolesnikova et al. 2000). The data altogether confirmed that mammalian cells have conserved the factors needed for a mitochondrial tRNA uptake pathway similar to that for $tRNA^{Lys}$(CUU) in *S. cerevisiae* (Entelis et al. 2001). Yeast $tRNA^{Lys}$ derivatives expressed in human cells from nuclear transgenes were indeed imported into mitochondria, correctly aminoacylated, and able to participate in organelle translation (Kolesnikova et al. 2004). Remarkably, $tRNA^{Lys}$ import partially rescued the mitochondrial functional deficiencies in cybrid cells and in patient-derived fibroblasts carrying the $tRNA^{Lys}$ m.8344A $>$ G MERRF mutation (Kolesnikova et al. 2004). In further experiments, the aminoacylation identity of the recombinant tRNAs was changed from lysine to leucine, so as to address the problem of mtDNA mutations in $tRNA^{Leu}$ genes, another major source of mitochondrial disease. Nuclear expression and mitochondrial transport of tRNA variants carrying the identity elements for mammalian organelle leucyl–tRNA synthetase in human cybrid cells allowed a significant rescue of the mtDNA $tRNA^{Leu}$ m.3243A $>$ G mutation, the major cause of the MELAS syndrome (Karicheva et al. 2011). These results establish the functionality of customized tRNAs imported into human mitochondria in vivo and open the way for exploiting allotopic expression of importable tRNAs to "complement" mtDNA mutations in tRNA genes (Karicheva et al. 2011; Kolesnikova et al. 2004).

9 Targeting Passenger RNAs into Mitochondria

An early work designed to demonstrate mitochondrial tRNA import in plants raised the idea that tRNAs might serve as vectors to drive extra RNAs into the organelles (Small et al. 1992). Mitochondrially imported tRNAs in higher plants include leucine tRNAs. A bean (*Phaseolus vulgaris*) tRNA leucine expressed from a nuclear transgene and carrying a four-base-pair insertion in the anticodon loop was recovered in the mitochondria of transformed potato (*Solanum tuberosum*) whole plants (Small et al. 1992). However, insertion of longer sequences was subjected to strong constraints and was not successful. On the other hand, tRNA leucine variants with an insertion in the variable region were poorly expressed and functionally defective (Small et al. 1992). The possibility was thus explored to graft the "passenger" RNA to the 5′-end of the tRNA as a trailor sequence.

Such an approach in turn had to face another difficulty, that is, escaping processing by RNase P, which generates the mature 5′-end of tRNAs (Schon 1995). Also, natural transcription of plant nuclear tRNA genes driven by the internal promoters for RNA polymerase III (A and B boxes) is believed to start from a CAA triplet generally located a few nucleotides upstream of the mature 5′-end of the tRNA and to be influenced by upstream regulation elements (Choisne et al. 1998; Geiduschek and Tocchini-Valentini 1988). Such a mechanism by itself would not allow the expression of a tRNA with a 5′-trailor sequence. All these

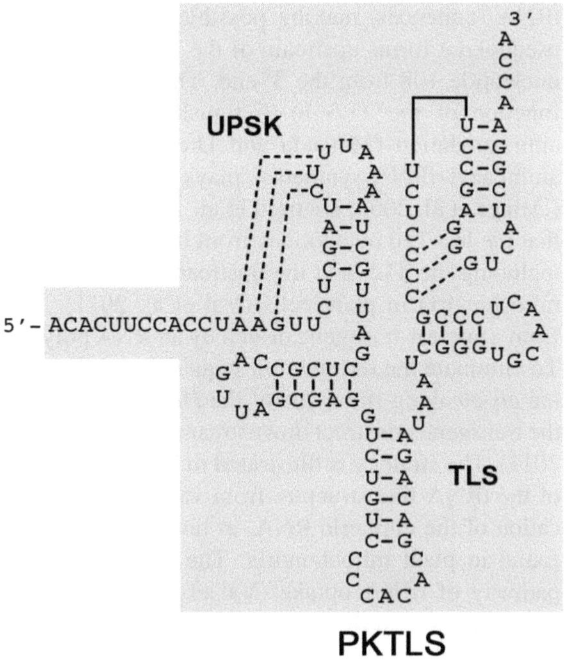

Fig. 1 A mitochondrial shuttle RNA (Val et al. 2011). The tRNA mimic (PKTLS) that serves as a shuttle to target passenger RNAs into mitochondria in plants corresponds to the last 120 nucleotides from the 3′-end of the *Turnip yellow mosaic virus* (TYMV) genomic RNA and comprises the tRNA-like structure (TLS) and the upstream pseudoknot (UPSK) (Matsuda and Dreher 2004)

constraints brought to a strategy that only retains the concept of a tRNA-like mitochondrial import shuttle and a tRNA mimic of viral origin was finally selected for that purpose. The genomic RNAs of many plant viruses possess a 3′ tRNA-like structure that can generally be aminoacylated (Dreher 2009; Fechter et al. 2001). Three groups of such tRNA mimics have been characterized on the basis of their aminoacylation identity, which can be histidine, tyrosine, or valine. Whereas the first two were not relevant, the latter group was of special interest as, besides leucine isoacceptors, mitochondrially imported tRNAs in higher plants also include valine tRNAs (Kumar et al. 1996; Maréchal-Drouard et al. 1990).

The tRNA-like structure (TLS) of the *Turnip yellow mosaic virus* (TYMV) is one of the best studied viral tRNA mimics (Fig. 1) and is a valine tRNA mimic (e.g., Matsuda and Dreher 2004). It is formed by the last 82 nucleotides at the 3′-end of the viral genomic RNA [recent structural considerations can be found in Hammond et al. (2009)] and is an excellent substrate for valyl–tRNA synthetase. Recognition of the tRNA by the relevant aminoacyl–tRNA synthetase has been shown to be required for in vivo mitochondrial import in plant cells (Delage et al. 2003; Dietrich et al. 1996). The TYMV TLS forms naturally at the 3′-end of the viral genomic and subgenomic RNAs, which thus by themselves represent a tRNA with a long upstream sequence. Although in some conditions it can be cleaved in vitro by the *Escherichia coli* enzyme at the level of the acceptor arm equivalent (Mans et al. 1990), the TYMV TLS does not seem to be processed by RNase P in vivo in plant cells. The D-arm and T-arm equivalents in the TLS do not fit the structure of the consensus A and B box RNA polymerase III internal promoters found in regular tRNA sequences, making possible an RNA polymerase II-driven expression. A pseudoknot forms upstream of the TLS in the TYMV genomic RNA, extending to nucleotide 108 from the 3′-end. This pseudoknot region contributes to optimized function of the TLS in 3′ translational enhancement, which is dependent on aminoacylation (Matsuda and Dreher 2004). As mentioned, recognition by the aminoacyl–tRNA synthetase plays a role in mitochondrial tRNA import in plants (Delage et al. 2003; Dietrich et al. 1996). On this basis, it was finally demonstrated that the last 120 nucleotides from the 3′-end of the TYMV genomic RNA (Fig. 1), including the TLS and the upstream pseudoknot, can drive a passenger RNA into mitochondria in plant cells (Val et al. 2011). The chimeric RNA was expressed from a nuclear transgene driven by an RNA polymerase II promoter and terminator. To eliminate the termination sequence and the poly(A) tail, the sequence encoding the *cis*-cleaving ribozyme of the *Hepatitis delta virus* (see below) was included in the transgene construct downstream of the tRNA mimic-encoding cDNA (Val et al. 2011). The strategy is illustrated in Fig. 2. Switching the aminoacylation specificity of the tRNA-like structure from valine to methionine abolished organelle translocation of the chimeric RNA, in line with the fact that cytosolic tRNAs^{Met} are not found in plant mitochondria. The process thus was likely to follow the natural pathway of tRNA uptake (Val et al. 2011). These data altogether established a multipurpose shuttle system for mitochondrial import of nuclearly expressed cargo RNA sequences in plant cells.

10 Targeting Catalytic RNAs into Mitochondria

Validation of a mitochondrial RNA shuttle system enabled to develop directed knockdown of organelle RNAs (Val et al. 2011). The strategy relied on the association of a *trans*-cleaving catalytic RNA, that is, a ribozyme, with the TYMV tRNA mimic. The advantage of ribozymes over other antisense strategies lies in that they carry their own specific RNA cleavage activity and do not rely on additional factors.

10.1 Trans-*Cleaving Ribozymes*

With the increasing amount of sequenced genomes, it has become abundantly clear that the complexity of life remains raveled in RNA. Although humans have more than three billion DNA base pairs in the genome, with most of these sequences being converted to RNA, only small portions represent genes (orfs) that are translated into proteins (Robertson 2010). Philip Sharp's and Richard Robert's initial findings that large noncoding RNA sequences or introns interrupt the coding sequences (exons) in genes led to the discovery of alternative splicing that can result in one gene producing more than one mRNA for protein synthesis (Witten and Ule 2011). The diverse functions of ribonucleic acids as information carriers

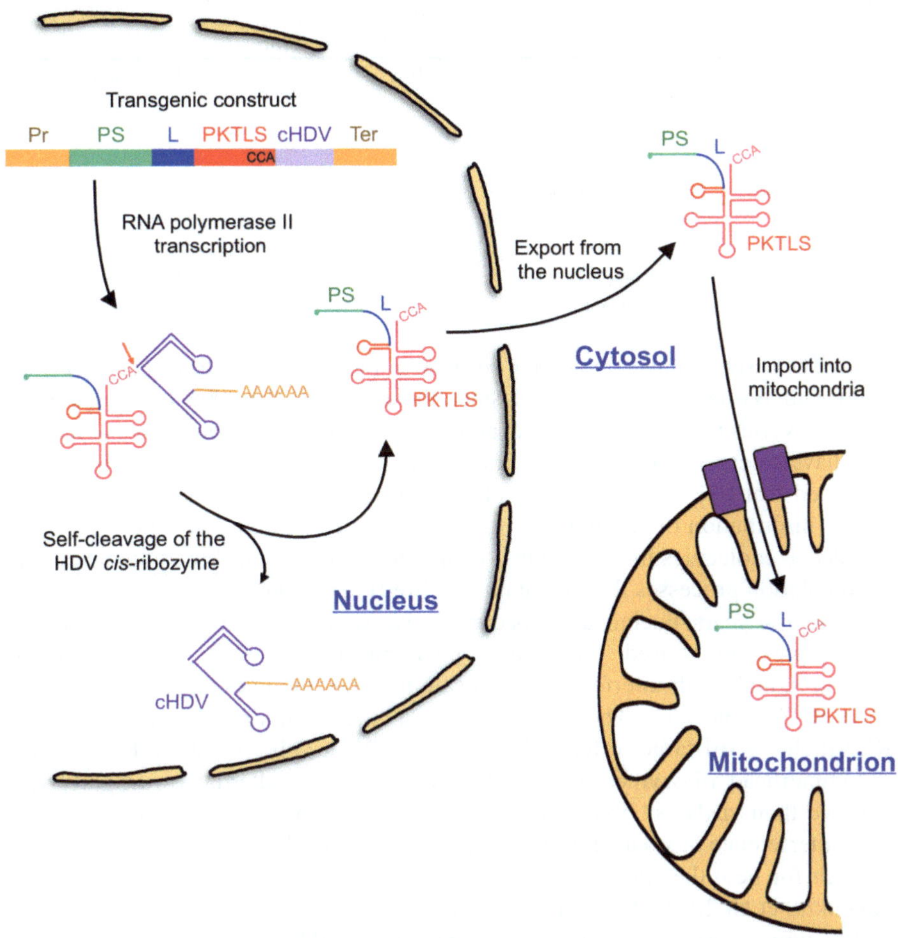

Fig. 2 Passenger RNA targeting into mitochondria in plants (Val et al. 2011). Plant cells are nuclearly transformed with constructs driven by an RNA polymerase II promoter (Pr) and terminator (Ter), the constructs code for the passenger sequence of interest (PS) combined with a linker (L), the last 120 nucleotides from the 3′-end of the TYMV genomic RNA (PKTLS), and the *Hepatitis delta virus cis*-cleaving ribozyme (cHDV). Following expression, self-cleavage of the cHDV eliminates the termination sequence and poly(A) tail. The passenger RNA associated with the PKTLS shuttle is then exported to the cytosol and imported into mitochondria

(mRNAs), adapters (tRNAs), or scaffolds and catalysts (rRNAs) reflect the crucial importance of RNA within the living cell and its significance as essential for life itself. A number of novel properties of RNA molecules make them useful in new applications in biotechnology and molecular medicine (Mulhbacher et al. 2010; Wan et al. 2011).

The biological role of both DNA and RNA as storage place and carrier of genetic information has taken a new leap in the early 1980s following the seminal discovery of RNA catalysts by Thomas Cech and Sidney Altman (Ferré-D'Amaré and Scott

Table 1 Characteristics of selected natural ribozymes

Group	Ribozyme	Molecule length (nt)	Reaction catalyzed
Small ribozymes	Hammerhead	~40	Transesterification
	HDV	~90	
	Hairpin	~70	
	Varkud satellite	~160	
Large ribozymes	Group I introns	~210	Transesterification
	Group II introns	<500	
	RNAse P	350–410	Hydrolysis
	snRNA (U2, U6)	180, 100	Transesterification
Ribosome	23S rRNA	~2,600	Peptidyl transferase

2010). They were named ribozymes because of their similarity to protein enzymes (Fedor and Williamson 2005). Ever since those reports, there has been a surge of new discoveries in that research area, and as a result, new functional roles of nucleic acids have emerged, including their involvement in gene regulation. Those findings confirmed that RNA has more facets than just being the carrier and transporter of genetic information or serving as a structural element. From that time, RNAs are regarded as molecules that have the potential to actively contribute to the catalytic and regulation processes that maintain cell life (Fedor 2009).

In the past years, nucleic acids became universal tools for selective silencing of gene expression. Based on natural phenomena, antisense oligonucleotides, ribozymes, and small interfering RNAs can influence the integrity and function of target RNA molecules. Though promising at the beginning, RNA interference (RNAi) shows off-target effects (Singh et al. 2011). Other catalytic RNAs, such as hammerhead ribozymes, represent a very attractive alternative. Slightly less effective than siRNAs, ribozymes exhibit fewer off-target effects than found in RNA interference mechanisms. These properties make them a subject of intensive research for new therapeutics against cancer and viral diseases, including AIDS (Hotchkiss et al. 2004; Weinberg et al. 2007).

Ribozymes possess a catalytic core composed of RNA only. Many are ribonucleoprotein complexes in which proteins do not take part in catalysis, but have structural or regulatory functions. Almost all of the naturally occurring ribozymes hydrolyze or ligate phosphodiester bonds in RNA. For the inactivation of their target RNA, some of them do not need any additional cell factors (Scherer and Rossi 2003). Up to today, nine types of natural catalytic RNA have been identified, from which viral-derived HDV, hairpin, and hammerhead ribozymes have the widest potential applications (Table 1, Fig. 3).

10.2 The Hairpin Ribozyme

The hairpin ribozyme is a small catalytic RNA found in satellite RNAs of plant viruses. It catalyzes self-cleavage and ligation reactions of the products of rolling

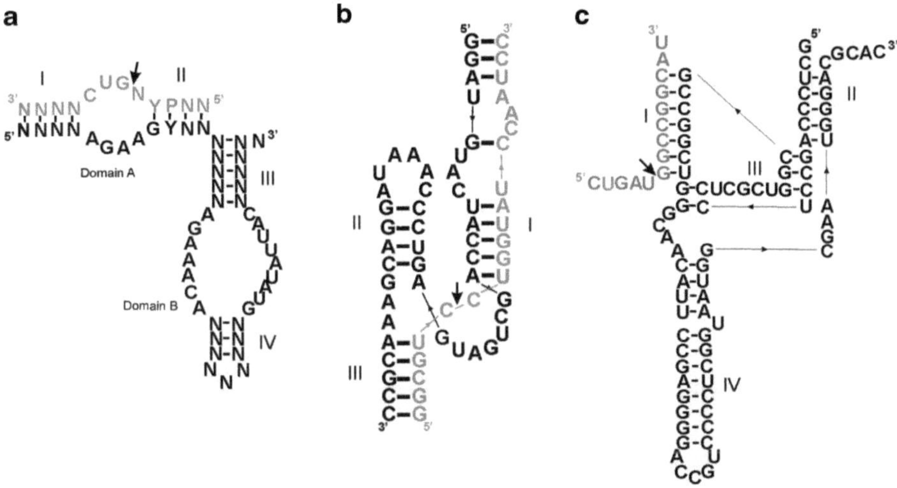

Fig. 3 Secondary structures of selected small catalytic RNAs acting in *trans*. (**a**) Hairpin ribozyme, (**b**) hammerhead ribozyme, (**c**) HDV ribozyme. *Arrows* indicate hydrolysis sites. *Gray*, target sequence; *black*, ribozyme sequence; I–IV, helices

circle virus replication. In contrast to hydrolysis by the hammerhead ribozyme, it does not require a metal ion for the reaction. The hairpin readily catalyzes both cleavage and ligation. The reaction can be driven in either direction by the appropriate choice of the construct and cleaved product sequence (Suydam et al. 2010).

Smaller artificial versions of the hairpin ribozyme have been developed to enable a more detailed experimental analysis of the molecule (Feldstein and Bruening 1993). A 50-nucleotide minimal catalytic domain and a 14-nucleotide substrate were identified (Hampel and Tritz 1989). A *trans*-acting ribozyme was developed for the catalysis of multiple substrates. The minimal hairpin ribozyme–substrate complex folds into a secondary structure that includes two domains, each consisting of two short helices separated by an internal loop (Fig. 3). Domain A contains the substrate and the primary substrate-recognition region of the ribozyme. Domain B is larger and contains the primary catalytic determinants of the ribozyme. The two domains are covalently joined via a phosphodiester linkage. They must interact with one another for catalysis to occur (Butcher et al. 1995). They can stack one atop the other, forming an inactive extended structure that resembles a hairpin, or, for catalysis to occur, lie parallel to one another in a fold that resembles a paperclip (Esteban et al. 1998).

The hairpin ribozyme accelerates the rate of phosphodiester transfer reactions by at least five orders of magnitude. To achieve this rate enhancement, the active site forms via a substrate helix docking event that constrains the scissile phosphate linkage and positions G8 and A38 for catalysis, both of which have been implicated as sites of proton transfer in general acid–base catalysis (Suydam et al. 2010). Structural and biochemical data indicate that the conserved A9 and A10 bases reside close to the scissile bond but make distinct contributions to catalysis (Spitale et al. 2009).

The hairpin ribozyme has been reengineered, resulting in a number of variants with extended or even new functions, such as activity control by external effectors (namely, oligonucleotides, flavine mononucleotide, and adenine), RNA fragment exchange reactions, self-processing hairpin ribozymes, as well as knockdown of specific RNA substrates by adapting the substrate-binding domain to the specific target sequence (Muller et al. 2011).

10.3 The Hepatitis Delta Virus Ribozyme (HDV Ribozyme)

Two variants of HDV ribozyme were discovered in the sense and antisense strands of the *Hepatitis delta virus* RNA. Of the known self-cleaving ribozymes, the HDV ribozyme is the only one that employs an active site divalent cation for catalysis (Ferré-D'Amaré and Scott 2010). A minimal natural ribozyme sequence with proficient in vitro self-cleavage activity is about 85 nucleotides long and adopts a secondary structure with four paired regions (P1–P4, I–IV, Fig. 3). The two pairings that define the 5′ and 3′ boundaries of the ribozyme, P1 and P2, form an atypical pseudoknot (Been and Wickham 1997). The cleavage site, located at the 5′-end of P1, is in close proximity to three single-stranded regions, consisting of a hairpin loop at the end of P3 and two sequences joining P1–P4 and P4–P2.

While the natural forms of the HDV ribozyme appear to be prone to misfolding, biochemical and mutagenesis studies from a number of laboratories have allowed the production of *trans*-acting ribozymes and smaller, more active *cis*-acting ribozymes (Been and Wickham 1997). The HDV ribozyme uses both metal ion and nucleobase catalysis in its cleavage mechanism. A reverse G·U wobble was observed in a recent crystal structure of the precleaved state. This unusual base-pair positions a Mg^{2+} ion to participate in catalysis (Veeraraghavan et al. 2011). The crystal structure of the precleaved HDV ribozyme, along with biochemical data, supports a mechanism for phosphodiester bond self-cleavage in which C75 acts as a general acid and the bound Mg^{2+} ion acts as a Lewis acid (Ganguly et al. 2011).

The reaction rate is sensitive to the identity of the upstream residue at position -1, with a nucleobase preference $U > C > A > G$, but no obvious explanation for these effects can be derived from existing experimental structures (Lee et al. 2011). In vitro selection of self-cleaving ribozymes from a human genomic library revealed an HDV-like ribozyme in the second intron of the human CPEB3 gene, and recent results suggest that this RNA affects episodic memory performance (Webb and Lupták 2011). Computational searches based on the secondary structure of the HDV/CPEB3 fold yielded numerous functional ribozymes in a wide variety of organisms. Genomic mapping of these RNAs suggested several biological roles, one of which is the 5′ processing of non-LTR retrotransposons (Webb and Lupták 2011).

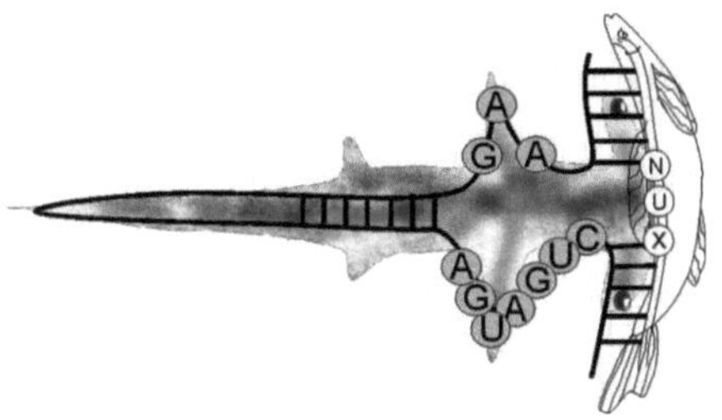

Fig. 4 Similarity of the hammerhead ribozyme to the hammerhead shark. GAA, CUGAUGA—catalytic core sequence; NUX—nucleotide target site sequence; N stands for any nucleotide and X cannot be G

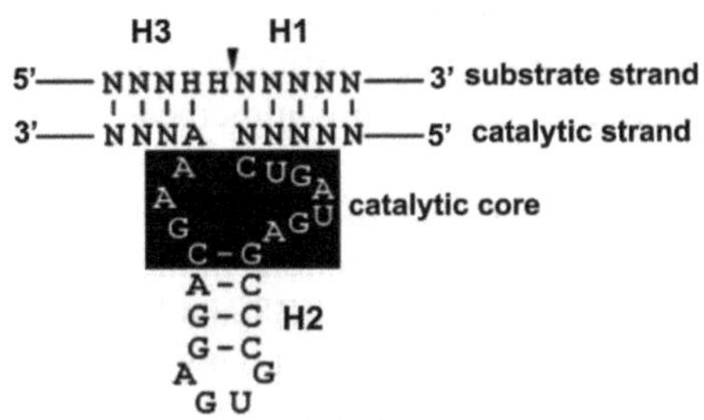

Fig. 5 Secondary structure of the hammehead ribozyme (HHRz). It is composed of three helices (H1–3). H1 and H3 are formed during substrate–enzyme binding; H2 is part of the catalytic core. Conservative nucleotides, required for hydrolysis, are highlighted on black background

The HDV ribozyme is another candidate for the future development of gene-inactivation systems. On top of this, a new generation of HDV ribozymes now exists that benefits from the addition of a specific on/off adaptor (specifically the SOFA-HDV ribozymes), which greatly increases both the ribozyme's specificity and its cleavage activity (Levesque et al. 2011).

10.4 The Hammerhead Ribozyme

One of the smallest known functional catalytic RNAs is the hammerhead ribozyme (HHRz), responsible for in *cis* viroid RNA hydrolysis during rolling circle replication (Buzayan et al. 1986). Such a ribozyme can be created from two different RNA molecules: one possessing the catalytic activity in *trans* and the second becoming the substrate of the reaction (Symons 1992). The specificity of the substrate–ribozyme recognition is embedded in the nucleotide complementarity (Fig. 4). Because of its small size and high activity, the hammerhead ribozyme has been a model of structure–function relations for 25 years (Perreault et al. 2011).

The hammerhead motif typically has a secondary structure composed of three helices (I, II, III) called arms, which surround an 11-nucleotide highly conservative catalytic domain (catalytic core) (Fig. 5). In a *trans*-acting configuration, helices I and III are formed upon hybridization of the ribozyme to the substrate complementary sequences.

Interaction of multihelical junctions, nonhelical elements (terminal loops and internal bulges), and pseudoknots position the substrate inside the catalytic pocket (Ferré-D'Amaré and Scott 2010). The tertiary structure of the minimal hammerhead ribozyme has been solved (Pley et al. 1994; Scott et al. 1995). The ribozyme secondary structure resembles the "Y" letter; helices II and III are coaxial, while helix I is at an acute angle to helix II. The catalytic pocket consists of the single-stranded sequence 5′-CUGA-3′, with a characteristic motif of U-turn between nucleotides U4–G5 (Pley et al. 1994). In 2006, the tertiary structure of the full-length hammerhead ribozyme was obtained (Martick and Scott 2006). This variant is twice as big as the minimal hammerhead and resembles the letter "γ," where helix II is coaxial to helix III. There are many differences in the secondary and tertiary structures of the minimal and full-length hammerhead ribozymes (Fig. 6). These differences concern also their activity in vitro and in vivo.

A comparison of the minimal and full-length hammerhead ribozyme crystal structures indicated that loop–bulge interactions in the full-length ribozyme induce changes in the active site that place residues G8 and U7 inside the catalytic pocket. Additionally, G8 forms a Watson–Crick bond with C3. This results in positioning C17 (hydrolysis site) in direct neighborhood of the catalytic center. Moreover, the structure suggests a potential participation of G8 and G12 in the general acid–base catalysis mechanism (Martick and Scott 2006).

Fundamental for efficient catalysis by the hammerhead ribozyme is a sequence 5′-NUX-3′ in the substrate strand, where N stands for any nucleotide and X cannot be G. Model experiments showed that the most effective hydrolysis occurs with a GUC or AUC target sequence (Kore et al. 1998). These relatively loose target sequence demandings and small sizes of the catalytic domain focused the attention on ribozymes acting as potential agents that can block gene expression via mRNA hydrolysis. Therefore, it is possible to design ribozymes active toward practically any RNA molecule. Active ribozymes demonstrate all features of an enzymatic reaction, that is, high specificity toward the substrate and possibility to catalyze many reaction cycles.

Hammerhead ribozymes catalyze sequence-specific degradation of target RNAs via internal transesterification (Ferré-D'Amaré and Scott 2010). In a first step, the ribozyme recognizes the target sequence, based on sequence complementarity, and the ribozyme–substrate complex adopts the catalytically active conformation. The 2′-OH-ribose group carries over a nucleophilic attack on the 3′-phosphorus according to an SN2 reaction model. The reaction products have a 2′,3′-cyclic phosphorus and a 5′-hydroxyl. The attacking group, hydrolyzed phosphorus, and leaving group are positioned in-line. The reaction products then dissociate, and the ribozyme is free to conduct another reaction, as protein enzymes do (Fedor and Uhlenbeck 1990). Hydrolysis is based on the general mechanism of acid–base catalysis in which electron pairs of heterocyclic bases act as general acid and general base (Westhof 1999).

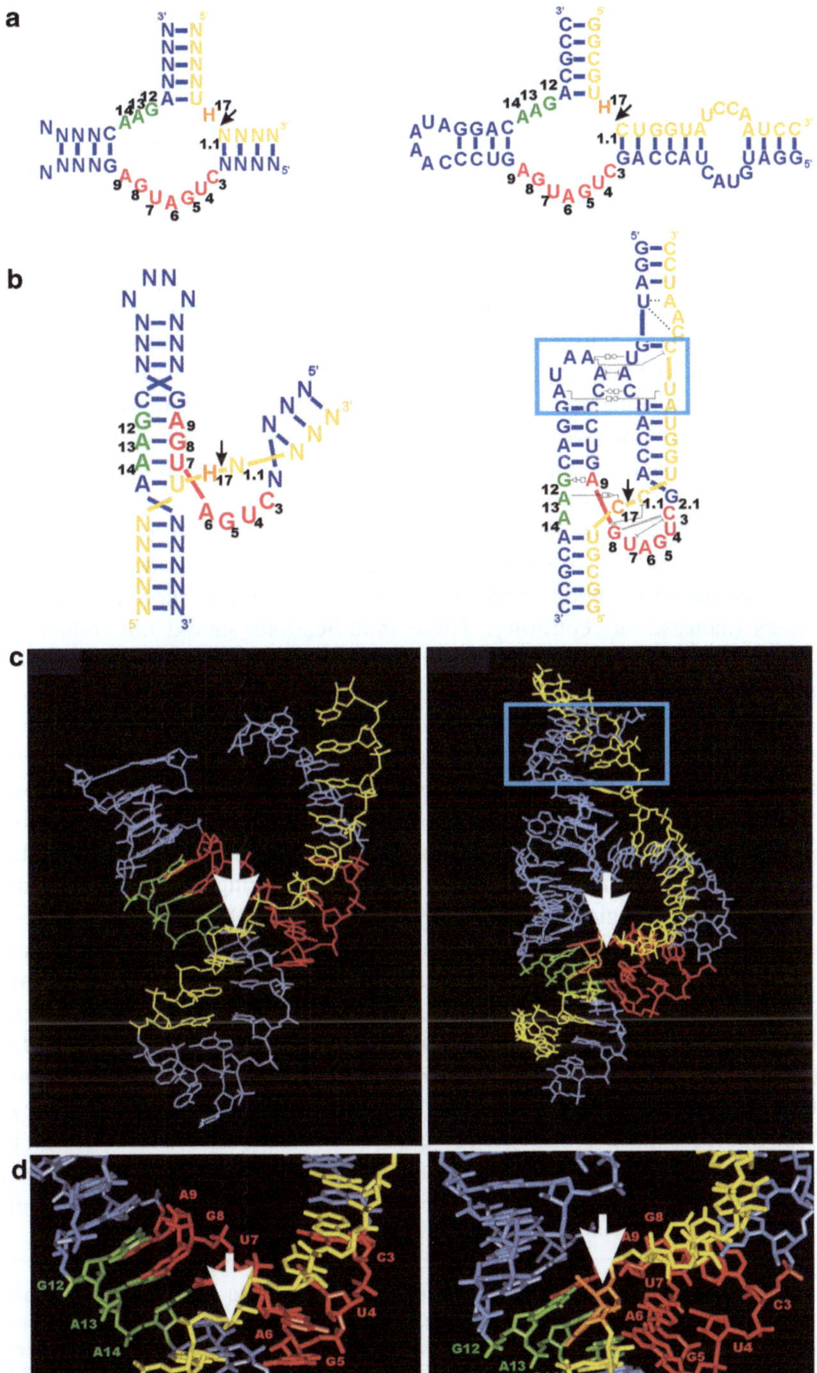

Fig. 6 Comparison of the minimal (*left*) and extended (*right*) hammerhead ribozyme structures. (**a**) Secondary structure, (**b**) secondary structure with tertiary interactions, (**c**) tertiary structure, and (**d**) stereo view of the catalytic core. The sequence and secondary structures are color-coded to

Usually, in vitro selection of active ribozymes preceded in vivo assays of gene silencing. However, in some cases, ribozyme variants that have the highest cleavage activity in vivo have decidedly lower cleavage activity in vitro (Crisell et al. 1993; Xu et al. 2001). Despite high specificity and many examples of ribozyme usefulness in vivo, there are still a few obstacles, such as low chemical RNA stability in physiological magnesium ion concentrations and cell nuclease sensitivity. The role of magnesium ions in catalysis is not ambiguous. The presence of bivalent metal ions is not conditional for hammerhead ribozyme catalytic activity. It was shown that the ribozyme can catalyze the reaction in the presence of a high concentration of monovalent metal ions (Li^+, Na^+, NH_4^+) (Curtis and Bartel 2001; O'Rear et al. 2001). Furthermore, research results indicate that magnesium and other ions most probably act only to stabilize the hammerhead ribozyme structure. The hammerhead ribozyme is active in high-pressure conditions with no magnesium ions (Fedoruk-Wyszomirska et al. 2007). Reportedly, a new mechanism of hydrolysis was proposed that rules out divalent ion participation.

The RNA durability can be much improved by introducing a recombinant vector construct that will allow the ribozyme to be constantly expressed in the cell (Weinberg et al. 2007). In the case of exogenous ribozymes, the most effective way of counteracting undesirable degradation is chemical modification, which decreases nuclease susceptibility. These modifications should not influence the ribozyme structure, cause a loss of catalytic properties, or be toxic to the cell.

Naturally occurring viroid ribozymes function in cell concentrations of divalent ions. Their activity is supported by additional regions outside the catalytic core (Khvorova et al. 2003). The hammerhead ribozyme crystal structure indicates that proper spatial orientation of helix I, one of the regions of ribozyme–substrate recognition, and helix II, element of the catalytic core, is necessary for hydrolysis (Fig. 6). It was shown that introduction of additional domains that create three-dimensional interactions with the catalytic core, stabilizing its structure, promotes hydrolysis in physiological conditions (Khvorova et al. 2003; Scott 2007). Tertiary interactions between the nonessential and peripheral loops capping helices I and II are responsible for conferring the intracellular activity of natural hammerhead ribozymes and are also required for artificial HHRz functionality in an intracellular environment (Khvorova et al. 2003). Different tertiary interactions modulate the cleavage rate by differentially stabilizing the core (Shepotinovskaya and Uhlenbeck 2010). Conformational dynamics analysis of full-length hammerhead ribozymes with the fluorescence resonance energy transfer method (FRET) indicated that helix I and II interaction acts similarly to magnesium ions for minimal ribozymes and allows the molecules to form a catalytically active conformation more frequently,

---

Fig. 6 (continued) match the structural features. *Blue,* ribozyme sequence; *red,* catalytic pocket; *green,* catalytic center; *yellow,* target sequence. Tertiary hydrogen-bonding contacts and tertiary stacking interactions are denoted as *thin black lines*. An *arrow* indicates the hydrolysis site. The *light-blue square* depicts loop–bulge contact. Tertiary structure source: RCSB PDB (1MME, 2OEU). Residue numbering and tertiary interaction representation as proposed by Martick and Scott (2006)

but it does not keep the ribozyme in a defined active state (McDowell et al. 2010). Thus, tertiary interactions taking place at a long distance from the ribozyme catalytic center indirectly influence its structure via changes in the geometry of helix interactions (Shepotinovskaya and Uhlenbeck 2010). Removal of these interactions leads to the formation of a structure characteristic for minimal variants in which the orientation of the helices remains unchanged, but the catalytic core rearranges to an open, inactive conformation. Tertiary interactions stabilize active, but thermodynamically less stable, conformations of the ribozyme catalytic core (Shepotinovskaya and Uhlenbeck 2010).

Since the original identification of hammerhead ribozymes, new reports on the occurrence of this motif in different organisms have emerged (de la Pena and Garcia-Robles 2010). A highly active hammerhead ribozyme was discovered in 2008 in the 3′-UTR of the human CLEC2 gene (Martick et al. 2008). It was shown that this molecule is active in *cis* both in vitro and in vivo, and a hypothesis on its role as a eukaryotic riboswitch was proposed (Scott et al. 2009). In the human microbiome, 20 candidates for new hammerhead ribozymes were identified (Jimenez et al. 2011). Bioinformatic research detected 284 unique motifs, from which only 124 examples were known ribozymes from plant pathogens and *A. thaliana* (Seehafer et al. 2011). The resembling 160 molecules were new examples identified in 50 different genomes, repeated regions, introns, or protein-coding regions (Table 2). They probably evolved independently (Seehafer et al. 2011). Some of the characterized ribozymes retain their activity in vitro. As to 2011, over 10,000 examples of hammerhead ribozymes are known (Perreault et al. 2011).

Ribozyme strategies inactivating mRNAs have been developed by different research groups for therapeutic applications, and several hammerhead ribozymes have been used in clinical trials. Successful decrease of HER-2, survivin, reverse transcriptase, or CCR5 expression has been reported, as well as application of ribozymes against prion diseases (Mulhbacher et al. 2010).

10.5 Ribozyme-Mediated Knockdown of Mitochondrial RNAs

In pioneering studies, a chimeric sequence made of a *trans*-cleaving hammerhead ribozyme attached to the 5′-end of the above-mentioned mitochondrial RNA shuttle (Sect. 9) through a specific linker was designed. The target of the ribozyme was the mitochondrial mRNA encoding the ATP9 subunit of the ATP synthase. The corresponding RNA was expressed under the control of an inducible promoter in stably transformed *Nicotiana tabacum* cells and efficiently imported into mitochondria (Val et al. 2011). The *trans*-cleaving ribozyme was active in the organelles and triggered a drastic knockdown of the *atp9* mRNA. Translation of the ATP9 protein was affected accordingly. The strategy was subsequently extended to *A. thaliana* whole plants (Val et al. 2011). Again, the level of the *atp9* mRNA decreased to a very low level upon expression of the *trans*-cleaving hammerhead ribozyme combined with the mitochondrial RNA shuttle. The approach thus allows

Table 2 Examples of *Eukaryotes* carrying hammerhead conservative sequences in their genomes, organized in order of decreasing number of identified unique motifs (Seehafer et al. 2011)

Number of unique motifs	Species	Common name
35	*Hydra magnipapillata*	Hydra
24	*Schistosoma mansoni*	Blood fluke
8	*Xenopus tropicalis*	Xenopus
5	*Drosophila pseudoobscura*	Fruit fly
4	*Tarsius syrichta*	Tarsier
4	*Canis familiaris*	Dog
4	*Echinops telfairi*	Lesser hedgehog tenrec
3	*Macaca mulatta*	Macaque
3	*Microcebus murinus*	Mouse lemur
3	*Mus musculus*	Mouse
3	*Equus caballus*	Horse
3	*Manodelphis domestica*	Opossum
3	*Drosophila persimilis*	Fruit fly
3	*Arabidopsis lyrata*	Lyre-leaved rock cress
3	*Arabidopsis thaliana*	Mouse-ear cress
2	*Homo sapiens*	Human
2	*Cavia porcellus*	Guinea pig
2	*Ochotona princes*	Pika
2	*Oryctolagus cuniculus*	Rabbit
2	*Rattus norvegicus*	Rat
2	*Bos Taurus*	Cow
2	*Vicugna pacos*	Alpaca
2	*Choloepus hoffmanni*	Sloth
2	*Dasypus novemcinctus*	Armadillo
2	*Taeniopygia guttata*	Zebra finch
2	*Danio rerio*	Zebra fish
2	*Culex quinquefasciatus*	Southern house mosquito
2	*Nasonia vitripennis*	Jewel wasp
2	*Tribolium castaneum*	Red floor beetle
2	*Vitis vinifera*	Wine
2	*Schistosoma japonicum*	Blood fluke
1	*Otolemur garnettii*	Bush baby
1	*Pan troglodytes*	Chimpanzee
1	*Spermophilus tridecemlineatus*	Squirrel
1	*Myotis lucifugus*	Microbat
1	*Sorex araneus*	Shrew
1	*Sus scrofa*	Pig
1	*Tursiops truncates*	Dolphin
1	*Loxodonta Africana*	Elephant
1	*Macropus eugenii*	Wallaby
1	*Gallus gallus*	Chicken
1	*Tetraodon nigroviridis*	Tetraodon
1	*Aedes aegypti*	Yellow fewer mosquito
1	*Physcomitrella patens*	Moss
1	*Aspergillus flavus*	Yellow mold

(continued)

Table 2 (continued)

Number of unique motifs	Species	Common name
1	*Aspergillus oryzae*	Koji mold
1	*Pichia stipitis*	Pichia
1	*Caenorhabditis briggsae*	Nematode worm
1	*Caenorhabditis remanei*	Nematode worm
1	*Ixodes scapularis*	Deer tick

directed functional manipulation of the organelles in living cells. Remarkably, the kinetics of target RNA knockdown was modulated according to the developmental stage and the need for mitochondrial function (Val et al. 2011). In fully developed plants cultivated under long-day conditions, expression of the ribozyme maintained a continuous low level of *atp9* mRNA. Conversely, in very young plantlets cultivated under short-day conditions, the initial knockdown was transiently released, suggesting that it had been sensed and tentatively compensated by an increased synthesis of the target RNA. The strategy developed thus appears to make a breakthrough in organelle genetics and to be especially suited for the investigation of mitochondrial RNA regulation processes and the identification of new functions in the large unassigned regions of the plant mitochondrial genomes.

11 Conclusion

Understanding the complexity of mitochondrial biogenesis and function in health and disease raised a number of genetic challenges that remained at the cutting edge of research, due to the failure of conventional methodologies in transforming the mtDNA. Progress in the assessment and deciphering of RNA trafficking processes has paved the way for unprecedented alternatives to investigate, manipulate, and rescue the mitochondrial genetic system. The first strategies of customized RNA targeting successfully developed in stable in vivo models highlight the potential of such approaches, and exploiting RNA technologies for mitochondrial genetics is likely to be only at its first steps. More shuttle systems for organelle targeting of nuclearly expressed RNAs are likely to become available in the next future. The different structural elements in human 5S rRNA were systematically analyzed in vitro and in vivo for a possible role in mitochondrial import. Domain beta proved to be dispensable, and a 5S rRNA derivative in which part of this domain was substituted for a 13-nucleotide piece of foreign sequence was efficiently transported into the organelles in human cells, opening the possibility of further customized substitutions (Smirnov et al. 2008). On the other hand, a randomized RNA library was generated from yeast $tRNA^{Lys}$(CUU) and submitted to a SELEX (systematic evolution of ligands by exponential enrichment) selection based on in vitro mitochondrial uptake and import factor binding (Kolesnikova et al. 2010). A set of new importable aptamers was characterized, with an improved efficiency relying on an alternative secondary structure versus the classical tRNA model. The information gathered lays the ground for the design of small RNA molecules able to target oligoribonucleotides into the organelles, especially to interfere with the replication of mutated mtDNA copies (Kolesnikova et al. 2010). In further in vitro experiments, stem-loop structures that function as targeting signals for PNPASE-mediated organelle uptake were identified in the mammalian RNase P RNA and MRP RNA (Wang et al. 2010). These may in turn be used to license PNPASE-dependent import of RNAs of interest. Other assays aimed at developing cargo RNA targeting driven by a customized shuttle protein. In this respect, it was established that the mammalian dihydrofolate reductase (DHFR) fused to an organelle-targeting signal can act as a carrier to translocate various cargo RNAs into isolated plant, yeast, or mammalian mitochondria (Sieber et al. 2011). Upon DHFR-mediated uptake of a tRNA precursor or a mitochondrial pre-mRNA, *in organello* processing and editing was observed, demonstrating the functional relevance of the approach and further suggesting a connection between RNA and protein import pathways. If validated in vivo following double nuclear transformation with transgenes encoding a cargo RNA and a specific shuttle protein, this strategy might become an alternative to the above-described tRNA-like shuttle-based system for the translocation of mRNAs into the organelles in plant or mammalian cells (Sieber et al. 2011). Finally, although this work remains controversial in the field, especially following an editorial expression of concern (Schekman 2010), it must be noted that further studies reported mammalian cell uptake and mitochondrial transport of various RNAs mediated by the above-mentioned RNA-specific, multisubunit import complex isolated from the protozoal parasite *L. tropica* (Adhya 2008; Adhya et al. 2011; Chatterjee et al. 2006; Mukherjee et al. 2007).

Acknowledgments Our projects are funded by the French Centre National de la Recherche Scientifique (CNRS, UPR2357), the Université de Strasbourg (UdS), the Agence Nationale de la Recherche (ANR-06-MRAR-037-02, ANR-09-BLAN-0240-01), and the Région Alsace. Support through grants from the Polish Ministry of Science and Higher Education and the National Science Center is acknowledged.

References

Adhya S (2008) *Leishmania* mitochondrial tRNA importers. Int J Biochem Cell Biol 40:2681–2685

Adhya S, Mahato B, Jash S et al (2011) Mitochondrial gene therapy: the tortuous path from bench to bedside. Mitochondrion 11:839–844

Ahmed AU, Fisher PR (2009) Import of nuclear-encoded mitochondrial proteins: a cotranslational perspective. Int Rev Cell Mol Biol 273:49–68

Allen JO, Fauron CM, Minx P et al (2007) Comparisons among two fertile and three male-sterile mitochondrial genomes of maize. Genetics 177:1173–1192

Alverson AJ, Rice DW, Dickinson S et al (2011) Origins and recombination of the bacterial-sized multichromosomal mitochondrial genome of cucumber. Plant Cell 23:2499–2513

Amiott EA, Jaehning JA (2006) Mitochondrial transcription is regulated via an ATP "sensing" mechanism that couples RNA abundance to respiration. Mol Cell 22:329–338
Ammerman ML, Fisk JC, Read LK (2008) gRNA/pre-mRNA annealing and RNA chaperone activities of RBP16. RNA 14:1069–1080
Aphasizhev R, Aphasizheva I (2011) Uridine insertion/deletion editing in trypanosomes: a playground for RNA-guided information transfer. Wiley Interdiscip Rev RNA 2:669–685
Aphasizheva I, Aphasizhev R (2010) RET1-catalyzed uridylylation shapes the mitochondrial transcriptome in *Trypanosoma brucei*. Mol Cell Biol 30:1555–1567
Asin-Cayuela J, Gustafsson CM (2007) Mitochondrial transcription and its regulation in mammalian cells. Trends Biochem Sci 32:111–117
Bandiera S, Hanein S, Lyonnet S (2011a) Mitochondria as novel players of the cellular RNA interference. J Biol Chem 286:le19
Bandiera S, Ruberg S, Girard M et al (2011b) Nuclear outsourcing of RNA interference components to human mitochondria. PLoS One 6:e20746
Barbrook AC, Howe CJ, Kurniawan DP et al (2010) Organization and expression of organellar genomes. Philos Trans R Soc Lond B Biol Sci 365:785–797
Barrey E, Saint-Auret G, Bonnamy B et al (2011) Pre-microRNA and mature microRNA in human mitochondria. PLoS One 6:e20220
Been MD, Wickham GS (1997) Self-cleaving ribozymes of hepatitis delta virus RNA. Eur J Biochem 247:741–753
Beitzinger M, Peters L, Zhu JY et al (2007) Identification of human microRNA targets from isolated argonaute protein complexes. RNA Biol 4:76–84
Bender A, Krishnan KJ, Morris CM et al (2006) High levels of mitochondrial DNA deletions in *substantia nigra* neurons in aging and Parkinson disease. Nat Genet 38:515–517
Bian Z, Li LM, Tang R et al (2010) Identification of mouse liver mitochondria-associated miRNAs and their potential biological functions. Cell Res 20:1076–1078
Binder S, Brennicke A (2003) Gene expression in plant mitochondria: transcriptional and post-transcriptional control. Philos Trans R Soc Lond B Biol Sci 358:181–189
Blum B, Simpson L (1990) Guide RNAs in kinetoplastid mitochondria have a nonencoded 3′ oligo (U) tail involved in recognition of the preedited region. Cell 62:391–397
Blum B, Bakalara N, Simpson L (1990) A model for RNA editing in kinetoplastid mitochondria: "guide" RNA molecules transcribed from maxicircle DNA provide the edited information. Cell 60:189–198
Bokori-Brown M, Holt IJ (2006) Expression of algal nuclear ATP synthase subunit 6 in human cells results in protein targeting to mitochondria but no assembly into ATP synthase. Rejuvenation Res 9:455–469
Bonen L (2008) *Cis-* and *trans-*splicing of group II introns in plant mitochondria. Mitochondrion 8:26–34
Bonnefoy N, Remacle C, Fox TD (2007) Genetic transformation of *Saccharomyces cerevisiae* and *Chlamydomonas reinhardtii* mitochondria. Methods Cell Biol 80:525–548
Bonnet C, Kaltimbacher V, Ellouze S et al (2007) Allotopic mRNA localization to the mitochondrial surface rescues respiratory chain defects in fibroblasts harboring mitochondrial DNA mutations affecting complex I or V subunits. Rejuvenation Res 10:127–144
Bonnet C, Augustin S, Ellouze S et al (2008) The optimized allotopic expression of ND1 or ND4 genes restores respiratory chain complex I activity in fibroblasts harboring mutations in these genes. Biochim Biophys Acta 1783:1707–1717
Börner GV, Mörl M, Janke A et al (1996) RNA editing changes the identity of a mitochondrial tRNA in marsupials. EMBO J 15:5949–5957
Bouzaidi-Tiali N, Aeby E, Charriere F et al (2007) Elongation factor 1a mediates the specificity of mitochondrial tRNA import in *T. brucei*. EMBO J 26:4302–4312
Brandina I, Smirnov A, Kolesnikova O et al (2007) tRNA import into yeast mitochondria is regulated by the ubiquitin-proteasome system. FEBS Lett 581:4248–4254

Brandon M, Baldi P, Wallace DC (2006) Mitochondrial mutations in cancer. Oncogene 25: 4647–4662
Brown GG, Zhang M (1995) Mitochondrial plasmids: DNA and RNA. In: Levings III CS, Vasil IK (ed) The molecular biology of plant mitochondria. Kluwer Academic Publishers, Dordrecht, the Netherlands, pp. 61-91
Bruske EI, Sendfeld F, Schneider A (2009) Thiolated tRNAs of *Trypanosoma brucei* are imported into mitochondria and dethiolated after import. J Biol Chem 284:36491–36499
Burzio VA, Villota C, Villegas J et al (2009) Expression of a family of noncoding mitochondrial RNAs distinguishes normal from cancer cells. Proc Natl Acad Sci USA 106:9430–9434
Butcher SE, Heckman JE, Burke JM (1995) Reconstitution of hairpin ribozyme activity following separation of functional domains. J Biol Chem 270:29648–29651
Buzayan JM, Gerlach WL, Bruening G (1986) Satellite tobacco ringspot virus RNA: A subset of the RNA sequence is sufficient for autolytic processing. Proc Natl Acad Sci USA 83:8859–8862
Campbell GR, Ziabreva I, Reeve AK et al (2011) Mitochondrial DNA deletions and neurodegeneration in multiple sclerosis. Ann Neurol 69:481–492
Chang DD, Clayton DA (1989) Mouse RNAase MRP RNA is encoded by a nuclear gene and contains a decamer sequence complementary to a conserved region of mitochondrial RNA substrate. Cell 56:131–139
Chase CD (2007) Cytoplasmic male sterility: a window to the world of plant mitochondrial-nuclear interactions. Trends Genet 23:81–90
Chateigner-Boutin AL, Small I (2010) Plant RNA editing. RNA Biol 7:213–219
Chateigner-Boutin AL, Small I (2011) Organellar RNA editing. WIREs RNA 2:493–506
Chatterjee S, Home P, Mukherjee S et al (2006) An RNA-binding respiratory component mediates import of type II tRNAs into *Leishmania* mitochondria. J Biol Chem 281:25270–25277
Choisne N, Carneiro VT, Pelletier G et al (1998) Implication of 5′-flanking sequence elements in expression of a plant tRNA(Leu) gene. Plant Mol Biol 36:113–123
Clement SL, Mingler MK, Koslowsky DJ (2004) An intragenic guide RNA location suggests a complex mechanism for mitochondrial gene expression in *Trypanosoma brucei*. Eukaryot Cell 3:862–869
Clifton SW, Minx P, Fauron CM et al (2004) Sequence and comparative analysis of the maize NB mitochondrial genome. Plant Physiol 136:3486–3503
Corell RA, Feagin JE, Riley GR et al (1993) *Trypanosoma brucei* minicircles encode multiple guide RNAs which can direct editing of extensively overlapping sequences. Nucleic Acids Res 21:4313–4320
Corral-Debrinski M (2007) mRNA specific subcellular localization represents a crucial step for fine-tuning of gene expression in mammalian cells. Biochim Biophys Acta 1773:473–475
Costa FF (2010) Non-coding RNAs: meet thy masters. Bioessays 32:599–608
Crausaz Esseiva A, Marechal-Drouard L, Cosset A et al (2004) The T-stem determines the cytosolic or mitochondrial localization of trypanosomal $tRNAs^{Met}$. Mol Biol Cell 15:2750–2757
Crisell P, Thompson S, James W (1993) Inhibition of HIV-1 replication by ribozymes that show poor activity *in vitro*. Nucleic Acids Res 21:5251–5255
Cruz-Reyes J, Zhelonkina A, Rusche L et al (2001) Trypanosome RNA editing: simple guide RNA features enhance U deletion 100-fold. Mol Cell Biol 21:884–892
Curtis EA, Bartel DP (2001) The hammerhead cleavage reaction in monovalent cations. RNA 7:546–552
de la Pena M, Garcia-Robles I (2010) Ubiquitous presence of the hammerhead ribozyme motif along the tree of life. RNA 16:1943–1950
Delage L, Duchêne AM, Zaepfel M et al (2003) The anticodon and the D-domain sequences are essential determinants for plant cytosolic tRNA(Val) import into mitochondria. Plant J 34: 623–633

Devaux F, Lelandais G, Garcia M et al (2010) Posttranscriptional control of mitochondrial biogenesis: spatio-temporal regulation of the protein import process. FEBS Lett 584:4273–4279

Dietrich A, Maréchal-Drouard L, Carneiro V et al (1996) A single base change prevents import of cytosolic tRNA(Ala) into mitochondria in transgenic plants. Plant J 10:913–918

Doersen CJ, Guerrier-Takada C, Altman S et al (1985) Characterization of an RNase P activity from HeLa cell mitochondria. Comparison with the cytosol RNase P activity. J Biol Chem 260:5942–5949

Dörner M, Altmann M, Pääbo S et al (2001) Evidence for import of a lysyl-tRNA into marsupial mitochondria. Mol Biol Cell 12:2688–2698

Dreher TW (2009) Role of tRNA-like structures in controlling plant virus replication. Virus Res 139:217–229

Duroc Y, Gaillard C, Hiard S et al (2005) Biochemical and functional characterization of ORF138, a mitochondrial protein responsible for Ogura cytoplasmic male sterility in *Brassicaceae*. Biochimie 87:1089–1100

Eliyahu E, Pnueli L, Melamed D et al (2010) Tom20 mediates localization of mRNAs to mitochondria in a translation-dependent manner. Mol Cell Biol 30:284–294

Eliyahu E, Lesnik C, Arava Y (2012) The protein chaperone Ssa1 affects mRNA localization to the mitochondria. FEBS Lett 586:64–69

Ellouze S, Augustin S, Bouaita A et al (2008) Optimized allotopic expression of the human mitochondrial ND4 prevents blindness in a rat model of mitochondrial dysfunction. Am J Hum Genet 83:373–387

Entelis NS, Kolesnikova OA, Dogan S et al (2001) 5 S rRNA and tRNA import into human mitochondria. Comparison of in vitro requirements. J Biol Chem 276:45642–45653

Entelis N, Brandina I, Kamenski P et al (2006) A glycolytic enzyme, enolase, is recruited as a cofactor of tRNA targeting toward mitochondria in *Saccharomyces cerevisiae*. Genes Dev 20:1609–1620

Esteban JA, Walter NG, Kotzorek G et al (1998) Structural basis for heterogeneous kinetics: reengineering the hairpin ribozyme. Proc Natl Acad Sci USA 95:6091–6096

Farajollahi S, Maas S (2010) Molecular diversity through RNA editing: a balancing act. Trends Genet 26:221–230

Fechter P, Rudinger-Thirion J, Florentz C et al (2001) Novel features in the tRNA-like world of plant viral RNAs. Cell Mol Life Sci 58:1547–1561

Fedor MJ (2009) Comparative enzymology and structural biology of RNA self-cleavage. Annu Rev Biophys Biomol Struct 38:271–299

Fedor MJ, Uhlenbeck OC (1990) Substrate sequence effects on "hammerhead" RNA catalytic efficiency. Proc Natl Acad Sci USA 87:1668–1672

Fedor MJ, Williamson JR (2005) The catalytic diversity of RNAs. Nat Rev Mol Cell Biol 6:399–412

Fedoruk-Wyszomirska A, Wyszko E, Giel-Pietraszuk M et al (2007) High hydrostatic pressure approach proves RNA catalytic activity without magnesium. Int J Biol Macromol 41:30–35

Feldstein PA, Bruening G (1993) Catalytically active geometry in the reversible circularization of 'mini-monomer' RNAs derived from the complementary strand of tobacco ringspot virus satellite RNA. Nucleic Acids Res 21:1991–1998

Fernandez-Silva P, Enriquez JA, Montoya J (2003) Replication and transcription of mammalian mitochondrial DNA. Exp Physiol 88:41–56

Ferré-D'Amaré AR, Scott WG (2010) Small self-cleaving ribozymes. Cold Spring Harb Perspect Biol 2:a003574

Fey J, Maréchal-Drouard L (1999) Compilation and analysis of plant mitochondrial promoter sequences: an illustration of a divergent evolution between monocot and dicot mitochondria. Biochem Biophys Res Commun 256:409–414

Frei U, Peiretti EG, Wenzel G (2004) Significance of cytoplasmic DNA in plant breeding. In: Janick J (ed) Plant Breeding Reviews. Wiley, Hoboken, pp 175–210

Fujii S, Toda T, Kikuchi S et al (2011) Transcriptome map of plant mitochondria reveals islands of unexpected transcribed regions. BMC Genomics 12:279
Gabay-Laughnan S, Kuzmin EV, Monroe J et al (2009) Characterization of a novel thermosensitive restorer of fertility for cytoplasmic male sterility in maize. Genetics 182:91–103
Gadir N, Haim-Vilmovsky L, Kraut-Cohen J et al (2011) Localization of mRNAs coding for mitochondrial proteins in the yeast *Saccharomyces cerevisiae*. RNA 17:1551–1565
Gallagher LJ, Betz SK, Chase CD (2002) Mitochondrial RNA editing truncates a chimeric open reading frame associated with S male-sterility in maize. Curr Genet 42:179–184
Ganguly A, Bevilacqua PC, Hammes-Schiffer S (2011) Quantum mechanical/molecular mechanical study of the HDV ribozyme: impact of the catalytic metal ion on the mechanism. J Phys Chem Lett 2:2906–2911
Garcia M, Delaveau T, Goussard S et al (2010) Mitochondrial presequence and open reading frame mediate asymmetric localization of messenger RNA. EMBO Rep 11:285–291
Gaspari M, Larsson NG, Gustafsson CM (2004) The transcription machinery in mammalian mitochondria. Biochim Biophys Acta 1659:148–152
Geiduschek EP, Tocchini-Valentini GP (1988) Transcription by RNA polymerase III. Annu Rev Biochem 57:873–914
Giegé P, Sweetlove LJ, Cognat V et al (2005) Coordination of nuclear and mitochondrial genome expression during mitochondrial biogenesis in *Arabidopsis*. Plant Cell 17:1497–1512
Giraud E, Van Aken O, Ho LH et al (2009) The transcription factor ABI4 is a regulator of mitochondrial retrograde expression of ALTERNATIVE OXIDASE1a. Plant Physiol 150:1286–1296
Golden DE, Hajduk SL (2005) The 3′-untranslated region of cytochrome oxidase II mRNA functions in RNA editing of African trypanosomes exclusively as a *cis* guide RNA. RNA 11:29–37
Golden DE, Hajduk SL (2006) The importance of RNA structure in RNA editing and a potential proofreading mechanism for correct guide RNA:pre-mRNA binary complex formation. J Mol Biol 359:585–596
Göringer HU, Katari VS, Böhm C (2011) The structural landscape of native editosomes in African trypanosomes. WIREs RNA 2:395–407
Grams J, McManus MT, Hajduk SL (2000) Processing of polycistronic guide RNAs is associated with RNA editing complexes in *Trypanosoma brucei*. EMBO J 19:5525–5532
Greaves LC, Reeve AK, Taylor RW et al (2012) Mitochondrial DNA and disease. J Pathol 226:274–286
Griffiths AJ (1995) Natural plasmids of filamentous fungi. Microbiol Rev 59:673–685
Guja KE, Garcia-Diaz M (2011) Hitting the brakes: termination of mitochondrial transcription. Biochim Biophys Acta. doi:10.1016/j.bbagrm.2011.1011.1004
Guy J, Qi X, Pallotti F et al (2002) Rescue of a mitochondrial deficiency causing Leber hereditary optic neuropathy. Ann Neurol 52:534–542
Hajduk S, Ochsenreiter T (2010) RNA editing in kinetoplastids. RNA Biol 7:229–236
Hammond JA, Rambo RP, Filbin ME et al (2009) Comparison and functional implications of the 3D architectures of viral tRNA-like structures. RNA 15:294–307
Hampel A, Tritz R (1989) RNA catalytic properties of the minimum (-)sTRSV sequence. Biochemistry 28:4929–4933
Handa H (2003) The complete nucleotide sequence and RNA editing content of the mitochondrial genome of rapeseed (*Brassica napus* L.): comparative analysis of the mitochondrial genomes of rapeseed and *Arabidopsis thaliana*. Nucleic Acids Res 31:5907–5916
Handa H (2008) Linear plasmids in plant mitochondria: peaceful coexistences or malicious invasions? Mitochondrion 8:15–25
Hashimi H, Cicova Z, Novotna L et al (2009) Kinetoplastid guide RNA biogenesis is dependent on subunits of the mitochondrial RNA binding complex 1 and mitochondrial RNA polymerase. RNA 15:588–599

Hernandez A, Panigrahi A, Cifuentes-Rojas C et al (2008) Determinants for association and guide RNA-directed endonuclease cleavage by purified RNA editing complexes from *Trypanosoma brucei*. J Mol Biol 381:35–48

Hernandez A, Madina BR, Ro K et al (2010) REH2 RNA helicase in kinetoplastid mitochondria: ribonucleoprotein complexes and essential motifs for unwinding and guide RNA (gRNA) binding. J Biol Chem 285:1220–1228

Heyne K, Mannebach S, Wuertz E et al (2004) Identification of a putative p53 binding sequence within the human mitochondrial genome. FEBS Lett 578:198–202

Hock MB, Kralli A (2009) Transcriptional control of mitochondrial biogenesis and function. Annu Rev Physiol 71:177–203

Holec S, Lange H, Kuhn K et al (2006) Relaxed transcription in *Arabidopsis* mitochondria is counterbalanced by RNA stability control mediated by polyadenylation and polynucleotide phosphorylase. Mol Cell Biol 26:2869–2876

Holec S, Lange H, Canaday J et al (2008) Coping with cryptic and defective transcripts in plant mitochondria. Biochim Biophys Acta 1779:566–573

Holzmann J, Frank P, Loffler E et al (2008) RNase P without RNA: identification and functional reconstitution of the human mitochondrial tRNA processing enzyme. Cell 135:462–474

Hong M, Simpson L (2003) Genomic organization of *Trypanosoma brucei* kinetoplast DNA minicircles. Protist 154:265–279

Horton TL, Landweber LF (2002) Rewriting the information in DNA: RNA editing in kinetoplastids and myxomycetes. Curr Opin Microbiol 5:620–626

Hotchkiss G, Maijgren-Steffensson C, Ahrlund-Richter L (2004) Efficacy and mode of action of hammerhead and hairpin ribozymes against various HIV-1 target sites. Mol Ther 10:172–180

Huang L, Mollet S, Souquere S et al (2011) Mitochondria associate with P-bodies and modulate microRNA-mediated RNA interference. J Biol Chem 286:24219–24230

Imanishi H, Hattori K, Wada R et al (2011) Mitochondrial DNA mutations regulate metastasis of human breast cancer cells. PLoS One 6:e23401

Janke A, Pääbo S (1993) Editing of a tRNA anticodon in marsupial mitochondria changes its codon recognition. Nucleic Acids Res 21:1523–1525

Jimenez RM, Delwart E, Luptak A (2011) Structure-based search reveals hammerhead ribozymes in the human microbiome. J Biol Chem 286:7737–7743

Jones AW, Yao Z, Vicencio JM et al (2011) PGC-1 family coactivators and cell fate: roles in cancer, neurodegeneration, cardiovascular disease and retrograde mitochondria-nucleus signalling. Mitochondrion. doi:10.1016/j.mito.2011.1009.1009

Kala S, Salavati R (2010) OB-fold domain of KREPA4 mediates high-affinity interaction with guide RNA and possesses annealing activity. RNA 16:1951–1967

Kaltimbacher V, Bonnet C, Lecoeuvre G et al (2006) mRNA localization to the mitochondrial surface allows the efficient translocation inside the organelle of a nuclear recoded ATP6 protein. RNA 12:1408–1417

Kaneko T, Suzuki T, Kapushoc ST et al (2003) Wobble modification differences and subcellular localization of tRNAs in *Leishmania tarentolae*: implication for tRNA sorting mechanism. EMBO J 22:657–667

Karicheva OZ, Kolesnikova OA, Schirtz T et al (2011) Correction of the consequences of mitochondrial 3243A > G mutation in the MT-TL1 gene causing the MELAS syndrome by tRNA import into mitochondria. Nucleic Acids Res 39:8173–8186

Khvorova A, Lescoute A, Westhof E et al (2003) Sequence elements outside the hammerhead ribozyme catalytic core enable intracellular activity. Nat Struct Biol 10:708–712

Kiss T, Filipowicz W (1992) Evidence against a mitochondrial location of the 7-2/MRP RNA in mammalian cells. Cell 70:11–16

Kloc M, Zearfoss NR, Etkin LD (2002) Mechanisms of subcellular mRNA localization. Cell 108:533–544

Kolesnikova OA, Entelis NS, Mireau H et al (2000) Suppression of mutations in mitochondrial DNA by tRNAs imported from the cytoplasm. Science 289:1931–1933

Kolesnikova OA, Entelis NS, Jacquin-Becker C et al (2004) Nuclear DNA-encoded tRNAs targeted into mitochondria can rescue a mitochondrial DNA mutation associated with the MERRF syndrome in cultured human cells. Hum Mol Genet 13:2519–2534

Kolesnikova O, Kazakova H, Comte C et al (2010) Selection of RNA aptamers imported into yeast and human mitochondria. RNA 16:926–941

Kore AR, Vaish NK, Kutzke U et al (1998) Sequence specificity of the hammerhead ribozyme revisited; the NHH rule. Nucleic Acids Res 26:4116–4120

Koslowsky DJ, Riley GR, Feagin JE et al (1992) Guide RNAs for transcripts with developmentally regulated RNA editing are present in both life cycle stages of *Trypanosoma brucei*. Mol Cell Biol 12:2043–2049

Koslowsky DJ, Reifur L, Yu LE et al (2004) Evidence for U-tail stabilization of gRNA/mRNA interactions in kinetoplastid RNA editing. RNA Biol 1:28–34

Kren BT, Wong PY, Sarver A et al (2009) MicroRNAs identified in highly purified liver-derived mitochondria may play a role in apoptosis. RNA Biol 6:65–72

Kubo T, Newton KJ (2008) Angiosperm mitochondrial genomes and mutations. Mitochondrion 8:5–14

Kuhn K, Weihe A, Borner T (2005) Multiple promoters are a common feature of mitochondrial genes in *Arabidopsis*. Nucleic Acids Res 33:337–346

Kuhn K, Richter U, Meyer EH et al (2009) Phage-type RNA polymerase RPOTmp performs gene-specific transcription in mitochondria of *Arabidopsis thaliana*. Plant Cell 21:2762–2779

Kumar R, Maréchal-Drouard L, Akama K et al (1996) Striking differences in mitochondrial tRNA import between different plant species. Mol Gen Genet 252:404–411

Landerer E, Villegas J, Burzio VA et al (2011) Nuclear localization of the mitochondrial ncRNAs in normal and cancer cells. Cell Oncol 34:297–305

Lee TS, Giambasu G, Harris ME et al (2011) Characterization of the structure and dynamics of the HDV ribozyme at different stages along the reaction path. J Phys Chem Lett 2:2538–2543

Leister D, Wang X, Haberer G et al (2011) Intracompartmental and intercompartmental transcriptional networks coordinate the expression of genes for organellar functions. Plant Physiol 157:386–404

Leung SS, Koslowsky DJ (1999) Mapping contacts between gRNA and mRNA in trypanosome RNA editing. Nucleic Acids Res 27:778–787

Leung SS, Koslowsky DJ (2001a) Interactions of mRNAs and gRNAs involved in trypanosome mitochondrial RNA editing: structure probing of an mRNA bound to its cognate gRNA. RNA 7:1803–1816

Leung SS, Koslowsky DJ (2001b) RNA editing in *Trypanosoma brucei*: characterization of gRNA U-tail interactions with partially edited mRNA substrates. Nucleic Acids Res 29:703–709

Levesque MV, Rouleau SG, Perreault JP (2011) Selection of the most potent specific on/off adaptor-hepatitis delta virus ribozymes for use in gene targeting. Nucleic Acid Ther 21:241–252

Li K, Smagula CS, Parsons WJ et al (1994) Subcellular partitioning of MRP RNA assessed by ultrastructural and biochemical analysis. J Cell Biol 124:871–882

Li F, Herrera J, Zhou S et al (2011a) Trypanosome REH1 is an RNA helicase involved with the 3′-5′ polarity of multiple gRNA-guided uridine insertion/deletion RNA editing. Proc Natl Acad Sci USA 108:3542–3547

Li P, Jiao J, Gao G et al (2011b) Control of mitochondrial activity by miRNAs. J Cell Biochem. doi:10.1002/jcb.24004

Liere K, Weihe A, Börner T (2011) The transcription machineries of plant mitochondria and chloroplasts: composition, function, and regulation. J Plant Physiol 168:1345–1360

Lightowlers RN (2011) Mitochondrial transformation: time for concerted action. EMBO Rep 12:480–481

Lipinski KA, Kaniak-Golik A, Golik P (2010) Maintenance and expression of the *S. cerevisiae* mitochondrial genome - from genetics to evolution and systems biology. Biochim Biophys Acta 1797:1086–1098

Litonin D, Sologub M, Shi Y et al (2010) Human mitochondrial transcription revisited: only TFAM and TFB2M are required for transcription of the mitochondrial genes *in vitro*. J Biol Chem 285:18129–18133
Liu Z, Butow RA (2006) Mitochondrial retrograde signaling. Annu Rev Genet 40:159–185
Ljubicic V, Joseph AM, Saleem A et al (2010) Transcriptional and post-transcriptional regulation of mitochondrial biogenesis in skeletal muscle: effects of exercise and aging. Biochim Biophys Acta 1800:223–234
Lorence A, Verpoorte R (2004) Gene transfer and expression in plants. Methods Mol Biol 267:329–350
Lu Q, Wierzbicki S, Krasilnikov AS et al (2010) Comparison of mitochondrial and nucleolar RNase MRP reveals identical RNA components with distinct enzymatic activities and protein components. RNA 16:529–537
Lukes J, Guilbride DL, Votypka J et al (2002) Kinetoplast DNA network: evolution of an improbable structure. Eukaryot Cell 1:495–502
Lukes J, Hashimi H, Zikova A (2005) Unexplained complexity of the mitochondrial genome and transcriptome in kinetoplastid flagellates. Curr Genet 48:277–299
Lung B, Zemann A, Madej MJ et al (2006) Identification of small non-coding RNAs from mitochondria and chloroplasts. Nucleic Acids Res 34:3842–3852
Madej MJ, Alfonzo JD, Huttenhofer A (2007) Small ncRNA transcriptome analysis from kinetoplast mitochondria of *Leishmania tarentolae*. Nucleic Acids Res 35:1544–1554
Madej MJ, Niemann M, Huttenhofer A et al (2008) Identification of novel guide RNAs from the mitochondria of *Trypanosoma brucei*. RNA Biol 5:84–91
Madina BR, Kuppan G, Vashisht AA et al (2011) Guide RNA biogenesis involves a novel RNase III family endoribonuclease in *Trypanosoma brucei*. RNA:10.1261/rna.2815911
Madison-Antenucci S, Grams J, Hajduk SL (2002) Editing machines: the complexities of trypanosome RNA editing. Cell 108:435–438
Magalhaes PJ, Andreu AL, Schon EA (1998) Evidence for the presence of 5 S rRNA in mammalian mitochondria. Mol Biol Cell 9:2375–2382
Mahapatra S, Ghosh S, Bera SK et al (1998) The D arm of $tRNA^{Tyr}$ is necessary and sufficient for import into *Leishmania* mitochondria *in vitro*. Nucleic Acids Res 26:2037–2041
Man PY, Turnbull DM, Chinnery PF (2002) Leber hereditary optic neuropathy. J Med Genet 39:162–169
Manfredi G, Fu J, Ojaimi J et al (2002) Rescue of a deficiency in ATP synthesis by transfer of MTATP6, a mitochondrial DNA-encoded gene, to the nucleus. Nat Genet 30:394–399
Mangus DA, Jang SH, Jaehning JA (1994) Release of the yeast mitochondrial RNA polymerase specificity factor from transcription complexes. J Biol Chem 269:26568–26574
Maniataki E, Mourelatos Z (2005) Human mitochondrial tRNAMet is exported to the cytoplasm and associates with the Argonaute 2 protein. RNA 11:849–852
Mans RM, Guerrier-Takada C, Altman S et al (1990) Interaction of RNase P from *Escherichia coli* with pseudoknotted structures in viral RNAs. Nucleic Acids Res 18:3479–3487
Marc P, Margeot A, Devaux F et al (2002) Genome-wide analysis of mRNAs targeted to yeast mitochondria. EMBO Rep 3:159–164
Maréchal-Drouard L, Guillemaut P, Cosset A et al (1990) Transfer RNAs of potato (*Solanum tuberosum*) mitochondria have different genetic origins. Nucleic Acids Res 18:3689–3696
Marker C, Zemann A, Terhorst T et al (2002) Experimental RNomics: identification of 140 candidates for small non-messenger RNAs in the plant *Arabidopsis thaliana*. Curr Biol 12:2002–2013
Martick M, Scott WG (2006) Tertiary contacts distant from the active site prime a ribozyme for catalysis. Cell 126:309–320
Martick M, Horan LH, Noller HF et al (2008) A discontinuous hammerhead ribozyme embedded in a mammalian messenger RNA. Nature 454:899–902
Maslov DA, Simpson L (1992) The polarity of editing within a multiple gRNA-mediated domain is due to formation of anchors for upstream gRNAs by downstream editing. Cell 70:459–467

Matsuda D, Dreher TW (2004) The tRNA-like structure of Turnip yellow mosaic virus RNA is a 3′-translational enhancer. Virology 321:36–46
McCauley DE, Olson MS (2008) Do recent findings in plant mitochondrial molecular and population genetics have implications for the study of gynodioecy and cytonuclear conflict? Evolution 62:1013–1025
McDowell SE, Jun JM, Walter NG (2010) Long-range tertiary interactions in single hammerhead ribozymes bias motional sampling toward catalytically active conformations. RNA 16:2414–2426
Mercer TR, Neph S, Dinger ME et al (2011) The human mitochondrial transcriptome. Cell 146:645–658
Michaud M, Maréchal-Drouard L, Duchêne AM (2010) RNA trafficking in plant cells: targeting of cytosolic mRNAs to the mitochondrial surface. Plant Mol Biol 73:697–704
Mileshina D, Ibrahim N, Boesch P et al (2011) Mitochondrial transfection for studying organellar DNA repair, genome maintenance and aging. Mech Ageing Dev 132:412–423
Millar AH, Whelan J, Soole KL et al (2011) Organization and regulation of mitochondrial respiration in plants. Annu Rev Plant Biol 62:79–104
Montoya J, Gaines GL, Attardi G (1983) The pattern of transcription of the human mitochondrial rRNA genes reveals two overlapping transcription units. Cell 34:151–159
Mukherjee S, Basu S, Home P et al (2007) Necessary and sufficient factors for the import of transfer RNA into the kinetoplast mitochondrion. EMBO Rep 8:589–595
Mulhbacher J, St-Pierre P, Lafontaine DA (2010) Therapeutic applications of ribozymes and riboswitches. Curr Opin Pharmacol 10:551–556
Muller S, Appel B, Krellenberg T et al (2011) The many faces of the hairpin ribozyme: structural and functional variants of a small catalytic RNA. IUBMB Life. doi:10.1002/iub.1575
Nishikura K (2010) Functions and regulation of RNA editing by ADAR deaminases. Annu Rev Biochem 79:321–349
Oca-Cossio J, Kenyon L, Hao H et al (2003) Limitations of allotopic expression of mitochondrial genes in mammalian cells. Genetics 165:707–720
Ochsenreiter T, Cipriano M, Hajduk SL (2007) KISS: the kinetoplastid RNA editing sequence search tool. RNA 13:1–4
Ochsenreiter T, Cipriano M, Hajduk SL (2008) Alternative mRNA editing in trypanosomes is extensive and may contribute to mitochondrial protein diversity. PLoS One 3:e1566
Ogihara Y, Yamazaki Y, Murai K et al (2005) Structural dynamics of cereal mitochondrial genomes as revealed by complete nucleotide sequencing of the wheat mitochondrial genome. Nucleic Acids Res 33:6235–6250
Ojaimi J, Pan J, Santra S et al (2002) An algal nucleus-encoded subunit of mitochondrial ATP synthase rescues a defect in the analogous human mitochondrial-encoded subunit. Mol Biol Cell 13:3836–3844
Olmedo-Monfil V, Cortes-Penagos C, Herrera-Estrella A (2004) Three decades of fungal transformation: key concepts and applications. Methods Mol Biol 267:297–313
O'Rear JL, Wang S, Feig AL et al (2001) Comparison of the hammerhead cleavage reactions stimulated by monovalent and divalent cations. RNA 7:537–545
Paquette J, Nicoghosian K, Qi GR et al (1990) The conformation of single-stranded nucleic acids tDNA versus tRNA. Eur J Biochem 189:259–265
Paris Z, Rubio MA, Lukes J et al (2009) Mitochondrial tRNA import in *Trypanosoma brucei* is independent of thiolation and the Rieske protein. RNA 15:1398–1406
Pelletier G, Budar F (2007) The molecular biology of cytoplasmically inherited male sterility and prospects for its engineering. Curr Opin Biotechnol 18:1–5
Penschow JL, Sleve DA, Ryan CM et al (2004) TbDSS-1, an essential *Trypanosoma brucei* exoribonuclease homolog that has pleiotropic effects on mitochondrial RNA metabolism. Eukaryot Cell 3:1206–1216

Perales-Clemente E, Fernandez-Silva P, Acin-Perez R et al (2010) Allotopic expression of mitochondrial-encoded genes in mammals: achieved goal, undemonstrated mechanism or impossible task? Nucleic Acids Res 39:225–234

Peralta S, Wang X, Moraes CT (2011) Mitochondrial transcription: lessons from mouse models. Biochim Biophys Acta. doi:10.1016/j.bbagrm.2011.1011.1001

Perreault J, Weinberg Z, Roth A et al (2011) Identification of hammerhead ribozymes in all domains of life reveals novel structural variations. PLoS Comput Biol 7:e1002031

Pfannschmidt T (2010) Plastidial retrograde signalling–a true "plastid factor" or just metabolite signatures? Trends Plant Sci 15:427–435

Pley HW, Flaherty KM, McKay DB (1994) Three-dimensional structure of a hammerhead ribozyme. Nature 372:68–74

Puranam RS, Attardi G (2001) The RNase P associated with HeLa cell mitochondria contains an essential RNA component identical in sequence to that of the nuclear RNase P. Mol Cell Biol 21:548–561

Quenault T, Lithgow T, Traven A (2011) PUF proteins: repression, activation and mRNA localization. Trends Cell Biol 21:104–112

Rackham O, Shearwood AM, Mercer TR et al (2011) Long noncoding RNAs are generated from the mitochondrial genome and regulated by nuclear-encoded proteins. RNA 17:2085–2093

Reichert A, Rothbauer U, Morl M (1998) Processing and editing of overlapping tRNAs in human mitochondria. J Biol Chem 273:31977–31984

Reifur L, Koslowsky DJ (2008) *Trypanosoma brucei* ATPase subunit 6 mRNA bound to gA6-14 forms a conserved three-helical structure. RNA 14:2195–2211

Reifur L, Yu LE, Cruz-Reyes J et al (2010) The impact of mRNA structure on guide RNA targeting in kinetoplastid RNA editing. PLoS One 5:e12235

Riley GR, Corell RA, Stuart K (1994) Multiple guide RNAs for identical editing of *Trypanosoma brucei* apocytochrome b mRNA have an unusual minicircle location and are developmentally regulated. J Biol Chem 269:6101–6108

Rinehart J, Krett B, Rubio MA et al (2005) Saccharomyces cerevisiae imports the cytosolic pathway for Gln-tRNA synthesis into the mitochondrion. Genes Dev 19:583–592

Robertson M (2010) The evolution of gene regulation, the RNA universe, and the vexed questions of artefact and noise. BMC Biol 8:97

Rossmanith W, Tullo A, Potuschak T et al (1995) Human mitochondrial tRNA processing. J Biol Chem 270:12885–12891

Rubio MA, Rinehart JJ, Krett B et al (2008) Mammalian mitochondria have the innate ability to import tRNAs by a mechanism distinct from protein import. Proc Natl Acad Sci USA 105:9186–9191

Ryan CM, Kao CY, Sleve DA et al (2006) Biphasic decay of guide RNAs in *Trypanosoma brucei.* Mol Biochem Parasitol 146:68–77

Salinas T, Duchene AM, Delage L et al (2006) The voltage-dependent anion channel, a major component of the tRNA import machinery in plant mitochondria. Proc Natl Acad Sci USA 103:18362–18367

Salinas T, Duchene AM, Marechal-Drouard L (2008) Recent advances in tRNA mitochondrial import. Trends Biochem Sci 33:320–329

Sanchez MI, Mercer TR, Davies SM et al (2011) RNA processing in human mitochondria. Cell Cycle 10:2904–2916

Scarpulla RC (2008) Transcriptional paradigms in mammalian mitochondrial biogenesis and function. Physiol Rev 88:611–638

Scarpulla RC (2011) Nucleus-encoded regulators of mitochondrial function: Integration of respiratory chain expression, nutrient sensing and metabolic stress. Biochim Biophys Acta. doi:10.1016/j.bbagrm.2011.1010.1011

Schekman R (2010) Editorial expression of concern: a bifunctional tRNA import receptor from Leishmania mitochondria. Proc Natl Acad Sci USA 107:9476

Scherer LJ, Rossi JJ (2003) Approaches for the sequence-specific knockdown of mRNA. Nat Biotechnol 21:1457–1465

Schmid B, Riley GR, Stuart K et al (1995) The secondary structure of guide RNA molecules from *Trypanosoma brucei*. Nucleic Acids Res 23:3093–3102

Schmitz-Linneweber C, Small I (2008) Pentatricopeptide repeat proteins: a socket set for organelle gene expression. Trends Plant Sci 13:663–670

Schnable PS, Wise RP (1998) The molecular basis of cytoplasmic male sterility and fertility restoration. Trends Plant Sci 3:175–180

Schneider A (2011) Mitochondrial tRNA import and its consequences for mitochondrial translation. Annu Rev Biochem 80:1033–1053

Schneider A, Maréchal-Drouard L (2000) Mitochondrial tRNA import: are there distinct mechanisms? Trends Cell Biol 10:509–513

Schon A (1995) Ribonuclease P from plant nuclei and photosynthetic organelles. Mol Biol Rep 22:139–145

Schon EA, DiMauro S (2007) Mitochondrial mutations: genotype to phenotype. Novartis Found Symp 287:214–233

Schumacher MA, Karamooz E, Zikova A et al (2006) Crystal structures of *T. brucei* MRP1/MRP2 guide-RNA binding complex reveal RNA matchmaking mechanism. Cell 126:701–711

Schwarzländer M, König AC, Sweetlove LJ et al (2011) The impact of impaired mitochondrial function on retrograde signalling: a meta-analysis of transcriptomic responses. J Exp Bot. doi:10.1093/jxb/err1374

Scott WG (2007) Morphing the minimal and full-length hammerhead ribozymes: implications for the cleavage mechanism. Biol Chem 388:727–735

Scott WG, Finch JT, Klug A (1995) The crystal structure of an all-RNA hammerhead ribozyme: a proposed mechanism for RNA catalytic cleavage. Cell 81:991–1002

Scott WG, Martick M, Chi YI (2009) Structure and function of regulatory RNA elements: ribozymes that regulate gene expression. Biochim Biophys Acta 1789:634–641

Seehafer C, Kalweit A, Steger G et al (2011) From alpaca to zebrafish: hammerhead ribozymes wherever you look. RNA 17:21–26

Seiwert SD, Stuart K (1994) RNA editing: transfer of genetic information from gRNA to precursor mRNA *in vitro*. Science 266:114–117

Seiwert SD, Heidmann S, Stuart K (1996) Direct visualization of uridylate deletion *in vitro* suggests a mechanism for kinetoplastid RNA editing. Cell 84:831–841

Seligmann H (2010) Undetected antisense tRNAs in mitochondrial genomes? Biol Direct 5:39

Shepotinovskaya I, Uhlenbeck OC (2010) Enhanced product stability in the hammerhead ribozyme. Biochemistry 49:4494–4500

Shlomai J (2004) The structure and replication of kinetoplast DNA. Curr Mol Med 4:623–647

Sieber F, Placido A, El Farouk-Ameqrane S et al (2011) A protein shuttle system to target RNA into mitochondria. Nucleic Acids Res 39:e96

Silvestri G, Ciafaloni E, Santorelli FM et al (1993) Clinical features associated with the A → G transition at nucleotide 8344 of mtDNA ("MERRF mutation"). Neurology 43:1200–1206

Singh S, Narang AS, Mahato RI (2011) Subcellular fate and off-target effects of siRNA, shRNA, and miRNA. Pharm Res 28:2996–3015

Small I, Maréchal-Drouard L, Masson J et al (1992) *In vivo* import of a normal or mutagenized heterologous transfer RNA into the mitochondria of transgenic plants: towards novel ways of influencing mitochondrial gene expression? EMBO J 11:1291–1296

Smeitink J, van den Heuvel L, DiMauro S (2001) The genetics and pathology of oxidative phosphorylation. Nat Rev Genet 2:342–352

Smirnov A, Tarassov I, Mager-Heckel AM et al (2008) Two distinct structural elements of 5 S rRNA are needed for its import into human mitochondria. RNA 14:749–759

Smirnov A, Comte C, Mager-Heckel AM et al (2010) Mitochondrial enzyme rhodanese is essential for 5 S ribosomal RNA import into human mitochondria. J Biol Chem 285:30792–30803

Smirnov A, Entelis N, Martin RP et al (2011) Biological significance of 5 S rRNA import into human mitochondria: role of ribosomal protein MRP-L18. Genes Dev 25:1289–1305

Spitale RC, Volpini R, Mungillo MV et al (2009) Single-atom imino substitutions at A9 and A10 reveal distinct effects on the fold and function of the hairpin ribozyme catalytic core. Biochemistry 48:7777–7779

St Johnston D (2005) Moving messages: the intracellular localization of mRNAs. Nat Rev Mol Cell Biol 6:363–375

Stuart KD, Schnaufer A, Ernst NL et al (2005) Complex management: RNA editing in trypanosomes. Trends Biochem Sci 30:97–105

Sugiyama Y, Watase Y, Nagase M et al (2005) The complete nucleotide sequence and multipartite organization of the tobacco mitochondrial genome: comparative analysis of mitochondrial genomes in higher plants. Mol Genet Genomics 272:603–615

Suydam IT, Levandoski SD, Strobel SA (2010) Catalytic importance of a protonated adenosine in the hairpin ribozyme active site. Biochemistry 49:3723–3732

Sylvestre J, Margeot A, Jacq C et al (2003) The role of the 3′ untranslated region in mRNA sorting to the vicinity of mitochondria is conserved from yeast to human cells. Mol Biol Cell 14:3848–3856

Symons RH (1992) Small catalytic RNAs. Annu Rev Biochem 61:641–671

Tarassov I, Entelis N (1992) Mitochondrially-imported cytoplasmic $tRNA^{Lys}$ (CUU) of *Saccharomyces cerevisiae*: in vivo and in vitro targetting systems. Nucleic Acids Res 20:1277–1281

Tuppen HA, Blakely EL, Turnbull DM et al (2010) Mitochondrial DNA mutations and human disease. Biochim Biophys Acta 1797:113–128

Unseld M, Marienfeld JR, Brandt P et al (1997) The mitochondrial genome of *Arabidopsis thaliana* contains 57 genes in 366,924 nucleotides. Nat Genet 15:57–61

Val R, Wyszko E, Valentin C et al (2011) Organelle trafficking of chimeric ribozymes and genetic manipulation of mitochondria. Nucleic Acids Res 39:9262–9274

Valkov VT, Scotti N, Kahlau S et al (2009) Genome-wide analysis of plastid gene expression in potato leaf chloroplasts and tuber amyloplasts: transcriptional and posttranscriptional control. Plant Physiol 150:2030–2044

Van Aken O, Zhang B, Carrie C et al (2009) Defining the mitochondrial stress response in *Arabidopsis thaliana*. Mol Plant 2:1310–1324

van der Spek H, Arts GJ, Zwaal RR et al (1991) Conserved genes encode guide RNAs in mitochondria of *Crithidia fasciculata*. EMBO J 10:1217–1224

Vanhamme L, Perez-Morga D, Marchal C et al (1998) *Trypanosoma brucei* TBRGG1, a mitochondrial oligo(U)-binding protein that co-localizes with an *in vitro* RNA editing activity. J Biol Chem 273:21825–21833

Veeraraghavan N, Ganguly A, Chen JH et al (2011) Metal binding motif in the active site of the HDV ribozyme binds divalent and monovalent ions. Biochemistry 50:2672–2682

Villegas J, Zarraga AM, Muller I et al (2000) A novel chimeric mitochondrial RNA localized in the nucleus of mouse sperm. DNA Cell Biol 19:579–588

Villegas J, Muller I, Arredondo J et al (2002) A putative RNA editing from U to C in a mouse mitochondrial transcript. Nucleic Acids Res 30:1895–1901

Villegas J, Burzio V, Villota C et al (2007) Expression of a novel non-coding mitochondrial RNA in human proliferating cells. Nucleic Acids Res 35:7336–7347

Wallace DC (1992) Diseases of the mitochondrial DNA. Annu Rev Biochem 61:1175–1212

Wan Y, Kertesz M, Spitale RC et al (2011) Understanding the transcriptome through RNA structure. Nat Rev Genet 12:641–655

Wang G, Chen HW, Oktay Y et al (2010) PNPASE regulates RNA import into mitochondria. Cell 142:456–467

Wang G, Shimada E, Koehler CM et al (2011) PNPASE and RNA trafficking into mitochondria. Biochim Biophys Acta. doi:10.1016/j.bbagrm.2011.1010.1001

Webb CH, Lupták A (2011) HDV-like self-cleaving ribozymes. RNA Biol 8:719–727

Weinberg MS, Ely A, Passman M et al (2007) Effective anti-hepatitis B virus hammerhead ribozymes derived from multimeric precursors. Oligonucleotides 17:104–112

Weng J, Aphasizheva I, Etheridge RD et al (2008) Guide RNA-binding complex from mitochondria of trypanosomatids. Mol Cell 32:198–209

Westhof E (1999) Chemical diversity in RNA cleavage. Science 286:61–62

Witten JT, Ule J (2011) Understanding splicing regulation through RNA splicing maps. Trends Genet 27:89–97

Wong LJ (2007) Diagnostic challenges of mitochondrial DNA disorders. Mitochondrion 7:45–52

Woo DK, Phang TL, Trawick JD et al (2009) Multiple pathways of mitochondrial-nuclear communication in yeast: intergenomic signaling involves ABF1 and affects a different set of genes than retrograde regulation. Biochim Biophys Acta 1789:135–145

Woodson JD, Chory J (2008) Coordination of gene expression between organellar and nuclear genomes. Nat Rev Genet 9:383–395

Xu R, Liu J, Xu F et al (2001) Activity identification of chimeric anti-caspase-3 mRNA hammerhead ribozyme *in vitro* and *in vivo*. Sci China C Life Sci 44:618–627

Yoshionari S, Koike T, Yokogawa T et al (1994) Existence of nuclear-encoded 5 S-rRNA in bovine mitochondria. FEBS Lett 338:137–142

Yu LE, Koslowsky DJ (2006) Interactions of mRNAs and gRNAs involved in trypanosome mitochondrial RNA editing: structure probing of a gRNA bound to its cognate mRNA. RNA 12:1050–1060

Zabala G, Gabay-Laughnan S, Laughnan JR (1997) The nuclear gene *Rf3* affects the expression of the mitochondrial chimeric sequence R implicated in S-type male sterility in maize. Genetics 147:847–860

Zhang SQ, Zhang FD, Xiao HL et al (2004) R region of S type of cytoplasmic male sterility in maize mitochondrial DNA is transcribed in both directions and may be associated with male sterility. Acta Botanica Sinica 46:337–341

Zhou H, Hu H, Lai M (2010a) Non-coding RNAs and their epigenetic regulatory mechanisms. Biol Cell 102:645–655

Zhou J, Liu L, Chen J (2010b) Mitochondrial DNA heteroplasmy in *Candida glabrata* after mitochondrial transformation. Eukaryot Cell 9:806–814

Zikova A, Kopecna J, Schumacher MA et al (2008) Structure and function of the native and recombinant mitochondrial MRP1/MRP2 complex from *Trypanosoma brucei*. Int J Parasitol 38:901–912

Zimmer SL, McEvoy SM, Li J et al (2011) A novel member of the RNase D exoribonuclease family functions in mitochondrial guide RNA metabolism in *Trypanosoma brucei*. J Biol Chem 286:10329–10340

RNA-Based Therapies for Inherited Metabolic Diseases

Belén Pérez, Magdalena Ugarte, and Lourdes R. Desviat

Contents

Abstract In many genetic diseases for which there is to date no effective treatment, research in novel therapeutical targets and treatment options is one of the main objectives. In this context, many inherited metabolic diseases (IMD) fulfill the above criteria, and significant research is being performed in mutation-specific therapies. Specifically, a special attention has been given to splicing defects that account for 15–30% of the mutant alleles in most IMD. Splicing mutations are most commonly found at the conserved 3′ or 5′ splice sites at intron–exon junctions, but other types of changes have been described, including mutations affecting the polypyrimidine tract, exonic mutations affecting splice enhancers or silencers, and deep intronic mutations that activate or create splice sites resulting in the aberrant inclusion of pseudoexons in the mRNA. The understanding of the precise mechanisms leading to the splicing defect has paved the way for the implementation of splicing-directed RNA-based therapies. In IMD, several studies have explored the therapeutical potential of targeting the aberrant pre-mRNA resulting from splicing mutations, using overexpression or silencing of splicing factors and splice blocking antisense oligonucleotides. In this chapter, we review the existing

B. Pérez • M. Ugarte (✉) • L.R. Desviat (✉)
Centro de Diagnóstico de Enfermedades Moleculares, Centro de Biología Molecular Severo Ochoa, UAM-CSIC, Universidad Autónoma de Madrid, 28049 Madrid, Spain

Centro de Investigación Biomédica en Red de Enfermedades Raras (CIBERER), Madrid, Spain
e-mail: mugarte@cbm.uam.es; lruiz@cbm.uam.es

V.A. Erdmann and J. Barciszewski (eds.), *From Nucleic Acids Sequences to Molecular Medicine*, RNA Technologies, DOI 10.1007/978-3-642-27426-8_14,

knowledge on the splicing defects in IMD and their associated pathological effect as revealed by functional studies and the use of possible therapeutical interventions that can be applied to modulate splicing.

Keywords Antisense therapy • Exonic splice enhancer • Exonic splice silencer • Pseudoexons • Splice factors • U1

1 Introduction

RNA splicing is a complex and tightly regulated process requiring *cis*- and *trans*-acting factors to be completed efficiently. The *cis*-acting factors include conserved motifs in or near the intron–exon boundaries that act as splicing signals: the 3′ and 5′ splice sites (3′ss, 5′ss), the branch point sequence 20–40 nucleotides upstream of the 3′ss, and the polypyrimidine tract between the branch point and the 3′ss (Cooper et al. 2009). A set of six different small nuclear ribonucleoproteins (snRNPs), U1, U2, U3, U4, U5, and U6, recognize these signals and mediate RNA splicing. In the first step, the 5′ss is recognized by base-pairing with the U1snRNA. In turn, the 3′ss is recognized by the U2 auxiliary factor (U2AF) heterodimer, formed by a small (U2AF35) and a large (U2AF65) subunit that recruit the U2 snRNP. Subsequently, the last three snRNPs (U4, U5, and U6), which are assembled into a trimeric complex, are recruited, and after specific interactions and conformational modifications, the catalytically active spliceosomal complex is formed (Tazi et al. 2005).

In addition to the conserved *cis*-acting sequences described, there is a battery of regulatory elements located in exons (ESE, exonic splice enhancers; ESS, exonic splice silencers) or introns (ISE, intronic splice enhancers; ISS, intronic splice silencers) that are required in some cases for correct exon definition or that mediate the use of alternative splice sites (Cartegni et al. 2002). Enhancers are frequently needed for those exons with weak 3′ and/or 5′ splicing signals, which, in turn, are more prone to splicing mutations (Eng et al. 2004). The regulatory elements are recognized by *trans*-acting splice factors that include serine/arginine-rich (SR) proteins, which bind enhancer sequences driving correct spliceosomal formation and selection of splice sites, and the members of the heterogeneous nuclear ribonucleoproteins (hnRNP) family that act as splicing repressors binding to silencer motifs (Long and Caceres 2009). SR proteins antagonize adjacent silencer elements and recruit the splicing machinery, resulting in the formation of bridge complexes across exons. Classical SR proteins include SF2/ASF (splicing factor 2/alternative splicing factor), SC35 (splicing component 35), and SRp20, SRp40, SRp55, and SRp75, named according to their apparent molecular mass in SDS-PAGE gels (Long and Caceres 2009). For each individual exon, splicing regulation may result from a complex interplay between different elements acting agonistically or antagonistically.

It has recently been proposed that splicing alterations constitute the most common (up to 60%) type of mutations directly causing genetic disease (Lopez-Bigas et al. 2005; Ward and Cooper 2009). Classically, splicing mutations have been described in the conserved 3′ or 5′ss with the majority of them affecting the invariant ag and gt dinucleotides (Krawczak et al. 2007), but in recent years, extensive evidence has accumulated of the presence of point mutations sometimes far away from these sites, affecting exonic or intronic nucleotides and causing a splicing defect. Such mutations alter splicing through creation or activation of cryptic splice sites or disruption of splicing regulatory signals. Formally and experimentally verified in many cases, coding mutations, which are standardly interpreted as nonsense, missense, or translationally silent or synonymous, can affect RNA processing and should be carefully evaluated (Cartegni et al. 2002). To this aim, besides the gold standard experimental RNA-based assays for aberrant transcript assessment, several bioinformatics analyses are to date freely available to predict splice site strength, the presence of *cis*-acting splicing regulatory sequences bound by splice factors and SR proteins, and the changes induced by a mutation (Hartmann et al. 2008). In addition, ex vivo analysis using minigenes is a reliable experimental tool to functionally analyze the individual effect of mutations on the splicing process, which is especially relevant when patient's RNA samples are not available (Baralle and Baralle 2005). Minigenes consist of a vector with exonic portions of a gene defined by functional 5′ss and 3′ss separated by intronic sequences where a polylinker is located and the exonic fragment of interest is cloned. The mutation under study is introduced by mutagenesis, and both the wild-type and mutant versions of the minigene are transfected in an established cell line (COS-7, HEK293, etc.), and 24 h later, RT-PCR is performed with vector-specific primers to selectively amplify processed transcripts from the minigene. If the mutation is involved in the splicing process, mutant minigenes will result in a different splicing pattern. Ideally, the wild-type minigene results in correct exon inclusion, while the mutant construct results in exon skipping or other aberrant transcripts. However, minigenes do not always recapitulate exactly the splicing defect occurring in vivo due to lack of full genomic context, so it is desirable to confirm the transcript profile in patients' cells when possible.

The diagnostic analysis of mRNA in genetic diseases has also resulted in the recognition of a novel class of disease-causing mutations in deep intronic regions distant from the invariant splice sites that cause missplicing by inducing inclusion of intronic sequences known as pseudoexons (Dhir and Buratti 2010). They frequently act by creating novel splice sites or by strengthening preexisting cryptic splice sites located in their proximity. Many of these pseudoexons correspond to transposable sequence elements derived from short or long interspersed nuclear elements (SINEs or LINES, respectively), primarily SINE-*Alu* sequences, which, in turn, are a prominent source of evolutionary new exons in the eukaryotic transcriptome (Vorechovsky 2010). Several examples of this pathogenic mechanism have been reported in human genetic disorders such as cystic fibrosis, ataxia telangiectasia, neurofibromatosis, Duchenne muscular dystrophy, and several inherited metabolic disorders (IMD) (Dhir and Buratti 2010; Perez et al. 2010).

The prevalence of splicing alterations in genetic disease and the fact that subtle changes in the transcript pattern can account for striking differences in disease phenotype (Ginjaar et al. 2000; Hefferon et al. 2004), milder in cases with detectable wild-type transcript due to partially penetrant or "leaky" splicing mutations (Vezain et al. 2011), have encouraged the research and development of new therapeutic approaches for what has been termed RNA-missplicing diseases (Hammond and Wood 2011). Splicing modulation can be achieved in a variety of ways, pharmacological and genetic, the latter involving interactions with the mutated pre-mRNA to correct the splicing defect. Herein we review current genetic therapies for missplicing, with a focus in enzyme deficiencies causing different IMD, most of them of autosomal recessive inheritance and currently lacking an effective treatment.

2 Splicing Defects in IMD

IMD are individually rare genetic diseases, although collectively they affect approximately 1/3,000 newborns (Lindner et al. 2011). Many IMD are potentially lethal, severe diseases and they are inadequately managed by current dietary and pharmacological therapies. The underlying cause of the disease is the functional deficiency of an enzyme or transport protein associated with a metabolic pathway and they are predominantly inherited in an autosomal recessive fashion. As in other genetic diseases, splicing defects account for 15–30% of the total mutant alleles, although as previously mentioned, without RNA analysis, the real frequency may be much higher.

Classical splicing mutations at 3′ and 5′ss have been described in all genes. The most common effects of mutations on the 3′ or 5′ss are the skipping of the corresponding exon and subsequent in-frame or out-frame deletions. In addition, in some cases, the changes activate cryptic splice sites inside the intron or the exon, generating a larger or shorter transcript, respectively. An example of activation of a cryptic exonic splice site has been reported in the MMAB gene responsible for methylmalonic aciduria cblB type. The variant change c.349-1G > C activates a cryptic exonic 3′ss producing the in-frame deletion of two amino acids, p. I117_Q118del (Jorge-Finnigan et al. 2010). Notably, this small in-frame deletion is pathogenic and no residual activity is rescued (unpublished results).

On the other hand, although considered infrequent, exonic mutations may affect splicing (Fig. 1) and may be currently misclassified, as genetic diagnosis is most frequently performed on genomic DNA samples without transcript analysis. Translationally silent or apparent "missense" mutations have been shown to affect splicing by disrupting or creating ESE or ESS or activating or creating 3′ and 5′ss. Most of the splicing mutations in the Human Gene Mutation database (HGMD Professional release 2011) correspond to 5′ss. Changes in the last exonic codon are included in this category, and even though these nucleotides are part of the conserved sequence involved in the binding to U1 snRNP, they are frequently

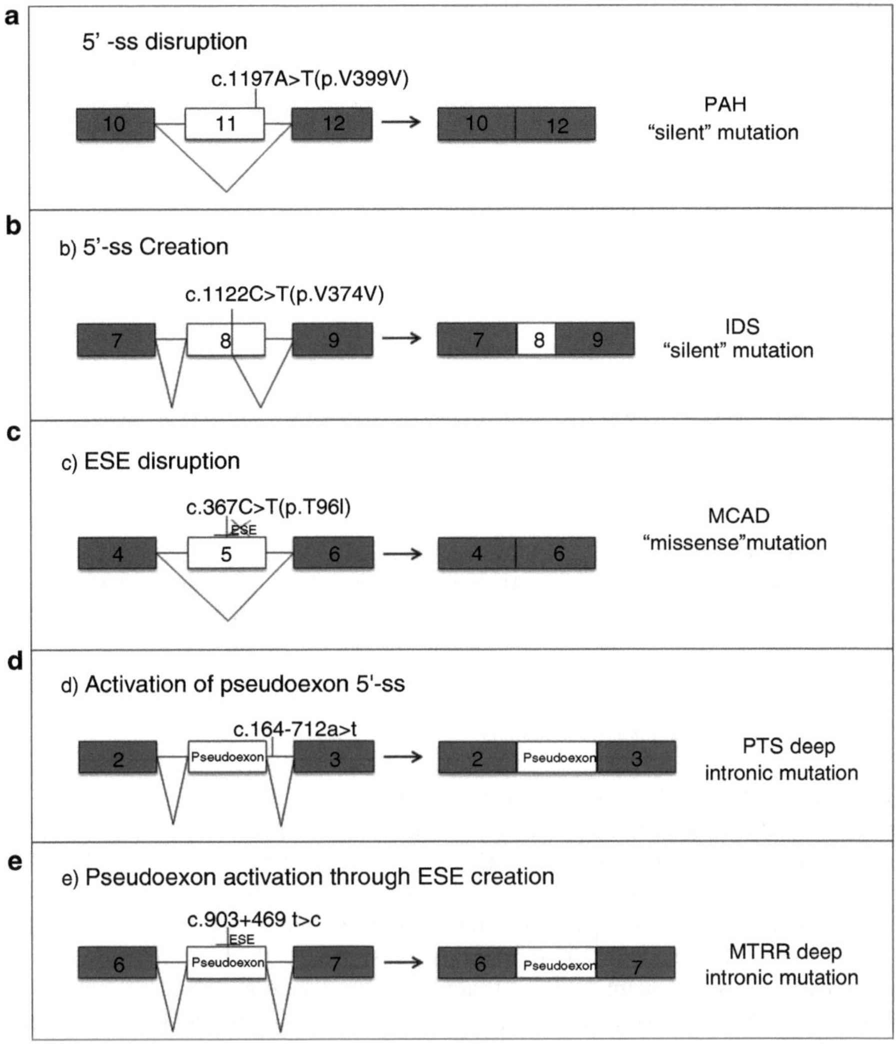

Fig. 1 Examples of mutations in exonic and deep intronic nucleotides in IMD genes causing splicing defects. A translationally silent mutation can have an effect on splicing through disruption of the 3′ss (**a**) (Chao et al. 2001) or through creation of a 5′ss (**b**) (Alves et al. 2006). In other cases, apparent "missense" mutations are really splicing defects due to disruption of ESE sequences (**c**) (Nielsen et al. 2007). Deep intronic mutations can induce aberrant pseudoexon inclusion by activation of its 5′ss (**d**) (Brasil et al. 2011) or by creation of an ESE (**e**) (Homolova et al. 2010)

wrongly predicted to produce an amino acid change. Table 1 shows examples of exonic mutations in IMD genes with a confirmed effect on splicing.

In addition, the number of cases with deep intronic mutations causing pseudoexon activation continues to grow in IMD and includes gene defects responsible for organic acidemias (Rincon et al. 2007; Tsuruta et al. 1998), lysosomal

Table 1 Examples of exonic mutations in IMD genes with an effect on splicing

Mutation	Gene	Effect	References
c.30C > G (silent)	PAH	Creation of an ESS	Chao et al. (2001)
c.1197A > T (silent)	PAH	Disruption of 5′ss	Dobrowolski et al. (2010)
c.951C > T (silent)	SCOT	Disruption of an ASF/SF2 binding site	Fukao et al. (2010)
c.1066G > C (silent)	IDS	Activation of cryptic ss	Alves et al. (2006)
c.1122C > T (silent)	IDS	Creation of 5′ss	Alves et al. (2006)
c.750G > A (silent)	ALDH7A1	Activation of cryptic ss	Salomons et al. (2007)
c.483C > T (silent)	PDHA1	Disruption of an SRp55 binding site	Boichard et al. (2008)
c.498C > T (silent)	PDHA1	Disruption of an SRp55 binding site	Boichard et al. (2008)
c.653A > G (p.K218R)[a]	PCCB	Disruption of 5′ss	Clavero et al. (2004)
c.257C > T (p.P86L)[a]	IDS	Activation of cryptic 3′ss	Alves et al. (2006)
c.362C > T (p.T96I)[a]	MCAD	Disruption of an ASF/SF2 binding site	Nielsen et al. (2007)
c.1054G > A (p.G352R)[a]	MCCB	Disruption of an SRp55 binding site	Stucki et al. (2009)
c.100G > A (p.G34S)[a]	BTD	Creation of cryptic 3′ss	Pomponio et al. (1997)
c.836A > G (p.Q279R)[a]	FAH	Disruption of 5′ss	Dreumont et al. (2001)
c.1808G > A	MUT	Creation of cryptic 5′ss	Martinez et al. (2005)
c.733G > A	MMAA	Disruption of 5′ss	Merinero et al. (2008)
c.584G > A	MMAB	Disruption of 5′ss	Jorge-Finnigan et al. (2010)
c.290G > A	MMAB	Disruption of 5′ss	Jorge-Finnigan et al. (2010)

[a]Initially misclassified as missense mutations

disorders (Rodriguez-Pascau et al. 2009; Vervoort et al. 1998), congenital disorders of glycosylation (Schollen et al. 2007), amino acid disorders (Ogino et al. 2007), and tetrahydrobiopterin deficiencies (Brasil et al. 2011; Ikeda et al. 1997; Meili et al. 2009). In the pseudoexon-activating mutations described, the change usually creates or activates an intronic 5′ss as has been described in MUT, PCCB, PMM2, or NPC1 genes (Perez et al. 2010). In these cases, usually the binding to U1snRNP is increased. Less frequent is the activation of the 3′ss although it has also been described (Table 2). In some cases, the change creates a binding site for an auxiliary splice factor as has been described in the PCCA gene where the intronic mutation creates a new SRp40 binding site in the middle of the pseudoexon, or in the MTRR gene where the c.903 + 469T > C mutation creates an SF2/ASF binding site (Fig. 1e) (Homolova et al. 2010; Rincon et al. 2007). In a particular case of 3-methylcrotonyl-CoA carboxylase deficiency, the inclusion of a pseudoexon in intron 10 was found to be mediated by the concomitant skipping of exon 11 due to a missense mutation disrupting an ESE (Stucki et al. 2009).

The frequency of pseudoexon-activating mutations ranges from 2 to 5% (Perez et al. 2010), although this may be an underestimation given that transcript analysis is not routinely performed for genotyping and full intron sequencing is almost never performed. In most cases, pseudoexon insertion results in a frameshift and a premature termination codon, eliciting the nonsense-mediated decay (NMD)

Table 2 Examples of pseudoexon-activating mutations in IMD successfully rescued by antisense therapy

Gene	gDNA change	RNA change	Predictable effect	References
MUT	c.1957-891C > A	r.1956_1057ins76	Increase in 5′ss score	Rincon et al. (2007)
MUT	c.1957-898A > G	r.1956_1057ins76	Increase in 5′ss score	Perez et al. (2009)
PCCA	c.1285-1416A > G	r.1284_1285ins84	Creates SRp40 binding site and eliminates SRp55 biding site	Rincon et al. (2007)
PCCB	c.654 + 462A > G	r.654_655ins72	Increase in 5′ss score	Rincon et al. (2007)
PMM2	c.640-15479C > T	r.639_640ins123	Increase in 5′ss score	Vega et al. (2009)
PTS	c.163 + 695_163 + 751del57	r.163_164ins45	Activation of branch point sequence	Brasil et al. (2011)
PTS	c.164-712A > T	r.163_164ins45/70	Increase in 5′ss score	Brasil et al. (2011)
PTS	c.84-322A > T	r.83_84ins79	Activation of 3′ss	Brasil et al. (2011)
NPC1	c.1554-1009G > A	r.1553_1554ins194	Creates a cryptic 5′ss	Rodriguez-Pascau et al. (2009)

mechanism which degrades the transcript. Treatment of patients' cells with cycloheximide, emetine, or other compounds to avoid NMD and subsequent cDNA analysis aids in the identification of these elusive mutations.

3 Genetic Therapies for Splicing Defects

The identification of splicing defects and of the molecular mechanism by which point mutations lead to the production of a specific aberrantly spliced transcript, along with the consideration of splicing variations as genetic modifiers of the phenotype, has opened an emerging field of study focused on splicing modulation as a novel therapeutical approach. In this sense, splice correction for mutations affecting 3′ or 5′ss at nucleotides other than the invariant ag and gt has been approached by overexpression of specific *trans*-acting factors binding to those sites (Nissim-Rafinia et al. 2000). Cystic fibrosis transmembrane conductance regulator (CFTR) function was restored to variable extents by overexpression of Htra2-β1 and SC35 in patient cells carrying a splicing mutation. Increased levels of these splice factors, also achieved by pharmacological treatment with the histone deacetylase inhibitor sodium butyrate, specifically modulated CFTR splicing by decreasing the level of aberrant CFTR transcripts, while overexpression of other splice factors had no effect (Nissim-Rafinia et al. 2004).

In congenital defects of glycosylation, this strategy has also been employed to evaluate the possibility of a novel therapy and to determine the importance of *trans*-acting factors in exon recognition especially in cases of weak 3′ss. The effect of overexpression of U2AF65, U2AF35, SC35, SRp40, SRp55, and ASF/SF2 was investigated with minigenes for mutations c.640-9T > G and c.256-1G > C in the PMM2 gene (Vega et al. 2009). In the first case, cotransfection with SC35 and ASF/

SF2 partially corrected exon skipping, although via the use of a cryptic splice site and in the second case, overexpression of U2AF65 and U2AF35 resulted in increased inclusion of the corresponding exon, demonstrating that effective splicing modulation can be achieved although by mechanistically different ways, which should be explored mutation by mutation (Vega et al. 2009).

Changes in the 5′ss constitute a good target for mutation-specific approaches. In this sense, U1snRNA overexpression has been assayed for several 5′ss mutations (Hartmann et al. 2010; Pinotti et al. 2008; Susani et al. 2004; Tanner et al. 2009). In propionic acidemia, one of the most common organic acidemias, the use of adapted U1snRNA complementary to the mutated site was investigated in patients' cells carriers of mutation c.1209 + 3A > G in the PCCA gene, a mutation that caused complete exon skipping. Functional analysis with minigenes indicated that in this case the lack of complementarity to U1snRNA at positions +5 and +6 influences the dependence of the 5′ss of an A at the +3 position for correct splicing. Overexpression of wild-type U1snRNA or with the corresponding matched nucleotide at position +3 does not rescue the splicing defect in patients' fibroblasts. This is only achieved with a mutated U1snRNA with perfect matches at positions +3, +5, and +6, in accordance with similar studies in other genes, where an extended U1snRNA-5′ss base-pairing is necessary for exon inclusion and that it not only the interaction between U1 snRNA and the 5′ss is involved in the splicing defect, as the mutation probably affects the affinity of the 5′ss for other splice factors (Sanchez-Alcudia et al. 2009). However, the efficient reversion of the splicing defect did not correlate with a functional enzymatic rescue. Further studies are necessary to investigate if the adapted U1snRNA also binds more stably to other 5′ss of key genes inhibiting the entry of other spliceosomal components due to delayed release of the U1snRNP, resulting in unforeseen side effects precluding a therapeutical defect.

In other cases, silencing the corresponding *trans*-binding splice factor by siRNA to prevent intronic cryptic splice site utilization may be useful. In pyruvate dehydrogenase deficiency, a mutation 26 nucleotides downstream of exon 7 was found to be responsible for activation of a cryptic splice site further downstream, leading to aberrant intron retention (Gabut et al. 2005). The mutation was demonstrated to result in an increased binding of the SR protein SC35, and siRNA-mediated reduction of the SC35 protein levels in primary fibroblasts almost completely corrected the aberrant splicing pattern (Gabut et al. 2005).

Another strategy for splicing modulation is the use of modified antisense oligonucleotides (AO) targeting splice sites that act by steric hindrance of the recognition and binding of the splicing apparatus to the selected sequences, thus forcing the machinery to use alternative sites. This was first reported for the beta-globin gene where AO were used to mask the activated intronic cryptic splice sites leading to abnormal splicing and causing beta-thalassemia (Dominski and Kole 1993). To date, splicing intervention with AO to induce the skipping of a specific exon allowing the restoration of the reading frame and restoring gene function is already in the clinical trial phase for Duchenne muscular dystrophy (Kinali et al. 2009; van Deutekom et al. 2007). Antisense therapy as personalized genetic

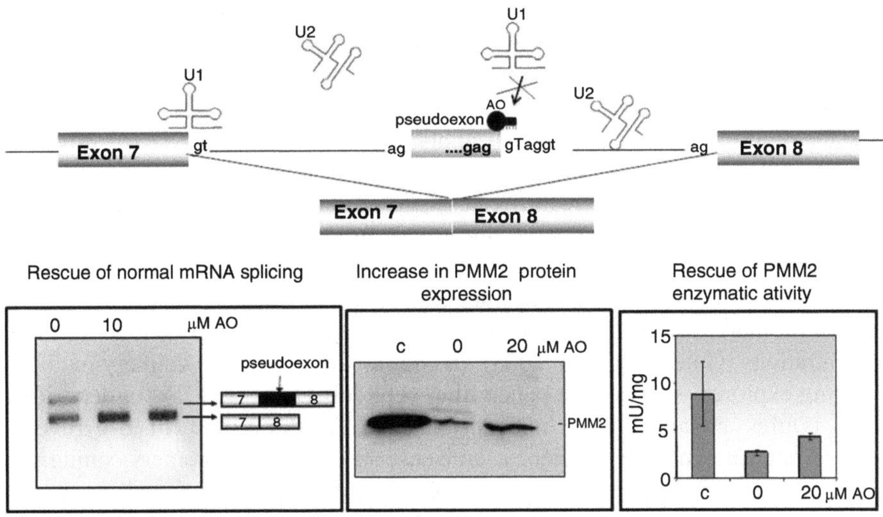

Fig. 2 Antisense therapy. Schematic figure and results of the use of morpholino antisense oligonucleotides to revert pathological pseudoexon inclusion in congenital defects of glycosylation (Vega et al. 2009)

medicine has also been widely investigated for deep intronic pseudoexon-activating mutations (Dhir and Buratti 2010; Perez et al. 2010) including several IMD (Perez et al. 2010) (Table 2). In propionic acidemia (MIM 606054), both *PCCA* and *PCCB* pseudoexons were targeted with AO rescuing normal transcript and enzyme protein and activity within the normal range (Rincon et al. 2007). In methylmalonic aciduria (MIM 251000) due to a defect in the *MUT* gene encoding methylmalonyl-CoA mutase, two different intronic mutations were shown to cause activation of the same pseudoexon, which was efficiently excluded from the mRNA by use of AO, recovering enzyme activity (Perez et al. 2009; Rincon et al. 2007). Similar results were obtained in cells from a patient with deficiency in phosphomannomutase 2 (MIM 212065) with a pseudoexon-activating mutation in heterozygous fashion in the PMM2 gene. In this case, the correction levels were 23% and 50%, for immunoreactive protein and activity, respectively (Fig. 2) (Vega et al. 2009). For Niemann–Pick type C (MIM 257220), AO treatment also reversed the aberrant pseudoexon inclusion (Rodriguez-Pascau et al. 2009). Finally, in 6-pyruvoyltetrahydropterin synthase deficiency, pathogenic inclusion of SINE- or LINE-derived cryptic exons in different patients due to three deep intronic mutations was reverted by use of AO directed to the pseudoexon 3′ or 5′ss. Transcriptional profile, enzyme activity, and pterin profile were normalized after antisense treatment (Brasil et al. 2011). In all the cases reported, antisense phosphorodiamidate morpholino oligonucleotides were used and splicing correction was shown to be dose and sequence specific, with no obvious cytotoxicity observed. The effect of AO transfection persists up to 15 days post transfection (Rincon et al. 2007). In the cases where enzymatic activity could be determined, the recovered levels were therapeutically significant even for heterozygous samples, taking into account that for recessive diseases residual activity is associated with milder phenotypes and that in animal models outcome can be improved with low expression levels of a transgene (Miyazaki et al. 2001).

A major limiting step for in vivo application of this therapy is the effective delivery of the AO into target cells. If the AO is to be administered systemically, it must be stable in the bloodstream, be resistant to degradation by nucleases, reach target cells, cross the cell membrane, and escape from cellular endosomes and reach the nucleus. To this aim, novel nucleotide chemistries have been developed with modified riboses and/or phosphate linkages, such as locked nucleic acids (LNA) that exhibit increased affinity to the target RNA, resulting in high efficiency at low concentrations (Guterstam et al. 2008). In addition, different AO delivery methods are being explored, such as cell-penetrating peptides (Du et al. 2011) or dendrimer based (Dufes et al. 2005). In methylmalonic aciduria and in 6-pyruvoyl-tetrahydropterin synthase deficiency, antisense morpholino oligomers conjugated to an octaguanidine dendrimer suitable for in vivo use were found to correct erroneous splicing in cell culture with similar efficiency to that obtained using a standard transfection reagent (Brasil et al. 2011; Perez et al. 2009).

For neurogenetic diseases, the inability of AO to cross the brain–blood barrier and efficiently target brain cells also limits clinical application of this therapeutic strategy. In this sense, both arginine-rich cell-penetrating peptides (Du et al. 2011) and carbosilane dendrimers (Jimenez et al. 2010) have been used for brain cell targeting with promising results.

One additional approach which could be used to correct splicing defects is spliceosome-mediated RNA *trans*-splicing (SMaRT) (Rodriguez-Martin et al. 2005). This technique involves naturally occurring *trans*-splicing between the target mRNA precursor and an externally added therapeutic pre-*trans*-splicing molecule (PTM) containing the correct exonic sequences. With some efficiency, the splicing machinery recognizes the PTM, and thus, wild-type transcript is produced. Although there are still few examples of applications of this technology to splicing defects, it has clear therapeutical potential and can be envisaged to be used for different missplicing diseases (Hammond and Wood 2011).

All these data represent the proof of concept of the applicability of genetic therapies for RNA-missplicing diseases and open the exciting prospect of modulating splicing as personalized genetic medicine for a subset of patients with IMD.

4 Concluding Remarks

The findings reviewed here support the hypothesis that genetic treatments that target a particular splicing mutation can be successfully developed. These include antisense oligonucleotides, splice factor overexpression, and trans-splicing approaches, which have all been tested with various outcomes in different gene defects. An advantage of these mutation-specific techniques over gene therapy strategies is that they correct endogenous transcripts expressed from their natural promoters. However, nonspecific side effects should be carefully evaluated in each case to investigate the possibility that nontarget transcripts are influenced by the treatment. Adequate in vivo models recapitulating the specific splicing defects are needed to confirm and extend the results obtained in cellular models of disease.

Acknowledgments This work was supported by grants from *Ministerio de Ciencia e Innovación* SAF2010-17272 (to LRD) and PI10/00455 (to BP). The authors also acknowledge the support received from *Centro de Investigación Biomédica en Red de Enfermedades Raras (CIBERER)* and the institutional grant from the *Fundación Ramón Areces* to the Centro de Biología Molecular Severo Ochoa.

References

Alves S, Mangas M, Prata MJ et al (2006) Molecular characterization of Portuguese patients with mucopolysaccharidosis type II shows evidence that the IDS gene is prone to splicing mutations. J Inherit Metab Dis 29:743–754

Baralle D, Baralle M (2005) Splicing in action: assessing disease causing sequence changes. J Med Genet 42:737–748

Boichard A, Venet L, Naas T et al (2008) Two silent substitutions in the PDHA1 gene cause exon 5 skipping by disruption of a putative exonic splicing enhancer. Mol Genet Metab 93:323–330

Brasil S, Viecelli HM, Meili D et al (2011) Pseudoexon exclusion by antisense therapy in 6-pyruvoyl-tetrahydropterin synthase deficiency. Hum Mutat 32:1019–1027

Cartegni L, Chew SL, Krainer AR (2002) Listening to silence and understanding nonsense: exonic mutations that affect splicing. Nat Rev Genet 3:285–298

Chao HK, Hsiao KJ, Su TS (2001) A silent mutation induces exon skipping in the phenylalanine hydroxylase gene in phenylketonuria. Hum Genet 108:14–19

Clavero S, Perez B, Rincon A et al (2004) Qualitative and quantitative analysis of the effect of splicing mutations in propionic acidemia underlying non-severe phenotypes. Hum Genet 115:239–247

Cooper TA, Wan L, Dreyfuss G (2009) RNA and disease. Cell 136:777–793

Dhir A, Buratti E (2010) Alternative splicing: role of pseudoexons in human disease and potential therapeutic strategies. FEBS J 277:841–855

Dobrowolski SF, Andersen HS, Doktor TK et al (2010) The phenylalanine hydroxylase c.30 C > G synonymous variation (p.G10G) creates a common exonic splicing silencer. Mol Genet Metab 100:316–323

Dominski Z, Kole R (1993) Restoration of correct splicing in thalassemic pre-mRNA by antisense oligonucleotides. Proc Natl Acad Sci USA 90:8673–8677

Dreumont N, Poudrier JA, Bergeron A et al (2001) A missense mutation (Q279R) in the fumarylacetoacetate hydrolase gene, responsible for hereditary tyrosinemia, acts as a splicing mutation. BMC Genet 2:9

Du L, Kayali R, Bertoni C et al (2011) Arginine-rich cell-penetrating peptide dramatically enhances AMO-mediated ATM aberrant splicing correction and enables delivery to brain and cerebellum. Hum Mol Genet 20:3151–3160

Dufes C, Uchegbu IF, Schatzlein AG (2005) Dendrimers in gene delivery. Adv Drug Deliv Rev 57:2177–2202

Eng L, Coutinho G, Nahas S et al (2004) Nonclassical splicing mutations in the coding and noncoding regions of the ATM Gene: maximum entropy estimates of splice junction strengths. Hum Mutat 23:67–76

Fukao T, Horikawa R, Naiki Y et al (2010) A novel mutation (c.951 C > T) in an exonic splicing enhancer results in exon 10 skipping in the human mitochondrial acetoacetyl-CoA thiolase gene. Mol Genet Metab 100:339–344

Gabut M, Mine M, Marsac C et al (2005) The SR protein SC35 is responsible for aberrant splicing of the E1alpha pyruvate dehydrogenase mRNA in a case of mental retardation with lactic acidosis. Mol Cell Biol 25:3286–3294

Ginjaar IB, Kneppers AL, v d Meulen JD et al (2000) Dystrophin nonsense mutation induces different levels of exon 29 skipping and leads to variable phenotypes within one BMD family. Eur J Hum Genet 8:793–796

Guterstam P, Lindgren M, Johansson H et al (2008) Splice-switching efficiency and specificity for oligonucleotides with locked nucleic acid monomers. Biochem J 412:307–313

Hammond SM, Wood MJ (2011) Genetic therapies for RNA mis-splicing diseases. Trends Genet 27:196–205

Hartmann L, Theiss S, Niederacher D et al (2008) Diagnostics of pathogenic splicing mutations: does bioinformatics cover all bases? Front Biosci 13:3252–3272

Hartmann L, Neveling K, Borkens S et al (2010) Correct mRNA processing at a mutant TT splice donor in FANCC ameliorates the clinical phenotype in patients and is enhanced by delivery of suppressor U1 snRNAs. Am J Hum Genet 87:480–493

Hefferon TW, Groman JD, Yurk CE et al (2004) A variable dinucleotide repeat in the CFTR gene contributes to phenotype diversity by forming RNA secondary structures that alter splicing. Proc Natl Acad Sci USA 101:3504–3509

Homolova K, Zavadakova P, Doktor TK et al (2010) The deep intronic c.903 + 469T > C mutation in the MTRR gene creates an SF2/ASF binding exonic splicing enhancer, which leads to pseudoexon activation and causes the cblE type of homocystinuria. Hum Mutat 31:437–444

Ikeda H, Matsubara Y, Mikami H et al (1997) Molecular analysis of dihydropteridine reductase deficiency: identification of two novel mutations in Japanese patients. Hum Genet 100:637–642

Jimenez JL, Clemente MI, Weber ND et al (2010) Carbosilane dendrimers to transfect human astrocytes with small interfering RNA targeting human immunodeficiency virus. BioDrugs 24:331–343

Jorge-Finnigan A, Aguado C, Sanchez-Alcudia R et al (2010) Functional and structural analysis of five mutations identified in methylmalonic aciduria cblB type. Hum Mutat 31:1033–1042

Kinali M, Arechavala-Gomeza V, Feng L et al (2009) Local restoration of dystrophin expression with the morpholino oligomer AVI-4658 in Duchenne muscular dystrophy: a single-blind, placebo-controlled, dose-escalation, proof-of-concept study. Lancet Neurol 8:918–928

Krawczak M, Thomas NS, Hundrieser B et al (2007) Single base-pair substitutions in exon-intron junctions of human genes: nature, distribution, and consequences for mRNA splicing. Hum Mutat 28:150–158

Lindner M, Gramer G, Haege G et al (2011) Efficacy and outcome of expanded newborn screening for metabolic diseases—report of 10 years from South-West Germany. Orphanet J Rare Dis 6:44

Long JC, Caceres JF (2009) The SR protein family of splicing factors: master regulators of gene expression. Biochem J 417:15–27

Lopez-Bigas N, Audit B, Ouzounis C et al (2005) Are splicing mutations the most frequent cause of hereditary disease? FEBS Lett 579:1900–1903

Martinez MA, Rincon A, Desviat LR et al (2005) Genetic analysis of three genes causing isolated methylmalonic acidemia: identification of 21 novel allelic variants. Mol Genet Metab 84:317–325

Meili D, Kralovicova J, Zagalak J et al (2009) Disease-causing mutations improving the branch site and polypyrimidine tract: pseudoexon activation of LINE-2 and antisense Alu lacking the poly(T)-tail. Hum Mutat 30:823–831

Merinero B, Perez B, Perez-Cerda C et al (2008) Methylmalonic acidaemia: examination of genotype and biochemical data in 32 patients belonging to mut, cblA or cblB complementation group. J Inherit Metab Dis 31:55–66

Miyazaki T, Ohura T, Kobayashi M et al (2001) Fatal propionic acidemia in mice lacking propionyl-CoA carboxylase and its rescue by postnatal, liver-specific supplementation via a transgene. J Biol Chem 276:35995–35999

Nielsen KB, Sorensen S, Cartegni L et al (2007) Seemingly neutral polymorphic variants may confer immunity to splicing-inactivating mutations: a synonymous SNP in exon 5 of MCAD protects from deleterious mutations in a flanking exonic splicing enhancer. Am J Hum Genet 80:416–432

Nissim-Rafinia M, Chiba-Falek O, Sharon G et al (2000) Cellular and viral splicing factors can modify the splicing pattern of CFTR transcripts carrying splicing mutations. Hum Mol Genet 9:1771–1778

Nissim-Rafinia M, Aviram M, Randell SH et al (2004) Restoration of the cystic fibrosis transmembrane conductance regulator function by splicing modulation. EMBO Rep 5:1071–1077

Ogino W, Takeshima Y, Nishiyama A et al (2007) Mutation analysis of the ornithine transcarbamylase (OTC) gene in five Japanese OTC deficiency patients revealed two known and three novel mutations including a deep intronic mutation. Kobe J Med Sci 53:229–240

Perez B, Rincon A, Jorge-Finnigan A et al (2009) Pseudoexon exclusion by antisense therapy in methylmalonic aciduria (MMAuria). Hum Mutat 30:1676–1682

Perez B, Rodriguez-Pascau L, Vilageliu L et al (2010) Present and future of antisense therapy for splicing modulation in inherited metabolic disease. J Inherit Metab Dis 33:397–403

Pinotti M, Rizzotto L, Balestra D et al (2008) U1-snRNA-mediated rescue of mRNA processing in severe factor VII deficiency. Blood 111:2681–2684

Pomponio RJ, Reynolds TR, Mandel H et al (1997) Profound biotinidase deficiency caused by a point mutation that creates a downstream cryptic 3′ splice acceptor site within an exon of the human biotinidase gene. Hum Mol Genet 6:739–745

Rincon A, Aguado C, Desviat LR et al (2007) Propionic and methylmalonic acidemia: antisense therapeutics for intronic variations causing aberrantly spliced messenger RNA. Am J Hum Genet 81:1262–1270

Rodriguez-Martin T, Garcia-Blanco MA, Mansfield SG et al (2005) Reprogramming of tau alternative splicing by spliceosome-mediated RNA trans-splicing: implications for tauopathies. Proc Natl Acad Sci USA 102:15659–15664

Rodriguez-Pascau L, Coll MJ, Vilageliu L et al (2009) Antisense oligonucleotide treatment for a pseudoexon-generating mutation in the NPC1 gene causing Niemann-Pick type C disease. Hum Mutat 30:E993–E1001

Salomons GS, Bok LA, Struys EA et al (2007) An intriguing "silent" mutation and a founder effect in antiquitin (ALDH7A1). Ann Neurol 62:414–418

Sanchez-Alcudia R, Perez B, Perez-Cerda C et al (2009) Rescue of a splicing defect in propionic acidemia with modified U1snRNA. Mol Genet Metab 98:123–124

Schollen E, Keldermans L, Foulquier F et al (2007) Characterization of two unusual truncating PMM2 mutations in two CDG-Ia patients. Mol Genet Metab 90:408–413

Stucki M, Suormala T, Fowler B et al (2009) Cryptic exon activation by disruption of exon splice enhancer: novel mechanism causing 3-methylcrotonyl-CoA carboxylase deficiency. J Biol Chem 284:28953–28957

Susani L, Pangrazio A, Sobacchi C et al (2004) TCIRG1-dependent recessive osteopetrosis: mutation analysis, functional identification of the splicing defects, and in vitro rescue by U1 snRNA. Hum Mutat 24:225–235

Tanner G, Glaus E, Barthelmes D et al (2009) Therapeutic strategy to rescue mutation-induced exon skipping in rhodopsin by adaptation of U1 snRNA. Hum Mutat 30:255–263

Tazi J, Durand S, Jeanteur P (2005) The spliceosome: a novel multi-faceted target for therapy. Trends Biochem Sci 30:469–478

Tsuruta M, Mitsubuchi H, Mardy S et al (1998) Molecular basis of intermittent maple syrup urine disease: novel mutations in the E2 gene of the branched-chain alpha-keto acid dehydrogenase complex. J Hum Genet 43:91–100 van Deutekom JC, Janson AA, Ginjaar IB et al (2007) Local dystrophin restoration with antisense oligonucleotide PRO051. N Engl J Med 357:2677–2686

Vega AI, Perez-Cerda C, Desviat LR et al (2009) Functional analysis of three splicing mutations identified in the PMM2 gene: toward a new therapy for congenital disorder of glycosylation type Ia. Hum Mutat 30:795–803

Vervoort R, Gitzelmann R, Lissens W et al (1998) A mutation (IVS8 + 0.6kbdelTC) creating a new donor splice site activates a cryptic exon in an Alu-element in intron 8 of the human beta-glucuronidase gene. Hum Genet 103:686–693

Vezain M, Gerard B, Drunat S et al (2011) A leaky splicing mutation affecting SMN1 exon 7 inclusion explains an unexpected mild case of spinal muscular atrophy. Hum Mutat 32:989–994

Vorechovsky I (2010) Transposable elements in disease-associated cryptic exons. Hum Genet 127:135–154

Ward AJ, Cooper TA (2009) The pathobiology of splicing. J Pathol 220:152–163

RNA-Cleaving DNA Enzymes and Their Potential Therapeutic Applications as Antibacterial and Antiviral Agents

P.I. Pradeepkumar and Claudia Höbartner

Contents

Abstract DNA catalysts are synthetic single-stranded DNA molecules that have been identified by in vitro selection from random sequence DNA pools. The most prominent representatives of DNA catalysts (also known as DNA enzymes, deoxyribozymes, or DNAzymes) catalyze the site-specific cleavage of RNA substrates. Two distinct groups of RNA-cleaving DNA enzymes are the 10-23 and 8-17 enzymes. A typical RNA-cleaving DNA enzyme consists of a catalytic core and two short binding arms which form Watson–Crick base pairs with the

---

P.I. Pradeepkumar (✉)
Department of Chemistry, Indian Institute of Technology Bombay, Powai, Mumbai 400076, India
e-mail: pradeep@chem.iitb.ac.in

C. Höbartner (✉)
Max Planck Institute for Biophysical Chemistry, Research Group Nucleic Acid Chemistry, Am Fassberg 11, 37077 Göttingen, Germany
e-mail: claudia.hoebartner@mpibpc.mpg.de

V.A. Erdmann and J. Barciszewski (eds.), *From Nucleic Acids Sequences to Molecular Medicine*, RNA Technologies, DOI 10.1007/978-3-642-27426-8_15,

RNA targets. RNA cleavage is usually achieved with the assistance of metal ions such as Mg^{2+}, Ca^{2+}, Mn^{2+}, Pb^{2+}, or Zn^{2+}, but several chemically modified DNA enzymes can cleave RNA in the absence of divalent metal ions. A number of studies have shown the use of 10-23 DNA enzymes for modest downregulation of therapeutically relevant RNA targets in cultured cells and in whole mammals. Here we focus on mechanistic aspects of RNA-cleaving DNA enzymes and their potential to silence therapeutically appealing viral and bacterial gene targets. We also discuss delivery options and challenges involved in DNA enzyme-based therapeutic strategies.

Keywords Antibacterials • Antivirals • Catalyst • Chemical modification • Delivery • Deoxyribozyme • DNA enzyme • Gene silencing • In vitro selection • Mutation • RNA cleavage

1 Introduction

The conventional biological role of both DNA and RNA as storage place and carrier of genetic information has taken a new leap in the early 1980s by the seminal discovery of RNA catalysts, called ribozymes, by Cech and Altman (Kruger et al. 1982; Guerrier-Takada et al. 1983). Ever since these reports, there has been a surge of new discoveries in this research area, and as a result, new functional roles of nucleic acids have emerged, including their involvement in gene regulation (Breaker 2004). Many ribozymes have been found in biological systems catalyzing important biochemical reactions, which include splicing of mRNAs (Doudna and Cech 2002) and peptide bond formation at the core of the ribosome (Leung et al. 2011). Moreover, a number of ribozymes with a wide variety of functions have been discovered in the laboratory using an elegant test tube evolution method called in vitro selection (Joyce 2007). These discoveries also shook the notion that not only proteins but also nucleic acids can function as catalysts with high precision and impressive rate enhancements. The ability of RNA to catalyze a particular reaction is aided by its propensity to fold and adopt defined 3D structures and thereby facilitate metal ion or cofactor-assisted catalysis (Lilley 2011). After the discovery of ribozymes, the possibility of the existence of DNA catalysts based on single-stranded DNAs was speculated the mid-1980s (Kruger et al. 1982), but no such catalysts have yet been found in any domains of life. In 1994, the first DNA catalyst was discovered in the laboratory using in vitro selection. Single-stranded DNA was shown to site-specifically cleave an RNA phosphodiester bond with the assistance of Pb^{2+} ions (Breaker and Joyce 1994).

The unique ability of single-stranded DNA to catalyze RNA cleavage prompted researchers to capitalize on the catalytic potential of DNA. Since DNA, unlike RNA, lacks a 2′-OH group and bears a 5-methyl group in uridine (i.e., thymidine), a question that remained to be answered was if DNA has the ability to achieve the catalytic diversity exhibited by ribozymes. Later in vitro selection experiments have shown ample evidence that DNA catalysis is not limited by the chemistry of the reaction but rather by the techniques involved to identify active DNA sequences. As a consequence, in addition to a series of RNA-cleaving enzymes, new DNA catalysts have been identified in a number of laboratories. These include DNA enzymes, which are capable of performing oxidative and hydrolytic cleavage of DNA (Carmi et al. 1998, Chandra et al. 2009), DNA phosphorylation (Li and Breaker 1999), DNA adenylation (Li et al. 2000b), phosphoramidate cleavage (Burmeister et al. 1997), peroxidation (Travascio et al. 1998), thymine dimer cleavage (Chinnapen and Sen 2004), porphyrin metalation (Li and Sen 1996), formation of phosphorothioester linkage (Levy and Ellington 2001), linear RNA ligation (formation both 2′–5′ and 3′–5′ linkages) (Purtha et al. 2005; Kost et al. 2008), synthesis of 2′,5′-branched nucleic acids (Coppins and Silverman 2004; Lee et al. 2011), lariat RNA synthesis (Wang and Silverman 2005), DNA depurination (Sheppard et al. 2000; Höbartner et al. 2007), formation of nucleopeptide linkages (Pradeepkumar et al. 2008) and nucleopeptides (Wong et al. 2011), and C–C bond formation (Diels–Alder reaction) (Chandra and Silverman 2008). Moreover, the search for new catalytic activities of DNA is currently actively being pursued in many laboratories. To gather detailed information on DNA catalysis and various reactions catalyzed by DNA, the reader is directed to excellent recent reviews published elsewhere (Silverman 2008, 2009, 2010; Schlosser and Li 2009a; Willner et al. 2008).

The focus of this chapter is the most widely studied class of deoxyribozymes, namely, RNA-cleaving DNA enzymes. These enzymes find broad applications not only in the conventional laboratory settings to manipulate large RNAs (Pyle et al. 2000; Silverman and Baum 2009) but also in the field of chemical sensing (Willner et al. 2008) and in the therapeutic arena, where they have shown the potential to downregulate gene expression (Baum and Silverman 2008). RNA-cleaving DNA enzymes offer specific advantages over RNA-cleaving ribozymes due to their chemical stability, easy accessibility by chemical solid-phase synthesis, and cost-effectiveness. The two major classes of RNA-cleaving enzymes are 10-23 and 8-17, both of which have been widely used in the above-mentioned applications. Both of these enzymes have a short catalytic core of 13–15 nt and recognize the target RNA sequence with the binding arms through Watson–Crick base pairing. Both enzymes require mostly divalent metal cations such as Mg^{2+}, Mn^{2+}, Zn^{2+}, Ca^{2+}, etc., for achieving efficient catalysis. Under optimized conditions, the 10-23 DNA enzyme exhibits k_{cat}/K_m of 10^9, which rivals the catalytic efficiency of ribozymes such as the hammerhead ribozyme and protein enzymes such as RNase A (Santoro and Joyce 1998). This fact underscores the robustness of DNA catalysis and provides opportunities to harness the catalytic activities of DNA for practical applications. Herein we review in vitro selection techniques used to identify RNA-cleaving DNA enzymes and summarize kinetic and mechanistic insights on the RNA cleavage reaction revealed by various biochemical and biophysical studies. Further, we give an overview on the potential use of RNA-cleaving DNA enzymes in the therapeutic area, where the focus is devoted to bacterial and viral targets.

2 In Vitro Selection Strategies

RNA-cleaving DNA enzymes are selected or evolved from a random library of DNA sequences (DNA pool) through repeated steps of selection and amplification (Joyce 2004). The subtle difference between in vitro *selection* and in vitro *evolution* techniques should be noted. If the process involves the identification of enzymes from the initial library of sequences without introducing genetic diversity between individual rounds of selective amplification, the procedure is called in vitro *selection.* However, if additional diversity is introduced, for example, using error-prone PCR during the selection process or a reselection is carried out using partially randomized libraries (based on the sequence information emerged from the initial selection), the process is called in vitro *evolution.* The random library is usually constructed by solid-phase DNA synthesis using a mixture of all four phosphoramidites in appropriate ratios based on their coupling efficiency (T > G > C > A). The DNA pool is generally designed with two well-defined primer binding sites, which enable PCR amplification after the selection step, and a random region, which usually varies from 20 to 80 nt in length. The length of the random region defines the sequence space (i.e., all possible sequences for a random region of length $n = 4^n$). For example, a random region of 40 nt has a sequence space of $4^{40} = 10^{24}$ sequences, which is impossible to sample experimentally due to practical limitations. To illustrate this, a 60-nt-long DNA pool containing 40 random nucleotides would require 1.6 mols of DNA, i.e., 30 kg, in order to achieve complete coverage of all 10^{24} possible sequences. In practice, typically 0.2–2 nanomoles of DNA pool is used to initiate the selection. This accounts to 10^{14}–10^{15} different molecules, which offer a sampling of only 10^{-9}–10^{-10} (e.g., $10^{14}/10^{24} = 10^{-10}$) of the sequence space. Increasing the length of the random region therefore leads to a decrease in the coverage of sequence space [for a detailed discussion, see Silverman (2008)]. A number of selection experiments have shown that a very large random region is not required to identify efficient DNA catalysts with a wide variety of functions (Silverman 2008; Schlosser and Li 2009a).

Two strategies for the in vitro selection process employed to identify RNA-cleaving DNA enzymes are shown in Fig. 1, termed "bead-based" and "gel-based" strategies (Santoro and Joyce 1997; Cruz et al. 2004). One of the crucial steps in this process is the physical separation of active DNA molecules from inactive ones. In the bead-based approach, this is usually accomplished by the help of a biotin tag which assists in retaining the inactive sequences on a streptavidin column after cleavage of the target RNA in the selection step (Fig. 1a). More stringent separations can be achieved in the gel-based approach, where the active enzymes are separated based on their different mobility in denaturing PAGE due to the size difference imparted during the cleavage step (Fig. 1b). These methods demand that the substrate and the DNA pool are covalently joined together at the beginning of the selection experiment. Therefore, in the first step of a selection cycle, the DNA pool is attached to an RNA substrate (or to a DNA containing an embedded ribonucleotide) either by performing PCR using a biotinylated primer or by enzymatic ligation (using T4 DNA or T4 RNA ligase). This is followed by the selection

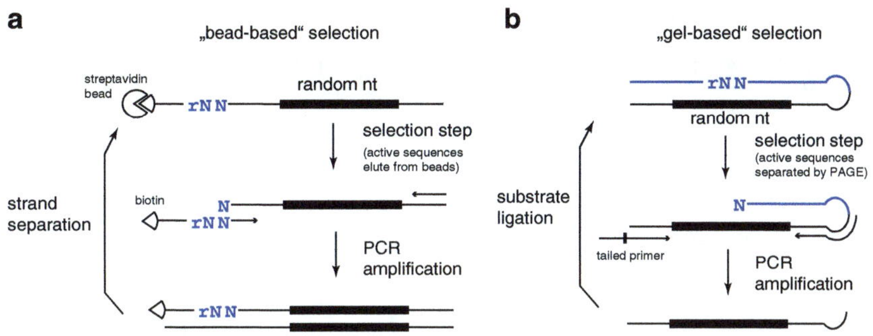

Fig. 1 In vitro selection strategies for RNA-cleaving deoxyribozymes. (**a**) "Bead-based" selection via biotin–streptavidin interaction. (**b**) "Gel-based" selection via PAGE separation. See text for details

step in which parameters such as buffer, pH, temperature, metal ion concentration, and incubation time are varied to set up different selection experiments in order to identify the most efficient DNA enzymes. The selection step is followed by fishing out the active DNA enzyme sequences using the methods described above. Then, the selected sequences are PCR amplified using appropriate primers. If the separation is carried out using the bead-based approach, the biotinylated primer will enable the capture of PCR product on the streptavidin column. A denaturation step is then performed to remove the undesired complementary strand of the double-stranded PCR product. If the separation involves a gel-based strategy, a reverse primer with an embedded ribonucleotide or a nonamplifiable linker is used. This provides the opportunity to generate a size difference in the amplified products required for PAGE-based separation of single-stranded DNA. This is achieved either by cleaving the ribonucleotide linkage or by utilizing the inherent difference in the size of the two strands. The selection and amplification steps are continued until no further increase in cleavage activity is observed. This typically requires between 5 and 15 round of in vitro selection. The individual sequences are then identified by the typical cloning and automated sequencing procedures. The cleavage activity of individual deoxyribozymes is analyzed by using solid-phase synthesized oligonucleotides that bind to the target RNA in a bimolecular format. This is usually referred to as "in trans" assay format, compared to the monomolecular "in cis" architecture used during selection.

3 RNA-Cleaving Deoxyribozymes: A Brief Overview

The chemical reaction catalyzed by all known RNA-cleaving deoxyribozymes is the formation of a 2′,3′-cyclic phosphate and a free 5′-hydroxyl terminus, originating from the nucleophilic attack of a 2′-hydroxyl group of the substrate

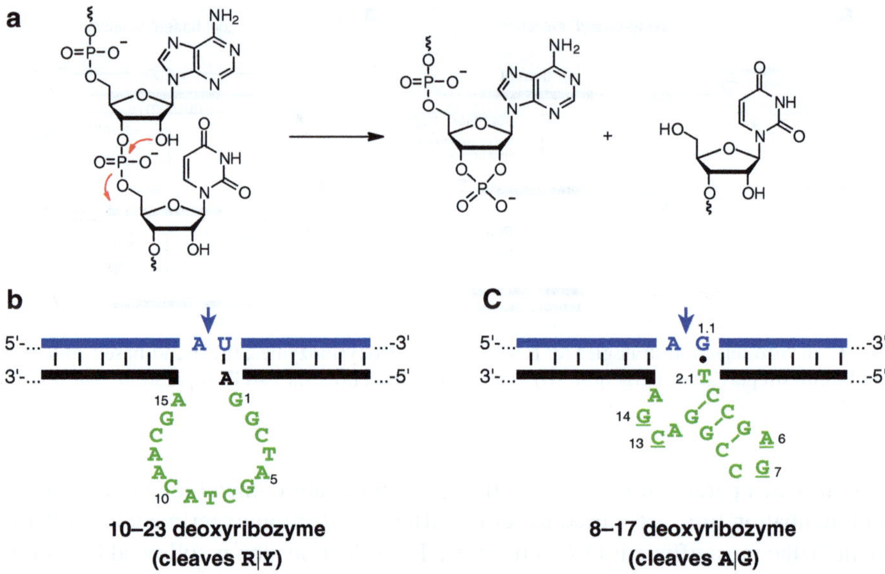

Fig. 2 (**a**) DNA-catalyzed RNA cleavage. Nucleophilic attack of the 2′-OH group onto the adjacent phosphodiester bond generates a 2′,3′-cyclic phosphate and a free 5′-OH terminus. (**b**) 10-23 deoxyribozyme. (**c**) 8-17 deoxyribozyme; conserved nucleotides most important for cleavage activity are *underlined*. The preferred cleavage sites are indicated by *blue arrows*

oligonucleotide on the adjacent phosphodiester bond (Fig. 2a). In principle, such an intramolecular cleavage reaction can occur randomly upon incubation of RNA under alkaline conditions. Deoxyribozymes, however, specifically cleave RNA only at predefined internal positions under mild conditions. The site specificity is programmed via the Watson–Crick base-paired binding arms, which generally leave 1 nt at the target site unpaired. The considerable rate enhancements and extremely high selectivities for particular sites render deoxyribozymes attractive tools for various applications.

The first reported deoxyribozyme catalyzed the cleavage of the phosphodiester bond on the 3′-side of single embedded ribonucleotide in a DNA substrate by using Pb^{2+} as a divalent metal ion cofactor (Breaker and Joyce 1994). At pH 7.0 and 23°C, in the presence of 1 mM $Pb(OAc)_2$, the DNA catalyst was able to cut its substrate in a bimolecular fashion and showed impressive Michaelis–Menten behavior, with k_{cat} of 1 min^{-1} and K_m of 2 μM, yielding ca. 30 turnovers per hour at a catalyst concentration of 0.01 μM and a substrate concentration of 1 μM. The rate enhancement over the uncatalyzed background reaction (10^{-4} min^{-1}) under comparable conditions amounts to 10^5. Although this initially reported deoxyribozyme was not capable of cleaving an all-RNA substrate, this prototype DNA catalyst spurred the development of successor deoxyribozymes that now offer highly useful properties.

Table 1 Representative examples of RNA phosphodiester-cleaving deoxyribozymes (consisting of unmodified DNA)

Deoxyribozyme name	Preferred cleavage site[a]	Random pool (nt)	Required cofactor (s)	max k_{obs} (min^{-1})	Rate enhancement	References
Pb^{2+}-depend.	a|G	50	Pb^{2+}	1	~10^5	Breaker and Joyce (1994)
E6	n.k.	40	Mg^{2+}	0.04	10^5	Breaker and Joyce (1995)
Na8	n.k.	40	Na^+	0.007	10^8	Geyer and Sen (1997)
HD2	a|G	40	L-histidine	0.2	10^6	Roth and Breaker (1998)
Mg5 (8-17 variant)	a|G	74	Ca^{2+}	0.1		Faulhammer and Famulok (1996)
17E (8-17 variant)	n|G	50	Zn^{2+}	1.35	10^8	Li et al. (2000a)
39E	a|G.	50	UO_2^{2+}	1.0		Liu et al. (2007)
10-23	r|y	50	Mg^{2+}	10	10^5	Santoro and Joyce (1997)
8-17	a|g	50	Mg^{2+}	10		Santoro and Joyce (1997)
Bipartite	a|a	40	Mn^{2+}	1.6		Feldman and Sen (2001)
8-17-like[b]	n|N	various	Mg^{2+}+Mn^{2+}	0.001 to 10		Cruz et al. (2004)

[a]Lowercase letters denote ribonucleotides, uppercase letters are used for DNA nucleotides; bold font: deoxyribozymes that cleave all-RNA substrates
[b]8-17 variants are known for all 16 possible dinucleotide junctions
n.k. = not known (not reported)

The most prominent and most widely studied classes of RNA-cleaving deoxyribozymes are the 10-23 and 8-17 DNA catalysts (Fig. 2b, c), which were first reported in 1997 (Santoro and Joyce 1997). A large variety of other RNA-cleaving DNA catalysts has been isolated by in vitro selection under various conditions, including specific metal ions or pH conditions. For example, DNA enzymes have been characterized that cleave (fluorophore-labeled) chimeric RNA–DNA substrates under acidic conditions, ranging from pH 3 to pH 6 (Liu et al. 2003). Others have been selected for the specific detection of uranium as uranyl ions UO_2^{2+} (Liu et al. 2007). An interesting RNA-cleaving deoxyribozyme has been generated from a binary nucleotide alphabet, which contains only guanosine and cytosine nucleotides, and still exhibits considerable Mn^{2+}-dependent cleavage activity (Schlosser and Li 2009b). An overview on selected examples of RNA-cleaving DNA enzymes and respective references can be found in Table 1. Excellent reviews are available that describe the details of in vitro selection and characterization of these DNA catalysts (Silverman 2005; Baum and Silverman 2008; Schlosser and Li 2010). In the following section, we focus on 10-23 and 8-17 DNA enzymes since most preclinical studies on DNA catalysts as therapeutic agents have been conducted with 10-23 deoxyribozymes, and 8-17 variants have found wide applications as components of biosensors.

4 On Mechanism and Folding of 10-23 and 8-17 Deoxyribozymes

The deoxyribozymes 10-23 and 8-17 were identified by in vitro selection with the goal to find general RNA-cleaving DNA catalysts that function under conditions resembling physiological environment (Santoro and Joyce 1997). The names are rather arbitrary and reflect the selection round and the clone number of the identified DNA. Both DNA enzymes turned out as useful catalysts to cleave all-RNA substrates with practically useful rate constants, ranging from 0.1 min^{-1} under simulated physiological conditions (2 mM Mg^{2+}) and up to 10 min^{-1} in the presence of high divalent metal ion concentrations (100 mM Mg^{2+}). Although the DNA enzymes have been biochemically characterized (Santoro and Joyce 1998), the exact catalytic mechanism of RNA cleavage is not yet understood in detail. Currently, no structure of a deoxyribozyme in an active conformation is known. A small number of reports describe attempts to solve the X-ray crystal structure of the10-23 DNA in complex with an uncleavable substrate (Nowakowski et al. 1999). However, the crystals revealed a dimeric DNA structure in an unproductive conformation that resembled a Holliday junction structure. A preliminary report on NMR structural studies confirmed secondary structure formation between substrate and deoxyribozyme, but no follow-up studies on more detailed NMR structural data have yet been reported (Choi et al. 2000). Below we summarize the reported results from biochemical and biophysical investigations on 10-23 and give a short overview on selected data on DNA folding studies of 8-17 deoxyribozymes [a recent comprehensive review on 8-17 DNA enzymes can be found in (Schlosser and Li 2010)].

4.1 *The 10-23 Deoxyribozyme*

The 10-23 deoxyribozyme generally cleaves an RNA substrate between a purine and a pyrimidine nucleotide (see Fig. 2b) and is active in the presence of Mg^{2+}, Mn^{2+}, or Ca^{2+} as divalent metal ion cofactor (Santoro and Joyce 1998). A linear increase in k_{obs} with a slope close to 1 was observed between pH 6.5 and 8.5, indicating that a single deprotonation event is part of the rate-determining step of RNA cleavage. Most likely, this reflects deprotonation of the 2′-OH group at the cleavage site, but it is not known if any of the core nucleotides could act as a general base to assists the deprotonation. In order to identify functionally relevant nucleotides in the catalytic core of the 10-23 DNA catalyst, a comprehensive mutation analysis was performed (Zaborowska et al. 2002). All 15 nt in the catalytic core were individually replaced by each of the other three natural nucleotides. The activity of the 45 mutants was evaluated by single time-point measurements of the cleaved RNA fraction under single turnover conditions (37°C, 20 min, in 50 mM Tris pH 7.5, 10 mM Mg^{2+}). Selected mutants were used for full kinetic measurements of k_{obs}. A number of

Table 2 Effects on catalytic activity of mutations and modifications in the catalytic core of the 10-23 deoxyribozyme

nt	G	G	C	T	A	G	C	T	A	C	A	A	C	G	A
Position	1	2	3	4	5	6	7	8	9	10	11	12	13	14	15
Mutation															
G			×	×	×		~	√	~	√	√	~	×		√
A	×	×	√	×		×	√	√		√			~	×	
C	×	×		×	√	×		√	×		~	√		×	×
T	×	×	~		×	×	√		√	√	√	√	×	×	~
Modification															
Inosine (I)	~	~	×	×	×	√	×	√	~	√	√	√	×	×	√
Purine (P)					√										
2AP						×								×	
c^7A					×						×	~			√
c^7G						×									
Deletion	×	×	×	×	×	×	√	√	~	×	~	×	×	×	×
Abasic	√	~	√	×	×	×	×	√	×	×	~	~	×	×	×
C3 linker	√	~	~	×			×		×	×	×	×			×
2′OMe	×	√	×	×	×	×	√	√	~	×	√	~	×	√	√
$2'(R)CH_3$				×				×							
$2'(S)CH_3$				×				√							
LNA				√				×							
PS	+	~	√	~	×	√	√	+	~	~	~	~	~	√	√

√ = modification is tolerated (no effect on catalytic activity), + = modification enhances catalysis, ~ = slightly reduced catalytic activity (less than 50% change in k_{obs} or final yield), × = modification is detrimental [no activity or trace amount of cleavage product (<10% of wt)]; empty fields indicate that this modification was not investigated on this position. For abbreviations and nucleotide structures see Fig. 3 and Fig. 4

nucleotides were found to be important (summarized in Table 2): nucleotides G1, G2, T4, G6, and G14 could not be replaced by any other nucleotide without complete loss of catalytic activity. In contrast, nucleotides at positions 7–12 could mostly be exchanged to any other of the natural nucleotides. In addition, the exocyclic functional groups of certain nucleotides were investigated. For example, inosine (Fig. 3) was comprehensively tested at all positions, and interesting results were found at positions C3 and C13. Based on the striking difference of adenosine and inosine substitution (adenosines were partially tolerated in place of cytidines, but inosines were detrimental), it was speculated that an exocyclic amino group is essential (originating from C or A). For position G6, the O^6-carbonyl group was found to be the most important feature because activity was abolished with adenine, but inosine did not have any effect. At position G14, both the N^2-amino and the O^6-carbonyl groups were essential, as indicated by the intolerance for inosine and 2-aminopurine. Interestingly, at position A5, cytidine showed some activity, but inosine was detrimental, again pointing toward an important amino group. However, when A5 was changed to purine, no change in activity was observed, indicating that another functional group must be responsible for the observed effects. It was speculated that a likely candidate is the lone pair of the N^1 ring nitrogen of adenine, which is also

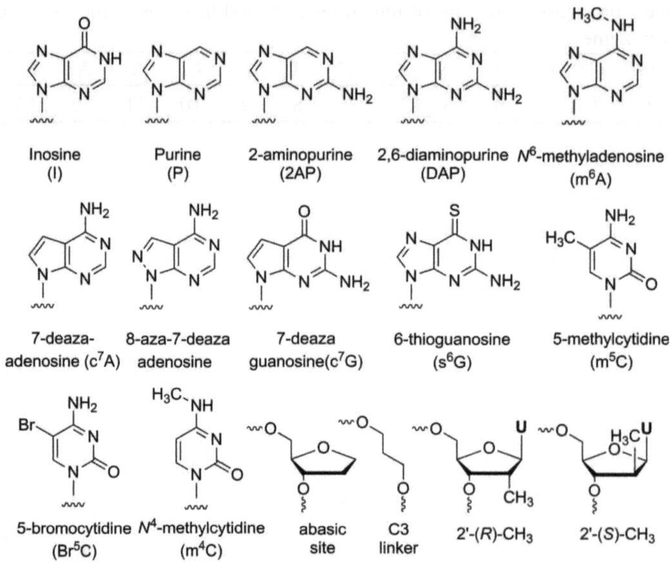

Fig. 3 Nucleoside modifications used to study mechanistic aspects of 10-23 and 8-17 deoxyribozymes (see also Fig. 4 for more examples)

found in purine, and analogously at N^3 of cytosine, but not in inosine and guanosine, which carry a protonated NH group as part of the lactam functionality. However, in a recent study, it was shown that c^7A (Fig. 3) at position A5 was detrimental, clearly pointing toward the importance of the N^7 atom (He et al. 2011). Interestingly, the activity of 10-23 was rescued by placing a nitrogen atom at position 8 of the purine ring, i.e., 8-aza-7-deaza-deoxyadenosine at position A5 supported phosphodiester cleavage. A similar effect was observed at position A11.

The mutation data revealed considerable flexibility at positions 7–12 but also suggested conserved nucleotides that are likely involved in formation of the catalytic structure. These results should prove helpful for interpretation of high-resolution deoxyribozyme structures; unfortunately, such structural data remained elusive up until today. Nevertheless, the mutation analysis highlighted essential nucleotides, and the results also give information on which positions might be amenable to chemical modifications or functionalization to protect the core against endonucleolytic attack or to modulate enzymatic activity.

The traditional enzymology approach, as described above, provides highly valuable data, but the process is rather laborious and time-consuming. Recently, combinatorial methods have been developed for the simultaneous assessment of all nucleotides in the catalytic core of DNA enzymes. Combinatorial mutation interference analysis (Wachowius et al. 2010) and nucleotide analog interference mapping of DNA (Wachowius and Höbartner 2011) were also applied to 10-23 and efficiently revealed essential residues that are largely comparable to the results from individual mutants (C. Höbartner, F. Wachowius, unpublished results).

To further narrow down the features of the catalyst, another study addressed the deletion of individual nucleotides from the 10-23 core (Zaborowska et al. 2005). It was found that C7 and T8 can be separately removed without severely affecting cleavage activity, but other tested deletions were detrimental. The double deletion of C7 and T8 resulted in about twofold reduced k_{obs} compared to the unmodified DNA, but the shortened deoxyribozyme was not able to cleave long RNA substrates. The effect of deleting individual nucleobases (instead of entire nucleotides) was systematically examined by replacing every nucleotide with an abasic site analog (a tetrahydrofuran derivative, Fig. 3 (Wang et al. 2010). Additionally, every nucleotide was replaced by an acyclic C3 linker (Fig. 3), which keeps the same number of atoms in the backbone chain, but confers much increased conformational flexibility to the DNA chain. The results from this study corroborated earlier findings on nucleotide requirements in the catalytic core and, in addition, defined the importance of structural preorganization of the DNA backbone by certain 2′-deoxyribose units. It is interesting to point out that substitution of T8 with an abasic site or the acyclic linker resulted in slightly increased deoxyribozyme activity compared to the parent DNA, whereas deletion of the entire nucleotide resulted in about twofold reduced k_{obs}. However, it should be noted that the activity tests (Wang et al. 2010) were done with a substrate containing a single embedded ribonucleotide, as opposed to the full RNA substrate used in the deletion and mutation analyses (Zaborowska et al. 2005).

In an earlier study, a 10-23 mutant was identified, which was 4 nt shorter than the parent DNA, and which showed high activity in the presence of Ca^{2+}, but had very low activity in the presence of Mg^{2+} (Sugimoto et al. 1999). A functional group mutation analysis in this variant identified the importance of carbonyl groups at G1 and G2 and showed that A7 and A8 could not be replaced by 2,6-diaminopurine (Fig. 3). These detrimental modifications affected the K_m, but did not influence k_{cat}; only one mutation, namely, G10-2AP, dramatically affected the k_{cat}. Moreover, 5-propinyl-C increased the catalytic activity by decreasing K_m (ca. fivefold) and increasing k_{cat} (almost twofold).

The tolerance of the 10-23 core nucleotides to ribose modifications was also investigated. In a comprehensive study reported in 2003, all loop nucleotides were systematically replaced by their 2′-OMe counterparts (Schubert et al. 2003). The activity assay revealed that the 2′-OMe group was detrimental at 9 out of 15 loop positions. When the remaining 6 nt simultaneously contained the 2′-OMe modification, the k_{obs} dropped by only twofold. A detailed discussion on the effect of 2′-OMe modifications can be found in Sect. 5.2. More recently, the influence of 2′-C-methyl modified uridine nucleotides (Fig. 3) at positions 4 and 8 of the 10-23 core sequence was investigated (Robaldo et al. 2010). The absolute configuration of the methyl group had a pronounced effect on the catalytic activity. This observation is most likely related to the different preferred sugar pucker conformations of 2′-(R)- and 2′-(S)-CH_3 modified nucleosides. Interestingly, incorporation of the 2′-(S)-CH_3 derivative at position 8 had little effect, whereas the 2′-(R)-CH_3 derivative reduced the cleavage rate by tenfold, at different Mg^{2+} concentrations between 0.5 and 10 mM. This is interesting with respect to the above-mentioned finding that

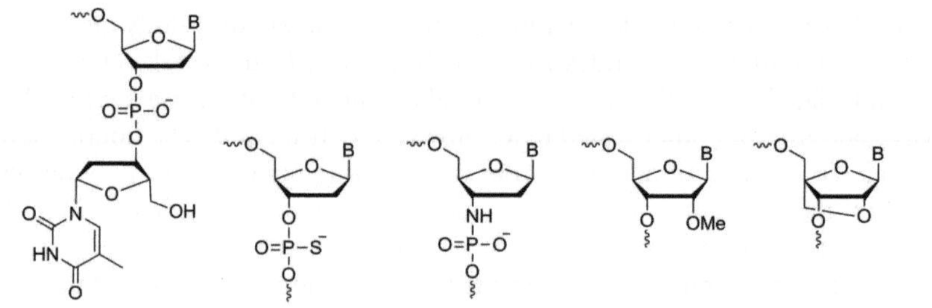

Fig. 4 Chemical modifications that have been introduced in RNA-cleaving DNA enzymes to impart drug-like properties

nucleotide T8 can be deleted without significantly affecting DNA activity. Introducing the modification at position 4 resulted in mostly inactive DNA variants at low Mg^{2+} concentrations. This 2′-C-methyl ribose modification is also appealing with respect to increased nuclease resistance of the modified DNA. Along the same lines, it should be mentioned that LNA nucleotides (Fig. 4) were tested at positions T4 and T8 in the catalytic core of the deoxyribozyme (Robaldo et al. 2010). Although LNA-T was more or less tolerated at T4, the T8-modified and double-modified derivatives exhibited drastically reduced catalytic activity, again indicating sensitivity to local structural changes induced by the conformationally restricted nucleotides. In contrast, LNA and 2′-OMe modifications were shown to be beneficial when introduced into the binding arms of 10-23 deoxyribozymes (Schubert et al. 2003; Donini et al. 2007). This will be discussed in more detail in the context of potential therapeutic applications in Sect. 5.

Besides nucleobases, the phosphate backbone can also be involved in the catalytic mechanism or in coordination of metal ions. Therefore, it is important to investigate functional and essential phosphate groups in the deoxyribozyme core region. In this respect, one nonbridging phosphate oxygen was systematically replaced by sulfur at every internucleotide phosphodiester bond of 10-23 (Nawrot et al. 2007). The cleavage kinetics were investigated in the presence of Mg^{2+} and of Mn^{2+}, in order to analyze the "thio" and "rescue" effects, following the hard and soft acids and bases (HSAB) principle. In the presence of Mg^{2+}, a 16-fold reduction in k_{obs} was observed when the phosphorothioate (PS) modification was placed at A5. This derivative also showed the largest rescue effect with Mn^{2+} ($k_{obs}Mn/k_{obs}Mg = 176$). By a combination with 6-thioguanosine substitution at position G6, it was suggested that a likely metal ion binding site is formed involving the carbonyl group of G6 and the phosphodiester between T4 and A5. Additional indication for a potential metal ion interaction was observed for the phosphodiester between T8 and A9. At nucleotides 3, 6, 7, and 14 of the loop, the sulfur substitution was well tolerated, and at nucleotides 1 and 8, the phosphorothioate even enhanced catalytic activity. It should also be noted as an advantage that the phosphorothioate modification confers increased nuclease stability to the DNA. Improved activity combined with increased stability provides interesting opportunities for biological or therapeutic applications (see also Sect. 5). In addition, the effect of phosphorothioate substitution at the scissile phosphate in the RNA substrate was investigated. Differential effects were found for the activity of 10-23 upon interaction with R_P- and S_P-phosphorothioate-containing substrates (Nawrot et al. 2008; He et al. 2002). While the S_P thio analog was efficiently cleaved with only slightly reduced rates compared to the unmodified RNA substrate, the R_P-configured phosphorothioate was resistant to cleavage by 10-23, even at high Mg^{2+} concentrations (up to 100 mM). Interestingly, this severe thio effect could not be rescued by the more thiophilic Mn^{2+} ions. These results suggest that the nonbridging oxygen atoms at the scissile phosphodiester bond are most likely not directly coordinated to metal ions, thus indicating that the catalytic mechanism does not involve metal-mediated charge neutralization at the nonbridging oxygen atoms. Rather, it is likely that metal ions directly coordinate to the 2′ oxygen and 5′ oxygen, facilitating deprotonation of the attacking 2′-OH group and neutralization of the developing charge at the 5′-phosphate terminus. Furthermore, it was hypothesized that the proton of the 2′-OH group migrates to the pro-R_P oxygen of the phosphate. This is consistent with the inactive nature of the R_P(S) substrate since the lower pK_A of the phosphorothioate compared to an unmodified phosphate would render H-migration less plausible. Further studies are expected in the near future, taking advantage of recently reported methods to produce 3′- and 5′-phosphorothiolate substrates, that could be helpful for disentangling the metal ion contributions to DNA-catalyzed RNA cleavage, similar to strategies employed for the investigation of ribozyme catalysis (Frederiksen and Piccirilli 2009).

Although there are some striking similarities between DNA-catalyzed RNA cleavage and hammerhead ribozyme activity, some experimental results do not fit into the picture of superimposable mechanisms. For example, the behavior of hammerhead and 10-23 against trivalent lanthanide ions is different. While the hammerhead ribozyme gives a bell-shaped curve of activity versus increasing La^{3+}-concentrations (in a background of constant Mg^{2+} concentration, i.e., activation at low [La^{3+}] and inhibition at high [La^{3+}]), the 10-23 DNA enzyme shows gradual inhibition (investigated between 0.5 and 10 μM $LaCl_3$) (He et al. 2002). Lanthanide probes were also used for studying the RNA cleavage by a Pb^{2+}-dependent deoxyribozyme (Geyer and Sen 1998), which was found to tolerate lanthanide ions for phosphodiester cleavage. A single lanthanide ion was most likely directly interacting with the 2′-OH group at the cleavage site. Different lanthanides were tested, showing that the smaller lanthanides (Tb^{3+}, Tm^{3+}, Lu^{3+}) more efficiently supported DNA-catalyzed phosphodiester cleavage compared to larger ones (Eu^{3+}, Nd^{3+}, Ce^{3+}, La^{3+}). These results were corroborated by Tb^{3+} luminescence experiments involving unmodified and phosphorothioate-containing substrates, which showed that the 2′-OH group was required for efficient Tb^{3+} binding, but the phosphate backbone at the cleavage site was likely not involved in direct metal ion contacts (Geyer and Sen 1998). Tb^{3+}-luminescence spectroscopy was used in a similar way to investigate metal ion binding in 8-17 deoxyribozymes [see Sect. 4.2; (Kim et al. 2008)].

Metal ions not only play a mechanistic role for DNA catalysis but probably possess an equally important structural role. By neutralizing negative charges, coordinating to phosphate groups and nucleobases, metal ions can induce proper folding of the DNA into a catalytically active conformation. Structural rearrangements upon metal ion binding were studied in the 10-23 DNA enzyme by fluorescence resonance energy transfer between two fluorophores, rhodamine and fluorescein, attached at the 5′- and 3′-termini of the binding arms (Cieslak et al. 2003). Fluorescence anisotropy experiments at different Mg^{2+} concentrations were also reported. The results showed that structural transitions in the DNA enzyme leading to a decreased end-to-end distance of the fluorophores occur at rather low Mg^{2+} concentrations, at which the deoxyribozyme does not yet have full activity (i.e., the folding midpoint observed by FRET was at ~2 mM Mg^{2+}). Since higher Mg^{2+} resulted in improved cleavage activity, but no further reduction of the FRET distances was observed, it was suggested that structural transitions at low Mg^{2+} concentrations are not sufficient for optimal catalytic activity. However, it is also conceivable that the single combination of fluorophores at terminal positions might not be ideal to observe subtle conformational changes required for activity. In this respect, more detailed studies (including single-molecule fluorescence experiments) were carried out for 8-17 deoxyribozymes, as discussed in Sect. 4.2 (Kim et al. 2007b; Lee et al. 2007, 2010).

As mentioned above, no 3D structure of 10-23 is available in an active conformation that would facilitate elucidation of structure–function relationships. Although a large amount of biochemical and biophysical data are available from the above-described reports, the relationship between tertiary structure and function remains unknown. Recently, a preliminary model for the tertiary structure of the 10-23 deoxyribozyme was investigated in silico using coarse-grained Brownian dynamics simulations (Kenward and Dorfman 2009). It was found that the DNA enzyme bends its substrate away from the cleavage point and a transition state was suggested with a buckled catalytic core. After the cleavage, a substantial extension of the end-to-end distance of the binding arms was observed. It is expected that this model provides the starting point for more detailed molecular dynamics simulations of deoxyribozymes and will be helpful to draw connections from experimental data to computational results.

4.2 *The 8-17 Family of Deoxyribozymes*

The "original" 8-17 deoxyribozyme (Fig. 2c) has a preference for cleavage of RNA substrates between adenosine and guanosine (Santoro and Joyce 1997). The catalytic core consists of 13–15 nt that present a typical 3-bp stem and a 3-nt loop as the characteristic features. Interestingly, this motif has been isolated more than five times from independent in vitro selection experiments in different laboratories (Li et al. 2000a; Peracchi 2000; Cruz et al. 2004; Schlosser et al. 2008a). The recurrence of this small motif makes it an interesting subject for fundamental studies on DNA catalysis but also has a practical benefit as a small, cheap, stable, and versatile endoribonuclease. Interestingly, out of the 15 nt organized into a defined three-way junction secondary structure, only 4 nt are absolutely conserved in more than 150 in vitro selected 8-17-like deoxyribozymes (Cruz et al. 2004). These are confined to A6, G7, C13, and G14 (underlined in Fig. 2c). Using deoxyribozyme analogs with modified nucleotides, the functional groups of these 4 nt essential for catalytic activity have been identified (Peracchi et al. 2005). It was found that A6 and G7 are sensitive to sterically demanding substitutions since m^6A (Fig. 3) had a larger impact on catalysis than purine at position A6, and s^6G supported catalysis even less efficiently than 2-aminopurine at position G7, which slowed down the reaction by about 1,000-fold. These results suggested the presence of a functionally important hydrogen-bonding network involving various H-bonding donors and acceptors of these two purine nucleotides. The most critical functional group of G14 was found to be the exocyclic amino group. Interestingly, s^6G (Fig. 3) was well tolerated under standard reaction conditions, but activity was strongly inhibited by submillimolar concentrations of thiophilic Cd^{2+} ions. This suggested that a strong direct metal coordination at this functional group does not support catalysis. On the contrary, the nucleotide at position 12 was found to modulate the metal ion cofactor requirements. While 8-17 with thymidine at position 12 is active with Mg^{2+}, Mn^{2+}, and Ca^{2+} to similar extents, a DNA enzyme with inosine at position 12 used Ca^{2+} fivefold more efficiently than the parent enzyme (activity with Mg^{2+} and Mn^{2+} was even slightly inhibited by the inosine mutation). Although such interesting observations cannot directly prove the participation of individual functional groups in the catalytic mechanism, they provide useful insights into potential tertiary interactions and may suggest structure–function relationships that can help model building and engineering of modified deoxyribozymes.

Sequence variations beyond these four conserved nucleotides were found to influence the efficiency of substrate cleavage and the nucleotide requirements and tolerance at the dinucleotide cleavage site (Schlosser et al. 2008b). Systematic characterization identified a subset of optimal 8-17 variants that can cleave all 16 possible dinucleotide junctions in chimeric RNA–DNA substrates (containing a single ribonucleotide). However, only a subset of dinucleotide junctions is efficiently cleaved in all-RNA substrates (Silverman and Baum 2009; Schlosser and Li 2010).

A variety of other biophysical studies have been conducted to collect insights into how individual residues contribute to folding and catalysis of 8-17 deoxyribozymes using different metal ion cofactors, including Mg^{2+}, Zn^{2+}, and Pb^{2+}. Metal-dependent folding of the 8-17 DNA enzyme was studied by bulk and single-molecule FRET experiments. The bulk study used a fluorophore-labeled DNA enzyme in complex with an inactivated substrate, in which the ribonucleotide at the cleavage site was replaced by a deoxyribonucleotide (Kim et al. 2007a). The DNA enzyme was found to fold into a pyramidal structure in the presence of Mg^{2+} and Zn^{2+}, but with Pb^{2+}, no global folding transition was observed (although Pb^{2+} is the most effective cofactor). A recent three-color bulk FRET study tried to correlate the metal-dependent folding with the identity of the dinucleotide junction in the substrate (Lam and Li 2010). It was found that DNA enzyme–substrate complexes with purine–purine junctions folded at lower Mn^{2+} concentrations than those containing pyrimidine–pyrimidine junctions, which is consistent with cleavage activity studies. Nevertheless, bulk studies with inactive substrates may be suboptimal to support strong conclusions on folding transitions. To exclude adverse effects of the inactive substrates which might change metal-binding properties, single-molecule FRET experiments were carried out which allow individual active deoxyribozymes to be studied with cleavable substrates. Moreover, smFRET can detect and distinguish less populated conformations that may be important for activity but that cannot be resolved in bulk FRET measurements. Using a total internal reflection (TIR) fluorescence microscope to observe fluorescently labeled, surface-bound deoxyribozymes, two distinct folding pathways that depend on the identity of the metal ion cofactor were confirmed (Kim et al. 2007b). In the presence of Mg^{2+} and Zn^{2+}, a structural change occurs before the cleavage step. In contrast, Pb^{2+} is accommodated in a prearranged conformation of the deoxyribozyme which directly leads to RNA cleavage. In another report, three-color alternating-laser excitation (3c-ALEX) was used to study the branched structure of 8-17 at the single-molecule level in solution (Lee et al. 2007). Again, different metal ion conditions resulted in different folding pathways of the DNA. Moreover, the important roles of A6 and G7 were confirmed, and it was shown that they are functionally involved in productive folding into an active conformation. Single-molecule experiments using ALEX-FRET were also demonstrated to provide accurate measurements of the cleavage rate of 8-17 deoxyribozymes and allowed the determination of apparent metal ion binding constants (Lee et al. 2010).

Metal ion binding in the 8-17 deoxyribozyme was also investigated by Tb^{3+} luminescence spectroscopy (Kim et al. 2008). These studies rely on the sensitized luminescence of Tb^{3+} by energy transfer from nucleic acids which can be detected at 543 nm. Tb^{3+} was found to reversibly inhibit 8-17 activity by competing for binding sites that are productively occupied by effective metal ion cofactors. Again, different effects were observed for Zn^{2+} and Pb^{2+}, and it was concluded that Pb^{2+} may bind to a different site than Zn^{2+} or Mg^{2+}. The differences in local and global folding induced by various metal ions were also studied by contact photo-cross-linking experiments using photoactivatable nucleotide analogs (Liu and Sen 2008, 2010). In particular, s^6G (Fig. 3) was incorporated into the substrate at the cleavage site and separately at various positions in the catalytic core. The thionucleotide in the enzyme substrate complex was excited by irradiation with UV light, and the cross-linked positions were identified by a combination of piperidine cleavage and primer extension experiments. These cross-linking studies resulted in a model for the active structure of the deoxyribozyme, in which the functionally essential nucleotides of the catalytic core, in particular G7 and G14, enclose the cleavage site in a sandwich-like topology (Liu and Sen 2008). Moreover, the results corroborated distinct properties of Pb^{2+}-activated deoxyribozymes, with respect to local conformation changes as opposed to global folding of the DNA enzyme–substrate complex (Liu and Sen 2010).

5 Potential Therapeutic Applications of RNA-Cleaving Deoxyribozymes Against Bacteria and Viruses

Ever since the discovery of RNA-cleaving 10-23 and 8-17 DNA enzymes, attempts have been embarked to exploit them as therapeutic agents by utilizing their ability to cleave disease-related mRNAs. Compared to other oligonucleotide-based gene silencing strategies, DNA enzymes offer several advantages (Bhindi et al. 2007). First of all, DNA enzymes do not require any cellular enzymes like RNase H (as for antisense strategies) or argonaute 2 (as for RNAi) to exhibit their cleavage activity on target RNAs. Compared to ribozymes, DNA enzymes are more target specific, and due to their versatility in target cleavage, identification of potential targets is relatively easy. Moreover, DNA is cheap and easy to synthesize compared to siRNAs or ribozymes. However, similar to other antisense technologies, the therapeutic utility of DNA enzyme is hampered by the present lack of efficient delivery agents. Therefore, although a plethora of reports is available in the literature on the use of DNA enzymes in in vitro settings and in animal models (Dass et al. 2008; Benson et al. 2008; Peracchi 2004), no clinical trials have so far been reported for an RNA-cleaving DNA enzyme. The therapeutic targets of RNA-cleaving DNA enzymes can be broadly classified into cancer, cardiovascular, allergy-related, bacterial, and viral targets. A detailed discussion on the DNA enzyme-mediated attenuation of all these targets is beyond the scope of this chapter. For a comprehensive outlook, readers are directed to the recent reviews available in the literature (Bhindi et al. 2007; Dass et al. 2008; Benson et al. 2008; Peracchi 2004; Achenbach et al. 2004). Here we focus on DNA enzyme-based strategies against bacteria and viruses. The therapeutic targets covered in this section and the observed effects of deoxyribozyme action are summarized in Table 3.

5.1 DNA Enzymes Targeting Bacteria

The utility of 10-23 DNA enzymes to cleave the mRNA of isocitrate lyase (*ICL*) of *Mycobacterium tuberculosis* has been investigated (Li et al. 2005). ICL is an enzyme which plays an important role in the metabolism of *M. tuberculosis* in the latent state of infection in macrophages. Therefore, *ICL* is one of the emerging targets for antituberculosis therapy. In vitro cleavage assays on a 1,871-nt-long ICL transcript in the presence of 15 mM Mg^{2+} (50 mM Tris–HCl, pH 7.0 at 37°C) revealed modest target attenuation by four out of five DNA enzymes tested. Under the same conditions, an analogous enzyme bearing mutations in the catalytic core was completely inactive. The DNA enzymes were tested to evaluate their ability to suppress the expression of *ICL* and thereby reduce the survival of *M. tuberculosis* in THP-1 cells. Since DNA enzymes were not capable of penetrating the cell wall of *M. tuberculosis*, suboptimal concentration of the antituberculosis drug isoniazid (INH) was used to soften the cell wall. After 4 and 7 days of treatment with DNA

Table 3 Targets and effects of antibacterial and antiviral RNA-cleaving deoxyribozymes

Disease/target	Strain/cell/animal/RNA	DNA enzyme/modification	Efficiency	References
Bacterial targets				
M. tuberculosis (*ICL*)	THP-1 cells	10-23	55–60% Growth reduction	Li et al. (2005)
M. tuberculosis (*TACO*)	RAW264.7 cells	10-23 (expr. vector)	78% Reduction in mRNA and 75% reduction in protein levels	Li et al. (2010)
ftsZ	DH5αpro cells	10-23 (expr. vector)	55–65% Growth reduction	Tan et al. (2004)
β-Lactamase	TEM-1 and TEM-3 strains	10-23 (and expr. vector)	58–63% Growth reduction (electroporated); 46–71% (vector based)	Chen et al. (2004a, b)
Methicillin-resistant *Staphylococcus aureus* (*blaR1*)	WHO-2 strain	10-23 (PS-mod.)	62% Reduction in colonies and 70% reduction in mRNA	Hou et al. (2007)
Viral targets				
HIV				
CCR5 and CXCR-4	HeLa and U87 cells	10-23	Inhibition of cell membrane fusion (based on luciferase reporter assay)	Goila and Banerjea (1998), Basu et al. (2000)
V3 loop	U87.CD4.CXCR4 and U87.CD4.CCR5 cells	10-23	68–81% of inhibition of viral replication	Zhang et al. (1999)
GAG	COS-1 cells	10-23	10-fold reduction of mRNA levels ; 95% reduction of HIV-1 gene expression (luciferase reporter assay)	Sriram and Banerjea (2000), Dash and Banerjea (2004), Sood et al. (2007a)
TAT and *TAT/REV*	THP-1 cells	10-23	8–10-fold reduction of mRNA levels	Unwalla and Banerjea (2001a, b), Sood et al. (2007b), Unwalla et al. (2006)
TAR	HeLa cells	10-23	80% Reduction of HIV indicator p24 antigen	Chakraborti and Banerjea (2003)
VPR-B, *VPR-C*	293-T cells	10-23	3.5- to 5-fold reduction of mRNA levels	Bano et al. (2007)
DIS and PBS at the 5′-UTR	HEK 293-T cells	10-23 (LNA mod.)	100% Reduction of viral antigen CA-p24	Jakobsen et al. (2007), Vester et al. (2006)

Hepatitis B and C				
5′-NCR	Huh7 cells	10-23 (PS-mod.)	45–67% Reduction of RNA (luciferase reporter assay)	Oketani et al. (1999)
5′-UTR	293rTA and Huh7 cells	10-23 (PS-mod.)	24–48% Reduction of viral RNA	Trepanier et al. (2006)
HCV 1b RNA	Huh7 cells	10-23 (2′-OMe mod.)	61% Reduction of viral RNA and 59% reduction of viral protein	Trepanier et al. (2008)
5′-UTR of IRES	Huh 7 cells	10-23	81% Inhibition of gene translation (luciferase reporter assay)	Roy et al. (2008)
X gene	HepG2 cells	10-23	4–6-fold reduction in X protein	Goila and Banerjea (2001)
X gene	AD293 cells	10-23	60% Reduction of protein, 80% reduction of RNA (GFP reporter assay)	Hou et al. (2006)
Other viral targets				
E6/E7 (HPV)	SiHa cells	10-23 (2′-OMe /LNA mod.)	50% Reduction of RNA	Reyes-Gutierrez and Alvarez-Salas (2009)
E6/E7 (HPV)	830-nt E6/E7 transcript (*in vitro*)	8-17 (LNA mod.)	$k_{obs} = 0.025–0.044\ min^{-1}$	Donini et al. (2007)
F gene (RSV)	Hep-2 cells and RSV-infected mice	10-23 (PS mod.)	Cells: 100% reduction of RNA; 87% reduction of protein; mice: 54% reduction of RNA	Xie et al. (2006), Yang et al. (2007)
3′-NCR (JEV)	J744E cells and JEV-infected mice	10-23 (PS/2′-OMe mod.)	Cells: 108-fold reductions of viral replicationmice: 99.99% inhibition of viral replication	Appaiahgari and Vrati (2007)
TIS (influenza virus A)	MDCK cells	10-23 (N3′-P5′ phosphoramidate linkages at ends)	99% Inhibition of viral replication	Takahashi et al. (2004)
5′-UTR (SARS-CoV)	Vero E6 cells	10-23	55–85% Reduction of protein levels (GFP reporter assay)	Wu et al. (2007)
5′-UTR (HRV-14)	835-nt HRV-14 transcript (in vitro)	10-23 (2′-OMe or LNA mod.)	$k_{obs} = 0.5\ min^{-1}$ and $v_{init} = 4.7\ nM\ min^{-1}$	Schubert et al. (2003)
5′-UTR (CAV-21)	797-nt CAV-21 transcript (in vitro)	10-23 (2′-OMe or LNA mod.)	100% RNA cleavage	Schubert et al. (2004)

expr. expression, *mod.* modification

enzyme at 5 μM concentrations, the bacterial growth (CFU levels) was reduced to 55–60% in comparison with the untreated cells. Since this initial work, no further studies have been reported on *ICL*-targeting DNA enzymes.

Tryptophan–aspartate-containing coat protein (*TACO*) plays a crucial role in the survival of *M. tuberculosis* in macrophages. Recently, researches have reported an efficient way to target this protein by generating 10-23 DNA enzymes in macrophages with the help of a single-stranded DNA expression vector based on Moloney mouse leukemia virus reverse transcriptase (*MMLV-RT*) (Li et al. 2010). To enhance the expression efficiency, the *MMLV-RT* gene and the oligonucleotide sequences for generating the DNA enzymes were cloned into a single plasmid under the control of two separate promoters. The vector was designed to generate the 10-23 DNA enzyme, which targets the coding region of *TACO*-mRNA. Control plasmids expressing mutant enzymes were also constructed. The efficiency of the active DNA containing vector was verified in murine macrophage cell line RAW264.7. Semiquantitative RT-PCR assays revealed a 78% reduction of *TACO*-mRNA levels at 48 h posttransfection. The corresponding reduction of the *TACO*-protein revealed by Western blot was 75%. Mutant plasmids did not induce any reduction of the mRNA or the protein levels. Though these results are encouraging in terms of DNA enzyme delivery, the need to coexpress viral transcriptase might lead to adverse effects. Moreover, the usefulness of this approach to reduce *M. tuberculosis* infection has not yet been reported.

A tetracycline-regulated expression vector based on *MMLV*-RT has been constructed to generate a 10-23 DNA enzyme which is able to target the *ftsZ* gene of bacteria (Tan et al. 2004). The *ftsZ* gene is essential for bacterial cell division and also for cell viability. Therefore, targeting the *ftsZ* mRNA is an attractive way to halt bacterial cell proliferation. When tested in DH5αpro cells, the single-stranded DNA expression vector was not only able to downregulate the *ftsZ* gene expression (revealed by Western blot) but was also able to repress bacterial cell proliferation. This was evident in the reduction (55–65%) of CFU levels and also in the formation of long filamentous cells. The inhibitory effect was found to depend on time as well as the amount of tetracycline used. Unfortunately, in these studies, a control vector expressing a mutant 10-23 enzyme was not investigated, and therefore, the contribution from antisense effect cannot be ruled out.

Targeting genes responsible for antibiotic resistance by RNA-cleaving DNA enzymes is an attractive way to sensitize drug-resistant bacteria to antibiotics. One of the important targets in this class is *β-lactamase* mRNA since *β-lactamase* is responsible for bacterial resistance against β-lactam antibiotics. Two 10-23 DNA enzymes targeting the start codon as well as the coding region of *β-lactamase* mRNA were designed (Chen et al. 2004a). A di-DNA enzyme was also designed by coupling two single DNA enzymes in tandem. In vitro cleavage kinetics using a short RNA substrate in the presence of 10 mM Mg^{2+} (50 mM Tris–HCl, pH 7.6 at 37°C) revealed that the DNA enzyme targeting the coding region had higher catalytic efficiency [$k_{cat}/K_m = 9.1 \times 10^7$ (mol/L)$^{-1}$ min^{-1}] than the enzyme targeting the start codon site [$k_{cat}/K_m = 6.4 \times 10^7$ (mol/L)$^{-1}$ min^{-1}] and the di-DNA enzyme [$k_{cat}/K_m = 3.1 \times 10^7$ (mol/L)$^{-1}$ min^{-1}]. This augmented catalytic efficiency of the enzyme against the coding region has been attributed to the increase in the target binding affinity (low K_m) due to high GC content in the binding arms. All the DNA enzymes were able to cleave a full-length *β-lactamase* RNA transcript (885 nt) with varying efficiencies (40–70% at 2 h). For in vivo studies, DNA enzymes have been delivered into ampicillin-resistant bacteria (TEM-1 and TEM-3) by electroporation. After 9 h, the growth inhibition assays and the *β-lactamase* activity measurement revealed that the di-DNA enzyme was most efficient in attenuating the drug-resistant mRNA. This is reflected in 58% growth inhibition for TEM-1 bacteria and 63% growth inhibition for TEM-3 bacteria. DNA enzymes bearing mutations in the catalytic core or in the binding arm did not show any activity. This underlines the specificity of the DNA enzyme-mediated approach to silence the *β-lactamase* mRNA. The enhanced in vivo activity of the di-DNA enzyme is intriguing, considering the complex structure of the DNA enzyme due to the presence of multiple catalytic loops and binding arms. To address the limitations of exogenous delivery of DNA enzymes to bacteria, researchers constructed a single-stranded DNA vector to express 10-23 DNA enzymes in vivo (Chen et al. 2004b). Using an M13mp18 phage vector, two circular DNA enzymes which target the same start codon and coding region of *β-lactamase* mRNA were generated. The catalytic efficiency of circular DNA enzymes was reduced to 25–30% of the corresponding linear DNA enzymes. The circular enzymes were shown to replicate inside the TEM-1 and TEM-3 bacteria and also were shown to inhibit bacterial growth. The enzyme targeted to the translation initiation site leads to 46% growth inhibition (53% reduction in *β-lactamase* activity), while the enzyme targeted to the start codon region leads to 71% growth inhibition (67% reduction in β-*lactamase* activity). Though these results are encouraging, the viability of these approaches in macrophages has not been reported so far. The efficient expression of DNA enzyme in such systems will be crucial to the potential success of these approaches in clinical settings.

An RNA-cleaving 10-23 DNA enzyme has also been used to target the *blaR1* gene of methicillin-resistant *Staphylococcus aureus* (MRSA) (Hou et al. 2007). The *blaR1* is involved in the sensor/transducer pathway of *blaR1-blaI-blaZ*, and therefore, targeting *blaR1* can lead to the silencing of *blaZ* gene, which codes for *β-lactamase*. A 10-23 enzyme modified with phosphorothioate linkages (PS) (see Fig. 4) in its binding arms has been introduced into MRSA strain WHO-2 by electroporation. A scrambled oligonucleotide sequence was used as a negative control in these experiments. The DNA enzyme at a concentration of 15 mg L^{-1} was able to reduce the number of WHO-2 colonies to 62% compared to the cells treated with the scrambled control. This leads to partial restoration of antibiotic susceptibility of the bacteria by reducing the minimum inhibitory concentration (MIC) of oxacillin from 32 μg mL^{-1} to 8 μg mL^{-1}. The poor retention of the DNA enzyme in the culture medium was responsible for the insufficient restoration of antibiotic susceptibility. RT-PCR revealed that the attenuation in the *blaR1* and *blaZ* mRNA was 70% and 59%, respectively.

5.2 DNA Enzymes Targeting Viral Diseases

5.2.1 Anti-HIV Strategies

In order to challenge HIV-1 entry and fusion into cells, DNA enzymes have been designed and targeted against the coding region of the chemokine receptor gene CCR5 (HIV coreceptor) (Goila and Banerjea 1998). In vitro cleavage assays on a 1,376-nt-long CCR5 transcript using a 10-23 DNA enzyme having 7-nt binding arms revealed more than 70% cleavage of the target under 2–10 mM Mg^{2+} concentrations. The membrane fusion activity assays using CCR5 reporter in HeLa cells demonstrated the ability of the DNA enzyme to reduce the membrane fusion activity significantly. Control experiments using unrelated DNA enzymes and a 14-mer oligonucleotide lacking the catalytic loop failed to induce any activity. This underlines the specificity of the DNA enzymes against CCR5. In a follow-up study, researchers designed DNA enzymes which target another important chemokine receptor CXCR-4 (Basu et al. 2000). In vitro cleavage assays in the presence of 10 mM Mg^{2+} on a 500-nt transcript revealed excellent cleavage by the DNA enzyme. A di-DNA enzyme in which 10-23 enzymes targeting CXCR-4 and CCR5 are arranged in tandem was able to cleave an engineered RNA target of CXCR-4 and CCR5 with extremely high efficiency. The ability of the CXCR-4 mono- and the di-DNA enzyme to alter the membrane fusion activity has been validated in U87 cells using luciferase-based reporter assays.

10-23 DNA enzymes were employed to target the V3 loop region of the HIV-1 genome (Zhang et al. 1999). Nine different DNA enzymes were tested for cleavage of a 414-nt RNA transcript under single turnover conditions. Four of these enzymes showed >50% cleavage. One enzyme was selected for further evaluation. This DNA enzyme exhibited modest catalytic efficiency [$k_{cat}/K_m = 3 \times 10^3$ (mol/L)$^{-1}$ min^{-1}], which was twofold higher than a ribozyme, targeting the same site. When tested to assess the inhibitory ability toward the replication of HIV-1(NL432) strains in U87.CD4.CXCR4 cells, the enzyme exhibited up to 81% inhibition of viral replication on day 2 of the treatment. However, the inhibitory ability of the enzyme against HIV-1(SF162) strain in U87.CD4.CCR5 cells was up to only 68%. The mutant enzyme, which lacks the catalytic ability, also exhibited 30% inhibition of SF162 replication. This may partially be attributed to the antisense effects of the DNA enzyme, and these results underscore the importance using appropriate control oligonucleotides while studying the DNA enzyme-mediated target cleavage in cellular systems. Single cycle infection assays using a luciferase-based reporter system showed that the DNA enzyme could inhibit the incoming HIV-1 virus to an appreciable extent (up to 74% inhibition after 36 h).

Since the target identification for DNA enzyme-mediated cleavage is challenging, 10-23 DNA enzymes with randomized binding arms (7 nt) were used to find the most efficient enzyme which could cleave the HIV-1 *GAG* RNA (Sriram and Banerjea 2000). Two combinatorial libraries were employed for this target selection approach. One was used to target any AU cleavage site in all AUG codons, and the other one was designed to target any purine–pyrimidine junction in the HIV 1-*GAG* RNA. After an extensive target screening (10 mM $MgCl_2$ in 50 mM Tris–HCl, pH 8 at 37°C) followed by primer extension analyses, the substrate binding sequences of two enzymes were identified which were able to cleave an AU junction (Dz-1836) and a GC junction (Dz-1810), respectively. However, in vitro cleavage assays using 220-nt transcripts revealed that the activity of both enzymes was suboptimal (<10% cleavage, 10 mM Mg^{2+}). Even under 50 mM Mg^{2+} conditions, the cleavage ability of the Dz1810 was only less than 20%. The cleavage assays using 740-nt-long *GAG* RNA transcript in the presence of 10 mM Mg^{2+} showed improved efficacy for Dz1810, while the Dz-1836 showed complete target cleavage. The RT-PCR analysis 48 h after transfection of COS-1 cells with *GAG* RNA expressing reporter and DNA enzymes revealed that Dz-1836 was able to suppress the *GAG* RNA levels tenfold, while the Dz-1810 enzyme was completely inactive. The anti-HIV activity of the Dz-1836 was further verified in COS-1 cells using a luciferase-based reporter assay in which ~95% reduction in the luciferase activity was observed 48 h after transfection. The specificity of inhibition was supported by testing a mutant enzyme and also an antisense oligonucleotide lacking the catalytic domain of the DNA enzyme. Both control experiments did not show any activity under these conditions. In a follow-up study, researchers investigated the cleavage ability of 10-23 enzymes targeting two different AU cleavage sites of HIV-1 *GAG* (p24) transcript which is 320 nt long (Dash and Banerjea 2004). Both enzymes exhibited different cleavage activities under varying Mg^{2+} concentrations. Though one of the enzymes did not induce any cleavage at 1 mM Mg^{2+}, the second enzyme was able to produce 80% cleavage. These varying cleavage potencies of 10-23 DNA enzymes underscore the importance of target accessibility. It has been recently demonstrated that use of 21-nt antisense oligos targeting both the upstream and downstream regions of DNA enzyme binding sites can dramatically increase the target cleavage (Sood et al. 2007a). Using this approach, a 220-nt HIV-1 Gag transcript was cleaved by 10-23 and 8-17 DNA enzymes which were found to be inefficient without the use of antisense oligos. Though the approach involving antisense oligos to augment DNA enzyme-mediated target cleavage works well in in vitro settings, the viability of such an approach in cellular systems has not been reported so far.

TAT and *REV* are regulatory proteins which play key roles in the transcription and replication of HIV-1. Therefore, targeting these genes provides an attractive way to halt the pathogenicity of HIV. The use of DNA enzymes to cleave *TAT* or *TAT/REV* RNAs has been reported (Unwalla and Banerjea 2001b). Several 10-23 enzymes with 7-nt binding arms were designed against the loop regions of *TAT* or *TAT/REV* RNA and tested for the cleavage efficiency under different Mg^{2+} concentrations. One 10-23 enzyme was able to cleave a 280-nt *TAT* RNA transcript efficiently in the presence of 1–5 mM Mg^{2+} (50 mM Tris–HCl pH 8 at 37°C). RT-PCR analysis of DNA enzyme-treated COS-1 cells revealed eight- to tenfold reduction in *TAT* expression. The mutant control failed to impart any target reduction, and this supports the specificity of the DNA enzyme-mediated approach to target important genes of HIV-1. It has also been demonstrated that many regions of *TAT* RNA which are predicted (by secondary structure analysis) to be single stranded are found to be completely inaccessible for DNA enzyme-mediated cleavage. The potency of RNA-cleaving DNA enzyme can also be increased by using them in combination with hammerhead ribozymes (Sood et al. 2007b; Unwalla et al. 2006). A DNA enzyme and a hammerhead ribozyme used in synergy achieved excellent cleavage (60%) of a 107-nt HIV-1 *TAT/REV/ENV* transcript under physiologically relevant conditions (50 mM Tris–HCl pH 7.5, 150 mM KCl, 1–2 mM $MgCl_2$ at 37°C). *TAT*-specific RT-PCR experiments of DNA enzyme- and ribozyme-treated THP-1 cells have shown 11-fold reduction in the mRNA levels compared to untreated cells. In order to enhance the delivery of 10-23 DNA enzyme to human macrophage cell line THP-1, 10 guanosine residues have been attached at the 3′-end of the DNA enzyme (Unwalla and Banerjea 2001a). A stretch of guanosine residues is known to interact with the scavenger receptor on the cell surface of macrophages (Pearson et al. 1993). In vitro cleavage assays in the presence of 10 mM Mg^{2+} of a 28-nt transcript showed that the guanosine-appended DNA enzymes were able to cleave the target, albeit at a lower efficiency than unmodified DNA enzymes. Cellular uptake assays using fluorescently labeled DNA enzymes demonstrated the ability of guanosine-tailed DNA enzymes to enter cells efficiently without the use of transfection agents. There was eightfold reduction of *TAT* RNA observed in RT-PCR assays, when guanosine-modified DNA enzymes were introduced into cells. This clearly demonstrates the power of guanosine-modified DNA enzymes to inhibit key proteins of HIV-1. Unfortunately, no further studies have been reported using this strategy in order to assess the clinical utility of guanosine tails on DNA enzymes.

The stem region of HIV-1-*TAR* RNA has been targeted using several RNA-cleaving DNA enzymes (Chakraborti and Banerjea 2003). Among the enzymes used, eight had 10-23 catalytic motifs and one contained the 8-17 motif. In vitro cleavage assays revealed that only two 10-23 variants were able to cleave an 80-nt full-length transcript at the single nucleotide bulge in the TAR RNA under 10 mM Mg^{2+} (50 mM Tris–HCl pH 8 at 37°C) conditions. The cleavage efficiency was modest, and it varied from 22% to 31%. A 10-23 mutant enzyme failed to cleave the target. One of the enzymes could induce the target cleavage even in the absence of Mg^{2+} albeit at very low efficiency. Nevertheless, this is one of the few reports of metal ion-independent cleavage by a 10-23 DNA enzyme. The reason for this unprecedented cleavage activity is not clear from the data available in the literature. The authors also demonstrated modest cleavage efficiency of the 10-23 DNA enzymes under simulated physiological conditions (2 mM Mg^{2+}). When both active 10-23 enzymes were cotransfected with the HIV1-encoded DNA into HeLa cells, >80% reduction of HIV indicator p24 antigen was observed. However, the mutant DNA enzyme also led to 40% reduction in p24 antigen which can be attributed to antisense effects. RT-PCR on RNA isolated from DNA enzyme-treated cells indicated 10- to 12-fold reduction in the HIV-1 *TAT* gene expression compared to the control cells which were not treated with DNA enzymes. The mutant DNA enzyme also induced fourfold reductions in *TAT* gene expression. This raises the question on the true catalytic efficiency of the DNA enzymes used in this study. Further studies on primary cells, human PMBCs, and cell lines infected with HIV-1 demonstrated fourfold reduction in the virus protection compared to the cells treated with mutant DNA enzymes.

Other important HIV-1 targets, which emerged recently for DNA enzyme-mediated gene silencing, are the *VpR*-B and *VpR*-C genes (Bano et al. 2007). *VpR* proteins are known to be responsible for controlling various cellular functions of HIV-1 and the G2 phase cell cycle arrest, thereby contributing to the overall pathology of the virus. In vitro cleavage assays of 315-nt-long *VpR* transcripts in the presence of 10 mM Mg^{2+} (50 mM Tris–HCl pH 8 at 37°C) identified two 10-23 enzymes with 7-nt binding arms which were capable of cleaving both *VpR-B* and *VpR-C* in a highly efficient manner. Mutated DNA enzymes could not impart any target cleavage under these conditions. RT-PCR assays of DNA enzyme-treated 293-T mammalian cells 48 h posttransfection revealed 3.5- to 5-fold reduction of *VpR* RNA. The corresponding reduction in the *VpR* protein was fourfold. It has also been demonstrated that in HeLa-CCR5 cells, the *VpR*-induced G2 cell cycle arrest can be reversed after 48 h by the action of *VpR*-specific DNA enzymes.

The ability of 10-23 DNA enzymes to cleave the 5′-UTR of the HIV-1 genome has been reported (Jakobsen et al. 2007). DNA enzymes having 9-nt binding arms were designed to cleave the accessible regions of the dimer initiation site (DIS) and the primer binding site (PBS) of the 5′-UTR. In vitro single turnover cleavage assays using a 355-nt transcript showed significant target cleavage under 10 mM Mg^{2+} (50 mM Tris–HCl pH 8 at 37°C) conditions. The cleavage efficiency of a DIS-targeting DNA enzyme was augmented to 100% by introducing two LNA (Fig. 4) modifications on each binding arm. It should be noted that the LNA modification could not enhance the cleavage efficiency of an enzyme targeting the PBS site. This underscores the fact that LNA modifications do not always enhance target accessibility and subsequent cleavage, as reported in previous studies (Vester et al. 2006). When tested for the ability to reduce the production of viral antigen CA-p24 in HEK 293-T cells, 72 h posttransfection, both LNA-modified enzymes have shown 100% target downregulation at 20–100 nM deoxyribozyme concentration. However, at these concentrations, significant cytotoxicity was observed. At 4 nM concentrations of DNA enzyme, the target silencing ability of the DIS- and PBS-targeting enzymes was only 3- and 18-fold, respectively.

5.2.2 Anti-hepatitis Virus Strategies

RNA-cleaving DNA enzymes have been utilized to target the hepatitis C virus (HCV) genome in the core and in the 5′-noncoding region (NCR) (Oketani et al. 1999). Under cell-free conditions and in the presence of 10 mM Mg^{2+} (10 mM Tris–HCl pH 7.4 at 37°C), a DNA enzyme having 11-nt binding arms cleaved a 731-nt HCV viral RNA at its core initiation site with high catalytic efficiency [$k_{cat}/K_m = 9.69 \times 10^4$ (mol/L)$^{-1}$ min^{-1}]. Increasing the binding arm length to 15 nt further enhanced the catalytic efficiency about 6.5-fold to [$k_{cat}/K_m = 6.29 \times 10^5$

(mol/L)$^{-1}$ min^{-1}]. The enzyme targeting 5′-NCR was found to be 60-fold less efficient in cleaving the target site than the one targeting core region of HCV RNA. When tested for the target RNA cleavage efficiency using luciferase-based reporter systems in HCV permissive hepatoma Huh7 cells, DNA enzymes modified with phosphorothioate linkages at the 5′- and 3′-ends showed 45–67% silencing of luciferase expression. To distinguish the antisense-mediated silencing of DNA enzymes, mutant enzymes having 15-nt binding arms were used. The results indicated 18–31% reduction of HCV RNA by these enzymes. Based on these results, authors concluded that both antisense and catalytic activities contributed to silencing of the target HCV RNA in cellular systems.

To target highly conserved regions of UTR of HCV, 10-23 DNA enzymes with 15-nt binding arms were modified with two PS linkages at both ends of binding arms (Trepanier et al. 2006). In vitro cleavage studies using a 976-nt-long transcript under 10 mM Mg^{2+} (50 mM Tris–HCl pH 7.5 at 37°C) conditions identified an enzyme with high catalytic efficiency [$k_{cat}/K_m = 5.7 \times 10^4$ (mol/L)$^{-1}$ min^{-1}]. When this enzyme was tested in human epithelial cell line 293rTA and Huh7 cells, the HCV RNA suppression after 24 h was 48% and 24%, respectively, as revealed by RT-PCR. In a follow-up study, researchers used a 10-23 DNA enzyme modified with four 2′-OMe nucleotides at both ends of the binding arms to target conserved regions of HCV 1b RNA (Trepanier et al. 2008). Cell-free cleavage assays using a 976-nt RNA transcript in the presence of 10 mM Mg^{2+} revealed fourfold increase in catalytic efficiency [$k_{cat}/K_m = 2.1 \times 10^5$ (mol/L)$^{-1}$ min^{-1}] compared to PS-modified DNA enzymes targeting the same target site. The RT-qPCR data revealed 61% target silencing in Huh7 cells after 6 h of treatment with the 2′-OMe-modified enzyme. The corresponding reduction of HCV protein was found to be 59% and HCV antigen was 84%. The better performance of 2′-OMe-modified enzyme was attributed to its enhanced target binding affinity.

10-23 DNA enzymes targeting various loop regions of the internal ribosomal entry site (IRES) located in the 5′-UTR of HCV have been reported (Roy et al. 2008). Since IRES RNA is closely associated with translation and viral replication, targeting IRES is an attractive strategy to halt HCV replication. In vitro cleavage assays using a 387-nt HCV-IRES transcript in the presence of 10 mM Mg^{2+} (50 mM Tris–HCl pH 7.5 at 37°C) identified several DNA enzymes with excellent efficiency to cleave IRES RNA of genotype b as well as of conserved sequences belonging to all six genotypes of HCV. A luciferase-based reporter system revealed that only one out of three DNA enzymes against genotype b that were active in vitro was able to inhibit HCV-IRES-mediated translation in Huh7 cells to a substantial level (81%). However, a mutant enzyme failed to show any RNA cleavage. These results were further verified using Northern blot hybridization assays. The ability of DNA enzyme to prevent HCV RNA replication has also been tested in Huh7.5 cells harboring a HCV1b replicon. Semiquantitative RT-PCR results revealed 70% inhibition of HCV1b genotype RNA synthesis by the DNA enzymes.

The versatility of 10-23 DNA enzymes to target the X gene of hepatitis B virus has been reported (Goila and Banerjea 2001). The X gene is a pleiotropic transactivator and augments the expression level of other hepatitis genes.

Two 10-23 DNA enzymes were used to target two highly conserved regions of X gene. A combination of both individual DNA enzymes into a tandem arrangement was also investigated. In vitro cleavage studies using a 465-nt X gene transcript under single turnover conditions at 10 mM of Mg^{2+} (50 mM Tris–HCl, pH 7.5 at 37°C) showed efficient target cleavage (~80% cleavage in 4 h) by the individual DNA enzymes. The cleavage ability of the tandem DNA enzyme was slightly lower (~55% cleavage in 4 h). DNA enzyme mutants failed to show any target cleavage activity. However, at 2 mM Mg^{2+} concentration, the cleavage efficiency was considerably lower for all tested DNA enzymes. The intracellular activity of these DNA enzymes was studied in HepG2 cells. The RT-PCR and Western blot analyses demonstrated more efficient X gene silencing (four- to sixfold reduction of X protein) by the individual DNA enzymes than by the tandem DNA enzyme. In a related study, DNA enzymes targeting the X gene using a HBx-EGFP reporter in AD293 cells have been explored (Hou et al. 2006). Different 10-23 enzymes having binding arms of 7, 8, and 9 nt were used, and their efficiency in target cleavage was monitored by flow cytometry and semiquantitative RT-PCR. Forty-eight hours after transfection, the DNA enzyme-treated cells showed ~60% reduction in fluorescence. The mRNA reduction was found to be ~80%. The control DNA enzymes, which were incapable of cleaving the target RNA, did not show any activity, ruling out antisense effects.

5.2.3 Other Viral RNA Targets

The polycistronic *E6*/*E7* mRNAs of human papillomavirus (HPV) have been an attractive target for nucleic acid-based therapy. The power of 10-23 and 8-17 DNA enzymes to cleave the *E6*/*E7* mRNAs at the bona fide antisense window has been recently demonstrated (Reyes-Gutierrez and Alvarez-Salas 2009). The in vitro cleavage of an mRNA transcript under single turnover conditions revealed that only the 10-23 enzyme was capable of cleaving the target efficiently ($k_{obs} = 0.013$ min^{-1}) compared to the 8-17 enzyme. To enhance the target accessibility and to fine-tune the nuclease stability of 10-23 DNA enzyme, chemical modifications such as PS or 2′-OMe or LNA were introduced at the binding arms and at positions 8 and 12 of the catalytic core. In vitro cleavage assays using a short *E6*/*E7* transcript in the presence of 10 mM Mg^{2+} (50 mM Tris–HCl pH 7.5, 150 mM NaCl, 2 mM spermidine at 37°C) revealed that the PS and 2′-OMe modifications in the catalytic core substantially reduced the cleavage activity compared to unmodified DNA enzymes. The DNA enzymes with 2′-OMe- or LNA-modified binding arms showed enhanced cleavage activity of the target compared to unmodified or PS-modified enzymes. The mutated DNA sequences which served as negative controls did not show any activity which validates the catalytic ability of the 10-23 DNA enzyme sequences. The 2′-OMe- and LNA-modified DNA enzymes were also tested to cleave full-length *E6*/*E7* transcript (>1,000 nt long) under 10 mM Mg^{2+} conditions. In these experiments, only LNA-modified DNA enzyme exhibited target cleavage activity. This can be attributed to the high target binding affinity of LNA-modified enzymes which help to access the target sites effectively. The LNA-modified DNA enzyme was further evaluated for the target mRNA cleavage in HPV-16-positive tumor cell line SiHa. RT-PCR analysis, 24 h posttransfection, revealed >50% E6/E7 mRNA reduction in cells treated with the LNA-modified DNA enzyme. Moreover, cell proliferation studies revealed dose-dependent killing (>50%) by the LNA-modified DNA enzyme 4 h posttransfection. The true catalytic ability and nontoxicity of LNA-modified DNA enzymes were evident in studies using proper mutant controls as well as in the HPV-16-negative cell line C33-A. This report clearly underscores the potential of LNA-modified 10-23 DNA enzymes to target HPV-16 viral mRNAs. The next logical step will be to test these modified DNA enzymes in mice bearing HPV-16.

In order to decipher the enhanced cleavage efficiency of LNA-modified DNA enzymes targeting long and structured mRNA targets such as E6/E7 mRNA, detailed kinetic and thermodynamic studies have been performed (Donini et al. 2007). In this study, four sets of 8-17 DNA enzymes and corresponding LNA-modified enzymes (4–5 modifications in the binding arms) were designed to target specific AG sites located within the first 200 nt of E6 coding sequence. All these cleavage sites were previously shown to be accessible for ribozyme and DNA enzyme-mediated cleavage. Kinetic studies using a short RNA substrate under single turnover conditions revealed that the LNA-modified DNA enzyme had a similar target association rate constant as the unmodified DNA enzyme. However, the dissociation rate constant of the substrate–enzyme complex was tenfold lower for the LNA-modified variant. Studies on the effect of Mg^{2+} have shown that LNA-modified 8-17 enzyme is capable of cleaving the target efficiently at $\leq$1.5 mM Mg^{2+} (50 mM Pipes-NaOH, pH 7.4 at 37°C) concentration. The enhanced efficiency of the LNA-modified enzyme is attributed to its capability to saturate the substrate at low enzyme concentration (100 nM) due to its high target binding affinity. Cleavage assays using prefolded E6 mRNA transcript (544 nt long) and E6/E7 transcript (830 nt long) have shown that LNA-modified DNA enzymes cleaved the target more efficiently (k_{obs} = 0.025–0.044 min^{-1}) compared to unmodified DNA enzymes (k_{obs} = 0.00007 min^{-1}).

Recently, 10-23 DNA enzymes have been designed to target respiratory syncytial virus (RSV) genomic RNA of subgroups A and B (Xie et al. 2006). In vitro cleavage assays using a 453-nt-long N gene RNA transcript in the presence of 10 mM Mg^{2+} (250 mM Tris–HCl pH 7.4 at 37°C) have shown 68% of the target cleavage after 4 h in a sequence-specific and time-dependent manner. Antisense oligonucleotides and a mutant DNA enzyme targeting the same site failed to produce any cleavage. When the DNA enzyme was tested in RSV-infected Hep-2 cells, 96% protection of RSV-infected cells was obtained after 5 days of treatment. Semiquantitative RT-PCR assays revealed 100% reduction of RSV F mRNA after 3 days of incubation with the DNA enzyme. These results were validated by Western blots in which the reduction of the RSV F protein level was up to 87%. The results obtained from these studies were in fact better than the therapeutic efficacy of anti-RSV drug ribavirin, which is currently in clinical use. It should be noted that to enhance the cell penetration and to increase the nuclease stability, the

DNA enzymes used in these studies were modified with PS units at both ends and a cholesterol moiety at the 3′-end. With these encouraging results in hand, the researchers tested the efficacy of the anti-RSV DNA enzyme in RSV-infected mice (Yang et al. 2007). Intranasal delivery of DNA enzyme resulted in a dose-dependent (0.2–0.8 mg) antiviral effect without producing any toxicity. RT-PCR assays have shown that the observed 54% mRNA reduction was a result of the DNA enzyme treatment. It would be interesting to see clinical tests of anti-RSV DNA enzymes because the results could be directly compared to an anti-RSV siRNA, which is currently in clinical trials (Meyers et al. 2009).

The ability of DNA enzymes to target the genomic RNA of Japanese encephalitis virus (JEV) has been demonstrated in in vitro and in vivo settings (Appaiahgari and Vrati 2007). Researchers designed a 10-23 DNA enzyme, which targets two direct repeat sequences located at the 3′-noncoding region (3′-NCR) of JEV virus. The 3′-NCR plays a crucial role in the replication of JEV. Both unmodified and completely PS-modified enzymes were used in these studies. Moreover, to enable the site-specific delivery to cells bearing scavenger receptors (ScRs), ten deoxyguanosine (poly-dG) residues were conjugated to the 3′-end of the unmodified and PS-modified enzymes. The in vitro cleavage assays under multiple turnover conditions in the presence of 2 mM Mg^{2+} (50 mM Tris–HCl, pH 7.5 at 37°C) revealed efficient cleavage of the 582-nt 3′-NCR sequence of JEV. It was also demonstrated that the poly-dG-modified DNA was taken up by murine macrophage cells, J744E. When tested for halting the JEV replication, the PS-modified DNA enzymes conjugated with poly-dG at the 3′-end showed 108-fold reductions of JEV replication (virus titers) at 5 μM enzyme concentrations. However, the randomized control DNA enzymes did not show any reduction in JEV titers supporting the specificity of DNA enzyme-mediated action. With these encouraging in vitro results in hand, researchers directly injected DNA enzymes (500 pmol) into JEV-infected mouse brains. After 72 h, the analysis of brain tissues revealed that the treatment with PS-modified DNA enzyme containing a PS-modified poly-dG tail resulted in 873-fold reduction in JEV titers. However, the unmodified DNA enzyme or a DNA enzyme control containing a randomized catalytic region did not show any substantial effect on JEV replication. Dose-dependent reduction of viral titers has also been reported in these studies. Injection of 1,000 pmol of completely PS-modified enzyme led to 99.99% of viral inhibition. The authors demonstrated the remarkable curing ability of DNA enzymes in mice infected with lethal JEV infection. These results clearly underscore the potential of RNA-cleaving DNA enzyme to treat viral infections of brain. Future studies are certainly warranted toward the realization of the clinical utility of these enzymes.

In order to target the translation initiation site (AUG) of influenza virus A, 10-23 DNA enzymes have been designed with varying binding arm lengths (7–9 nt) (Takahashi et al. 2004). To impart nuclease resistance, one or two N3′–P5′ phosphoramidate linkages (Fig. 4) have been introduced at the 3′- and 5′-ends of the enzymes. The kinetic evaluation of in vitro target cleavage using a 25-nt short RNA substrate under 25 mM Mg^{2+} (50 mM Tris–HCl, pH 7.5 at 37°C) has been carried under multiple turnover conditions. The DNA enzyme with 9-nt binding arms and two N3′–P5′ phosphoramidate linkages at the 3′- and 5′-ends exhibited high catalytic efficiency [$k_{cat}/K_m = 1 \times 10^6$ (mol/L)$^{-1}$ min^{-1}], which was comparable to that of unmodified enzyme. When tested for the effect on viral replication in MDCK cells infected with influenza virus, the end-modified enzyme has shown 99% inhibition of virus. However, the in vitro inactive DNA enzyme controls also exhibited 49–58% inhibitory activity which points toward the contribution from the antisense mode of action.

An attempt to find viable strategies to address deadly severe acute respiratory syndrome-associated coronavirus (SARS-CoV), 10-23 DNA enzymes have been designed to target the 5′-UTR of a highly conserved fragment in the SARS genome (Wu et al. 2007). DNA enzyme having 9-nt binding arms was used to test the target cleavage efficacy on a 1-kb RNA transcript in in vitro settings. Under single turnover conditions and in the presence of 10 mM Mg^{2+}(50 mM Tris–HCl, pH 7.5 at 37°C), the 10-23 enzyme exhibited a k_{obs} of 0.064 min^{-1}. Under multiple turnover conditions, the initial velocity of the cleavage reaction was found to be 0.4 nM min^{-1}. Mutant enzymes did not show any significant cleavage under these conditions. Cellular efficacy of the DNA enzyme was evaluated using a SARS-CoV 5′-UTR-eGFP fusion plasmid in Vero E6 cells. FACS analysis revealed dose-dependent target attenuation (55–85%) of DNA enzyme-treated cells, while the mutant enzyme was found to be completely inactive. These results were further confirmed by RT-PCR analysis.

The effects of chemical modifications (2′-OMe, LNA, PS, etc.) in the binding arms and in the catalytic core of 10-23 DNA enzymes were investigated on deoxyribozymes targeting the 5′-UTR of human rhinovirus 14 (HRV14) (Schubert et al. 2003). The HRV14 is one of the viruses responsible for common cold. In vitro cleavage studies have been carried out using a 835-nt RNA transcript, which encodes the HRV14 genome. Both single turnover and multiple turnover experiments were performed in the presence of 10 mM Mg^{2+} (50 mM Tris–HCl, pH 7.5 at 37°C), and the respective rate constant (k_{obs}) and initial velocity (v_{init}) were determined to assess the catalytic potential. The cleavage assays revealed that DNA enzymes with 9-nt binding arms and four 2′-OMe or four LNA modifications on each binding arms were able to enhance the k_{obs} by factors of 2 and 4, respectively. Under multiple turnover conditions, the 2′-OMe-modified enzyme exhibited a sixfold enhancement in initial velocity, while the LNA-modified enzyme had shown a severe deceleration. The adverse effects of LNA-modified enzyme have been attributed to the high target binding, which influences the product release. The presence of PS modification in the binding arms severely affected the target cleavage under single and multiple turnover conditions. This can be attributed to the poor target recognition capabilities of PS modification. When the binding arms were shortened to 7 nt, the k_{obs} and v_{init} were drastically increased to several fold for 2′-OMe- and LNA-modified enzymes. For example, an enzyme with 7-nt binding arms having five 2′-OMe modifications on each arm enhanced the cleavage activity by 20-fold ($k_{obs} = 0.5$ min^{-1} and $v_{init} = 4.7$ nM min^{-1}) compared to the unmodified counterpart. Researchers also tried to evaluate the effect of

2′-OMe modification in the 15-nt catalytic core of the 10-23 DNA enzyme. Systematic replacement of each nucleotide in the catalytic core with 2′-OMe modification led to the identification of six positions (2, 7, 8, 11, 14, and 15), which are amenable to the modifications without losing the cleavage efficiency. This is one of the very few reports in which chemical modifications have been successfully utilized in the catalytic core of 10-23 enzyme. To address the nuclease vulnerability of DNA enzymes in cellular environment, an optimized DNA enzyme with 7-nt binding arms containing five 2′-OMe modifications on each binding arm and six 2′-OMe modifications in the catalytic core has been designed. This highly engineered DNA enzyme had an impressive cleavage potential, which is indicated by a tenfold increase in the k_{obs} (0.57 min^{-1}) and v_{init} (2 nM min^{-1}) compared to the unmodified enzyme. The enzyme exhibited a half-life of 25 h in 10% fetal calf serum. Though these results are promising from the therapeutic point of view, unfortunately, the potential of this modified enzyme to cleave the RNA target of HRV14 in cellular systems is not reported in the literature. In a related study, researchers demonstrated the potential of using 2′-OMe or LNA modification on the binding arms of DNA enzymes to cleave inaccessible targets present in the 5′-UTR of coxsackievirus (CAV-21) (Schubert et al. 2004). In spite of the fact that the target site in the CAV-21 shares sequence homology with the 5′-UTR of HRV-14, unmodified DNA enzyme in the presence of 10 mM Mg^{2+} (50 mM Tris–HCl, pH 7.0 at 37°C) could not impart the cleavage of a 797-nt CAV-21 transcript. This shows that target sequences in HRV-14 and CAV-21 do not share similar accessibility features for DNA enzyme-promoted cleavage. However, when the 7-nt binding arms of the enzyme were modified with five 2′-OMe or three LNA moieties, complete cleavage of the target was observed under single turnover conditions. These studies point toward the need to perform thorough target accessibility studies while designing DNA enzymes against the conserved regions of mRNAs belonging to different organisms.

5.3 Delivery Strategies for DNA Enzymes

In order to capitalize the full therapeutic potential of RNA-cleaving DNA enzymes, the use of in vivo delivery agents is warranted. Unfortunately, only a handful of reports exist in the literature dealing with the DNA enzyme delivery issue (Tan et al. 2009b). The first study on deoxyribozyme delivery used DNA enzymes encapsulated in microspheres made of poly(lactic acid) and poly (glycolic acid) (PLGA) to target EGFR mRNA (Khan et al. 2004). Although these studies demonstrated the ability of sustained release of DNA enzymes in vitro, the effect on the target gene downregulation and the amount of DNA enzyme encapsulated in microspheres were not reported. To deliver anti-*c-MYC* DNA enzymes to subcutaneous tumors in mice, nanoparticles based on linear polymers of cyclodextrin complexed with DNA enzymes (termed “polyplexes”) have been used (Pun et al. 2004). In order to enhance the surface properties and to harness receptor-mediated delivery into tumor cells, polyplexes were modified with polyethylene glycol and transferrin (Tf-PEG-polyplex). Transferrin is an iron-containing protein whose receptor is highly expressed in tumor cells. Intravenous and intraperitoneal injection of Tf-PEG-polyplexes containing fluorescently tagged DNA enzymes resulted in the maximum accumulation of the DNA enzymes in tumor sites. However, the toxicity profiles of the Tf-PEG-polyplexes and the capability of the delivered DNA enzymes to reduce the tumor volume have not been reported in this study.

Poly(propylene imine) dendrimers have also been used for the in vitro and in vivo delivery of DNA enzymes. DNA enzymes complexed with cationic dendrimers showed >80% transfection efficiency in carcinoma A2780 cells (Tack et al. 2006). Moreover, intravenous injection of fluorescently labeled dendrimer–DNA enzyme complexes into ovarian carcinoma-bearing mice resulted in the accumulation of DNA enzymes in the tumor and also in the cell nucleus. When tested for toxicity, the DNA enzyme–dendrimer complexes showed minimal toxicity. It should be noted that toxicity profiles of these complexes in noncancerous cell lines have not been reported so far. Moreover, the gene silencing activity of the DNA enzymes delivered using cationic dendrimers has not been confirmed. This demands more studies on these delivery agents before they can make their way to clinical trials. The use of gold nanoparticles linked with cationic polymers (polylysine/polyethylamine) and transferrin has been reported for the delivery of DNA enzymes targeting *c-MYC* in HT-29 colon carcinoma cells (Tack et al. 2008). Varying transfection efficiencies (43–77%) were reported based on the nature of the cationic polymers used. No further studies have so far been reported on the toxicity profile of this gold nanoparticle-based delivery system as well as the in vivo efficacy of mRNA downregulation.

The most promising DNA enzyme delivery strategy reported in the literature is nanoparticles made of the carbohydrate-based biopolymer chitosan and DNA enzymes (Tan et al. 2009a). These chitosan-based nanoparticles were used to deliver the *c-JUN*-targeting DNA enzyme Dz13 to osteosarcoma cells SaOS-2 (Tan et al. 2010). Dz13-mediated apoptosis was evident in these studies. Further studies in tumor-bearing mice models demonstrated nontoxicity and the power of the chitosan-Dz13 nanoparticles to treat the osteosarcoma efficiently. Though these results are highly encouraging, it is difficult to predict whether chitosan-based delivery systems may once be used for the delivery of DNA enzymes targeting viral, cardiovascular, and other cancer targets. Moreover, the true clinical utility of this delivery system has yet to be investigated.

5.4 Challenges and Opportunities for DNA Enzyme-Based Therapeutics

In the last 12 years, there has been considerable progress in DNA enzyme research, which has been devoted to the development of deoxyribozymes as therapeutic agents against various deadly diseases as well as against viral and bacterial pathogens. These efforts include attempts to impart nuclease stability and enhance the target accessibility, specificity, and target binding affinity by introducing various chemical modifications in the binding arms and in the catalytic core of DNA enzymes. The chemical modifications are more or less well tolerated in the binding arms, but they are found to be less tolerated in the catalytic core (see above Sect. 4). To impart endonucleolytic resistance, incorporation of modifications in the catalytic core will be beneficial (Schubert et al. 2003). The search for an ideal chemical modification which is capable of imparting all desired therapeutic properties to DNA enzyme is far from over. It should also be emphasized that most of the studies in this direction have been directed to chemically fine-tune 10-23 DNA enzyme perhaps owing to its early known high in vitro cleavage efficacy on any purine–pyrimidine RNA junction. The discovery of the large family of 8-17 DNA enzymes which are capable of practically cleaving any dinucleotide RNA junction (Cruz et al. 2004) should encourage researchers to evaluate the therapeutic utility of these short DNA catalysts as well.

One of the important aspects which was not explored completely with respect to therapeutic development is the in vitro selection of new chemically modified RNA-cleaving DNA enzymes, which are tailored with functionalities that are capable of imparting drug-like properties. The nonavailability of a universal DNA polymerase which can incorporate sugar-, base-, and/or backbone-modified nucleoside triphosphates into a growing DNA chain poses major challenges to such efforts. Though attempts to select metal ion-independent DNA enzymes using chemically modified triphosphates have been carried out, the evolved enzymes were found to be less superior in their cleavage efficiency compared to 10-23 and 8-17 enzymes (Santoro et al. 2000; Sidorov et al. 2004; Ting et al. 2004; Hollenstein et al. 2009). The rationale behind these studies has been the fact that both 10-23 and 8-17 require high Mg^{2+} (>5 mM) for very efficient in vitro RNA cleavage. This concentration is far from the physiological Mg^{2+} concentration (0.2–0.8 mM). However, the available data on the excellent cleavage activity of 10-23 enzymes in cellular systems and in animal models demonstrate that therapeutically demanding target cleavage can indeed be achieved with the help of known enzymes. If one wants to really improve the in vivo efficiency, it may be desirable to evolve new RNA-cleaving enzymes under physiologically relevant metal ion concentrations. Another possibility is to perform in vitro selection experiments in conditions mimicking the cellular environment analogous to efforts which have been undertaken for the in vitro selection of cell-specific aptamers (Raddatz et al. 2008).

Off-target effects of DNA enzymes have not been widely investigated unlike in the case of antisense oligonucleotides and siRNAs. Recently, it has been reported that some DNA enzymes such as Dz13, which was earlier reported for therapeutic action on a wide range of tumors, are actually promoting apoptosis due to the activation of inhibitor of caspase-activated deoxyribonuclease and protein kinase C delta (Dass et al. 2010). The sequence composition of the DNA enzyme can also impart unwanted toxicity. DNA enzymes which are rich in Gs have been shown to enhance cytotoxic effects (Goodchild et al. 2007). These results are alarming and demand for more rigorous off-target/toxicity profiling of any DNA enzyme which is slated for in vivo use. At the delivery front, DNA enzyme therapeutics may benefit from the major research efforts which are currently under way for antisense oligonucleotides and siRNAs (Stanton and Colletti 2010).

6 Conclusions and Future Perspectives

In summary, artificial single-stranded DNA enzymes can serve as useful tools for the site-specific cleavage of RNA for various applications. Seventeen years after the initial report, this intriguing property of DNA is still of conceptual interest with respect to exploring the general catalytic ability and potential limitations of DNA enzymes. The basic mechanistic principles of DNA-catalyzed scission of RNA phosphodiester bonds have been studied by biochemical methods, and the DNA folding processes have been explored by biophysical approaches. Nevertheless, the molecular details of how such impressive rate enhancements of up to 10^8-fold are achieved are far from being fully understood. Clearly, more biochemical and structural data are needed to disentangle the contributions of individual nucleotides, metal ions, and water molecules in the active sites of DNA catalysts.

With respect to applications as therapeutic agents, the efficiency of DNA catalysts compares well with other nucleic acid-based strategies to downregulate protein expression, but, at the same time, faces similar difficulties as well. However, unlike in the case of other DNA- or RNA-based therapeutic agents, detailed pharmacokinetic and pharmacodynamic studies are yet to be conducted for RNA-cleaving DNA enzymes. To increase the in vivo efficiency, promising developments can currently be found in recent in vitro selection attempts that strive to improve the tolerance for low divalent metal ion concentrations by expanding the chemical diversity of DNA nucleotides. Modified nucleotide modifications will also serve effectively for enhancing the lifetime of DNA enzymes under the demanding conditions encountered in biological environments. Together with modern strategies for delivery of modified DNA enzymes, it can be expected that deoxyribozymes may find their way to clinic in the near future. Even though RNA-cleaving DNA enzymes have not yet made it to real-life therapeutic applications, they have been serving the scientific community as valuable research tools to address biochemical, biomedical, and bioanalytical problems.

Acknowledgments Financial support from Department of Science and Technology (DST), Government of India (FAST track scheme, SR/FT/LS-133/2008) to P.I.P. is gratefully acknowledged. C.H. is supported by the Max Planck Society, and P.I.P. is a recipient of a Max-Planck India Fellowship (MPG-DST scheme).

References

Achenbach JC, Chiuman W, Cruz RP et al (2004) DNAzymes: from creation in vitro to application in vivo. Curr Pharm Biotechnol 5:321–336

Appaiahgari MB, Vrati S (2007) DNAzyme-mediated inhibition of Japanese encephalitis virus replication in mouse brain. Mol Ther 15:1593–1599

Bano AS, Gupta N, Sharma Y et al (2007) HIV-1 VprB and C RNA cleavage by potent 10-23 DNAzymes that also cause reversal of G2 cell cycle arrest mediated by Vpr genes. Oligonucleotides 17:465–472

Basu S, Sriram B, Goila R et al (2000) Targeted cleavage of HIV-1 coreceptor-CXCR-4 by RNA-cleaving DNA-enzyme: inhibition of coreceptor function. Antiviral Res 46:125–134

Baum DA, Silverman SK (2008) Deoxyribozymes: useful DNA catalysts in vitro and in vivo. Cell Mol Life Sci 65:2156–2174

Benson VL, Khachigian LM, Lowe HC (2008) DNAzymes and cardiovascular disease. Br J Pharmacol 154:741–748

Bhindi R, Fahmy RG, Lowe HC et al (2007) Brothers in arms: DNA enzymes, short interfering RNA, and the emerging wave of small-molecule nucleic acid-based gene-silencing strategies. Am J Pathol 171:1079–1088

Breaker RR (2004) Natural and engineered nucleic acids as tools to explore biology. Nature 432:838–845

Breaker RR, Joyce GF (1994) A DNA enzyme that cleaves RNA. Chem Biol 1:223–229

Breaker RR, Joyce GF (1995) A DNA enzyme with Mg^{2+}-dependent RNA phosphodiesterase activity. Chem Biol 2:655–660

Burmeister J, von Kiedrowski G, Ellington AD (1997) Cofactor-assisted self-cleavage in DNA Libraries with a 3′-5′ phosphoramidate bond. Angew Chem Int Ed 36:1321–1324

Carmi N, Balkhi SR, Breaker RR (1998) Cleaving DNA with DNA. Proc Natl Acad Sci USA 95:2233–2237

Chakraborti S, Banerjea AC (2003) Inhibition of HIV-1 gene expression by novel DNA enzymes targeted to cleave HIV-1 TAR RNA: potential effectiveness against all HIV-1 isolates. Mol Ther 7:817–826

Chandra M, Silverman SK (2008) DNA and RNA can be equally efficient catalysts for carbon-carbon bond formation. J Am Chem Soc 130:2936–2937

Chandra M, Sachdeva A, Silverman SK (2009) DNA-catalyzed sequence-specific hydrolysis of DNA. Nat Chem Biol 5:718–720

Chen F, Li Z, Wang R et al (2004a) Inhibition of ampicillin-resistant bacteria by novel mono-DNAzymes and di-DNAzyme targeted to beta-lactamase mRNA. Oligonucleotides 14:80–89

Chen F, Wang R, Li Z et al (2004b) A novel replicating circular DNAzyme. Nucleic Acids Res 32:2336–2341

Chinnapen DJ, Sen D (2004) A deoxyribozyme that harnesses light to repair thymine dimers in DNA. Proc Natl Acad Sci U S A 101:65–69

Choi YJ, Han HJ, Lee JH et al (2000) Synthesis and NMR studies of the RNA-cleaving DNA enzyme. B Kor Chem Soc 21:955–956

Cieslak M, Szymanski J, Adamiak RW et al (2003) Structural rearrangements of the 10-23 DNAzyme to beta 3 integrin subunit mRNA induced by cations and their relations to the catalytic activity. J Biol Chem 278:47987–47996

Coppins RL, Silverman SK (2004) A DNA enzyme that mimics the first step of RNA splicing. Nat Struct Mol Biol 11:270–274

Cruz RPG, Withers JB, Li Y (2004) Dinucleotide junction cleavage versatility of 8-17 deoxyribozyme. Chem Biol 11:57–67

Dash BC, Banerjea AC (2004) Sequence-specific cleavage activities of DNA enzymes targeted against HIV-1 Gag and Nef regions. Oligonucleotides 14:41–47

Dass CR, Choong PF, Khachigian LM (2008) DNAzyme technology and cancer therapy: cleave and let die. Mol Cancer Ther 7:243–251

Dass CR, Tan ML, Galloway SJ et al (2010) Dz13 induces a cytotoxic stress response with upregulation of E2F1 in tumor cells metastasizing to or from bone. Oligonucleotides 20:79–91
Donini S, Clerici M, Wengel J et al (2007) The advantages of being locked. Assessing the cleavage of short and long RNAs by locked nucleic acid-containing 8-17 deoxyribozymes. J Biol Chem 282:35510–35518
Doudna JA, Cech TR (2002) The chemical repertoire of natural ribozymes. Nature 418:222–228
Faulhammer D, Famulok M (1996) The Ca^{2+} ion as a cofactor for a novel RNA-cleaving deoxyribozyme. Angew Chem Int Ed Engl 35:2837–2841
Feldman AR, Sen D (2001) A new and efficient DNA enzyme for the sequence-specific cleavage of RNA. J Mol Biol 313:283–294
Frederiksen JK, Piccirilli JA (2009) Identification of catalytic metal ion ligands in ribozymes. Methods 49:146–166
Geyer CR, Sen D (1997) Evidence for the metal-cofactor independence of an RNA phosphodiester-cleaving DNA enzyme. Chem Biol 4:579–593
Geyer CR, Sen D (1998) Lanthanide probes for a phosphodiester-cleaving, lead-dependent, DNAzyme. J Mol Biol 275:483–489
Goila R, Banerjea AC (1998) Sequence specific cleavage of the HIV-1 coreceptor CCR5 gene by a hammer-head ribozyme and a DNA-enzyme: inhibition of the coreceptor function by DNA-enzyme. FEBS Lett 436:233–238
Goila R, Banerjea AC (2001) Inhibition of hepatitis B virus X gene expression by novel DNA enzymes. Biochem J 353:701–708
Goodchild A, King A, Gozar MM et al (2007) Cytotoxic G-rich oligodeoxynucleotides: putative protein targets and required sequence motif. Nucleic Acids Res 35:4562–4572
Guerrier-Takada C, Gardiner K, Marsh T et al (1983) The RNA moiety of ribonuclease P is the catalytic subunit of the enzyme. Cell 35:849–857
He QC, Zhou JM, Zhou DM et al (2002) Comparison of metal-ion-dependent cleavages of RNA by a DNA enzyme and a hammerhead ribozyme. Biomacromolecules 3:69–83
He J, Zhang D, Wang Q et al (2011) A novel strategy of chemical modification for rate enhancement of 10-23 DNAzyme: a combination of A9 position and 8-aza-7-deaza-2′-deoxyadenosine analogs. Org Biomol Chem 9:5728–5736
Höbartner C, Pradeepkumar PI, Silverman SK (2007) Site-selective depurination by a periodate-dependent deoxyribozyme. Chem Commun:2255–2257
Hollenstein M, Hipolito CJ, Lam CH et al (2009) A self-cleaving DNA enzyme modified with amines, guanidines and imidazoles operates independently of divalent metal cations (M2+). Nucleic Acids Res 37:1638–1649
Hou W, Ni Q, Wo J et al (2006) Inhibition of hepatitis B virus X gene expression by 10-23 DNAzymes. Antiviral Res 72:190–196
Hou Z, Meng JR, Zhao JR et al (2007) Inhibition of beta-lactamase-mediated oxacillin resistance in *Staphylococcus aureus* by a deoxyribozyme. Acta Pharmacol Sin 28:1775–1782
Jakobsen MR, Haasnoot J, Wengel J et al (2007) Efficient inhibition of HIV-1 expression by LNA modified antisense oligonucleotides and DNAzymes targeted to functionally selected binding sites. Retrovirology 4:29
Joyce GF (2004) Directed evolution of nucleic acid enzymes. Annu Rev Biochem 73:791–836
Joyce GF (2007) Forty years of in vitro evolution. Angew Chem Int Ed Engl 46:6420–6436
Kenward M, Dorfman KD (2009) Coarse-grained brownian dynamics simulations of the 10-23 DNAzyme. Biophys J 97:2785–2793
Khan A, Benboubetra M, Sayyed PZ et al (2004) Sustained polymeric delivery of gene silencing antisense ODNs, siRNA, DNAzymes and ribozymes: in vitro and in vivo studies. J Drug Target 12:393–404
Kim HK, Liu J, Li J et al (2007a) Metal-dependent global folding and activity of the 8-17 DNAzyme studied by fluorescence resonance energy transfer. J Am Chem Soc 129:6896–6902
Kim HK, Rasnik I, Liu J et al (2007b) Dissecting metal ion-dependent folding and catalysis of a single DNAzyme. Nat Chem Biol 3:763–768

Kim HK, Li J, Nagraj N et al (2008) Probing metal binding in the 8-17 DNAzyme by TbIII luminescence spectroscopy. Chemistry 14:8696–8703

Kost DM, Gerdt JP, Pradeepkumar PI et al (2008) Controlling the direction of site-selectivity and regioselectivity in RNA ligation by Zn2+-dependent deoxyribozymes that use 2′,3′-cyclic phosphate RNA substrates. Org Biomol Chem 6:4391–4398

Kruger K, Grabowski PJ, Zaug AJ et al (1982) Self-splicing RNA: autoexcision and autocyclization of the ribosomal RNA intervening sequence of Tetrahymena. Cell 31:147–157

Lam JC, Li Y (2010) Influence of cleavage site on global folding of an RNA-cleaving DNAzyme. Chembiochem 11:1710–1719

Lee NK, Koh HR, Han KY et al (2007) Folding of 8-17 deoxyribozyme studied by three-color alternating-laser excitation of single molecules. J Am Chem Soc 129:15526–15534

Lee NK, Koh HR, Han KY et al (2010) Single-molecule, real-time measurement of enzyme kinetics by alternating-laser excitation fluorescence resonance energy transfer. Chem Commun 46:4683–4685

Lee CS, Mui TP, Silverman SK (2011) Improved deoxyribozymes for synthesis of covalently branched DNA and RNA. Nucleic Acids Res 39:269–279

Leung EK, Suslov N, Tuttle N et al (2011) The mechanism of peptidyl transfer catalysis by the ribosome. Annu Rev Biochem 80:527–555

Levy M, Ellington AD (2001) Selection of deoxyribozyme ligases that catalyze the formation of an unnatural internucleotide linkage. Bioorg Med Chem 9:2581–2587

Li Y, Breaker RR (1999) Phosphorylating DNA with DNA. Proc Natl Acad Sci USA 96:2746–2751

Li Y, Sen D (1996) A catalytic DNA for porphyrin metallation. Nat Struct Biol 3:743–747

Li J, Zheng W, Kwon AH, Lu Y (2000a) In vitro selection and characterization of a highly efficient Zn(II)-dependent RNA-cleaving deoxyribozyme. Nucleic Acids Res 28:481–488

Li Y, Liu Y, Breaker RR (2000b) Capping DNA with DNA. Biochemistry 39:3106–3114

Li J, Zhu D, Yi Z et al (2005) DNAzymes targeting the icl gene inhibit ICL expression and decrease *Mycobacterium tuberculosis* survival in macrophages. Oligonucleotides 15:215–222

Li J, Wang N, Luo Q et al (2010) The 10-23 DNA enzyme generated by a novel expression vector mediate inhibition of taco expression in macrophage. Oligonucleotides 20:61–68

Lilley DM (2011) Catalysis by the nucleolytic ribozymes. Biochem Soc Trans 39:641–646

Liu Y, Sen D (2008) A contact photo-cross-linking investigation of the active site of the 8-17 deoxyribozyme. J Mol Biol 381:845–859

Liu Y, Sen D (2010) Local rather than global folding enables the lead-dependent activity of the 8-17 deoxyribozyme: evidence from contact photo-crosslinking. J Mol Biol 395:234–241

Liu Z, Mei SH, Brennan JD et al (2003) Assemblage of signaling DNA enzymes with intriguing metal-ion specificities and pH dependences. J Am Chem Soc 125:7539–7545

Liu J, Brown AK, Meng X et al (2007) A catalytic beacon sensor for uranium with parts-per-trillion sensitivity and millionfold selectivity. Proc Natl Acad Sci U S A 104:2056–2061

Meyers R, Alvarez R, Elbashir S et al (2009) RNA interference-mediated silencing of the respiratory syncytial virus nucleocapsid defines a potent antiviral strategy. Antimicrob Agents Chemother 53:3952–3962

Nawrot B, Widera K, Wojcik M et al (2007) Mapping of the functional phosphate groups in the catalytic core of deoxyribozyme 10-23. FEBS J 274:1062–1072

Nawrot B, Widera K, Sobczak M et al (2008) Effect of R-P and S(P)phosphorothioate substitution at the scissile site on the cleavage activity of deoxyribozyme 10-23. Curr Org Chem 12:1004–1009

Nowakowski J, Shim PJ, Prasad GS et al (1999) Crystal structure of an 82-nucleotide RNA-DNA complex formed by the 10-23 DNA enzyme. Nat Struct Biol 6:151–156

Oketani M, Asahina Y, Wu CH et al (1999) Inhibition of hepatitis C virus-directed gene expression by a DNA ribonuclease. J Hepatol 31:628–634

Pearson AM, Rich A, Krieger M (1993) Polynucleotide binding to macrophage scavenger receptors depends on the formation of base-quartet-stabilized four-stranded helices. J Biol Chem 268:3546–3554
Peracchi A (2000) Preferential activation of the 8-17 deoxyribozyme by Ca^{2+} ions. Evidence for the identity of 8-17 with the catalytic domain of the Mg5 deoxyribozyme. J Biol Chem 275:11693–11697
Peracchi A (2004) Prospects for antiviral ribozymes and deoxyribozymes. Rev Med Virol 14:47–64
Peracchi A, Bonaccio M, Clerici M (2005) A mutational analysis of the 8-17 deoxyribozyme core. J Mol Biol 352:783–794
Pradeepkumar PI, Höbartner C, Baum DA et al (2008) DNA-catalyzed formation of nucleopeptide linkages. Angew Chem Int Ed 47:1753–1757
Pun SH, Tack F, Bellocq NC et al (2004) Targeted delivery of RNA-cleaving DNA enzyme (DNAzyme) to tumor tissue by transferrin-modified, cyclodextrin-based particles. Cancer Biol Ther 3:641–650
Purtha WE, Coppins RL, Smalley MK et al (2005) General deoxyribozyme-catalyzed synthesis of native 3′-5′ RNA linkages. J Am Chem Soc 127:13124–13125
Pyle AM, Chu VT, Jankowsky E et al (2000) Using DNAzymes to cut, process, and map RNA molecules for structural studies or modification. Methods Enzymol 317:140–146
Raddatz MS, Dolf A, Endl E et al (2008) Enrichment of cell-targeting and population-specific aptamers by fluorescence-activated cell sorting. Angew Chem Int Ed Engl 47:5190–5193
Reyes-Gutierrez P, Alvarez-Salas LM (2009) Cleavage of HPV-16 E6/E7 mRNA mediated by modified 10-23 deoxyribozymes. Oligonucleotides 19:233–242
Robaldo L, Montserrat JM, Iribarren AM (2010) 10-23 DNAzyme modified with (2′R)- and (2′S)-2′-deoxy-2′-C-methyluridine in the catalytic core. Bioorg Med Chem Lett 20:4367–4370
Roth A, Breaker RR (1998) An amino acid as a cofactor for a catalytic polynucleotide. Proc Natl Acad Sci U S A 95:6027–6031
Roy S, Gupta N, Subramanian N et al (2008) Sequence-specific cleavage of hepatitis C virus RNA by DNAzymes: inhibition of viral RNA translation and replication. J Gen Virol 89:1579–1586
Santoro SW, Joyce GF (1997) A general purpose RNA-cleaving DNA enzyme. Proc Natl Acad Sci U S A 94:4262–4266
Santoro SW, Joyce GF (1998) Mechanism and utility of an RNA-cleaving DNA enzyme. Biochemistry 37:13330–13342
Santoro SW, Joyce GF, Sakthivel K et al (2000) RNA cleavage by a DNA enzyme with extended chemical functionality. J Am Chem Soc 122:2433–2439
Schlosser K, Li Y (2009a) Biologically inspired synthetic enzymes made from DNA. Chem Biol 16:311–322
Schlosser K, Li Y (2009b) DNAzyme-mediated catalysis with only guanosine and cytidine nucleotides. Nucleic Acids Res 37:413–420
Schlosser K, Li Y (2010) A versatile endoribonuclease mimic made of DNA: characteristics and applications of the 8-17 RNA-cleaving DNAzyme. Chembiochem 11:866–879
Schlosser K, Gu J, Lam JC et al (2008a) In vitro selection of small RNA-cleaving deoxyribozymes that cleave pyrimidine-pyrimidine junctions. Nucleic Acids Res 36:4768–4777
Schlosser K, Gu J, Sule L et al (2008b) Sequence-function relationships provide new insight into the cleavage site selectivity of the 8-17 RNA-cleaving deoxyribozyme. Nucleic Acids Res 36:1472–1481
Schubert S, Gul DC, Grunert HP et al (2003) RNA cleaving '10-23' DNAzymes with enhanced stability and activity. Nucleic Acids Res 31:5982–5992
Schubert S, Furste JP, Werk D et al (2004) Gaining target access for deoxyribozymes. J Mol Biol 339:355–363
Sheppard TL, Ordoukhanian P, Joyce GF (2000) A DNA enzyme with *N*-glycosylase activity. Proc Natl Acad Sci U S A 97:7802–7807

Sidorov AV, Grasby JA, Williams DM (2004) Sequence-specific cleavage of RNA in the absence of divalent metal ions by a DNAzyme incorporating imidazolyl and amino functionalities. Nucleic Acids Res 32:1591–1601

Silverman SK (2005) In vitro selection, characterization, and application of deoxyribozymes that cleave RNA. Nucleic Acids Res 33:6151–6163

Silverman SK (2008) Catalytic DNA (deoxyribozymes) for synthetic applications-current abilities and future prospects. Chem Commun:3467–3485

Silverman SK (2009) Deoxyribozymes: selection design and serendipity in the development of DNA catalysts. Acc Chem Res 42:1521–1531

Silverman SK (2010) DNA as a versatile chemical component for catalysis, encoding, and stereocontrol. Angew Chem Int Ed 49:7180–7201

Silverman SK, Baum DA (2009) Use of deoxyribozymes in RNA research. Methods Enzymol 469:95–117

Sood V, Gupta N, Bano AS et al (2007a) DNA-enzyme-mediated cleavage of human immunodeficiency virus type 1 Gag RNA is significantly augmented by antisense-DNA molecules targeted to hybridize close to the cleavage site. Oligonucleotides 17:113–121

Sood V, Unwalla H, Gupta N et al (2007b) Potent knock down of HIV-1 replication by targeting HIV-1 Tat/Rev RNA sequences synergistically with catalytic RNA and DNA. AIDS 21:31–40

Sriram B, Banerjea AC (2000) In vitro-selected RNA cleaving DNA enzymes from a combinatorial library are potent inhibitors of HIV-1 gene expression. Biochem J 352(Pt 3):667–673

Stanton MG, Colletti SL (2010) Medicinal chemistry of siRNA delivery. J Med Chem 53:7887–7901

Sugimoto N, Okumoto Y, Ohmichi T (1999) Effect of metal ions and sequence of deoxyribozymes on their RNA cleavage activity. J Chem Soc Perkin Trans 2:1381–1386

Tack F, Bakker A, Maes S et al (2006) Modified poly(propylene imine) dendrimers as effective transfection agents for catalytic DNA enzymes (DNAzymes). J Drug Target 14:69–86

Tack F, Noppe M, Van Dijck A et al (2008) Delivery of a DNAzyme targeting c-myc to HT29 colon carcinoma cells using a gold nanoparticulate approach. Pharmazie 63:221–225

Takahashi H, Hamazaki H, Habu Y et al (2004) A new modified DNA enzyme that targets influenza virus A mRNA inhibits viral infection in cultured cells. FEBS Lett 560:69–74

Tan XX, Rose K, Margolin W et al (2004) DNA enzyme generated by a novel single-stranded DNA expression vector inhibits expression of the essential bacterial cell division gene ftsZ. Biochemistry 43:1111–1117

Tan ML, Choong PF, Dass CR (2009a) Cancer, chitosan nanoparticles and catalytic nucleic acids. J Pharm Pharmacol 61:3–12

Tan ML, Choong PF, Dass CR (2009b) DNAzyme delivery systems: getting past first base. Expert Opin Drug Deliv 6:127–138

Tan ML, Dunstan DE, Friedhuber AM et al (2010) A nanoparticulate system that enhances the efficacy of the tumoricide Dz13 when administered proximal to the lesion site. J Control Release 144:196–202

Ting R, Thomas JM, Lermer L et al (2004) Substrate specificity and kinetic framework of a DNAzyme with an expanded chemical repertoire: a putative RNaseA mimic that catalyzes RNA hydrolysis independent of a divalent metal cation. Nucleic Acids Res 32:6660–6672

Travascio P, Li Y, Sen D (1998) DNA-enhanced peroxidase activity of a DNA-aptamer-hemin complex. Chem Biol 5:505–517

Trepanier J, Tanner JE, Momparler RL et al (2006) Cleavage of intracellular hepatitis C RNA in the virus core protein coding region by deoxyribozymes. J Viral Hepat 13:131–138

Trepanier JB, Tanner JE, Alfieri C (2008) Reduction in intracellular HCV RNA and virus protein expression in human hepatoma cells following treatment with 2′-O-methyl-modified anti-core deoxyribozyme. Virology 377:339–344

Unwalla H, Banerjea AC (2001a) Inhibition of HIV-1 gene expression by novel macrophage-tropic DNA enzymes targeted to cleave HIV-1 TAT/Rev RNA. Biochem J 357:147–155

Unwalla H, Banerjea AC (2001b) Novel mono- and di-DNA-enzymes targeted to cleave TAT or TAT-REV RNA inhibit HIV-1 gene expression. Antiviral Res 51:127–139
Unwalla H, Chakraborti S, Sood V et al (2006) Potent inhibition of HIV-1 gene expression and TAT-mediated apoptosis in human T cells by novel mono- and multitarget anti-TAT/Rev/Env ribozymes and a general purpose RNA-cleaving DNA-enzyme. Antiviral Res 72:134–144
Vester B, Hansen LH, Lundberg LB et al (2006) Locked nucleoside analogues expand the potential of DNAzymes to cleave structured RNA targets. BMC Mol Biol 7:19
Wachowius F, Höbartner C (2011) Probing essential nucleobase functional groups in aptamers and deoxyribozymes by nucleotide analogue interference mapping of DNA. J Am Chem Soc: ASAP
Wachowius F, Javadi-Zarnaghi F, Höbartner C (2010) Combinatorial mutation interference analysis reveals functional nucleotides required for DNA catalysis. Angew Chem Int Ed 49:8504–8508
Wang Y, Silverman SK (2005) Efficient one-step synthesis of biologically related lariat RNAs by a deoxyribozyme. Angew Chem Int Ed 44:5863–5866
Wang B, Cao L, Chiuman W et al (2010) Probing the function of nucleotides in the catalytic cores of the 8-17 and 10-23 DNAzymes by abasic nucleotide and C3 spacer substitutions. Biochemistry 49:7553–7562
Willner I, Shlyahovsky B, Zayats M et al (2008) DNAzymes for sensing, nanobiotechnology and logic gate applications. Chem Soc Rev 37:1153–1165
Wong OY, Pradeepkumar PI, Silverman SK (2011) DNA-catalyzed covalent modification of amino acid side chains in tethered and free peptide substrates. Biochemistry 50:4741–4749
Wu S, Xu J, Liu J et al (2007) An efficient RNA-cleaving DNA enzyme can specifically target the 5′-untranslated region of severe acute respiratory syndrome associated coronavirus (SARS-CoV). J Gene Med 9:1080–1086
Xie YY, Zhao XD, Jiang LP et al (2006) Inhibition of respiratory syncytial virus in cultured cells by nucleocapsid gene targeted deoxyribozyme (DNAzyme). Antiviral Res 71:31–41
Yang XQ, Zhou J, Xie YY et al (2007) Inhibition of respiratory syncytial virus of subgroups A and B using deoxyribozyme DZ1133 in mice. Virus Res 130:241–248
Zaborowska Z, Furste JP, Erdmann VA et al (2002) Sequence requirements in the catalytic core of the "10-23" DNA enzyme. J Biol Chem 277:40617–40622
Zaborowska Z, Schubert S, Kurreck J et al (2005) Deletion analysis in the catalytic region of the 10-23 DNA enzyme. FEBS Lett 579:554–558
Zhang X, Xu Y, Ling H et al (1999) Inhibition of infection of incoming HIV-1 virus by RNA-cleaving DNA enzyme. FEBS Lett 458:151–156

Hammerhead Ribozymes Against Virus and Viroid RNAs

Alberto Carbonell, Ricardo Flores, and Selma Gago

Contents

Abstract The hammerhead ribozyme, a small catalytic motif that promotes self-cleavage of the RNAs in which it is found naturally embedded, can be manipulated to recognize and cleave specifically in *trans* other RNAs in the presence of Mg^{2+}. To be really effective, hammerheads need to operate at the low concentration of Mg^{2+} existing in vivo. Evidence has been gathered along the last years showing that tertiary stabilizing motifs (TSMs), particularly interactions between peripheral loops, are critical for the catalytic activity of hammerheads at physiological levels of Mg^{2+}. These TSMs, in two alternative formats, have been incorporated into a

A. Carbonell • R. Flores (✉) • S. Gago (✉)
Instituto de Biología Molecular y Celular de Plantas (UPV-CSIC), Universidad Politécnica de Valencia, Avenida de los Naranjos, Valencia 46022, Spain
e-mail: rflores@ibmcp.upv.es; selmag@ibmcp.upv.es

V.A. Erdmann and J. Barciszewski (eds.), *From Nucleic Acids Sequences to Molecular Medicine*, RNA Technologies, DOI 10.1007/978-3-642-27426-8_16,

new generation of more efficient *trans*-cleaving hammerheads, some of which are active in vitro and *in planta* when targeted against the highly structured RNA of a viroid (a small plant pathogen). This strategy has potential to confer protection against other RNA replicons, like RNA viruses infecting plants and animals.

Keywords Argonaute • Avocado sunblotch viroid • Catalytic RNAs • Central conserved region • Complementary DNA • Delta antigen • Deoxyribonuclease • Dicer-like • Double-stranded RNA • Hepatitis B virus • Hepatitis delta virus • Large isoform delta antigen • MicroRNA • Messenger RNA • Non-protein-coding RNAs • Nuclear-encoded RNA polymerase • Peach latent mosaic viroid • Plastid-encoded RNA polymerase • Polyacrylamide gel electrophoresis • Potato spindle tuber • Potato spindle tuber viroid • RNA polymerase I • RNA polymerase II • RNA polymerase III • Ribonuclease • RNA-induced silencing complex • RNA viruses • Short isoform delta antigen • Small interfering RNA • Tobacco mosaic virus • Transfer RNA • Viroids • Viroid-derived small RN

1 A Historical Overview: Hammerhead Ribozymes in Their Natural Context

To defend themselves against DNA viruses, bacteria have evolved a quite effective strategy based on restriction enzymes, which are DNases that recognize short nucleotide stretches of the virus DNA and proceed to degrade it specifically, while the cell DNA is protected because of previous chemical modifications catalyzed by other enzymes (Arber and Linn 1969). Although bacteria are also invaded by RNA viruses, for reasons that we do not know, they have not evolved restriction RNases, which neither have been described in eukaryotes. Evolution has endowed the latter with alternative defensive barriers: the immunitary system in higher vertebrates, whose final effectors are proteins that specifically bind certain chemical signatures present in the invading virus or cellular pathogen, and the RNA interference (RNAi) system, whose final effectors are small RNAs that in animals and, particularly, in plants also play a defensive role (in addition to other functions) (Fire et al. 1998). A key component of the RNAi system is RISC (*R*NA-*i*nduced *s*ilencing *c*omplex), which somehow acts as a restriction RNase, although with a mechanism quite different from that of restriction DNases from bacteria (Hammond et al. 2000). Intriguingly, small RNAs derived from clustered, regularly interspaced, short palindromic repeats (CRISPR) involved in an adaptable and heritable RNAi-like pathway conferring resistance to phage infection have been recently reported in bacteria (and archaea) [see for a review Gesner et al. (2011)].

Therefore, restriction RNases (using this term sensu lato because such enzymes do not exist sensu stricto) have potential to become a kind of "magic bullet" that could promote the selective degradation of pathogenic RNAs, including those of viruses and viroids. These latter pathogens incite diseases in plants similar to those caused by viruses, but they differ from viruses in structure, function, and evolutionary origin. Viroids are exclusively composed of a small, circular, non-protein-coding RNA and, consequently, they are "naked genomes" not protected by a capsid protein as in viruses (Diener 2003; Tabler and Tsagris 2004; Flores et al. 2005; Ding 2009; see also Chapter "Viroids: the Smallest Know Infectious Agents Cause Accumulation of Viroid-Specific Small RNAs*"). Moreover, certain viroids properly manipulated can serve as a source of restriction RNases. A brief overview on how viroids replicate will illustrate this point.

Viroid replication occurs through a rolling-circle mechanism (Branch and Robertson 1984). The infecting most abundant viroid RNA, to which the (+) polarity is arbitrarily assigned, hijacks a cellular RNA polymerase for generating oligomeric (−) strands by reiterative transcription of the circular template. These oligomeric (−) RNAs may (1) serve directly as template for producing oligomeric (+) RNAs that are cleaved to unit-length segments by an RNase and circularized by an RNA ligase or (2) be previously cleaved to unit-length segments and ligated to the monomeric (−) forms, which serve as the template for synthesizing oligomeric (+) RNAs, as indicated previously. Therefore, the two pathways of this rolling-circle mechanism, termed asymmetric and symmetric, respectively, demand three RNA catalytic activities: polymerase, nuclease, and ligase, which initially were thought to be of host origin considering that viroids are non-protein-coding RNAs. The polymerase involved in transcription of *Potato spindle tuber viroid* (PSTVd) and other members of the family *Pospiviroidae* that replicate in the nucleus through the asymmetric pathway is RNA polymerase II (Mühlbach and Sänger 1979; Flores and Semancik 1982; Schindler and Mühlbach 1992), while a nuclear-encoded chloroplastic RNA polymerase mediates transcription of *Avocado sunblotch viroid* (ASBVd) and other members of the family *Avsunviroidae* that replicate in the chloroplast through the symmetric pathway (Navarro et al. 2000; Rodio et al. 2007). Intriguingly, although DNA is the physiological template of these nuclear and chloroplastic RNA polymerases, viroids manage to redirect them to transcribe RNA. The information regarding the third enzymatic activity (RNA ligase) involved in viroid replication is limited, but enzymes of this class with nuclear and chloroplastic localization most likely mediate circularization in the families *Pospiviroidae* and *Avsunviroidae*, respectively (Flores et al. 2011).

Going back to the second step of the replication cycle, cleavage of the oligomeric RNAs into their unit-length counterparts is not catalyzed by a host enzyme in the family *Avsunviroidae* but, remarkably, by ribozymes (of the hammerhead class) embedded in both polarity strands. This is a finding with deep functional, evolutionary and applied implications. Of the several ribozyme classes described, hammerhead ribozymes (hereafter abbreviated as hammerheads) initially discovered in viroid and certain satellite RNAs (structurally similar to viroids but functionally dependent on a helper virus) (Prody et al. 1986; Hutchins et al. 1986; Forster and Symons 1987) are the simplest and most studied. Hammerheads are small RNA motifs that at room temperature, physiological pH, and the presence of a divalent cation (generally Mg^{2+}) self-cleave in a specific phosphodiester bond, generating 5′-hydroxyl and 2′,3′ cyclic phosphodiester termini through a transesterification mechanism. The name hammerhead derives from its initial representation in two dimensions recalling the shape of this tool (Fig. 1a), and it is still used despite its tridimensional conformation being different (see below).

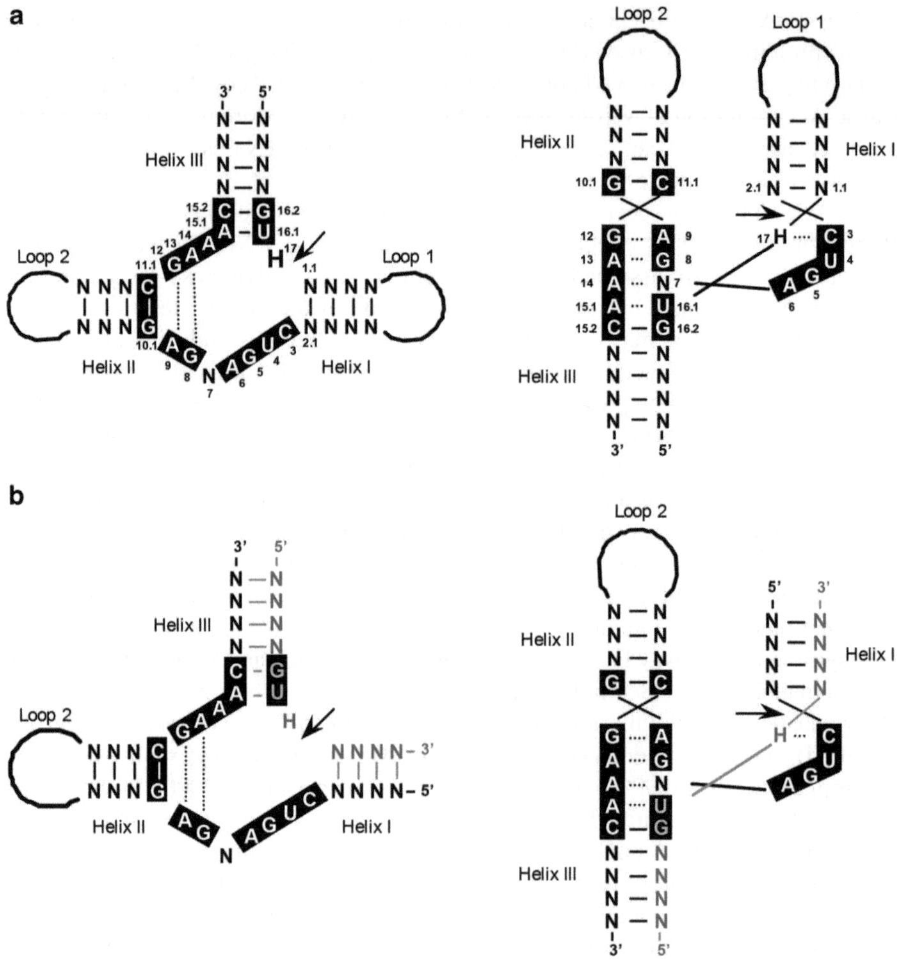

Fig. 1 Structure of the hammerhead ribozyme. (**a**) Schematic representation of the consensus ribozyme and its numbering as originally proposed (*left*) and according to crystallographic data obtained with an artificial hammerhead (*right*). (**b**) Transformation of the natural *cis*-acting format into a *trans*-acting format by disrupting loop 1; the ribozyme and the substrate are denoted with *black* and *gray* fonts, respectively. Nucleotides strictly or highly conserved in most natural hammerheads are on a *black background*, and the self-cleavage site is marked with an *arrow*. N represents any nucleotide, and H represents any nucleotide except G. *Dashes* and *discontinuous lines* indicate Watson–Crick pairs and noncanonical interactions, respectively

2 Manipulating *Cis*-Acting Hammerheads to Act in *Trans*

The consensus hammerhead resulting from a comparative analysis of natural hammerheads present in viroid and satellite RNAs (Flores et al. 2001) (Fig. 1a) reveals that these ribozymes have a strictly conserved central core flanked by three helices (I, II, and III) with minimal sequence restrictions. Site-directed mutagenesis of an artificial hammerhead has shown that changes in the nucleotides forming the central core are accompanied by severe drops of the catalytic cleavage constant (Ruffner et al. 1990), and crystallographic studies with other artificial hammerheads have unveiled a complex array of interactions between these nucleotides (Pley et al. 1994; Scott et al. 1995). Overall, the 3D structure of the hammerhead resembles a misshapen Y with its basis formed by helix III and the two arms by helices I and II (with the latter being almost colinear with helix III) (Fig. 1a).

It is important to highlight that during viroid replication hammerheads act in *cis* (self-cleaving the RNA in which they are embedded through a single turnover mechanism) (Hutchins et al. 1986). However, a simple experiment with deep implications revealed that a ribozyme of this class can be manipulated to act in *trans* by just splitting it into the ribozyme itself and the substrate (Fig. 1b); to this aim, loop 1, which caps helix I and seemed dispensable, was opened (Uhlenbeck 1987). Two corollaries were inferred from this experiment: (1) it should be possible to design hammerheads targeting specific RNAs for degradation (restriction RNases), the only requisite for which was to identify in the target RNA a GUH [the trinucleotide preceding the self-cleavage site in most natural hammerheads, where H represents any nucleotide except G; see for a review Flores et al. (2001)] and synthesize ribozyme arms complementary to the sequences of the substrate flanking the cleavage site (forming helices I and III) (Fig. 1b) and (2) one molecule of ribozyme could act on several molecules of substrate, through a multiple turnover mechanism, thus increasing the catalytic efficiency. Both predictions were confirmed experimentally (Haseloff and Gerlach 1988), paving the way for a new biotechnological tool ("the genetic scissors"), a kind of antisense RNA with the advantage that a molecule of ribozyme can process more than one molecule of substrate, thus functioning as a true catalyst, in other words, a catalytic antisense RNA.

A suite of additional studies better defined the hammerhead features for efficient and versatile cleavage in *trans*. Illustrative examples include that a helix I of only 3 bp is sufficient for the catalytic activity, showing some advantages of an asymmetric design with helices I and III of different lengths (Tabler et al. 1994); that helix II could be reduced to a few nucleotides, giving rise to what were termed minizymes (McCall et al. 1992); and that there were alternatives to the GUX trinucleotide preceding the cleavage site, like AUA, that were also catalytically active (Kore et al. 1998).

One of the first experiments with this new tool consisted in embedding a hammerhead into an antisense RNA against the RNA of *Human immunodeficiency virus* (HIV): when both RNAs were cotransfected in human cells, the catalytic domain exhibited an inhibitory effect on HIV replication four- to sevenfold higher than a catalytically inactive mutant or the antisense RNA itself (Homann et al. 1993). Similar experiments were performed in plants of *Nicotiana clevelandii* in which an antisense RNA against *Plum pox virus* (PPV), or a variant with an embedded hammerhead, was expressed episomically: following PPV inoculation, plants displayed some delay in the onset of symptoms, which was more pronounced with the catalytic variant (Liu et al. 2000). These results were stimulating but revealed a limited efficiency. Additional studies with antisense RNAs—containing or not hammerheads against both polarity strands of *Citrus exocortis viroid* (CEVd), a close relative of PSTVd—expressed transgenically in tomato (an experimental host for CEVd), showed a moderate decrease in viroid titer when the plants were challenged-inoculated; moreover, this effect was only detected with the construct against the (−) polarity strand (which accumulates in vivo considerably less than its complementary counterpart), and the catalytic domain did not provide any additional effect (Atkins et al. 1995). In contrast, in a parallel work with PSTVd and potato, a significant resistance was observed in transgenic lines expressing the construct against the viroid (−) strand, with the effect being associated with the catalytic domain because control lines in which this domain was inactivated did not display resistance (Yang et al. 1997). A feasible explanation for these somewhat contradictory results is that PSTVd accumulates to lower levels than CEVd, as also occurs with (−) with respect to (+) strands.

3 A Critical Issue: Colocalization of Ribozyme and Substrate

Although hammerheads offer a promising approach to target and degrade specific RNAs in vivo, important hurdles remain to be overcome for the practical implementation of this technology (Castanotto et al. 2002). To be effective, a hammerhead requires (1) stability against RNases, which for ribozymes that are externally synthesized and then delivered can be enhanced by substituting the 2′-OH of the riboses by other groups; (2) high expression, which for ribozymes synthesized intracellularly is achieved by putting the corresponding constructs under the control of strong promoters; (3) easy access of the ribozyme to the complementary region of its cognate RNA, a question difficult to evaluate considering that this region may adopt stable secondary structures or be associated with proteins; and (4) colocalization of ribozyme and substrate in the same subcellular compartment.

To diminish the effects resulting from the secondary structure of the target RNA around the cleavage site, hammerheads produced intracellularly that are able to unwind double-stranded structures have been designed; to this end, the ribozyme is fused to an RNA motif that interacts and recruits RNA helicases, thus creating a hybrid ribozyme with significantly higher catalytic activity in vivo (Kawasaki and Taira 2002). The issue of colocalizing ribozyme and substrate has been experimentally tackled by coexpressing within the same animal cells two retroviral vectors, one coding for a hammerhead targeting an artificial RNA substrate and the other this same RNA. Due to the common packaging signal, the anticipated outcome was the colocalization of the ribozyme with the substrate, but not with a control RNA not fused to the viral vector. The observed results confirmed this prediction: while the ribozyme reduced more than 90% the titer of the infectious virus fused to the target RNA, no effects were observed with the control RNA (Sullenger and Cech 1993). In a second study using yeast as a model system, a hammerhead was directed to the nucleolus (by fusing it to a small nucleolar RNA), wherein its target RNA (another nucleolar RNA) was located. The colocalization produced an efficiency close to 100% (Samarsky et al. 1999). Incidentally, this approach could be in principle extrapolated for controlling PSTVd and related viroids that accumulate in the nucleolus.

4 An Unanticipated Participant: Interactions Between Peripheral Loops of Natural Hammerheads Greatly Increase Their Self-Cleavage Activity

Despite extensive research efforts on hammerheads, a key aspect of their mechanism remained unnoticed up to 2003. Until then, most data were obtained with artificial hammerheads operating in *trans* (the format required for ribozymes targeting specific RNAs, which additionally facilitates kinetic analyses in a protein-free medium) and indicated that for their effective functioning in vitro, they needed Mg^{2+} at 5–10 mM, while the concentration in vivo of this cation is about 0.5 mM. As in other similar situations, this paradox pointed to an incomplete understanding of the question. Subsequent results obtained in vitro and in vivo have demonstrated that hammerheads acting in *cis* (the natural format) self-cleave much faster than their *trans*-acting derivatives and that modifications of peripheral loops 1 and 2 of natural hammerheads induce a drastic reduction of their self-cleavage catalytic constant (De la Peña et al. 2003; Khvorova et al. 2003). These data show that, in contrast with the accepted view, regions external to the central conserved core play a key catalytic role in natural hammerheads, suggesting that tertiary interactions between peripheral loops 1 and 2 facilitate sampling of the catalytic-active conformation and permit effective functioning at the low physiological concentrations of Mg^{2+} (Fig. 2a). Other studies support this view and explain why minimal *trans*-cleaving hammerheads require higher concentrations of this cation for adopting the active folding (Rueda et al. 2003; Canny et al. 2004; Penedo et al. 2004). Moreover, the interactions between peripheral loops might be stabilized by proteins, as suggested by the observation that a chloroplastic protein stimulates in vitro, and presumably in vivo, the hammerhead-mediated self-cleavage of a viroid RNA (Daròs and Flores 2002).

Crystallographic analysis of a natural hammerhead present in transcripts of *Schistosoma mansoni* has shown that tertiary interactions between loops far apart from the active site do exist and prime this ribozyme for catalysis at submillimolar Mg^{2+}. More specifically, the new structure confirms previous suggested roles for some conserved nucleotides and offers an explanation for the roles of others that are also conserved (Martick and Scott 2006). Similar analyses extended to the natural hammerhead of satellite RNA of *Tobacco ring spot virus* (sTRSV) (Prody et al. 1986), which is a typical hammerhead with loops 1 and 2 (the former is substituted by an internal loop in the hammerhead from *S. mansoni*), reinforce this view by providing the precatalytic (enzyme–substrate complex) and postcatalytic

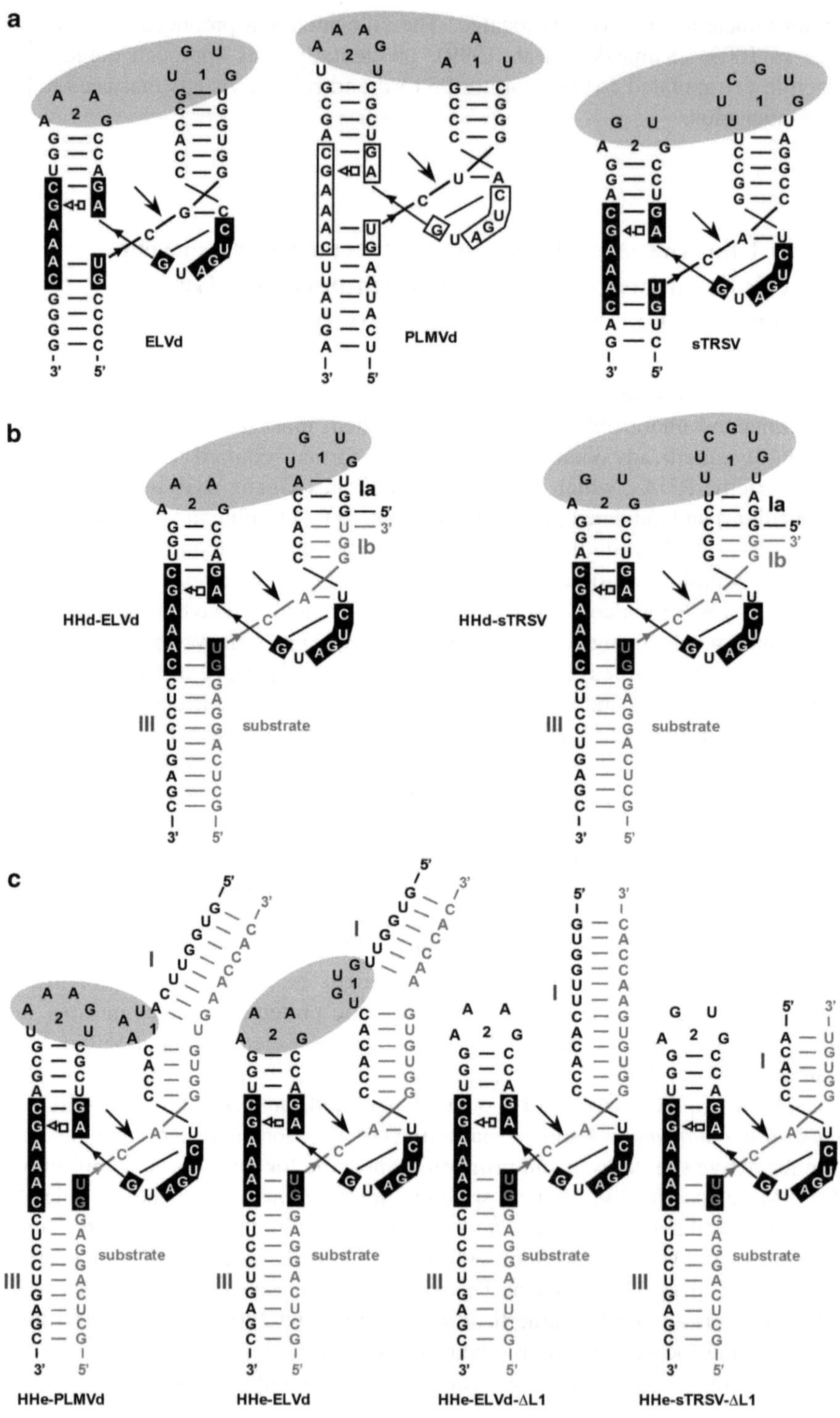

Fig. 2 (continued)

(enzyme–product complex) crystal structures of an active full-length hammerhead RNA that self-cleaves in the crystal (Chi et al. 2008). This same work has also revealed a Hoogsteen pair between an A in stem-loop II and a U in a nonhelical region of stem I that is apparently conserved in most natural hammerheads possibly due to its functional relevance (Chi et al. 2008). Furthermore, examination by NMR spectroscopy of loops 1 and 2 of the hammerheads from *Chrysanthemum chlorotic mottle viroid* (CChMVd) (Navarro and Flores 1997) and dissection of their roles by site-directed mutagenesis, self-cleavage kinetics, and infectivity bioassays have shown that specific nucleotides mediate the interactions between loops 1 and 2; the relevance of these nucleotides is evidenced by their conservation in most natural hammerheads (Dufour et al. 2009).

5 A New Generation of *Trans*-Acting Hammerheads Operating In Vitro and In Vivo at Physiological Concentrations of Magnesium

Besides explaining the low efficiency in vivo of hammerheads with *trans* format, these findings provided momentum for resuming the study of hammerheads under a new perspective. Soon afterward the interactions between peripheral loops became evident, these tertiary stabilizing motifs (TSMs) were incorporated into a new generation of more efficient *trans*-cleaving hammerheads with two distinct designs: (1) extending stem I and including loop 1 as a bulge in the hybridizing arm of this stem (extended format) (Saksmerprome et al. 2004; Weinberg and Rossi 2005) and (2) embedding within stem I the 5′ and 3′ termini of the ribozyme and substrate, respectively (discontinuous format) (Burke and Greathouse 2005) (Fig. 2b, c). Some of these hammerheads are active in vitro at low Mg^{2+} concentration against short RNA substrates. In particular, discontinuous hammerheads (HHd) derived from sTRSV and extended hammerheads (HHe) derived from *Peach latent mosaic viroid* (PLMVd) (Hernández and Flores 1992) are the most efficient when compared with other *trans*-cleaving hammerheads (Saksmerprome et al. 2004; Weinberg and Rossi 2005; Burke and Greathouse 2005).

Fig. 2 (**a**) Self-cleaving hammerheads derived from ELVd, PLMVd, and sTRSV. Motifs conserved in most natural hammerheads are within *boxes*, and self-cleavage sites are marked by *arrows*. *Black* and *white backgrounds* refer to (+) and (−) polarities, respectively. *Dashes* denote Watson–Crick (and wobble) pairs and the *open square–triangle* a Hoogsteen/sugar edge interaction. *Ovals* represent the proposed tertiary interactions between loops 1 and 2. (**b**) Schematic representation of the complex formed by discontinuous hammerheads (HHd) derived from ELVd and sTRSV and their short RNA substrates. Hammerhead and substrate nucleotides are shown with *black* and *gray* fonts, respectively. Ia and Ib refer to the distal and proximal halves of stem I, respectively. (**c**) Schematic representation of the complex formed by extended hammerheads (HHe) derived from PLMVd, ELVd, and sTRSV and their short RNA substrates. HHe-PLMVd and HHe-ELVd hammerheads contain TSMs, while the minimal HHe-ELVd-ΔL1 and HHe-sTRSV-ΔL1 hammerheads lack these motifs. Adapted from Carbonell et al. (2011)

However, other natural hammerheads like those of *Eggplant latent viroid* (ELVd) (Fadda et al. 2003) have not been adapted to a *trans* design, despite displaying higher self-cleavage rates than other natural hammerheads at very low Mg^{2+} concentrations (Carbonell et al. 2006) and despite appearing particularly convenient for the discontinuous format (because their long stem I of seven base pairs should facilitate substrate binding and folding of loop 1) (Fig. 2a).

In the remaining part of this chapter, we will focus on our recent efforts aimed at designing discontinuous and extended hammerheads derived from ELVd, PLMVd, and sTRSV with TSMs (Fig. 2b, c) (Carbonell et al. 2011). We have first examined the ability of these hammerheads to catalyze the *trans*-cleavage in vitro of short RNA substrates and of a long and highly structured RNA containing the complete sequence of the pathogen PSTVd (Diener 1971; Gross et al. 1978). Then we have selected an extended PLMVd-derived hammerhead with natural TSMs, which displays in vitro the highest cleavage rate, for transient expression bioassays in *Nicotiana benthamiana* plants to examine its activity in vivo against PSTVd infection.

6 *Trans*-Cleavage In Vitro of Short RNA Substrates by Discontinuous and Extended Hammerheads

Previous studies have shown that *cis*-acting ELVd hammerheads with a GUC (or AUC) trinucleotide preceding the self-cleavage site are the most active at very low Mg^{2+} concentration (Carbonell et al. 2006). Therefore, we took these hammerheads as a starting platform for designing HHd with preserved TSMs, for *trans*-cleaving a short fragment of the PSTVd minus (−) strand with a GUC target site. Analyses of several HHd revealed that the most efficient at low Mg^{2+} were derived from natural hammerheads in which the sequences of loops 1 and 2 remain unaltered. Unexpectedly, another ELVd-derived ribozyme with an artificial loop 1 (AAAA) was also active, suggesting that alternative tertiary interactions between artificial loops 1 and the wild-type loop 2 might promote cleavage at submillimolar Mg^{2+}. PLMVd-derived hammerheads were not tested because they are inappropriate for the discontinuous format (their stem I has only five base pairs) (Fig. 2a).

Because experiments with short RNA substrates have reported that extended hammerheads derived from PLMVd are particularly efficient (Saksmerprome et al. 2004; Weinberg and Rossi 2005), we designed a variant (HHe-PLMVd) targeting a short RNA fragment of the PSTVd (−) strand. This HHe-PLMVd, with natural loops 1 and 2 to preserve their interactions, displayed a high cleavage constant even at 0.1 mM Mg^{2+}. We next examined two extended hammerheads, HHe-ELVd (also with loops 1 and 2 as in the natural ribozyme) and HHe-ELVd-ΔL1 (a minimal hammerhead in which loop 1 was deleted); the higher catalytic constant of the former highlighted the need of TSMs for activity at submillimolar Mg^{2+}. An additional control with HHe-sTRSV-ΔL1, a modified version of the sTRSV ribozyme without TSMs but with some activity in vivo when stably expressed in transgenic potato plants (Yang et al. 1997), confirmed this view.

Altogether these experiments showed that some ELVd-derived hammerheads in discontinuous format target for cleavage short RNA substrates. However, their lower efficiency when compared with a sTRSV-derived hammerhead suggests that a hybridizing stem of only three base pairs ensures substrate binding. On the other hand, extended versions of the ELVd hammerhead were active against a short RNA substrate, but their catalytic constant, lower than that derived from the PLMVd hammerhead, indicated that the simple transposition of loop 1 to adapt a hammerhead to the *trans* format may not preserve a high catalytic activity (Weinberg and Rossi 2005).

7 *Trans*-Cleavage In Vitro of a Highly Structured RNA by Discontinuous and Extended Hammerheads

Early attempts to target long RNA substrates with minimal hammerheads met with limited success: the in vitro *trans*-cleavage constants were 100-fold lower than those obtained for short RNA substrates, a likely consequence of catalytically inactive complexes being formed between the substrate and the ribozyme (Hormes and Sczakiel 2002), or of folded substrates restricting access of the ribozyme to the vicinity of the cleavage site (Campbell et al. 1997). As indicated above, discovery of TSMs in natural hammerheads (De la Peña et al. 2003; Khvorova et al. 2003) prompted their incorporation into ribozymes with discontinuous and extended formats for enhanced activity at low Mg^{2+} concentration. However, these studies have been performed in vitro and against short RNA substrates, with only one extended PLMVd-derived hammerhead having been tested against a long RNA (a 258-nt fragment of the HIV genome) (Saksmerprome et al. 2004).

To provide additional data in this context, the hammerheads described in the previous section were next examined in vitro against a full-length PSTVd (−) RNA (359 nt) with a GUC target site and two vector tails of 17 and 88 nt; this RNA adopts a rodlike conformation according to previous studies and our own RNase T1 probing experiments. In contrast with data obtained for the short RNA substrates, most of the HHd derived from ELVd catalyzed cleavage of the long substrate more efficiently than their sTRSV-derived counterpart. Moreover, the higher cleavage rate of a variant in which the U–A base pair closing loop 1 was substituted by a stronger C–G base pair supports that the stability of stem Ib is critical for preserving the TSMs. Also worth noting is that two of the ELVd-derived hammerheads with artificial loops 1 (UACG and AAAA) were active at submillimolar Mg^{2+}, thus suggesting that alternative TSMs between artificial sequences of loop 1 and the natural loop 2 might promote cleavage as reported for a sTRSV-derived hammerhead with an artificial tetraloop (Burke and Greathouse 2005). Because the sequences of natural hammerheads have been selected as a compromise between self-cleavage and other functions that they additionally mediate (De la Peña and

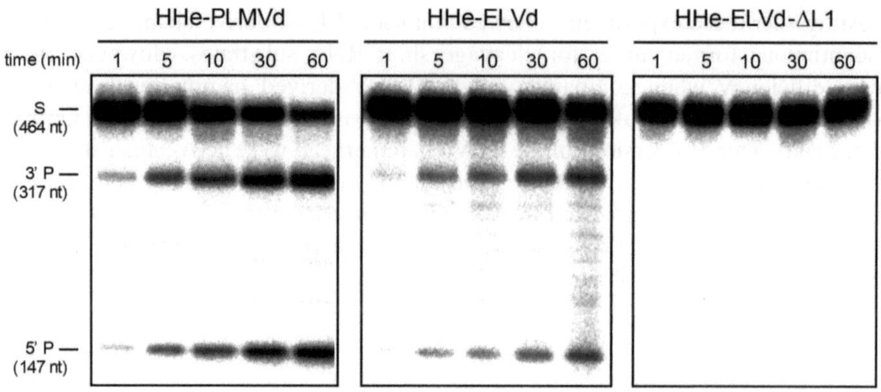

Fig. 3 Analysis by denaturing PAGE (5%) and autoradiography of *trans*-cleavage in vitro (at 0.1 mM Mg^{2+}) of PSTVd (−) RNA catalyzed by three extended hammerheads (HHe) derived from PLMVd and ELVd (see Fig. 2c). The positions and size of the substrate (S) and of the resulting 3′ and 5′ cleavage products (P) are indicated. HHe-ELVd-ΔL1 refers to the minimal hammerhead without TSMs. Adapted from Carbonell et al. (2011)

Flores 2001; Carbonell et al. 2006), it is not surprising that artificial sequences forming part of the TSMs (Khvorova et al. 2003; Saksmerprome et al. 2004) or the catalytic core (Carbonell et al. 2006) could even enhance cleavage in a *trans* format.

The most efficient extended hammerhead against the long RNA substrate, especially at submillimolar Mg^{2+}, was derived from PLMVd (Fig. 3), in line with previous in vitro selection studies at low Mg^{2+} concentration in which a PLMVd-derived hammerhead with only two transitions in loop 2 with respect to the natural variant was selected for the fastest self-cleavage (Saksmerprome et al. 2004). The nucleotides forming the asymmetric bulging loop of the HHe-PLMVd most likely generate TSMs resembling those existing in the natural hammerhead, because a smaller bulging loop (of three nucleotides) permits less alternative interactions than in the HHe-ELVd (with a bulging loop of four nucleotides). Supporting this view, extended hammerheads derived from sTRSV and CChMVd, with bulging loops of seven nucleotides, also display low catalytic efficiency (Weinberg and Rossi 2005).

8 *Trans*-Cleavage In Vivo of a Viroid RNA by an Extended PLMVd-Derived Hammerhead

Considering that at low Mg^{2+} concentration the HHe-PLMVd displayed the best activity against short and long RNAs in vitro, this hammerhead was chosen for further evaluation *in planta*. For this purpose, two cultures of *Agrobacterium tumefaciens* transformed with constructs expressing the ribozyme and the

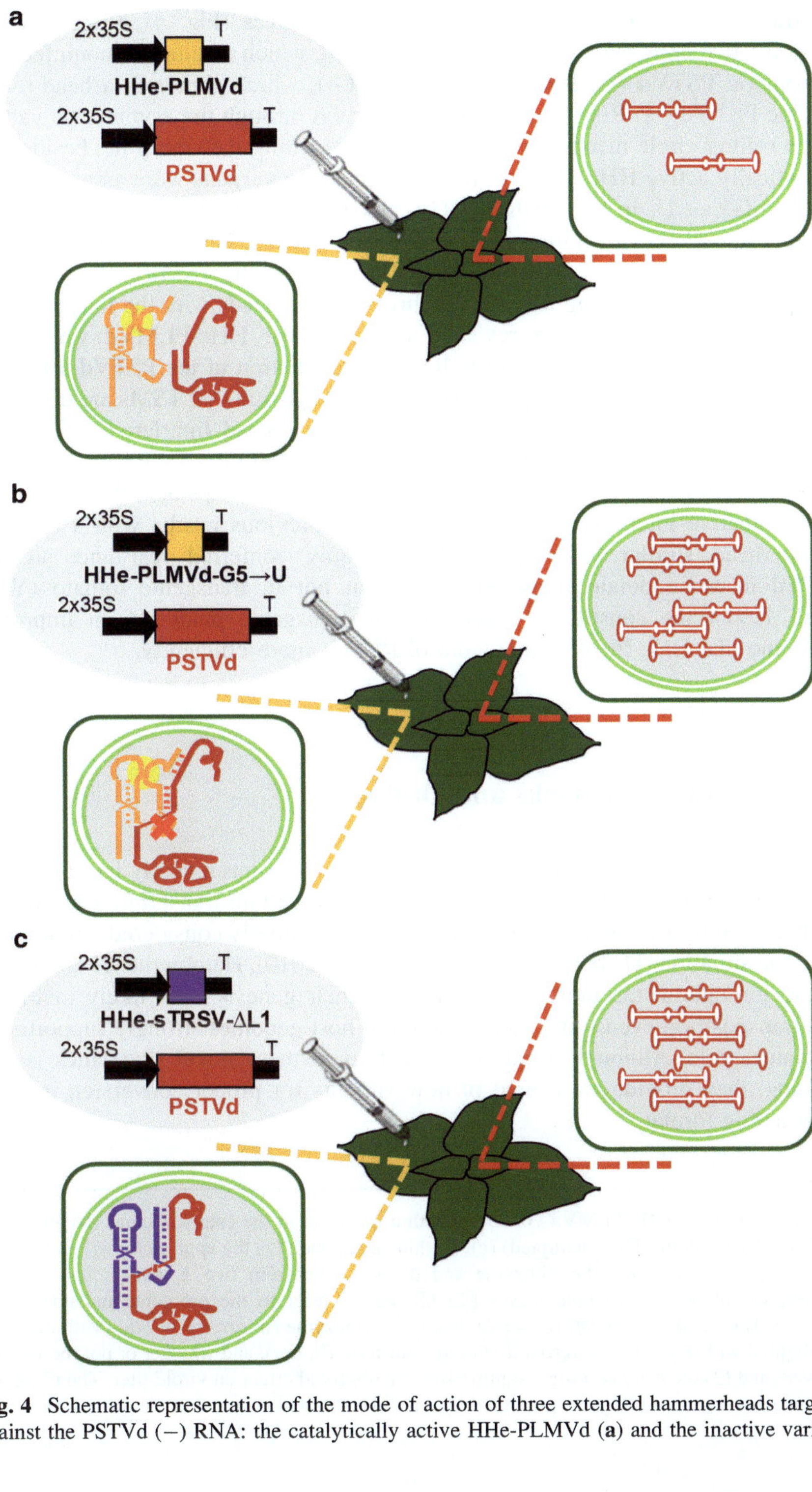

Fig. 4 Schematic representation of the mode of action of three extended hammerheads targeted against the PSTVd (−) RNA: the catalytically active HHe-PLMVd (**a**) and the inactive variants

substrate were coinfiltrated in *N. benthamiana* leaves (Fig. 4). Regarding the substrate, two constructs were used: mPSTVd(−), which results in a noninfectious monomeric PSTVd (−) RNA, and dPSTVd (−), which generates a head-to-tail dimeric PSTVd (−) RNA that triggers replication through the asymmetric variant of the rolling-circle mechanism (see above). Regarding the ribozyme, besides the catalytically active HHe-PLMVd, two other inactive variants were assayed: HHe-PLMVd-G5 → U, in which the CUGA box of the central conserved core was mutated into CUUA (resulting in a catalytically inactive hammerhead), and HHe-sTRSV-ΔL1, in which loop 1 was deleted (disrupting the TSM).

Bioassays coexpressing each of the three hammerheads with the noninfectious mPSTVd(−) RNA substrate revealed that only the HHe-PLMVd was active in vivo, strongly suggesting that the lower accumulation of the PSTVd transcript most likely results from ribozyme-mediated cleavage and that TSMs are critical in this respect (Fig. 4). Moreover, this same hammerhead interfered with viroid infection when coexpressed with the infectious dPSTVd (−) RNA, indicating that it may target the primary dimeric transcript and perhaps also the oligomeric (−) replicative intermediates (Fig. 4). Because previous results with a minimal hammerhead similar to HHe-sTRSV-ΔL1 only conferred resistance against PSTVd in some potato transgenic lines, but not in transgenic tomato (Yang et al. 1997), the constitutive expression in transgenic plants of an improved ribozyme like HHe-PLMVd may control PSTVd more efficiently.

9 Concluding Remarks and Outlooks

The use of bioinformatic tools for searching databases has recently led to another unexpected turn: the presence of hammerheads in certain transcripts from newts, crickets, and schistosomes is not the exception, as initially considered, but a rather common situation (De la Peña and Garcia-Robles 2010). Hammerheads are widely spread throughout the biological scale, with their genesis most likely involving retrotransposition events and their location in host genomes strongly supporting a regulatory role. Although a careful dissection of these novel ribozymes is still pending, they provide an arsenal of new variants for proper conversion into the *trans*-acting format.

Fig. 4 (continued) HHe-PLMVd-G5 → U (with a mutation in the catalytic core) (**b**) and HHe-sTRSV-ΔL1 (with the TSM disrupted) (**c**). Within each panel: (1) the *upper left oval outlines* the two constructs expressing the ribozyme and the substrate from two *A. tumefaciens* cultures coinfiltrated in *N. benthamiana* leaves (2×35S and T refer to the promoter and terminator, respectively); (2) the *lower left rectangle* details the ribozyme (in *orange* or *blue*, with the TSM highlighted with a *yellow background*) and the substrate (in *red*), and whether or not the latter is cleaved; and (3) the *right rectangle* summarizes the observed effect on viroid titer. The ribozyme and the genomic viroid RNA (represented by a rodlike structure) are not at the same scale

Acknowledgments Our research is currently being supported by grants BFU2008-03154 and BFU2011-28443 from the Ministerio de Ciencia e Innovación of Spain and by the Generalitat Valenciana (ACOMP/2010/278).

References

Arber W, Linn S (1969) DNA modification and restriction. Annu Rev Biochem 38:467–500

Atkins D, Young M, Uzzell S et al (1995) The expression of antisense and ribozyme genes targeting citrus exocortis viroid in transgenic plants. J Gen Virol 76:1781–1790

Branch AD, Robertson HD (1984) A replication cycle for viroids and other small infectious RNAs. Science 223:450–454

Burke DH, Greathouse ST (2005) Low-magnesium, trans-cleavage activity by type III, tertiary stabilized hammerhead ribozymes with stem I discontinuities. BMC Biochem 6:14

Campbell TB, McDonald CK, Hagen M (1997) The effect of structure in a long target RNA on ribozyme cleavage efficiency. Nucleic Acids Res 25:4985–4993

Canny M, Jucker F, Kellogg E et al (2004) Fast cleavage kinetics of a natural hammerhead ribozyme. J Am Chem Soc 126:10848–10849

Carbonell A, De la Peña M, Flores R et al (2006) Effects of the trinucleotide preceding the self-cleavage site on eggplant latent viroid hammerheads: differences in co- and post-transcriptional self-cleavage may explain the lack of AUC in most natural hammerheads. Nucleic Acids Res 34:5613–5622

Carbonell A, Flores R, Gago S (2011) Trans-cleaving hammerhead ribozymes with tertiary stabilizing motifs: in vitro and in vivo activity against a structured viroid RNA. Nucleic Acids Res 39:2432–2444

Castanotto D, Li JR, Michienzi A et al (2002) Intracellular ribozyme applications. Biochem Soc Trans 30:1140–1145

Chi Y, Martick M, Lares M et al (2008) Capturing hammerhead ribozyme structures in action by modulating general base catalysis. PLoS Biol 6:2060–2068

Daròs JA, Flores R (2002) A chloroplast protein binds a viroid RNA in vivo and facilitates its hammerhead-mediated self-cleavage. EMBO J 21:749–759

De la Peña M, Flores R (2001) An extra nucleotide in the consensus catalytic core of a viroid hammerhead ribozyme: implications for the design of more efficient ribozymes. J Biol Chem 276:34586–34593

De la Peña M, Garcia-Robles I (2010) Ubiquitous presence of the hammerhead ribozyme motif along the tree of life. RNA 16:1943–1950

De la Peña M, Gago S, Flores R (2003) Peripheral regions of natural hammerhead ribozymes greatly increase their self-cleavage activity. EMBO J 22:5561–5570

Diener TO (1971) Potato spindle tuber virus. IV. Replicating, low molecular weight RNA. Virology 45:411–428

Diener TO (2003) Discovering viroids—a personal perspective. Nat Rev Microbiol 1:75–80

Ding B (2009) The biology of viroid-host interactions. Annu Rev Phytopathol 47:105–131

Dufour D, De la Peña M, Gago S et al (2009) Structure-function analysis of the ribozymes of chrysanthemum chlorotic mottle viroid: a loop-loop interaction motif conserved in most natural hammerheads. Nucleic Acids Res 37:368–381

Fadda Z, Daròs JA, Fagoaga C et al (2003) Eggplant latent viroid, the candidate type species for a new genus within the family *avsunviroidae*. J Virol 77:6528–6532

Fire A, Xu S, Montgomery MK et al (1998) Potent and specific genetic interference by double-stranded RNA in *caenorhabditis elegans*. Nature 391:806–811

Flores R, Semancik JS (1982) Properties of a cell-free system for synthesis of citrus exocortis viroid. Proc Natl Acad Sci USA 79:6285–6288

Flores R, Hernández C, De la Peña M et al (2001) Hammerhead ribozyme structure and function in plant RNA replication. Meth Enzymol 341:540–552
Flores R, Hernández C, Martínez de Alba AE et al (2005) Viroids and viroid-host interactions. Annu Rev Phytopathol 43:117–139
Flores R, Grubb D, Elleuch A et al (2011) Rolling-circle replication of viroids, viroid-like satellite RNAs and hepatitis delta virus: variations on a theme. RNA Biol 8:200–206
Forster AC, Symons RH (1987) Self-cleavage of plus and minus RNAs of a virusoid and a structural model for the active sites. Cell 49:211–220
Gesner EM, Schellenberg MJ, Garside EL et al (2011) Recognition and maturation of effector RNAs in a CRISPR interference pathway. Nat Struct Mol Biol 18:688–692
Gross HJ, Domdey H, Lossow C et al (1978) Nucleotide sequence and secondary structure of potato spindle tuber viroid. Nature 273:203–208
Hammond SM, Bernstein E, Beach D et al (2000) An RNA-directed nuclease mediates post-transcriptional gene silencing in *Drosophila* cells. Nature 404:293–296
Haseloff J, Gerlach WL (1988) Simple RNA enzymes with new and highly specific endoribonuclease activity. Nature 334:585–591
Hernández C, Flores R (1992) Plus and minus RNAs of peach latent mosaic viroid self cleave in vitro via hammerhead structures. Proc Natl Acad Sci USA 89:3711–3715
Homann M, Tzortzakaki S, Rittner K et al (1993) Incorporation of the catalytic domain of a hammerhead ribozyme into antisense RNA enhances its inhibitory effect on the replication of human immunodeficiency virus type 1. Nucleic Acids Res 21:2809–2814
Hormes R, Sczakiel G (2002) The size of hammerhead ribozymes is related to cleavage kinetics: the role of substrate length. Biochimie 84:897–903
Hutchins CJ, Rathjen PD, Forster AC et al (1986) Self-cleavage of plus and minus RNA transcripts of avocado sunblotch viroid. Nucleic Acids Res 14:3627–3640
Kawasaki H, Taira K (2002) Identification of genes by hybrid ribozymes that couple cleavage activity with the unwinding activity of an endogenous RNA helicase. EMBO Rep 3:443–450
Khvorova A, Lescoute A, Westhof E et al (2003) Sequence elements outside the hammerhead ribozyme catalytic core enable intracellular activity. Nat Struct Biol 10:708–712
Kore A, Vaish NK, Kutzke U et al (1998) Sequence specificity of the hammerhead ribozyme revisited; the NHH rule. Nucleic Acids Res 26:4116–4120
Liu B, Tabler M, Tsagris M (2000) Episomal expression of a hammerhead ribozyme directed against plum pox virus. Virus Res 68:15–23
Martick M, Scott WG (2006) Tertiary contacts distant from the active site prime a ribozyme for catalysis. Cell 126:1–12
McCall MJ, Hendry P, Jennings PA (1992) Minimal sequence requirements for ribozyme activity. Proc Natl Acad Sci USA 89:5710–5714
Mühlbach HP, Sänger HL (1979) Viroid replication is inhibited by α-amanitin. Nature 278:185–188
Navarro B, Flores R (1997) Chrysanthemum chlorotic mottle viroid: unusual structural properties of a subgroup of self-cleaving viroids with hammerhead ribozymes. Proc Natl Acad Sci USA 14:11262–11267
Navarro JA, Vera A, Flores R (2000) A chloroplastic RNA polymerase resistant to tagetitoxin is involved in replication of avocado sunblotch viroid. Virology 268:218–225
Penedo J, Wilson T, Jayasena S et al (2004) Folding of the natural hammerhead ribozyme is enhanced by interaction of auxiliary elements. RNA 10:880–888
Pley HW, Flaherty KM, McKay DB (1994) Three-dimensional structure of a hammerhead ribozyme. Nature 372:68–74
Prody GA, Bakos JT, Buzayan JM et al (1986) Autolytic processing of dimeric plant virus satellite RNA. Science 231:1577–1580
Rodio ME, Delgado S, De Stradis A et al (2007) A viroid RNA with a specific structural motif inhibits chloroplast development. Plant Cell 19:3610–3616

Rueda D, Wick K, McDowell SE et al (2003) Diffusely bound Mg^{2+} ions slightly reorient stems I and II of the hammerhead ribozyme to increase the probability of formation of the catalytic core. Biochemistry 42:9924–9936

Ruffner DE, Stormo GD, Uhlenbeck OC (1990) Sequence requirements of the hammerhead RNA self-cleavage reaction. Biochemistry 29:10695–10702

Saksmerprome V, Roychowdhury-Saha M, Jayasena S et al (2004) Artificial tertiary motifs stabilize trans-cleaving hammerhead ribozymes under conditions of submillimolar divalent ions and high temperatures. RNA 10:1916–1924

Samarsky DA, Ferbeyre G, Bertrand E et al (1999) A small nucleolar RNA:ribozyme hybrid cleaves a nucleolar RNA target in vivo with near-perfect efficiency. Proc Natl Acad Sci USA 96:6609–6614

Schindler IM, Mühlbach HP (1992) Involvement of nuclear DNA-dependent RNA polymerases in potato spindle tuber viroid replication: a reevaluation. Plant Sci 84:221–229

Scott WG, Finch JT, Klug A (1995) The crystal structure of an all-RNA hammerhead ribozyme: a proposed mechanism for RNA catalytic cleavage. Cell 81:991–1002

Sullenger BA, Cech TR (1993) Tethering ribozymes to a retroviral packaging signal for destruction of viral RNA. Science 262:1566–1569

Tabler M, Tsagris M (2004) Viroids: petite RNA pathogens with distinguished talents. Trends Plant Sci 9:339–348

Tabler M, Homann M, Tzortzakaki S et al (1994) A three-nucleotide helix I is sufficient for full activity of a hammerhead ribozyme: advantages of an asymmetric design. Nucleic Acids Res 22:3958–3965

Uhlenbeck OC (1987) A small catalytic oligonucleotide. Nature 328:596–600

Weinberg MS, Rossi JJ (2005) Comparative single-turnover kinetic analyses of trans-cleaving hammerhead ribozymes with naturally derived non-conserved sequence motifs. FEBS Lett 579:1619–1624

Yang X, Yie Y, Zhu F et al (1997) Ribozyme-mediated high resistance against potato spindle tuber viroid in transgenic potatoes. Proc Natl Acad Sci USA 94:4861–4865

Suppression of Hepatitis C Viral Genome Replication with RNA-Cleaving Deoxyribozyme

Dal-Hee Min and Dong-Eun Kim

Contents

Abstract Downregulation of viral genes via oligonucleotide-based gene therapy is a potential strategy for the treatment of virus infection such as hepatitis C. Hepatitis C virus (HCV) is a small-sized, enveloped, positive-sense single-stranded RNA virus. As HCV has highly mutative properties and strong drug resistance, effective antiviral drug for HCV infection is currently unavailable. One of the potential therapeutic strategies for hepatitis C treatment is to cleave HCV RNA genome with proper antisense nucleic acids, thereby inhibiting virus replication in host. RNA-cleaving antisense oligodeoxyribozyme, known as DNAzyme, is an attractive therapeutic oligonucleotide which enables cleavage of mRNA in a sequence-specific

D.-H. Min
Department of Chemistry, Seoul National University, Seoul 151-747, Republic of Korea

D.-E. Kim (✉)
WCU Program and Basic Research Laboratory, Department of Bioscience and Biotechnology, Konkuk University, Seoul 143-701, Republic of Korea
e-mail: kimde@konkuk.ac.kr

V.A. Erdmann and J. Barciszewski (eds.), *From Nucleic Acids Sequences to Molecular Medicine*, RNA Technologies, DOI 10.1007/978-3-642-27426-8_17,

manner and thus silencing target gene. In this chapter, we discuss current status of functional antisense oligonucleotides that have been applied to inhibit HCV replication in vitro and in vivo. In particular, the DNAzyme and the DNAzyme conjugated nanoparticle system are discussed in detail to demonstrate a successful usage of functional oligonucleotide and its delivery in vivo for further therapeutic application of functional oligonucleotides in the treatment of hepatitis C.

Keywords Antisense oligonucleotide • DNAzyme • Hepatitis C virus • Nanomaterial-based gene delivery • RNA cleavage

1 Introduction

Infection with hepatitis C virus (HCV) causes chronic hepatitis, if untreated, which can eventually lead to liver cirrhosis and hepatocellular carcinoma (Hoofnagle 2002). Although a combinational therapy of interferon-α and the nucleoside analog (e.g., ribavirin) brought encouraging results, relatively poor efficacy and significant side effects are shown in over 50% of treated patients, not achieving stable virus clearance (McHutchison and Fried 2003; Shepherd et al. 2000). To date, intensive efforts have been directed to develop novel drugs against HCV; effective anti-HCV drug is not, however, available because of HCV's high rate of mutation and drug resistance (Zein 2000). Thus, alternative HCV therapeutics in conjunction with the current therapy regimens will be needed in near future to treat HCV-infected patients (Walker et al. 2003).

HCV is an enveloped virus with a positive-sense, single-stranded ~9,500 nucleotides (nt) RNA genome that encodes a single long open reading frame, which is translated into a polyprotein including the core (C), envelope (E1, E2), and nonstructural (NS2 to NS5b) groups of proteins (Choo et al. 1991; Kato et al. 1990) (Fig. 1a). Among different isolates of HCV with considerable nucleotide variability, HCV genotypes 1a and 1b are most clinically relevant (Takamizawa et al. 1991). Translation in the host cytoplasm is initiated under control of the internal ribosomal entry site (IRES) of 340 nt, which is located at the viral 5′ untranslated region (5′ UTR) (Tsukiyama-Kohara et al. 1992). HCV IRES is folded into a stable secondary structure and highly conserved among all HCV genotypes, which contains three distinct stem loops (II–IV) and a pseudoknot (Honda et al. 1999) (Fig. 1b). IRES directs the translational machinery to the initiator AUG codon, and mutations in various regions of the IRES cause deleterious effects in translation, which was proved in many in vitro studies. Since the IRES is unique in HCV RNA genome, which is distinguished from the cap-dependent translation of host cell mRNA, blocking of the IRES region with antisense oligonucleotides (AS-ODNs) could be exploited to achieve selective suppression of HCV gene expression (Honda et al. 1996).

The IRES-mediated translation synthesizes HCV polyprotein, which is subsequently processed into mature viral structural and nonstructural proteins by a series

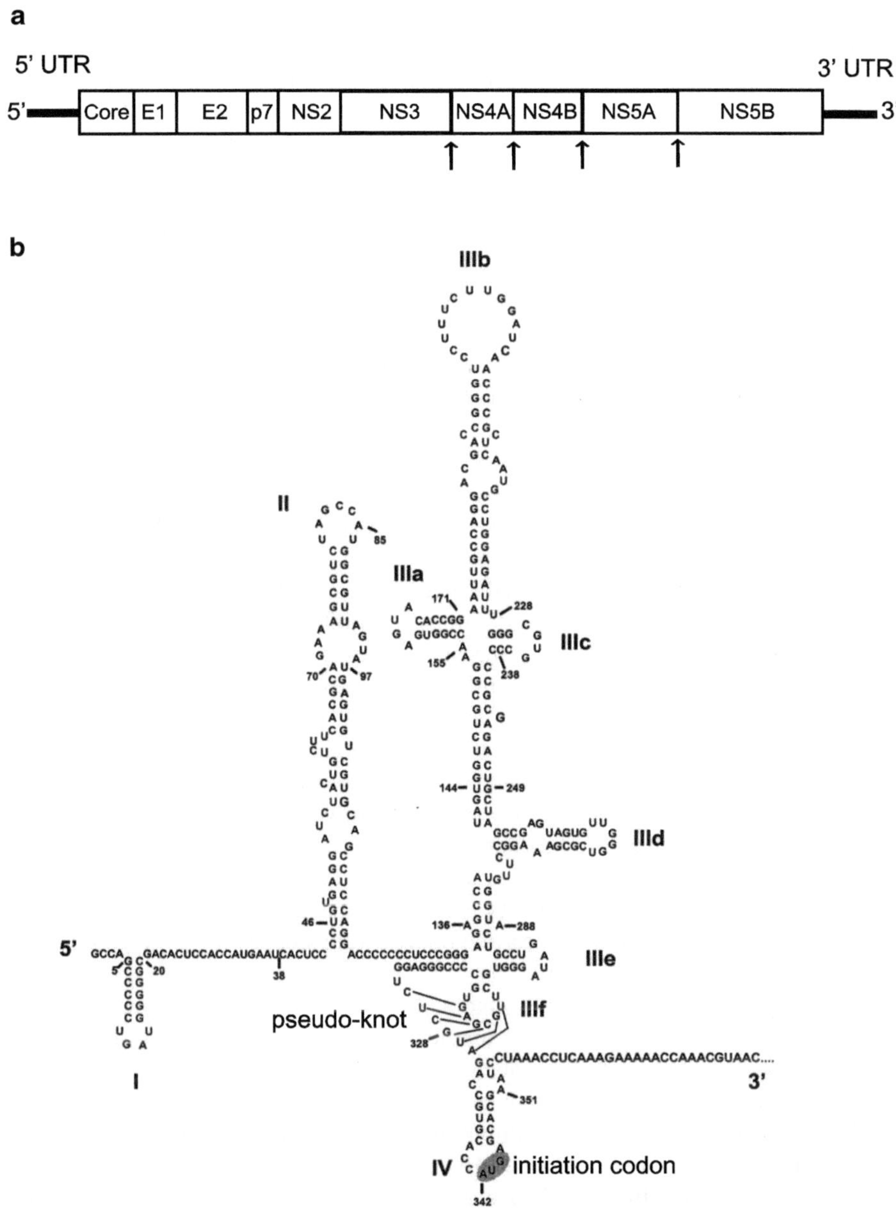

Fig. 1 (**a**) HCV RNA genome structure. Sites of proteolytic cleavage by NS3 are indicated by *arrows*. (**b**) Sequence and secondary structure of internal ribosomal entry site (IRES) located in the 5′ UTR of HCV RNA genome

of cotranslational and posttranslational cleavages by host signal peptidases (Hijikata et al. 1993; Mizushima et al. 1994) and two viral proteases: NS2-3 (Grakoui et al. 1993) and NS3 (Bartenschlager et al. 1993; Manabe et al. 1994; Tomei et al. 1993). Among the HCV nonstructural gene products, NS3 contains a trypsin-like serine protease activity (see Fig. 1a for cleavage by NS3) and a helicase activity in the N-terminal and C-terminal, respectively (Yao et al. 1995). The viral protease NS3 has been identified as an attractive target for anti-HCV drugs, because its activity is indispensable for processing many of the nonstructural proteins of HCV.

As an effective modality to treat the HCV, selective attenuation of the expression of viral genes is counted as one of the appealing antiviral strategies. As such, specific knockdown of the viral gene expression with functional nucleic acids has attracted considerable attention, which is regarded as gene therapy. Gene therapy generally refers to one of the therapeutic options to treat diseases caused by defects in gene expression by regulating gene expression at a posttranscriptional level (Dobson 2006; El-Aneed 2004; Labhasetwar 2005). To date, there have been tremendous efforts to utilize small functional oligonucleotides to specifically inhibit aberrant target genes, including short interfering RNA (siRNA) (McManus and Sharp 2002), antisense oligonucleotide (McMahon et al. 2011), ribozyme (Lewin and Hauswirth 2001), and deoxyribozyme (DNAzyme) (Dass et al. 2008; Isaka 2007). One of the most significant advantages of using these agents for disease treatment is that almost all of the diseases caused by unregulated gene expression—including cancer and viral diseases—may be "treatable" by blocking protein synthesis through degradation of the related mRNAs. Traditional drugs based on small organic compounds are limited in their function because they inhibit enzyme activities and/or protein–protein interactions by exerting on "already expressed proteins."

As a potential target for oligonucleotide-based gene therapy, pathogenic HCV RNA genome has been targeted by various antisense oligodeoxynucleotides (AS-ODNs) to selectively inactivate replication of the viral genome. Several groups have tested AS-ODNs that inhibit HCV viral genome replication and viral polyprotein synthesis both in vitro and in mice models (Alt et al. 1999; Brown-Driver et al. 1999; Hanecak et al. 1996; Lima et al. 1997; Mizutani et al. 1995; Seki and Honda 1995; Wakita and Wands 1994; Yao et al. 1995; Zhang et al. 1999). Phase I/II clinical trials on chronically HCV-infected patients with a phosphorothioate-modified antisense oligonucleotide (ISIS14803) were carried out but were stopped for reasons of lack of efficacy (McHutchison et al. 2006).

Despite great potential of small oligonucleotides for disease treatment, delivery of the oligonucleotides remains the major obstacle to its therapeutic application because of its fast degradation by nucleases in physiological condition, inefficient cellular uptake, and lack of targeting capability. If AS-ODNs that were sensitive to nuclease degradation were used without appropriate delivery vehicle, low-affinity profiles toward their target and side effects were often observed in vivo (Crooke 2004). In addition, the therapeutic use of oligonucleotides is mainly challenged by tissue delivery after systemic administration. The administered oligonucleotides need to travel through the bloodstream, out of the circulation, and act against the target cells. Then, it must find its target mRNA and be knocking out the message. Therefore, development of effective delivery vehicles is essential for successful oligonucleotide-based therapy. Notable delivery methods for small oligonucleotides among the ones developed to date will be discussed in the following section by showing successful demonstration in their strategy. Despite the drawbacks of the

AS-ODNs and delivery issues, catalytic oligonucleotides such as DNAzymes, ribozymes, and small interfering RNAs have attracted a particular attention at the present time (De Francesco and Migliaccio 2005). This review will focus exclusively on the recent DNAzyme strategies against HCV to give a practical insight toward functional AS-ODNs. Detailed study of isolation of DNAzymes that specifically cleave HCV RNA genome and suppress HCV genome replication in cells and mice will be discussed.

2 Antiviral Antisense Oligonucleotides

2.1 RNA-Cleaving Antisense Oligonucleotides: DNAzymes

One of the strategies to inhibit HCV replication in cells is to cleave HCV RNA genome with either ribozyme or deoxyribozyme (Bartolome et al. 2004; Oketani et al. 1999; Shippy et al. 1999; Trepanier et al. 2006). For example, natural hammerhead ribozyme self-cleaving motif has been modified to bind and cleave the passenger RNA strand (Fig. 2A). The modified version of natural hammerhead ribozyme (*trans*-cleaving hammerhead ribozyme) was applied to cleave the target RNA in vitro (Choi et al. 2008; Ludwig et al. 1998). Ribozymes perform catalytic reactions with great precision through Watson–Crick base pairing with complementary sequences. Because ribozymes can be encoded and transcribed from DNA, administration of ribozyme has been accomplished by elaborate vector designs used for overexpression of the ribozyme. Alternatively, efforts have also been made in the synthesis and delivery of ribozymes, not using expressible gene vectors. However, ribozymes are not effective due to their short half-life and difficulty in large-scale synthesis. A decade later, similar but different class of RNA-cleaving oligonucleotides, deoxyribozymes, has entered into the stage of gene therapy tools because of their small size, ease of synthesis, and moderate resistance to chemical and enzymatic degradation (Steele et al. 2003).

Deoxyribozymes, the RNA-cleaving short DNA molecules, also named as DNA enzymes (DNAzymes), were derived by in vitro selection from a combinatorial library of DNA sequences that are capable of cleaving a short target RNA molecule in a sequence-specific manner (Santoro and Joyce 1997). DNAzyme possessing a catalytic motif of "10-23" has a potential to bind and cleave any target RNA that contains a purine–pyrimidine junction (Santoro and Joyce 1997) (Fig. 2A), allowing far greater flexibility in choosing the target sites than hammerhead and hairpin ribozymes. The Watson–Crick base pairing of binding arms confers substrate recognition and binding of DNAzyme to any desired target site of RNA, which makes DNAzyme a powerful tool for gene inactivation strategies. DNAzymes offer several advantages, including cost-effectiveness, straightforward chemical modification, and relatively high stability in serum compared to RNA (Appaiahgari and Vrati 2007).

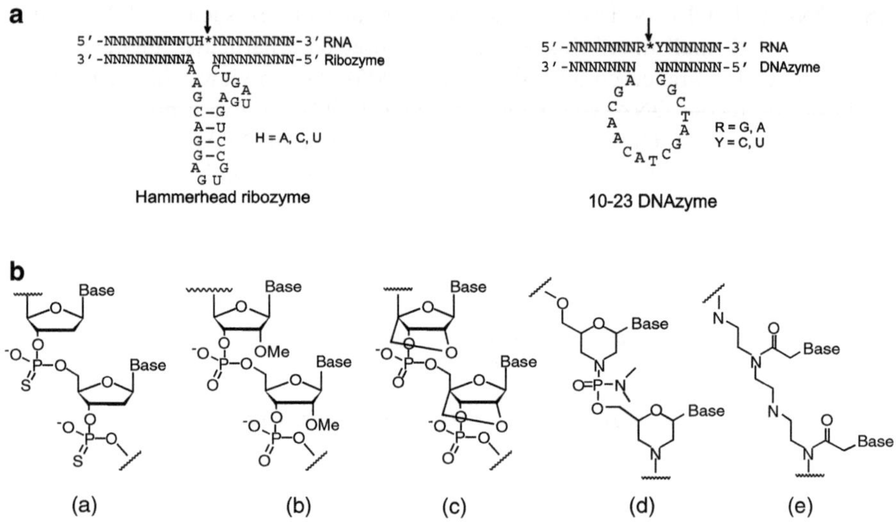

Fig. 2 (**A**) RNA-cleaving antisense oligonucleotides: hammerhead ribozyme and 10-23 DNAzyme. (**B**) Chemically modified antisense oligonucleotides: (a) phosphorothioate, (b) 2′-*O*-methyl oligoribonucleotides, (c) locked nucleic acids (LNA), (d) morpholino nucleic acid, (e) peptide nucleic acid (PNA)

The potential utility of DNAzymes as effective gene therapy agents has been demonstrated by experiments in which pathogenic gene expression was modulated at the posttranscriptional level in various clinically relevant model systems in vitro and in vivo (Achenbach et al. 2004; Dass 2004). A number of research groups have utilized DNAzymes to selectively cleave and disrupt the function of target genes (Lu et al. 2005; Santiago et al. 1999; Wu et al. 1999). Despite the flexibility in potential cleavage sites for DNAzyme, selection and identification of DNAzymes that might be effective and optimal in binding and cleavage of the target mRNA have to be determined empirically. Efforts to predict RNA secondary structure and its influence on the binding of other nucleic acids through free-energy minimization algorithm were not accurate, particularly in long RNAs (Heale et al. 2005; Kretschmer-Kazemi Far and Sczakiel 2003; Luo and Chang 2004; Schubert et al. 2005; Westerhout and Berkhout 2007). Clearly, finding accessible regions for DNAzyme binding in the long stretch of target RNAs gives a major limitation to the effectiveness of DNAzymes. In addition to these challenges, like other oligonucleotides, the lack of a safe, efficient system for the delivery of DNAzymes to target disease sites and tissues remains a major obstacle to the clinical use of DNAzymes.

2.2 *Chemical Modifications of Antisense Oligonucleotides*

Oligonucleotides without chemical modifications are generally vulnerable to nuclease degradation in vivo. To overcome this problem, AS-ODNs have been chemically modified on the sugar and the backbone. As the first and second generation, AS-ODNs were chemically modified on backbone phosphate linkage with replacement of oxygen atom to sulfur (phosphorothioates) [Fig. 2B(a)], methyl group (methylphosphonates), or amines (phosphoramidates). Among these, the phosphorothioates have been the most successful in gene silencing due to their sufficient resistance to nucleases. Alternatively, chemical modification has been also applied to the position 2′ of ribose with an alkoxy group (e.g., methyl group) [Fig. 2B(b)]. For example, antisense 2′-*O*-methyl oligoribonucleotides targeting the HCV IRES were shown to inhibit translation of a HCV RNA genome in rabbit reticulocyte lysate (Tallet-Lopez et al. 2003). DNAzymes have been also modified using the phosphorothioate linkages to render the oligonucleotide more resistant to endogenous nucleases. However, such alterations resulted in several deleterious sequence-independent effects such as toxicity (Wahlestedt et al, 2000), immunologic responsiveness (Fluiter et al. 2003), and increased affinity for cellular proteins (Rockwell et al. 1997).

The third generation of chemically modified oligonucleotides containing structural elements has shown an efficient control of HCV replication in vitro, including locked nucleic acids (LNAs) (Laxton et al. 2011), peptide nucleic acids (PNAs) (Alotte et al. 2008), and morpholino ODN (McCaffrey et al. 2003). All of the modifications enhanced nuclease resistance without hampering specific binding to the target RNA sequence. As shown in Fig. 2B(c), LNA nucleotide has the ribose moiety modified with an extra bridge connecting the 2′ oxygen and 4′ carbon, which effectively increases affinity for complementary sequences (Braasch and Corey 2001). The LNA-incorporated oligonucleotides showed increased thermal stability of duplexes toward complementary DNA or RNA, stability toward 3′ exonuclease degradation (Petersen et al. 2000). Recently, LNA-based ODN targeting the host microRNA miR-122 entered to the phase II clinical test as of late 2010 (Lanford et al. 2010). LNAs have been also applied to modify DNAzymes (Vester et al. 2002; Schubert et al. 2003) to increase binding affinity. LNA incorporation into DNAzymes, however, has been found to diminish both catalytic activity under single turnover conditions (Vester et al. 2002) and biological potency (Fluiter et al. 2005).

Alternatively, morpholinos [Fig. 2B(d)] are another promising ODNs that possess favorable hybridization, nuclease stability, and toxicity profiles. Morpholino ODNs designed to target the HCV IRES prevented HCV IRES translation in a preclinical mouse model (McCaffrey et al. 2003). In addition, PNAs are nucleic acid analogs containing natural nucleoside bases on a pseudo-peptide backbone with strong affinity to DNA and RNA sequences (Nielsen 1997) [Fig. 2B(e)]. PNA ODN targeting HCV IRES sequence showed a strong inhibition of HCV IRES-driven translation in a rabbit reticulocyte lysate (Alotte et al. 2008). Despite its specific affinity to and strong knockout of the target RNA, PNA needs an appropriate transfection reagent to be delivered into the cell because of its charge neutrality and insolubility.

3 Various Strategies for Oligonucleotide Delivery

An efficient and biocompatible delivery system for oligonucleotides is required for its successful clinical application. The carrier type can be classified into two categories—viral and nonviral approaches (Park et al. 2006). Viral vectors have proven their efficiency in the transfection, but there remains safety issue as a major concern of long standing. Nonviral approaches include various cationic lipids, polymers, proteins (e.g., protamine), and various inorganic nanomaterials (Fig. 3). This section will focus on nonviral approaches for oligonucleotide delivery with more in-depth discussion for nanomaterial-based delivery strategies which are recently explosively studied.

3.1 Oligonucleotide Delivery with Functional Polymers

Cationic Lipids. To date, various drugs have been formulated into liposomes which consist of cationic phospholipids to improve cellular uptake and pharmacokinetics (Fig. 3a). Liposomes can fuse with cell membranes and thereby induce efficient cellular entry of drugs. Commercially available phospholipids for oligonucleotide transfection (i.e., Lipofectamine 2000) are typically in the form of mixtures of cationic, fusogenic, and other types of lipids. The cationic lipid-based delivery systems have been extensively used to deliver oligonucleotides into mammalian cells (Li and Szoka 2007). Many successful demonstrations have been reported for oligonucleotide delivery assisted with cationic lipids in vitro and in vivo (Zimmermann et al. 2006; Khoury et al. 2006; Morrissey et al. 2005). However, cytotoxicity of cationic lipid itself should be overcome to be more widely useful as a therapeutic formula.

Polymers. Polyethylenimine (PEI; Fig. 3b) is one of the most widely used cationic polymers for oligonucleotide delivery (Ge et al. 2004; Schiffelers et al. 2004; Grzelinski et al. 2006). PEI–oligonucleotide complexes are thought to electrostatically interact with cell surface and be internalized by endocytosis. There are a number of reports showing therapeutic applicability of PEI for oligonucleotide delivery in many disease models. One of the concerns on PEI is the high toxicity at raised concentrations. Several studies are reported on improving biocompatibility and reducing cytotoxicity of PEI by chemical modification. Other types of cationic polymers based on polyhistidine and polylysines are also demonstrated for their capability as oligonucleotide delivery vehicles.

Proteins. Proteins presenting high cationic charge densities are also harnessed for oligonucleotide delivery (Choi et al. 2010). Protamine-antibody conjugate protein (Fig. 3c) was used for targeted siRNA delivery showing high efficacy (Song et al. 2005). Antibody served as a targeting moiety for receptor-specific binding to cells and protamine noncovalently adsorbed siRNA. This approach minimized adopting synthetic or artificial components in the system.

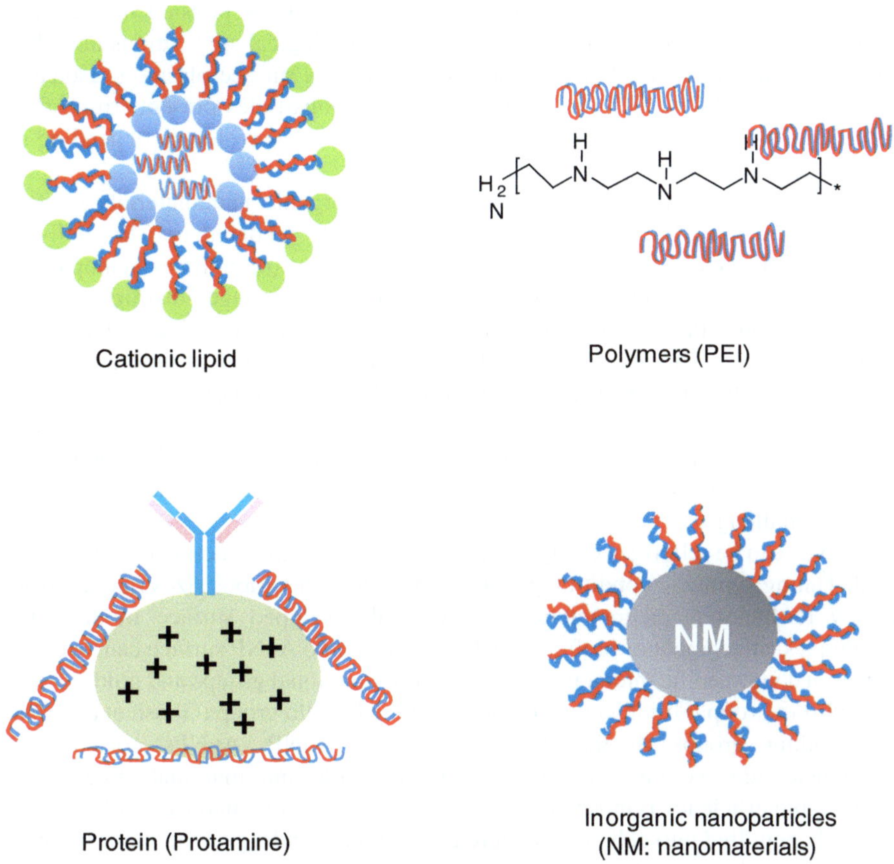

Fig. 3 Various strategies for oligonucleotide delivery designed for therapeutic applications

3.2 *Oligonucleotides Delivery with Inorganic Nanomaterials*

Recent rapid progress in nanotechnology accelerated the development of new interdisciplinary research areas. Nanomedicine refers to the use of nanomaterials to devise medical innovations in a wide range of medical applications from drug discovery and delivery to biomarker discovery and diagnostics (Sekhon and Kamboj 2010). Specifically, inorganic biomedicine bridges inorganic nanomaterials and biomedicine by actively utilizing unique physical and chemical properties of proper inorganic nanomaterials to biomedical applications. Commonly used inorganic nanomaterials (Fig. 3) include gold nanoparticle (AuNP) (Patel et al. 2010), quantum dot (QD) (Han et al. 2001), and magnetic nanoparticle (MNP) (Brown et al. 2005). When gold, silver, and even semiconductors are made in very small sizes, usually below hundreds of nanometers, they no longer behave in ways they did in bulk. These nanomaterials possess unique physicochemical and optical properties which are beneficial for various biological applications including oligonucleotide delivery. For example, AuNP induces surface plasmon resonance for facile optical detection, and semiconductor nanoparticles present strong fluorescence with very narrow emission spectra at different wavelength depending on their size.

One of the important advantages of using inorganic nanomaterials in drug delivery is that multifunctionality can be modularly rendered into the delivery system by introducing targeting ligands, imaging probes (often times, nanoparticle itself serves as an imaging agent), multiple drugs, and triggered drug releasing system into/onto the pertinent nanomaterials while controlling size and shape of delivery vehicle. To date, DNA has been readily conjugated to inorganic nanomaterials which can form structures of a limited number of particles or aggregates of these particles (Deng et al. 2005; Zhao et al. 2006). Since DNA can act as a scaffold due to its potential to form organized microstructures, the development of DNA delivery vehicles with various inorganic nanoparticles presents a facile and feasible challenge.

AuNPs can be relatively easily synthesized by chemical reduction of chloroauric acid. Tuning synthetic condition generates AuNPs with various sizes ranging from a few to hundreds nanometers. Due to well-developed surface modification techniques, attachment of biofunctional molecules onto AuNP surface can be easily performed via specific interaction between thiol functional groups and gold surface. The Mirkin group (Giljohann et al. 2009) and the Rotello group (Ghosh et al. 2008) are leading groups in the bioapplication of AuNP. AuNPs coated with oligonucleotides were used to deliver oligonucleotides into mammalian cells with appreciable efficiency (Ghosh et al. 2008). It is believed that nanosized objects can be easily engulfed into cells with different degrees depending on the shape, relative size, surface charge, and coating material. Low cytotoxicity of AuNP made it very attractive nanomaterial for in vivo application.

QD is one of the semiconductor nanomaterials made of cadmium selenide or cadmium telluride (Xing and Rao 2008). Nowadays, QD is gradually replacing conventional organic dyes in some bioprobing experiments due to bright fluorescence, narrow emission spectra, broad excitation wavelength, and lack of photobleaching even under prolonged irradiation (Smith et al. 2009). QD is reported to show gene silencing when introduced to cells in the form of QD–oligonucleotide complex whereby location of the drug delivery vehicle can be monitored by imaging of QD fluorescence in situ (Derfus et al. 2007). However, safety of QD is an issue to be solved since UV irradiation is reported to induce the release of extremely toxic cadmium metal ion from QD (Derfus et al. 2004).

MNP has been popularly used for MRI contrast agent. MNP is considered to be much safer than other nanomaterials such as QD for in vivo application. Dextran-coated iron oxide MNP is one of the commonly used MNPs for MRI, which is approved by United States Food and Drug Administration (Sekhon and Kamboj 2010). The MNP exhibits long circulation time and enhances MRI visualization of tumors. There have been efforts to use MNP–oligonucleotide complex to achieve

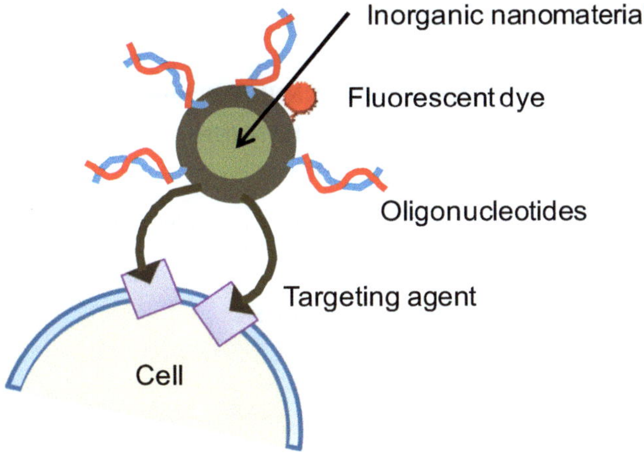

Fig. 4 A structure of multifunctional nanoparticle

target gene knockdown in cultured cells and in vivo (Agrawal et al. 2009). This approach allows "noninvasive imaging" of the delivery vehicles in live objects to track localization, accumulation, and secretion of the MNP–oligonucleotide complexes. Scientists have developed MRI probes with better performance as a contrast agent by tuning sizes or composition of MNPs (Jang et al. 2009). Multi-functional MNPs have been well explored to present targeting moiety, drug, and imaging agent in one nanoparticle (Huh et al. 2005; McCarthy and Weissleder 2008) (Fig. 4). Chemical functionalization of MNP with antibodies, peptides, proteins, organic fluorescent dyes, and oligonucleotides could allow the strategy for the multifunctional MNP.

4 Suppression of HCV Genome Replication with DNAzyme

4.1 In Vitro Selection of DNAzymes that Cleave HCV RNA

AS-ODNs that direct their target by Watson–Crick base pairing need to be designed by avoiding the underlying secondary structure of target RNA, which hampers accessibility of AS-ODNs. Because of unpredictable steric and topological constraints of long stretch of RNA, knowledge of the location of unpaired loop does not guarantee effective hybridization sites (Lima et al. 1997; Matveeva et al. 1997). To overcome this problem, pools of 10-23 DNAzymes were constructed, which possess randomized annealing arm sequence (Dz pools I and II in Fig. 5), and screened for accessible sequences in full-length transcript of HCV NS3 RNA as a target RNA (Lee et al. 2010). The Dz pool I contains oligonucleotides (58 nt) consisting of a central 10-23 catalytic core motif (15 nt) flanked by two binding

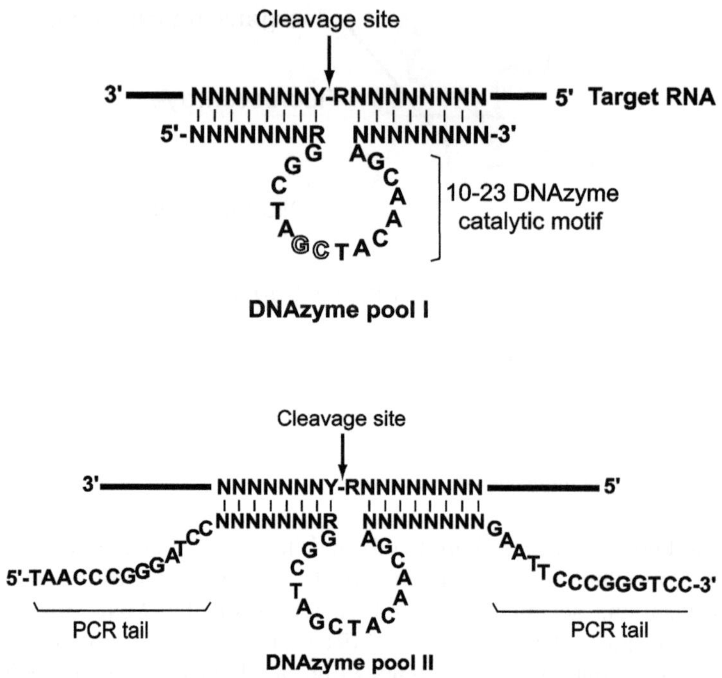

Fig. 5 DNAzymes used as random sequences pool for in vitro selection of DNAzymes that cleave target RNA

arms of randomized sequences, and the Dz pool II contains defined sequences for PCR amplification at both 5′/3′ ends of Dz pool I sequence. Sequence diversity was created by randomizing sequences of two arms totaling 16 nt (8 and 7 Ns plus 1 purine). The fixed terminal sequences of Dz pool II were designed to use PCR amplification of the DNAzymes that are bound to the target RNA substrate in the absence of magnesium.

Selection of accessible cleavage sites by DNAzyme was performed using combinatorial approach as shown in Fig. 6, which is precisely described elsewhere (Lee et al. 2010). The cleaved RNA products from either Dz pool I or II were subjected to primer extension and analyzed on sequencing gels in comparison with the corresponding dideoxy sequencing ladder, and the cleavage sites were identified. Eighteen cleavage sites were identified in the HCV NS3 target RNA (Table 1). The positions of effective DNAzymes in the full-length target RNA tended to be clustered into groups separated by regions containing unreactive sites. In many cases, relatively active target sites such as cleavage sites 1,476–1,488, 904–911, 670–681, and 468–475 nt (Table 1) were very close or even overlapping. This result indicates that these DNAzyme target sites may form an accessible single-stranded loop in the RNA secondary structure (Lee et al. 2010).

Eight DNAzyme-working sites were chosen out of eighteen identified cleavage sites based on the band intensity of cleavage products in the sequencing gel, and the

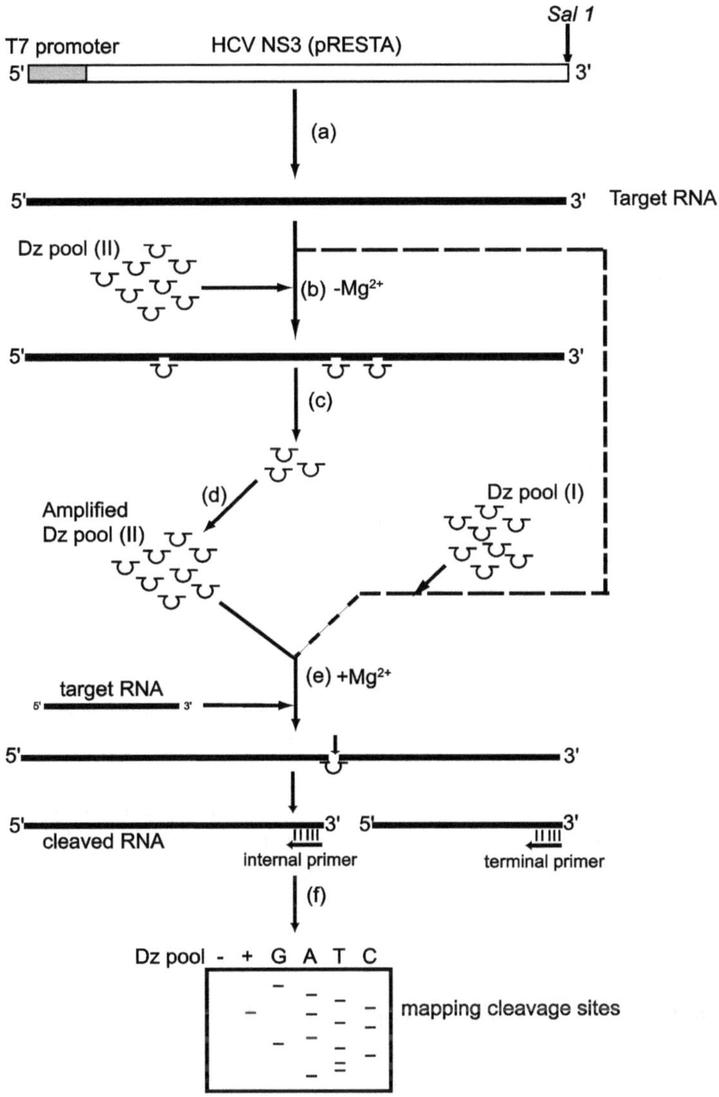

Fig. 6 Schematic of DNAzyme selection procedure by using the randomized DNAzyme pools [see the reference for detail; Lee et al. (2010)]

corresponding DNAzymes were synthesized (Table 1). Cleavage of the target RNA by these synthesized DNAzymes was examined individually. Products of RNA cleavage reaction were observed with the Dz681 and Dz904 targeting on 681st and 904th base of the NS3 RNA, respectively (Fig. 7a). The RNA cleavage activity was more prominent with a shorter RNA substrate (500 nt) of HCV NS3 RNA fragment (501st base to 1,000th base, Fig. 7b). Potential cleavage sites for DNAzymes identified after the screening process mainly reside at a localized portion of the

Table 1 Cleavage sites in the HCV NS3 RNA screened with the DNAzyme library [adapted from Lee et al. (2010)]

RNA sequence	Cleavage site (base)	DNAzyme sequence[a]
5′UGGCAUGC**AIU**GUCAGCU 3′	1,488	–
5′AUCAUGGC**AIU**GCAUGUC 3′	1,484	(Dz1488)5′GACATGCAggctagctacaacgaGCCATGAT 3′
5′CCAAAUUC**AIU**CAUGGCA 3′	1,476	–
5′UAGGAGCC**GIU**CCAAAAU 3′	1,434	(Dz1434)5′ATTTTGGAggctagctacaacgaGGCTCCTA 3′
5′ACGGGCCA**AIC**ACCCCUG 3′	1,407	(Dz1407)5′CAGGGGTGggctagctacaacgaTGGCCCGT 3′
5′GGUUGCCC**GIU**CUGCCAG 3′	1,191	–
5′GCGGGCUU**AIC**CUGAAUA 3′	1,168	–
5′UGUAACAC**AIU**GUGUCAC 3′	911	–
5′ACUGUAAC**AIC**AUGUGUC 3′	909	–
5′GAUUGACU**GIU**AACACAU 3′	904	(Dz904)5′ATGTGTTAggctagctacaacgaAGTCAATC 3′
5′GACAUCUC**AIU**UUUCUGC 3′	717	–
5′GCAAAGCC**AIU**CCCCAUU 3′	681	(Dz681)5′AATGGGGAggctagctacaacgaGGCTTTGC 3′
5′CUUCUACG**GIC**AAAGCCA 3′	673	–
5′CCCCUUCU**AIC**GGCAAAG 3′	670	(Dz670)5′TTTTGCCGggctagctacaacgaAGAAGGGG 3′
5′ACCCCAAU**AIU**CGAGGAG 3′	627	–
5′CGCCCUAC**AIC**AUCAUAA 3′	475	(Dz475)5′TTATGATGggctagctacaacgaGTAGGCGC 3′
5′CCGGGGGG**GIC**CUACGAC 3′	468	(Dz468)5′GTCGTAGGggctagctacaacgaGCCCCCGG 3′
5′ACCCCAAC**AIU**CAGAACU 3′	375	–
NA	NA	(DzI)5′GCATCAAAggctagctacaacgaTGTGTAGA 3′
5′GCAAAGCC**AIU**CCCCAUU 3′	681	(mtDz681)5′AATGGGGAggcta**tg**tacaacgaGGCTTTGC 3′
5′GCAAAGCC**AIU**CCCCAUU 3′	681	(asDz681)5′AATGGGGAGGCTTTGC 3′

[a]DNAzyme sequences (name in parenthesis) are designed to be complementary to sequences flanking the selected target cleavage site. The 10-23 catalytic motif is indicated in lower case and the sequence for the target binding arms is in upper case

RNA stretch. Therefore, a strategy to identify DNAzyme target sites together with in vitro RNA cleavage assay demonstrates its usefulness in design of RNA-cleaving oligonucleotides targeted to a long-stretched RNA (Lee et al. 2010).

4.2 Inhibitory Effect of DNAzymes on HCV Replication in Hepatic Cells

The selected DNAzymes were examined for inhibition of HCV genome replication in a cell culture system, which mimics persistent infection by harboring HCV genome (Lee et al. 2010). HCV permissive hepatoma cell line Huh-7 was transfected with HCV subgenomic replicon RNA that was transcribed from HCV replicon plasmid (pFK-I389neo/NS3-3′/5.1) (Lohmann et al. 2001) and placed under antibiotic selective pressure. Cell viability under G418 selection pressure reflects ongoing replication of HCV replicon in these cells. Each DNAzyme was transfected into the HCV replicon-transfected cells, and the cell viability was assessed 48 h after transfection in G418-containing media. As shown in Fig. 8, cell viability was significantly reduced in HCV replicon cells treated with four

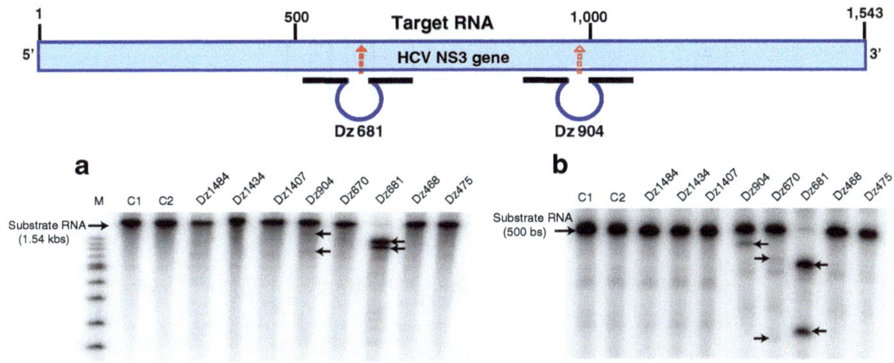

Fig. 7 HCV RNA cleavage by the selected DNAzymes in vitro. Cleavage of (**a**) the full-length HCV NS3 RNA and (**b**) 500-nt fragment of NS3 RNA

DNAzymes such as Dz904, 681, 670, and 468 ($P < 0.005$). As a control, a DNAzyme with irrelevant sequence that does not match NS3 target gene (DzI) was also tested. DzI did not show a significant efficacy against HCV replicon cells. The other control DNAzyme (mtDz681), which abrogated RNA cleavage activity by changing two nucleotides in the catalytic motif (see Fig. 5) but retaining the same binding arm sequences as of Dz681, was tested for cell viability of the HCV replicon cells. In addition, another variant of the most efficient DNAzyme (Dz681), which contains only the binding arm sequences of Dz681 without the catalytic motif (asDz681), was assayed to address if the target binding site is also available for conventional antisense oligodeoxynucleotide (ODN). Both mtDz681 and asDz681 significantly decreased the cell viability of the HCV replicon-transfected cells by 23% and 40%, respectively (Fig. 8a), suggesting that antisense ODN against the same target site as for Dz 681 is also effective.

In order to determine whether the selected DNAzymes can inhibit the propagation of the HCV replicon by eliminating the HCV RNA and protein expression, Western blot analysis and RT-PCR were performed (Lee et al. 2010). In the Western blot analysis for the NS3 expression, Dz681 showed the most potent inhibition of NS3 expression among 10 ODNs tested including the Dz681 variants (mtDz681 and asDz681) (Fig. 8b). At 48 h posttransfection with Dz681, levels of HCV NS3 protein were substantially decreased, when compared with the constitutive expression of β-actin (internal control), which was also observed with its antisense version (asDz681). Other DNAzymes including the control DNAzyme (DzI) that was ineffective in the cell viability decrease such as Dz475, 1407, 1434, and 1484 have failed to show significant decrease in the level of HCV NS3 protein in cells. Overall, the extent of inhibition of NS3 protein expression by each DNAzyme is correlated with the result of the cell viability test. Semiquantitative RT-PCR was also performed with HCV NS3-specific primers to further examine the downregulation of viral RNA by HCV NS3-specific DNAzymes. As shown in Fig. 8b, the HCV NS3 RNA expression was decreased by DNAzymes, Dz904, 670,

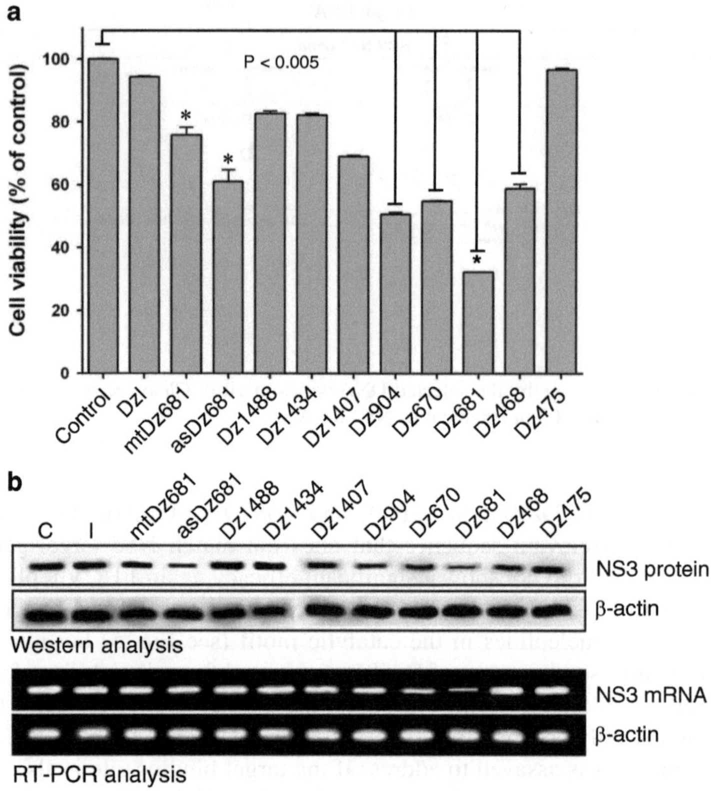

Fig. 8 (**a**) Cell viability assay to assess the efficacy of DNAzymes on suppression of HCV genome replication in Huh-7 cells. (**b**) Effect of DNAzymes on inhibition of target gene NS3 expression, protein, and mRNA [Recap from Lee et al. (2010)]

and 681, which is consistent with the results of Western blot analysis and cell viability test. Interestingly, antisense ODN (asDz681) targeting the Dz681 binding site less inhibited an expression of viral NS3 RNA than that of NS3 protein. We speculate that inhibition of target protein expression was more pronounced than that of the corresponding RNA due to a working mechanism of antisense ODN in cells, in which antisense ODN reduces protein expression by hindering of ribosomal assembly and translation in addition to RNase H degradation of RNA.

4.3 *Delivery of DNAzyme with Iron Oxide Nanoparticles for HCV Gene Knockdown*

After the RNA-cleaving DNAzyme targeting the HCV NS3 RNA genome was identified, an iron oxide nanoparticle-based system has been recently developed for

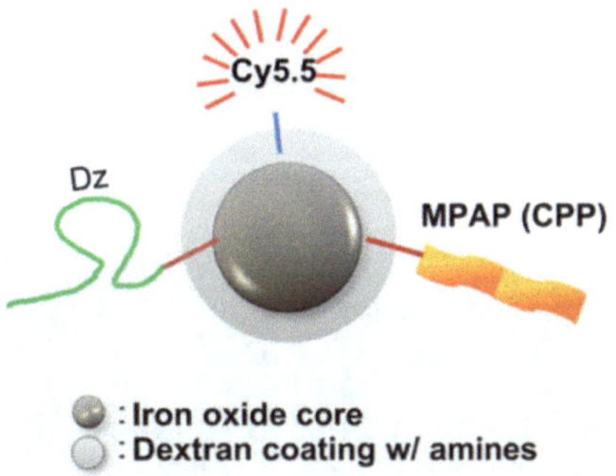

Fig. 9 Schematic of the multifunctional iron oxide nanoparticle for DNAzyme delivery (*Dz* DNAzyme, *MPAP* myristoylated polyarginine peptide, *Cy5.5* fluorescent dye, *CPP* cell-penetrating peptide) [Recap from Ryoo et al. (2012)]

the delivery of DNAzymes to treat hepatitis C (Ryoo et al. 2012). In this study, the iron oxide nanoparticles (MNs) were designed to have a magnetic core, coated with dextran, and are conjugated to the DNAzyme (Dz681) targeting the HCV NS3 RNA, a near-infrared fluorescent dye (Cy5.5), and a cell-penetrating peptide (CPP) that aids in membrane translocation (Fig. 9). The conjugated Cy5.5 dye enables tracking of the therapeutic nanoparticle in vitro and in vivo using fluorescence imaging, and the iron oxide core can be used for tracking via noninvasive magnetic resonance imaging. The fluorescent dye Cy5.5 was conjugated to the aminated MNs using Cy5.5 *N*-hydroxysuccinimide ester. Cy5.5–MNs were conjugated to thiolated Dz and a myristoylated polyarginine peptide (MPAP; myristic acid-ARRRRRRRC), using a cross-linker, sulfosuccinimidyl-6-(30-(2-pyridyldithio)-propionamido)-hexanoate (sulfo-LC-SPDP). MPAP which acts as a cell-penetrating peptide was used to enhance the efficiency of cellular uptake (Nelson et al. 2007). In the MN, linked to both Dz and MPAP (Dz–MPAP–MN), Dz and MPAP were attached via a disulfide linkage which can be cleaved in reducing environments, such as that in the cytoplasm (Saito et al. 2003).

Uptake of MN, Dz–MN, and Dz–MPAP–MN by the HCV replicon cells was evaluated by Cy5.5 fluorescence microscopy 12 h after MN treatment. Cells transfected with the HCV subgenomic replicon RNA containing luciferase reporter gene (pFK-I_{389}neo/NS3-3′/5.1) (Lohmann et al. 2001) were used for assessing inhibition of HCV genome replication in cells by luciferase activity measurement and Western blot analysis. NS3-silencing efficiency of Dz-conjugated nanoparticles transfected in Huh-7 Luc-Neo cells was analyzed in vitro by Western blot analysis of HCV NS3 expression. The Dz-conjugated nanoparticles efficiently inhibited the HCV NS3 replication in cultured Huh-7 Luc-Neo cells. In addition, luciferase (reporter gene) assays of Huh-7 Luc-Neo cells treated with Dz–MPAP–MN indicate dose-dependent downregulation of the NS3 target gene. As shown in Fig. 10, Dz–MPAP–MN was most efficiently internalized and accumulated, primarily at the perinuclear region in the Huh-7 cells, whereas MN and Dz–MN were rarely observed inside the cells. Thus, the addition of the MPAP cell-penetrating peptide to Dz–MN greatly increases the extent of cellular uptake and NS3 knockdown.

To examine the efficacy of the Dz-conjugated nanoparticles in vivo, mice were injected subcutaneously with Dz–MPAP–MN-treated Huh-7 replicon cells

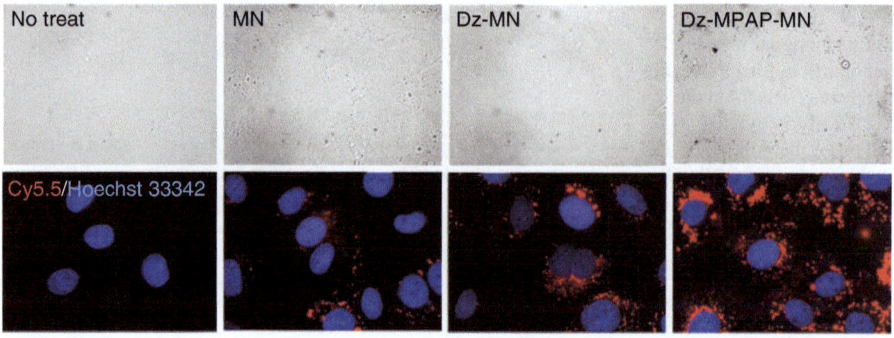

Fig. 10 Fluorescence images of Huh-7 Luc-Neo cells after treatment with various magnetic iron oxide nanoparticle (MN) species (*Blue* Hoechst 33342-stained nuclei, *red* Cy5.5-conjugated MNs). Cellular uptake efficiency was much higher for the MPAP-conjugated MN than for the MN species lacking MPAP. Scale bar is 20 μm

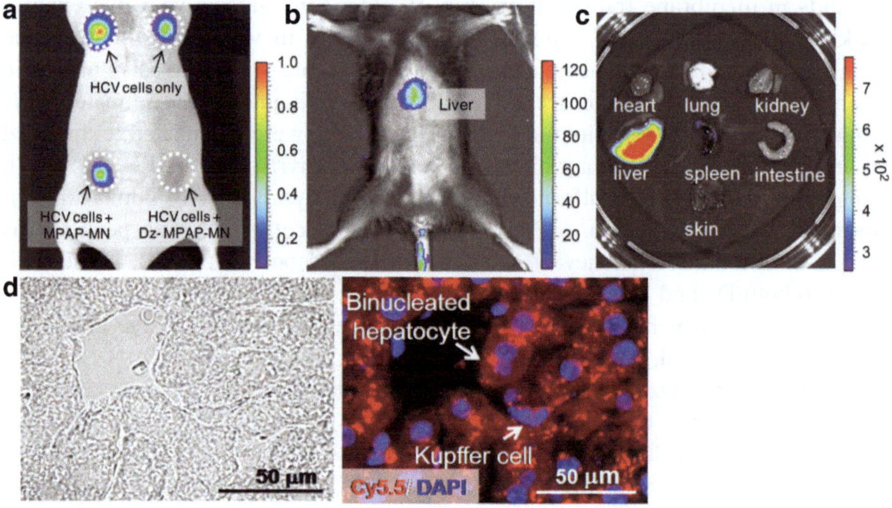

Fig. 11 Analysis of Dz–MPAP–MN efficacy in vivo and biodistribution of Dz–MPAP–MN. (**a**) Subcutaneous injection of the HCV replicon cells treated with Dz–MPAP–MN into mice reduced the luciferase signal, indicating diminished HCV replication in vivo. (**b–d**) Biodistribution of Dz–MPAP–MN after intravenous administration to mice; (**b**) a whole-body image showing intense Cy5.5 fluorescence in the area of the liver, (**c**) fluorescence images of the extracted organs show that the particles accumulate primarily in the liver, (**d**) bright field and fluorescence images of the extracted liver tissue slice are shown. Cellular uptake and accumulation of the intravenously injected Dz–MPAP–MNs were observed in both hepatocytes and Kupffer cells [Recap from Ryoo et al. (2012)]

(Ryoo et al. 2012). Although the ideal test for the in vivo efficacy of drug-loaded nanoparticles would use a virus-infected animal model, small animals (e.g., mice) are not susceptible to infection by human viruses, such as HIV-1 and HCV. Indeed,

HCV infects only humans and chimpanzees (Bukh 2004). In fact, generating xenograft tumors of the HCV replicon Huh-7 cells was not straightforward; the subcutaneously injected cells rapidly lost all of their luciferase activity, within 4 days, because the host rejection process rapidly eliminated the replicon RNA (Zhu et al. 2006). As shown in the whole mouse image in Fig. 11a, the cells co-injected with Dz–MPAP–MN exhibited a lower luciferase signal than the controls, suggesting that Dz–MPAP–MN may effectively downregulate its target gene in vivo. For practical application of systemically introduced nanoparticles, the particles should accumulate in the liver, the site of Dz function. When mice were injected with Dz–MPAP–MN via their tail veins, the particles were found exclusively in the liver within 30 min after injection, as shown by Cy5.5 fluorescence imaging of the nanoparticles and ex vivo imaging after organ extraction (Fig. 11b, c). Examination of liver tissue sections by fluorescence microscopy showed that the injected particles were taken up by both hepatocytes and Kupffer cells (liver macrophages) in the liver. The accumulation of the particles in both types of cells, rather than just in the Kupffer cells, is an encouraging sign for their potential use for hepatitis C treatment (Fig. 11d).

5 Conclusions

RNA-cleaving DNAzymes are attractive functional AS-ODNs as therapeutic candidates, because (1) their ability to cleave RNA substrates in a sequence-specific manner can be used to silence target genes, (2) they are much more stable than ribozymes, and (3) they are less expensive than siRNA molecules. Since a key challenge for clinical application of functional AS-ODNs as an antiviral agent is to develop an efficient and safe delivery system, several nonviral delivery systems for oligonucleotides have been developed. Among them, inorganic nanoparticles are well characterized as DNA delivery and imaging agents. As an example for a usage of DNAzyme against RNA virus such as HCV, DNAzymes possessing randomized annealing arm sequence has been used to identify accessible cleavage sites in the target RNA, HCV NS3 RNA that encodes viral helicase and protease. When the selected DNAzyme was transfected into the hepatoma cells harboring the HCV subgenomic replicon, the DNAzyme efficiently inhibited HCV RNA replication by reducing the expression of HCV NS3. Therefore, the HCV genomic RNA-cleaving DNAzyme is a new tool for viral gene modulation, which can mitigate the chronic HCV replication in vivo, if combined with an appropriate delivery vehicle. To this end, the DNAzyme delivery system that harnesses well-characterized iron oxide nanoparticles was developed and exhibited effective silencing of expression of the HCV NS3 gene in mice. Thus, the presented study is an updated information regarding a demonstration of an effective design and validation of functional DNAzyme against viral RNA genome and its delivery system with a high potential for future clinical use in the treatment of hepatitis C.

Acknowledgments This work was supported by NRF grants (2010-0019306, 2011-0016385) and the WCU project (R33-10128) funded by the MEST, Republic of Korea.

References

Achenbach JC, Chiuman W, Cruz RP et al (2004) DNAzymes: from creation in vitro to application in vivo. Curr Pharm Biotechnol 5:321–336

Agrawal A, Min DH, Singh N et al (2009) Functional delivery of siRNA in mice using dendriworms. ACS Nano 3:2495–2504

Alotte C, Martin A, Caldarelli SA et al (2008) Short peptide nucleic acids (PNA) inhibit hepatitis C virus internal ribosome entry site (IRES) dependent translation in vitro. Antiviral Res 80:280–287

Alt M, Eisenhardt S, Serwe M et al (1999) Comparative inhibitory potential of differently modified antisense oligodeoxynucleotides on hepatitis C virus translation. Eur J Clin Invest 29:868–876

Appaiahgari MB, Vrati S (2007) DNAzyme-mediated inhibition of Japanese encephalitis virus replication in mouse brain. Mol Ther 15:1593–1599

Bartenschlager R, Ahlborn-Laake L, Mous J et al (1993) Nonstructural protein 3 of the hepatitis C virus encodes a serine-type proteinase required for cleavage at the NS3/4 and NS4/5 junctions. J Virol 67:3835–3844

Bartolome J, Castillo I, Carreno V (2004) Ribozymes as antiviral agents. Minerva Med 95:11–24

Braasch DA, Corey DR (2001) Locked nucleic acid (LNA): fine-tuning the recognition of DNA and RNA. Chem Biol 8:1–7

Brown AB, Mahmood U, Cortes ML et al (2005) Magnetic resonance imaging and characterization of spontaneous lesions in a transgenic mouse model of tuberous sclerosis as a model for endothelial cell-based transgene delivery. Hum Gene Ther 16:1367–1376

Brown-Driver V, Eto T, Lesnik E et al (1999) Inhibition of translation of hepatitis C virus RNA by 2-modified antisense oligonucleotides. Antisense Nucleic Acid Drug Dev 9:145–154

Bukh J (2004) A critical role for the chimpanzee model in the study of hepatitis C. Hepatology 39:1469–1475

Choi WH, Choi BR, Kim JH et al (2008) Design and kinetic analysis of hammerhead ribozyme and DNAzyme that specifically cleave TEL-AML1 chimeric mRNA. Biochem Biophys Res Commun 374:169–174

Choi YS, Lee JY, Suh JS et al (2010) The systemic delivery of siRNAs by a cell penetrating peptide, low molecular weight protamine. Biomaterials 31:1429–1443

Choo QL, Richman KH, Han JH et al (1991) Genetic organization and diversity of the hepatitis C virus. Proc Natl Acad Sci USA 88:2451–2455

Crooke ST (2004) Progress in antisense technology. Annu Rev Med 55:61–95

Dass CR (2004) Deoxyribozymes: cleaving a path to clinical trials. Trends Pharmacol Sci 25:395–397

Dass CR, Choong PF, Khachigian LM (2008) DNAzyme technology and cancer therapy: cleave and let die. Mol Cancer Ther 7:243–251

De Francesco R, Migliaccio G (2005) Challenges and successes in developing new therapies for hepatitis C. Nature 436:953–960

Deng Z, Tian Y, Lee SH et al (2005) DNA-encoded self-assembly of gold nanoparticles into one-dimensional arrays. Angew Chem Int Ed 44:3582–3585

Derfus AM, Chan WCW, Bhatia SN (2004) Probing the cytotoxicity of semiconductor quantum dots. Nano Lett 4:11–18

Derfus AM, Chen AA, Min DH et al (2007) Targeted quantum dot conjugates for siRNA delivery. Bioconjug Chem 18:1391–1396

Dobson J (2006) Gene therapy progress and prospects: magnetic nanoparticle-based gene delivery. Gene Ther 13:283–287

El-Aneed A (2004) An overview of current delivery systems in cancer gene therapy. J Control Release 94:1–14

Fluiter K, ten Asbroek AL, de Wissel MB et al (2003) In vivo tumour growth inhibition and biodistribution studies of locked nucleic acid (LNA) antisense oligonucleotides. Nucleic Acids Res 31:953–962

Fluiter K, Frieden M, Vreijling J et al (2005) Evaluation of LNA modified DNAzymes targeting a single nucleotide polymorphism in the large subunit of RNA polymerase II. Oligonucleotides 15:246–254

Ge Q, Filip L, Bai A et al (2004) Inhibition of influenza virus production in virus-infected mice by RNA interference. Proc Natl Acad Sci USA 101:8676–8681

Ghosh P, Han G, De M et al (2008) Gold nanoparticles in delivery applications. Adv Drug Deliv Rev 60:1307–1315

Giljohann DA, Seferos DS, Prigodich AE et al (2009) Gene regulation with polyvalent siRNA-nanoparticle conjugates. J Am Chem Soc 131:2072–2073

Grakoui A, McCourt DW, Wychowski C et al (1993) A second hepatitis C virus-encoded proteinase. Proc Natl Acad Sci USA 90:10583–10587

Grzelinski M, Urban-Klein B, Martens T et al (2006) RNA interference-mediated gene silencing of pleiotrophin through polyethylenimine-complexed small interfering RNAs in vivo exerts antitumoral effects in glioblastoma xenografts. Hum Gene Ther 17:751–766

Han M, Gao X, Su JZ et al (2001) Quantum-dot-tagged microbeads for multiplexed optical coding of biomolecules. Nat Biotechnol 19:631–635

Hanecak R, Brown-Driver V, Fox MC et al (1996) Antisense oligonucleotide inhibition of hepatitis C virus gene expression in transformed hepatocytes. J Virol 70:5203–5212

Heale BS, Soifer HS, Bowers C et al (2005) siRNA target site secondary structure predictions using local stable substructures. Nucleic Acids Res 33:e30

Hijikata M, Mizushima H, Akagi T et al (1993) Two distinct proteinase activities required for the processing of a putative nonstructural precursor protein of hepatitis C virus. J Virol 67:4665–4675

Honda M, Brown EA, Lemon SM (1996) Stability of a stem-loop involving the initiator AUG controls the efficiency of internal initiation of translation on hepatitis C virus RNA. RNA 2:955–968

Honda M, Beard MR, Ping LH et al (1999) A phylogenetically conserved stem-loop structure at the 5′ border of the internal ribosome entry site of hepatitis C virus is required for cap-independent viral translation. J Virol 73:1165–1174

Hoofnagle JH (2002) Course and outcome of hepatitis C. Hepatology 36:S21–S29

Huh YM, Jun YW, Song HT et al (2005) In vivo magnetic resonance detection of cancer by using multifunctional magnetic nanocrystals. J Am Chem Soc 127:12387–12391

Isaka Y (2007) DNAzymes as potential therapeutic molecules. Curr Opin Mol Ther 9:132–136

Jang JT, Nah H, Lee JH et al (2009) Critical enhancements of MRI contrast and hyperthermic effects by dopant-controlled magnetic nanoparticles. Angew Chem Int Ed 48:1234–1238

Kato N, Hijikata M, Ootsuyama Y et al (1990) Molecular cloning of the human hepatitis C virus genome from Japanese patients with non-A, non-B hepatitis. Proc Natl Acad Sci USA 87:9524–9528

Khoury M, Louis-Plence P, Escriou V et al (2006) Efficient new cationic liposome formulation for systemic delivery of small interfering RNA silencing tumor necrosis factor alpha in experimental arthritis. Arthritis Rheum 54:1867–1877

Kretschmer-Kazemi Far R, Sczakiel G (2003) The activity of siRNA in mammalian cells is related to structural target accessibility: a comparison with antisense oligonucleotides. Nucleic Acids Res 31:4417–4424

Labhasetwar V (2005) Nanotechnology for drug and gene therapy: the importance of understanding molecular mechanisms of delivery. Curr Opin Biotechnol 16:674–680

Lanford RE, Hildebrandt-Eriksen ES, Petri A et al (2010) Therapeutic silencing of microRNA-122 in primates with chronic hepatitis C virus infection. Science 327:198–201

Laxton C, Brady K, Moschos S et al (2011) Selection, optimization, and pharmacokinetic properties of a novel, potent antiviral locked nucleic acid-based antisense oligomer targeting hepatitis C virus internal ribosome entry site. Antimicrob Agents Chemother 55:3105–3114

Lee B, Kim KB, Oh S et al (2010) Suppression of hepatitis C virus genome replication in cells with RNA-cleaving DNA enzymes and short-hairpin RNA. Oligonucleotides 20:285–296

Lewin AS, Hauswirth WW (2001) Ribozyme gene therapy: applications for molecular medicine. Trends Mol Med 7:221–228

Li W, Szoka FC Jr (2007) Lipid-based nanoparticles for nucleic acid delivery. Pharm Res 24:438–449

Lima WF, Brown-Driver V, Fox M et al (1997) Combinatorial screening and rational optimization for hybridization to folded hepatitis C virus RNA of oligonucleotides with biological antisense activity. J Biol Chem 272:626–638

Lohmann V, Korner F, Dobierzewska A et al (2001) Mutations in hepatitis C virus RNAs conferring cell culture adaptation. J Virol 75:1437–1449

Lu ZX, Ye M, Yan GR et al (2005) Effect of EBV LMP1 targeted DNAzymes on cell proliferation and apoptosis. Cancer Gene Ther 12:647–654

Ludwig J, Blaschke M, Sproat BS (1998) Extending the cleavage rules for the hammerhead ribozyme: mutating adenosine15.1 to inosine15.1 changes the cleavage site specificity from N16.2U16.1H17 to N16.2C16.1H17. Nucleic Acids Res 26:2279–2285

Luo KQ, Chang DC (2004) The gene-silencing efficiency of siRNA is strongly dependent on the local structure of mRNA at the targeted region. Biochem Biophys Res Commun 318:303–310

Manabe S, Fuke I, Tanishita O et al (1994) Production of nonstructural proteins of hepatitis C virus requires a putative viral protease encoded by NS3. Virology 198:636–644

Matveeva O, Felden B, Audlin S et al (1997) A rapid in vitro method for obtaining RNA accessibility patterns for complementary DNA probes: correlation with an intracellular pattern and known RNA structures. Nucleic Acids Res 25:5010–5016

McCaffrey AP, Meuse L, Karimi M et al (2003) A potent and specific morpholino antisense inhibitor of hepatitis C translation in mice. Hepatology 38:503–508

McCarthy JR, Weissleder R (2008) Multifunctional magnetic nanoparticles for targeted imaging and therapy. Adv Drug Deliv Rev 60:1241–1251

McHutchison JG, Fried MW (2003) Current therapy for hepatitis C: pegylated interferon and ribavirin. Clin Liver Dis 7:149–161

McHutchison JG, Patel K, Pockros P et al (2006) A phase I trial of an antisense inhibitor of hepatitis C virus (ISIS 14803), administered to chronic hepatitis C patients. J Hepatol 44:88–96

McMahon KM, Mutharasan RK, Tripathy S et al (2011) Biomimetic high density lipoprotein nanoparticles for nucleic acid delivery. Nano Lett 11:1208–1214

McManus MT, Sharp PA (2002) Gene silencing in mammals by small interfering RNAs. Nat Rev Genet 3:737–747

Mizushima H, Hijikata M, Tanji Y et al (1994) Analysis of N-terminal processing of hepatitis C virus nonstructural protein 2. J Virol 68:2731–2734

Mizutani T, Kato N, Hirota M et al (1995) Inhibition of hepatitis C virus replication by antisense oligonucleotide in culture cells. Biochem Biophys Res Commun 212:906–911

Morrissey DV, Lockridge JA, Shaw L et al (2005) Potent and persistent in vivo anti-HBV activity of chemically modified siRNAs. Nat Biotechnol 23:1002–1007

Nelson AR, Borland L, Allbritton NL et al (2007) Myristoyl-based transport of peptides into living cells. Biochemistry 46:14771–14781

Nielsen PE (1997) Peptide nucleic acid (PNA) from DNA recognition to antisense and DNA structure. Biophys Chem 68:103–108

Oketani M, Asahina Y, Wu CH et al (1999) Inhibition of hepatitis C virus-directed gene expression by a DNA ribonuclease. J Hepatol 31:628–634

Park TG, Jeong JH, Kim SW (2006) Current status of polymeric gene delivery systems. Adv Drug Deliv Rev 58:467–486

Patel PC, Giljohann DA, Daniel WL et al (2010) Scavenger receptors mediate cellular uptake of polyvalent oligonucleotide-functionalized gold nanoparticles. Bioconjug Chem 21:2250–2256

Petersen M, Nielsen CB, Nielsen KE et al (2000) The conformations of locked nucleic acids (LNA). J Mol Recognit 13:44–53
Rockwell P, O'Connor WJ, King K et al (1997) Cell-surface perturbations of the epidermal growth factor and vascular endothelial growth factor receptors by phosphorothioate oligodeoxynucleotides. Proc Natl Acad Sci USA 94:6523–6528
Ryoo SR, Jang H, Kim KS et al (2012) Functional delivery of DNAzyme with iron oxide nanoparticles for hepatitis C virus gene knockdown. Biomaterials 33: 2754–2761 doi:10.1016/j.biomaterials.2011.12.015
Saito G, Swanson JA, Lee KD (2003) Drug delivery strategy utilizing conjugation via reversible disulfide linkages: role and site of cellular reducing activities. Adv Drug Deliv Rev 55:199–215
Santiago FS, Lowe HC, Kavurma MM et al (1999) New DNA enzyme targeting Egr-1 mRNA inhibits vascular smooth muscle proliferation and regrowth after injury. Nat Med 5:1264–1269
Santoro SW, Joyce GF (1997) A general purpose RNA-cleaving DNA enzyme. Proc Natl Acad Sci USA 94:4262–4266
Schiffelers RM, Ansari A, Xu J et al (2004) Cancer siRNA therapy by tumor selective delivery with ligand-targeted sterically stabilized nanoparticle. Nucleic Acids Res 32:e149
Schubert S, Gul DC, Grunert HP et al (2003) RNA cleaving "10-23" DNAzymes with enhanced stability and activity. Nucleic Acids Res 31:5982–5992
Schubert S, Grunweller A, Erdmann VA et al (2005) Local RNA target structure influences siRNA efficacy: systematic analysis of intentionally designed binding regions. J Mol Biol 348:883–893
Sekhon BS, Kamboj SR (2010) Inorganic nanomedicine–part 1. Nanomedicine 6:516–522
Seki M, Honda Y (1995) Phosphorothioate antisense oligodeoxynucleotides capable of inhibiting hepatitis C virus gene expression: in vitro translation assay. J Biochem 118:1199–1204
Shepherd J, Waugh N, Hewitson P (2000) Combination therapy (interferon alfa and ribavirin) in the treatment of chronic hepatitis C: a rapid and systematic review. Health Technol Assess 4:1–67
Shippy R, Lockner R, Farnsworth M et al (1999) The hairpin ribozyme. Discovery, mechanism, and development for gene therapy. Mol Biotechnol 12:117–129
Smith AM, Mohs AM, Nie S (2009) Tuning the optical and electronic properties of colloidal nanocrystals by lattice strain. Nat Nanotechnol 4:56–63
Song E, Zhu P, Lee SK et al (2005) Antibody mediated in vivo delivery of small interfering RNAs via cell-surface receptors. Nat Biotechnol 23:709–717
Steele D, Kertsburg A, Soukup GA (2003) Engineered catalytic RNA and DNA: new biochemical tools for drug discovery and design. Am J Pharmacogenomics 3:131–144
Takamizawa A, Mori C, Fuke I et al (1991) Structure and organization of the hepatitis C virus genome isolated from human carriers. J Virol 65:1105–1113
Tallet-Lopez B, Aldaz-Carroll L, Chabas S et al (2003) Antisense oligonucleotides targeted to the domain IIId of the hepatitis C virus IRES compete with 40S ribosomal subunit binding and prevent in vitro translation. Nucleic Acids Res 31:734–742
Tomei L, Failla C, Santolini E et al (1993) NS3 is a serine protease required for processing of hepatitis C virus polyprotein. J Virol 67:4017–4026
Trepanier J, Tanner JE, Momparler RL et al (2006) Cleavage of intracellular hepatitis C RNA in the virus core protein coding region by deoxyribozymes. J Viral Hepat 13:131–138
Tsukiyama-Kohara K, Iizuka N, Kohara M et al (1992) Internal ribosome entry site within hepatitis C virus RNA. J Virol 66:1476–1483
Vester B, Lundberg LB, Sorensen MD et al (2002) LNAzymes: incorporation of LNA-type monomers into DNAzymes markedly increases RNA cleavage. J Am Chem Soc 124:13682–13683
Wahlestedt C, Salmi P, Good L et al (2000) Potent and nontoxic antisense oligonucleotides containing locked nucleic acids. Proc Natl Acad Sci USA 97:5633–5638
Wakita T, Wands JR (1994) Specific inhibition of hepatitis C virus expression by antisense oligodeoxynucleotides. In vitro model for selection of target sequence. J Biol Chem 269:14205–14210

Walker MP, Appleby TC, Zhong W et al (2003) Hepatitis C virus therapies: current treatments, targets and future perspectives. Antivir Chem Chemother 14:1–21

Westerhout EM, Berkhout B (2007) A systematic analysis of the effect of target RNA structure on RNA interference. Nucleic Acids Res 35:4322–4330

Wu Y, Yu L, McMahon R et al (1999) Inhibition of bcr-abl oncogene expression by novel deoxyribozymes (DNAzymes). Hum Gene Ther 10:2847–2857

Xing Y, Rao J (2008) Quantum dot bioconjugates for in vitro diagnostics & in vivo imaging. Cancer Biomark 4:307–319

Yao ZQ, Zhou YX, Feng XM et al (1995) Specific inhibition of hepatitis b virus gene expression by an antisense oligonucleotide in vitro. Acta Virol 39:227–230

Zein NN (2000) Clinical significance of hepatitis C virus genotypes. Clin Microbiol Rev 13:223–235

Zhang H, Hanecak R, Brown-Driver V et al (1999) Antisense oligonucleotide inhibition of hepatitis C virus (HCV) gene expression in livers of mice infected with an HCV-vaccinia virus recombinant. Antimicrob Agents Chemother 43:347–353

Zhao W, Gao Y, Kandadai SA et al (2006) DNA polymerization on gold nanoparticles through rolling circle amplification: towards novel scaffolds for three-dimensional periodic nanoassemblies. Angew Chem Int Ed 45:2409–2413

Zhu Q, Oei Y, Mendel DB et al (2006) Novel robust hepatitis C virus mouse efficacy model. Antimicrob Agents Chemother 50:3260–3268

Zimmermann TS, Lee AC, Akinc A et al (2006) RNAi-mediated gene silencing in non-human primates. Nature 441:111–114

Design of Synthetic shRNAs for Targeting Hepatitis C: A New Approach to Antiviral Therapeutics

Brian H. Johnston and Qing Ge

Contents

Abstract Small hairpin RNAs (shRNAs) are widely used as gene silencing tools and typically consist of a duplex stem of 19–29 bp, a loop, and often a dinucleotide overhang at the 3′ end. Like siRNAs, shRNAs show promise as potential therapeutic agents due to their high level of specificity and potency, although effective delivery to target tissues remains a challenge. Algorithms used to predict siRNA performance are frequently used to design shRNAs as well. However, the differences between these two kinds of RNAi mediators indicate that the factors affecting target gene silencing will not be the same for siRNAs and shRNAs. Stem and loop lengths, structures of the termini, the identity of nucleotides

B.H. Johnston (✉)
SomaGenics, Inc., 2161 Delaware Avenue, Santa Cruz, CA 95060, USA

Department of Immunology, Peking University Health Science Center, Beijing 100191, PR China
e-mail: bjohnston@somagenics.com

Q. Ge
Department of Immunology, Peking University Health Science Center, Beijing 100191, PR China

V.A. Erdmann and J. Barciszewski (eds.), *From Nucleic Acids Sequences to Molecular Medicine*, RNA Technologies, DOI 10.1007/978-3-642-27426-8_18,

adjacent to and near the loop, and the position of the guide (antisense) strand all affect the efficacy of shRNAs. In addition, shRNAs with 19-bp or shorter stem lengths are processed and function differently than those with longer stems. In this review, we describe studies of targeting the hepatitis C virus that have provided guidelines for an optimal design for short (19 bp) shRNAs (sshRNAs) that are highly potent, stable in biological fluids, and have minimal immunostimulatory properties.

Keywords Antivirals • Hepatitis C • RNAi • shRNAs • sshRNAs

1 Introduction

While antibiotics have greatly reduced the mortality and morbidity associated with bacterial infections, the development of effective antivirals has proceeded much more slowly. Hepatitis C virus (HCV) infections remain a worldwide health problem, and no vaccine is currently available. While a number of small molecule inhibitors of virally encoded enzymes are showing promise in clinical trials and some have now been approved by the US FDA, cures of HCV generally require multiple agents with different mechanisms of action due to the development of viral resistance. Such viral escape results from the high error rate of the viral RNA-dependent RNA polymerase, which allows rapid exploration of sequence space, creating myriad quasispecies that may include mutations conferring resistance to any given drug. Small interfering RNAs (siRNAs) and their cousins small hairpin RNAs (shRNAs) have two advantages over small molecule drugs that suggest a potential solution to the viral resistance problem: first, they can potentially target conserved but "undruggable" sequences anywhere in the viral genome, including noncoding sequences, and second, multiple small siRNAs or shRNAs can be combined to foil the ability of the virus to escape through single-site mutations. Chronic HCV infection, which represents the most common form of this disease, is also attractive as a disease target for RNAi because the course of the infection is slow. Thus, there is plenty of time for the biology of RNA interference to take place, in contrast to some very rapid viral infections in which viral replication may outrun treatments that require some hours to take full effect.

In this review, we describe the development of a new class of RNA interference (RNAi) effectors, short shRNAs, or sshRNAs, as HCV drugs. sshRNAs have been designed to target the conserved internal ribosome entry site (IRES) element of the HCV genome, one of the most highly conserved regions of the viral genome. Their short length (around 40 nt) distinguishes them from more traditional shRNAs in both potency and mechanism of action.

2 Hepatitis C Virus

HCV, a single-stranded, positive-sense RNA virus, is the most common blood-borne RNA virus, with an estimated 4.1 million persons currently infected in the United States and 180 million worldwide. A total of 3–4 million people are newly infected each year (Armstrong et al. 2000; Alter et al. 1999; Shepard et al. 2005; Ray Kim 2002). Approximately 250,000 human immunodeficiency virus (HIV)-infected persons in the United States are coinfected with HCV (http://www.cdc.gov/ncidod/diseases/hepatitis/c/). HCV is recognized as a major cause of end-stage liver disease, such as liver cancer (with 1–5% of HCV cases leading to this outcome) and cirrhosis (10–20% of cases), and is the leading indication for liver transplantation in the Western world (Shepard et al. 2005). Chronic liver disease ranks as the tenth leading cause of death in the United States, and HCV is estimated to account for 40–60% of these cases. Mortality related to HCV infection (death from liver failure or hepatocellular carcinoma) is expected to increase over the next two decades (Deuffic-Burban et al. 2007).

There are six major genotypes of HCV. The standard treatment is a combination of pegylated interferon (IFN)-α and ribavirin, which is effective in 40–80% of patients, depending on the genotype. Efficacy has been around 50% in patients infected with HCV genotype 1, which comprises nearly 70% of cases in the western hemisphere, although the success rate is improving with the addition of newly approved protease inhibitors to the standard treatment. The response rate for HIV–HCV coinfected patients is lower, estimated at 30–40% (Torriani et al. 2004). Because interferon is associated with severe adverse effects, including flu-like symptoms, hematologic abnormalities, and depression, patient compliance has been poor with an estimated 30% of patients refusing treatment. Considering that the incidence of HCV is increasing worldwide and that within 10 years more deaths from HCV than HIV are predicted by the Centers for Disease Control, safer and more efficacious HCV drugs are urgently needed (Deuffic-Burban et al. 2004, 2006; Law et al. 2003; Salomon et al. 2002).

More than 40 HCV drug candidates are under development. They can be divided into those that target the virus directly [direct-acting antiviral (DAA) agents] and those that affect host targets (Ronn and Sandstrom 2008; Liu-Young and Kozal 2008; Beaulieu 2007). DAAs include small molecules as well as ribozymes, antisense oligonucleotides (ASOs), decoy RNAs, RNA aptamers, siRNAs, and shRNAs. Host-targeting agents include small molecules such as cyclophilin inhibitors, antifibrotic agents, antibodies, modified interferons with improved pharmacokinetics, and oligonucleotides complementary to the required host factor microRNA-122 (miR-122) (Lanford et al. 2010). As mentioned above, the high genetic diversity and rapid mutation and turnover rates of HCV (10^{10}–10^{12} new particles produced per day with an error rate of 10^{-3}–10^{-5} mutations per nucleotide per genomic replication) result in the rapid emergence of viral resistance with many single DAAs (Okamoto et al. 1992; Neumann et al. 1998; Ogata et al. 1991; Sarrazin and Zeuzem 2010). For instance, although clinical studies showed that the use of protease inhibitors in combination with pegylated interferon-α (peg-IFN) and ribavirin increased the rate of sustained viral response (SVR) by at least 20% compared with peg-IFN and ribavirin alone in patients infected with HCV genotype 1, the existence of viral variants with reduced susceptibility has been observed (Deuffic-Burban et al. 2004, 2009, 2012; Sarrazin and Zeuzem 2010; Robinson et al. 2011).

Of the oligonucleotide-based DAAs, siRNAs and shRNA generally have the best efficacy and potency in vitro as well as in animals (Ronn and Sandstrom 2008; McHutchison et al. 2006). In addition, the ability of RNAi to efficiently limit viral replication, and to target multiple genes and/or sequences simultaneously, makes this an attractive therapeutic approach for limiting the emergence of resistant mutants. Both siRNAs and vector-expressed shRNAs have been shown to significantly decrease HCV RNA replication and protein expression in cell culture as well as in animal systems (Watanabe et al. 2007; Randall and Rice 2004; Kapadia et al. 2003; Wilson et al. 2003; Yokota et al. 2003; Seo et al. 2003; Takigawa et al. 2004; Prabhu et al. 2005; Sen et al. 2003; Kanda et al. 2007). More recently, synthetic shRNAs have gained attention, as discussed below.

3 RNA Interference

RNA interference (RNAi) plays a central role in the regulation of eukaryotic gene expressions associated with various biological processes ranging from development to cell homeostasis. Diseases, particularly cancers, are often associated with the dysregulation of particular miRNAs (Shenouda and Alahari 2009).

The mechanism of RNAi is complex and can involve alternative pathways (Filipowicz 2005; Siomi and Siomi 2009). Double-stranded RNA (dsRNA) molecules are recognized and processed by one or more RNase III-family enzymes (Hammond et al. 2000; Zamore et al. 2000; Bernstein et al. 2001). Primary transcripts encoding microRNAs (pri-miRs) are processed in the nucleus by a "microprocessor" complex to individual hairpins before export to the cytoplasm. In the cytoplasm, dsRNAs longer than ~23 bp, including hairpins, are processed by Dicer into double-stranded siRNAs whose strands are each 21–23 nucleotides (nt) in length. Dicer contains a PAZ domain that binds specifically to the 3′ end of single-stranded RNA and two RNase III domains that possess the catalytic cleavage activity. The distance between the 3′-overhang-binding PAZ domain and the active site of the RNase III domains provides a molecular ruler corresponding to the length of an siRNA duplex (Jinek and Doudna 2009). The cleavage products of Dicer have characteristic termini, a monophosphate group at the 5′ end, and a two-nucleotide overhang with a 3′-hydroxyl at the 3′ end (MacRae and Doudna 2007). sRNAs, whether generated by dicing of a precursor RNA or introduced from outside the cell, are loaded into an RNA-induced silencing complex (RISC) containing a protein of the Argonaute (Ago) family (Elbashir et al. 2001; Lee et al. 2004; Pham et al. 2004; Tomari et al. 2004a, b; Jinek and Doudna 2009). Ago proteins contain PAZ, middle (MID), and PIWI domains. The binding of an siRNA to an

Ago is aided by the presence of a 5′ phosphate group and a 3′ dinucleotide overhang at the termini. As with Dicer, the 3′ overhang binds to the PAZ domain of Ago. The 5′ phosphate group binds in a pocket at the interface between the MID domain and the PIWI domain (Jinek and Doudna 2009).

Of the four human Agos, Ago2 is thought to be the only one able to mediate cleavage of a target mRNA (Hammond et al. 2001; Okamura et al. 2004; Meister et al. 2004; Rand et al. 2004; Liu et al. 2004; Song et al. 2004; Rivas et al. 2005). To render the "guide" (antisense) strand of an siRNA available to pair with its target, the passenger strand must be removed. The selection of which strand is to be removed and which retained is thought to be largely governed by asymmetry in the thermodynamic profile of the siRNA duplex termini (Schwarz et al. 2003; Khvorova et al. 2003). After the RISC-loading complex (which contains the dsRNA binding protein TRBP along with Dicer and Ago2) loads an siRNA duplex into Ago2, Ago2 cleaves (or "slices") the "passenger" (sense) strand at a position opposite 10 nt from the 5′-phosphate of the guide strand, facilitating the dissociation and/or degradation by C3PO (Ye et al. 2011) of the resulting passenger-strand fragments. The loss of the passenger strand from the complex produces the active RISC. The anchoring of the 5′ phosphate group into the binding pocket of Ago is essential, and the distance from it determines the position of the cleavage site in the passenger strand (this Ago2-mediated passenger-strand cleavage was also found in the processing of some microRNAs [miRNAs] (Diederichs and Haber 2007).

When passenger-strand slicing is blocked by chemical modification or by mismatches between the two strands of the siRNA, a slower, alternative pathway dissociates and destroys the passenger strand, possibly via an ATP-dependent helicase (RNA helicase A), yielding the active RISC (Matranga et al. 2005; Leuschner et al. 2006; Miyoshi et al. 2005; Kraynack and Baker 2006; Robb and Rana 2007; Rand et al. 2005). RISC uses the bound single-stranded RNA molecule as a guide to "search" the resident population of messenger RNAs (mRNAs) for complementary sequences, eventually cleaving these transcripts and thereby downregulating the expression of the targeted gene (Hammond et al. 2000; Dorsett and Tuschl 2004; Kim and Rossi 2008; Elbashir et al. 2001; Nykanen et al. 2001; Martinez et al. 2002). The currently favored model for target recognition and cleavage by Ago2 is as follows: The target binds to the seed region of the 5′ half of the guide sequence and then base pairing proceeds toward its 3′ end. This results in the 3′ end of the guide strand dissociating from the PAZ domain, leading to a conformational change that positions the active site of Ago2 at the cleavage site on the target (Filipowicz 2005; Tomari and Zamore 2005). The position of the scissile phosphate group of the target mRNA is similar to that in the passenger strand, i.e., 10 nt from the 5′-phosphate group of the guide strand. Mismatches at the 10th and 11th nucleotides prevent the slicing activity (Jinek and Doudna 2009).

The endogenous RNAi machinery has been exploited to advance a wide range of studies involving gene function analysis, pathway mapping, drug target validation, and host–pathogen interactions (Natt 2007; Dorsett and Tuschl 2004; Iorns et al. 2007). New understanding of how RNAi regulates gene expression is also leading to the rapid development of RNAi-based therapeutics, especially in the area of viral disease. For example, RNAi approaches targeting viral genes, including those of the HIV, influenza, and hepatitis A, B, and C viruses (Table 1), respiratory syncytial virus, polio virus, the SARS coronavirus, alphaviruses, and the Marburg and dengue fever viruses, have been shown to limit viral replication in cell culture and, in some cases, in animals (Rossi et al. 2007; Barik and Bitko 2006; Watanabe et al. 2007; Arbuthnot et al. 2007; Seyhan et al. 2007; Gitlin et al. 2002; Fowler et al. 2005; Haasnoot et al. 2003). A number of siRNA drug candidates are in clinical trials (Castanotto and Rossi 2009). Moreover, mechanistic studies of dsRNA/siRNA/miRNA processing by Dicer and Ago2 have also led to the development of algorithms for efficient target selection; chemical modification for improved specificity, functionality, and longevity; and new designs for RNAi triggers such as Dicer substrate, ssiRNA, etc., that benefit both basic and applied research (Bolcato-Bellemin et al. 2007; Collingwood et al. 2008).

4 Structure–Activity Relationships of shRNAs

RNAi effectors can be generated by chemical synthesis of siRNAs and sshRNAs, Dicer cleavage of longer synthetic dsRNAs and shRNAs, or processing of shRNAs and long dsRNAs transcribed from DNA or viral vectors (Dorsett and Tuschl 2004; Chang et al. 2006; Bernards et al. 2006; Fewell and Schmitt 2006; Vlassov et al. 2006; Amarzguioui et al. 2006; Ge et al. 2010a, b). The transcription of shRNAs or dsRNAs can be driven by Pol II promoters or alternatively by Pol III promoters such as the H1 promoter of RNase P or the U6 snRNA promoter. shRNA has received considerable attention due to its widespread use in DNA vector-based shRNA libraries for various loss-of-function screens, generation of cell lines or transgenic animals that express silencing triggers against targets of interest, and therapeutic approaches (Grimm et al. 2006; Li et al. 2005a, b).

4.1 Parameters Involved

An shRNA consists of largely paired antisense and sense sequences connected by a loop of unpaired nucleotides. A duplex stem of typically 24–29 bp, either fully paired or with miRNA-style internal mismatches or loops, is commonly used in vector-expressed shRNAs (Silva et al. 2005; Stegmeier et al. 2005; Boudreau et al. 2008). Although target site selection is critical to silencing activity, the structural design of the shRNA also plays a significant role. With appropriate design, shRNAs can be at least as active as siRNAs targeting the same site. Although the structure–activity relationship of siRNAs has been extensively examined and a 19-bp RNA duplex with 2-nucleotide overhangs at the 3′ ends of each strand is widely used, there have been only a handful of studies on the effect of hairpin structures on efficacy of target silencing. The factors involved in shRNA design are loop size and

Table 1 Hepatitis C-specific siRNAs and shRNAs

Target region in HCV	Model	Biological effect	References
NS5B	Hydrodynamic injection of si/shRNA and the reporter plasmid of NS5B-luciferase fusion	Reduced luciferase expression in mouse liver by 75% (siRNA) to 98% (expressed shRNA)	McCaffrey et al. (2002)
NS3, NS5B	Selectable subgenomic HCV replicon cell culture	HCV RNA replication and protein expression were inhibited more than 30-fold	Kapadia et al. (2003)
Core	HCV replicon cell culture	80-fold decrease in HCV RNA	Randall et al. (2003)
NS3-1, NS5B	HCV replicon cell culture	Effectively suppressed replication of the HCV replicon without suppressing host gene expression	Takigawa et al. (2004)
NS3, NS5B	HCV subgenomic replicon cell culture	HCV RNA synthesis reduced by 90%	Wilson et al. (2003), Wilson and Richardson (2005)
E2, NS3, NS5B	Transient HCV1a replication model	Effectively suppressed HCV RNA replication and protein expression	Prabhu et al. (2005)
Core, E2	EGFP reporter in cell culture	Suppressed EGFP expression	Liu et al. (2006)
Core, NS3, NS4A, NS4B	Genomic HCV replicon cell culture and hydrodynamic injection mouse model	Effectively suppressed viral replication in a dose-dependent manner	Kim et al. (2006), Shin et al. (2009)
NS5A	Cell culture-grown HCV genotype 1a	Effectively inhibited NS5A and core protein expression	Sen et al. (2003)
Stem–loop II of 5′ UTR	HCV subgenomic and full-length infectious replicon cell culture	Suppressed GFP expression and IRES mRNA in the case of six different HCV genotypes	Prabhu t al. (2006)
5′ UTR	HCV IRES-reporter and HCV subgenomic replicon cell culture	~80% suppression of HCV replication with concentrations of siRNA as low as 2.5 nM	Yokota et al. (2003)
5′ UTR	HCV subgenomic replicon with the luciferase gene	Suppressed the luciferase reporter expression	Seo et al. (2003)
5′ UTR	Cell culture-grown HCV genotype 2a, 1a, and 1b replicon system	Inhibited viral genome replication and infectivity titers	Kanda et al. (2007)
5′ UTR	HCV serum infected Huh-7 cells that supports genotype-4 replication	Suppressed HCV RNA by ~25-fold	Zekri et al. (2009)

(continued)

Table 1 (continued)

Target region in HCV	Model	Biological effect	References
5′ UTR	Cell culture-grown HCV and replicon system	Suppressed HCV replications	Ray and Kanda (2009)
5′ UTR	HCV replicon cell culture	Additive HCV inhibitory effects for combinations of ribozymes and siRNAs	Jarczak et al. (2005)
5′ UTR	HCV subgenomic replicon, HCV-luciferase reporter in cell culture	Effectively suppressed RNA replication in replicon; up to 98% knockdown of HCV-luciferase reporter	Ge et al. (2010a, b), Ilves et al. (2006), Vlassov et al. (2007)
5′ UTR	Hydrodynamic injection of HCV-luciferase reporter with shRNA in mice	Up to 99% knockdown of HCV-luciferase reporter in mouse liver	Wang et al. (2005)
5′ UTR	HCV-luciferase reporter expressed in mouse liver with lipid nanoparticle formulation	90% suppression of IRES-luciferase expression in mouse liver	Dallas, Ma et al. (submitted)
5′ UTR	HCV-infected chimeric uPA/SCID mice, lipid nanoparticle formulation	2.5 $\log_{10}$ viral load reduction	Dallas, Ma et al. (submitted)
5′ UTR NS3, NS4A, NS4B, NS5B	HCV subgenomic replicon cell culture	Suppressed HCV RNA and NS5B protein levels up to 75% with single siRNA and 90% with siRNA combinations	Korf et al. (2007)

sequence, stem length, presence of internal mismatches or single-stranded overhangs, and whether it is expressed or synthetic and directly delivered. As with siRNAs, stem length determines whether the molecule is a substrate for Dicer. In the context of vector expression, fully matched shRNAs have been compared with shRNAs containing internal mismatches from sequence alterations in the passenger arm, and the fully matched shRNAs were found to be more potent (Li et al. 2007; Boudreau et al. 2008). The loop can be almost any size, from 2 to 10 or more nucleotides. Brummelkamp et al. found that an expressed 19-bp shRNA with a 9-nt loop provided better target knockdown than similar shRNAs with 5- or 7-nt loops (Brummelkamp et al. 2002). The efficacy seen with expressed shRNAs depends on expression level as well as design parameters (Hinton et al. 2008; Kawasaki et al. 2003; Zhou et al. 2009). To investigate design parameters alone, several groups have examined the impact of structural changes to *synthetic* hairpins and found that the lengths of both stems and loops can affect efficacy. Li et al. reported that in the context of 4-nt loops, shRNAs with 29-bp stems silenced target gene expression more efficiently than those with 19-bp stems, but 19-bp shRNAs

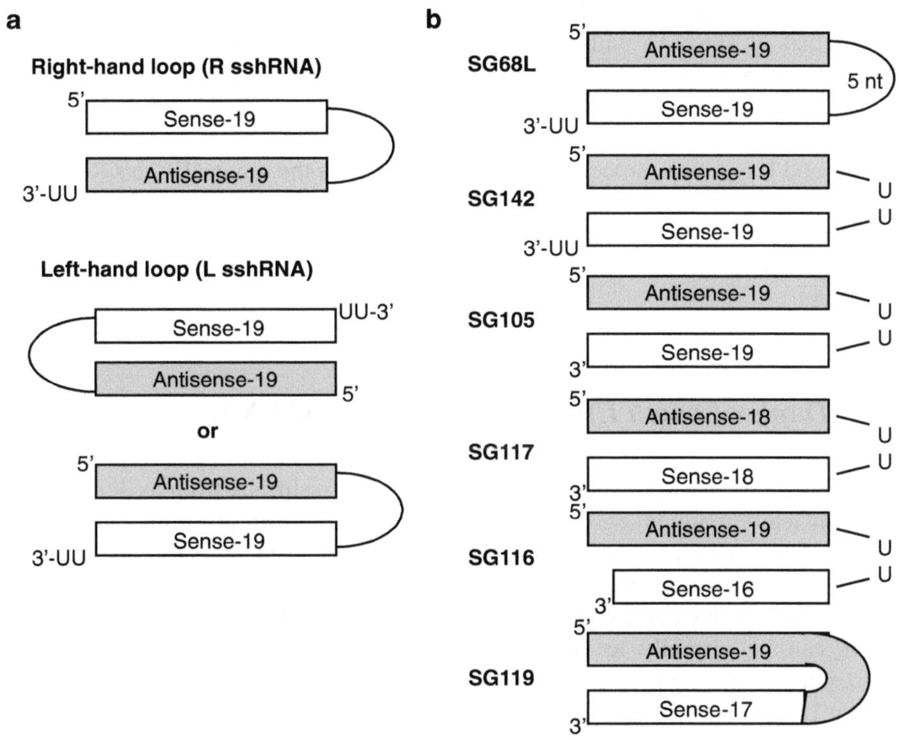

Fig. 1 Structure and activity of R- and L-sshRNAs. (**a**) General structures of R- and L-sshRNAs. Antisense and sense are relative to target sequence. The antisense strand normally becomes the guide strand in RISC. (**b**) Representative structures of L-sshRNAs, with sense and antisense strands of equal or unequal length. Adapted from Ge et al. (2010b)

with 9-nt loops outperformed shRNAs with longer stems, including 29-bp shRNAs with 4-nt loops (Li et al. 2007). Our group found that, in the context of 10-nt loops, 19-bp shRNAs were somewhat more potent than similar 19-bp and 25-bp siRNAs and 25-bp shRNAs were less potent than any of the 19-bp shRNAs or siRNAs tested (Vlassov et al. 2007).

4.2 Loop Position

The position of the antisense sequence within the hairpin also affects shRNA efficacy. shRNAs have often been designed with the sense sequence at the 5′ end of the hairpin (right-hand loop, R-type shRNAs) (Vlassov et al. 2007; Li et al. 2007) (Fig. 1). Harborth et al. reported that an shRNA with its antisense sequence at the 5′ end of the hairpin (left-hand loop, L-type shRNA) showed comparable silencing efficacy to an R shRNA if the stem length was 21–29 nt (Harborth et al. 2003). However, when the stem length was shortened to 19 bp (with a 4-nt loop), much less potency was found with R shRNA, whereas L shRNA retained a potency comparable to that of shRNAs with 21–29-bp stems. Similar results were obtained by another group with a CD8-specific shRNA (McManus et al. 2002). Together, these results suggested that 19-bp shRNAs may be processed differently from shRNAs of 24 bp or longer. A major difference is that, of the longer shRNAs, both R and L types are processed by Dicer to generate the same ~19-bp siRNAs. In contrast, 19-bp shRNAs were found not to be Dicer substrates (Siolas et al. 2005; McManus et al. 2002). To distinguish shRNAs with 19 or fewer base pairs from longer, Dicer substrate shRNAs, we have designated the former as short shRNAs or sshRNAs.

4.3 Duplex Length Effects on Activity of sshRNAs

Shortening the sense sequence of a synthetic 19-bp L-sshRNA from its 3′ end to 17 or 16 nt while maintaining the length of the antisense arm at 19 nt significantly reduced gene silencing activity, suggesting that having duplex structure at the 5′ end of the antisense sequence is important(Fig. 2) (Ge et al. 2010b). However, the overall length of the duplex can be shorter than 19 bp. sshRNAs having 17- or 18-bp stems can be virtually as potent as similar 19-bp versions (Fig. 2). Further shortening of the stem to 16 nt in each strand (connected by UU, which might form a 15-bp stem with a GUUC loop) resulted in somewhat lower activity, and an sshRNA with 15 nt in each arm of the duplex had very little activity. This indicates that potent silencing activity requires a hairpin with a duplex length of at least 16 bp. Interestingly, an sshRNA with a stem of 16 bp consisting of a 19-nt antisense sequence connected directly to a 17-nt sense sequence showed similar potency compared to its parent molecule with 19-bp stem and UU linker.

4.4 Mechanistic Differences of L- and R-sshRNAs

To further examine the structure–function relationships of sshRNAs and how they may differ from Dicer-substrate shRNAs, our group undertook an extensive structure–activity and mechanistic study of sshRNAs targeting three partially overlapping sequences within the internal ribosome entry site (IRES) of HCV (Ge et al. 2010a, b; Dallas et al. 2012). For two of these target sites, L-type sshRNAs showed significantly higher potency than R-type; for the third, there was little difference between the two hairpin types. Unlike R-sshRNAs, where loop sizes of 5 nt or greater were optimal (Vlassov et al. 2007; Li et al. 2007), the L-sshRNAs were more potent when the loop size was very small (1 or 2 nt) than when it was larger (5 or 10 nt) (Fig. 3). The IC_{50} of an L-sshRNA of 19 bp and a UU loop was slightly more potent than a corresponding siRNA targeting the same region. The loop sequence appeared not to affect the sshRNA activity since three

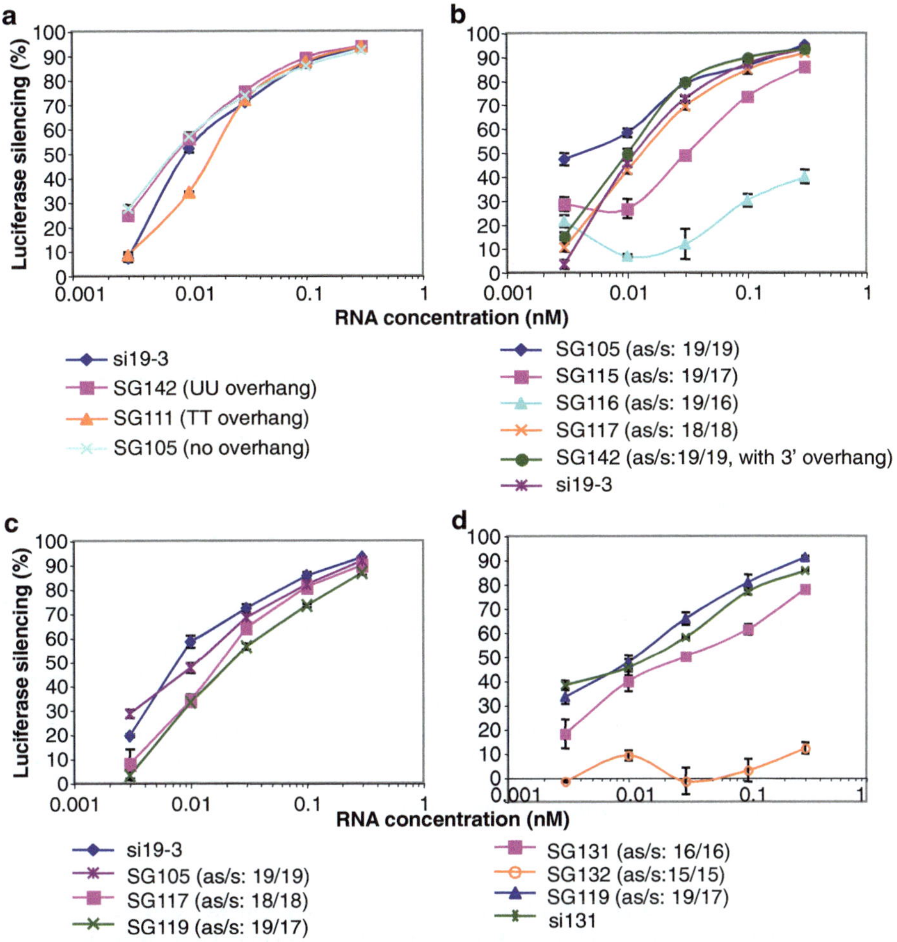

Fig. 2 Comparison of the activities of L-sshRNAs differing in 3′ overhang and stem length. (**a**) Overhang comparison. L-sshRNAs with and without a 3′ overhang as well as an siRNA, si19-3, all targeting the same sequence were chemically synthesized, and their ability to inhibit HCV IRES-dependent luciferase expression was compared in 293FT cells. (**b**) Stem length comparison. L-sshRNAs (UU loop) against the same target sequence but with different stem lengths were compared for their inhibitory activity in 293FT cells. SG105 differs from SG142 in lacking a 3′ UU overhang. (**c**) Effect of using part of the antisense sequence as the loop. SG119 has a 19-nt antisense sequence directly linked with a 17-nt sense sequence, probably forming a 16-bp duplex and 4-nt loop (UGCA). (**d**) Stem length comparison with SG119 derivatives. SG131 and SG132 have UU to connect the 3′ end of the antisense and 5′ end of the sense strands. They were compared with SG119 for target knockdown in 293FT cells. si131 has two complementary strands 16 nt in length with UU overhangs at their 3′ ends. as/s values represent the nucleotide lengths of antisense and sense strands. Adapted from Ge et al. (2010b)

different loop sequences, including one derived from the microRNA miR-23, gave the same results. However, nucleotides adjacent to the loop may affect the activity as they can affect the actual size of the loop. For example, a CG base pair next to a

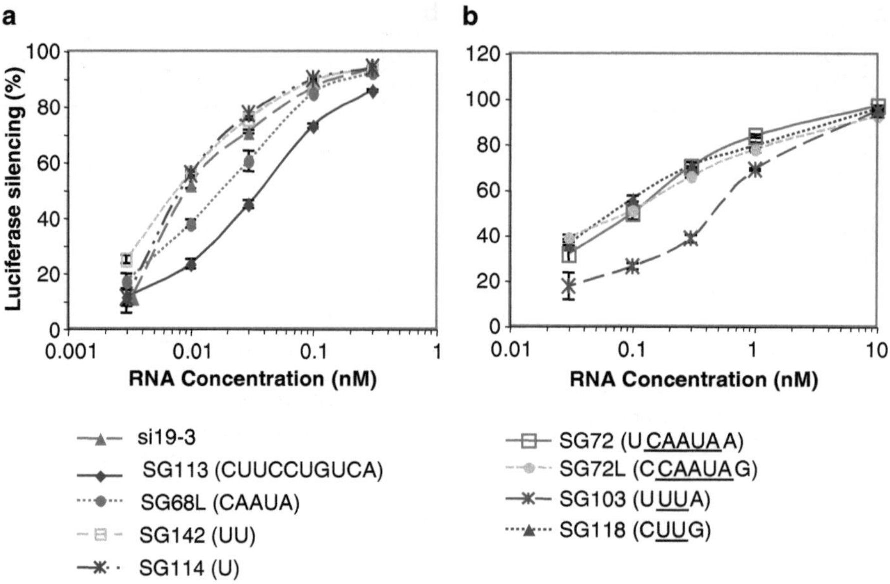

Fig. 3 Comparison of L-sshRNA activity with various loop structures and base pairs adjacent to the loop. (**a**) Loop length comparison, using L-sshRNAs against the same target region as sh68 and si19-3 but with various lengths and sequences of loops. sshRNAs were chemically synthesized, and their abilities to inhibit HCV IRES-dependent luciferase expression were compared in 293FT cells. (**b**) Loop-adjacent base pair comparison. sshRNAs against the same target region as si72 but with loops of 5 nt (SG72 and SG72L) and 2 nt (SG118 and SG103) and left- (SG72L and SG118) and right-loop (SG72 and SG103) orientation were compared for their inhibitory activity in 293FT cells. Loop sequences are *underlined*. Adapted from Ge et al. (2010b)

UU loop can be paired and result in an actual 2-nt loop (Jucker and Pardi 1995), whereas if that base pair is AU or UA, strain in a 2-nt loop may keep them unpaired, resulting in a 4-nt loop of AUUU or UUUA. Based on studies involving chemical modification, conditional dicer-knockout cells, and Ago immunoprecipitation, we concluded (Dallas et al. 2012) that L-sshRNAs having UU loops [which are naturally quite resistant to RNase cleavage (Ge et al. 2010a)] require passenger-arm cleavage to be maximally active, but the loops remain intact in the active RISC, with the guide strand breaking its base pairs with the truncated passenger strand to allow pairing with the target. Passenger-strand "slicing" is needed apparently to facilitate opening of the guide-passenger duplex. L-sshRNAs have their guide sequence at the 5′ side of the loop, so the 5′ end of the guide sequence is immediately available for phosphorylation and binding to the MID pocket of Ago2. R-sshRNAs, on the other hand, require a larger loop for full activity because the loop must be cleaved by some nuclease to create a terminal phosphate at the 5′ end of the guide strand (see Fig. 1). In this case, passenger-arm slicing is less important because the cleavage of the loop itself facilitates removal of the passenger strand in active RISC.

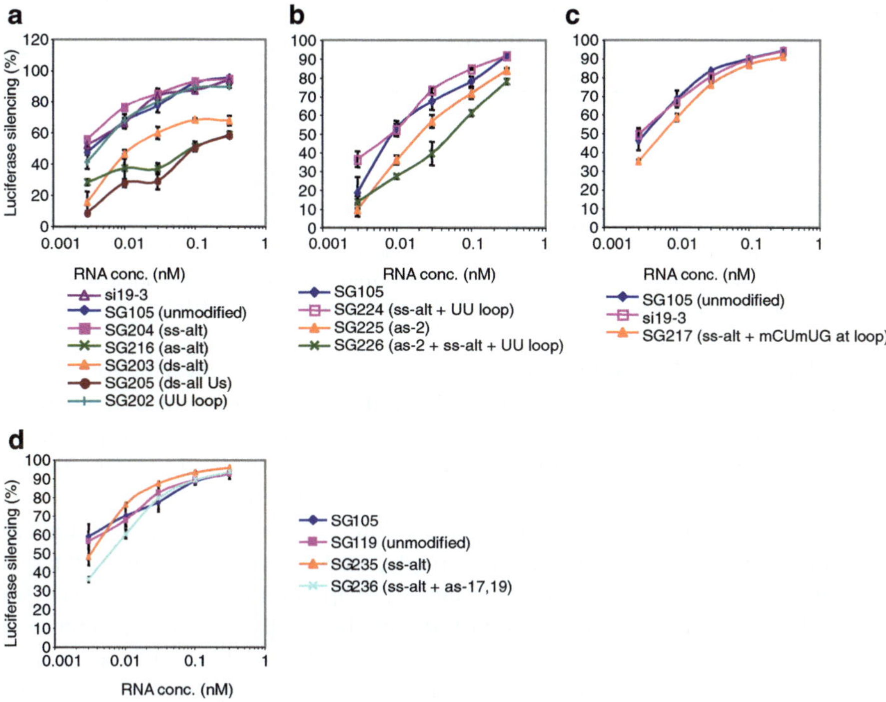

Fig. 4 Effect of 2′-*O*-Me modifications in the stem and/or loop regions on activities of L-sshRNAs. Activity was determined by potency in suppressing the expression of an HCV IRES-fLuc reporter in 293FT cells, by cotransfecting them in triplicate with the reporter DNA. An unmodified siRNA specific for the same target, si19-3, was used as a positive control. (**a–c**) Comparison of an unmodified sshRNA (SG105) and its derivatives containing 2′-*O*-Me modifications in the stem and/or loop. (**d**) Comparison of SG119 and its derivatives containing 2′-*O*-Me modifications in the stem and loop. Each construct is identified by sequence number and a shorthand notation for the region modified, e.g., as-2 means 2′-*O*-methylated at position 2 of the antisense strand, ss-alt means 2′-*O*-methylated at every other nucleotide of the sense strand except for the slicer site. Adapted from Ge et al. (2010a)

Whereas the presence of short flanking sequences such as a 3′-overhang generally enhance the efficiency of gene knockdown for R shRNAs (Siolas et al. 2005; Vlassov et al. 2007), the presence or absence of a 3′ overhang has relatively little effect on the silencing ability of L-sshRNAs (Fig. 2a). This appears to be because the loop is generally not cleaved in L-sshRNAs, and when the stem opens up, the loop is available to bind to the PAZ domain of Ago2 (Dallas et al. 2012).

5 Monomeric Versus Oligomeric shRNAs

Synthetic shRNAs are usually simply dissolved in H_2O or a buffer prior to in vitro or in vivo use. Whether these shRNAs form the expected hairpin structures in solution is frequently ignored. We have found that synthetic shRNAs, irrespective of stem

Table 2 Heating and snap cooling reduces immune stimulation by unmodified sshRNAs

	IFN-β		TNF-α	
Heating/snap cooling	−	+	−	+
SG105	47.5 ± 0.4	5.0 ± 0.6	6,192 ± 2,422	319 ± 81
SG117	23.1 ± 2.7	4.6 ± 0.7	2,552 ± 73	195 ± 37
SG119	7.3 ± 1.8	2.6 ± 0.6	5.9 ± 2.8	5.4 ± 2.1

Three different unmodified sshRNAs (100 nM), either untreated or subjected to 95°C heating (4 min) and snap cooling, were transfected into MRC5 cells in triplicate. RNA was extracted from cells 24 h post-transfection, and IFN-β and TNF-α mRNAs were quantified by RT-PCR. The mean values and standard errors of the relative RNA levels (fold differences) of cytokine genes were calculated and normalized to levels of GAPDH. Adapted from Ge et al. (2010b).

length, loop size, or L vs. R loop orientation, can form dimers or even oligomers (Ge et al. 2010b). Dimerization of hairpin RNAs has been also documented in retroviral RNAs, tRNAs, and some artificial RNA hairpins (Sun et al. 2007). The propensity of hairpin RNAs to dimerize depends on their loop size, sequence, and concentration as well as how they are handled (Bernacchi et al. 2005; Liu et al. 2005). When shRNAs are heated to 95°C and quickly cooled in an ice bath (snap cooling), dimerization can be eliminated (Ge et al. 2010a). Interestingly, sshRNAs showed efficient target knockdown both before and after the heating/snap cooling procedure, even for some sshRNAs that were predominantly dimers before the treatment, suggesting that the dimers of certain sshRNAs are functional molecules that can be processed and utilized by the RNAi machinery with similar efficiency as the monomers. However, the presence of long duplex regions in dimers can provoke immune stimulation. The heating/cooling procedure greatly reduces this immune stimulation (Table 2).

6 Chemical Modification to Improve Pharmacological Properties

Like all RNAs, shRNAs are nonideal as drugs due to their susceptibility to degradation by nucleases and the tendency of some sequences and structural features to cause unwanted immune stimulation. In a study of the effects of 2′-*O*-methyl, 2′-deoxy (DNA), and phosphorothioate modifications at various positions in the stem, loop, and overhangs, we found that placing a 2′-*O*-Me on each nucleotide of the loop and alternate nucleotides of the passenger arm, but leaving an unmodified window of 4 nt at the slicer site, provided significantly greater stability in 10% human serum (Fig. 5) while abrogating induction of the innate immune system (Ge et al. 2010a). It can be seen from Table 3 that two sshRNAs of the same structure but different sequences can have very different immunostimulatory properties. Blunt sshRNAs are particularly efficient at inducing RIG-I. In each case, however, 2′-*O*-methylation renders them non-stimulatory, by blocking recognition by pattern recognition sensors such as RIG-I. These modifications had essentially no effect on potency, but placing modifications in most positions of the guide arm or the slicer

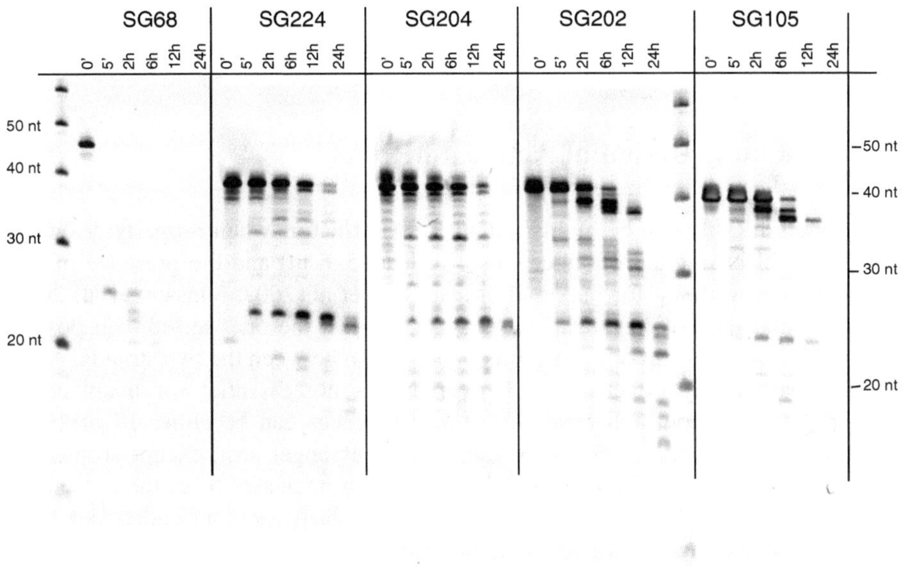

Fig. 5 Effect of 2′-*O*-Me modifications and loop size on serum stability of sshRNAs. sshRNAs were incubated with 10% human serum at 37°C for the times shown. Aliquots were analyzed by denaturing 12% PAGE. Adapted from Ge et al. (2010a)

Table 3 Effect of 2′-*O*-Me modification on immune stimulation by sshRNAs

2′-*O*-Me	SG142 (3′-overhang)		SG118 (3′-overhang)		T7 transcribed shRNA	Blunt-ended version of SG142	
	–	+	–	+	–	–	+
IFN-b	1.5 ± 0.2	1.3 ± 0.1	192 ± 7	2.2 ± 0.5	393 ± 35	7.3 ± 1.8	0.6 ± 0.1
IL-6	2.8 ± 0.1	1.3 ± 0.1	21.5 ± 1.7	1.5 ± 0.2	83 ± 27	4.6 ± 1.4	0.9 ± 0.1
TNF-a	7.7 ± 0.4	1.5 ± 0.4	548 ± 92	1.8 ± 0.6	319 ± 49	5.9 ± 2.8	1.3 ± 0.2
TLR3	4.3 ± 0.2	1.3 ± 0.1	67.6 ± 3.7	1.5 ± 0.1	95 ± 5	8.0 ± 0.9	1.1 ± 0.1
TLR7	1.5 ± 0.4	7.0 ± 3.0	6.5 ± 0.8	2.6 ± 1.2	7.8 ± 4.0	1.2 ± 0.4	0.3 ± 0.04
TLR8	1.4 ± 0.1	4.5 ± 1.7	5.3 ± 0.4	N.D.	4.1 ± 1.1	2.1 ± 0.4	0.5 ± 0.1
RIG-I	28.4 ± 10.2	2.8 ± 1.5	18.9 ± 8.5	3.9 ± 1.6	141 ± 35	289 ± 73	13.5 ± 6.1
PKR	3.6 ± 0.4	1.2 ± 0.1	12.1 ± 2.2	1.2 ± 0.1	6.2 ± 2.2	4.3 ± 0.4	1.1 ± 0.2

Shown are mean values and standard errors of the mRNA levels (relative to an untreated control and normalized to GAPDH) of genes of interest. 100-nM sshRNAs with (+) and without (–) 2′-*O*-Me modifications on the loop and alternate nucleotides of the passenger strand were transfected into human MRC-5 cells in triplicate (without heating and snap cooling). SG142 and SG118 are 19-bp sshRNAs with UU loops and 3′-UU overhangs targeting different sequences on the HCV viral RNA. RNA was extracted from cells 24 h post-transfection, and quantitative RT-PCR was performed. Cells that received the transfection reagent Lipofectamine 2000, alone showed no change in levels of the tested genes. A T7-transcribed shRNA, used as positive control, was transfected into cells in equivalent amounts (on a mononucleotide basis). N.D., not done. Adapted from Ge et al. (2010a).

site of the passenger arm reduced their efficacy. Phosphorothioate modifications were found to induce interferon-beta (IFN-β) and TNF-α in MRC-5 cells.

7 Summary: Design of Active sshRNAs

Synthetic sshRNAs can be highly potent RNAi effectors when properly designed. 19-bp R-sshRNAs require longer loops (at least 5–6 nt) and the presence of a 3′ dinucleotide overhang for maximal efficacy (Li et al. 2007; Vlassov et al. 2007; Siolas et al. 2005). In contrast, L-sshRNAs, at least the ones tested, can possess a loop as short as 1–2 nt or even a direct connection between the two strands. A UU loop is particularly effective. A 3′ overhang is not essential for many active L-sshRNAs, and both antisense and sense sequences can be either 18 or 19 nt. Modification of every other nucleotide on the passenger arm, except around the slicer site, is helpful for increasing stability against nucleases (e.g., those found in serum) and minimizing immune stimulation, particularly for blunt-ended sshRNAs. A heat–snap cooling step immediately prior to use is advisable to eliminate dimers.

Acknowledgment This work was supported by National Institutes of Health grant numbers R44AI056611, R44AI074256, and R43AI074214 (B.H.J.).

References

Alter MJ, Kruszon-Moran D, Nainan OV et al (1999) The prevalence of hepatitis C virus infection in the United States, 1988 through 1994. N Engl J Med 341:556–562

Amarzguioui M, Lundberg P, Cantin E et al (2006) Rational design and in vitro and in vivo delivery of Dicer substrate siRNA. Nat Protoc 1:508–517

Arbuthnot P, Longshaw V, Naidoo T et al (2007) Opportunities for treating chronic hepatitis B and C virus infection using RNA interference. J Viral Hepat 14:447–459

Armstrong GL, Alter MJ, McQuillan GM et al (2000) The past incidence of hepatitis C virus infection: implications for the future burden of chronic liver disease in the United States. Hepatology 31:777–782

Barik S, Bitko V (2006) Prospects of RNA interference therapy in respiratory viral diseases: update 2006. Expert Opin Biol Ther 6:1151–1160

Beaulieu PL (2007) Non-nucleoside inhibitors of the HCV NS5B polymerase: progress in the discovery and development of novel agents for the treatment of HCV infections. Curr Opin Investig Drugs 8:614–634

Bernacchi S, Ennifar E, Toth K et al (2005) Mechanism of hairpin-duplex conversion for the HIV-1 dimerization initiation site. J Biol Chem 280:40112–40121

Bernards R, Brummelkamp TR, Beijersbergen RL (2006) shRNA libraries and their use in cancer genetics. Nat Methods 3:701–706

Bernstein E, Caudy AA, Hammond SM et al (2001) Role for a bidentate ribonuclease in the initiation step of RNA interference. Nature 409:363–366

Bolcato-Bellemin AL, Bonnet ME, Creusat G et al (2007) Sticky overhangs enhance siRNA-mediated gene silencing. Proc Natl Acad Sci USA 104:16050–16055

Boudreau RL, Monteys AM, Davidson BL (2008) Minimizing variables among hairpin-based RNAi vectors reveals the potency of shRNAs. RNA 14:1834–1844

Brummelkamp TR, Bernards R, Agami R (2002) A system for stable expression of short interfering RNAs in mammalian cells. Science 296:550–553

Castanotto D, Rossi JJ (2009) The promises and pitfalls of RNA-interference-based therapeutics. Nature 457:426–433

Chang K, Elledge SJ, Hannon GJ (2006) Lessons from nature: microRNA-based shRNA libraries. Nat Methods 3:707–714

Collingwood MA, Rose SD, Huang L et al (2008) Chemical modification patterns compatible with high potency dicer-substrate small interfering RNAs. Oligonucleotides 18:187–200

Dallas A, Ge Q, Ilves H et al (2012) Right- and Left-loop short shRNAs have distinct and unusual mechanisms of silencing. Submitted

Deuffic-Burban S, Wong JB, Valleron AJ et al (2004) Comparing the public health burden of chronic hepatitis C and HIV infection in France. J Hepatol 40:319–326

Deuffic-Burban S, Mohamed MK, Larouze B et al (2006) Expected increase in hepatitis C-related mortality in Egypt due to pre-2000 infections. J Hepatol 44:455–461

Deuffic-Burban S, Poynard T, Sulkowski MS et al (2007) Estimating the future health burden of chronic hepatitis C and human immunodeficiency virus infections in the United States. J Viral Hepat 14:107–115

Deuffic-Burban S, Abiteboul D, Lot F et al (2009) Costs and cost-effectiveness of different follow-up schedules for detection of occupational hepatitis C virus infection. Gut 58:105–110

Deuffic-Burban S, Mathurin P, Pol S et al (2012) Impact of hepatitis C triple therapy availability upon the number of patients to be treated and associated costs in France: a model-based analysis. Gut 61:290–296

Diederichs S, Haber DA (2007) Dual role for argonautes in microRNA processing and posttranscriptional regulation of microRNA expression. Cell 131:1097–1108

Dorsett Y, Tuschl T (2004) siRNAs: applications in functional genomics and potential as therapeutics. Nat Rev Drug Discov 3:318–329

Elbashir SM, Lendeckel W, Tuschl T (2001) RNA interference is mediated by 21- and 22-nucleotide RNAs. Genes Dev 1:188–200

Fewell GD, Schmitt K (2006) Vector-based RNAi approaches for stable, inducible and genome-wide screens. Drug Discov Today 11:975–982

Filipowicz W (2005) RNAi: the nuts and bolts of the RISC machine. Cell 122:17–20

Fowler T, Bamberg S, Moller P et al (2005) Inhibition of Marburg virus protein expression and viral release by RNA interference. J Gen Virol 86:1181–1188

Ge Q, Dallas A, Ilves H et al (2010a) Effects of chemical modification on the potency, serum stability, and immunostimulatory properties of short shRNAs. RNA 16:118–130

Ge Q, Ilves H, Dallas A et al (2010b) Minimal-length short hairpin RNAs: the relationship of structure and RNAi activity. RNA 16:106–117

Gitlin L, Karelsky S, Andino R (2002) Short interfering RNA confers intracellular antiviral immunity in human cells. Nature 418:430–434

Grimm D, Streetz KL, Jopling CL et al (2006) Fatality in mice due to oversaturation of cellular microRNA/short hairpin RNA pathways. Nature 441:537–541

Haasnoot PC, Cupac D, Berkhout B (2003) Inhibition of virus replication by RNA interference. J Biomed Sci 10:607–616

Hammond SM, Bernstein E, Beach D et al (2000) An RNA-directed nuclease mediates post-transcriptional gene silencing in Drosophila cells. Nature 404:293–296

Hammond SM, Boettcher S, Caudy AA et al (2001) Argonaute2, a link between genetic and biochemical analyses of RNAi. Science 293:1146–1150

Harborth J, Elbashir SM, Vandenburgh K et al (2003) Sequence, chemical, and structural variation of small interfering RNAs and short hairpin RNAs and the effect on mammalian gene silencing. Antisense Nucleic Acid Drug Dev 13:83–105

Hinton TM, Wise TG, Cottee PA et al (2008) Native microRNA loop sequences can improve short hairpin RNA processing for virus gene silencing in animal cells. JRNAi Gene Silencing 4:295–301

Ilves H, Kaspar RL, Wang Q et al (2006) Inhibition of hepatitis C IRES-mediated gene expression by small hairpin RNAs in human hepatocytes and mice. Ann NY Acad Sci 1082:52–55

Iorns E, Lord CJ, Turner N et al (2007) Utilizing RNA interference to enhance cancer drug discovery. Nat Rev Drug Discov 6:556–568

Jarczak D, Korf M, Beger C et al (2005) Hairpin ribozymes in combination with siRNAs against highly conserved hepatitis C virus sequence inhibit RNA replication and protein translation from hepatitis C virus subgenomic replicons. FEBS J 272:5910–5922

Jinek M, Doudna JA (2009) A three-dimensional view of the molecular machinery of RNA interference. Nature 457:405–412

Jucker FM, Pardi A (1995) Solution structure of the CUUG hairpin loop: a novel RNA tetraloop motif. Biochemistry 34:14416–14427

Kanda T, Steele R, Ray R et al (2007) Small interfering RNA targeted to hepatitis C virus 5′-nontranslated region exerts potent antiviral effect. J Virol 81:669–676

Kapadia SB, Brideau-Andersen A, Chisari FV (2003) Interference of hepatitis C virus RNA replication by short interfering RNAs. Proc Natl Acad Sci USA 100:2014–2018

Kawasaki H, Suyama E, Iyo M et al (2003) siRNAs generated by recombinant human Dicer induce specific and significant but target site-independent gene silencing in human cells. Nucleic Acids Res 31:981–987

Khvorova A, Reynolds A, Jayasena SD (2003) Functional siRNAs and miRNAs exhibit strand bias. Cell 115:209–216

Kim D, Rossi J (2008) RNAi mechanisms and applications. Biotechniques 44:613–616

Kim M, Shin D, Kim SI et al (2006) Inhibition of hepatitis C virus gene expression by small interfering RNAs using a tri-cistronic full-length viral replicon and a transient mouse model. Virus Res 122:1–10

Korf M, Meyer A, Jarczak D et al (2007) Inhibition of HCV subgenomic replicons by siRNAs derived from plasmids with opposing U6 and H1 promoters. J Viral Hepat 14:122–132

Kraynack BA, Baker BF (2006) Small interfering RNAs containing full 2′-*O*-methylribonucleotide-modified sense strands display Argonaute2/eIF2C2-dependent activity. RNA 12:163–176

Lanford RE, Hildebrandt-Eriksen ES, Petri A et al (2010) Therapeutic silencing of microRNA-122 in primates with chronic hepatitis C virus infection. Science 327:198–201

Law MG, Dore GJ, Bath N et al (2003) Modelling hepatitis C virus incidence, prevalence and long-term sequelae in Australia, 2001. Int J Epidemiol 32:717–724

Lee YS, Nakahara K, Pham JW et al (2004) Distinct roles for Drosophila Dicer-1 and Dicer-2 in the siRNA/miRNA silencing pathways. Cell 117:69–81

Leuschner PJ, Ameres SL, Kueng S et al (2006) Cleavage of the siRNA passenger strand during RISC assembly in human cells. EMBO Rep 7:314–320

Li L, Lin X, Staver M et al (2005a) Evaluating hypoxia-inducible factor-1alpha as a cancer therapeutic target via inducible RNA interference in vivo. Cancer Res 65:7249–7258

Li MJ, Kim J, Li S et al (2005b) Long-term inhibition of HIV-1 infection in primary hematopoietic cells by lentiviral vector delivery of a triple combination of anti-HIV shRNA, anti-CCR5 ribozyme, and a nucleolar-localizing TAR decoy. Mol Ther 12:900–909

Li L, Lin X, Khvorova A et al (2007) Defining the optimal parameters for hairpin-based knockdown constructs. RNA 13:1765–1774

Liu J, Carmell MA, Rivas FV et al (2004) Argonaute2 is the catalytic engine of mammalian RNAi. Science 305:1437–1441

Liu HW, Cosa G, Landes CF et al (2005) Single-molecule FRET studies of important intermediates in the nucleocapsid-protein-chaperoned minus-strand transfer step in HIV-1 reverse transcription. Biophys J 89:3470–3479

Liu M, Ding H, Zhao P et al (2006) RNA interference effectively inhibits mRNA accumulation and protein expression of hepatitis C virus core and E2 genes in human cells. Biosci Biotechnol Biochem 70:2049–2055

Liu-Young G, Kozal MJ (2008) Review: hepatitis C protease and polymerase inhibitors in development. AIDS Patient Care STDS 22:449–457
MacRae IJ, Doudna JA (2007) Ribonuclease revisited: structural insights into ribonuclease III family enzymes. Curr Opin Struct Biol 17:138–145
Martinez J, Patkaniowska A, Urlaub H et al (2002) Single-stranded antisense siRNAs guide target RNA cleavage in RNAi. Cell 110:563
Matranga C, Tomari Y, Shin C et al (2005) Passenger-strand cleavage facilitates assembly of siRNA into Ago2-containing RNAi enzyme complexes. Cell 123:607–620
McCaffrey AP, Meuse L, Pham TT et al (2002) RNA interference in adult mice. Nature 418:38–39
McHutchison JG, Patel K, Pockros P et al (2006) A phase I trial of an antisense inhibitor of hepatitis C virus (ISIS 14803), administered to chronic hepatitis C patients. J Hepatol 44:88–96
McManus MT, Petersen CP, Haines BB et al (2002) Gene silencing using micro-RNA designed hairpins. RNA 8:842–850
Meister G, Landthaler M, Patkaniowska A et al (2004) Human Argonaute2 mediates RNA cleavage targeted by miRNAs and siRNAs. Mol Cell 15:185–197
Miyoshi K, Tsukumo H, Nagami T et al (2005) Slicer function of Drosophila Argonautes and its involvement in RISC formation. Genes Dev 19:2837–2848
Natt F (2007) siRNAs in drug discovery: target validation and beyond. Curr Opin Mol Ther 9:242–247
Neumann AU, Lam NP, Dahari H et al (1998) Hepatitis C viral dynamics in vivo and the antiviral efficacy of interferon-alpha therapy. Science 282:103–107
Nykanen A, Haley B, Zamore PD (2001) ATP requirements and small interfering RNA structure in the RNA interference pathway. Cell 107:309–321
Ogata N, Alter HJ, Miller RH et al (1991) Nucleotide sequence and mutation rate of the H strain of hepatitis C virus. Proc Natl Acad Sci USA 88:3392–3396
Okamoto H, Kojima M, Okada S et al (1992) Genetic drift of hepatitis C virus during an 8.2-year infection in a chimpanzee: variability and stability. Virology 190:894–899
Okamura K, Ishizuka A, Siomi H et al (2004) Distinct roles for Argonaute proteins in small RNA-directed RNA cleavage pathways. Genes Dev 18:1655–1666
Pham JW, Pellino JL, Lee YS et al (2004) A Dicer-2-dependent 80s complex cleaves targeted mRNAs during RNAi in Drosophila. Cell 117:83–94
Prabhu R, Vittal P, Yin Q et al (2005) Small interfering RNA effectively inhibits protein expression and negative strand RNA synthesis from a full-length hepatitis C virus clone. J Med Virol 76:511–519
Prabhu R, Garry RF, Dash S (2006) Small interfering RNA targeted to stem-loop II of the 5′ untranslated region effectively inhibits expression of six HCV genotypes. Virol J 3:100
Rand TA, Ginalski K, Grishin NV et al (2004) Biochemical identification of Argonaute 2 as the sole protein required for RNA-induced silencing complex activity. Proc Natl Acad Sci USA 101:14385–14389
Rand TA, Petersen S, Du F et al (2005) Argonaute2 cleaves the anti-guide strand of siRNA during RISC activation. Cell 123:621–629
Randall G, Rice CM (2004) Interfering with hepatitis C virus RNA replication. Virus Res 102:19–25
Randall G, Grakoui A, Rice CM (2003) Clearance of replicating hepatitis C virus replicon RNAs in cell culture by small interfering RNAs. Proc Natl Acad Sci USA 100:235–240
Ray Kim W (2002) Global epidemiology and burden of hepatitis C. Microbes Infect 4:1219–1225
Ray RB, Kanda T (2009) Inhibition of HCV replication by small interfering RNA. Methods Mol Biol 510:251–262
Rivas FV, Tolia NH, Song JJ et al (2005) Purified Argonaute2 and an siRNA form recombinant human RISC. Nat Struct Mol Biol 12:340–349
Robb GB, Rana TM (2007) RNA helicase A interacts with RISC in human cells and functions in RISC loading. Mol Cell 26:523–537

Robinson M, Tian Y, Delaney WEt et al (2011) Preexisting drug-resistance mutations reveal unique barriers to resistance for distinct antivirals. Proc Natl Acad Sci USA 108:10290–10295

Ronn R, Sandstrom A (2008) New developments in the discovery of agents to treat hepatitis C. Curr Top Med Chem 8:533–562

Rossi JJ, June CH, Kohn DB (2007) Genetic therapies against HIV. Nat Biotechnol 25:1444–1454

Salomon JA, Weinstein MC, Hammitt JK et al (2002) Empirically calibrated model of hepatitis C virus infection in the United States. Am J Epidemiol 156:761–773

Sarrazin C, Zeuzem S (2010) Resistance to direct antiviral agents in patients with hepatitis C virus infection. Gastroenterology 138:447–462

Schwarz DS, Hutvagner G, Du T et al (2003) Asymmetry in the assembly of the RNAi enzyme complex. Cell 115:199–208

Sen A, Steele R, Ghosh AK et al (2003) Inhibition of hepatitis C virus protein expression by RNA interference. Virus Res 96:27–35

Seo MY, Abrignani S, Houghton M et al (2003) Small interfering RNA-mediated inhibition of hepatitis C virus replication in the human hepatoma cell line Huh-7. J Virol 77:810–812

Seyhan AA, Alizadeh BN, Lundstrom K et al (2007) RNA interference-mediated inhibition of Semliki Forest Virus replication in mammalian cells. Oligonucleotides 17:473–484

Shenouda SK, Alahari SK (2009) MicroRNA function in cancer: oncogene or a tumor suppressor? Cancer Metastasis Rev 28:369–378

Shepard CW, Finelli L, Alter MJ (2005) Global epidemiology of hepatitis C virus infection. Lancet Infect Dis 5:558–567

Shin D, Lee H, Kim SI et al (2009) Optimization of linear double-stranded RNA for the production of multiple siRNAs targeting hepatitis C virus. RNA 15:898–910

Silva JM, Li MZ, Chang K et al (2005) Second-generation shRNA libraries covering the mouse and human genomes. Nat Genet 37:1281–1288

Siolas D, Lerner C, Burchard J et al (2005) Synthetic shRNAs as potent RNAi triggers. Nat Biotechnol 23:227–231

Siomi H, Siomi MC (2009) On the road to reading the RNA-interference code. Nature 457:396–404

Song JJ, Smith SK, Hannon GJ et al (2004) Crystal structure of Argonaute and its implications for RISC slicer activity. Science 305:1434–1437

Stegmeier F, Hu G, Rickles RJ et al (2005) A lentiviral microRNA-based system for single-copy polymerase II-regulated RNA interference in mammalian cells. Proc Natl Acad Sci USA 102:13212–13217

Sun X, Li JM, Wartell RM (2007) Conversion of stable RNA hairpin to a metastable dimer in frozen solution. RNA 13:2277–2286

Takigawa Y, Nagano-Fujii M, Deng L et al (2004) Suppression of hepatitis C virus replicon by RNA interference directed against the NS3 and NS5B regions of the viral genome. Microbiol Immunol 48:591–598

Tomari Y, Zamore PD (2005) Perspective: machines for RNAi. Genes Dev 19:517–529

Tomari Y, Du T, Haley B et al (2004a) RISC assembly defects in the Drosophila RNAi mutant armitage. Cell 116:831–841

Tomari Y, Matranga C, Haley B et al (2004b) A protein sensor for siRNA asymmetry. Science 306:1377–1380

Torriani FJ, Rodriguez-Torres M, Rockstroh JK et al (2004) Peginterferon Alfa-2a plus ribavirin for chronic hepatitis C virus infection in HIV-infected patients. N Engl J Med 351:438–450

Vlassov AV, Ilves H, Johnston BH (2006) Inhibition of hepatitis C IRES-mediated gene expression by 8-17 deoxyribozymes in human tissue culture cells. Dokl Biochem Biophys 410:257–259

Vlassov AV, Korba B, Farrar K et al (2007) shRNAs targeting hepatitis C: effects of sequence and structural features, and comparison with siRNA. Oligonucleotides 17:223–236

Wang Q, Contag CH, Ilves H et al (2005) Small hairpin RNAs efficiently inhibit hepatitis C IRES-mediated gene expression in human tissue culture cells and a mouse model. Mol Ther 12:562–568

Watanabe T, Umehara T, Kohara M (2007) Therapeutic application of RNA interference for hepatitis C virus. Adv Drug Deliv Rev 59:1263–1276
Wilson JA, Richardson CD (2005) Hepatitis C virus replicons escape RNA interference induced by a short interfering RNA directed against the NS5b coding region. J Virol 79:7050–7058
Wilson JA, Jayasena S, Khvorova A et al (2003) RNA interference blocks gene expression and RNA synthesis from hepatitis C replicons propagated in human liver cells. Proc Natl Acad Sci USA 100:2783–2788
Ye X, Huang N, Liu Y et al (2011) Structure of C3PO and mechanism of human RISC activation. Nat Struct Mol Biol 18:650–657
Yokota T, Sakamoto N, Enomoto N et al (2003) Inhibition of intracellular hepatitis C virus replication by synthetic and vector-derived small interfering RNAs. EMBO Rep 4:602–608
Zamore PD, Tuschl T, Sharp PA et al (2000) RNAi: double-stranded RNA directs the ATP-dependent cleavage of mRNA at 21 to 23 nucleotide intervals. Cell 101:25–33
Zekri AR, Bahnassy AA, El-Din HM et al (2009) Consensus siRNA for inhibition of HCV genotype-4 replication. Virol J 6:13
Zhou D, Zhang J, Wang C et al (2009) A method for detecting and preventing negative RNA interference in preparation of lentiviral vectors for siRNA delivery. RNA 15:732–740

The Diverse Active Sites in Splicing, Debranching, and MicroRNA Processing Around RNA Phosphodiester Bonds

Sourav Kumar Dey, Eduardo Paredes, Molly Evans, and Subha R. Das

Contents

Abstract The cleavage and ligation reactions at RNA phosphodiester bonds are the central reactions catalyzed by enzymes in critical cellular regulatory pathways. In pre-mRNA splicing, two phospho-transesterifications result in the right mRNA for protein synthesis with the intervening intron removed as a lariat structure. The lariat RNA is then debranched by an enzyme that specifically acts on this 2′-5′-branched RNA. Following debranching, some of these introns that include pre-microRNA sequences can be processed by Dicer that cleaves the RNA to provide microRNAs. Dicer and Drosha, enzymes that act on much bigger primary

S.K. Dey • E. Paredes • M. Evans • S.R. Das (✉)
Department of Chemistry and Center for Nucleic Acids Science and Technology,
Carnegie Mellon University, 4400 Fifth Avenue, Pittsburgh, PA 15213, USA
e-mail: srdas@andrew.cmu.edu

V.A. Erdmann and J. Barciszewski (eds.), *From Nucleic Acids Sequences to Molecular Medicine*, RNA Technologies, DOI 10.1007/978-3-642-27426-8_19,

transcripts, are both RNase III-like enzymes that cleave the RNA phosphodiester linkage. All these reactions are in related pathways, and the RNA phosphodiester bonds are most likely cleaved with the aid of two metal ions, yet the active sites that host these could be composed entirely of RNA or entirely of protein, or possibly a hybrid of the two. Where unknown, it is possible to estimate some of these active site architectures through homology to closely related enzymes. Better insight into these related process and active sites will play a key role in leveraging these important RNA regulatory processes for molecular medicine.

Keywords 2'-5'-branched RNA • Dicer • Debranching • Drosha • Group II intron • Lariat intron • Mirtron • 2'-5'-phosphodiesterase • Phospho-transesterification • Ribozyme • RNAse III • Spliceosome • Two metal ion mechanism

Abbreviations

3′-SS	3′-splice site
5′-SS	5′-splice site
Aa-RNase III	*Aquifex aeolicus* Ribonuclease III
BP	Branch point
BPS	Branch point sequence
Dbr	Debranching enzyme
ISL	Intramolecular stem loop
mRNA	Messenger RNA
miRNA	MicroRNA
RISC	RNA-induced silencing complex
RNAi	RNA interference
RNase	Ribonuclease
RNP	Ribonucleoprotein
snRNA	Small nuclear RNA
snoRNA	Small nucleolar RNA
U2AF	U2 auxiliary factor

1 Introduction

With the rapid accrual of sequenced genomes, it has become abundantly clear that the complexity of life remains raveled in RNA. Humans have the most DNA, which is converted to RNA, yet only small portions that represent "genes" are translated into proteins (Eddy 2001; Mattick 2007; Pertea and Salzberg 2010). Sharps and Robert's initial finding that large noncoding RNA sequences or introns interrupt the coding sequences or exons in genes led to the discovery of alternative splicing that can result in one gene producing more than one messenger RNA (mRNA) for protein synthesis (Sharp 1987; Black 2003). During the process of pre-mRNA

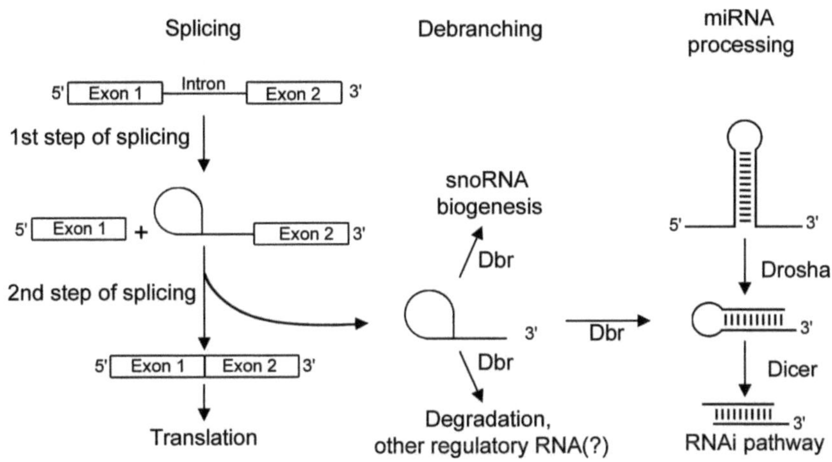

Fig. 1 Splicing, debranching, and microRNA (miRNA) pathways. The coding sequences (exons) in pre-mRNA are joined together through two phospho-transesterification steps in splicing. The noncoding intron sequence is removed as a "lariat" RNA. This 2′-5′-branched lariat RNA is then debranched by the lariat debranching enzyme (Dbr) before degradation or incorporation into other regulatory pathways. Some debranched introns are cleaved by Dicer to generate short double-stranded RNAs which can take part in RNA interference (RNAi)

splicing an intron that is flanked by two exons is removed from the pre-mRNA (Fig. 1) and this removal is mediated by the macromolecular spliceosomal apparatus that consists of five small nuclear RNAs (snRNA) and more than a hundred proteins (Wahl et al. 2009). The specific details of the two chemical steps of splicing that involve the phospho-transesterification reactions at the phosphate backbone remain under scrutiny. However, a wealth of information on how RNA active sites can catalyze the same type of reaction has been gleaned over the years from the intensive investigation of self-splicing introns—the group I and group II ribozymes (Gesteland et al. 2006). Recent crystallographic analyses of these continue to inform how the spliceosomal apparatus could operate. As in the group II intron, the spliceosome also removes the noncoding intron sequence as a lariat (Fig. 1).

The lariat RNA is a unique RNA that in addition to the normal 3′-5′-phosphodiester-linked backbone includes a branch point (BP) adenosine residue that is linked at the 2′-position to the 5′-end of the RNA sequence (Figs. 1 and 2) at the start of the intron (Ruskin et al. 1984). This 2′-5′-phosphodiester linkage in the lariat RNA backbone undergoes debranching by a unique lariat debranching enzyme (Dbr) that cleaves that particular bond (Chapman and Boeke 1991). Although the cleaved 2′-5′-phosphodiester is an unusual linkage, Dbr is related to well-described metallo-phosphatases such as calcineurin and serine/tyrosine phosphatases that are involved in removal of phosphates from proteins and Mre11, a nuclease involved in DNA repair (Khalid et al. 2005). Following debranching, some introns are processed further in the biogenesis of C/D box small nucleolar

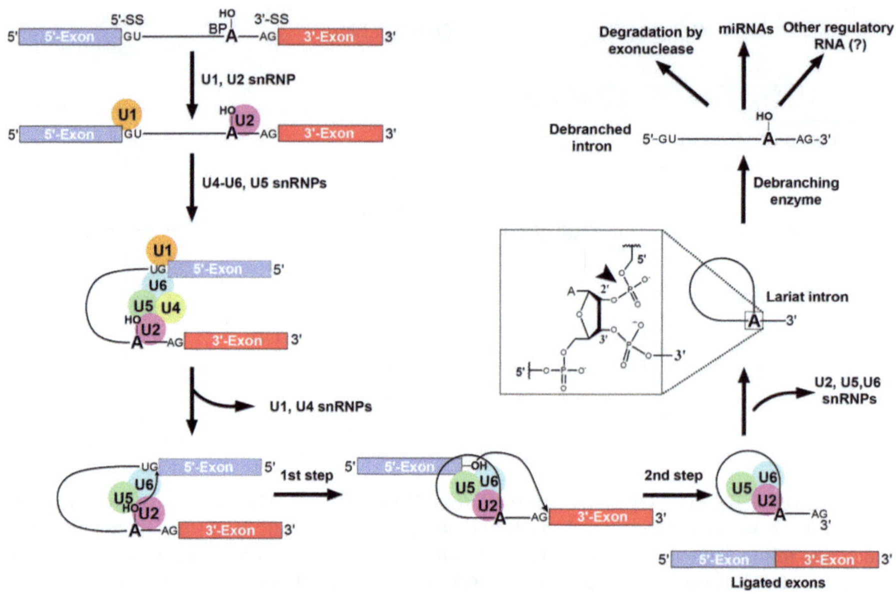

Fig. 2 Splicing of pre-mRNA by the spliceosome and debranching of the lariat intron. The most conserved residues at the 5′-splice site, 3′-splice site, and branch point and only the snRNPs at different stages of spliceosomal assembly and catalysis are shown. The two phospho-transesterification steps in splicing start by attack of the 2′-OH of the branch point adenosine (1st step) and then the 3′-OH of the 5′-exon (2nd step). The *inset* shows the connectivity of the lariat intron at the branch point. The lariat intron is debranched after the removal of the spliceosomal snRNPs by Dbr by cleavage at the phosphate indicated by the arrowhead

RNAs (snoRNAs) (Ooi et al. 1998). Very recently, it was discovered that some introns following debranching (Fig. 1) are processed into microRNAs (Bartel et al. 2007; Lai et al. 2007).

MicroRNAs (miRNAs) and RNA interference (RNAi) are among the most prominent discoveries in gene regulation in recent times (van den Berg et al. 2010). The regulatory microRNAs (miRNAs) are derived from longer precursor RNAs after the cleavage by endonucleases like Drosha, which generates the pre-miRNA stem loop structure, and Dicer, which generates short double-stranded RNAs (Figs. 1 and 3). Both Drosha and Dicer are endoribonucleases of the ribonuclease III (RNase III) family and cleave the RNA phosphodiester bond.

All these phosphodiester bond cleavages are critical in gene regulatory pathways. Thus, the mechanism(s) by which these phosphodiester bonds are cleaved is an area of significant research. It is also instructive to examine the various postulated and elucidated mechanisms for the phospho-transesterification reaction in the related contexts of splicing, debranching, and miRNA biogenesis (Fig. 1) as all of these processes are potential targets for molecular medicine. The similarities and distinctions between the active sites and mechanisms for reactions in these pathways provide important clues to understanding and identifying features

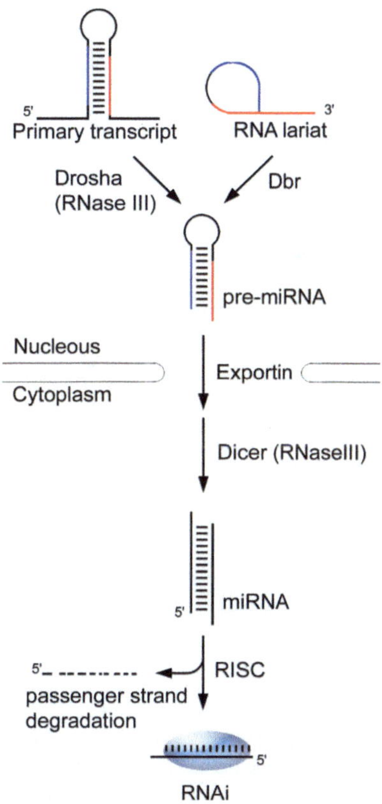

Fig. 3 Mirtron and miRNA pathways. Debranching of lariat introns generates an RNA hairpin similar to a Drosha cleaved pre-miRNA. After nuclear export by exportin, and Dicer cleavage, the mature miRNA is loaded into a RISC complex for RNAi. Both Drosha and Dicer include RNase III domains that catalyze the RNA cleavage

that may succumb to therapeutic intervention. These active site architectures are examined in this chapter with respect to the phospho-transesterification reaction. In all these, an RNA phosphodiester bond is cleaved within an active site that maybe composed with residues from protein, RNA, or both (Fig. 4). The mechanism of these or closely related enzymes involves the phosphoryl transfer reaction mediated with the aid of metal ion cofactors.

2 Metal Ion Catalysis of Transesterification Reactions

2.1 Two-Metal Ion Mechanism of Phosphoryl Transfer Reactions

A two-metal ion mechanism for phosphoryl transfer was proposed by Steitz et al. for the 3′-5′-exonuclease activity of *Escherichia coli* DNA polymerase I (Beese and Steitz 1991). In this mechanism, two divalent metal ions separated by approximately 4 Å form inner-sphere complexes with the nucleophilic water and the

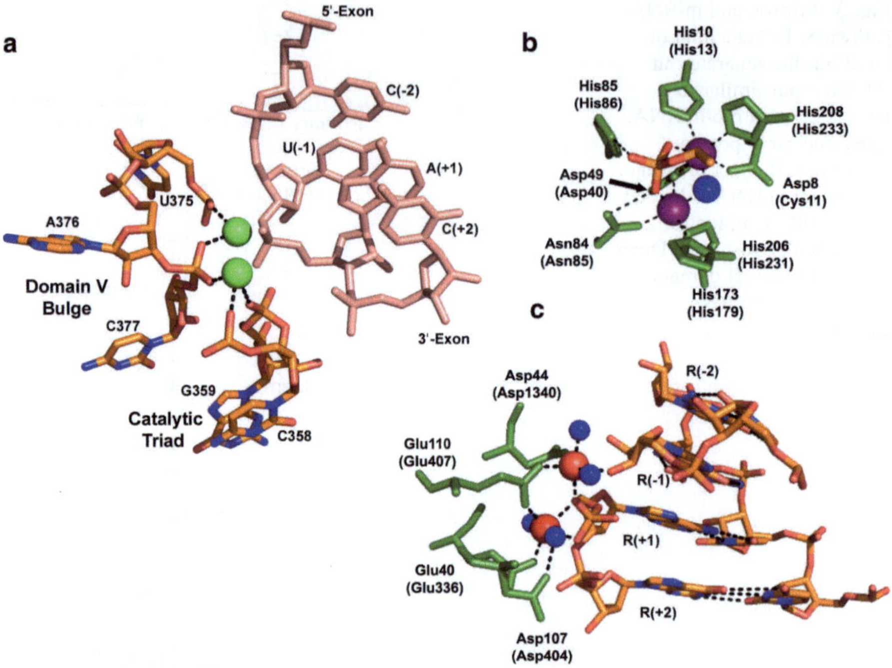

Fig. 4 Active sites observed in crystal structures of self-splicing group II intron, Mre11, and RNase III. (**a**) Crystal structure of the group IIC intron from *O. iheyensi* with an oligonucleotide mimicking the ligated exons (Purple) (PDB 1IGI). The catalytically important Mg^{2+} ions shown as green spheres are separated by 4.6 Å. The invariant domain V bulge and the catalytic triad residues are shown in *red*, *orange*, and *blue*. (**b**) Crystal structure of the manganese and 5′-dAMP-bound *Pyrococcus furiosus* Mre11 (PDB 1II7). Manganese ions are in *purple*; the bridging water is *blue*. Corresponding yeast Dbr1 residues based on amino acid similarities are in parentheses. (**c**) Post-cleavage crystal structure of Aa-RNase III (PDB 2NUG) shows the two magnesium ions (*red*) bound by four acidic residues (*green*). Four bound water molecules (*blue*) and the cleaved RNA duplex (orange) are also bound after RNA cleavage. The numbers in the parenthesis show the corresponding residues in Dicer

scissile phosphate (Fig. 5). One of the metal ions interacts with an attacking water molecule to facilitate nucleophilic attack on the scissile phosphate bond. The other metal ion coordinates with the 3′-oxygen of the leaving group and also with the nonbridging oxygen of the phosphate group to stabilize the negative charge on the leaving group and the pentacoordinated transition state.

Such a two-metal ion mechanism is also postulated for several other nucleic acid-processing enzymes like alkaline phosphatase, RNase H domain of HIV reverse transcriptase, and T7 DNA polymerase (Kim and Wyckoff 1991; Davies et al. 1991; Doublie et al. 1998). Interestingly, although these enzymes do not have any structural similarities, the distance between the two metal ions is maintained at about 4 Å, which suggests an important role of the positioning of the two metal ions within the active site. Based on the two-metal ion mechanism of nucleic

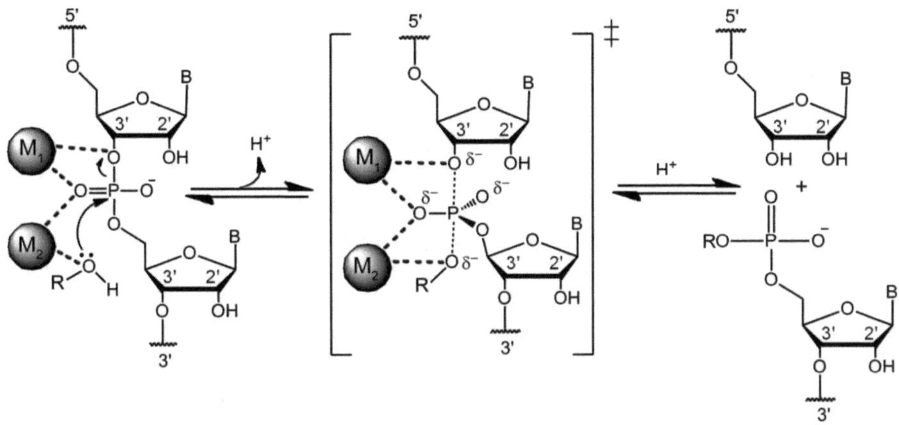

Fig. 5 A two-metal ion mechanism for phospho-transesterifications in RNA. Two divalent metal ions separated by ~4 Å form an inner-sphere complex with the scissile phosphate and the attacking nucleophile. Metal ion M_1 stabilizes the negative charge developed on the 3′-oxyanion of the leaving group. Metal ion M_2 acts as a Lewis acid and activates the nucleophile for attack at the scissile phosphate. The nucleophile could be a water molecule (R = H) or a hydroxyl group of RNA (R = 3′- or 2′-RNA), depending on the reaction. Both the metal ions also stabilize the pentacoordinated transition state by interacting with the bridging and nonbridging oxygen atoms of the scissile phosphate bond. By microscopic reversibility, the transition state for the reverse reaction would be identical. Varied active sites architectures in numerous enzymes—protein or RNA based—enable the catalysis of this reaction

acid-processing enzymes, Steiz and Steitz proposed an analogous two-metal ion mechanism for ribozymes (Steitz and Steitz 1993).

2.2 *Difference and Similarities Between Protein and RNA Active Sites and the Role of Metal Ions*

The discovery of ribozymes fuelled prolific research into how an active site composed of RNA could catalyze reactions. The large ribozymes—the group I and group II introns—that autocatalyze phosphoryl transfer reactions involved in their splicing do so with the aid of metal ion cofactors similar to many protein enzymes. While both protein and RNA can use metal ions for catalysis, there are differences in the binding mode of the metal ions. Proteins have a backbone that is conformationally less flexible than RNA. Additionally, in proteins, a large number of hydrophobic amino acid side chains cause dense packing which allows for metal binding sites with high specificity and affinity. The main metal binding groups of proteins are the sulfide residue of cysteine, the imidazole ring of histidine, and the carboxylate side chains of glutamic and aspartic acid that can form highly selective and tight metal ion binding pockets. Occasionally other side chains like the phenolate of tyrosine, the hydroxyl group of serine and threonine, the thioether of methionine, etc., are used with lower binding affinity toward metal ions (Bertini et al. 2001).

In RNA, the negatively charged phosphate backbone can bind nonspecifically with a large number of metal ions with low binding constants. This charge screening is important for the three-dimensional folding of RNA. Along with these nonspecific metal ion binding sites, there are few specific metal ion binding sites that bind with relatively higher binding constants. Although the main metal ion binding ligands for RNA are the negatively charged nonbridging oxygen atoms of the phosphate group, other potential metal binding sites include the N7 and N3 of purine nucleobases, the N1 of adenine, the N3 of cytosine, the carbonyl oxygen of the C6 in guanosine, and the C2 and C4 oxygens in pyrimidines (Kapinos et al. 1999). In RNA, all the nitrogen atoms involved in metal ion binding are less basic than imidazole, which is one of the most used residues for metal ion binding in proteins, and therefore have lower affinity (Sigel and Pyle 2007).

A common strategy utilized by the enzymes is to stabilize the charges that develop in the transition state (Warshel 1981). Bound metal ions like Mg^{2+} typically used as the metal ion cofactors are redox inert, have a high positive charge density due to their lower size, and tend to bind water molecules rather than bulky ligands. In some cases, Mn^{2+} or Zn^{2+} can even be used as the metal ion cofactors with softer ligands like thiolate to bind these metal ions. A divalent metal ion can polarize the carbonyl or phosphoryl group of the substrate making it more electrophilic and can stabilize the negative charge buildup during the transition state (Figs. 5 and 6). To achieve such feats, the metal ions have to be in inner-sphere contact with the ligands. In case of ribozymes, such tightly bound metal ions can be identified within the constellation of other nonspecific metal ions that stud the ribozyme by "metal ion specificity switch" experiments. In these, an important molecular contact, such as the 3′-oxygen of the leaving group or the nonbridging oxygen of the scissile phosphate group, is replaced with sulfur. If this atomically substituted ribozyme is inactive in the presence of Mg^{2+}, but the activity is rescued by thiophilic metal ions such as Mn^{2+}, Zn^{2+}, or Cd^{2+}, then a direct inner-sphere contact can be inferred between the metal ion and the active site functional group (Piccirilli et al. 1993; Weinstein et al. 1997; Yoshida et al. 1999).

Metal ions can even alter the acid–base properties of the residues adjacent to it. For example, an adjacent metal ion can make the protonation of neighboring residues difficult, resulting in a pK_a shift (Oda et al. 1994). In some cases, positively charged residues of the protein side chains can play the same role (Highbarger et al. 1996). Additionally, water molecules or an alcohol moiety can serve as a nucleophile or a general base during the reaction when coordinated to a metal ion. Indeed greater nucleophilic capabilities of metal-coordinated hydroxides and alkoxides are well known (Bevilacqua 2003; Adams et al. 2004; Walter 2007). Thus, metal ions play several key roles in both ribozyme and protein enzyme chemistry. While the roles of the metal ions are similar, the ligands which bind to the metal ion to create the active site are different.

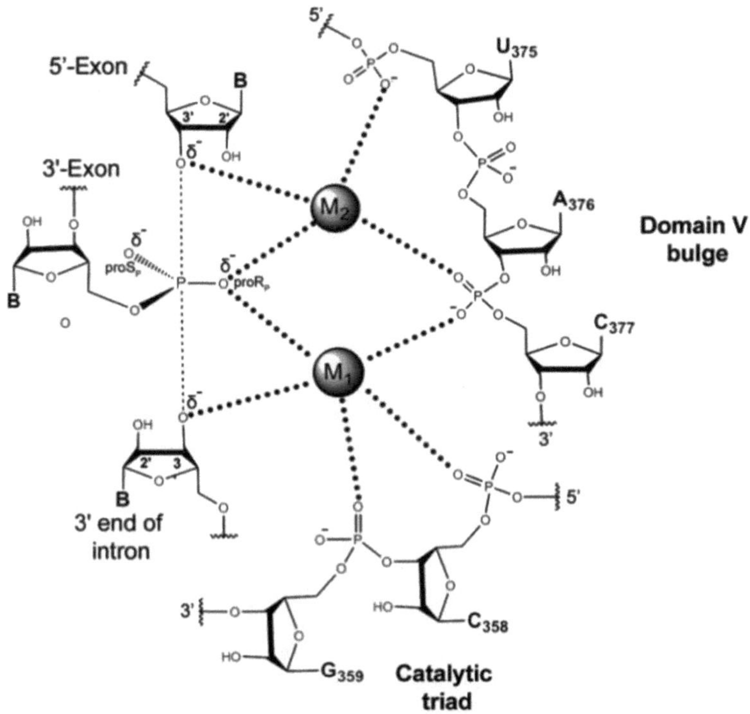

Fig. 6 Proposed mechanism for the transition state stabilization by two metal ions in the second step of splicing by self-cleaving group II intron of *O. iheyensi*. The highly conserved metal ion binding residues from the catalytic CGC triad and domain V bulge found in the postcleavage crystal structures are shown here. The two metal ions separated by 3.7 Å are proposed to activate the nucleophile and the leaving group and also stabilize the pentacoordinated transition state. This active site architecture may share significant similar features with that of the spliceosome

3 Splicing: Reactions Catalyzed by the Group II Intron and the Spliceosome

All intronic sequences in the pre-mRNA are removed via two transesterification reactions. In the first step of splicing, the 2′-hydroxyl group from a conserved adenosine residue attacks the phosphodiester bond at 5′-splice site (5′-SS), resulting in a lariat intermediate (Fig. 2). The second step of splicing involves nucleophilic attack of the 3′-hydroxyl of the 5′-exon to the phosphodiester bond at 3′-splice site (3′-SS) (Sharp 1987).

These two phosphoryl transfer reactions in the self-splicing group II intron have been studied extensively using atom-specific chemical modification as well as mutagenesis. Substitution of the nonbridging prochiral oxygen of the reactive phosphate group by sulfur shows that both steps of splicing undergo an inversion of configuration suggesting a normal S_N2-type mechanism (Padgett et al. 1994;

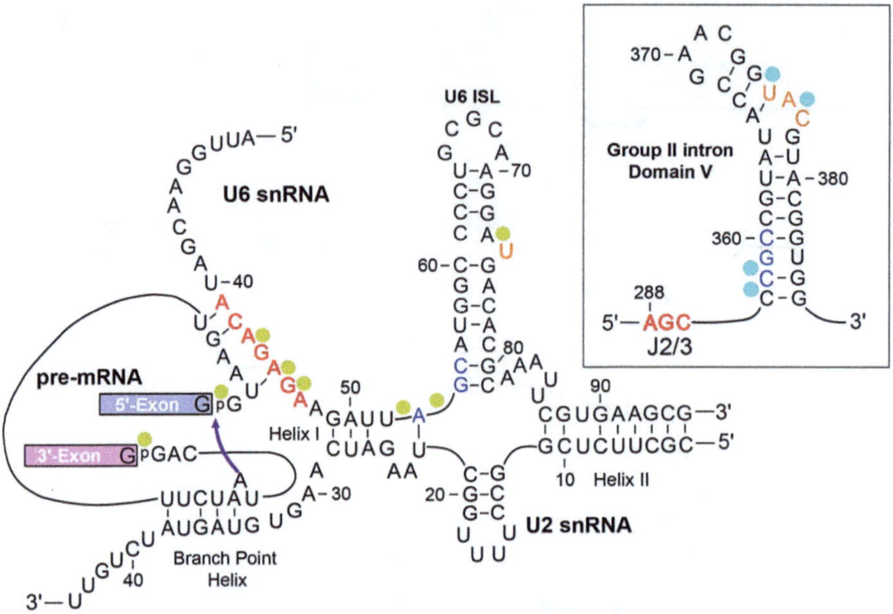

Fig. 7 A proposed model of the secondary structure of catalytically important spliceosomal U2–U6 snRNAs structure with the pre-mRNA in the catalytic spliceosome (Sun and Manley 1995). The sequence and numbering refer to human U2 and U6 snRNA. The nucleotide upstream of 5′-SS base pairs with the U6 snRNA and the branch site nucleotides base pair with U2 snRNA. The *arrow* shows the attack of the conserved branch point adenosine 2′-hydroxyl at the 5′-SS (first step of splicing). *Colors* highlight the highly conserved "ACAGAGA" box (*red*), AGC triad (*blue*), and bulged U (*orange*) of the ISL. *Yellow circles* indicate the metal ion binding phosphate residues found by metal ion specificity switch experiments or other functional and structural studies. The *inset* shows the secondary structure of the *O. iheyensi* of group IIC intron domain V along with the invariant J2/3 from crystal structure. The residues that play similar roles as those in U6 snRNA are highlighted in the same color. The *cyan circles* indicate the metal ion binding phosphate residues observed in the crystal structure

Podar et al. 1995). Incorporation of an R_P phosphorothioate both at 5′- and 3′-splice site completely blocks the reaction, whereas S_P phosphorothioate at the 5′-SS splice site only slightly inhibits the reaction, but this has almost no effect on the second step. These features suggest that the two steps of splicing are somewhat mechanistically different in having different active sites or that at least there is significant conformational rearrangement after the first step of splicing.

Group II introns can be divided into three major families: IIA, IIB, and IIC (Toor et al. 2001). Although there is very little sequence conservation in group II introns, all of them consist of six domains. Out of these domains, domain V contains the important residues for binding of catalytically important metal ions—an almost invariant AGC triad (CGC in group IIC introns) and a two nucleotide AC bulge (Figs. 4a, 6, and 7) also essential for catalytic activity (Pyle 2010). Domain VI contains a bulged adenosine that attacks the 5′-SS during the first step of splicing resulting in the 2′-5′-branched RNA lariat (Chu et al. 2001). Metal ion specificity switch experiments reveal the metal ion binding residues of the 5′-and 3′-splice sites and other catalytically important residues. Replacement of the 3′-oxygen of the scissile phosphate at the 5′- and 3′-splice sites with sulfur results in complete blockage of both the steps of splicing. However, replacement of magnesium with a more thiophilic ion like Mn^{2+} or Cd^{2+} results in recovery in the activity, providing evidence for the direct binding of the metal ions with those oxygen atoms (Sontheimer et al. 1999; Gordon et al. 2000a). A two-metal ion mechanism for phosphoryl transfer reactions in the group II intron was proposed based on the atom-specific chemical modification and mutagenesis studies (Figs. 4a and 6). Recent crystallographic analyses of a group II intron (type IIC) from eubacterium *Oceanobacillus iheyensis* depict structures after the second step of splicing (Toor et al. 2008a, b). Tertiary interactions between different domains organize the catalytic site around domain V. A highly interesting feature of the group II intron crystal structure is the triple helix formed between the CGC triad, the 5′-AC bulge, and the J2/3 region that acts as a scaffold for the catalytic center (Fig. 4a). The proximity of the backbone phosphate groups in this region results in a negatively charged metal binding pocket, and the group IIC intron crystal structure shows two Mg^{2+} ions in this pocket separated by ~4 Å. These metal ions are directly bound to the nonbridging oxygen of the phosphate groups from the domain V bulge and the catalytic triad (Figs. 4a and 6). Although the first crystal structure did not include any exons, a second structure includes a nucleic acid fragment mimicking the ligated exons and depicts inner-sphere coordination of the scissile phosphate (Fig. 4a) with the two divalent metal ions (Toor et al. 2008b, 2010). The distance between these two divalent metal ions and the relative positioning between the important residues strongly suggest that in the group II intron the operative mechanism is a two-metal ion-catalyzed reaction for the phosphoryl transfer steps of splicing (Toor et al. 2009).

These two steps of intron splicing and exon ligation are the same ones catalyzed by the spliceosome, a huge ribonucleoprotein (RNP) complex (Will and Lührmann 2006). The spliceosome is a highly dynamic and complex RNP apparatus which undergoes several conformational changes during its assembly and catalytic activity (Wahl et al. 2009). The modern eukaryotic spliceosome comprises five small nuclear RNAs (U1, U2, U4, U5, and U6) (Fig. 2) and more than a hundred proteins. The mechanism of the intron removal reaction catalyzed by the spliceosome is similar to the self-splicing reaction of group II introns (Padgett and Dayie 2008; Keating et al. 2010). Phosphorothioate analysis has shown that both splicing reactions occur via a normal S_N2 mechanism resulting in inversion of configuration of the reactive phosphate group (Moore and Sharp 1993; Maschhoff and Padgett 1993). Substitution of the nonbridging phosphate oxygen with sulfur at the 5′-SS shows a metal ion specificity switch, though the sulfur substitution 3′-SS shows no similar effect (Sontheimer et al. 1997; Gordon et al. 2000b). Still, in both these cases, a bimolecular reaction mimicking the second step of splicing shows metal ion binding by the 3′-oxyanion leaving group. The stereochemical course of the reactions and the metal ion requirement in the mRNA splicing also suggest that the two steps of splicing occur at different active sites or at least a conformational change occurs after the first step, similar to the group II intron (Padgett et al. 1994). A recent single molecule study of spliceosomal assembly suggests that a stepwise association of the different snRNA is required (Fig. 2) for catalytic activation of the spliceosome, rather than a preformed active complex (Hoskins et al. 2011).

3.1 Role of the snRNAs in Pre-mRNA Splicing

The U1, U2, U4, U5, and U6 RNAs are the main building blocks of spliceosome (Fig. 2) along with about 80 (in yeast) to 170 (in human) proteins depending on the species (Wahl et al. 2009). Interestingly, some metazoan species and plants contain a second type of spliceosome consisting of snRNAs analogous to U1, U2, U4, and U6, while sharing U5 between both types of spliceosome (Patel and Steitz 2003). All the snRNAs are associated with six Sm proteins except U6, which contains seven Sm-like (LSm) proteins (Veretnik et al. 2009). None of these snRNPs have a preformed active site and are remodeled throughout the course of the spliceosomal activation by a cascade of allosteric interactions (Brow 2002).

The first event of spliceosomal activation is the ATP-independent binding of U1 snRNP to the 5′-SS (Fig. 2). Base pairing interactions of U1 with the 5′-SS is also stabilized by Sm proteins and a U1-specific RNA recognition motif (Pomeranz Krummel et al. 2009; Hoskins et al. 2011). All further events in spliceosomal assembly require ATP hydrolysis. After the binding of U1, the U2 snRNP binds to the 3′-SS and branch point sequence (BPS). In addition to the proteins associated with U2 snRNP, a non-snRNP protein, the U2 auxiliary factor (U2AF) plays a crucial role in 3′-SS and BPS recognition. U2AF binds to the polypyrimidine tract upstream of 3′-SS and recruits the U2 snRNP for base pairing interactions with the BPS resulting in the spliceosomal A complex (Valcarcel et al. 1996). The next event in the spliceosomal assembly is ATP-dependent binding of the U4–U6/U5 tri-snRNP forming the B complex (Fig. 2). The U4 snRNA, which is initially base-paired with U6 snRNA in the tri-snRNP complex, is not a part of the active spliceosome complex. The extensive base pairing between U4 and U6 is disrupted during the formation of active spliceosome (the B^* complex), and the U4 and U1 snRNPs are discarded from the spliceosome (Will and Lührmann 2006). The U5 snRNA along with its associated proteins is believed to be responsible for aligning the exons for optimal ligation during the second step of splicing, though U5 which was initially thought to be essential for splicing is dispensable (Segault et al. 1999; Will and Lührmann 2006). This strongly suggests that U2 and U6 are the only snRNAs required for catalytic activity. After the removal of U1 and U4, U6 snRNA forms extensive base pair with U2 which is bound to the BPS. The U6 snRNA also forms base pairing interaction with the pre-mRNA (Fig. 7) using the invariant "ACAGAGA" sequence, juxtaposing the substrates to the catalytically important AGC triad and the asymmetric bulge of the U6 intramolecular stem loop (ISL) (Rhode et al. 2006). After the first step of the reaction, the spliceosome undergoes another rearrangement in the RNP networks before forming the C complex (Konarska et al. 2006). After the second catalytic step, the spliceosome undergoes Prp43-mediated dissociation (Fig. 2) which releases the spliced mRNA, lariat intron, and the U2, U5, and U6 snRNPs (Tsai et al. 2005).

Along with RNA–RNA interactions, RNA–protein and protein–protein interactions play crucial roles in spliceosomal assembly and potentially also in catalysis (Abelson 2008; Wahl et al. 2009). Recent studies indicate that the human spliceosome contains approximately 170 protein factors which exchange binding sites during stepwise assembly (Bessonov et al. 2008). The spliceosome undergoes a multitude of dynamic remodeling events mediated by protein factors during each stage of its assembly. Enzymatic activities such as DExD/H-type RNA-dependent ATPase/helicase and peptidyl-prolyl cis/trans isomerase are indispensable for accurate function of the spliceosome. Such remodeling events are necessary for (1) recognition of multiple splice sites for precise splicing reactions, which is especially important for alternative splicing; (2) flexibility, as the interactions between the spliceosomal components are weak, though a number of such interactions make the spliceosome stable; and (3) the change in the active site component from the first step to the second step of splicing.

3.2 *An RNA-Centric View of the Spliceosomal Active Site Architecture*

The chemistry of the two steps of splicing by the spliceosome displays enormous similarities with the self-splicing group II introns. The active site of the group II intron is formed by the catalytic AGC triad, the asymmetric bulge of domain V, and the joining sequence J2/3, which hold the two catalytic metal ions (Figs. 4a and 6). The U6 RNA which is part of the active spliceosome has all these elements suggesting that it can be responsible for the active site of the spliceosome (Fig. 7) (Keating et al. 2010). The evolutionary conserved elements of the U6 snRNA are the ACAGAGA box (which is believed to play the same role as J2/3 of group II intron), an AGC triad, and a conserved bulged U residue in the intramolecular stem loop (Fig. 7). Mutation in these sequences resulted in either complete or partial block of splicing, while mutations in other residues have a nominal effect (Datta and Weiner 1993; Fabrizio and Abelson 1990; Madhani et al. 1990). Hydroxyl radical probing experiments have demonstrated that the ACAGAGA box resides close to the ISL, both before and after the first step of splicing (Rhode et al. 2006). Strengthening the base pairing interaction of between ACAGAGA box and the 5′-SS in yeast by restoring the base complementarity resulted in favored first step of splicing, while it reduced the efficiency of the second step indicating functional association between U6 and 5′-SS and a conformational change between the two steps of splicing (Konarska et al. 2006).

Another important snRNA for the spliceosomal activity is U2. In the activated spliceosome, U2 forms a number of catalytically important base-paired helices with the U6 snRNA (Fig. 7). Although the exact nature of base pairing between U2 and U6 snRNAs is not clear, several models have been proposed based on mutational studies. An early model of U2–U6 base pairing in yeast suggested a three-helix junction where the AGC triad forms base pair with U2 (Madhani and Guthrie 1992). This model is believed to be important for both steps of splicing (Hilliker and Staley 2004; Mefford and Staley 2009). Although an NMR study of yeast U2–U6 snRNA in the absence of Mg^{2+} suggested a four-helix junction structure (Sashital et al. 2004), another model was proposed for human U2–U6 base pairing. Here the AGC triad forms intramolecular base pair within the U6, which extends the ISL (Fig. 7) creating a four-way junction (Sun and Manley 1995). A recent single molecule FRET study of protein-free yeast U2–U6 snRNA indicated a Mg^{2+}-dependent conformational change from a four-helix junction to three-helix junction with increasing concentration of Mg^{2+}, where the bulged U residue is important for folding (Guo et al. 2009). Another important role of U2 is recognition of the BPS via base pairing (Wu and Manley 1989). Finally, similar to domain VI of group II intron, a bulged adenosine of the U2–BPS base pair acts as a nucleophile for the first step of splicing.

Divalent metal ions are required for the folding and catalytic activity in the spliceosome. Important metal ion binding pockets have been identified using chemical modification and structural elucidation (Fig. 7). One of the most important metal ion binding pockets of the U2–U6 structure is a bulged U residue within the U6 ISL. A metal ion specificity switch experiment suggests that the pro-S_P phosphate oxygen of the bulged U is involved in metal ion coordination during the first step of splicing, although the role of this oxygen for the second step is not clear (Yean et al. 2000). NMR studies also provide evidence for metal ion binding by U6 ISL (Huppler et al. 2002). Interestingly, similar evidence for metal ion binding by the asymmetric bulge of domain V in a group II intron (Fig. 7) had also been obtained (Sigel et al. 2000). Apart from the ISL, phosphorothioate interference analysis has also detected metal ion binding by the ACAGAGA box and the AGC triad (Yu et al. 1995; Fabrizio and Abelson 1992). In a lanthanide ion (Tb^{3+})-based fluorescence assay, metal ion binding by all the three important domains of U2–U6 complex has been observed (Yuan et al. 2007). Taken together, these suggest that functionally assembled U2–U6 snRNA forms important tertiary interactions, which helps in defining the metal ion binding pockets that are essential for catalysis.

3.3 *Spliceosome: Catalysis by RNA or Protein?*

Together with its RNA and protein components, the spliceosome is a highly efficient and dynamic ribonucleoprotein machine. But the question of whether the active site of the spliceosome is composed of RNA, protein, or both of them remains to be answered. Although it is very clear that the base-paired U2–U6 snRNAs are part of the catalytically active spliceosome, their exact role in catalysis remains unresolved as there are also many proteins involved in spliceosomal assembly with unknown functions (Wahl et al. 2009). A few studies have emerged of protein-free splicing-like reactions by in vitro-assembled U2–U6 snRNAs. An initial study suggested catalysis by human U2–U6 complex, where the branch point adenosine was joined to the G residue of the AGC triad (Valadkhan and Manley 2001). However, the reaction was highly inefficient. In a more recent study, human U2–U6 snRNA is shown to catalyze a two-step *trans*-splicing reaction, where the first step is hydrolysis of a phosphate bond (Valadkhan et al. 2009). Such hydrolysis-mediated splicing reactions are also observed in group II introns (Podar et al. 1998). Cross-linking analysis and mutational complementation show that the reaction occurs in the vicinity of the ACAGAGA box. Further, Tb^{3+}-mediated cleavage also shows metal ion binding in the evolutionary conserved regions of U2–U6 (Lee et al. 2010).

Although protein-free splicing-like reactions are mediated by U2–U6 RNAs, there are several protein factors which are present in close vicinity of the active site in the spliceosome. Among these is a phylogenetically conserved U5-specific protein, Prp8 (Grainger and Beggs 2005). In addition to U6 and U5 snRNAs, Prp8 is shown to cross-link to the 5′-SS, 3′-SS, and BPS, indicating its proximity to the active site in spliceosome (Turner et al. 2006). Prp8 is also known to interact with two other helicases, Snu114 and Brr2, that are responsible for dynamic rearrangement of the spliceosome (Liu et al. 2006). A recent crystal structure of Prp8 reveals an RNase H-like domain which interacts with the 5′-SS, suggesting that Prp8 may directly participate in catalysis (Pena et al. 2008; Ritchie et al. 2008; Yang et al. 2008). Another important protein is the U2-specific SF3b14a/p14 protein that interacts with the BPS (Will et al. 2001). Although the role of several proteins in spliceosome is well understood, the extent to which the proteins are involved is still unclear.

4 Debranching of the Lariat Intron

After the lariat intron is excised from the pre-mRNA, the helicase Prp43 removes U2, U5, and U6 snRNPs to be recycled to a new splicesome (Fig. 2) (Wahl et al. 2009). The Dbr debranches the lariat that is then often degraded from the 5′-end by the 5′→3′-exonucleases Rat1 and Xrn1 and from the 3′-end by the exosome with its 3′→5′-exonuclease activity (Coller and Parker 2004; Garneau et al. 2007). Although certain mRNAs can be cleaved in the branch by the endonuclease Rnt1 to allow for complete degradation without debranching, the majority of the lariats can only be degraded from the 3′-end by the exosome (Danin-Kreiselman et al. 2003). As the exosome cannot cleave the 2′-5′-branched structure, a lack of debranching activity results in the accumulation of "nibbled" lariat introns that lack their 3′-tail (Chapman and Boeke 1991). Accumulation of lariat introns in Dbr null mutants has different phenotypes in different species; while there is no noticeable change in *Saccharomyces cerevisiae*, *Schizosaccharomyces pombe* null mutants have a severe growth defect, with aberrant elongated cell shape (Nam et al. 1997). This could be due to the amount of introns found within the genome—*S. pombe* is estimated to contain approximately 40 times the number of introns than *S. cerevisiae*.

4.1 Active Site Similarities of Debranching Enzyme with Mre11

1. Debranching activity was first isolated from HeLa cells, and the yeast Dbr was subsequently identified in a genetic screen to identify mutants that were deficient in retrotransposition (Ruskin and Green 1985; Chapman and Boeke 1991). Dbr is a Mn^{2+}-dependent 2′-5′-phosphodiesterase that is essential for lariat degradation; hydrolysis of the branch point by Dbr yields a 5′-phosphate end and a 2′-hydroxyl at the branch attachment site (Fig. 2). Dbr is well conserved among eukaryotes; the *S. cerevisiae* null mutant phenotype can be alleviated by expression of *S. pombe*, *Caenorhabditis elegans*, human, and mouse Dbr (Nam et al. 1997; Kim et al. 2000, 2001). Although no structural information of Dbr is available, it has significant sequence homology to other metallo-phosphodiesterases such as λ phosphoprotein phosphatase, calcineurin, and Mre11 that belong to the metallo-phosphoesterase superfamily (Koonin 1994; Zhuo et al. 1994). Specifically, this superfamily is characterized by a dinuclear metal center with the three signature motifs DIH(X)$_{\sim 25}$GDYVDR(X)$_{\sim 27}$GNHE important in metal ion binding and catalysis (Zhuo et al. 1994; Rusnak and Mertz 2000; Goldberg et al. 1995; Voegtli et al. 2000; Hopfner et al. 2001). Structural studies of this superfamily show that conserved His, Asp, and Asn side chains of the signature motif bind the two metal ions while conserved His and Asn side chains of the signature motif contact the phosphate backbone of the substrate (Fig. 8) (Goldberg et al. 1995; Voegtli et al. 2000; Hopfner et al. 2001; Kissinger et al. 1995). *S. cerevisiae*, *M. musculus*, *Drosophila melanogaster*, *Arabidopsis thaliana*, and *S. pombe* debranching enzymes show similar conservation of this motif, suggesting that the structure of the active site and the coordination of the manganese ions may be similar in Dbr (Khalid et al. 2005).

Crystallized Mre11 shows coordination of the manganese ions in the active site by seven amino acid residues (Asp8, His10, Asp49, Asn84, His173, His206, His208) and also a bridging water molecule (Figs. 4b and 8) (Hopfner et al. 2001). Mre11 was also crystallized with multiple DNA substrates (Hopfner et al. 2001; Williams et al. 2008). Based on the structures of Mre11 with these DNA molecules, Asn84 and His85 of the GNHE motif and the manganese ions are believed to ensure proper positioning of the reactive groups and activate the phosphate for nucleophilic attack through coordination of the phosphate (Hopfner et al. 2001). Manganese ion M_2 (Fig. 8) is proposed to bind either a hydroxide ion or the bridging water molecule

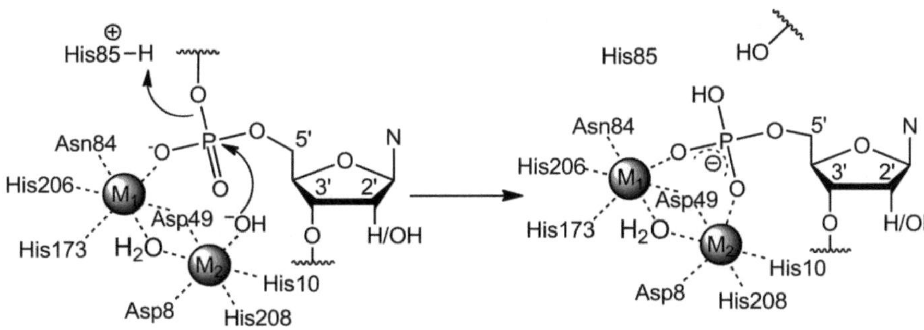

Fig. 8 A proposed mechanism of Mre11 cleavage of DNA that could resemble that of Dbr cleavage of branched RNA. The binuclear manganese cluster is coordinated by seven amino acids and a bridging water molecule. The two manganese ions, Ans84 and His85, coordinate the scissile phosphate to position the reactive group and activate it for nucleophilic attack. Collinear nucleophilic attack either by a hydroxide ion or by a bridging water molecule is coordinated by manganese ion M_2. His85 is proposed to act in transition state stabilization and then as a general acid to donate a proton to the 3′-oxygen of the leaving group. In Dbr, the leaving group would be the 2′-oxygen of the branch point adenosine

for an S_N2 nucleophilic attack (Hopfner et al. 2001; Khalid et al. 2005). After attack by the nucleophile, His85 of Mre11 is proposed to act in the pentacoordinate transition state stabilization and then act as a general acid to donate a proton to the leaving 3′-oxyanion of the DNA, resulting in the 3′-hydroxyl and 5′-phosphate products (Hopfner et al. 2001).

From the homology between Mre11 and Dbr (Figs. 4b and 8), a mechanism for Dbr can be proposed based on the crystal structure of Mre11 (Khalid et al. 2005; Mourani and Damha 2006). For metal ion coordination, six of the seven corresponding amino acids in Dbr (His13, Asp40, Asn85, His179, His231, and His223) are conserved (Fig 4b), essential for debranching activity, and intolerant of conservative substitutions (Khalid et al. 2005). Thus, the binuclear manganese cluster and coordination is likely similar between Mre11 and Dbr. It has also been proposed that Dbr residues Asn85 and His86 of the GNHE motif along with the manganese ions coordinate the 2′-5′-scissile phosphate similar to Asn84, His86, and manganese ion coordination of the 3′-5′-scissile phosphate in Mre11. This proposed coordination is supported by the fact that Asn85 and His86 are intolerant of conservative substitutions (Khalid et al. 2005). Additional support is the lack of catalytic activity on the R_P 2′-5′-phosphorothioate substrate; the larger sulfur atom would disrupt coordination by Asn85 and manganese ion M_1, leading to nonoptimal enzyme–substrate interactions and lack of catalytic activity (Mourani and Damha 2006). If His86 of Dbr indeed corresponds to His85 of Mre11, His86 could also act in transition state stabilization and then as a general acid as His85 does in Mre11. The function of additional conserved Dbr residues is yet to be determined.

4.2 Substrate Specificity of Debranching Enzyme

To recognize the DNA substrate, Mre11 depends only on the DNA phosphate backbone, as expected for a nonspecific nuclease; the Mre11 active site was shown to bind to the phosphate moiety of the molecule through Asn84, His85, and both Mn^{2+} ions (Figs. 4b and 8) (Hopfner et al. 2001). However, Dbr-specific cleavage of only the 2′-5′-phosphodiester bond in a branched RNA likely requires a different mechanism of recognition that it could use to distinguish the RNA-specific 2′-5′-linkage.

Requirements of the stem and 2′-branch residues for efficient cleavage suggest that the RNA substrate either acts to stabilize the active site or even acts in catalysis. In the stem, it has been shown that nonconserved branch points of C and G are not efficiently debranched by Dbr (Jacquier and Rosbash 1986; Arenas and Hurwitz 1987). Synthetic branches and lariats with nonnatural branch residues show differing cleavage efficiencies (Pratico et al. 2005; Paredes et al. 2011). Additionally, a nucleotide 3′ of the branch point is required for efficient cleavage; a 3′-hydroxyl and 3′-phosphate do not support efficient cleavage (Ruskin and Green 1985; Ooi et al. 2001). This suggests that Dbr recognizes both the 2′- and 3′-residues at the branch point. Conversely, the identity of 5′-nucleotide contributes little to efficient cleavage; in fact, the nucleotide is not even required for efficient cleavage (Ganeshan et al. 1995).

Features of the 2′-branch sequence of the RNA substrate are essential for debranching activity (Fig. 2). Initial indication of the essential nature of the 2′-branch sequence arises from studies of Dbr acting on a trinucleotide substrate but not a branched structure with only the 2′-phosphate (Arenas and Hurwitz 1987). Other studies show that a purine rather than a pyrimidine nucleobase is better tolerated at the 2′-residue (Nam et al. 1994). Additionally, the pro-R_P oxygen of the 2′-5′-phosphate linkage is essential for debranching activity, while the pro-S_P oxygen is not (Mourani and Damha 2006). Thus, these portions of the RNA substrate are required for efficient catalysis; however, the role each of the elements play (structural or catalytic) is yet to be determined.

5 Alternate Lariat Intron Functions

Lariat introns generated and released in the process of splicing have recently been associated with additional regulatory RNA pathways (Figs. 1 and 2). Processing of small nucleolar RNAs (snoRNAs) found within introns requires the activity of Dbr for maturation (Tycowski and Steitz 2001; Brameier et al. 2011; Boeke et al. 1998). Additionally, in *S. cerevisiae*, Dbr mutants are defective in Ty retrotransposition, though the role of Dbr in this process is yet unclear (Karst et al. 2000; Salem et al. 2003; Ye et al. 2005; Coombes and Boeke 2005). Finally, lariat introns that code for miRNAs (mirtrons) require debranching for their processing within the RNA interference (RNAi) pathway.

The RNAi pathway uses small ~22-nt regulatory microRNAs which constitute one of the largest gene families with over 600 members in animal genomes (Bartel 2004; Kim 2005; Carthew and Sontheimer 2009; van den Berg et al. 2010; Winter and Diederichs 2011). These miRNAs associate with Argonaute proteins to post-transcriptionally repress target transcripts. Although they can regulate perfectly matched transcripts, the majority of miRNAs recognize their targets with as little as 7 nts from positions 2 to 8 known as the miRNA seed. The biogenesis of miRNAs proceeds in a stepwise and cell-compartmentalized manner (Fig. 3) (Bartel 2004; Kim 2005; van den Berg et al. 2010). Traditional RNA hairpins that enter the miRNA pathway are generally transcribed from noncoding genes by RNA polymerase II as a primary transcript containing one or more pre-miRNA hairpins. These pre-miRNA hairpins are first cleaved from the main RNA transcript into ~65 nt hairpins by Drosha (Figs. 1 and 3), a ribonuclease (RNase) (Wu et al. 2000; Lee et al. 2003; Denli et al. 2004). Drosha is mainly found in the nucleus and exists as a part of a protein complex called the microprocessor complex that also contains a double-stranded RNA-binding protein, Pasha (Denli et al. 2004; Lai et al. 2009). Pasha recognizes the basal hairpin duplex and then binds Drosha which cleaves ~10 nt away from the hairpin's single-stranded flanking sequences. Double-stranded RNA cleavage is achieved by two successive single-strand breaks at two RNase III domains: RNase IIIa and RNase IIIb (Lee et al. 2003). As the cleavage sites are located two nucleotides away from each other, Drosha generates a 3′-overhang important in downstream RNA processing events. The pre-miRNA hairpins are exported from the nucleus by a protein called exportin (Bartel 2004; Kim 2005; van den Berg et al. 2010). Following nuclear export, pre-miRNA hairpins are processed into mature miRNAs by Dicer, an RNase that is only found in the cytoplasm (Bernstein et al. 2001; MacRae et al. 2006; Ji 2008; Strould et al. 2008; Chakravarthy et al. 2010). Dicer recognizes the 3′-overhang and cuts the RNA hairpin ~20–22 nt away toward the hairpin loop also at RNase III domains, similar to Drosha (MacRae et al. 2006). Mature miRNAs are then loaded into the RNA-induced silencing complex (RISC) that is guided by the miRNA to cleave their mRNA targets.

Although most miRNAs derive from noncoding genes, about one third of them are mirtrons coming from the introns of mRNA-coding genes (Lin et al. 2006; Bartel et al. 2007; Lai et al. 2007). These mirtron sequences are located as isolated sequences flanked by exons (Fig. 2) and thus require efficient splicing for their excision from the main transcript sequence. The highly conserved splice site residues, once spliced and debranched, generate the 3′-overhang hairpin end similar to those generated by Drosha (Figs. 2 and 3). Therefore, these pre-miRNA transcripts require Dbr activity but do not require Drosha activity for maturation. Following Dbr debranching, mirtron-derived pre-miRNAs are exported and processed like traditional miRNAs.

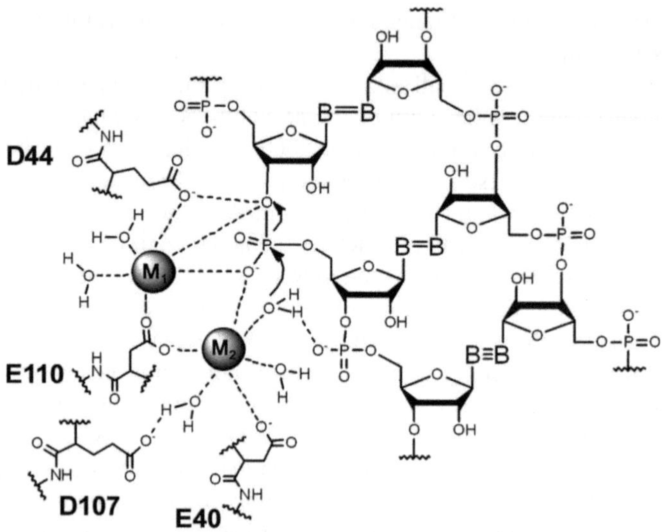

Fig. 9 A two-metal ion mechanism for RNA cleavage in the RNase III domain of Aa-RNase III. Drosha and Dicer active sites are structurally similar. The catalytic metal activates a bound water molecule for nucleophilic attack on the scissile phosphate. Following attack the 3′-hydroxyl leaves behind the 5′-phosphate. RNase III cleavage is not sequence specific, and interactions with the double-stranded RNA substrate distal to the cleaved bond confer specificity in Drosha and Dicer

5.1 Cleavage Mechanism in RNase III Active Sites

All mirtron and miRNA processing cleavage steps by Drosha and Dicer in the RNAi pathways are carried out at RNase III domains. There are four classes of RNase III enzymes classified by increasing molecular weight and complexity of the polypeptide chain, and these four classes are exemplified by bacterial RNase III, *S. cerevisiae* Rnt1p, *D. melanogaster* Drosha, and the *H. sapiens* Dicer, respectively (Drider and Condon 2004). Although the crystal structures for Drosha and Dicer have been solved, the most well-studied RNase III-like enzyme is bacterial RNase III. Even though all of the RNase III enzymes have a dsRNA binding domain and at least one endonuclease domain, the difference within the classes arises from extensions at the N terminus and in the case of Drosha and Dicer, an additional endonuclease domain (Ji 2006; Strould et al. 2008). The *Aquifex aeolicus* RNase III (Aa-RNase III), a bacterial RNase, acts as a homodimer with a F41 residue acting as a ball and socket joint, while Drosha and Dicer exhibit an intramolecular dimer joined similarly by a lysine (L41 in *Giardia intestinalis* Dicer-Gi Dicer) (Ji et al. 2008). These junctions appear to be responsible for the proper positioning of the two catalytic motifs for RNA cleavage (Ji et al. 2008; Ji 2008).

RNA cleavage in Aa-RNase III is magnesium dependent and is thought to proceed in a single step with a two-metal ion-catalyzed S_N2-type mechanism (Fig. 9) for cleavage of the scissile phosphate (Ji 2006; Sun et al. 2005; Blaszczyk et al. 2001; Ji et al. 2008). Magnesium ions that are hexacoordinate are held by four acidic residues at the active site with the rest of the coordination sphere binding the substrate and a suspected nucleophilic water molecule (Figs. 4c and 9). Once bound, the metal ion activates a water nucleophile that attacks the scissile phosphate releasing a 5′-phosphate and a 3′-hydroxyl. The requirement of E40 and D107 side-chain residues in both size and charge together with other requirements in and around the active site suggests the involvement of the second metal ion in the mechanism (Ji et al. 2008; Blaszczyk et al. 2001).

Although several crystal structures have been solved for members of the RNase III classes crystallized with metal ions, only the Giardia Dicer crystal structure (PDB ID 2FFL) shows the position of the two magnesium ions (MacRae et al. 2006). The position was determined by using the anomalous difference electron density map from a crystal derivatized with Er^{3+} normally used to identify any transient Mn^{2+} ions in proteins (Sundaramoorthy et al. 2005). In this crystal structure, the first metal ion is located at metal site M_1 in both endonuclease domains. The second magnesium ion is located at slightly different positions with respect to the active site residues in the two different RNase III domains and suggests two possible occupancy states of the second metal ion, namely, metal sites B and C. The crystal structure of the *Mycobacterium tuberculosis* RNase III with Ca^{2+} (PDB ID 2A11) shows two calcium ions at sites A and C (5.4 Å apart), suggesting that C might be the occupied site in the native enzyme. This distance is similar to other enzymes that are thought to use a two-metal ion mechanism (MacRae et al. 2006). However, the recent structure of bacterial RNase III with its nicked dsRNA product (Fig. 4c) reveals that the four acidic residues bind the two metal ions that are separated by 3.9 Å (Ji et al. 2008). These metal ions also coordinate the 3′-hydroxyl and 5′-phosphate products and four inner-sphere water molecules. Therefore, although the requirement of two metal ions for catalysis has been established, the exact positioning of the metal ions and the residues that coordinate them before, during, and after phosphodiester cleavage bear further investigation.

Thus, the RNA phosphodiester bond cleavages mediated in the spliceosome could be at an RNA active site like that of the group II intron, or in a hybrid RNA–protein active site or in a protein-only active site. Dbr mediates the RNA 2′-5′-phosphodiester cleavage within an enzyme active site though the involvement of the RNA substrate could be critical in distinguishing its specific function from other metallo-phosphatase reactions that closely resemble it. Unlike the highly sequence or linkage-specific reactions of splicing and debranching, RNase III-mediated RNA cleavage within Drosha and Dicer is within a protein active site that is not designed for sequence specificity. Rather, the specificity of Drosha and Dicer in miRNA processing arises from the structure and length of the substrate sequence. Though these enzymes that are critical components in cellular regulatory pathways have diverse active sites, they share a common phosphodiester cleavage function. It is therefore not surprising to see that these enzymes have substrates with shared origins in their pathways. As these regulatory networks are related, identifying specific and distinguishing characteristics of each will become critical in enabling molecular interventions.

Acknowledgment The authors thank the DSF Charitable Foundation for financial support of the Das laboratory.

References

Abelson J (2008) Is the spliceosome a ribonucleoprotein enzyme? Nat Struct Mol Biol 15:1235–1237

Adams PL, Stahley MR, Kosek AB et al (2004) Crystal structure of a self-splicing group I intron with both exons. Nature 430:45–50

Arenas J, Hurwitz J (1987) Purification of a RNA debranching activity from HeLa cells. J Biol Chem 262:4274–4279

Bartel DP (2004) MicroRNAs: genomics, biogenesis, mechanism, and function. Cell 116:281–297

Bartel DP, Ruby JG, Jan CH (2007) Intronic microRNA precursors that bypass Drosha processing. Nature 448:83–86

Beese LS, Steitz TA (1991) Structural basis for the 3′-5′ exonuclease activity of Escherichia coli DNA polymerase I: a two metal ion mechanism. EMBO J 10:25–33

Bernstein E, Caudy AA, Hammond SM et al (2001) Role for a bidentate ribonuclease in the initiation step of RNA interference. Nature 409:363–366

Bertini I, Sigel A, Sigel H (eds) (2001) Handbook on metalloproteins. Marcel Dekker, New York

Bessonov S, Anokhina M, Will CL et al (2008) Isolation of an active step I spliceosome and composition of its RNP core. Nature 452:846–850

Bevilacqua PC (2003) Mechanistic considerations for general acid–base catalysis by RNA: revisiting the mechanism of the hairpin ribozyme. Biochemistry 42:2259–2265

Black DL (2003) Mechanisms of alternative pre-messenger RNA splicing. Annu Rev Biochem 72:291–336

Blaszczyk J, Tropea JE, Bubunenko M et al (2001) Crystallographic and modeling studies of RNase III suggest a mechanism for double-stranded RNA cleavage. Structure 9:1225–1236

Boeke JD, Ooi SL, Samarsky DA et al (1998) Intronic snoRNA biosynthesis in Saccharomyces cerevisiae depends on the lariat-debranching enzyme: intron length effects and activity of a precursor snoRNA. RNA 4:1096–1110

Brameier M, Herwig A, Reinhardt R et al (2011) Human box C/D snoRNAs with miRNA like functions: expanding the range of regulatory RNAs. Nucleic Acids Res 39:675–686

Brow DA (2002) Allosteric cascade of spliceosome activation. Annu Rev Genet 36:333–360

Carthew RW, Sontheimer EJ (2009) Origins and mechanisms of miRNAs and siRNAs. Cell 136:642–655

Chakravarthy S, Sternberg SH, Kellenberger CA et al (2010) Substrate-specific kinetics of dicer-catalyzed RNA processing. J Mol Biol 404:392–402

Chapman KB, Boeke JD (1991) Isolation and characterization of the gene encoding yeast debranching enzyme. Cell 65:483–492

Chu VT, Adamidi C, Liu Q et al (2001) Control of branch-site choice by a group II intron. EMBO J 20:6866–6876

Coller J, Parker R (2004) Eukaryotic mRNA decapping. Annu Rev Biochem 73:861–890

Coombes CE, Boeke JD (2005) An evaluation of detection methods for large lariat RNAs. RNA 11:323–331

Danin-Kreiselman M, Lee CY, Chanfreau G (2003) RNAse III-mediated degradation of unspliced pre-mRNAs and lariat introns. Mol Cell 11:1279–1289

Datta B, Weiner AM (1993) The phylogenetically invariant ACAGAGA and AGC sequences of U6 small nuclear RNA are more tolerant of mutation in human cells than in Saccharomyces cerevisiae. Mol Cell Biol 13:5377–5382

Davies JF, Hostomska Z, Hostomsky Z et al (1991) Crystal-structure of the ribonuclease-H domain of HIV-1 reverse-transcriptase. Science 252:88–95

Denli AM, Tops BBJ, Plasterk RHA et al (2004) Processing of primary microRNAs by the microprocessor complex. Nature 432:231–235

Doublie S, Tabor S, Long AM et al (1998) Crystal structure of a bacteriophage T7 DNA replication complex at 2.2 Å resolution. Nature 391:251–258

Drider D, Condon C (2004) The continuing story of endoribonuclease III. J Mol Microb Biotech 8:195–200

Eddy SR (2001) Non-coding RNA genes and the modern RNA world. Nat Rev Genet 2:919–929

Fabrizio P, Abelson J (1990) Two domains of yeast U6 small nuclear RNA required for both steps of nuclear precursor messenger RNA splicing. Science 250:404–409

Fabrizio P, Abelson J (1992) Thiophosphates in yeast U6 snRNA specifically affect pre-mRNA splicing in vitro. Nucleic Acids Res 20:3659–3664

Ganeshan K, Tadey T, Nam K et al (1995) Novel approaches to the synthesis and analysis of branched RNA. Nucleos Nucleot 14:1009–1013

Garneau NL, Wilusz J, Wilusz CJ (2007) The highways and byways of mRNA decay. Nat Rev Mol Cell Biol 8:113–126

Gesteland RF, Cech TR, Atkins JF (eds) (2006) The RNA world, 3rd edn. Cold Spring Harbor Lab Press, New York

Goldberg J, Huang HB, Kwon YG et al (1995) Three-dimensional structure of the catalytic subunit of protein serine/threonine phosphatase-1. Nature 376:745–753

Gordon PM, Sontheimer EJ, Piccirilli JA (2000a) Kinetic characterization of the second step of group II intron splicing: role of metal ions and the cleavage site 2′-OH in catalysis. Biochemistry 39:12939–12952

Gordon PM, Sontheimer EJ, Piccirilli JA (2000b) Metal ion catalysis during the exon-ligation step of nuclear pre-mRNA splicing: extending the parallels between the spliceosome and group II introns. RNA 6:199–205

Grainger RJ, Beggs JD (2005) Prp8 protein: at the heart of the spliceosome. RNA 11:533–557

Guo Z, Karunatilaka KS, Rueda D (2009) Single-molecule analysis of protein-free U2-U6 snRNAs. Nat Struct Mol Biol 16:1154–1159

Highbarger LA, Gerlt JA, Kenyon GL (1996) Mechanism of the reaction catalyzed by acetoacetate decarboxylase. Importance of lysine 116 in determining the pK(a) of active-site lysine 115. Biochemistry 35:41–46

Hilliker AK, Staley JP (2004) Multiple functions for the invariant AGC triad of U6 snRNA. RNA 10:921–928

Hopfner KP, Karcher A, Craig L et al (2001) Structural biochemistry and interaction architecture of the DNA double-strand break repair Mre11 nuclease and Rad50-ATPase. Cell 105:473–485

Hoskins AA, Friedman LJ, Gallagher SS et al (2011) Ordered and dynamic assembly of single spliceosomes. Science 331:1289–1295

Huppler A, Nikstad LJ, Allmann AM et al (2002) Metal binding and base ionization in the U6 RNA intramolecular stem-loop structure. Nat Struct Biol 9:431–435

Jacquier A, Rosbash M (1986) RNA splicing and intron turnover are greatly diminished by amutant yeast branch point. Proc Natl Acad Sci USA 83:5835–5839

Ji XH (2006) Structural basis for non-catalytic and catalytic activities of ribonuclease III. Acta Crystallogr D 62:933–940

Ji X (2008) The mechanism of RNase III Action: how dicer dices. In: Paddison PJ, Vogt PK (eds) RNA interference, vol 320, Current Topics in Microbiology and Immunology. Springer, Berlin, pp 99–116

Ji X, Gan J, Shaw G et al (2008) A stepwise model for double-stranded RNA processing by ribonuclease III. Mol Microbiol 67:143–154

Kapinos LE, Song B, Sigel H (1999) Acid-base and metal-ion-coordinating properties of benzimidazole and derivatives (=1,3-dideazapurines) in aqueous solution: interrelation between complex stability and ligand basicity. Chem Eur J 5:1794–1802

Karst SM, Rutz NL, Menees TM (2000) The yeast retrotransposons Ty1 and Ty3 require the RNA lariat debranching enzyme, Dbr1p for efficient accumulation of reverse transcripts. Biochem Biophys Res Commun 268:112–117

Keating KS, Toor N, Perlman PS et al (2010) A structural analysis of the group II intron active site and implications for the spliceosome. RNA 16:1–9

Khalid MF, Damha MJ, Shuman S et al (2005) Structure-function analysis of yeast RNA debranching enzyme (Dbr1), a manganese-dependent phosphodiesterase. Nucleic Acids Res 33:6349–6360

Kim VN (2005) MicroRNA biogenesis: coordinated cropping and dicing. Nat Rev Mol Cell Biol 6:376–385

Kim EE, Wyckoff HW (1991) Reaction mechanism of alkaline phosphatase based on crystal structures two-metal ion catalysis. J Mol Biol 218:449–464

Kim JW, Kim HC, Kim GM et al (2000) Human RNA lariat debranching enzyme cDNA complements the phenotypes of Saccharomyces cerevisiae dbr1 and Schizosaccharomyces pombe dbr1 mutants. Nucleic Acids Res 28:3666–3673

Kim HC, Kim GM, Yang JM et al (2001) Cloning, expression, and complementation test of the RNA lariat debranching enzyme cDNA from mouse. Mol Cells 11:198–203

Kissinger CR, Parge HE, Knighton DR et al (1995) Crystal structures of human calcineurin and the human FKBP12–FK506–calcineurin complex. Nature 378:641–644

Konarska MM, Vilardell J, Query CC (2006) Repositioning of the reaction intermediate within the catalytic center of the spliceosome. Mol Cell 21:543–553

Koonin EV (1994) Conserved sequence pattern in a wide variety of phosphoesterases. Protein Sci 3:356–358

Lai EC, Okamura K, Hagen JW et al (2007) The mirtron pathway generates microRNA-class regulatory RNAs in Drosophila. Cell 130:89–100

Lai EC, Martin R, Smibert P et al (2009) A drosophila pasha mutant distinguishes the canonical microRNA and mirtron pathways. Mol Cell Biol 29:861–870

Lee Y, Ahn C, Han JJ et al (2003) The nuclear RNase III Drosha initiates microRNA processing. Nature 425:415–419

Lee C, Jaladat Y, Mohammadi A et al (2010) Metal binding and substrate positioning by evolutionarily invariant U6 sequences in catalytically active protein-free snRNAs. RNA 16:2226–2238

Lin SL, Miller JD, Ying SY (2006) Intronic microRNA (miRNA). J Biomed Biotechnol 2006: 26818

Liu S, Rauhut R, Vornlocher HP et al (2006) The network of protein-protein interactions within the human U4/U6.U5 tri-snRNP. RNA 12:1418–1430

MacRae IJ, Zhou KH, Li F et al (2006) Structural basis for double-stranded RNA processing by dicer. Science 311:195–198

Madhani HD, Guthrie C (1992) A novel base-pairing interaction between U2 and U6 snRNAs suggests a mechanism for the catalytic activation of the spliceosome. Cell 71:803–817

Madhani HD, Bordonne R, Guthrie C (1990) Multiple roles for U6 snRNA in the splicing pathway. Genes Dev 4:2264–2277

Maschhoff KL, Padgett RA (1993) The stereochemical course of the first step of pre-mRNA splicing. Nucleic Acids Res 21:5456–5462

Mattick JS (2007) A new paradigm for developmental biology. J Exp Biol 210:1526–1547

Mefford MA, Staley JP (2009) Evidence that U2/U6 helix I promotes both catalytic steps of pre-mRNA splicing and rearranges in between these steps. RNA 15:1386–1397

Moore MJ, Sharp PA (1993) Evidence for two active sites in the spliceosome provided by stereochemistry of pre-mRNA splicing. Nature 365:364–368

Mourani R, Damha M (2006) Synthesis, characterization, and biological properties of small branched RNA fragments containing chiral (R-P and S-P) 2′,5′-phosphorothioate linkages. Nucleos Nucleot Nucleic Acids 25:203–229

Nam KB, Hudson RHE, Chapman KB et al (1994) Yeast lariat debranching enzyme–substrate and sequence specificity. J Biol Chem 269:20613–20621

Nam K, Lee G, Trambley J et al (1997) Severe growth defect in a Schizosaccharomyces pombe mutant defective in intron lariat degradation. Mol Cell Biol 17:809–818

Oda Y, Yamazaki T, Nagayama K et al (1994) Individual ionization constants of all the carboxyl groups in Ribonuclease HI from Escherichia coli determined by NMR. Biochemistry 33: 5275–5284

Ooi SL, Samarsky DA, Fournier MJ et al (1998) Intronic snoRNA biosynthesis in Saccharomyces cerevisiae depends on the lariat-debranching enzyme: Intron length effects and activity of a precursor snoRNA. RNA 4:1096–1110

Ooi SL, Dann C, Nam K et al (2001) RNA lariat debranching enzyme. Methods Enzymol 342:233–248

Padgett RA, Dayie KT (2008) A glimpse into the active site of a group II intron and maybe the spliceosome, too. RNA 14:1697–1703

Padgett RA, Podar M, Boulanger SC et al (1994) The stereochemical course of group-Ii Intron self-splicing. Science 266:1685–1688

Paredes E, Grahacharya D, Evans M, Raney E, Dey SK, Macbeth M, Das SR (2011) Unpublished results

Patel AA, Steitz JA (2003) Splicing double: insights from the second spliceosome. Nat Rev Mol Cell Biol 4:960–970

Pena V, Rozov A, Fabrizio P et al (2008) Structure and function of an RNase H domain at the heart of the spliceosome. EMBO J 27:2929–2940

Pertea M, Salzberg SL (2010) Between a chicken and a grape: estimating the number of human genes. Gen Biol 11:206

Piccirilli JA, Vyle JS, Caruthers MH et al (1993) Metal-ion catalysis in the tetrahymena ribozyme reaction. Nature 361:85–88

Podar M, Perlman PS, Padgett RA (1995) Stereochemical selectivity of group II intron splicing, reverse splicing, and hydrolysis reactions. Mol Cell Biol 15:4466–4478

Podar M, Chu VT, Pyle AM et al (1998) Group II intron splicing in vivo by first-step hydrolysis. Nature 391:915–918

Pomeranz Krummel DA, Oubridge C, Leung AK et al (2009) Crystal structure of human spliceosomal U1 snRNP at 5.5 A resolution. Nature 458:475–480

Pratico ED, Wang Y, Silverman SK (2005) A deoxyribozyme that synthesizes 2′,5′-branched RNA with any branch-site nucleotide. Nucleic Acids Res 33:3503–3512

Pyle AM (2010) The tertiary structure of group II introns: implications for biological function and evolution. Crit Rev Biochem Mol 45:215–232

Rhode BM, Hartmuth K, Westhof E, Lührmann R (2006) Proximity of conserved U6 and U2 snRNA elements to the 5′ splice site region in activated spliceosomes. EMBO J 25: 2475–2486

Ritchie DB, Schellenberg MJ, Gesner EM et al (2008) Structural elucidation of a PRP8 core domain from the heart of the spliceosome. Nat Struct Mol Biol 15:1199–1205

Ruskin B, Green MR (1985) An RNA processing activity that debranches RNA lariats. Science 229:135–140

Ruskin B, Krainer AR, Maniatis T et al (1984) Excision of an intact intron as a novel lariat structure during pre-mRNA splicing in vitro. Cell 38:317–331

Rusnak F, Mertz P (2000) Calcineurin: form and function. Physiol Rev 80:1483–1521

Salem LA, Boucher CL, Menees TM (2003) Relationship between RNA lariat debranching and Ty1 element retrotransposition. J Virol 77:12795–12806

Sashital DG, Cornilescu G, McManus CJ et al (2004) U2-U6 RNA folding reveals a group II intron-like domain and a four-helix junction. Nat Struct Mol Biol 11:1237–1242

Segault V, Will CL, Polycarpou-Schwarz M et al (1999) Conserved loop I of U5 small nuclear RNA is dispensable for both catalytic steps of pre-mRNA splicing in HeLa nuclear extracts. Mol Cell Biol 19:2782–2790

Sharp PA (1987) Splicing of messenger RNA precursors. Science 235:766–771

Sigel RKO, Pyle AM (2007) Alternative roles for metal ions in enzyme catalysis and the implications for ribozyme chemistry. Chem Rev 107:97–113

Sigel RKO, Vaidya A, Pyle AM (2000) Metal ion binding sites in a group II intron core. Nat Struct Biol 7:1111–1116

Sontheimer EJ, Sun S, Piccirilli JA (1997) Metal ion catalysis during splicing of premessenger RNA. Nature 388:801–805

Sontheimer EJ, Gordon PM, Piccirilli JA (1999) Metal ion catalysis during group II intron self-splicing: parallels with the spliceosome. Genes Dev 13:1729–1741

Steitz TA, Steitz JA (1993) A general two-metal-ion mechanism for catalytic RNA. Proc Natl Acad Sci USA 90:6498–6502

Strould RM, Du Z, Lee JK et al (2008) Structural and biochemical insights into the dicing mechanism of mouse dicer: a conserved lysine is critical for dsRNA cleavage. Proc Natl Acad Sci USA 105:2391–2396

Sun JS, Manley JL (1995) A novel U2-U6 snRNA structure is necessary for mammalian messenger-RNA splicing. Genes Dev 9:843–854

Sun W, Pertzev A, Nicholson AW (2005) Catalytic mechanism of Escherichia coli ribonuclease III: kinetic and inhibitor evidence for the involvement of two magnesium ions in RNA phosphodiester hydrolysis. Nucleic Acids Res 33:807–815

Sundaramoorthy M, Youngs HL, Gold MH et al (2005) High-resolution crystal structure of manganese peroxidase: substrate and inhibitor complexes. Biochemistry 44:6463–6470

Toor N, Hausner G, Zimmerly S (2001) Coevolution of group II intron RNA structures with their intron-encoded reverse transcriptases. RNA 7:1142–1152

Toor N, Keating KS, Taylor SD et al (2008a) Crystal structure of a self-spliced group II intron. Science 320:77–82

Toor N, Rajashankar K, Keating KS et al (2008b) Structural basis for exon recognition by a group II intron. Nat Struct Mol Biol 15:1221–1222

Toor N, Keating KS, Pyle AM (2009) Structural insights into RNA splicing. Curr Opin Struc Biol 19:260–266

Toor N, Keating KS, Fedorova O et al (2010) Tertiary architecture of the Oceanobacillus iheyensis group II intron. RNA 16:57–69

Tsai RT, Fu RH, Yeh FL et al (2005) Spliceosome disassembly catalyzed by Prp43 and its associated components Ntr1 and Ntr2. Genes Dev 19:2991–3003

Turner IA, Norman CM, Churcher MJ et al (2006) Dissection of Prp8 protein defines multiple interactions with crucial RNA sequences in the catalytic core of the spliceosome. RNA 12:375–386

Tycowski KT, Steitz JA (2001) Non-coding snoRNA host genes in Drosophila: expression strategies for modification guide snoRNAs. Eur J Cell Biol 80:119–125

Valadkhan S, Manley JL (2001) Splicing-related catalysis by protein-free snRNAs. Nature 413:701–707

Valadkhan S, Mohammadi A, Jaladat Y et al (2009) Protein-free small nuclear RNAs catalyze a two-step splicing reaction. Proc Natl Acad Sci USA 106:11901–11906

Valcarcel J, Gaur RK, Singh R et al (1996) Interaction of U2AF65 RS region with pre-mRNA branch point and promotion of base pairing with U2 snRNA. Science 273:1706–1709

van den Berg A, Slezak-Prochazka I, Durmus S et al (2010) MicroRNAs, macrocontrol: regulation of miRNA processing. RNA 16:1087–1095

Veretnik S, Wills C, Youkharibache P et al (2009) Sm/Lsm genes provide a glimpse into the early evolution of the spliceosome. PLoS Comput Biol 5:e1000315

Voegtli WC, White DJ, Reiter NJ et al (2000) Structure of the bacteriophage lambda Ser/Thr protein phosphatase with sulfate ion bound in two coordination modes. Biochemistry 39:15365–15374

Wahl MC, Will CL, Lührmann R (2009) The spliceosome: design principles of a dynamic RNP machine. Cell 136:701–718

Walter NG (2007) Ribozyme catalysis revisited: is water involved? Mol Cell 28:923–929

Warshel A (1981) Electrostatic basis of structure-function correlation in proteins. Acc Chem Res 14:284–290

Weinstein LB, Jones BCNM, Cosstick R et al (1997) A second catalytic metal ion in a group I ribozyme. Nature 388:805–808

Will CL, Lührmann R (2006) Spliceosome structure and function. In: Gesteland RF, Cech TR, Atkins JF (eds) The RNA world, 3rd edn. Cold Spring Harbor Lab Press, New York, pp 369–400

Will CL, Schneider C, MacMillan AM et al (2001) A novel U2 and U11/U12 snRNP protein that associates with the pre-mRNA branch site. EMBO J 20:4536–4546

Williams RS, Moncalian G, Williams JS et al (2008) Mre11 dimers coordinate DNA end bridging and nuclease processing in double-strand-break repair. Cell 135:97–109

Winter J, Diederichs S (2011) MicroRNA biogenesis and cancer. In: Wu W (ed) MicroRNA and cancer methods and protocols, vol 676, Methods in molecular biology. Humana, New York, pp 3–22

Wu J, Manley JL (1989) Mammalian pre-mRNA branch site selection by U2 snRNP involves base pairing. Genes Dev 3:1553–1561

Wu HJ, Xu H, Miraglia LJ et al (2000) Human RNase III is a 160-kDa protein involved in preribosomal RNA processing. J Biol Chem 275:36957–36965

Yang K, Zhang L, Xu T et al (2008) Crystal structure of the beta-finger domain of Prp8 reveals analogy to ribosomal proteins. Proc Natl Acad Sci USA 105:13817–13822

Ye Y, De Leon J, Yokoyama N et al (2005) DBR1 siRNA inhibition of HIV-1 replication. Retrovirology 2:9

Yean SL, Wuenschell G, Termini J et al (2000) Metal-ion coordination by U6 small nuclear RNA contributes to catalysis in the spliceosome. Nature 408:881–884

Yoshida A, Sun SG, Piccirilli JA (1999) A new metal ion interaction in the Tetrahymena ribozyme reaction revealed by double sulfur substitution. Nat Struct Biol 6:318–321

Yu YT, Maroney PA, Darzynkiwicz E et al (1995) U6 snRNA function in nuclear pre-mRNA splicing: a phosphorothioate interference analysis of the U6 phosphate backbone. RNA 1:46–54

Yuan F, Griffin L, Phelps L et al (2007) Use of a novel Forster resonance energy transfer method to identify locations of site-bound metal ions in the U2-U6 snRNA complex. Nucleic Acids Res 35:2833–2845

Zhuo SQ, Clemens JC, Stone RL et al (1994) Mutational analysis of a Ser/Thr phosphatase - identification of residues important in phosphoesterase substrate binding and catalysis. J Biol Chem 269:26234–26238

MicroRNAs and Their Antagonists as Novel Therapeutics

Jeppe Vinther, Jakob Lewin Rukov, and Noam Shomron

Contents

Abstract During the past decade, it has been realized that microRNAs (miRNAs) regulate expression of about half of human genes and thereby act as key mediators of most pathways involved in human health and disease. Given that miRNAs are often deregulated in disease processes, it comes as no surprise that therapeutic

J. Vinther
Department of Biology, University of Copenhagen, 2200 Copenhagen N, Denmark

J.L. Rukov
Department of Biology, University of Copenhagen, 2200 Copenhagen N, Denmark

Sackler Faculty of Medicine, Tel Aviv University, Tel Aviv 69978, Israel

N. Shomron (✉)
Sackler Faculty of Medicine, Tel Aviv University, Tel Aviv 69978, Israel
e-mail: nshomron@post.tau.ac.il

V.A. Erdmann and J. Barciszewski (eds.), *From Nucleic Acids Sequences to Molecular Medicine*, RNA Technologies, DOI 10.1007/978-3-642-27426-8_20,

targeting of miRNA expression and function is a subject of intense investigation. Recently, the field has seen considerable progress, and the first miRNA drugs, in the form of miRNA antagonists, have entered clinical trials. In this chapter we will (1) explore how miRNAs function in health and how they change during disease development, focusing on the mode of miRNA regulation (often, this involves moderate control of many targets); (2) describe the strategies used to manipulate miRNA expression and the challenges associated in a clinical setting; (3) provide an overview of miRNAs and their antagonists as therapeutics; and (4) finally, discuss the therapeutic potential of miRNA manipulation and the caveats that come with it.

Keywords MicroRNA • MicroRNA antagonism • MicroRNA delivery • RNA drugs • Therapeutics

1 Introduction

The large majority of currently approved drugs target receptors or enzymes. The number of proteins or genes that have been successfully targeted by an approved drug is in the hundreds. However, despite extensive research, the therapeutic gain of continuing traditional methods of drug development is somewhat limited. In order to overcome this, alternatives to targeting proteins are of essence, and the pharmaceutical industry is exploring new routes for novel therapeutics. One major avenue is the research on microRNAs (miRNAs). miRNAs are small (~22 nucleotides) regulatory noncoding RNAs that downregulate gene expression. The miRNAs are often involved in disease etiology and function as regulators of many genes and pathways, which make them candidate molecules for development of therapeutics. Research performed during the past decade, since the discovery of miRNAs, has clearly demonstrated that miRNA function can be successfully modulated both in cell culture and in vivo. In cell culture, miRNAs' function can be enhanced or antagonized using standard transfections of different types of miRNA mimics or inhibitors. These experiments have paved the road for much of our current understanding of miRNA function. In vivo, manipulation of miRNA performance is more challenging, but there has been large progress. In particular, therapeutic miRNA antagonism now seems to be a feasible and novel option to treat various human diseases.

2 miRNA Biogenesis and Function

miRNAs were originally described as rare developmental regulators (Lee et al. 1993; Wightman et al. 1993) but have emerged as essential players in a wide range of cellular processes (Chi et al. 2009; Friedman et al. 2009). miRNAs downregulate gene expression at the posttranscriptional level, most likely by a combination of

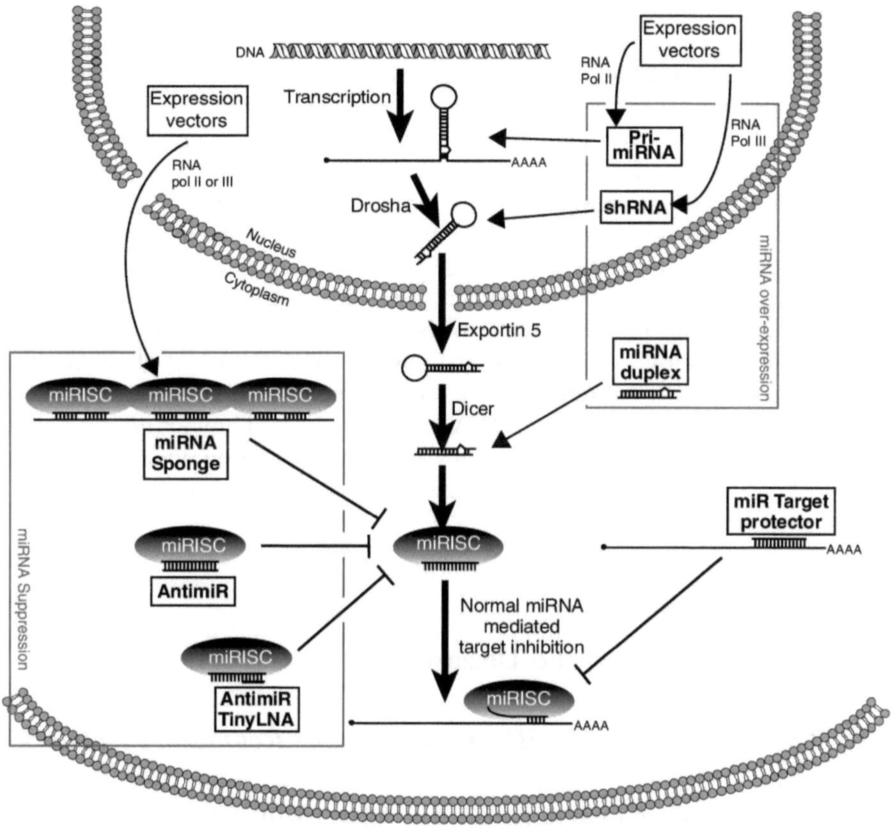

Fig. 1 miRNA biogenesis, function, and therapeutic interference. The miRNA biogenesis pathway, commencing with miRNA gene transcription and ending with binding to a miRNA target gene, is schematically drawn centrally in this image. On both sides of this pathway, miRNA manipulations are shown. miRNA therapeutics may utilize and interfere with miRNA biogenesis and function at several different stages. By introducing miRNAs into the miRNA processing pathway, miRNAs may be overexpressed (*top right-hand side*). RNA polymerase II-based expression vectors produce pri-miRNAs that necessitate both Drosha and Dicer processing, while miRNA hairpins that bypass Drosha processing can be produced from RNA polymerase III-based expression vectors. miRNA duplexes may be directly introduced and incorporated into RISC (all three levels are indicated in the *top right-hand side corner* of the image). Endogenous miRNA levels may be sequestered by several methods (*bottom left-hand side*). Among these are miRNA sponges which are long, artificial RNA transcripts that contain multiple miRNA binding sites and therefore dominate competition for miRISC. miRNAs may be targeted and neutralized by full length or tiny antimiRs (all three levels of regulation are indicated in the *bottom left-hand side corner* of the image). Finally, to only influence the targeting of a single gene, miRNA target protectors may be used to selectively cover miRNA target sites (or surrounding sequences)

inhibition of translation and degradation of the mRNA (Djuranovic et al. 2011; Filipowicz et al. 2005; Guo et al. 2010). Mature miRNAs reside in hairpins that are expressed as parts of long primary transcripts (pri-miRNAs), often as fractions of an intron in protein coding genes or as longer individual transcripts (Fig. 1). miRNA biogenesis includes two subsequent cleavage reactions. First, the miRNA hairpin is recognized and excised by the RNase Drosha and its cofactor DGCR8, leading to a ~70 nucleotide pre-miRNA, which is transported from the nucleus to the cytoplasm by the export protein exportin 5 (Gregory et al. 2004). In the cytosol, a second RNase named Dicer cleaves the pre-miRNA, which produces a duplex consisting of a miRNA guide strand and a mainly nonfunctional miRNA* passenger strand (Gregory et al. 2005) (miRNA* are also termed -5p or -3p depending on the pre-miRNA strand they originate from). The mature miRNA is hereafter incorporated into the RNA-induced silencing complex (RISC), while the miRNA* is degraded or incorporated into RISC at a lower frequency (Gregory et al. 2005). Partial complementarity between the miRNA and the mRNA 3′ untranslated region (UTR) directs the loaded RISC (termed miRISC) to the mRNA target (Lewis et al. 2005). The primary determinant of this recognition is target complementarity in the nucleotides residing in positions 2–8 on the miRNA (termed the "seed" region) (Lewis et al. 2003). Inhibition of expression occurs by blocking translation and, according to recent data primarily, by degradation of the target mRNA (Djuranovic et al. 2011; Guo et al. 2010).

3 miRNAs in Health and Disease

3.1 *The Importance of miRNAs in Cellular Regulatory Networks*

In the miRNA database (miRbase) (Griffiths-Jones 2010), 1,049 different human mature miRNAs are registered, although the actual number of active miRNAs is more likely to be around two-thirds this number (Chiang et al. 2010). Since each miRNA potentially targets hundreds of genes (Baek et al. 2008; Lim et al. 2005; Selbach et al. 2008; Shomron et al. 2009), around 60% of human genes are estimated to be targets of evolutionarily conserved miRNA regulation (Friedman et al. 2009). miRNA targeting is therefore widespread, making these small molecules pivotal managers of gene expression. Importantly, for miRNA-based therapeutics, many essential cellular processes, including differentiation, apoptosis, and stress response, are controlled by varying miRNA levels (Stefani and Slack 2008).

3.2 *miRNAs in Disease: A Case Story*

Consistent with their important roles in cellular regulatory networks, miRNAs have been shown to be deregulated in a long list of diseases, including diabetes (Ling et al. 2009), cancer (Calin and Croce 2006), and neurological (Provost 2010) and cardiovascular diseases (Cheng and Zhang 2010), to name a few.

An example of how a miRNA plays a central role in physiological processes associated with pathological conditions is the importance of the miR-208 family in cardiac contractility. The primary determinants of heart muscle cell contraction are the myosin heavy chain (MHC) proteins αMHC (encoded by *Myh6*), a fast ATPase delivering strong contraction activity, and βMHC (encoded by *Myh7*), a slow ATPase delivering lower contractile efficiency (Dorn 2011; Liu and Olson 2010). During development, βMHC is highly expressed, while αMHC is expressed at lower levels. However, in the adult heart expression shifts in favor of αMHC, a process induced by thyroid hormone (T3) (Dorn 2011; Liu and Olson 2010). Interestingly, the only heart specific miRNA, miR-208a (Liu and Olson 2010), resides within an intron of αMHC, while the related miR-208b (which is also found in slow skeletal muscle cells) similarly resides within an intron of *Myh7*. Another related gene, *Myh7b*, encodes a protein of unknown function but carries the miRNA, miR-499, which is also related in sequence to miR-208a and miR-208b. The expressions of miR-208a, miR-208b, and miR-499 are tightly coupled to their host genes (Callis et al. 2009; van Rooij et al. 2009), and the three miRNAs share six bases of their seed region which suggests that they regulate the same proteins (Liu and Olson 2010).

Following mechanical stress, αMHC levels drop [but not miR-208a levels, due to its long half-life (van Rooij et al. 2007)] while βMHC/miR-208b and MYH7B/miR-499 levels increase (van Rooij et al. 2009). This shift is associated with cardiac hypertrophy, the heart's primary response to stress (Callis et al. 2009). Cardiac hypertrophy is considered an adaptive mechanism to sustain cardiac output during stress; however, prolonged periods of pathological hypertrophy may lead to heart failure (Callis et al. 2009). In both humans and rodents, a part of this stress response is mediated by first miR-208a and later miR-208b and miR-499 through targeting of the T3 cofactor thyroid hormone receptor-associated protein 1 (THRAP1) (Callis et al. 2009; van Rooij et al. 2007).

Collectively, miR-208a/b and miR-499 are referred to as myomiRs (van Rooij et al. 2009). In animal models, myomiR activity has also been shown to lead to fibrillation (van Rooij et al. 2007). Furthermore, overexpression of miR-208a leads to failure in ventricular contraction, and depletion leads to deficiencies in cardiac conduction in the form of atrial fibrillation (Callis et al. 2009). Importantly and consistent with their functions in animal models, all three myomiRs are upregulated in patients with dilated cardiomyopathy (Satoh et al. 2010), and miR-208a is consistently upregulated in myocardial infarct patients (Boštjančič et al. 2010).

This example illustrates how miRNAs perform central functions in cardiac contractility and are thus intrinsic to certain disease states. It does, however, also exemplify some of the caveats of miRNA therapeutics, since pathological conditions may follow both miRNA overexpression (cardiac hypertrophy) and depletion (aberrant cardiac conduction). This means that it would be crucial to avoid excessive perturbation of miRNA levels during treatment. That said, the intricate role of myomiRs in cardiac disease make them appealing drug targets, and the anti-myomiR drugs under development (Table 1) are promising potential agents against chronic heart failure.

Table 1 miRNAs currently being pursued as drug targets and having passed the initial discovery phase

miRNA	Indication	Drug design	Company	Phase	Relevant publications
Let-7	Non-small cell lung cancer	miRNA mimic	miRNA Therapeutics	Preclinical testing	Trang et al. (2011)
miR-15/195	Post-MI remodeling	LNA/DNA antimiR	miRagen, Santaris	Preclinical testing	van Rooij et al. (2006)
miR-16	Prostate cancer	miRNA mimic	miRNA Therapeutics	Preclinical testing	Takeshita et al. (2010)
miR-21	Fibrosis	Antagomir	Regulus, sanofi-aventis	Preclinical testing	Thum et al. (2008)
miR-29	Cardiac fibrosis	miRNA mimic	miRagen	Lead optimization	van Rooij et al. (2008)
miR-33a/b	Atherosclerosis	F/MOE antimiR	Regulus	Preclinical testing	Rayner et al. (2011)
miR-34	Non-small cell lung cancer	miRNA mimic	miRNA Therapeutics	Preclinical testing	Trang et al. (2011)
miR-122	HCV	F/MOE antimiR	Regulus, GlaxoSmithKline	Preclinical testing	Esau et al. (2006), Jopling et al. (2005)
miR-122	HCV	LNA/DNA antimiR	Santaris	Phase 2	Elmen et al. (2008b), Lanford et al. (2010)
miR-132	Angiogenesis	Antagomir	Regulus	Preclinical testing	Anand et al. (2010)
miR-155	Inflammatory diseases	N/A	Regulus, Glaxo SmithKline	Preclinical testing	O'Connell et al. (2007, 2008)
miR-208/499	Chronic heart failure	LNA/DNA antimiR	miRagen, Santaris	Preclinical testing	van Rooij et al. (2007, 2009)
miR-451	Polycythemia vera	LNA/DNA antimiR	miRagen, Santaris	Preclinical testing	Patrick et al. (2010)

Note: Information is from company websites, press releases, and papers that are published coauthored or licensed by the mentioned companies. It is likely that the information is not complete. Moreover, the drug design mentioned here may not correspond to the actual drugs that are being tested by the companies, because the actual design in most cases is undisclosed (N/A: Not available, MI: Myocardial infarction)

3.3 *miRNAs in Cancer*

miRNAs are frequently deregulated in cancer. For instance, miR-21 overexpression in an in vivo mouse model leads to a pre-B cell lymphoma phenotype (Selcuklu et al. 2009), and the miR-15a/miR-16 cluster is frequently deleted in B cell chronic lymphocytic leukemia (CLL) (Calin et al. 2002). Many miRNAs are consistently and uniquely upregulated or downregulated in multiple different cancer types. miR-21 is, for instance, also upregulated in colon cancer, prostate cancer, glioma, and lung cancer, to name a few (Selcuklu et al. 2009). miR-15a/miR-16 are also downregulated in prostate cancer and pituitary cancer (Aqeilan et al. 2009). These far-reaching miRNAs are therefore termed oncomirs or tumor suppressor miRNAs, respectively. The deregulated miRNA has a downstream effect on many relevant targets. Oncomirs typically target tumor suppressor genes and are overexpressed in cancer, and tumor suppressor miRNAs typically target oncogenes and are depleted in cancer.

4 Identifying miRNA Therapeutic Targets

4.1 *The Appeal of miRNAs as Therapeutic Targets*

An important aspect of miRNA functionality and their related use as therapeutic agents is that they may regulate cellular processes by affecting levels of many proteins involved in the same cellular pathway. For instance, the response of T cells upon activation depends on an intracellular signaling cascade involving a series of phosphatases which results in increased intracellular Ca^{2+} levels (Li et al. 2007). miR-181a participates in regulating this sensitivity in T cells by targeting four genes encoding phosphatases: *SHP-2*, *PTPN22*, *DUSP5*, and *DUSP6*, all active in the T cell signal transduction pathway. miR-181a thus targets the pathway at several different levels. Importantly, modulating levels of these four proteins individually is not sufficient to reproduce the effect of miR-181a (Li et al. 2007). Thus, miRNAs allow for modifying pathway efficacy even in cases where it may be ineffective to modify levels of single proteins, ascertained by their often modest level.

Furthermore, diseases such as cancer are characterized by global changes in a series of different regulatory systems, such as apoptosis, DNA repair, and cell adhesion (Garzon et al. 2010). Since miRNAs may potentially target dozens or even hundreds of genes, one or a few miRNAs may at the same time affect different target genes within several regulatory systems (O'Day and Lal 2010). miRNAs can therefore be regarded as centrally located connectors affecting multiple pathways at the same time. By reverting miRNA levels, it may be possible to rewire multiple divergent systems to the desired phenotype (Garzon et al. 2010).

Another advantage of miRNA therapeutics is the relatively simple, sequence-based nature of oligonucleotide drugs. Whereas de novo development of more traditional protein-based drugs is challenging due to three-dimensional protein structure prediction and protein–drug interactions, sequence-based miRNA drugs are essentially one-dimensional. Knowledge on miRNA sequence automatically confers knowledge on the design of potential associated drugs. This may (in the long run) entail simplicity in the drug design and development pipeline.

4.2 Considerations for Therapeutic Targeting of miRNAs

miRNAs have characteristics that make them uniquely suited for therapeutic targets. However, current research on their specific role in disease raises a few concerns that should be taken into account when determining the overall therapeutic value of a miRNA. In most cases, for instance, the exact role of the miRNA in disease etiology is not entirely defined. As a consequence, it might be difficult to assess whether the miRNA takes part in pathogenesis or is simply deregulated as a secondary effect of other regulatory changes in the cell. This distinction is important for miRNA therapeutics since disease regulatory miRNAs will in general be more suitable drug targets than follower miRNAs.

Studies have furthermore shown that miRNA-affected gene targets may be context dependent and therefore might vary between tissues. For instance, the otherwise well-established miR-21 target phosphatase and tensin homolog (PTEN) does not seem to be a target of miR-21 in MCF-7 breast cancer cells (Qi et al. 2009; Terao et al. 2011). Thus, the disease-related potential of a miRNA (and therefore also its therapeutic suitability) depends on the actions/target profile of the miRNA in the specific cellular context.

Another point is that the interaction between a miRNA and a potential target depends on (among other things) the stoichiometric relationship between the two molecules. When miRNA levels are increased (such as when treating with miRNA mimics; see below), previously untargeted mRNAs with weak target sites may be expected to become targeted by the miRNA. Conversely, some targets will be expected to react more strongly than others when miRNAs are inhibited. The miRNA target profiles in cells treated with miRNA therapeutics will therefore not necessarily reflect the target profile in untreated cells and should be closely monitored.

In many diseases, the underlying genetic changes may vary even though the pathological condition is the same. As such, miRNA deregulation in a certain disease may be frequent but not universal. The miR-15a/16 cluster is, for instance, not deleted or downregulated in all cases of CLL (Calin et al. 2002). In cases where miRNA deregulation is not a source of pathogenesis, it will of course be less productive to therapeutically target the miRNA. In these cases, miRNA pharmacogenomic approaches may help determine whether the miRNA is suited for therapeutic targeting (Rukov and Shomron 2011).

Whereas the examples presented above mainly deal with how miRNAs are directly involved in pathogenesis, some miRNAs have been described as performing protective functions and may present an alternative type of miRNA therapeutics. For instance, several studies have demonstrated that a protective function against ischemia, which is induced by mild ischemic stress and known as ischemic preconditioning (IP), is mediated by miR-21 (Cheng et al. 2010; Dong et al. 2009; Yin et al. 2009). miR-21 is upregulated during IP, and increased miR-21 level is in itself sufficient to induce protection against apoptosis (an indicator of IP) in a cell line model (Dong et al. 2009). Inhibition of miR-21 abolished the protective effect of IP in a rat in vivo model (Cheng et al. 2010). Such protective miRNAs (and, conversely, miRNAs that increase predisposition to disease) may be relevant therapeutic targets, for instance (in the above example), in prophylactic treatment of ischemia.

4.3 *Mechanisms of miRNA Deregulation*

A separate issue is how miRNA function initially becomes deregulated, since understanding the background of deregulation may inform on the best strategy for developing therapeutics. Several underlying reasons can be envisioned:

1. The miRNA may be mutated. Particularly, mutations in the seed sequence may significantly modify the regulatory impact of a miRNA. In a related scenario, the miRNA sequence may be preserved, but miRNA gene copy number may vary between healthy and diseased tissue, due to deletion or duplication of the miRNA gene [as exemplified with the miR-15a/16 cluster in cancer (Calin et al. 2002)].
2. The biogenesis of the miRNA may be affected. A minority of miRNAs require specific factors for their expression which, if deregulated, will affect the expression of such subgroups of miRNAs. For instance, pri-miR-18a requires binding of the protein hnRNP A1 for Drosha recognition to occur (Guil and Caceres 2007; Michlewski et al. 2008). Global changes in miRNA expression can also be seen as a result of failure in canonical miRNA processing factors, as exemplified by Drosha overexpression in cervical squamous cell carcinoma, which leads to major changes in miRNA output (Muralidhar et al. 2011). In these cases, a pathological condition may arise due to perturbations of many miRNAs, and even though the disease is rooted in miRNA deregulation, the miRNAs themselves may not be suitable therapeutic targets. Rather, the treatment should focus on the underlying causes of the miRNA deregulation.
3. Transcription of the miRNA gene may vary. This may occur, for instance, due to aberrant transcription factor expression or changes in methylation patterns. The transcription factors E12/E47, for example, have been found to be overexpressed in breast cancer stem cells and drive miR-495 expression, which in turn promotes oncogenesis (Hwang-Verslues et al. 2011).

4. miRNA target sites may change. miRNA target site gain/loss can take place following mutations in the 3′ UTR sequence of an mRNA. A single-nucleotide polymorphism in a miR-187 target in transforming growth factor, beta 1 (TGFB1), has, for instance, been associated with an increased risk of breast cancer (Nicoloso et al. 2010). In this case, miRNA profiles will remain unchanged, only the target profile varies.

5 Strategies for the Manipulation of miRNA Function

5.1 miRNA Manipulation in Cell Culture

Several different strategies for manipulation miRNA function have been developed (Fig. 1). Early approaches focused on cultured cells, and a variety of different strategies have now been developed. One of the earliest and still most used methods to introduce miRNAs into cells is the use of miRNA duplexes (Lim et al. 2005; Nielsen et al. 2007; Zeng et al. 2003). Like siRNA duplexes, miRNA duplexes consist of ~21 bp double-stranded RNA (dsRNA) with 2 bp overhangs at the 3′ ends but are designed so that the guide strand is identical to the mature miRNA strand rather than having complete complimentarity to a target gene. In miRNA-introduced cells, after miRNA transfection, the action of the microprocessor complex and dicer is bypassed, and the mature miRNA strand is incorporated into RISC. miRNAs can also be expressed in cell culture by transfection with plasmids that express the miRNA hairpin from RNA polymerase III promoters, such as the H1 or U6 promoters (Paddison et al. 2002). Using this strategy, Drosha cleavage is bypassed while Dicer cleavage is still needed to produce the active miRNA. Alternatively, the miRNA hairpin and its immediate genomic surroundings can be expressed from RNA polymerase II promoters to allow the mature miRNA to be produced through the canonical miRNA pathway (Zeng et al. 2002). The use of miRNA expression constructs makes it possible to create cells lines that stably express miRNAs and create transgenic animals that overexpress miRNAs.

Transfection with miRNA duplexes or miRNA expression vectors typically results in a very high level of mature miRNA in the cells, and even though the repressive effect on the individual miRNA target gene is relatively modest, a strong signal for the miRNA seed match in the 3′ UTR of target mRNAs can be observed (Grimson et al. 2007). However, such experiments produce a miRNA level much higher than what is observed for most endogenously expressed miRNAs. Thus, the miRNA may be expressed out of context, meaning that the observed miRNA targets may not be physiologically relevant.

Inhibition of miRNAs has the large advantage over miRNA overexpression since the native endogenous levels of miRNAs are modified. Interestingly, the expression and/or function of most miRNAs are repressed in tumor-derived cells (Lu et al. 2005). An effective method to antagonize miRNA function is to use oligonucleotides that are complementary to the miRNA (called antimiRs). When

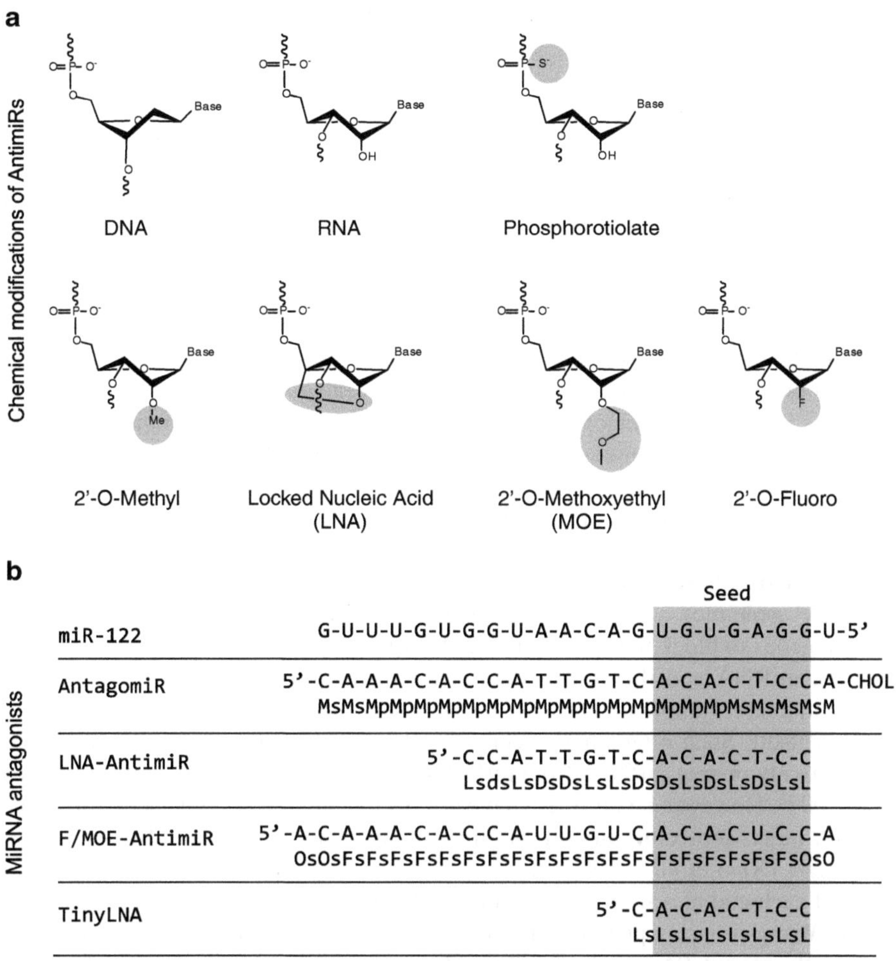

Fig. 2 Chemically modified oligonucleotides for miRNA inhibition. (**a**) Nucleotide modifications. First panel shows an unmodified dinucleotide, and in the following panels, modifications are highlighted in *gray* and labeled underneath. (**b**) A miRNA (exemplified by miR-122) and aligned antagonist sequence and modified elements. *D* DNA, *M* 2′-OMe, *L* LNA, *O* MOE, *F* 2′-OF, *CHOL* cholesterol, *p* phosphodiester, *s* phosphorothioate

transfected into cells, antimiRs sterically block the miRNA by hybridization and prevents the miRNA from binding to the miRNA targets (Fig. 1). Importantly, DNA and RNA oligos cannot resist degradation by intracellular nucleases. In contrast, oligos that have 2′-OMe (Hutvagner et al. 2004; Meister 2004), 2′-OF, or 2′-*O*-(2-methoxyethyl) (MOE) (Davis et al. 2006) substitutions or a methylene bridge connecting the 2′-O atom and the 4′-C (LNA) (Orom et al. 2006) are stable and can inhibit miRNA function (Fig. 2). AntimiR oligos can be further stabilized by substituting a nonbridging oxygen group of the oligo phosphodiester backbone with a sulfur group (phosphorothioate modification) (Davis et al. 2006). The different modifications can be combined in many ways to produce effective antimiRs, but oligos containing the LNA and 2′-OF modifications, which increase their affinity to the miRNA target, are most potent (Lennox and Behlke 2010). An alternative to the use of oligos to antagonize miRNAs is transfection with expression constructs that produce transcripts containing binding sites for miRNA. Such constructs are termed "miRNA sponges," because they titrate the miRNA away from their endogenous targets (Fig. 1) (Ebert et al. 2007). miRNA sponges can be designed with RNA polymerase II or III promoter and be delivered by transfection or with viral vectors (Ebert et al. 2007; Ebert and Sharp 2010). In general, the miRNA sponge construct contains 4–10 miRNA binding sites that are designed to be either perfectly complementary to the miRNA or to have a few central mismatches to avoid miRISC induced cleavage of the construct. The sponge strategy is not as effective as the antimiRs for miRNA antagonism but allows the creation of stably transfected cells that permanently antagonize a miRNA (Ebert and Sharp 2010).

Each miRNA has many target genes, and antimiRs and miRNA sponges will affect the expression of all of these genes. This may be an advantage and required for a potential therapeutic effect, but in a minority of cases, it is possible that the therapeutic benefit stems from only one dominant miRNA target gene (for instance, if a SNP generates a target in the gene's 3′ UTR). In this case, it would be advantageous to block the specific miRNA target in question rather than to antagonize the miRNA and thereby its entire target repertoire. This can be done with oligos that are complementary to the miRNA target sequence rather than the miRNA itself. Such oligos have been named miRNA protectors and were shown to effectively protect mRNAs from miR-430 regulation in zebrafish embryos (Choi et al. 2007). However, so far, there have not been many reports of successful use of miRNA target protectors in cell culture. One exception is a study by Wang and coworkers (Xiao ct al. 2007), who used a protector oligo consisting of five LNA bases at each end of a stretch of central DNA bases. However, the results obtained by the Wang group is puzzling, because the target protectors they applied could potentially recruit RNase H upon hybridization to the target mRNA and induce degradation of the target (Kurreck et al. 2002). More studies are therefore needed to determine the optimal design and usefulness of miRNA protectors.

5.2 miRNA Manipulation In Vivo

Compared to experiments performed in cell culture, therapeutic manipulation of miRNA presents some additional challenges, in particular regarding delivery, specificity, and tolerability of the drug. However, the efforts to target miRNAs therapeutically have previous experimental predecessors. Single-stranded antisense oligonucleotides (ASOs) are a class of drugs that have been under development for decades, and significant progress has been made. With the use of nucleotide modifications such as LNA, 2′-MOE, 2′-F, and phosphorothioate, most of the delivery and safety issues have been solved, so far leading to one approved drug and many more in the pipeline (Bennett and Swayze 2010). Likewise, there have been considerable efforts to extend the successful use of siRNAs in cell culture to therapeutic settings, and some advancement has been made in delivering siRNA to the relevant tissues (Lares et al. 2010).

5.2.1 miRNA Antagonism

Antagomir molecules are a type of ASOs that target a miRNA rather than an mRNA. The first demonstration that miRNA antagonism had the potential to become a new productive area of drug development was published in 2005 by Stoffel and coworkers (Krutzfeldt et al. 2005). This groundbreaking work demonstrated that the function of liver-specific miR-122 could be antagonized by treating mice with a 2′-OMe oligo containing phosphorothioate backbone modifications in the 3′ and 5′ ends and a cholesterol conjugation (Fig. 2). This oligo design is known as the antagomir and solves many of the problems related to inhibition of miRNAs in vivo. First, the cholesterol and the phosphorothioates improve the biodistribution of the antagomir by mediating binding to the abundant plasma proteins, which is essential to avoid rapid clearance via filtration in the kidneys. Second, the phosphorothioates and the 2′-OMe modification protect the oligo from nucleases, and third, the cholesterol promotes cellular uptake by an unknown mechanism. The antagomir-122 was administered intravenously and was shown to antagonize miR-122 function by derepressing miR-122 target genes. Moreover, antagomirs targeted against miR-16 were shown to be active in liver, kidney, lung, heart, skeletal muscle, skin, adrenal gland, ovary, intestine, colon, fat, and bone marrow as judged by northern blotting for miR-16 under increased denaturing conditions (Krutzfeldt et al. 2005). Northern blotting of miRNA levels after antisense experiments can be difficult to interpret, because the antisense oligo can interfere with probing. Thus, it is likely that not all the tissues mentioned in the original paper are amendable to antagomir-mediated miRNA inhibition. This was later shown to be the case for the skin miRNA miR-203, which could not be targeted by systemic delivery of antagomirs (Yi et al. 2008), whereas the fat miRNAs miR-103 and miR-107 have been successfully targeted (Trajkovski et al. 2011).

An alternative approach to the antagomirs is the use of antimiR molecules without cholesterol conjugation. Instead, the antimiRs have a backbone that is completely phosphorothioate substituted to promote binding of the oligo to plasma proteins, thereby avoiding filtration in the kidneys. The antimiR design is identical to the ASOs that have been shown to be effectively delivered to many different tissues after intravenous administration (Greenberger et al. 2008; Yu et al. 2007). Both 2′-MOE, chimeric 2′-MOE/2′-F, and chimeric LNA/DNA phosphorothioate-modified antimiR-122 oligos were shown to antagonize miR-122 after intravenous injection, leading to a decrease in plasma cholesterol (Davis et al. 2009; Elmen et al. 2008b; Esau et al. 2006) and derepression of miR-122 target genes (Elmen et al. 2008b). Because of the high binding affinity of LNA-modified bases, the LNA/DNA antimiR can be shortened to 15 bases and still regain the ability to antagonize miR-122. An advantage of antimiRs based on the high-affinity oligos (LNA and 2′-F/MOE) is that they require lower concentration to antagonize their target miRNA. Antagomir silencing of miR-122 in mouse liver was shown to be effective after a dosing of somewhere between 40 and 80 mg/kg bodyweight on three consecutive days (Krutzfeldt et al. 2007). In contrast, 2′-F/MOE and LNA/DNA antimiRs were shown to effectively antagonize mouse miR-122 and derepress a miR-122 target gene after a single dose of 11 mg/kg and 25 mg/kg, respectively (Davis et al. 2009). Moreover, three intravenous injections of 10 mg/kg with the LNA/DNA antimiR-122 were shown to induce an effective and long-lasting reduction in plasma cholesterol levels in African green monkeys without any apparent toxic effects (Elmen et al. 2008a). These concentrations are in a range that is applicable for human therapeutics. miR-122 inhibition is an interesting target because of the (1) cholesterol-lowering effect of miR-122 antagonism and (2) requirement of miR-122 for the replication of hepatitis C virus (HCV) (Jopling et al. 2005). The latter indication is of particular notice, as there is a medical need for an improved treatment of HCV infections and since experiments in a nonhuman primate HCV model indicate that the LNA/DNA-based antimiR-122 might become an effective therapeutics (Lanford et al. 2010). Following these experiments, a phase I clinical trial showed that the LNA/DNA-based antimiR-122 is well tolerated in healthy human subjects. At the time of writing, the antimiR-122 is in the finishing stages of a phase II clinical trial on HCV-infected patients. The result of this trial is eagerly anticipated and will probably give a valuable indication whether an antimiR strategy will be an effective treatment for HCV infections. Because of the relative ease of designing an effective antimiR, several different miRNA antagonists are being investigated in preclinical studies (see Table 1). One example is a miR-208 antimiR, which is being tested for efficacy against chronic heart failure. In this case, both miR-208a and miR-208b contribute to disease progression (as explained above), and a successful miRNA therapy would probably have to antagonize both family members, while the third myomiRs, miR-499, would probably be more challenging to target with the same oligo (van Rooij et al. 2007, 2009). Recently, Kauppinen and coworkers showed that an antimiR-122 shortened to eight bases, which are complementary to the miR-122 seed (consisting entirely of LNA bases with phosphorothioate substitutions), could robustly derepress mRNAs containing miR-122 seed matches. This occurred without affecting mRNA containing sequences that were complementary to the antimiR itself (Obad et al. 2011). These results suggest that a tinyLNA targeted against the miR-208 seed region could be an effective antagonist against both miR-208 family members.

5.2.2 miRNA Mimics

miRNA repression contributes to many human diseases (Cheng et al. 2010; Ling et al. 2009; Provost 2010), and in these cases, reestablishment of the miRNA expression could potentially have therapeutic benefit. There are two major strategies for introducing miRNA mimics into cells. Either miRNA duplexes are used in combination with different kinds of lipid formulations or the miRNA is delivered using a viral vector, which encodes a miRNA hairpin (Garzon et al. 2010). When miRNAs are repressed in cancer, reintroduction of miRNA by injection of miRNA hairpin or viral vector expressing the miRNA leads to an inhibition of tumor growth (Wang et al. 2008). However, injections are not a feasible treatment strategy and would be ineffective toward metastases. Thus, there has been much effort to develop methods to deliver miRNA mimics to different types of cancer cells. Slack and coworkers recently published a promising example of the use of nanoparticles formed by a neutral lipid formulation to deliver Let-7 or miR-34 to non-small cell lung cancer cells (NSCLC) in a K-Ras mouse model, which develops these tumors (Trang et al. 2011). Both Let-7 and miR-34 are downregulated in NSCLC, but after eight intravenous injections of the neutral lipid complexed Let-7 or miR-34 duplexes over 16 days each at a concentration of 1 mg/kg per injection, the lung tumors had decreased in size compared to the control. This study demonstrates that systemic delivery of miRNA duplexes to tumor tissue in therapeutically relevant concentrations is possible. Another example of systemic administration of a miRNA mimic was published by Mendell and coworkers (Kota et al. 2009). In this study, a mouse model of hepatocellular carcinoma (HCC) was used to investigate the effect of reintroduction of miR-26a, which is repressed in these cells. An adeno-associated virus (AAV) vector encoding a miR-26 hairpin was administered by intravenous injection at a dose of 10^{12} viral genomes. Expression of miR-26a in the cancer cell dramatically reduced cell proliferation and induced a tumor-specific apoptosis without any apparent toxicity.

5.3 *Prospects and Potential Pitfalls of miRNA Manipulation*

As described in the previous sections, miRNAs are involved in many different disease states, and Table 1 lists some of the different miRNAs that are currently being pursued as potential drug targets. Humans have at least several hundred miRNA genes (Chiang et al. 2010), and even though some of these will not be therapeutically relevant, there will still be many currently unexplored potential drug targets. Importantly, many of these targets can be antagonized simply by synthesizing a complementary oligo with the right chemistry and injecting this intravenously. On a longer timescale, upregulating miRNA levels with miRNA mimics may also become an option.

Despite the growing optimism surrounding miRNA therapeutics and the fact that antimiR therapy seems to be effective, there are still some unknowns regarding usefulness of miRNA therapies.

5.3.1 Multiple Targets of miRNAs

miRNAs have hundreds of targets and are therefore likely to affect several different pathways (O'Day and Lal 2010). In general, these pathways must be involved in important functions, because both miRNA and their targets have been conserved over long evolutionary distances. The importance of these functions is the major reason that miRNA therapeutics is so promising, but at the same time means that miRNA manipulation could affect important functions that cause severe adverse effect. For example, besides affecting cholesterol metabolism and being involved in HCV replication, there is evidence suggesting that miRNA-122 functions as a tumor suppressor in liver (Kutay et al. 2006), which at face value makes miR-122 antagonism somewhat less attractive. It is therefore likely that therapies based on miR-122 will have a relatively short duration and perhaps be used in combination with other types of treatment to eradicate the HCV infection completely. For other miRNAs, it is possible that similar problems with inherent miRNA function causing adverse effects will arise. In some cases, it may become possible to use miRNA target protector instead of antimiRs, thereby selectively inhibiting only some of the miRNA functions, but it is still unclear whether target protectors will work in a therapeutic setting.

5.3.2 Tissue Delivery

A vital issue in any therapeutics is tissue delivery, a key for all effective drugs. For miRNA mimics, delivery remains a major hurdle, and it is likely that a unique solution will have to be developed in each individual case. AAV-mediated delivery does show some promise and could in some cases be used to deliver miRNA mimics. In contrast, phosphorothioated antimiR molecules can be delivered subcutaneously or intravenously and do not require any sort of formulation. The antimiR oligos probably have a similar tissue distribution as other types of antisense phosphorothioated oligonucleotide drugs, which can enter cells in many different tissues. The highest concentration of oligo is obtained in kidney, liver, intestine, bone marrow, lymph nodes, and spleen with lower concentrations found in heart, lung, ovaries, uterus, and testes (Greenberger et al. 2008; Yu et al. 2007). However, recent work indicates that phosphorothioated oligos enter cells by two different mechanisms: (1) a productive mechanism leading to effective targeting of RNA and (2) an unproductive route where the oligo is sequestered away from the target (Koller et al. 2011). Since the unproductive uptake mechanism is responsible for the bulk of oligo uptake, it remains unclear exactly which tissues will have a therapeutically relevant concentration of antimiRs after systemic administration. Two studies recently showed that phosphorothioated oligos can enter some cell lines without the use of any transfection reagents (Koller et al. 2011; Stein et al. 2010), suggesting that productive oligo uptake is an inherent ability of human cells, which might have been lost by most cultured cells as a part of the immortalization process. The cell lines that have retained the productive oligo uptake should greatly facilitate the investigation of this mechanism and have already helped determine that productive uptake of phosphorothioated oligos depends on a clathrin- and caveolin-independent endocytotic process involving AP2M1 (Koller et al. 2011).

5.3.3 Specificity of miRNA and Antagonist Drugs

For any potential drug, specificity is a concern that cannot be ignored. For miRNA mimics, tissue specific delivery is very important in order to avoid the inhibition of genes that are not true targets for the miRNA. Once delivered to the designated tissue, miRNA mimics will probably not have detrimental off-target effects since potential off-target genes expressed in the tissue would have been under constant evolutionary selection to avoid being targeted by the specific miRNA. For antimiRs, off-target effects can result from hybridization to other RNA sequences than the miRNA molecules. It was demonstrated that two mismatches in a 15 base LNA/DNA antimiR are enough to completely cancel the miRNA antagonism (Elmen et al. 2008b). Although the sequence of a 15-base-long antimiR is relatively unique in the human genome, subparts of the oligo will have many matches in the genome. Additionally, at body temperature, annealing of these subparts to complementary RNAs may be thermodynamically favorable. Specificity issues might also rise from hybridization of antimiRs to proteins, which may also cause adverse effects. Such interactions could be dependent both on the chemistry and the specific sequence of the oligo (Bennett and Swayze 2010) and are difficult to predict in advance. Thus, further studies are needed to determine how much specific and unspecific hybridization of this nature occurs in cells, and if such spurious hybridization of antimiRs to off targets causes any unwanted effects.

6 Outlook and Conclusion

In a relatively short period of time, miRNAs have been established as master regulators of gene expression and as central players in disease development, thereby becoming the focus of efforts to develop a new generation of therapeutics. The results of phase II trials of the first miRNA antagonist drug are being eagerly anticipated, and a series of other drugs may reach a similar stage of development shortly. That said, challenges still remain, particularly related to specificity under physiological conditions and associated adverse effects. Other problems pertain to drug delivery, which, if solved, may greatly increase the range of malignancies treatable by miRNA therapeutics. Recent developments in applied miRNA research are in parallel establishing miRNAs as effective diagnostic markers (Hu et al. 2010) and predictors of optimal individual treatment (Rukov and Shomron 2011). The ongoing research emanating from the central role of miRNAs in disease may thus result in a combined tool, where miRNAs potentially facilitate diagnosis as well as both treatment planning and execution.

References

Anand S, Majeti BK, Acevedo LM et al (2010) MicroRNA-132-mediated loss of p120RasGAP activates the endothelium to facilitate pathological angiogenesis. Nat Med 16:909–914
Aqeilan RI, Calin GA, Croce CM (2009) miR-15a and miR-16-1 in cancer: discovery, function and future perspectives. Cell Death Differ 17:215–220
Baek D, Villen J, Shin C et al (2008) The impact of microRNAs on protein output. Nature 455:64–71
Bennett CF, Swayze EE (2010) RNA targeting therapeutics: molecular mechanisms of antisense oligonucleotides as a therapeutic platform. Annu Rev Pharmacol Toxicol 50:259–293
Boštjančič E, Zidar N, Štajer D et al (2010) MicroRNAs miR-1, miR-133a, miR-133b and miR-208 are dysregulated in human myocardial infarction. Cardiology 115:163–169
Calin GA, Croce CM (2006) MicroRNA signatures in human cancers. Nat Rev Cancer 6:857–866
Calin GA, Dumitru CD, Shimizu M et al (2002) Frequent deletions and down-regulation of micro-RNA genes miR15 and miR16 at 13q14 in chronic lymphocytic leukemia. Proc Natl Acad Sci USA 99:15524–15529
Callis TE, Pandya K, Seok HY et al (2009) MicroRNA-208a is a regulator of cardiac hypertrophy and conduction in mice. J Clin Invest 119:2772–2786
Cheng Y, Zhang C (2010) MicroRNA-21 in cardiovascular disease. J Cardiovasc Transl Res 3:251–255
Cheng Y, Zhu P, Yang J et al (2010) Ischaemic preconditioning-regulated miR-21 protects heart against ischaemia/reperfusion injury via anti-apoptosis through its target PDCD4. Cardiovasc Res 87:431–439
Chi SW, Zang JB, Mele A et al (2009) Argonaute HITS-CLIP decodes microRNA-mRNA interaction maps. Nature 460:479–486
Chiang HR, Schoenfeld LW, Ruby JG et al (2010) Mammalian microRNAs: experimental evaluation of novel and previously annotated genes. Genes Dev 24:992–1009
Choi WY, Giraldez AJ, Schier AF (2007) Target protectors reveal dampening and balancing of Nodal agonist and antagonist by miR-430. Science 318:271–274
Davis S, Lollo B, Freier S et al (2006) Improved targeting of miRNA with antisense oligonucleotides. Nucleic Acids Res 34:2294–2304
Davis S, Propp S, Freier SM et al (2009) Potent inhibition of microRNA in vivo without degradation. Nucleic Acids Res 37:70–77
Djuranovic S, Nahvi A, Green R (2011) A parsimonious model for gene regulation by miRNAs. Science 331:550–553
Dong S, Cheng Y, Yang J et al (2009) MicroRNA expression signature and the role of microRNA-21 in the early phase of acute myocardial infarction. J Biol Chem 284:29514–29525
Dorn GW II (2011) MicroRNAs in cardiac disease. Transl Res 157:226–235
Ebert MS, Neilson JR, Sharp PA (2007) MicroRNA sponges: competitive inhibitors of small RNAs in mammalian cells. Nat Met 4:721–726
Ebert MS, Sharp PA (2010) MicroRNA sponges: progress and possibilities. RNA 16:2043–2050
Elmen J, Lindow M, Schutz S et al (2008a) LNA-mediated microRNA silencing in non-human primates. Nature 452:896–899
Elmen J, Lindow M, Silahtaroglu A et al (2008b) Antagonism of microRNA-122 in mice by systemically administered LNA-antimiR leads to up-regulation of a large set of predicted target mRNAs in the liver. Nucleic Acids Res 36:1153–1162
Esau C, Davis S, Murray SF et al (2006) miR-122 regulation of lipid metabolism revealed by in vivo antisense targeting. Cell Metab 3:87–98
Filipowicz W, Jaskiewicz L, Kolb FA et al (2005) Post-transcriptional gene silencing by siRNAs and miRNAs. Curr Opin Struct Biol 15:331–341
Friedman RC, Farh KK, Burge CB et al (2009) Most mammalian mRNAs are conserved targets of microRNAs. Genome Res 19:92–105

Garzon R, Marcucci G, Croce CM (2010) Targeting microRNAs in cancer: rationale, strategies and challenges. Nat Rev Drug Discov 9:775–789

Greenberger LM, Horak ID, Filpula D et al (2008) A RNA antagonist of hypoxia-inducible factor-1alpha, EZN-2968, inhibits tumor cell growth. Mol Cancer Ther 7:3598–3608

Gregory RI, Chendrimada TP, Cooch N et al (2005) Human RISC couples microRNA biogenesis and posttranscriptional gene silencing. Cell 123:631–640

Gregory RI, K-p Y, Amuthan G et al (2004) The microprocessor complex mediates the genesis of microRNAs. Nature 432:235–240

Griffiths-Jones S (2010) miRBase: microRNA sequences and annotation. Current Protocols in Bioinformatics DOI: 10.1002/0471250953.bi1209s29

Grimson A, Farh KK-H, Johnston WK et al (2007) MicroRNA targeting specificity in mammals: determinants beyond seed pairing. Mol Cell 27:91–105

Guil S, Caceres JF (2007) The multifunctional RNA-binding protein hnRNP A1 is required for processing of miR-18a. Nat Struct Mol Biol 14:591–596

Guo H, Ingolia NT, Weissman JS et al (2010) Mammalian microRNAs predominantly act to decrease target mRNA levels. Nature 466:835–840

Hu Z, Chen X, Zhao Y et al (2010) Serum microRNA signatures identified in a genome-wide serum microRNA expression profiling predict survival of non-small-cell lung cancer. J Clin Oncol 28:1721–1726

Hutvagner G, Simard MJ, Mello CC et al (2004) Sequence-specific inhibition of small RNA function. PLoS Biol 2:E98

Hwang-Verslues WW, Chang PH, Wei PC et al (2011) miR-495 is upregulated by E12/E47 in breast cancer stem cells, and promotes oncogenesis and hypoxia resistance via downregulation of E-cadherin and REDD1. Oncogene 30:2463–2474

Jopling CL, Yi M, Lancaster AM et al (2005) Modulation of hepatitis C virus RNA abundance by a liver-specific MicroRNA. Science 309:1577–1581

Koller E, Vincent TM, Chappell A et al (2011) Mechanisms of single-stranded phosphorothioate modified antisense oligonucleotide accumulation in hepatocytes. Nucleic Acids Res 39:4795–4807

Kota J, Chivukula RR, O'Donnell KA et al (2009) Therapeutic microRNA delivery suppresses tumorigenesis in a murine liver cancer model. Cell 137:1005–1017

Krutzfeldt J, Kuwajima S, Braich R et al (2007) Specificity, duplex degradation and subcellular localization of antagomirs. Nucleic Acids Res 35:2885–2892

Krutzfeldt J, Rajewsky N, Braich R et al (2005) Silencing of microRNAs in vivo with 'antagomirs'. Nature 438:685–689

Kurreck J, Wyszko E, Gillen C et al (2002) Design of antisense oligonucleotides stabilized by locked nucleic acids. Nucleic Acids Res 30:1911–1918

Kutay H, Bai S, Datta J et al (2006) Downregulation of miR-122 in the rodent and human hepatocellular carcinomas. J Cell Biochem 99:671–678

Lanford RE, Hildebrandt-Eriksen ES, Petri A et al (2010) Therapeutic silencing of microRNA-122 in primates with chronic hepatitis C virus infection. Science 327:198–201

Lares MR, Rossi JJ, Ouellet DL (2010) RNAi and small interfering RNAs in human disease therapeutic applications. Trends Biotechnol 28:570–579

Lee RC, Feinbaum RL, Ambros V (1993) The *C. elegans* heterochronic gene lin-4 encodes small RNAs with antisense complementarity to lin-14. Cell 75:843–854

Lennox KA, Behlke MA (2010) A direct comparison of anti-microRNA oligonucleotide potency. Pharm Res 27:1788–1799

Lewis BP, Burge CB, Bartel DP (2005) Conserved seed pairing, often flanked by adenosines, indicates that thousands of human genes are microRNA targets. Cell 120:15–20

Lewis BP, Shih IH, Jones-Rhoades MW et al (2003) Prediction of mammalian microRNA targets. Cell 115:787–798

Li Q-J, Chau J, Ebert PJR et al (2007) miR-181a is an intrinsic modulator of T cell sensitivity and selection. Cell 129:147–161

Lim LP, Lau NC, Garrett-Engele P et al (2005) Microarray analysis shows that some microRNAs downregulate large numbers of target mRNAs. Nature 433:769–773
Ling HY, Ou HS, Feng SD et al (2009) Changes in microRNA profile and effects of miR-320 in insulin-resistant 3T3-L1 adipocytes. Clin Exp Pharmacol Physiol 36:E32–E39
Liu N, Olson EN (2010) MicroRNA regulatory networks in cardiovascular development. Dev Cell 18:510–525
Lu J, Getz G, Miska EA et al (2005) MicroRNA expression profiles classify human cancers. Nature 435:834–838
Meister G (2004) Sequence-specific inhibition of microRNA- and siRNA-induced RNA silencing. RNA 10:544–550
Michlewski G, Guil S, Semple CA et al (2008) Posttranscriptional regulation of miRNAs harboring conserved terminal loops. Mol Cell 32:383–393
Muralidhar B, Winder D, Murray M et al (2011) Functional evidence that Drosha overexpression in cervical squamous cell carcinoma affects cell phenotype and microRNA profiles. J Pathol 224:496–507
Nicoloso MS, Sun H, Spizzo R et al (2010) Single-nucleotide polymorphisms inside microRNA target sites influence tumor susceptibility. Cancer Res 70:2789–2798
Nielsen CB, Shomron N, Sandberg R et al (2007) Determinants of targeting by endogenous and exogenous microRNAs and siRNAs. RNA 13:1894–1910
O'Connell RM, Rao DS, Chaudhuri AA et al (2008) Sustained expression of microRNA-155 in hematopoietic stem cells causes a myeloproliferative disorder. J Exp Med 205:585–594
O'Connell RM, Taganov KD, Boldin MP et al (2007) MicroRNA-155 is induced during the macrophage inflammatory response. Proc Natl Acad Sci USA 104:1604–1609
O'Day E, Lal A (2010) MicroRNAs and their target gene networks in breast cancer. Breast Cancer Res 12:201
Obad S, dos Santos CO, Petri A et al (2011) Silencing of microRNA families by seed-targeting tiny LNAs. Nat Genet 43:371–378
Orom UA, Kauppinen S, Lund AH (2006) LNA-modified oligonucleotides mediate specific inhibition of microRNA function. Gene 372:137–141
Paddison PJ, Caudy AA, Bernstein E et al (2002) Short hairpin RNAs (shRNAs) induce sequence-specific silencing in mammalian cells. Genes Dev 16:948–958
Patrick DM, Zhang CC, Tao Y et al (2010) Defective erythroid differentiation in miR-451 mutant mice mediated by 14-3-3zeta. Genes Dev 24:1614–1619
Provost P (2010) MicroRNAs as a molecular basis for mental retardation, Alzheimer's and prion diseases. Brain Res 1338:58–66
Qi L, Bart J, Tan L et al (2009) Expression of miR-21 and its targets (PTEN, PDCD4, TM1) in flat epithelial atypia of the breast in relation to ductal carcinoma in situ and invasive carcinoma. BMC Cancer 9:163
Rayner KJ, Sheedy FJ, Esau CC et al (2011) Antagonism of miR-33 in mice promotes reverse cholesterol transport and regression of atherosclerosis. J Clin Invest 121:2921–2931
Rukov JL, Shomron N (2011) MicroRNA pharmacogenomics: post-transcriptional regulation of drug response. Trends Mol Med 17:412–423
Satoh M, Minami Y, Takahashi Y et al (2010) Expression of microRNA-208 is associated with adverse clinical outcomes in human dilated cardiomyopathy. J Card Fail 16:404–410
Selbach M, Schwanhausser B, Thierfelder N et al (2008) Widespread changes in protein synthesis induced by microRNAs. Nature 455:58–63
Selcuklu SD, Donoghue MT, Spillane C (2009) miR-21 as a key regulator of oncogenic processes. Biochem Soc Trans 37:918–925
Shomron N, Golan D, Hornstein E (2009) An evolutionary perspective of animal microRNAs and their targets. J Biomed Biotechnol 2009:594738
Stefani G, Slack FJ (2008) Small non-coding RNAs in animal development. Nat Rev Mol Cell Biol 9:219–230

Stein CA, Hansen JB, Lai J et al (2010) Efficient gene silencing by delivery of locked nucleic acid antisense oligonucleotides, unassisted by transfection reagents. Nucleic Acids Res 38:e3

Takeshita F, Patrawala L, Osaki M et al (2010) Systemic delivery of synthetic microRNA-16 inhibits the growth of metastatic prostate tumors via downregulation of multiple cell-cycle genes. Mol Ther 18:181–187

Terao M, Fratelli M, Kurosaki M et al (2011) Induction of miR-21 by retinoic acid in estrogen receptor-positive breast carcinoma cells. J Biol Chem 286:4027–4042

Thum T, Gross C, Fiedler J et al (2008) MicroRNA-21 contributes to myocardial disease by stimulating MAP kinase signalling in fibroblasts. Nature 456:980–984

Trajkovski M, Hausser J, Soutschek J et al (2011) MicroRNAs 103 and 107 regulate insulin sensitivity. Nature 474:649–653

Trang P, Wiggins JF, Daige CL et al (2011) Systemic delivery of tumor suppressor microRNA mimics using a neutral lipid emulsion inhibits lung tumors in mice. Mol Ther 19:1116–1122

van Rooij E, Quiat D, Johnson BA et al (2009) A family of microRNAs encoded by myosin genes governs myosin expression and muscle performance. Dev Cell 17:662–673

van Rooij E, Sutherland LB, Liu N et al (2006) A signature pattern of stress-responsive microRNAs that can evoke cardiac hypertrophy and heart failure. Proc Natl Acad Sci USA 103:18255–18260

van Rooij E, Sutherland LB, Qi X et al (2007) Control of stress-dependent cardiac growth and gene expression by a microRNA. Science 316:575–579

van Rooij E, Sutherland LB, Thatcher JE et al (2008) Dysregulation of microRNAs after myocardial infarction reveals a role of miR-29 in cardiac fibrosis. Proc Natl Acad Sci USA 105:13027–13032

Wang H, Garzon R, Sun H et al (2008) NF-kappaB-YY1-miR-29 regulatory circuitry in skeletal myogenesis and rhabdomyosarcoma. Cancer Cell 14:369–381

Wightman B, Ha I, Ruvkun G (1993) Posttranscriptional regulation of the heterochronic gene lin-14 by lin-4 mediates temporal pattern formation in *C. elegans*. Cell 75:855–862

Xiao J, Yang B, Lin H et al (2007) Novel approaches for gene-specific interference via manipulating actions of microRNAs: examination on the pacemaker channel genes HCN2 and HCN4. J Cell Phys 212:285–292

Yi R, Poy MN, Stoffel M, Fuchs E (2008) A skin microRNA promotes differentiation by repressing 'stemness'. Nature 452:225–229

Yin C, Salloum FN, Kukreja RC (2009) A novel role of microRNA in late preconditioning. Circ Res 104:572–575

Yu RZ, Kim TW, Hong A et al (2007) Cross-species pharmacokinetic comparison from mouse to man of a second-generation antisense oligonucleotide, ISIS 301012, targeting human apolipoprotein B-100. Drug Metab Dispos 35:460–468

Zeng Y, Wagner EJ, Cullen BR (2002) Both natural and designed micro RNAs can inhibit the expression of cognate mRNAs when expressed in human cells. Mol Cell 9:1327–1333

Zeng Y, Yi R, Cullen BR (2003) MicroRNAs and small interfering RNAs can inhibit mRNA expression by similar mechanisms. Proc Natl Acad Sci USA 100:9779–9784

microRNAs in Human Diseases and Viral Infections

Alessandra Mescalchin and Tobias Restle

Contents

Abstract microRNAs (miRNAs) are a class of short noncoding RNAs derived from either cellular or viral transcripts that act posttranscriptionally to regulate messenger RNA stability and translation. In humans, several hundred different miRNA species have been identified to fine-tune gene expression. Among them, an increasing number has been shown to be involved in the development and progression of a variety of diseases such as cancer, heart failure, asthma, and even allergies. Besides, several pathogenic DNA viruses, such as adeno-, herpes-, and polyomaviruses, encode for own viral miRNAs. These miRNAs are involved in the modulation of viral pathogenicity and replication properties. Moreover, the hepatitis C virus (HCV) exploits a cellular miRNA to boost viral replication and mRNA translation.

As it requires no more than partial target recognition for inhibitory effects to occur, it is perspicuous that in principle, a single miRNA can regulate the expression of several hundred mRNAs. Thus, the dysregulated expression or mutation of just one miRNA species might be sufficient to cause an onset of pathological

A. Mescalchin • T. Restle (✉)
Institute of Molecular Medicine, University of Lübeck, Center for Structural and Cell Biology in Medicine (CSCM), Ratzeburger Allee 160, 23538 Lübeck, Germany
e-mail: restle@imm.uni-luebeck.de

V.A. Erdmann and J. Barciszewski (eds.), *From Nucleic Acids Sequences to Molecular Medicine*, RNA Technologies, DOI 10.1007/978-3-642-27426-8_21,

conditions. Hence, miRNAs are very interesting and highly promising targets for novel therapeutic approaches, and their functional modulation may represent an effective strategy to fight disease progression and viral infection.

In this chapter, we describe the biology and mechanisms of action of miRNAs with particular emphasis on pathologically relevant species and current strategies for therapeutic intervention.

Keywords Disease • microRNA • RNAi • Virus

Abbreviations

ANP32A	Acidic nuclear phosphoprotein 32 family member A
AML	Acute myeloid leukemia
AAV	Adeno-associated virus
AD	Alzheimer's disease
API5	Apoptosis inhibitor 5
Ago	Argonaute
BACE1	Beta-site amyloid precursor protein-cleaving enzyme 1
CLL	B-cell chronic lymphocytic leukemia
ChIP	Chromatin immunoprecipitation
DGCR8	DiGeorge syndrome critical region gene 8
DNMT	DNA methyltransferase
ds	Double strand
EBV	Epstein-Barr virus
HvAV	Heliothis virescens ascovirus
HCV	Hepatitis C virus
HCC	Hepatocellular carcinoma
HSV	Herpes simplex virus
HVT	Herpesvirus of turkeys
hAV	Human adenovirus
hCMV	Human cytomegalovirus
HHV	Human herpesvirus
HMDD	Human miRNA-associated disease database
IP07	Importin 7
ILTV	Infectious laryngotracheitis virus
IL-8	Interleukin-8
IRES	Internal ribosome entry site
KSHV	Kaposi's sarcoma-associated herpesvirus
KCNJ2	K(+) channel subunit Kir2.1
LMP2A	Latent membrane protein 2A
MICB	Major histocompatibility complex class I-related chain B
MDV	Marek's disease virus
miRNAs	microRNAs

mCMV	Mouse cytomegalovirus
MHV-68	Murine herpesvirus 68
MyoD	Myogenic differentiation 1
nt	Nucleotides
2′*O*-Me	2′*O*-methyl
PUMA	p53 upregulated modulator of apoptosis
PB	processing bodies
PS	Phophorothioate
PTEN	Phosphatase and tensin homolog
PI3K	Phosphoinositide 3-kinase
PTGS	Posttranscriptional gene silencing
Rbl2	Rb-like protein 2
rLCV	Rhesus lymphocryptovirus
RRV	Rhesus monkey rhadinovirus
RLC	RISC loading complex
RISC	RNA-induced silencing complex
RNAi	RNA interference
siRNAs	Short interfering RNAs
SV40	Simian virus 40
SOCS1	Suppressor of cytokine signaling 1
TRBP	TAR RNA-binding protein
TH2	T helper-2
THBS1	Thrombospondin-1
TOMM22	Translocase of outer mitochondrial membrane 22
UTR	Untranslated region
HHV-3/VZV	Varicella zoster virus

1 Introduction

miRNAs are ubiquitous small noncoding RNAs, ~21–22 nucleotides (nt) long, which play an important role in the fine tuning of gene expression [for a comprehensive review, see Bartel (2009)]. These tiny RNAs typically act as negative regulators of mRNA expression at the posttranscriptional level triggering translational repression or even cleavage of their corresponding target transcripts [reviewed by Huntzinger and Izaurralde (2011)]. This occurs upon recruitment of miRNAs by cellular proteins involved in the RNA interference (RNAi) pathway followed by binding of the resulting complexes to miRNA-matching sequences within the 3′-untranslated region (UTR) of target mRNAs. Interestingly, certain cellular miRNAs, such as miR-369-3 and let-7, were shown to activate mRNA translation during cell cycle arrest (Vasudevan et al. 2007), suggesting that regulation of gene expression by miRNAs might change during the course of the cell cycle.

As yet, in humans, more than 1,000 miRNA precursors have been identified. They play critical roles in several key cellular processes, such as development, cell growth, tissue differentiation, and even apoptosis (Bartel 2009; Ghildiyal and

Zamore 2009). However, under pathological conditions, dysregulation of miRNA biogenesis and function may result in a variety of diseases [reviewed by Cho (2007) and Kerr et al. (2011)]. Several miRNAs have been associated to human pathological states, such as cardiovascular disorders (Small and Olson 2011), viral infections (Norman and Sarnow 2010; Carnero et al. 2011), cancer (Akao et al. 2010), and even asthma (Chiba and Misawa 2010) and allergies (Lu et al. 2009b). Modulation of individual pathogenic miRNA species may represent an effective strategy to fight disease progression and by the same token offer the chance to gain deeper insight into the biology of action of these small RNAs. Despite the large number of existing data concerning connections between miRNAs and diseases, further studies are essential to shed light on the underlying pathomechanisms.

2 The RNAi Process

The phenomenon of RNAi was first conclusively described 1998 in the nematode *Caenorhabditis elegans* (Fire et al. 1998). Back then, injection of dsRNA complementary to the worm encoded *unc-2* mRNA into the nematode led to suppression of this particular transcript. Moreover, shortly after, it could be shown that the observed specific inhibition of gene expression affected the entire organism and even was heritable (Grishok et al. 2000). At about the same time, antisense RNAs (~25 nt) were found in plants (Hamilton and Baulcombe 1999), thereby providing evidence for a role of endogenous RNA in posttranscriptional gene silencing (PTGS). RNAi is mediated by short RNAs (~21–22 nt) which are produced from long dsRNA molecules by an RNase III-like enzyme (Elbashir et al. 2001b), meanwhile identified as Dicer (Hutvagner et al. 2001). A further study using *Drosophila* embryo lysate demonstrated that for an efficient target downregulation, these short RNAs, termed short interfering RNAs (siRNAs), need to fulfill certain requirements, such as to possess a length of 21 nt, 2 nt 3′-overhangs, and no mismatch in the center region of the oligonucleotide duplex which would interfere with target RNA cleavage (Elbashir et al. 2001c). Notably, such siRNAs have also been described to efficiently repress expression of target genes upon transfection into mammalian cells (Elbashir et al. 2001a).

The RNAi process was originally already described in 1993 for *C. elegans* and called posttranscriptional regulation mechanism (Lee et al. 1993; Wightman et al. 1993), albeit its relevance was not fully recognized until 1998 owing to the work of Fire and Mello, who received the Nobel Prize in Physiology or Medicine in 2006 for their groundbreaking discovery (Fire et al. 1998). Already in the early 1990s, the *lin-4* locus was described to encode short RNAs (22 and 61 nt) partially complementary to the 3′-UTR of *lin-14* mRNA (Lee et al. 1993). The LIN-14 protein controls the timing of developmental events in *C. elegans*, and its translation is regulated by an antisense mechanism. Hence, lin-4 was the first miRNA to be identified. A few years later, a second miRNA, called let-7, was characterized in nematodes (Reinhart et al. 2000) and found to be highly conserved across species (Pasquinelli et al. 2000). Since then, many more miRNAs have been identified which are envisaged to target up to 30% of all human genes (Krek et al. 2005; Lewis et al. 2005).

Both siRNAs and miRNAs participate in the RNAi process and thus in the posttranscriptional regulation of gene expression. Generally speaking, the process is initiated upon cleavage of long dsRNA molecules into short siRNA or miRNA duplexes by Dicer. However, it should be kept in mind that even though both pathways, i.e., generation of siRNAs or miRNAs, are highly related there are distinct differences (Fig. 1).

The short dsRNA molecules are then recognized by the RNA-induced silencing complex (RISC). Normally, the strand with lower 5′-thermodynamic stability (termed guide strand) is loaded into RISC, whereas the other (termed passenger strand) gets degraded (Khvorova et al. 2003; Schwarz et al. 2003). Silencing of target genes occurs either via mRNA degradation or prevention of mRNA translation. In case of siRNAs, guide stands are usually perfectly complementary to target regions within the corresponding mRNA, and this full complementarity triggers the cleavage of target transcripts by the Argonaute 2 (Ago2) protein (Fig. 2), a key component of RISC (Liu et al. 2004; Meister et al. 2004). miRNAs, in contrast, present guide strands which are only partially complementary (ranging from the 2nd to the 7–8th nt calculated from the 5′-end of the guide strand, also referred to as "seed" region) to target mRNAs and as a result trigger repression of mRNA translation by RISC (Filipowicz et al. 2008; Castanotto and Rossi 2009).

Commonly, a particular miRNA can interfere with the translation of up to some hundred different mRNAs. There are three principally different scenarios how this may take place. Accordingly, translational repression by miRNAs can either occur during initiation or elongation of protein synthesis or by unrelated processes like triggering of RNA degradation [reviewed by Huntzinger and Izaurralde (2011)].

First evidence of miRNA-mediated translational inhibition during the initiation phase was derived from experiments performed analyzing target RNA/miRNA complexes (Pillai et al. 2005). Such complexes did not co-sediment upon sucrose density gradient centrifugation with the polysomal fraction but rather with fractions of lower mass containing fewer ribosomes. Moreover, while mRNAs without a cap at the 5′-end were not subject to repression (Pillai et al. 2005), replacing the cap structure with an internal ribosome entry site (IRES) impaired miRNA-mediated inhibition (Humphreys et al. 2005), indicative of a cap-dependent mechanism. In support of this interpretation, a motif within the MID domain of Ago2 (Fig. 2) was identified, which shows close homology to the cap-binding domain of an essential initiation factor, eIF4E (Kiriakidou et al. 2007). Based on these results, one can speculate Ago2 may compete with eIF4E for the binding to the cap, thereby hampering initiation of translation. Furthermore, depletion of eIF6, a protein which inhibits association of the two ribosomal subunits, leads to decreased miRNA activity in human cells, indicating that miRISC may interfere with eIF6 to reduce association of polysomes with the target mRNA, exposing it and favoring its degradation (Chendrimada et al. 2007).

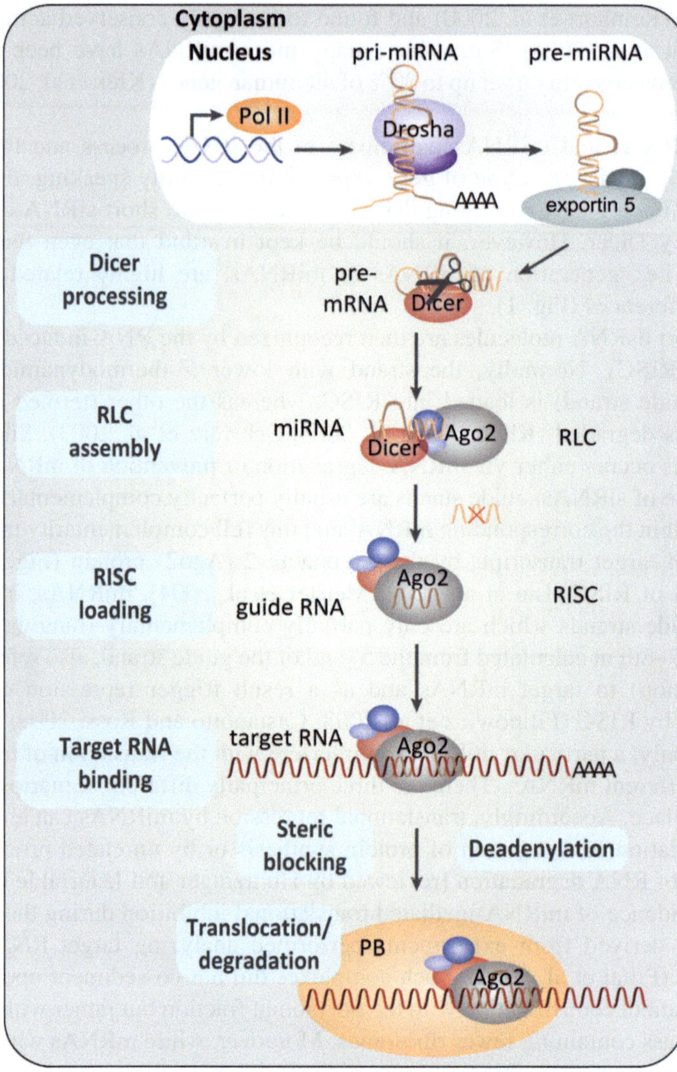

Fig. 1 Mechanism of miRNA-mediated RNAi. Primary RNA precursors (pri-miRNA) of up to several hundred nt in length are transcribed by RNA polymerase II followed by processing through Drosha into pre-miRNAs (~ 60 nt). Subsequently, pre-miRNAs are exported from the nucleus to the cytoplasm by exportin 5 and further processed by Dicer to generate double-stranded miRNA intermediates. These miRNA intermediates are loaded into the RISC loading complex (RLC) where strand selection takes place based on the relative thermodynamic stability of the 5′-ends. One strand is incorporated to form miRISC, whereas the other one gets degraded in specialized processing bodies (PB). Upon formation of a ternary complex with an at least partially complementary target, mRNA translational inhibition occurs. Typically, a single miRNA molecule can interfere with the translation of up to several hundred different mRNAs by several mechanisms, such as interfering with initiation of translation by competing for binding to the cap, promoting "ribosome drop-off," or degradation of mRNA. Adapted from Jinek and Doudna (2009)

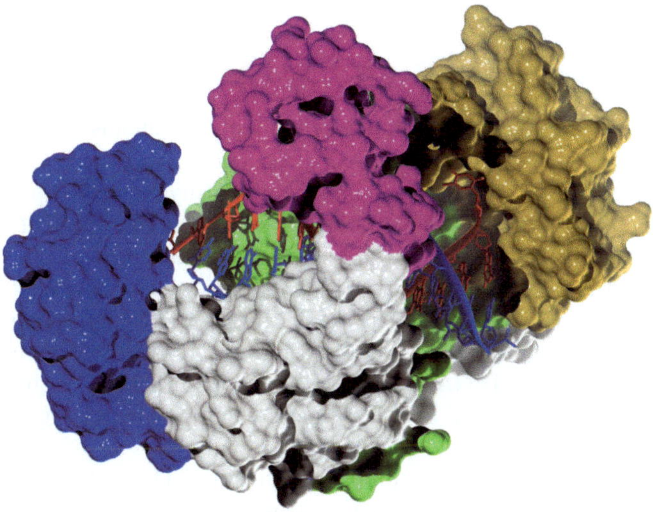

Fig. 2 X-ray structure of the ternary complex of *Thermus thermophilus* Argonaute bound to a 21-nt guide DNA and a 20-nt target RNA (pdb file: 3F73). The protein contains four defined domains, N-terminal, PAZ, MID, and PIWI which are color coded *blue*, *magenta*, *gold*, and *green*, respectively. Additionally, two linker regions are shown in *gray*. Guide DNA is shown in *red* and target RNA in *blue*

On the other hand, there is evidence that miRNAs may be able to inhibit translation at a step subsequent to initiation. A possible mechanism of repression during the elongation step was described as "ribosome drop-off" (Petersen et al. 2006). In this study, the authors compared the association of polysomes upon blockage of initiation of translation using mRNAs containing miRNA binding sites and thus are repressible and control mRNAs without such binding sites. They found that the miRNA-repressed mRNA dissociated more rapidly from the polysomes when compared to the control, suggesting that the miRNA-mediated inhibition may lead to a premature dissociation of the ribosomes from the transcript. Moreover, mRNAs containing viral IRES at the 5′-UTR and no cap were also silenced, further supporting that miRNAs mediate repression following initiation of translation.

Additionally, miRNAs can apparently trigger mRNA degradation, which in this case is not a result of Ago2-catalyzed cleavage but rather mediated through other mechanisms including target deadenylation by the CAF1-CCR4-NOT complex and decapping by the DCP2 enzyme (Behm-Ansmant et al. 2006; Eulalio et al. 2009). Other studies describe the ability of miRNAs to promote deadenylation of capped (Wu et al. 2006) as well as of IRES-containing mRNAs (Wakiyama et al. 2007), further supporting this alternative mode of action. Whatever pathway, i.e., blockage of translational initiation, translational elongation, or mRNA degradation, finally turns out to be of fundamental importance, the result stays the same, namely, inhibition of protein synthesis.

3 miRNAs

miRNAs play vital roles in the regulation of gene expression, i.e., during the cell cycle (Vasudevan et al. 2008), animal development (Wienholds and Plasterk 2005), and cell fate specification (Lu et al. 2008a; Gangaraju and Lin 2009). In humans (Fig. 1), they are derived from long, capped, polyadenylated RNA precursors (Cai et al. 2004) transcribed by RNA polymerase II or in the case of Alu repeat-associated miRNAs by the RNA polymerase III (Borchert et al. 2006). The precursors fold into well-defined hairpin structures which are recognized by the nuclear RNase III-type protein Drosha and its cofactor DiGeorge syndrome critical region gene 8 (DGCR8) and are processed into pre-miRNAs (~60 nt) (Lee et al. 2003). For "mirtrons," miRNA transcripts embedded in intronic regions, cleavage into pre-miRNA is performed by the splicing machinery bypassing processing by Drosha (Ruby et al. 2007).

In the nucleus, exportin 5 in complex with Ran-GTP binds the pre-miRNAs, exports them into the cytoplasm, and releases them upon hydrolysis of GTP (Bohnsack et al. 2004). Dicer then recognizes the 3′-protruding ends of the hairpin pre-miRNAs via its PAZ domain and cleaves them close to their terminal loop generating ~21–22 nt long ds miRNA intermediates [reviewed by Winter et al. (2009) and Kim et al. (2009)]. Upon Dicer processing, these miRNA intermediates are loaded into the RISC which is composed of Ago proteins, TAR RNA-binding protein (TRBP), and Dicer (Chendrimada et al. 2005). Here, strand selection takes place, and the antisense strand is incorporated to form the miRISC guiding the complex to complementary mRNA species (Khvorova et al. 2003; Schwarz et al. 2003). The passenger strand is normally degraded, except when both strands of the miRNA intermediate are loaded to form active miRISCs (Okamura et al. 2008).

The expression and activity of miRNAs are highly regulated by transcriptional and/or posttranscriptional control mechanisms which depend on tissue- or developmental-specific factors [reviewed by Winter et al. (2009) and Gangaraju and Lin (2009)]. An example is the highly conserved let-7 miRNA. In humans, the expression of this miRNA was shown to be controlled by the LIN28 protein (Heo et al. 2008). This protein acts as a negative regulator of let-7 maturation via recruitment of the uridylyl transferase TUT4, which adds a poly-U tail to the precursor let-7, labeling it for degradation. While uridylation seems to decrease miRNA stability, adenylation was described to interfere with their activity. In fact, a recent study in human cells has shown that mature miRNAs carrying a poly-A tail at the 3′-end are depleted in Ago2 and Ago3 complexes, indicating that this modification may interfere with RISC loading (Burroughs et al. 2010).

Drosha contributes to the regulation of miRNA expression by processing different primary miRNA sequences with different efficiencies and thus leading to a biased expression of the corresponding mature miRNAs (Feng et al. 2011). At the transcriptional level, miRNA expression can be regulated by several transcription factors, for example, the oncogenic protein c-Myc which is involved in the control of cell growth and apoptosis. Dysregulated expression of c-Myc is one of the most common traits in cancer and was shown to activate transcription of the miR-17-92 cluster (O'Donnell et al. 2005). Two miRNAs in this cluster, miR-17-5p and miR-20a, negatively regulate expression of the transcription factor E2F1 which promotes cell cycle progression. Apparently, the expression of this gene is also activated by c-Myc, suggesting that c-Myc tightly controls proliferative signals at both mRNA and miRNA levels. Other examples are miR-1, miR-133, and miR-206 which are highly conserved muscle-specific miRNAs and preferentially transcribed during the myoblast-myotube transition (Rao et al. 2006). By chromatin immunoprecipitation (ChIP) analysis, the myogenic transcription factor myogenin and MyoD (myogenic differentiation 1) were identified to bind to sequences upstream of the coding region of these miRNAs. Both proteins are myogenic transcriptional regulatory factors which activate the expression of a number of muscle-specific genes during myogenesis, and likely also of miR-1, miR-133, and miR-206. Furthermore, cellular proteins such as Importin-8 and Pumilio-1, which is an RNA-binding protein, can as well modulate the activity of the miRISC. Importin-8 was described to be involved in the efficient loading of Ago/miRNA complexes onto target mRNAs probably via direct interaction with Ago proteins (Weinmann et al. 2009), whereas Pumilio-1 can bind to the *p27* tumor suppressor 3′-UTR in response to growth factor-based stimulation leading to local structural changes within the mRNA (Kedde et al. 2010). This in turn facilitates the binding of miRISCs loaded with miR-221 or miR-222 to the target *p27* mRNA, thereby promoting cell cycle entry (Kedde et al. 2010). In sum, the modulation of the expression and activity of these tiny regulators of gene expression is very complex and requires the coordinated activity of many cellular factors. Changes leading to perturbations of the existing equilibrium, for example, overexpression or depletion of certain miRNA species, may lead to pathological states (Lu et al. 2008b) or even to cell death (Cheng et al. 2005).

3.1 miRNAs and Cancer

Many types of cancer are caused by altered expression of genes, such as tumor suppressor genes or oncogenes, which may lead to uncontrolled cell proliferation, to cell death avoidance, and to cell migration when the tumor progresses to malignancy [reviewed by Kelly and Strasser (2011)]. Such a dysregulation of gene expression can occur through several mechanisms, for instance, genomic alterations (Chin et al. 2011) or aberrant expression of miRNAs [reviewed by White et al. (2011)].

Genes encoding human miRNAs are often located both in cancer-associated genomic regions and/or at fragile sites in the human genome (Calin et al. 2004), suggesting that miRNAs might play a role in cancer (Table 1).

A pioneering study on miRNA expression in B-cell chronic lymphocytic leukemia (CLL) has shown that deletions and downregulation of two clustered miRNA genes, i.e., miR-15 and miR-16, are frequent in patients with this type of cancer, suggesting a correlation between changes in miRNA expression and CLL (Calin et al. 2004). In support of these findings, it was reported that both miRNAs can downregulate BCL-2,

Table 1 Selected examples of cancer-associated miRNAs

miRNAs	Target mRNAs	Dysregulation in cancer	References
miR-15, miR-16	BCL-2	Downregulated in CCL	Calin et al. (2004), Cimmino et al. (2005)
miR-29a	Ski	Downregulated in AML	Teichler et al. (2011)
miR-143	BCL-2, TOP2A, and PLK1	Downregulated in liposarcoma	Ugras et al. (2011)
miR-198	c-MET	Downregulated in HCC	Tan et al. (2011)
miR-17-92 cluster	Several	Upregulated in solid tumors and lymphomas	Ota et al. (2004)
miR-21	ANP32A and SMARCA4	Upregulated in a variety of different cancers	Krichevsky and Gabriely (2009), Schramedei et al. (2011)
miR-26a	PTEN	Upregulated in human glioma	Huse et al. (2009)
miR-155	SOCS1	Upregulated in a variety of different cancers	Iorio et al. (2005), Gironella et al. (2007), Greither et al. (2010)
miR-372, miR-373	LATS2	Upregulated in human testicular germ cell tumors	Li et al. (2003)

which is an antiapoptotic protein and overexpressed in CLL (Cimmino et al. 2005). Upon loss of miR-15 and miR-16, an increased cell survival and the development of a more aggressive form of CLL were observed, indicating a role for these miRNAs as tumor suppressors. Differential gene expression profiling of normal adipose tissue and well-differentiated liposarcoma has allowed the identification of miR-143 as additional tumor suppressor miRNA (Ugras et al. 2011). Interestingly, restoring miR-143 expression in liposarcoma cells led to reduced cell proliferation, increased apoptosis, and decreased expression of genes, such as BCL-2, TOP2A, and PLK1 which are usually overexpressed in cancer cells. Further evidence that miRNAs play an important role in cancer has risen from the observation that miR-29a expression appears to be downregulated in acute myeloid leukemia (AML) (Teichler et al. 2011). This miRNA is located at a cancer-specific fragile site in the human genome and when getting lost in AML causes upregulation of the nuclear oncogene Ski. In hepatocellular carcinoma (HCC), the decreased expression of miR-198 was shown to be related to an enhanced c-MET activity (Tan et al. 2011). This proto-oncogene is a target of miR-198, and overexpression in HCC triggers cell migration and invasion. Additional examples of tumor suppressor miRNAs in HCC include the above-described miR-143 and miR-145 which are transcribed at very low levels in colorectal tumors (Slaby et al. 2007).

Apart from the downregulation of particular miRNAs, there are other mechanisms pointing toward a clear correlation between these small RNAs and cancer development. Chromosomal translocation, for example, can cause highly increased expression levels of miRNAs which are involved in oncogenesis. An example is miR-21 which controls cell growth by inhibiting apoptosis and was found to be upregulated in a variety of different cancers (Krichevsky and Gabriely 2009). This miRNA targets the tumor suppressor genes acidic nuclear phosphoprotein 32 family member A (ANP32A) as well as SMARCA4 which are both involved in chromatin remodeling processes, causing a reduced apoptosis and an increased invasiveness of cancer cells (Schramedei et al. 2011). Another study described the involvement of miR-155 in oncogenesis (Iorio et al. 2005; Gironella et al. 2007; Greither et al. 2010). This miRNA is highly expressed in many human cancers, leading to a downregulation of the suppressor of cytokine signaling 1 (SOCS1) which is a tumor suppressor gene. In breast cells, ectopic expression of miR-155 induced cell proliferation and transformation, while recovery of SOCS1 expression reduced the pro-tumorigenic function of miR-155 (Jiang et al. 2010). Amplification of the miR-26a-2 locus leads to overexpression of miR-26a in human glioma (Huse et al. 2009). This miRNA is a downregulator of the tumor suppressor phosphatase and tensin homolog (PTEN) which modulates cell cycle progression and cell survival. A further study described overexpression of the miR-17-92 cluster in lung cancer, particularly visible in small-cell lung cancer histology samples (Hayashita et al. 2005). This cluster which consists of seven miRNAs (miR-17-5p, -17-3p, -18a, -19a, -19b, -20a, and -92) is located in the intron 3 of the *C13orf25* gene on chromosome 13q31 which is a region often amplified in solid tumors and lymphomas (Ota et al. 2004). Upregulation of the miR-17-92 cluster increases cell proliferation, implying an oncogenic function of these miRNAs, but the role of the individual components of the cluster is still largely unknown [reviewed by van Haaften and Agami (2010)]. Moreover, genetic screens aimed to study possible cooperation between miRNAs and oncogenes in the context of cellular transformation have allowed the identification miR-372 and miR-373, which led to the development of human testicular germ cell tumors. These miRNAs act most likely via direct inhibition of the expression of the tumor suppressor LATS2 an inhibitor of G1/S cell cycle transition (Li et al. 2003).

In addition, aberrant expression of miRNAs can also be caused by DNA methylation, leading to reduced miRNA transcription (Han et al. 2007). Here, miRNA expression profiles of a colon cancer cell line versus a DNA methyltransferase knockout cell line were compared. Other examples of aberrant hypermethylation include miR-9-1, miR-124a3, miR-148, miR-152, and miR-663 in human breast cancer tissue (Lehmann et al. 2008). Furthermore, for miR-9-1, a direct correlation between the methylation status and a decreased expression could be shown, which clearly indicates epigenetic inactivation of this miRNA.

Notwithstanding the rapidly increasing knowledge concerning the involvement of miRNAs in cancer development, many more studies are required to fully recognize the underlying mechanisms. We just started to get a glimpse of the tip of the iceberg.

3.2 *Virally Encoded miRNAs*

While in higher vertebrates RNAi merely plays an inferior role concerning antiviral defense mechanisms, various dsDNA viruses abuse this fundamental principle of gene regulation by producing own regulatory RNAs turning a basic host defense system into a viral offense strategy [for a review, see Umbach and Cullen (2009), Boss et al. (2009), Skalsky and Cullen (2010), and Grundhoff and Sullivan (2011)]. The first experimental evidence of virally encoded miRNAs (Pfeffer et al. 2004) was found for the human herpesvirus 4/Epstein-Barr virus (HHV-4/EBV). Subsequently, applying advanced computational prediction methods, many more miRNAs were identified, supported by experimental evidence like array analyses, cloning, and northern blotting (Grundhoff et al. 2006; Cai and Cullen 2006). The reasons for miRNA exploitation by viruses seem quite obvious: negligible need of coding capacity, fast response as no proteins have to be translated, low immunogenicity, and potentially large host target range make miRNAs ideal candidates for the modulation of host defense mechanisms like immune evasion and apoptosis as well as viral replication strategies. So far, all viruses shown to encode for miRNAs belong to the group of DNA viruses, presumably owing to their replication strategy requiring nuclear localization which in turn is a prerequisite for Drosha function (Fig. 1). There have been miRNAs described for adeno-, herpes-, and polyomaviruses (Table 2).

Adeno- and *polyomaviridae* only code for few or even just one miRNA. In contrast, over 150 transcripts were identified for the *herpesviridae*. Even more intriguing than the sheer number is the fact that miRNAs were found in almost every single herpesvirus looked at so far. Human herpesvirus 8/Kaposi's sarcoma-associated herpesvirus (HHV-8/KSHV) alone encodes for 12 pre-miRNAs, organized in a cluster within the latency-associated region (Fig. 3). In total, at least 25 KSHV mature miRNAs have been detected by multiple groups. For some of the viral pre-miRNAs, both strands are processed and might readily be incorporated into the RISC. Apart from this ambiguity in strand selection, there are posttranscriptional modifications in case of miRK10, where a single nucleotide is exchanged within the seed region, leading to two distinct and active isoforms, 10a and l0b. On the other hand, a cell line-specific single-nucleotide polymorphism in case of pre-miR-K5 and pre-miR-K9 greatly diminishes the processing of these pre-miRNAs. Still, viruses with this variability are able to conduct both infective cycles, i.e., lytic as well as latent, without apparent impairments (Gottwein et al. 2006; Umbach and Cullen 2010). Since all KSHV miRNAs are expressed to high levels during latency, they are suspected to be of importance for the establishment or maintenance of the latent state (Murphy et al. 2008). The other human pathogenic member of the rhadinoviruses, EBV, makes even more extensive use of miRNAs with 25 encoded pre-miRNAs identified up to now (Zhu et al. 2009).

Interestingly, thus far no miRNAs were either predicted or identified for hepatitis B virus, human papillomavirus, human parvovirus, or human herpesvirus-3/varicella zoster virus (HHV-3/VZV). For poxviruses, miRNAs have been predicted; however, they still await experimental proof.

3.2.1 Evolution of Viral miRNAs

On first view, it might seem plausible to assume that viral miRNAs have been captured from the host via horizontal gene transfer. Yet based on the low level of sequence homology with host miRNAs and even to those of closely related viruses

Table 2 Virus-encoded miRNAs

Virus family	Virus	Host	Number of pre-miRNAs	References
Adenovirus	hAV	Human	2	Andersson et al. (2005), Aparicio et al. (2006), Sano et al. (2006), Xu et al. (2007)
Ascovirus	HvAV	Insect	1	Hussain et al. (2008)
α-Herpesvirus	HSV-1	Human	8	Cui et al. (2006), Umbach et al. (2008, 2009)
	HSV-2	Human	6	Tang et al. (2008, 2009), Umbach et al. (2010)
	MDV-1	Avian	14	Burnside et al. (2006), Yao et al. (2008)
	MDV-2	Avian	18	Waidner et al. (2009), Yao et al. (2007)
	HVT	Avian	17	Waidner et al. (2009)
	ILTV	Avian	7	Waidner et al. (2009)
β-Herpesvirus	hCMV	Human	11	Dunn et al. (2005), Grey et al. (2005), Pfeffer et al. (2005)
	mCMV	Mouse	18	Buck et al. (2007), Dölken et al. (2007)
γ-Herpesvirus	EBV	Human	25	Cai et al. (2006), Pfeffer et al. (2004), Zhu et al. (2009)
	KSHV	Human	12	Cai et al. (2005), Pfeffer et al. (2005), Samols et al. (2005), Grundhoff et al. (2006)
	MHV-68	Mouse	9	Pfeffer et al. (2005)
	rLCV	Primate	32	Cai et al. (2006), Walz et al. (2010)
	RRV	Primate	15	Schäfer et al. (2007)
Polyomavirus	BKV	Human	1	Seo et al. (2008)
	JCV	Human	1	Seo et al. (2008)
	MCV	Human	1	Seo et al. (2009)
	SV40	Primate	1	Sullivan et al. (2005)

EBV Epstein-Barr virus, *hCMV* human cytomegalovirus, *mCMV* mouse cytomegalovirus, *HSV-1 and HSV-2* herpes simplex virus 1 and 2, *KSHV* Kaposi's sarcoma-associated herpesvirus, *MDV-1 and MDV-2* Marek's disease virus types 1 and 2, *HVT* herpesvirus of turkeys, *ILTV* infectious laryngotracheitis virus, *rLCV* rhesus lymphocryptovirus, *RRV* rhesus monkey rhadinovirus, *MHV-68* murine herpesvirus 68, *hAV* human adenovirus, *HvAV* Heliothis virescens ascovirus [adopted from Skalsky and Cullen (2010)]. As briefly outlined in the text the actual number of mature miRNAs might be significantly higher. Please note: the list of polyomavirus miRNAs is not complete

(Schäfer et al. 2007; Walz et al. 2010), there is little evidence in direct support of this presumption. Merely miR-K11 of KSHV and M4 from the chicken α-herpesvirus Marek's disease virus serotype 1 (MDV-1) share identical seed regions with the human miR-155 (Skalsky et al. 2007), while only few bases match in the respective 3′-regions. Strikingly, miR-155 is involved in crucial cellular differentiation pathways and has been implicated in lymphomagenesis (Costinean et al. 2006). Moreover, MDV-1 causes tumors in its natural host, while its relative MDV-2, which lacks a miR-155 homolog, cannot. Accordingly, it is assumed that miR-155 functions are critical for herpesviral infection and tumorigenesis. To an even lower extent, there are additional viral miRNAs showing partial homology to host

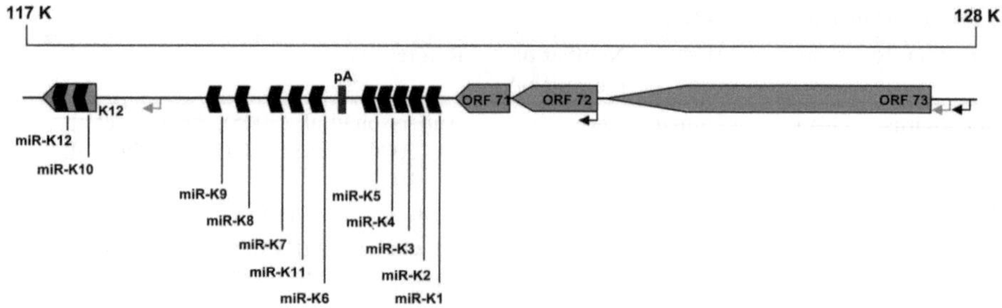

Fig. 3 The latency-associated region and miRNA cluster of KSHV. The miRNAs (*black directional arrows*) miR-K1-9 and miR-K11 are clustered within one intergenic region, flanked by latently expressed proteins (*gray directional arrows*). MiR-K10 and -12 are located in the coding region of *K12*. Transcription of miRNAs including the *K12* locus occurs during latent as well as productive infection. All latent mRNAs start off from four promoters, two latent and two lytic ones (*black and gray arrows*)

miRNAs (Gottwein and Cullen 2008). However, due to the fast evolution of viruses, it is difficult to conclusively resolve the question of the origin of viral miRNAs, as already single-nucleotide mutations, especially if they occur in the seed region, significantly alter miRNA target range and thus make them prone to further rapid selection. This is supported by the finding that even related strains of the same virus display considerable degrees of sequence variability in particular miRNAs while conservation in others remains high (Marshall et al. 2007). Apart from sequence variations, genomic location and organization of miRNA coding regions are at best preserved among members of a given subfamily.

3.2.2 Targets of Viral miRNAs

The intrinsically low sequence complementarity combined with the low conservation of viral miRNAs renders precise target prediction quite difficult (Lewis et al. 2003). This accounts for the relatively small number of validated target genes reported so far. A further complication represents the lack of appropriate in vivo model systems. Thus, most of the targets listed below still await in vivo validation. Among them are a number of lytic viral transcripts as well as various cellular mRNAs involved in cell proliferation and survival, antiviral defense, and stress response.

The polyomavirus SV40 (Simian virus 40) codes for a single *cis*-acting miRNA targeting the virally encoded large T-antigen. Downregulation of this viral protein especially during later stages of the replication cycle has been shown in vitro to reduce activation of cytotoxic T cells (Sullivan et al. 2005). However, if this indeed resembles, the natural function of this miRNA remains to be seen (Sullivan et al. 2009). Alike, a comparable mechanism has been proposed for miR-BART22 of EBV which targets the viral latent membrane protein 2A (LMP2A) (Lung et al. 2009). Furthermore, miR-BART2, another EBV miRNA, can induce cleavage of its target BALF5 (viral polymerase), even though the biological relevance is not fully understood (Barth et al. 2008). Finally, EBV LMP1 is targeted by three EBV miRNAs, miR-BART1-5p, miRBART16, and miRBART17-5p (Lo et al. 2007). LMP1 is a constitutively active viral mimic of the tumor necrosis factor receptor family and induces cell proliferation during latency.

For human herpesvirus 5/cytomegalovirus (HHV-5/CMV), it could be shown that downregulation of the major histocompatibility complex class I-related chain B (MICB) via hcmv-miR-UL112 reduces susceptibility to cytotoxic T cells (Stern-Ginossar et al. 2007). MICB was also identified as target of EBV- and KSHV-encoded miRNAs (Nachmani et al. 2009). Additional cellular targets of EBV miRNAs include CXCL-11 (Xia et al. 2008), which can attract T cells and the p53 upregulated modulator of apoptosis (PUMA) (Choy et al. 2008). Recent findings also suggest the translocase of outer mitochondrial membrane 22 (TOMM22) and importin 7 (IP07) as EBV miRNA-regulated genes (Dölken et al. 2010). In this context, KSHV represents the virus with the most fully characterized cellular targets of viral miRNAs. In *cis*, effects are conferred through regulation of Rta by KSHV miR-K4-5p and miR-K9* (Lu et al. 2009a, Bellare and Ganem 2009). This might be a factor to establish and maintain latency. In *trans*, KSHV can introduce an anti-apoptotic state by repression of the apoptosis mediators BCLAF1 by miR-K5 (Ziegelbauer et al. 2009), the TNF-associated TWEAK receptor by miR-K3 (Abend et al. 2010), and thrombospondin-1 (THBS1) by miR-K6-5p (and others) (Samols et al. 2007). In addition, the latter plays a role in neo-angiogenesis. Cell cycle and transcriptional activity could be targeted through inhibition of the transcription factor MAF by miR-K6 and -K11 (Hansen et al. 2010), which also leads to cellular reprogramming. Moreover, transcriptional repression via DNA methylation is induced, because the DNA methyltransferase (DNMT) inhibitor Rb-like protein 2 (Rbl2) is a target of miR-K4-5p (Lu et al. 2009a), whereas miR-K1 apparently activates transcription, and progression of cell cycle, by decreasing protein levels of the NFKB-inhibitor IKB (Lei et al. 2010) and p21 (Gottwein and Cullen 2010), respectively. Potential targets that might affect B-cell transformation include BACH1 (Skalsky et al. 2007) and presumably others. However, hitherto the in vivo effects have not been investigated since no adequate animal model for latent KSHV infection has yet been established.

3.2.3 Host miRNAs and Viral Replication

Viruses are able not only to encode miRNAs but also to exploit cellular miRNAs for their own purposes (Chang et al. 2008; Nachmani et al. 2010). A prominent example is HCV which uses a highly conserved liver-specific miRNA, termed miR-122, to facilitate its replication and translation by a so far unknown mechanism (Jopling et al. 2005; Chang et al. 2008; Henke et al. 2008). HCV is a single-stranded positive-sense RNA flavivirus, and its genome contains two miR-122 binding sites in the 5′-UTR (Fig. 4). Interaction of miR-122 with the HCV genome has been

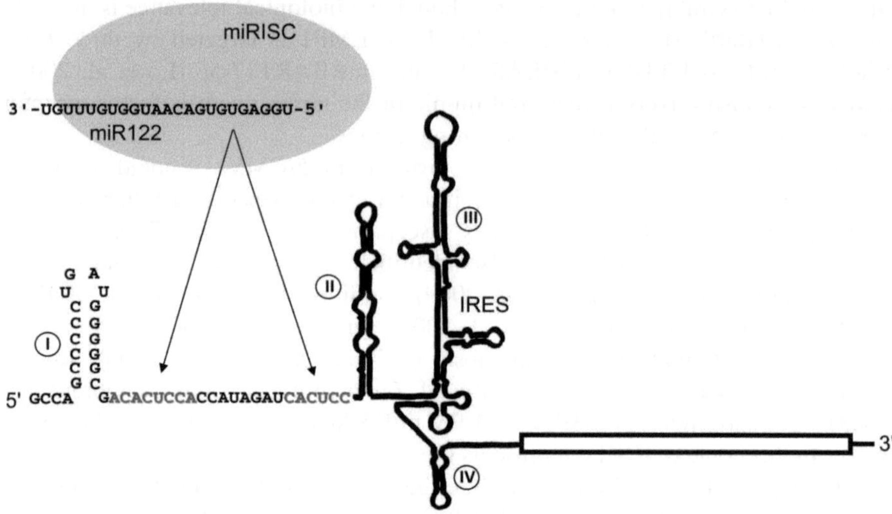

Fig. 4 Secondary structure of the 5′-UTR of the HCV genome. The secondary structure of the 5′-UTR of the HCV genome consists of four domains labeled I–IV of which the IRES is composed of domains II–IV. The predicted miR-122 binding sites are colored in *gray*. On top, miR-122 is shown bound to RISC [adopted from Fraser and Doudna (2007)]

shown to increase accumulation of viral RNA in cultured liver cells, modulating HCV abundance (Jopling et al. 2005; Randall et al. 2007).

3.3 miRNAs in Other Diseases

miRNAs are by far not just involved in cancer and viral infections (Lu et al. 2008b). Indeed, a steadily increasing number of studies report about miRNA-disease associations. Since this area of research is still rather young, many more associations are to be expected.

Examples are miR-1, miR-29, miR-21, miR-133, and miR-208a which are involved in heart disease [reviewed by Small and Olson (2011)]. Acute myocardial infarction results in a dysregulation of miR-29 family members causing reduced expression levels in the region of the heart adjacent to infarction. Normally, these miRNAs inhibit the expression of proteins, such as collagens, fibrillins, and elastin, and thus their dysregulation contributes to the formation of cardiac fibrosis (van Rooij et al. 2008). Another example represents miR-1, which is highly expressed in mammalian hearts [reviewed by Mishima et al. (2007)]. This miRNA is overexpressed in humans with coronary artery disease, and it was described that the high level of expression causes heart arrhythmia in normal and infarcted hearts, possibly by targeting KCNJ2 (encoding the K(+) channel subunit Kir2.1) and GJA1 (encoding connexin 43) which are important for cardiac conductance (Yang et al. 2007).

Recently, miR-143 and miR-145 have been found to be significantly downregulated in colonic epithelium in patients with ulcerative colitis when compared to normal colonic mucosa (Pekow et al. 2011). Transfection of colorectal cancer cells with miR-143 and miR-145 followed by analysis of putative target protein expression profiles allowed the identification of their target transcripts. Among them, there are the proto-oncogene K-RAS and the apoptosis inhibitor 5 (API5), suggesting these miRNAs function as tumor suppressors.

Another recent study could show that the level of miR-155 was fivefold increased in cystic fibrosis lung epithelial cells compared to control cells (Bhattacharyya et al. 2011). This miRNA specifically reduces the level of SHIP1, a phosphatase involved in inhibiting the PI3K (phosphoinositide 3-kinase) pathway. The activation of PI3K upon miR-155 overexpression results in an increased interleukin-8 (IL-8) expression and consequently leads to inflammatory processes. Moreover, a correlation between allergic asthma and upregulation of miR-126 in the airway wall tissue of a mouse model was described (Mattes et al. 2009). The disease is characterized by abnormal T helper-2 (TH2) lymphocyte responses to inhaled antigens which eventually causes inflammation. Inhibition of miR-126 reduced inflammation via diminishing TH2 responses, thereby suppressing the asthmatic phenotype.

Interestingly, even the Alzheimer's disease (AD) has recently been associated with dysregulation of miRNAs (Wang et al. 2008). One of them is miR-127 which was found to be significantly downregulated even in patients at the earliest clinical stage of disease. This miRNA targets the beta-site amyloid precursor protein-cleaving enzyme 1 (BACE1) which is essential for the generation of beta-amyloid plaques that characterize AD. Thus, the low levels of miR-127 may be involved in accelerated disease progression caused by a derepression of BACE1.

Further information on miRNAs and their association with diseases is listed in the human miRNA-associated disease database (HMDD) which as of July 2011 contains 3,017 miRNA-disease associations (http://cmbi.bjmu.edu.cn/hmdd) (Lu et al. 2008b).

4 Tools for the Modulation of miRNA Expression Levels and Activity

Dysregulation of miRNA expression has been described to cause or at least facilitate the development of several pathological states (Lu et al. 2008b). Accordingly, controlling miRNA expression or activity can be considered a promising therapeutic approach for the treatment of various diseases (Brown and Naldini 2009; Garzon et al. 2010). As a result, in the last years, miRNA inhibitors, so-called anti-miRs, anti-miRNAs, antagomirs, target protectors, or miRmasks, have been developed which consist of chemically modified, single-stranded oligonucleotides as a rule fully complementary to the guide strand of the corresponding target miRNA

(Krützfeldt et al. 2005; Weiler et al. 2006; Choi et al. 2007; Xiao et al. 2007; Stenvang and Kauppinen 2008). They are designed to specifically bind endogenous miRNAs, thus blocking their activity.

Morpholino-modified target protectors have been applied in zebrafish for the disruption of interactions between miR-430 and the corresponding target mRNAs, i.e., *squint* and *lefty* which are involved in transforming growth factor-β signaling pathways (Choi et al. 2007). Those results showed that injection of miR-430-directed target protectors into fish embryos led to an increased *squint* and *lefty* expression and activity, indicating that this approach is suited to inhibit specific miRNA-mRNA interactions in vivo. An analogous strategy was used in rats, in which 2′*O*-Me-modified oligonucleotides were applied to abrogate interactions between cardiac pacemaker channel-encoding genes (HCN2 and HCN4) and their cognate miRNAs miR-1 and miR-133 (Xiao et al. 2007). The switching off of these individual miRNA activities caused an acceleration of the rat's heart rate and increased pacemaker activities.

A particularly promising example is miR-122, a highly conserved liver-specific miRNA, which is exploited by HCV to facilitate viral replication and translation (Jopling et al. 2005; Chang et al. 2008; Henke et al. 2008; Roberts et al. 2011). Binding of miR-122 to the 5′-UTR of the HCV genome (Fig. 4) has been shown to increase accumulation of viral RNA in cultured liver cells (Jopling et al. 2005; Randall et al. 2007), and hence, repression of miR-122 by an antisense approach seemed a possible antiviral strategy for this potentially life-threatening infectious disease. To test this assumption, either Huh7 cells or mice were initially transfected with RNA oligonucleotides fully complementary to miR-122 modified either with 2′*O*-methyl (2′*O*-Me) (Jopling et al. 2005) or 2′*O*-Me including phosphorothioate (PS) backbone chemistry and a cholesterol group at the 3′-end to increase bioavailability (Krützfeldt et al. 2005). The latter was injected into the tail veins of mice and led to an efficient and specific suppression of endogenous miR-122. Advanced studies in mice and primates showed that locked nucleic acid (LNA) modifications are as well suited to silence miR-122 (Elmen et al. 2008a, b). The observed effects could be further improved by combining LNAs with DNA bases and a PS backbone generating a new inhibitor termed SPC3649 which led to a specific and stabile downregulation of the target while showing a low toxicity profile (Elmen et al. 2008a). Thereupon, the therapeutic potential of SPC3649 was tested in chimpanzees chronically infected with HCV and resulted in a reduction of viral load by up to 2.6-fold for HCV genome equivalents in serum and to an improved liver histology (Lanford et al. 2010). Currently, phase 2a clinical trials are being conducted to test safety and tolerability of SPC3649 in HCV-infected patients. Recently, powerful downregulation of miRNAs by seed-targeting 8-mer tiny LNAs could be shown, an approach capable of drastically cutting the price tag for a potential nucleic acid-based HCV therapy (Obad et al. 2011).

Another approach, which permits interference with the activity of upregulated miRNAs, is based on endogenously transcribed RNAs containing several binding sites for a given miRNA. Such molecules, so-called decoys (Haraguchi et al. 2009), sponges (Ebert et al. 2007), or erasers (Sayed et al. 2008), are suited to specifically trap a particular miRNA, eventually preventing interaction with the natural target. There are decoys which either carry perfectly or imperfectly matching target sites, though to antagonize miRNAs imperfect binding sites seem to be more efficient than fully complementary ones [reviewed by Brown and Naldini (2009)].

Besides aberrant overexpression of miRNAs, the opposite might as well cause pathological conditions (Cimmino et al. 2005; Slaby et al. 2007). In this case, it is essential to restore expression levels of affected miRNAs rather than knocking them down or out [reviewed by Garzon et al. (2010)]. This can, for example, be achieved by transfection of synthetic miRNA mimics, resembling the natural counterpart or alternative by endogenous expression of miRNA precursors. These precursors can be expressed under the control of tissue-specific promoters. For several reasons like efficient transduction of target cells in combination with low immunogenicity and long-lasting effects, adeno-associated viral (AAV) vectors have become the vehicle of choice in recent years (Nirenberg 2008). The first approach was applied to restore miR-29 expression in HCC (Xiong et al. 2010). This miRNA targets the mRNA of the antiapoptotic proteins BCL-2 and Mcl-1. As proposed, introduction of miR-29 dramatically repressed the ability of HCC cells to form tumors in nude mice. The second procedure was used, for example, to overexpress miR-26a via AAV transduction in a liver cancer mouse model (Kota et al. 2009). miR-26a can be considered as tumor suppressor and is downregulated in HCC. Consistent with these previous findings, restoration of the miRNA led to apoptosis of cancer cells, to inhibition of cell proliferation, and, in sum, to a decreased disease progression.

Taken together, these results suggest that modulation of aberrant miRNA expression or activity applying above-mentioned strategies is a highly promising starting point for the development of novel therapies fighting disease progression.

5 Conclusions

miRNAs are a fast progressing fascinating topic in the ever growing field of RNA biology. Just in recent years, we started to learn about their fundamental roles in the origin and development of a variety of diseases. While the observation that aberrant expression of a tiny single RNA species can cause such devastation outcomes like cancer is frightening on the one hand, it might offer a fair chance to specifically interfere with this pathological dysfunction, thus providing new therapeutic avenues. Especially the approach to antagonize miRNAs by chemically modified antisense oligonucleotides seems promising. Here, knocking down miR-122 to challenge HCV infections is currently one of most advanced approaches. However, as common to all nucleic acid-based strategies, the problem of delivery has to be overcome before such therapies might become available to the general public. In this regard, the perspectives are somewhat disillusioning. Moreover, as we just began to understand some of the underlying principles, more basic research will be necessary to comprehend this complicated interconnected network. There is no doubt the years to come will remain thrilling.

Acknowledgments We thank Andrea Deerberg for providing Fig. 1 and Sven Müller-Loennies (Institute of Microbiology and Infection Biology, Research Center Borstel, Leibniz Center for Medicine and Biosciences, Germany) for critical reading of the manuscript. We would like to apologize to those authors whose work was not cited directly owing to space limitations. This work was financially supported by a grant of the "Medizinische Fakultät, Lübeck" (FKZ: E25: 2011).

References

Abend JR, Uldrick T, Ziegelbauer JM (2010) Regulation of tumor necrosis factor-like weak inducer of apoptosis receptor protein (TWEAKR) expression by Kaposi's sarcoma-associated herpesvirus microRNA prevents TWEAK-induced apoptosis and inflammatory cytokine expression. J Virol 84:12139–12151

Akao Y, Nakagawa Y, Hirata I et al (2010) Role of anti-oncomirs miR-143 and -145 in human colorectal tumors. Cancer Gene Ther 17:398–408

Andersson MG, Haasnoot PC, Xu N et al (2005) Suppression of RNA interference by adenovirus virus-associated RNA. J Virol 79:9556–9565

Aparicio O, Razquin N, Zaratiegui M et al (2006) Adenovirus virus-associated RNA is processed to functional interfering RNAs involved in virus production. J Virol 80:1376–1384

Bartel DP (2009) MicroRNAs: target recognition and regulatory functions. Cell 136:215–233

Barth S, Pfuhl T, Mamiani A et al (2008) Epstein-Barr virus-encoded microRNA miR-BART2 down-regulates the viral DNA polymerase BALF5. Nucleic Acids Res 36:666–675

Behm-Ansmant I, Rehwinkel J, Doerks T et al (2006) mRNA degradation by miRNAs and GW182 requires both CCR4:NOT deadenylase and DCP1:DCP2 decapping complexes. Genes Dev 20:1885–1898

Bellare P, Ganem D (2009) Regulation of KSHV lytic switch protein expression by a virus-encoded microRNA: an evolutionary adaptation that fine-tunes lytic reactivation. Cell Host Microbe 6:570–575

Bhattacharyya S, Balakathiresan NS, Dalgard C et al (2011) Elevated miR-155 promotes inflammation in cystic fibrosis by driving hyperexpression of interleukin-8. J Biol Chem 286:11604–11615

Bohnsack MT, Czaplinski K, Gorlich D (2004) Exportin 5 is a RanGTP-dependent dsRNA-binding protein that mediates nuclear export of pre-miRNAs. RNA 10:185–191

Borchert GM, Lanier W, Davidson BL (2006) RNA polymerase III transcribes human microRNAs. Nat Struct Mol Biol 13:1097–1101

Boss IW, Plaisance KB, Renne R (2009) Role of virus-encoded microRNAs in herpesvirus biology. Trends Microbiol 17:544–553

Brown BD, Naldini L (2009) Exploiting and antagonizing microRNA regulation for therapeutic and experimental applications. Nat Rev Genet 10:578–585

Buck AH, Santoyo-Lopez J, Robertson KA et al (2007) Discrete clusters of virus-encoded micrornas are associated with complementary strands of the genome and the 7.2-kilobase stable intron in murine cytomegalovirus. J Virol 81:13761–13770

Burnside J, Bernberg E, Anderson A et al (2006) Marek's disease virus encodes MicroRNAs that map to meq and the latency-associated transcript. J Virol 80:8778–8786

Burroughs AM, Ando Y, de Hoon MJ et al (2010) A comprehensive survey of 3′ animal miRNA modification events and a possible role for 3′ adenylation in modulating miRNA targeting effectiveness. Genome Res 20:1398–1410

Cai X, Cullen BR (2006) Transcriptional origin of Kaposi's sarcoma-associated herpesvirus microRNAs. J Virol 80:2234–2242

Cai X, Hagedorn CH, Cullen BR (2004) Human microRNAs are processed from capped, polyadenylated transcripts that can also function as mRNAs. RNA 10:1957–1966

Cai X, Lu S, Zhang Z et al (2005) Kaposi's sarcoma-associated herpesvirus expresses an array of viral microRNAs in latently infected cells. Proc Natl Acad Sci U S A 102:5570–5575

Cai X, Schafer A, Lu S et al (2006) Epstein-Barr virus microRNAs are evolutionarily conserved and differentially expressed. PLoS Pathog 2:e23

Calin GA, Sevignani C, Dumitru CD et al (2004) Human microRNA genes are frequently located at fragile sites and genomic regions involved in cancers. Proc Natl Acad Sci U S A 101:2999–3004

Carnero E, Sutherland JD, Fortes P (2011) Adenovirus and miRNAs. Biochim Biophys Acta 25: E406–414

Castanotto D, Rossi JJ (2009) The promises and pitfalls of RNA-interference-based therapeutics. Nature 457:426–433

Chang J, Guo JT, Jiang D et al (2008) Liver-specific microRNA miR-122 enhances the replication of hepatitis C virus in nonhepatic cells. J Virol 82:8215–8223

Chendrimada TP, Gregory RI, Kumaraswamy E et al (2005) TRBP recruits the Dicer complex to Ago2 for microRNA processing and gene silencing. Nature 436:740–744

Chendrimada TP, Finn KJ, Ji X et al (2007) MicroRNA silencing through RISC recruitment of eIF6. Nature 447:823–828

Cheng AM, Byrom MW, Shelton J et al (2005) Antisense inhibition of human miRNAs and indications for an involvement of miRNA in cell growth and apoptosis. Nucleic Acids Res 33:1290–1297

Chiba Y, Misawa M (2010) MicroRNAs and their therapeutic potential for human diseases: MiR-133a and bronchial smooth muscle hyperresponsiveness in asthma. J Pharmacol Sci 114:264–268

Chin L, Hahn WC, Getz G et al (2011) Making sense of cancer genomic data. Genes Dev 25:534–555

Cho WC (2007) OncomiRs: the discovery and progress of microRNAs in cancers. Mol Cancer 6:60

Choi WY, Giraldez AJ, Schier AF (2007) Target protectors reveal dampening and balancing of Nodal agonist and antagonist by miR-430. Science 318:271–274

Choy EY, Siu KL, Kok KH et al (2008) An Epstein-Barr virus-encoded microRNA targets PUMA to promote host cell survival. J Exp Med 205:2551–2560

Cimmino A, Calin GA, Fabbri M et al (2005) miR-15 and miR-16 induce apoptosis by targeting BCL2. Proc Natl Acad Sci U S A 102:13944–13949

Costinean S, Zanesi N, Pekarsky Y et al (2006) Pre-B cell proliferation and lymphoblastic leukemia/high-grade lymphoma in E(mu)-miR155 transgenic mice. Proc Natl Acad Sci U S A 103:7024–7029

Cui C, Griffiths A, Li G et al (2006) Prediction and identification of herpes simplex virus 1-encoded microRNAs. J Virol 80:5499–5508

Dölken L, Perot J, Cognat V et al (2007) Mouse cytomegalovirus microRNAs dominate the cellular small RNA profile during lytic infection and show features of posttranscriptional regulation. J Virol 81:13771–13782

Dölken L, Malterer G, Erhard F et al (2010) Systematic analysis of viral and cellular microRNA targets in cells latently infected with human gamma-herpesviruses by RISC immunoprecipitation assay. Cell Host Microbe 7:324–334

Dunn W, Trang P, Zhong Q et al (2005) Human cytomegalovirus expresses novel microRNAs during productive viral infection. Cell Microbiol 7:1684–1695

Ebert MS, Neilson JR, Sharp PA (2007) MicroRNA sponges: competitive inhibitors of small RNAs in mammalian cells. Nat Methods 4:721–726

Elbashir SM, Harborth J, Lendeckel W et al (2001a) Duplexes of 21-nucleotide RNAs mediate RNA interference in cultured mammalian cells. Nature 411:494–498

Elbashir SM, Lendeckel W, Tuschl T (2001b) RNA interference is mediated by 21- and 22-nucleotide RNAs. Genes Dev 15:188–200

Elbashir SM, Martinez J, Patkaniowska A et al (2001c) Functional anatomy of siRNAs for mediating efficient RNAi in Drosophila melanogaster embryo lysate. EMBO J 20:6877–6888

Elmen J, Lindow M, Schutz S et al (2008a) LNA-mediated microRNA silencing in non-human primates. Nature 452:896–899

Elmen J, Lindow M, Silahtaroglu A et al (2008b) Antagonism of microRNA-122 in mice by systemically administered LNA-antimiR leads to up-regulation of a large set of predicted target mRNAs in the liver. Nucleic Acids Res 36:1153–1162

Eulalio A, Huntzinger E, Nishihara T et al (2009) Deadenylation is a widespread effect of miRNA regulation. RNA 15:21–32

Feng Y, Zhang X, Song Q et al (2011) Drosha processing controls the specificity and efficiency of global microRNA expression. Biochim Biophys Acta 1809:700–707

Filipowicz W, Bhattacharyya SN, Sonenberg N (2008) Mechanisms of post-transcriptional regulation by microRNAs: are the answers in sight? Nat Rev Genet 9:102–114

Fire A, Xu S, Montgomery MK et al (1998) Potent and specific genetic interference by double-stranded RNA in Caenorhabditis elegans. Nature 391:806–811

Fraser CS, Doudna JA (2007) Structural and mechanistic insights into hepatitis C viral translation initiation. Nat Rev Microbiol 5:29–38

Gangaraju VK, Lin H (2009) MicroRNAs: key regulators of stem cells. Nat Rev Mol Cell Biol 10:116–125

Garzon R, Marcucci G, Croce CM (2010) Targeting microRNAs in cancer: rationale, strategies and challenges. Nat Rev Drug Discov 9:775–789

Ghildiyal M, Zamore PD (2009) Small silencing RNAs: an expanding universe. Nat Rev Genet 10:94–108

Gironella M, Seux M, Xie MJ et al (2007) Tumor protein 53-induced nuclear protein 1 expression is repressed by miR-155, and its restoration inhibits pancreatic tumor development. Proc Natl Acad Sci U S A 104:16170–16175

Gottwein E, Cullen BR (2008) Viral and cellular microRNAs as determinants of viral pathogenesis and immunity. Cell Host Microbe 3:375–387

Gottwein E, Cullen BR (2010) A human herpesvirus microRNA inhibits p21 expression and attenuates p21-mediated cell cycle arrest. J Virol 84:5229–5237

Gottwein E, Cai X, Cullen BR (2006) A novel assay for viral microRNA function identifies a single nucleotide polymorphism that affects Drosha processing. J Virol 80:5321–5326

Greither T, Grochola LF, Udelnow A et al (2010) Elevated expression of microRNAs 155, 203, 210 and 222 in pancreatic tumors is associated with poorer survival. Int J Cancer 126:73–80

Grey F, Antoniewicz A, Allen E et al (2005) Identification and characterization of human cytomegalovirus-encoded microRNAs. J Virol 79:12095–12099

Grishok A, Tabara H, Mello CC (2000) Genetic requirements for inheritance of RNAi in *C. elegans*. Science 287:2494–2497

Grundhoff A, Sullivan CS (2011) Virus-encoded microRNAs. Virology 411:325–343

Grundhoff A, Sullivan CS, Ganem D (2006) A combined computational and microarray-based approach identifies novel microRNAs encoded by human gamma-herpesviruses. RNA 12:733–750

Hamilton AJ, Baulcombe DC (1999) A species of small antisense RNA in posttranscriptional gene silencing in plants. Science 286:950–952

Han L, Witmer PD, Casey E et al (2007) DNA methylation regulates MicroRNA expression. Cancer Biol Ther 6:1284–1288

Hansen A, Henderson S, Lagos D et al (2010) KSHV-encoded miRNAs target MAF to induce endothelial cell reprogramming. Genes Dev 24:195–205

Haraguchi T, Ozaki Y, Iba H (2009) Vectors expressing efficient RNA decoys achieve the long-term suppression of specific microRNA activity in mammalian cells. Nucleic Acids Res 37:e43

Hayashita Y, Osada H, Tatematsu Y et al (2005) A polycistronic microRNA cluster, miR-17-92, is overexpressed in human lung cancers and enhances cell proliferation. Cancer Res 65:9628–9632

Henke JI, Goergen D, Zheng J et al (2008) microRNA-122 stimulates translation of hepatitis C virus RNA. EMBO J 27:3300–3310

Heo I, Joo C, Cho J et al (2008) Lin28 mediates the terminal uridylation of let-7 precursor MicroRNA. Mol Cell 32:276–284
Humphreys DT, Westman BJ, Martin DI et al (2005) MicroRNAs control translation initiation by inhibiting eukaryotic initiation factor 4E/cap and poly(A) tail function. Proc Natl Acad Sci U S A 102:16961–16966
Huntzinger E, Izaurralde E (2011) Gene silencing by microRNAs: contributions of translational repression and mRNA decay. Nat Rev Genet 12:99–110
Huse JT, Brennan C, Hambardzumyan D et al (2009) The PTEN-regulating microRNA miR-26a is amplified in high-grade glioma and facilitates gliomagenesis in vivo. Genes Dev 23:1327–1337
Hussain M, Taft RJ, Asgari S (2008) An insect virus-encoded microRNA regulates viral replication. J Virol 82:9164–9170
Hutvagner G, McLachlan J, Pasquinelli AE et al (2001) A cellular function for the RNA-interference enzyme Dicer in the maturation of the let-7 small temporal RNA. Science 293:834–838
Iorio MV, Ferracin M, Liu CG et al (2005) MicroRNA gene expression deregulation in human breast cancer. Cancer Res 65:7065–7070
Jiang S, Zhang HW, Lu MH et al (2010) MicroRNA-155 functions as an OncomiR in breast cancer by targeting the suppressor of cytokine signaling 1 gene. Cancer Res 70:3119–3127
Jinek M, Doudna JA (2009) A three-dimensional view of the molecular machinery of RNA interference. Nature 457:405–412
Jopling CL, Yi M, Lancaster AM et al (2005) Modulation of hepatitis C virus RNA abundance by a liver-specific MicroRNA. Science 309:1577–1581
Kedde M, van Kouwenhove M, Zwart W et al (2010) A Pumilio-induced RNA structure switch in p27-3' UTR controls miR-221 and miR-222 accessibility. Nat Cell Biol 12:1014–1020
Kelly GL, Strasser A (2011) The essential role of evasion from cell death in cancer. Adv Cancer Res 111:39–96
Kerr TA, Korenblat KM, Davidson NO (2011) MicroRNAs and liver disease. Transl Res 157:241–252
Khvorova A, Reynolds A, Jayasena SD (2003) Functional siRNAs and miRNAs exhibit strand bias. Cell 115:209–216
Kim VN, Han J, Siomi MC (2009) Biogenesis of small RNAs in animals. Nat Rev Mol Cell Biol 10:126–139
Kiriakidou M, Tan GS, Lamprinaki S et al (2007) An mRNA m7G cap binding-like motif within human Ago2 represses translation. Cell 129:1141–1151
Kota J, Chivukula RR, O'Donnell KA et al (2009) Therapeutic microRNA delivery suppresses tumorigenesis in a murine liver cancer model. Cell 137:1005–1017
Krek A, Grun D, Poy MN et al (2005) Combinatorial microRNA target predictions. Nat Genet 37:495–500
Krichevsky AM, Gabriely G (2009) miR-21: a small multi-faceted RNA. J Cell Mol Med 13:39–53
Krützfeldt J, Rajewsky N, Braich R et al (2005) Silencing of microRNAs in vivo with 'antagomirs'. Nature 438:685–689
Lanford RE, Hildebrandt-Eriksen ES, Petri A et al (2010) Therapeutic silencing of microRNA-122 in primates with chronic hepatitis C virus infection. Science 327:198–201
Lee RC, Feinbaum RL, Ambros V (1993) The *C. elegans* heterochronic gene lin-4 encodes small RNAs with antisense complementarity to lin-14. Cell 75:843–854
Lee Y, Ahn C, Han J et al (2003) The nuclear RNase III Drosha initiates microRNA processing. Nature 425:415–419
Lehmann U, Hasemeier B, Christgen M et al (2008) Epigenetic inactivation of microRNA gene hsa-mir-9-1 in human breast cancer. J Pathol 214:17–24
Lei X, Bai Z, Ye F et al (2010) Regulation of NF-kappaB inhibitor IkappaBalpha and viral replication by a KSHV microRNA. Nat Cell Biol 12:193–199

Lewis BP, Shih IH, Jones-Rhoades MW et al (2003) Prediction of mammalian microRNA targets. Cell 115:787–798

Lewis BP, Burge CB, Bartel DP (2005) Conserved seed pairing, often flanked by adenosines, indicates that thousands of human genes are microRNA targets. Cell 120:15–20

Li Y, Pei J, Xia H et al (2003) Lats2, a putative tumor suppressor, inhibits G1/S transition. Oncogene 22:4398–4405

Liu J, Carmell MA, Rivas FV et al (2004) Argonaute2 is the catalytic engine of mammalian RNAi. Science 305:1437–1441

Lo AK, To KF, Lo KW et al (2007) Modulation of LMP1 protein expression by EBV-encoded microRNAs. Proc Natl Acad Sci U S A 104:16164–16169

Lu J, Guo S, Ebert BL et al (2008a) MicroRNA-mediated control of cell fate in megakaryocyte-erythrocyte progenitors. Dev Cell 14:843–853

Lu M, Zhang Q, Deng M et al (2008b) An analysis of human microRNA and disease associations. PLoS One 3:e3420

Lu J, Verma SC, Murakami M et al (2009a) Latency-associated nuclear antigen of Kaposi's sarcoma-associated herpesvirus (KSHV) upregulates survivin expression in KSHV-associated B-lymphoma cells and contributes to their proliferation. J Virol 83:7129–7141

Lu TX, Munitz A, Rothenberg ME (2009b) MicroRNA-21 is up-regulated in allergic airway inflammation and regulates IL-12p35 expression. J Immunol 182:4994–5002

Lung RW, Tong JH, Sung YM et al (2009) Modulation of LMP2A expression by a newly identified Epstein-Barr virus-encoded microRNA miR-BART22. Neoplasia 11:1174–1184

Marshall V, Parks T, Bagni R et al (2007) Conservation of virally encoded microRNAs in Kaposi sarcoma-associated herpesvirus in primary effusion lymphoma cell lines and in patients with Kaposi sarcoma or multicentric Castleman disease. J Infect Dis 195:645–659

Mattes J, Collison A, Plank M et al (2009) Antagonism of microRNA-126 suppresses the effector function of TH2 cells and the development of allergic airways disease. Proc Natl Acad Sci U S A 106:18704–18709

Meister G, Landthaler M, Patkaniowska A et al (2004) Human Argonaute2 mediates RNA cleavage targeted by miRNAs and siRNAs. Mol Cell 15:185–197

Mishima Y, Stahlhut C, Giraldez AJ (2007) miR-1-2 gets to the heart of the matter. Cell 129:247–249

Murphy E, Vanicek J, Robins H et al (2008) Suppression of immediate-early viral gene expression by herpesvirus-coded microRNAs: implications for latency. Proc Natl Acad Sci U S A 105:5453–5458

Nachmani D, Stern-Ginossar N, Sarid R et al (2009) Diverse herpesvirus microRNAs target the stress-induced immune ligand MICB to escape recognition by natural killer cells. Cell Host Microbe 5:376–385

Nachmani D, Lankry D, Wolf DG et al (2010) The human cytomegalovirus microRNA miR-UL112 acts synergistically with a cellular microRNA to escape immune elimination. Nat Immunol 11:806–813

Nirenberg M (2008) Introduction to AAV Vector special issue. Gene Ther 15:807

Norman KL, Sarnow P (2010) Hepatitis C virus 'Achilles' heel-dependence on liver-specific microRNA miR-122. Cell Res 20:247–249

O'Donnell KA, Wentzel EA, Zeller KI et al (2005) c-Myc-regulated microRNAs modulate E2F1 expression. Nature 435:839–843

Obad S, dos Santos CO, Petri A et al (2011) Silencing of microRNA families by seed-targeting tiny LNAs. Nat Genet 43:371–378

Okamura K, Phillips MD, Tyler DM et al (2008) The regulatory activity of microRNA* species has substantial influence on microRNA and 3' UTR evolution. Nat Struct Mol Biol 15:354–363

Ota A, Tagawa H, Karnan S et al (2004) Identification and characterization of a novel gene, C13orf25, as a target for 13q31-q32 amplification in malignant lymphoma. Cancer Res 64:3087–3095

Pasquinelli AE, Reinhart BJ, Slack F et al (2000) Conservation of the sequence and temporal expression of let-7 heterochronic regulatory RNA. Nature 408:86–89

Pekow JR, Dougherty U, Mustafi R et al (2011) miR-143 and miR-145 are downregulated in ulcerative colitis: putative regulators of inflammation and protooncogenes. Inflamm Bowel Dis 18:94–100

Petersen CP, Bordeleau ME, Pelletier J et al (2006) Short RNAs repress translation after initiation in mammalian cells. Mol Cell 21:533–542

Pfeffer S, Zavolan M, Grasser FA et al (2004) Identification of virus-encoded microRNAs. Science 304:734–736

Pfeffer S, Sewer A, Lagos-Quintana M et al (2005) Identification of microRNAs of the herpesvirus family. Nat Methods 2:269–276

Pillai RS, Bhattacharyya SN, Artus CG et al (2005) Inhibition of translational initiation by Let-7 MicroRNA in human cells. Science 309:1573–1576

Randall G, Panis M, Cooper JD et al (2007) Cellular cofactors affecting hepatitis C virus infection and replication. Proc Natl Acad Sci U S A 104:12884–12889

Rao PK, Kumar RM, Farkhondeh M et al (2006) Myogenic factors that regulate expression of muscle-specific microRNAs. Proc Natl Acad Sci U S A 103:8721–8726

Reinhart BJ, Slack FJ, Basson M et al (2000) The 21-nucleotide let-7 RNA regulates developmental timing in *Caenorhabditis elegans*. Nature 403:901–906

Roberts AP, Lewis AP, Jopling CL (2011) miR-122 activates hepatitis C virus translation by a specialized mechanism requiring particular RNA components. Nucleic Acids Res 39:7716–7729

Ruby JG, Jan CH, Bartel DP (2007) Intronic microRNA precursors that bypass Drosha processing. Nature 448:83–86

Samols MA, Hu J, Skalsky RL et al (2005) Cloning and identification of a microRNA cluster within the latency-associated region of Kaposi's sarcoma-associated herpesvirus. J Virol 79:9301–9305

Samols MA, Skalsky RL, Maldonado AM et al (2007) Identification of cellular genes targeted by KSHV-encoded microRNAs. PLoS Pathog 3:e65

Sano M, Kato Y, Taira K (2006) Sequence-specific interference by small RNAs derived from adenovirus VAI RNA. FEBS Lett 580:1553–1564

Sayed D, Rane S, Lypowy J et al (2008) MicroRNA-21 targets Sprouty2 and promotes cellular outgrowths. Mol Biol Cell 19:3272–3282

Schäfer A, Cai X, Bilello JP et al (2007) Cloning and analysis of microRNAs encoded by the primate gamma-herpesvirus rhesus monkey rhadinovirus. Virology 364:21–27

Schramedei K, Morbt N, Pfeifer G et al (2011) MicroRNA-21 targets tumor suppressor genes ANP32A and SMARCA4. Oncogene 30:2975–2985

Schwarz DS, Hutvagner G, Du T et al (2003) Asymmetry in the assembly of the RNAi enzyme complex. Cell 115:199–208

Seo GJ, Fink LH, O'Hara B et al (2008) Evolutionarily conserved function of a viral microRNA. J Virol 82:9823–9828

Seo GJ, Chen CJ, Sullivan CS (2009) Merkel cell polyomavirus encodes a microRNA with the ability to autoregulate viral gene expression. Virology 383:183–187

Skalsky RL, Cullen BR (2010) Viruses, microRNAs, and host interactions. Annu Rev Microbiol 64:123–141

Skalsky RL, Samols MA, Plaisance KB et al (2007) Kaposi's sarcoma-associated herpesvirus encodes an ortholog of miR-155. J Virol 81:12836–12845

Slaby O, Svoboda M, Fabian P et al (2007) Altered expression of miR-21, miR-31, miR-143 and miR-145 is related to clinicopathologic features of colorectal cancer. Oncology 72:397–402

Small EM, Olson EN (2011) Pervasive roles of microRNAs in cardiovascular biology. Nature 469:336–342

Stenvang J, Kauppinen S (2008) MicroRNAs as targets for antisense-based therapeutics. Expert Opin Biol Ther 8:59–81

Stern-Ginossar N, Elefant N, Zimmermann A et al (2007) Host immune system gene targeting by a viral miRNA. Science 317:376–381
Sullivan CS, Grundhoff AT, Tevethia S et al (2005) SV40-encoded microRNAs regulate viral gene expression and reduce susceptibility to cytotoxic T cells. Nature 435:682–686
Sullivan CS, Sung CK, Pack CD et al (2009) Murine Polyomavirus encodes a microRNA that cleaves early RNA transcripts but is not essential for experimental infection. Virology 387:157–167
Tan S, Li R, Ding K et al (2011) miR-198 inhibits migration and invasion of hepatocellular carcinoma cells by targeting the HGF/c-MET pathway. FEBS Lett 585:2229–2234
Tang S, Bertke AS, Patel A et al (2008) An acutely and latently expressed herpes simplex virus 2 viral microRNA inhibits expression of ICP34.5, a viral neurovirulence factor. Proc Natl Acad Sci U S A 105:10931–10936
Tang S, Patel A, Krause PR (2009) Novel less-abundant viral microRNAs encoded by herpes simplex virus 2 latency-associated transcript and their roles in regulating ICP34.5 and ICP0 mRNAs. J Virol 83:1433–1442
Teichler S, Illmer T, Roemhild J et al (2011) MicroRNA29a regulates the expression of the nuclear oncogene Ski. Blood 118:1899–1902
Ugras S, Brill ER, Jacobsen A et al (2011) Small RNA sequencing and functional characterization reveals microRNA-143 tumor suppressor activity in liposarcoma. Cancer Res 71:5659–5669
Umbach JL, Cullen BR (2009) The role of RNAi and microRNAs in animal virus replication and antiviral immunity. Genes Dev 23:1151–1164
Umbach JL, Cullen BR (2010) In-depth analysis of Kaposi's sarcoma-associated herpesvirus microRNA expression provides insights into the mammalian microRNA-processing machinery. J Virol 84:695–703
Umbach JL, Kramer MF, Jurak I et al (2008) MicroRNAs expressed by herpes simplex virus 1 during latent infection regulate viral mRNAs. Nature 454:780–783
Umbach JL, Nagel MA, Cohrs RJ et al (2009) Analysis of human alphaherpesvirus microRNA expression in latently infected human trigeminal ganglia. J Virol 83:10677–10683
Umbach JL, Wang K, Tang S et al (2010) Identification of viral microRNAs expressed in human sacral ganglia latently infected with herpes simplex virus 2. J Virol 84:1189–1192
van Haaften G, Agami R (2010) Tumorigenicity of the miR-17-92 cluster distilled. Genes Dev 24:1–4
van Rooij E, Sutherland LB, Thatcher JE et al (2008) Dysregulation of microRNAs after myocardial infarction reveals a role of miR-29 in cardiac fibrosis. Proc Natl Acad Sci U S A 105:13027–13032
Vasudevan S, Tong Y, Steitz JA (2007) Switching from repression to activation: microRNAs can up-regulate translation. Science 318:1931–1934
Vasudevan S, Tong Y, Steitz JA (2008) Cell-cycle control of microRNA-mediated translation regulation. Cell Cycle 7:1545–1549
Waidner LA, Morgan RW, Anderson AS et al (2009) MicroRNAs of Gallid and Meleagrid herpesviruses show generally conserved genomic locations and are virus-specific. Virology 388:128–136
Wakiyama M, Takimoto K, Ohara O et al (2007) Let-7 microRNA-mediated mRNA deadenylation and translational repression in a mammalian cell-free system. Genes Dev 21:1857–1862
Walz N, Christalla T, Tessmer U et al (2010) A global analysis of evolutionary conservation among known and predicted gammaherpesvirus microRNAs. J Virol 84:716–728
Wang WX, Rajeev BW, Stromberg AJ et al (2008) The expression of microRNA miR-107 decreases early in Alzheimer's disease and may accelerate disease progression through regulation of beta-site amyloid precursor protein-cleaving enzyme 1. J Neurosci 28:1213–1223
Weiler J, Hunziker J, Hall J (2006) Anti-miRNA oligonucleotides (AMOs): ammunition to target miRNAs implicated in human disease? Gene Ther 13:496–502

Weinmann L, Hock J, Ivacevic T et al (2009) Importin 8 is a gene silencing factor that targets argonaute proteins to distinct mRNAs. Cell 136:496–507
White NM, Fatoohi E, Metias M et al (2011) Metastamirs: a stepping stone towards improved cancer management. Nat Rev Clin Oncol 8:75–84
Wienholds E, Plasterk RH (2005) MicroRNA function in animal development. FEBS Lett 579:5911–5922
Wightman B, Ha I, Ruvkun G (1993) Posttranscriptional regulation of the heterochronic gene lin-14 by lin-4 mediates temporal pattern formation in *C. elegans*. Cell 75:855–862
Winter J, Jung S, Keller S et al (2009) Many roads to maturity: microRNA biogenesis pathways and their regulation. Nat Cell Biol 11:228–234
Wu L, Fan J, Belasco JG (2006) MicroRNAs direct rapid deadenylation of mRNA. Proc Natl Acad Sci U S A 103:4034–4039
Xia T, O'Hara A, Araujo I et al (2008) EBV microRNAs in primary lymphomas and targeting of CXCL-11 by ebv-mir-BHRF1-3. Cancer Res 68:1436–1442
Xiao J, Yang B, Lin H et al (2007) Novel approaches for gene-specific interference via manipulating actions of microRNAs: examination on the pacemaker channel genes HCN2 and HCN4. J Cell Physiol 212:285–292
Xiong Y, Fang JH, Yun JP et al (2010) Effects of microRNA-29 on apoptosis, tumorigenicity, and prognosis of hepatocellular carcinoma. Hepatology 51:836–845
Xu N, Segerman B, Zhou X et al (2007) Adenovirus virus-associated RNAII-derived small RNAs are efficiently incorporated into the rna-induced silencing complex and associate with polyribosomes. J Virol 81:10540–10549
Yang B, Lin H, Xiao J et al (2007) The muscle-specific microRNA miR-1 regulates cardiac arrhythmogenic potential by targeting GJA1 and KCNJ2. Nat Med 13:486–491
Yao Y, Zhao Y, Xu H et al (2007) Marek's disease virus type 2 (MDV-2)-encoded microRNAs show no sequence conservation with those encoded by MDV-1. J Virol 81:7164–7170
Yao Y, Zhao Y, Xu H et al (2008) MicroRNA profile of Marek's disease virus-transformed T-cell line MSB-1: predominance of virus-encoded microRNAs. J Virol 82:4007–4015
Zhu JY, Pfuhl T, Motsch N et al (2009) Identification of novel Epstein-Barr virus microRNA genes from nasopharyngeal carcinomas. J Virol 83:3333–3341
Ziegelbauer JM, Sullivan CS, Ganem D (2009) Tandem array-based expression screens identify host mRNA targets of virus-encoded microRNAs. Nat Genet 41:130–134

Dysregulation of MicroRNA Expression and Human Diseases?

To-Ha Thai

Contents

Abstract Data from in vivo gain- and loss-of-function studies in mouse models demonstrate without a doubt that microRNAs (miRNAs), alongside coding genes, control diverse biological processes in mammalians. The regulation of miRNA expression is tightly controlled, and often the same rules and regulations that govern coding gene expression apply also to miRNAs. Similar to coding genes, altering the levels and the temporal expression of a specific miRNA clearly affects the proper development and function of the tissue where it is expressed. Therefore, it is reasonable to argue that the dysregulated control of miRNA expression would give rise to diseases as has been well established for coding genes. Along the same line of reasoning, there is an impetus to use miRNAs as diagnostic biomarkers and to develop therapeutic agents to target miRNAs for the treatment of various diseases. In this chapter, we will summarize studies establishing the physiological roles of miRNAs in mammalians and discuss whether there is a link between dysregulated

T.-H. Thai (✉)
Division of Rheumatology, Department of Medicine, Beth Israel Deaconess Medical Center, Harvard Medical School, 330 Brookline Avenue, CLS-936, Boston, MA 02115, USA
e-mail: tthai@bidmc.harvard.edu

V.A. Erdmann and J. Barciszewski (eds.), *From Nucleic Acids Sequences to Molecular Medicine*, RNA Technologies, DOI 10.1007/978-3-642-27426-8_22,

miRNA expression and diseases. In addition, we will review progresses toward targeting miRNAs as therapeutic agents.

Keywords Adaptive immunity • Cancer • Innate immunity • Lymphoproliferative diseases • MicroRNAs • Therapeutic targets

1 Introduction

MicroRNAs are small, ~21–22 nucleotide (nt) long, noncoding regulatory RNAs, first discovered in *Caenorhabditis elegans* for its role in regulating the expression of coding genes (Fig. 1) (Lee et al. 1993; Moss et al. 1997; Wightman et al. 1993). Most mammalian miRNAs are transcribed by RNA polymerase II (pol II), the same polymerase that directs the transcription of coding genes. miRNAs are encoded in intergenic regions, in sense or antisense orientation within introns of specific genes, or by noncoding transcripts. In the nucleus, miRNAs are derived from larger precursors called primary (pri)-miRNAs that are processed into 75-nt long precursor (pre)-miRNA hairpins by the RNAse III enzyme Drosha (Denli et al. 2004; Gregory et al. 2004). The hairpins are exported into the cytoplasm by exportin-5. In the cytoplasm, Dicer, an RNAse III-like enzyme, processes the pre-miRNA hairpins to generate small RNA duplexes (Grishok et al. 2001; Hutvagner et al. 2001; Ketting et al. 2001; Knight and Bass 2001). One and sometimes both of the strands will be incorporated into the RNA-induced silencing complex (RISC) or microribonucleoprotein (miRNP) complex containing Argonaute 2 (Ago 2) (Lingel et al. 2003; Song et al. 2003; Yan et al. 2003), among other proteins. The RISC/miRNP complex executes miRNA functions (Pillai et al. 2007), and miRNAs perform their functions by forming a duplex with the target gene(s) in the 3′ untranslated (UTR) region of its messenger RNA (mRNA). Depending on the degree of complementarity between the miRNA/mRNA duplex, this interaction usually leads to the downregulation of protein expression by translational repression, mRNA cleavage, or promotion

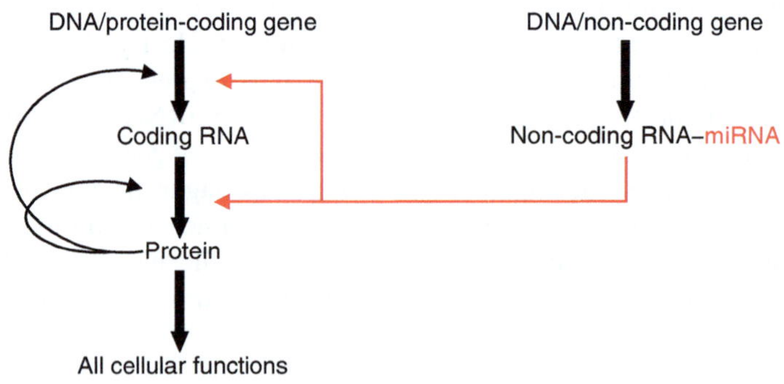

Fig. 1 miRNA functional mechanism. miRNAs are encoded by noncoding genes in the nucleus and are then exported into the cytoplasm where they regulate coding gene expression by translational repression or transcript degradation. Thus, miRNAs control complex biological processes

of mRNA decay (Fabian et al. 2010; Kim 2005). Interestingly, the interaction between mRNAs and miRNAs affects not only the translation of the target mRNA but also the miRNA stability, and this reciprocal relationship depends on the extent of base pairing between the two molecules (Ameres and Fukunaga 2010; Cazalla et al. 2010).

To date, hundreds of miRNAs have been identified in species including viruses, plants, nematodes, mice, and humans (Bartel and Bartel 2003; Berezikov and Plasterk 2005; Cullen 2007), and the number is still increasing (miRBase, http://www.mirbase.org/). Most human miRNAs are conserved in the mouse, and about one-third of *C. elegans* miRNAs have vertebrate homologues suggesting that a large fraction of miRNAs play evolutionary conserved developmental and/or functional roles.

2 MicroRNAs and the Immune System

2.1 *Effects of Global Reduction of miRNA Expression*

The first hint that miRNAs play an important role in the immune system comes from mouse studies in which the *dicer-1* or *Ago2* gene is conditionally deleted. Although the ablation of *dicer* leads to a global reduction of miRNAs and other small RNAs, it is clear that inactivation of Dicer activity, thus the expression of most if not all miRNAs, impacts the development and differentiation of lymphocytes. The effect depends on the lymphoid developmental stage at which Dicer activity is abrogated. The inactivation of Dicer activity at the double negative (DN) stage of T cell development results in a significant reduction of total thymic cellularity, including the T cell receptor (TCR) $\alpha\beta$ lineage (Cobb et al. 2005). On the other hand, the loss of Dicer activity at the double positive (DP) stage results in a reduction in $CD8^+$ and T regulatory (T_{reg}) cells. The effect on peripheral $CD4^+$ helper T (T_H) cells is small, and the resulting $CD4^+$ T_H cells preferentially express interferon-γ, an essential effector cytokine of the T_H1 lineage (Chong et al. 2008; Cobb et al. 2006; Muljo et al. 2005). It is noteworthy to point out that cell death and defective proliferation occur regardless of the stages at which Dicer activity is deleted.

Ablation of Dicer activity at the earliest stage of B cell development results in a severe block from pro- to pre-B cell transition (Koralov et al. 2008), and deletion of Ago2 from the miRNP complex results in the dysregulation of B cell and erythrocyte development (O'Carroll et al. 2007).

Hence, the question arises as to why Dicer and Ago2 activity is important for early B cell development but less so for early T cell development. One explanation is that the developmental stage at which Dicer is deleted in the B cell case may not be equivalent to that of the T cell case. It should be noted that not all miRNAs are equally ablated in both cases. It is also evident that not all T cell lineage commitment programs are subjected to Dicer control. Irrespective of the cell type examined, Dicer activity is required for cell survival and proliferation, suggesting that the defects in lymphoid development may be partially due to the reduction of a group of miRNAs controlling cell survival and proliferation. Could all these phenotypes be ascribed to miRNAs, and if so, which subsets of miRNAs are involved in each cell type and at which stage of lineage commitment? The answer awaits further studies.

Although Dicer and Ago2 are among the major factors involved in miRNA biogenesis and function, they are also involved in small interfering RNA (siRNA) and other small RNA biology. Therefore, the defects observed in the above mutant mice may not be solely due to the loss of miRNAs.

2.2 *Functional Repression of Altering Individual miRNAs*

2.2.1 Lymphocyte Development

Through several seminal genetic studies, it is apparent by 2004 that individual miRNAs exert control on the immune system (Fig. 2 and Table 1). First, it has been shown that lethally irradiated mice reconstituted with bone marrow cells overexpressing the thymus-enriched miRNA, miR-181, whose expression is dynamically regulated during T cell development, have a higher and lower percentage of B and T cells in the periphery, respectively (Chen et al. 2004). Furthermore, in vitro studies suggest that miR-181a regulates T cell development and function by modulating TCR signaling strength rather than turning on or off the signal. miR-181a appears to regulate T cell function by controlling the levels of T cell receptor (TCR)-related phosphatases (Li et al. 2007). The generation and analysis of miR-181 loss-of-function mice would be valuable to provide a more physiological flavor to this intriguing and tantalizing hypothesis.

Subsequently, works from two different groups demonstrate the important role of miRNAs in B cell development (Xiao et al. 2007; Zhou et al. 2007). Loss- and gain-of-function studies show that miR-150, a mature lymphocyte-specific miRNA, regulates B cell differentiation by controlling the transcription factor c-myb in a dose-dependent manner (Xiao et al. 2007). Therefore, miR-150 may also act as a dimmer rather than an on/off switch in the modulation of B cell development, and it achieves this delicate balance by fine-tuning the expression levels of its targets, such as c-myb, in B cells. Although c-myb plays an important role in both T and B cell development, ectopic expression or deletion of miR-150 has a stronger effect on B than T cell development. Thus, other miR-150 targets may be more crucial to T cell development.

The developmental programs in B and T cells are controlled by many transcription factors, among which some are lineage and stage specifically expressed and some are more ubiquitously expressed. It is not surprising that it is the case with miRNAs such as the miR-17~92 cluster, and mature miRNAs from this cluster are broadly but differentially expressed in various mouse tissues (Xiao et al. 2008). The miR-17~92 cluster consists of seven individual miRNAs,

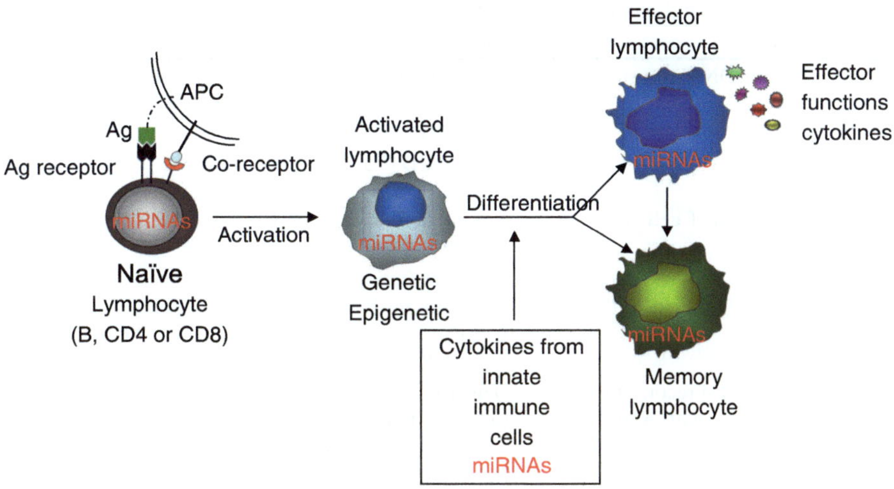

Fig. 2 miRNA control of the immune system. Several seminal studies have shown that miRNAs play important roles in the control of both the innate and adaptive immunity

miR-17-5p, miR-17-3p, miR-18, miR-19a, miR-19b1, miR-20, and miR-92, and they have been implicated in the development of lymphomas (He et al. 2005). Although born at Mendelian ratio, miR-17~92$^{-/-}$ pups die within minutes after birth with lung hypoplasia and a ventricular septal defect (Ventura et al. 2008). In addition, there is a drastic reduction in fetal pre-B cells in miR-17~92$^{-/-}$ embryos. Adult B cell development is also impaired, resulting in a reduction in all B cell subsets. Interestingly, T cell development is not affected. The pre-B cell compartment is severely compromised, while the pro-B cell population is mildly reduced even though both populations express Bim at a similar level; *Bim* is a predicted miR-17~92 target. These results imply that pre-B cell survival is more sensitive to miR-17~92 control compared to that of pro-B, and the extent to which Bim contributes to the defect observed in miR-17~92$^{-/-}$ B cell development remains to be determined.

2.2.2 Lymphocyte Function

Adaptive Immunity

Many miRNAs seem to participate in both hematopoiesis and immune responses, and the same mechanisms may be employed in both processes. In addition to the dysregulated B cell development observed, miR-150$^{-/-}$ mice also displayed augmented steady-state levels of serum immunoglobulin (Ig) A, IgG1, IgG2b, and IgM, as well as an enhanced T-dependent immune response, perhaps due to the increased c-myb levels. It has been shown that c-myb promotes lymphocyte survival by controlling the prosurvival factor Bcl2 (Taylor et al. 1996). The data suggest that

Table 1 miRNA control on the immune system in vivo

Mouse model		Phenotype	References
$Dicer^{-/-}$	DN T cell stage	↓ Thymic cellularity Substantial ↓ in TCRαβ lineage ↑ % of TCRγδ thymocytes Normal $CD4^{+}$ and $CD8^{+}$ development	Cobb et al. (2005)
	DP T cell stage	Normal thymic cellularity Slightly ↓ T cell numbers T_H1 skew (unsuppressible IFN-γ)	Cobb et al. (2006)
	Earliest B cell stage	Defect in pro-B → pre-B Rescued by transgenic expression of Ig^{HEL} BCR	Koralov et al. (2008)
$Ago2^{-/-}$	BM reconstitution with $Ago2^{-/-}$	Defective pre-B → mature B cell Erythroid hyperplasia with high-reticulocyte anemia Heinz body inclusions and splenomegaly	O'Carroll et al. (2007)
miR-181	Viral-induced overexpression	Substantial ↑ in peripheral B cell numbers Significant ↓ in $CD8^{+}$ T cells	Chen et al. (2004)
$miR\text{-}150^{-/-}$	Early embryonic deletion	Substantial ↑ in B1 cell numbers ↑ IgA, IgG1, IgG2b, IgM ↑ c-myb induction upon IgM cross-linking	Xiao et al. (2007)
$miR\text{-}17\sim92^{-/-}$	β-actin-promoted Cre-induced deletion	Drastic ↓ in fetal pre-B cells Neonatal lethality with pulmonary hypoplasia Ventricular and atrial septal defect	Ventura et al. (2008)
miR-17~92	hCD2-iCre transgene ectopic/ overexpression	Widespread lymphocytosis Lymphadenopathy/splenomegaly with follicular hyperplasia ↑ B1-a and ↓ marginal zone B cells Disruption of T-cell peripheral tolerance Hyperimmunoglobulinemia (IgM, IgG2a, IgG2b, IgG3) ↑ Titers of anti-ssDNA and anti-dsDNA	Xiao et al. (2008)
$miR\text{-}155^{-/-}$	BAC-based embryonic exon deletion	Defects in T-dependent and T-independent responses Disrupted germinal center formation containing less B cells ↑ Susceptibility to infections No defects in B cell proliferation Slight Th2 skew and disturbed Treg homeostasis	Thai et al. (2007)

(continued)

Table 1 (continued)

Mouse model		Phenotype	References
miR-155	CD21-Cre-induced B cell overexpression	↑ % of GC B cells ↑ T-dependent response	
	Retroviral induced BM reconstitution	Expansion of granulocyte/ monocyte compartment Pathologic features of acute myeloid leukemia (AML)	O'Connell et al. (2008)
	Eμ enhancer pro-B cell overexpression	High-grade polyclonal lymphoma resembling ALL	Costinean et al. (2006)
miR-223$^{-/y}$	Germline deletion	Hyperresponsive granulocytes Enhanced granulocyte fungicidal activity Hyperproliferation of granulocyte progenitor	Johnnidis et al. (2008)
miR-146a$^{-/-}$	Germline deletion	Autoimmunity Myeloproliferative manifestations Tumor development in secondary lymphoid organs	Boldin et al. (2011)
DLEU2/miR-15a/16-1$^{-/-}$	Germline and conditional deletion	B cell autonomous lymphoproliferative disorders Recapitulates the spectrum of CLL-associated phenotypes	Klein et al. (2010)

DN double negative, *DP* double positive, *BM* bone marrow, *TCR* T cell receptor, *IFN-γ* interferon gamma, *Ig* immunoglobulin, *BCR* B cell receptor, *Ago2* Argonaute 2, *ss* single stranded, *ds* double stranded, *GC* germinal center, *AML* acute myelogenous leukemia, *ALL* acute lymphoblastic leukemia, *HEL* hen egg lysozyme

miR-150 also regulates the immune response; doing so whether through c-myb or other targets remains to be determined.

One miRNA that participates in both innate and adaptive immunity is miR-155, and it does not appear to regulate lymphocyte development. *Bic*, the gene encoding miR-155, was discovered as a recurrent integration site of avian leukosis virus in chicken lymphoma cells (Tam et al. 1997). miR-155 expression is induced in macrophages and lymphocytes after stimulation (O'Connell et al. 2007; Thai et al. 2007). Dysregulated miR-155 expression results in impaired immune responses to T-dependent and T-independent antigens (Rodriguez et al. 2007; Thai et al. 2007). Both *tnf* and *lt-α* gene expression are compromised in the mutant cells, suggesting that miR-155 controls the germinal center (GC) response at least in part at the level of cytokine production. More recently, the transcription factor PU.1, which is highly expressed in GC but downregulated in post-GC B cells (Cattoretti et al. 2006), is found to be expressed in in vitro activated miR-155-deficient spleen B cells (Vigorito et al. 2007). PU.1 is also one of the predicted targets of human miR-155 (John et al. 2004). However, PU.1-deficient mice do not exhibit any defects in GC reaction nor plasma cell differentiation (Polli et al. 2005). Therefore, it is not clear what role PU.1 plays in GC reaction and how it is connected to miR-155.

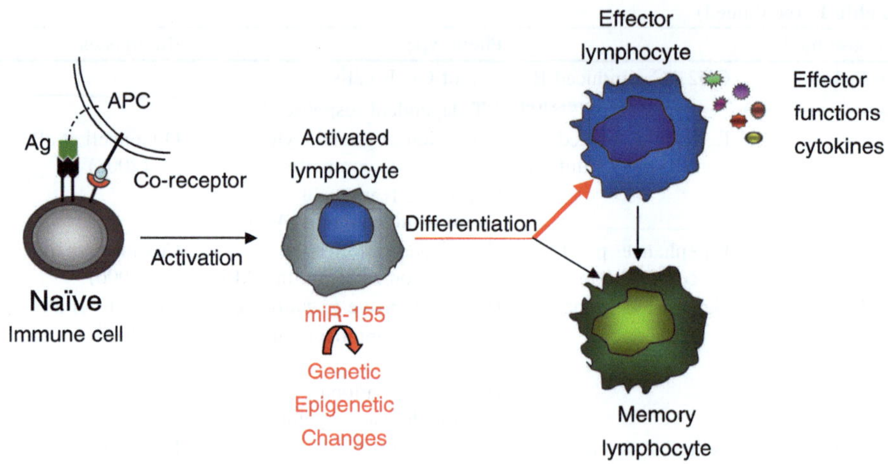

Fig. 3 Regulation of lymphocyte effector functions by miR-155. Although many genes have been identified to be putative miR-155 targets, it is conceivable that miR-155 controls a pathway(s) downstream of receptor activation, which might modulate genetic/epigenetic modifications in lymphoid effector genes

Being a true lymphoid effector miRNA, miR-155 regulates, in addition, T_H2 differentiation (Rodriguez et al. 2007; Thai et al. 2007) and T regulatory (Treg) cell homeostasis (Kohlhaas et al. 2009; Lu et al. 2009). Interestingly, the forkhead box P3 (FoxP3) transcription factor appears to regulate the expression of miR-155 in Treg cells (Marson et al. 2007; Zheng et al. 2007).

One might wonder how does miR-155 regulate B and T cell effector functions. Does it do so by controlling the expression of a yet to be identified master regulator or by modulating different targets as discussed above? Another plausible scenario would be that miR-155 controls a pathway(s) downstream of receptor activation, which might modulate genetic/epigenetic modifications in lymphoid effector genes (Fig. 3).

Innate Immunity

The role of miR-155 in innate immunity is illustrated by a recent report showing that miR-155 is induced in bone marrow (BM) cells from $Rag1^{-/-}$ mice after lipopolysaccharide (LPS) injection (O'Connell et al. 2007). Moreover, mice receiving BM cells overexpressing miR-155 have an expanded granulocyte/monocyte compartment (O'Connell et al. 2008). Additionally PU.1, among other proteins, was downregulated in the mouse macrophage cell line RAW 264.7 transduced with a miR-155-overexpressing retrovirus. It is not clear what role PU.1 plays in adaptive and innate immunity, but it is evident that miR-155 would provide the clue to decipher this previously unknown connection.

The importance of miRNAs in innate immunity is also underscored by work demonstrating that miR-223, a myeloid-specific miRNA encoded in the X chromosome, negatively regulates granulocyte lineage specification and inflammatory response (Johnnidis et al. 2008). In the absence of miR-223 (miR-223$^{-/y}$), granulocytes are hypermature and hypersensitive to stimuli, and they display enhanced fungicidal activity. As a result, miR-223$^{-/y}$ mice develop spontaneous inflammatory lung pathology and exhibit an exaggerated tissue destruction following endotoxin challenge. The mutant mice display an expanded granulocytic compartment as a result of progenitor hyperproliferation. Ablation of *Mef2c*, a transcription factor that promotes progenitor proliferation and a miR-223 target predicted with the highest confidence in miR-223$^{-/y}$ mice, does not rectify the functional defects, while the progenitor proliferation and granulocyte differentiation defects are rescued. These results suggest that miR-223 is a negative regulator of progenitor proliferation, and granulocyte differentiation and activation; in addition, the data imply that miR-223 may regulate distinct targets during granulocyte development and function.

Similar to miR-155, miR-146a is induced in human monocytes after in vitro activation by LPS, suggesting that this miRNA may be involved in mammalian microbial infection (Taganov et al. 2006). In a recent publication, the same group demonstrates that miR-146a null mice manifest an exaggerated inflammatory response to endotoxin challenge (Boldin et al. 2011). In addition, old miR-146a$^{-/-}$ mice develop spontaneous autoimmune symptoms characterized by splenomegaly, lymphadenopathy, and multiorgan inflammation that ultimately lead to premature death. miR-146a also regulates immunological tolerance by controlling the IFN-γ signaling pathway in Treg cells (Lu et al. 2010). Thus, miR-155 and miR-146a appear to govern both the innate and adaptive immune responses to foreign and endogenous antigens. The dysregulated expression of either miRNA will lead to the breakdown of immune regulation and tolerance in mice. Because of their inducible nature, miR-155 and miR-146a would be excellent targets for the treatment of inflammatory and autoimmune diseases without interfering with normal immune response.

3 MicroRNAs and Other Biological Systems

In the past few years, several in vivo gain- and loss-of-function studies have been carried out that definitively demonstrate the spectrum of biological systems controlled by miRNAs (Table 2). In the cardiovascular system, in the in vivo mouse model, the muscle-specific miR-1-2 controls cardiogenesis, myocyte cell cycle, and electrical conduction (Zhao et al. 2007). Similarly, the cardiovascular-specific miR-143/145 cluster controls arterial smooth muscle cell functions as shown in the miR-143/145 null mouse (Boettger et al. 2009; Elia et al. 2009; Xin et al. 2009).

Type II diabetes has become the world most prevalent chronic disease, and miRNAs would likely contribute to its pathogenesis. Indeed, it has been shown that miR-375 controls pancreatic α- and β-cell numbers, and genetic deletion of this miRNA in the obese mouse leads to the development of severe diabetes (Poy et al. 2009). The importance of miRNAs in diabetes is underscored by a study carried out by Jordan and colleagues who show that miR-143 regulates glucose metabolism in vivo (Jordan et al. 2011). Although these mouse studies are crucial, we have not yet identified genetic mutations in the human genome that definitively link these miRNAs in human diseases.

Table 2 miRNA control on complex biological processes in vivo

Mouse model		Phenotype	References
miR-1-2$^{-/-}$	Germline deletion	Regulates cardiogenesis Cardiac conduction Myocyte cell cycle	Zhao et al. (2007)
miR-143/145$^{-/-}$	Germline deletion	Modulates cytoskeletal dynamics and responsiveness of smooth muscle cells injury	Xin et al. (2009)
		Arterial smooth muscle contractile	Boettger et al. (2009)
		Smooth muscle cell maintenance and vascular homeostasis	Elia et al. (2009)
miR-375$^{-/-}$	Germline deletion	Regulates normal pancreatic α- and β-cell mass Severe diabetic state in obese mice	Poy et al. (2009)
miR-143 overexpression	Rosa26 targeted transgenic	Inhibits insulin-induced AKT activation Impairs glucose metabolism	Jordan et al. (2011)
miR-206$^{-/-}$	Germline deletion	Promotes neuromuscular synapses Delayed ALS progression	Williams et al. (2009)
miR-144/451$^{-/-}$	Germline deletion	Impairs late erythroblast maturation Erythroid hyperplasia Splenomegaly and mild anemia	Rasmussen et al. (2010)
miR-33$^{-/-}$	Germline deletion	Dysregulated serum HDL levels	Horie et al. (2010)
miR-290~295$^{-/-}$	Germline deletion	Embryonic lethality	Medeiros et al. (2011)

ALS amyotrophic lateral sclerosis, *HDL* high-density lipoprotein

In the nervous system, miR-206 promotes the regeneration of neuromuscular synapses in mice. The absence of this miRNA in amyotrophic lateral sclerosis (ALS) mouse model accelerates ALS progression (Williams et al. 2009). However, the role of miR-206 in the pathogenesis of ALS has not been confirmed since genetic deletion of miR-206 has not been identified in ALS patients.

4 MicroRNAs and Diseases

4.1 Cancer

There appears to be a connection between the deregulation of developmental programs and cancer. Because miRNAs play an important role in tissue development, it is of no surprise that altered expression of miRNAs has been linked to cancer. As with protein-coding genes, miRNAs can be classified as tumor suppressors or oncogenes/oncomirs.

One of the earliest indications that dysregulated miRNA expression patterns may contribute to hematopoietic malignancies comes from observations made by Calin and colleagues (Calin et al. 2002, 2004). The authors noted that a 30-kb region within human chromosome 13q14 is deleted in more than half the cases of chronic lymphocytic leukemia (CLL). This is the region from which human miR-15a/16-1 cluster is derived, and it is proposed that the miR-15a/16-1 cluster acts as a tumor suppressor, and the loss of this cluster may contribute to the development of CLL. Although intriguing, the validity of the data is called into question (Borkhardt et al. 2006; Chng 2006). Recently however, the debate as to whether miR-15a/16-1 cluster is involved in the CLL manifestation has been partially resolved with the generation of *DELU2/miR-15a/16-1* null mouse lines (Klein et al. 2010). In this study, the authors generate two mouse models. In one model, the genomic region analogous to that of human and contains several genes including the miR-15a/16-1 cluster is deleted, thus creating a mutant mouse that harbors similar genetic deletion (*ΔDELU2/miR-15a/16-1*) found in a subset of CLL patients. In another mouse, only the miR-15a/16-1 cluster is deleted. Interestingly, the *ΔDELU2/miR-15a/16-1* mouse displays a more aggressive disease phenotype compared to the miR-15a/16-1 null mouse. Thus, the results suggest that the *DELU2/miR-15a/16-1* locus contains additional tumor-suppressive genetic elements that are distinct from the miR-15a/16-1 cluster. The study also indicates that deletion of the *DELU2/miR-15a/16-1* region causes CLL in a subset of patients and that the miR-15a/16-1 cluster contributes partially to the pathogenesis of CLL. This is the first genetic evidence to directly link the dysregulated expression of miRNAs in the pathogenesis of a human disease.

The most studied presumptive oncogenic miRNA is miR-155. As discussed above, constitutive and high expression of miR-155 in collaboration with c-myc leads to an increase in a subtype of chicken B cell lymphoma (Tam et al. 1997). Studies in humans showed that miR-155 is highly expressed in most diffuse large B cell lymphomas (DLBCL), Hodgkin's lymphoma, and primary mediastinal DLBCL (Eis et al. 2005; Kluiver et al. 2005; van den Berg et al. 2003). Interestingly, miR-155 transgenic (Eμ-miR-155) mice develop a preleukemic pre-B cell proliferation evident in the spleen and the bone marrow, followed by frank B cell malignancy with no involvement of the mature B cell compartment (Costinean et al. 2006). It is not clear why these mice do not develop mature B cell lymphoma. One interesting finding that arises in Teng and Dorsett studies is that the enzyme activation-induced cytidine deaminase (AID), pivotal in class switch recombination (CSR) and somatic hypermutation (SHM) (Chaudhuri and Alt 2004; Di Noia and Neuberger 2007; Honjo et al. 2004), is a miR-155 target (Dorsett et al. 2008; Teng et al. 2008). AID is implicated in oncogenic reciprocal translocations between the *IgH* and *c-myc* loci (*c-myc/IgH*), a hazardous misstep during CSR (Muramatsu et al. 2000; Ramiro et al. 2004). Intriguingly, *c-myc/IgH* translocations increase 15-fold in activated miR-155$^{-/-}$ B cells compared to controls (Dorsett et al. 2008). These results suggest that the expression of miR-155 must be tightly and timely regulated during the GC reaction to ensure proper CSR and SHM without generating oncogenic *c-myc/IgH* translocations. They also imply that dysregulated miR-155 expression might be one of the first hits that initiate B cell lymphomagenesis during the GC reaction in the mouse. Unfortunately, to date, we have no human genetic studies that directly indicate miR-155 in a human disease, e.g., abnormal activation of the human *bic* promoter resulting in B cell lymphoma. All existing data are correlative and suggestive.

Another group of miRNAs often amplified in a subset of B cell lymphomas is the miR-17~92 cluster (Ota et al. 2004). Mice with ectopic expression of the human miR-17~92 cluster develop a generalized lymphoproliferative disease characterized by an increase in activated $CD4^+$ T cells, antigen-experienced $CD4^+$ and $CD8^+$ T cells, B1a B cells, and GC B cells (Xiao et al. 2008). These mice also develop autoimmune symptoms with elevated anti-single- or anti-double-stranded antibodies. Both *Pten* and *Bim* are predicted miR-17~92 targets, and their levels are reduced in the mutant $CD4^+$ T cells, suggesting that the expansion of lymphoid cells in these mice could in part be due to the downregulation of these two proapoptotic factors. Although one could not rule out the role of E2F1 and E2F2, both of which have been shown to be direct miR-17~92 targets (O'Donnell et al. 2005; Sylvestre et al. 2007). The miR-17~92 mutant mice do not develop a true lymphoma disease as seen in a previously published miR-17~92 transgenic mouse model generated with *Eμ-myc* bone marrow infected with retrovirus expressing a subset of miRNAs of the miR-17~92 cluster (He et al. 2005). Results from these two mouse models suggest that the level of miR-17~92 expression and the cooperation with c-myc might be important to lymphomagenesis. Therefore, the sole amplification of the miR-17~92 cluster might not be sufficient to induce B cell lymphomas; as suggested instead, a second genetic abnormality must be occurring to induce malignancy (Ota et al., 2004).

Another mouse model in which the dysregulation of a miRNA leads to the development of malignancy is the miR-146a$^{-/-}$ mouse discussed above. As a result of macrophage hyperresponsiveness to bacterial LPS, aged miR-146a$^{-/-}$ mice display myeloproliferative disorder with frank tumor infiltration in their secondary lymphoid organs. Genetic deletion of miR-146a has not been identified in human; thus, it is not clear whether the absence of miR-146a would lead to myeloid malignancy in human.

4.2 Lymphoproliferative/Autoimmune Diseases

To date, the only two miRNA mouse models in which autoimmune manifestations are observed are the miR-17~92 transgenic and the miR-146a null mice, discussed above. Although it is suggested that miR-101 might be involved in the pathogenesis of lupus-like diseases in the *sanroque* mouse, however, the authors do not provide direct genetic evidence for such a case (Yu et al. 2007). While the observations are intriguing, the final proof implicating miR-101 in lupus pathogenesis awaits the generation of a miR-101 mutant mouse model. The paucity of available miRNA mouse models for the study of autoimmune diseases such as systemic lupus erythematosus (SLE) is not due to a lack of interest in linking miRNAs to autoimmune diseases. Indeed, a few groups have identified several miRNAs that are differentially expressed in lupus patients by miRNA array assays using mostly peripheral blood mononuclear cells (PBMCs) (Dai et al. 2007; Tang et al. 2009; Pan et al. 2010; Zhao et al. 2010; Te et al. 2010). The type of miRNAs and their putative targets identified in these studies range from those associated with innate immunity (Tang et al. 2009) to inflammation (Zhao et al. 2010) and DNA methylation (Pan et al. 2010). The suggestions from the data are tantalizing, yet again the final proof waits the generation of mouse models for each of these miRNAs.

5 MicroRNAs as Potential Drug Targets

Due to their small size and potential roles in disease processes, miRNAs have been intensively exploited as potential drug targets by both academic institutions and biopharmaceutical companies. As discussed above, the fluctuation of miRNA levels is often associated with disease states. If the overexpression of a particular miRNA proved to be undesirable, then it is reasonable to propose that by reducing the level of that miRNA, one would be able to alleviate the conditions. If, however, the repression of a miRNA induced a disease state, then by replacing the same miRNA into affected cells, one would be able to rectify the defects.

Several studies have demonstrated that the introduction of a small inhibitory nucleic acid molecule, either RNA or DNA, into animal models of disease abrogates molecular pathology and, in some cases, the disease itself. To date, only three miRNAs have been effectively targeted in vivo: miR-122, miR-155, and miR-21. The injection of an anti-miRNA modified oligonucleotide (locked nucleic acids, LNA) against the liver-specific miR-122 into mice and nonhuman primates selectively and effectively lowers the level of this miRNA in the liver, reduces serum cholesterol, and resolves hepatitis C infection, and the effects are long lasting (Krutzfeldt et al. 2005; Elmen et al. 2008a, b; Lanford et al. 2010). The US Food and Drug Administration (FDA) approves a clinical phase 2a trial to assess the efficacy and tolerability of the anti-miR-122 LNA (miravirsen-SPC3649) in treatment-naïve patients with chronic hepatitis C infection. This represents the first miRNA-targeted drug to enter clinical trials. At this point, no data are available to indicate if miravirsen-SPC3649 is efficacious in humans. As discussed above, the dysregulated expression of miR-155 and miR-146a has been implicated in inflammation and cancer in mice. However, to date, no targeted drugs for these miRNAs have moved into in vivo animal models or clinical trials. One group though demonstrated that it is possible to inhibit miR-155 in vivo; however, the authors fell short in providing evidence as to whether this inhibition recapitulates the phenotype observed in the miR-155 knockout mouse (Fabani et al. 2010). Recently, miR-21 has been shown to be effectively targeted in vivo in the lupus-prone mouse model B6-*sle123* with miR-21 LNA (Garchow et al. 2011). Since miR-21 is ubiquitously and constitutively expressed, it is not clear how targeting miR-21 would ameliorate lupus-like symptoms in their study. In addition, there is no mouse or human genetic data to support the results.

In vivo miRNA replacement therapy has not been fully explored. However, one group is successful in inhibiting cancer cell proliferation, inducing tumor-specific apoptosis, and providing protection from disease progression by systemically administering miR-26a into a hepatocellular carcinoma (HCC) mouse model (Kota et al. 2009). The replacement protocol described here requires the use of adeno-associated virus to express miR-26a in vivo. It might be more advantageous to develop means to introduce miRNA mimics, thus avoiding unwanted adjuvant effects inherent in adenovirus vector.

Although results in animal models proved promising, many hurdles must be overcome before anti-miRNAs or miRNA replacement therapy could be routinely used as therapeutic agents in humans: first, the mode of delivery to achieve specificity in vivo; second, the bioavailability of these reagents; third, the potential off-target effects of these RNA molecules; and finally, the cost-effectiveness of developing large-scale RNA agents.

6 Perspective

We can conclude without any doubt that miRNAs participate, along with coding genes, in the regulation of complex biological processes in mammalians including the hematopoietic/immune, the cardiovascular, the endocrine, the nervous, and the reproductive systems.

However, there is a dearth of human genetic studies in humans to implicate the dysregulation of miRNA expression in the pathogenesis of human diseases. To date, only one study in which targeted deletion of the *LEU2/miR-15a/16-1* region faithfully recapitulates disease states found in a subset of CLL patients. However, the specific deletion of the *miR-15a/16-1* cluster only partially reproduces the CLL phenotype. What accounts for the paucity of human genetic evidence to implicate miRNAs in human diseases? Is it because most miRNAs do not act as master regulators of gene expression, but rather as modifiers of gene levels? Or most miRNAs reside in more stable genomic regions that are protected from agents that induce genetic instability? Or perhaps the direction of research carried out by human biologists precludes the identification of genetic evidence implicating miRNAs in human diseases? The answer is probably all of the above. While the first two issues are beyond our ability to address since we could not change the nature of biology, we could redirect our effort in searching for a link between miRNAs and human diseases. At this stage, the miRNA biology field does not need any more in vitro studies documenting the expression pattern of miRNAs in human diseases; what it needs are in-depth genomic and in vivo functional studies, such as the one described for the *DLEU2/miR15a/16-1* in CLL, to definitively draw a connection between miRNA expression/function and human diseases.

It is apparent from in vivo gain- and loss-of-function data that ablation or overexpression of one miRNA affects a spectrum of biological functions in both function types. Therefore, when interpreting gain- and loss-of-function results, one must take into account the fact that the observed phenotype undoubtedly results from the dysregulated expression of a network of target genes. These considerations must be included in the design of anti-miRNAs or miRNA mimics as therapeutic agents.

References

Ameres SL, Fukunaga R (2010) Riding in silence: a little snowboarding, a lot of small RNAs. Silence 1:8

Bartel B, Bartel DP (2003) MicroRNAs: at the root of plant development? Plant Physiol 132:709–717

Berezikov E, Plasterk RH (2005) Camels and zebrafish, viruses and cancer: a microRNA update. Hum Mol Genet 14(Spec No. 2):R183–R190

Boettger T, Beetz N, Kostin S et al (2009) Acquisition of the contractile phenotype by murine arterial smooth muscle cells depends on the Mir143/145 gene cluster. J Clin Invest 119:2634–2647

Boldin MP, Taganov KD, Rao DS et al (2011) miR-146a is a significant brake on autoimmunity, myeloproliferation, and cancer in mice. J Exp Med 208:1189–1201

Borkhardt A, Fuchs U, Tuschl T (2006) MicroRNA in chronic lymphocytic leukemia. N Engl J Med 354:524–525

Calin GA, Dumitru CD, Shimizu M et al (2002) Frequent deletions and down-regulation of micro-RNA genes miR15 and miR16 at 13q14 in chronic lymphocytic leukemia. Proc Natl Acad Sci USA 99:15524–15529

Calin GA, Sevignani C, Dumitru CD et al (2004) Human microRNA genes are frequently located at fragile sites and genomic regions involved in cancers. Proc Natl Acad Sci USA 101:2999–3004

Cattoretti G, Shaknovich R, Smith PM et al (2006) Stages of germinal center transit are defined by B cell transcription factor coexpression and relative abundance. J Immunol 177:6930–6939

Cazalla D, Yario T, Steitz J (2010) Down-regulation of a host microRNA by a Herpesvirus saimiri noncoding RNA. Science 328:1563–1566

Chaudhuri J, Alt FW (2004) Class-switch recombination: interplay of transcription, DNA deamination and DNA repair. Nat Rev Immunol 4:541–552

Chen CZ, Li L, Lodish HF et al (2004) MicroRNAs modulate hematopoietic lineage differentiation. Science 303:83–86

Chng WJ (2006) MicroRNA in chronic lymphocytic leukemia. N Engl J Med 354:524–525

Chong MM, Rasmussen JP, Rudensky AY et al (2008) The RNAseIII enzyme Drosha is critical in T cells for preventing lethal inflammatory disease. J Exp Med 205:2005–2017

Cobb BS, Hertweck A, Smith J et al (2006) A role for Dicer in immune regulation. J Exp Med 203:2519–2527

Cobb BS, Nesterova TB, Thompson E et al (2005) T cell lineage choice and differentiation in the absence of the RNase III enzyme Dicer. J Exp Med 201:1367–1373

Costinean S, Zanesi N, Pekarsky Y et al (2006) Pre-B cell proliferation and lymphoblastic leukemia/high-grade lymphoma in E(mu)-miR155 transgenic mice. Proc Natl Acad Sci USA 103:7024–7029

Cullen BR (2007) Immunology. Outwitted by viral RNAs. Science 317:329–330

Dai Y, Huang YS, Tang M et al (2007) Microarray analysis of microRNA expression in peripheral blood cells of systemic lupus erythematosus patients. Lupus 16:939–946
Denli AM, Tops BB, Plasterk RH et al (2004) Processing of primary microRNAs by the Microprocessor complex. Nature 432:231–235
Di Noia JM, Neuberger MS (2007) Molecular mechanisms of antibody somatic hypermutation. Annu Rev Biochem 76:1–22
Dorsett Y, McBride KM, Jankovic M et al (2008) MicroRNA-155 suppresses activation-induced cytidine deaminase-mediated Myc-Igh translocation. Immunity 28:630–638
Eis PS, Tam W, Sun L et al (2005) Accumulation of miR-155 and BIC RNA in human B cell lymphomas. Proc Natl Acad Sci USA 102:3627–3632
Elia L, Quintavalle M, Zhang J et al (2009) The knockout of miR-143 and -145 alters smooth muscle cell maintenance and vascular homeostasis in mice: correlates with human disease. Cell Death Differ 16:1590–1598
Elmen J, Lindow M, Schutz S et al (2008a) LNA-mediated microRNA silencing in non-human primates. Nature 452:896–899
Elmen J, Lindow M, Silahtaroglu A et al (2008b) Antagonism of microRNA-122 in mice by systemically administered LNA-antimiR leads to up-regulation of a large set of predicted target mRNAs in the liver. Nucleic Acids Res 36:1153–1162
Fabani MM, Abreu-Goodger C, Williams D et al (2010) Efficient inhibition of miR-155 function in vivo by peptide nucleic acids. Nucleic Acids Res 38:4466–4475
Fabian MR, Sonenberg N, Filipowicz W (2010) Regulation of mRNA translation and stability by microRNAs. Annu Rev Biochem 79:351–379
Garchow BG, Encinas OB, Leung YT et al (2011) Silencing of microRNA-21 in vivo ameliorates autoimmune splenomegaly in lupus mice. EMBO Mol Med 3:605–615
Gregory RI, Yan KP, Amuthan G et al (2004) The Microprocessor complex mediates the genesis of microRNAs. Nature 432:235–240
Grishok A, Pasquinelli AE, Conte D et al (2001) Genes and mechanisms related to RNA interference regulate expression of the small temporal RNAs that control *C. elegans* developmental timing. Cell 106:23–34
He L, Thomson JM, Hemann MT et al (2005) A microRNA polycistron as a potential human oncogene. Nature 435:828–833
Honjo T, Muramatsu M, Fagarasan S (2004) AID: how does it aid antibody diversity? Immunity 20:659–668
Horie T, Ono K, Horiguchi M et al (2010) MicroRNA-33 encoded by an intron of sterol regulatory element-binding protein 2 (Srebp2) regulates HDL in vivo. Proc Natl Acad Sci USA 107:17321–17326
Hutvagner G, McLachlan J, Pasquinelli AE et al (2001) A cellular function for the RNA-interference enzyme Dicer in the maturation of the let-7 small temporal RNA. Science 293:834–838
John B, Enright AJ, Aravin A et al (2004) Human MicroRNA targets. PLoS Biol 2:e363
Johnnidis JB, Harris MH, Wheeler RT et al (2008) Regulation of progenitor cell proliferation and granulocyte function by microRNA-223. Nature 451:1125–1129
Jordan SD, Kruger M, Willmes DM et al (2011) Obesity-induced overexpression of miRNA-143 inhibits insulin-stimulated AKT activation and impairs glucose metabolism. Nat Cell Biol 13:434–446
Ketting RF, Fischer SE, Bernstein E et al (2001) Dicer functions in RNA interference and in synthesis of small RNA involved in developmental timing in *C. elegans*. Genes Dev 15:2654–2659
Kim VN (2005) MicroRNA biogenesis: coordinated cropping and dicing. Nat Rev Mol Cell Biol 6:376–385
Klein U, Lia M, Crespo M et al (2010) The DLEU2/miR-15a/16-1 cluster controls B cell proliferation and its deletion leads to chronic lymphocytic leukemia. Cancer Cell 17:28–40

Kluiver J, Poppema S, de Jong D et al (2005) BIC and miR-155 are highly expressed in Hodgkin, primary mediastinal and diffuse large B cell lymphomas. J Pathol 207:243–249

Knight SW, Bass BL (2001) A role for the RNase III enzyme DCR-1 in RNA interference and germ line development in *Caenorhabditis elegans*. Science 293:2269–2271

Kohlhaas S, Garden OA, Scudamore C et al (2009) Cutting edge: the Foxp3 target miR-155 contributes to the development of regulatory T cells. J Immunol 182:2578–2582

Koralov SB, Muljo SA, Galler GR et al (2008) Dicer ablation affects antibody diversity and cell survival in the B lymphocyte lineage. Cell 132:860–874

Kota J, Chivukula RR, O'Donnell KA et al (2009) Therapeutic microRNA delivery suppresses tumorigenesis in a murine liver cancer model. Cell 137:1005–1017

Krutzfeldt J, Rajewsky N, Braich R et al (2005) Silencing of microRNAs in vivo with 'antagomirs'. Nature 438:685–689

Lanford RE, Hildebrandt-Eriksen ES, Petri A et al (2010) Therapeutic silencing of microRNA-122 in primates with chronic hepatitis C virus infection. Science 327:198–201

Lee RC, Feinbaum RL, Ambros V (1993) The *C. elegans* heterochronic gene lin-4 encodes small RNAs with antisense complementarity to lin-14. Cell 75:843–854

Li QJ, Chau J, Ebert PJ et al (2007) miR-181a is an intrinsic modulator of T cell sensitivity and selection. Cell 129:147–161

Lingel A, Simon B, Izaurralde E et al (2003) Structure and nucleic-acid binding of the Drosophila Argonaute 2 PAZ domain. Nature 426:465–469

Lu LF, Boldin MP, Chaudhry A et al (2010) Function of miR-146a in controlling treg cell-mediated regulation of Th1 responses. Cell 142:914–929

Lu LF, Thai TH, Calado DP et al (2009) Foxp3-dependent microRNA155 confers competitive fitness to regulatory T cells by targeting SOCS1 protein. Immunity 30:80–91

Marson A, Kretschmer K, Frampton GM et al (2007) Foxp3 occupancy and regulation of key target genes during T-cell stimulation. Nature 445:931–935

Medeiros LA, Dennis LM, Gill ME et al (2011) Mir-290–295 deficiency in mice results in partially penetrant embryonic lethality and germ cell defects. Proc Natl Acad Sci USA 108:14163–14168

Moss EG, Lee RC, Ambros V (1997) The cold shock domain protein LIN-28 controls developmental timing in *C. elegans* and is regulated by the lin-4 RNA. Cell 88:637–646

Muljo SA, Ansel KM, Kanellopoulou C et al (2005) Aberrant T cell differentiation in the absence of Dicer. J Exp Med 202:261–269

Muramatsu M, Kinoshita K, Fagarasan S et al (2000) Class switch recombination and hypermutation require activation-induced cytidine deaminase (AID), a potential RNA editing enzyme. Cell 102:553–563

O'Carroll D, Mecklenbrauker I, Das PP et al (2007) A slicer-independent role for Argonaute 2 in hematopoiesis and the microRNA pathway. Genes Dev 21:1999–2004

O'Connell RM, Rao DS, Chaudhuri AA et al (2008) Sustained expression of microRNA-155 in hematopoietic stem cells causes a myeloproliferative disorder. J Exp Med 205:585–594

O'Connell RM, Taganov KD, Boldin MP et al (2007) MicroRNA-155 is induced during the macrophage inflammatory response. Proc Natl Acad Sci USA 104:1604–1609

O'Donnell KA, Wentzel EA, Zeller KI et al (2005) c-Myc-regulated microRNAs modulate E2F1 expression. Nature 435:839–843

Ota A, Tagawa H, Karnan S et al (2004) Identification and characterization of a novel gene, C13orf25, as a target for 13q31-q32 amplification in malignant lymphoma. Cancer Res 64:3087–3095

Pan W, Zhu S, Yuan M et al (2010) MicroRNA-21 and microRNA-148a contribute to DNA hypomethylation in lupus CD4+ T cells by directly and indirectly targeting DNA methyltransferase 1. J Immunol 184:6773–6781

Pillai RS, Bhattacharyya SN, Filipowicz W (2007) Repression of protein synthesis by miRNAs: how many mechanisms? Trends Cell Biol 17:118–126

Polli M, Dakic A, Light A et al (2005) The development of functional B lymphocytes in conditional PU.1 knock-out mice. Blood 106:2083–2090
Poy MN, Hausser J, Trajkovski M et al (2009) miR-375 maintains normal pancreatic alpha- and beta-cell mass. Proc Natl Acad Sci USA 106:5813–5818
Ramiro AR, Jankovic M, Eisenreich T et al (2004) AID is required for c-myc/IgH chromosome translocations in vivo. Cell 118:431–438
Rasmussen KD, Simmini S, Abreu-Goodger C et al (2010) The miR-144/451 locus is required for erythroid homeostasis. J Exp Med 207:1351–1358
Rodriguez A, Vigorito E, Clare S et al (2007) Requirement of bic/microRNA-155 for normal immune function. Science 316:608–611
Song JJ, Liu J, Tolia NH et al (2003) The crystal structure of the Argonaute2 PAZ domain reveals an RNA binding motif in RNAi effector complexes. Nat Struct Biol 10:1026–1032
Sylvestre Y, De Guire V, Querido E et al (2007) An E2F/miR-20a autoregulatory feedback loop. J Biol Chem 282:2135–2143
Taganov KD, Boldin MP, Chang KJ et al (2006) NF-kappaB-dependent induction of microRNA miR-146, an inhibitor targeted to signaling proteins of innate immune responses. Proc Natl Acad Sci USA 103:12481–12486
Tam W, Ben-Yehuda D, Hayward WS (1997) bic, a novel gene activated by proviral insertions in avian leukosis virus-induced lymphomas, is likely to function through its noncoding RNA. Mol Cell Biol 17:1490–1502
Tang Y, Luo X, Cui H et al (2009) MicroRNA-146A contributes to abnormal activation of the type I interferon pathway in human lupus by targeting the key signaling proteins. Arthritis Rheum 60:1065–1075
Taylor D, Badiani P, Weston K (1996) A dominant interfering Myb mutant causes apoptosis in T cells. Genes Dev 10:2732–2744
Te JL, Dozmorov IM, Guthridge JM et al (2010) Identification of unique microRNA signature associated with lupus nephritis. PLoS One 5:e10344
Teng G, Hakimpour P, Landgraf P et al (2008) MicroRNA-155 is a negative regulator of activation-induced cytidine deaminase. Immunity 28:621–629
Thai TH, Calado DP, Casola S et al (2007) Regulation of the germinal center response by microRNA-155. Science 316:604–608
van den Berg A, Kroesen BJ, Kooistra K et al (2003) High expression of B-cell receptor inducible gene BIC in all subtypes of Hodgkin lymphoma. Genes Chromosomes Cancer 37:20–28
Ventura A, Young AG, Winslow MM et al (2008) Targeted deletion reveals essential and overlapping functions of the miR-17 through 92 family of miRNA clusters. Cell 132:875–886
Vigorito E, Perks KL, Abreu-Goodger C et al (2007) MicroRNA-155 regulates the generation of immunoglobulin class-switched plasma cells. Immunity 27:847–859
Wightman B, Ha I, Ruvkun G (1993) Posttranscriptional regulation of the heterochronic gene lin-14 by lin-4 mediates temporal pattern formation in *C. elegans*. Cell 75:855–862
Williams AH, Valdez G, Moresi V et al (2009) MicroRNA-206 delays ALS progression and promotes regeneration of neuromuscular synapses in mice. Science 326:1549–1554
Xiao C, Calado DP, Galler G et al (2007) MiR-150 controls B cell differentiation by targeting the transcription factor c-Myb. Cell 131:146–159
Xiao C, Srinivasan L, Calado DP et al (2008) Lymphoproliferative disease and autoimmunity in mice with increased miR-17-92 expression in lymphocytes. Nat Immunol 9:405–414
Xin M, Small EM, Sutherland LB et al (2009) MicroRNAs miR-143 and miR-145 modulate cytoskeletal dynamics and responsiveness of smooth muscle cells to injury. Genes Dev 23:2166–2178
Yan KS, Yan S, Farooq A et al (2003) Structure and conserved RNA binding of the PAZ domain. Nature 426:468–474
Yu D, Tan AH, Hu X et al (2007) Roquin represses autoimmunity by limiting inducible T-cell co-stimulator messenger RNA. Nature 450:299–303

Zhao X, Tang Y, Qu B et al (2010) MicroRNA-125a contributes to elevated inflammatory chemokine RANTES via targeting KLF13 in systemic lupus erythematosus. Arthritis Rheum 62:3425–3435

Zhao Y, Ransom JF, Li A et al (2007) Dysregulation of cardiogenesis, cardiac conduction, and cell cycle in mice lacking miRNA-1-2. Cell 129:303–317

Zheng Y, Josefowicz SZ, Kas A et al (2007) Genome-wide analysis of Foxp3 target genes in developing and mature regulatory T cells. Nature 445:936–940

Zhou B, Wang S, Mayr C et al (2007) miR-150, a microRNA expressed in mature B and T cells, blocks early B cell development when expressed prematurely. Proc Natl Acad Sci USA 104:7080–7085

Noncoding RNAs: Identification of Cancer-Associated MicroRNAs

Muller Fabbri and George A. Calin

Contents

Abstract Naturally occurring microRNAs (miRNAs) are small noncoding RNAs 19–24 nucleotides (nt) long, typically cleaved from 60- to 110-nt hairpin precursors (pre-miRNAs) that are produced from large precursors (pri-miRNAs) and are encoded in the genomes of invertebrates, vertebrates, and plants. miRNAs act as

M. Fabbri (✉)
Department of Molecular Virology, Immunology, and Medical Genetics, The Ohio State University Comprehensive Cancer Center, 1092 Biomedical Research Tower, 460 West 12th Avenue, Columbus, OH 43210, USA
e-mail: muller.fabbri@osumc.edu

G.A. Calin (✉)
Center for RNA Interference and Non-Coding RNAs, The University of Texas MD Anderson Cancer Center, Houston, TX, USA

Department of Experimental Therapeutics, The University of Texas MD Anderson Cancer Center, 1881 East Road, Houston, TX 77054, USA
e-mail: gcalin@mdanderson.org

V.A. Erdmann and J. Barciszewski (eds.), *From Nucleic Acids Sequences to Molecular Medicine*, RNA Technologies, DOI 10.1007/978-3-642-27426-8_23,

regulators of gene expression during development and differentiation at the transcriptional, posttranscriptional, and/or translational level, although most target genes have yet to be identified. Many miRNAs are conserved in sequence between distantly related organisms, suggesting that these molecules participate in essential processes. In this chapter, we present principles related to the basic and translational research that has emerged in the last decade, a period that can be truly considered the "miRNA revolution" in molecular oncology. Furthermore, we present a compendium of information about the main miRNAs that have been identified in the last several years as playing important roles in cancer. Also, we alert the reader to several additional reviews that may provide a deeper understanding of this new and exciting field of research.

Keywords Biomarker • Cancer • MicroRNA • Mutation • Noncoding RNA

1 What Are MicroRNAs?

Structurally, microRNAs (miRNAs) are 19- to 24-nucleotide (nt) long RNAs processed from much longer primary transcripts (hundreds to thousands of nucleotides) that arise from hairpin loop structures (60–110 nt) after successive enzymatic maturation steps by the ribonucleases Drosha in the nucleus and Dicer in the cytoplasm (Ambros 2008).

Functionally, miRNAs regulate gene expression in a sequence-specific fashion. Initially transcribed by RNA polymerase II as long, capped, polyadenylated precursor miRNAs (pri-miRNAs), miRNAs undergo a complex processing sequence. First, the double-stranded RNA-specific ribonuclease Drosha, in conjunction with its binding partner DiGeorge syndrome critical region gene 8 (DGCR8, or Pasha), processes pri-miRNAs into hairpin RNAs of 60–110 nt known as pre-miRNAs. Translocated from the cell nucleus to the cytoplasm by means of Exportin 5, pre-miRNA is processed by a ribonuclease III (Dicer) and transactivating response RNA-binding protein (TRBP, which binds human immunodeficiency virus 1) into a 19- to 24-nt duplex. Finally, the duplex interacts with a large protein, RNA-induced silencing complex (RISC), which includes argonaute proteins (AGO1-4 in humans). One strand of the miRNA duplex remains stably associated with RISC and becomes the mature miRNA, which guides the RISC complex mainly (but not exclusively) to the 3′-untranslated region (UTR) of the target messenger RNAs (mRNAs). Consequently, the translation and/or stability of targeted mRNAs is impaired, causing a reduction in protein expression levels (Filipowicz et al. 2008). Evidence is emerging that miRNAs' effects on gene expression may be more varied than initially proposed. For example, miRNAs can activate rather than suppress mRNA expression in particular cell-cycle conditions. Upon cell-cycle exit, *miR-363-3* and *let-7*, by recruiting and modifying specific ribonucleoproteins (RNPs) such as AGO2 and fragile X mental retardation autosomal homolog 1 (FXR1) to AU (adenosine–uridine)-rich elements inside ribonucleoproteins

(RNPs) 3′-UTRs, can activate the translation of proteins they normally repress during cell proliferation (Vatolin et al. 2006; Duursma et al. 2008; Vasudevan et al. 2007). Furthermore, miRNA-induced mRNA repression was also found to occur via binding sites located inside mRNA coding sequences, as shown for miRNAs regulating fundamental processes such as embryonic stem cell differentiation (Tay et al. 2008). Additionally, specific miRNAs that carry a distinct hexanucleotide terminal motif, such as *miR-29b*, were found to be enriched in the cell nucleus, suggesting that miRNA has additional functions in different subcellular components (Hwang et al. 2007). Some evidence bears this theory out, as it has been shown that miRNAs in the nucleus may act at the promoter level, affecting transcription; for example, *miR-373* binds to the *CDH1* promoter and stimulates transcription (Place et al. 2008).

Finally, miRNAs can directly interact with proteins; for example, the interaction of miR-368 with hnRNP E2 (heterogeneous ribonucleoprotein type E2) is independent of the miRNA seed sequence, and this interaction leads to the release of *CCAAT/enhancer binding protein alpha* (*CEBPA*) mRNA from hnRNP E2-mediated translational inhibition (Eiring et al. 2010). Each miRNA has hundreds or thousands of target genes. We have demonstrated that a specific cluster of two miRNAs (namely, the miR-15a/16-1 cluster) is able to regulate, directly and indirectly, about 14% of the whole genome in a leukemic cell model (Calin et al. 2008). Therefore, it is likely that the full coding genome is under the control of miRNAs, as suggested by evidence that these miRNA genes may be involved in any type of physiological process and pathway, including B-cell lineage fate (*miR-181*), B-cell survival (*miR-15a* and *miR-16-1*) (Cimmino et al. 2005), cell proliferation control (*miR-125b* and *let-7*), brain patterning (*miR-430*), pancreatic cell insulin secretion (*miR-375*), and adipocyte development (*miR-145*) (Negrini et al. 2009).

In this chapter, we present principles related to the basic and translational research that has emerged in the last decade, a period that can be truly considered the "miRNA revolution" in molecular oncology. Furthermore, we present a compendium of information about the main miRNAs that have been identified in the last several years as playing important roles in cancer. Also, we alert the reader to several additional reviews that may provide a deeper understanding of this new and exciting field of research.

2 Principles of miRNA Involvement in Human Cancers

The main principles of miRNA involvement in human cancers are summarized below. Such findings are important not only for scientists in general but also for oncologists, as the field of noncoding RNAs (ncRNAs) has already been shown to touch every aspect of human oncology.

2.1 miRNA Alterations Are Ubiquitous

miRNAs are involved in the pathophysiology of all types of human tumors, both benign and malignant. miRNAs differentially expressed between tumors and normal tissues have been identified in lymphomas, breast cancers, lung cancers, papillary thyroid carcinomas, glioblastomas, hepatocellular carcinomas, pancreatic tumors, pituitary adenomas, cervical cancers, prostate cancers, kidney and bladder cancers, and colorectal cancers (Calin and Croce 2006). Furthermore, miRNA alterations have been identified in other human diseases, including cardiac and autoimmune disorders and psychiatric conditions such as schizophrenia (Small et al. 2010; Edwards et al. 2010). The development of high-throughput profiling methods to detect miRNA expression in human tissues has provided invaluable tools to investigate the role of miRNAs in both physiological and pathological conditions (Liu et al. 2008). The accumulation of data has increased exponentially in the last 9 years, and those data clearly show that alterations in miRNA genes play a critical role in cancer initiation and progression (Table 1) (Spizzo et al. 2009a, b).

Recently, investigators using high-throughput profiling techniques observed a link in cytogenetically well-defined chronic lymphocytic leukemia (CLL) samples between the level of expression of *miR-34a*, a member of a miRNA family positively regulated by P53, and the level of response to DNA damage, the P53 status, and, significantly, the degree of response to fludarabine-based treatment. Low *miR-34a* expression levels were statistically significantly associated with impaired response to DNA damage, mutations in P53, and resistance to fludarabine, either with or without P53 deletion. Upregulation of *miR-34a* expression after irradiation was associated with induction of B-cell lymphoma-associated X protein (Bax) and p21 expression but not P53 upregulated modulator of apoptosis (Puma) expression (Zenz et al. 2009). These findings are straightforward and provide a new piece to the recently identified puzzle of miRNA involvement in drug resistance and sensitivity in patients with CLL.

2.2 Multiple Mechanisms Causing Abnormal miRNA Expression

The main mechanism responsible for miRNome (a term we coined for the full complement of miRNAs present in a genome) alterations in cancer cells is aberrant gene expression, which is characterized by abnormal expression of mature and/or precursor miRNA sequences in cancer cells compared with expression in cells from the corresponding normal tissues. This aberrant expression is caused by several independent or combined mechanisms, such as localization of miRNAs at cancer-associated genomic regions (CAGR) (Calin et al. 2004), epigenetic regulation of miRNA expression (Esteller 2008), or development of abnormalities in miRNA-processing genes and proteins, including mutations in Dicer, TRBP, or Exportin 5 (Melo and Esteller 2011). In various types of tumors, sometimes multiple mechanisms account for the deregulation of a specific miRNA. For example, the tumor suppressor *miR-34a* is positively controlled by TP53 (Chang et al. 2007; Fabbri et al. 2011), is regulated by MYC (Chang et al. 2008), is silenced by aberrant cytosine–guanine (CpG) methylation (Lodygin et al. 2008), and is located at 1p36 (Wei et al. 2008), a chromosomal region frequently lost in neuroblastomas.

Table 1 The most frequently dysregulated miRNAs in human cancer

MicroRNA	Dysregulation in cancer	Validated miRNA targets	Function
miR-15a and miR-16-1 cluster	Downregulated in CLL and prostate cancer	BCL2, CCND1, WNT3A, P53, MCL1	TSG/OG
let-7 family	Downregulated in lung and breast cancers	C-MYC, HMGA2, MYCN	TSG/OG
miR-34 family (miR-34a, miR-34b, miR-34c)	Downregulated in lung and pancreatic cancers	BCL2 MYCN	TSG
miR-29 family (miR-29a, miR-29b, miR-29c)	Downregulated in lung cancer, CLL, AML, breast cancer, and cholangiocarcinoma	TCL1, MCL1, DNMT3s	TSG
miR-143 and miR-145 cluster	Downregulated in colorectal cancer	ERK5, C-MYC	TSG
miR-21	Upregulated in glioblastoma, CLL, and AML and in prostate, pancreatic, gastric, colon, breast, lung, and liver cancers	PTEN, PCDC4, TPM1	OG
miR-155	Upregulated in Burkitt lymphoma, Hodgkin disease, primary mediastinal non-Hodgkin lymphoma, CLL, and AML and in lung, breast, pancreatic cancers	c-maf, C/EBPbeta, SHIP	OG
miR-17-92 cluster	Upregulated in breast, colon, lung, pancreatic, prostate, and gastric cancers and lymphomas	PTEN, BIM	OG
miR-221/222	Upregulated in glioblastoma and in thyroid, prostate, colon, pancreas, and stomach cancers	P27Kip1	OG
miR-10b	Upregulated in breast cancer	HOXD10	OG
miR-372/373	Upregulated in testicular cancer	LATS2	OG

CLL chronic lymphocytic leukemia, *AML* acute myeloid leukemia, *TSG* tumor-suppressor gene, *OG* oncogene. Gene symbol as in NCBI database

Accordingly, numerous genetic studies have identified miRNA abnormalities in human cancers by dissecting their transcriptional regulators (Chang et al. 2008; Wei et al. 2008; He et al. 2007a, b). Cancer-associated miRNAs have been located downstream of major oncogenes and tumor-suppressor genes (TSGs) with transcription factor activities; for example, P53 promotes tumorigenesis via the transactivation of all members of the *miR-34* family (He et al. 2007a, b), and MYC promotes tumorigenesis by both positively and negatively regulating the transcription of different miRNAs (e.g., the miR-17-92 cluster and let-7 family, respectively) (Chang et al. 2008). Similarly, transcription factors that govern the programming of metastatic gene expression have been found to control miRNAs also involved in metastasis, such as the pleiotropic transcription factor TWIST1 (Ma et al. 2007), which transactivates the prometastatic miR-10b, and SMAD4, which activates miR-155 downstream of transforming growth factor beta (TGF-β) signaling (Nicoloso et al. 2009).

2.3 miRNAs as Both Oncogenes and TSGs

Various mechanisms allow miRNAs to act either as oncogenes (such as miR-21 or miR-155, which causes acute B-cell leukemia in transgenic mice models) (Costinean et al. 2006) or as TSGs (such as the miR-15a/16-1 cluster, deletion of which causes CLL in knockout mice) (Klein et al. 2010). In particular, while some miRNAs act mainly as TSGs, other miRNAs are frequently overexpressed in human cancers and target TSGs, thereby exerting a tumorigenic function (Negrini et al. 2009; Fabbri et al. 2007a, b). The miR-15a/16-1 cluster, for instance, controls the expression of about 14% of all the genes in the human genome (Calin et al. 2008), and these miRNAs act as TSGs in CLL by targeting the antiapoptotic gene *BCL2* (Cimmino et al. 2005). More recently, we showed that the same miR-15a/16-1 cluster directly targets the TSG P53 in CLL (Fabbri et al. 2011), evidence that explains the indolent course of CLL in patients carrying the 13q chromosomal deletion, which affects the locus of this miRNA cluster (Calin et al. 2002). miRNAs with well-established roles as oncogenes include the miR-17-92 cluster, which is transactivated by the *c-MYC* oncogene and dramatically accelerates lymphomagenesis in murine models (He et al. 2005; Mendell 2008); miR-155, which induces leukemia in transgenic murine models (Costinean et al. 2006) and has an important function as a regulator of inflammation and immune response (Thai et al. 2007; Rodriguez et al. 2007; Tili et al. 2007); and miR-21, which targets important TSGs such as *PTEN1* (Cully et al. 2006) and *PDCD4* in several neoplasms (Meng et al. 2007; Asangani et al. 2008; Frankel et al. 2008).

In some instances, the same miRNA acts as an oncogene in one type of cancer cells and as a TSG in others because the targets and mechanisms of action differ; for example, miR-221 is hyperexpressed in liver cancers because the TSG *PTEN* is targeted (Pineau et al. 2010), but miR-221 is downregulated in erythroblastic leukemias because the c-KIT oncogene is targeted (Felli et al. 2005) (for other examples, see Table 1). Consequently, it has become evident that miRNAs can affect all the hallmarks of malignant cells: (1) self-sufficiency in growth signals (let-7 family), (2) insensitivity to antigrowth signals (miR-17-92 cluster), (3) evasion of apoptosis (miR-34a), (4) limitless replicative potential (miR-372/373 cluster), (5) angiogenesis (miR-210), and (6) invasion and metastasis (miR-10b) (Hanahan and Weinberg 2011).

In addition to the described mechanism of action wherein miRNAs can act as oncogenes by targeting TSGs or as TSGs by targeting oncogenes, miRNAs can

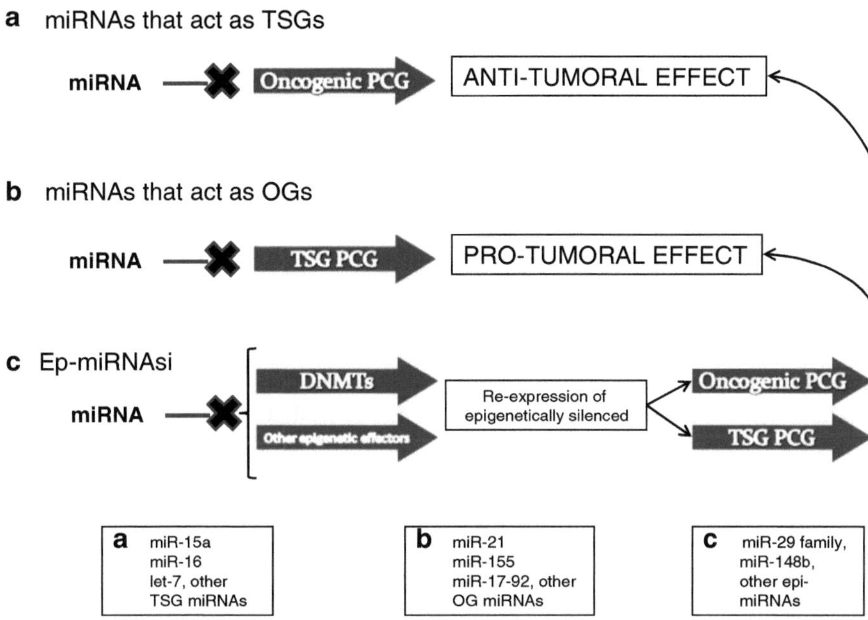

Fig. 1 Validated mechanisms of action of miRNAs. (**a**) A miRNA acts as a tumor-suppressor gene (TSG) and directly targets oncogenes (OGs); this results in an antitumoral effect. (**b**) A miRNA acts as an OG and directly targets TSGs; this results in a protumoral effect. (**c**) An epi-miRNA directly targets effectors of the epigenetic machinery [such as DNA methyltransferases (DNMTs) or other epigenetic effectors] and indirectly affects the expression of epigenetically regulated OGs and TSGs. As a consequence, both an overall antitumoral and an overall protumoral effect can occur. The three *boxes* at the *bottom* show examples of miRNAs that belong to each of the three described mechanisms of action. *PCG* protein-coding gene

actually affect gene expression through another indirect epigenetic-mediated mechanism. We have demonstrated that a family of miRNAs (namely, the miR-29 family) can directly target and reduce the expression of two key enzymes—DNA methyltransferases (DNMT) 3A and 3B—that are responsible for gene silencing by promoter hypermethylation (Fabbri et al. 2007a, b). Also, this family of miRNAs can indirectly silence DNMT1 by targeting its activator Sp1 (Garzon et al. 2009), thereby leading to a global inhibitory effect on all the main DNMTs and to a global hypomethylating effect. Consequently, we observed reexpression of the TSGs whose expression in cancer was silenced by epigenetic promoter hypermethylation, revealing that miRNAs regulating the expression of effectors of the epigenetic machinery (such as DNMTs), called "epi-miRNAs," indirectly affect the expression of epigenetically regulated genes. Several other epi-miRNAs have been identified since then, and the number continues to grow, including miRNAs that target other effectors of the epigenetic machinery (Fabbri 2008). Figure 1 summarizes the mechanisms of action of miRNAs.

2.4 Mutations Affecting the miRNA–Target Complex

Germline and somatic mutations in active precursor or primary miRNA molecules may contribute to cancer predisposition and initiation (such as the miR-15a/16 cluster mutations that occur in rare families with high incidences of both CLL and breast cancer). In the initial report of sequence variations in miRNAs, we reported on two patients diagnosed with CLL, one of whom had a family history of CLL and breast cancer, a C-T homozygous substitution in the pri-miR-16-1 at 7 nt in the 3′ direction after the end of the pre-miRNA (Calin et al. 2005). We found that this substitution was associated with lower levels of mature miR-16 production, revealing a functional impact on the processing of this miRNA. Mutation in a nearly identical location in the 3′ flanking region of miR-16-1 was described in the New Zealand black (NZB) mouse, a model for human CLL that spontaneously develops the disease when it ages (Raveche et al. 2007). In another example, scientists screened for genetic variants in 17 selected miRNA genes, which were predicted to regulate key breast cancer genes, in 42 patients with familial breast cancer (Shen et al. 2009). They identified seven new variants, two in pre-miRNAs (pre-miR-30c-1 and pre-miR-21) and five in pri-miRNA transcripts (pri-miR-17, pri-miR-24-1, pri-miR-125a, pri-miR-191, and pri-miR-125b-1). Interestingly, the variants in pre-miR-30c-1 and pri-miR-17 were rare and were only observed in noncarriers of BRCA1/2 mutations. Since miR-17 can target BRCA1 and the described variant affects the processing of miR-17 (Shen et al. 2009), miRNA genomic variations can potentially alter the regulation of key breast cancer genes.

Furthermore, polymorphisms in mRNAs targeted by miRNAs also influence cancer risk, such as the let-7 complementary single-nucleotide polymorphism (SNP) site in the KRAS 3′-UTR that was found to be significantly associated with an increased risk of non-small-cell lung carcinoma in moderate smokers (Pelletier and Weidhaas 2010; Song and Chen 2011). Genome-wide bioinformatics analysis has predicted that approximately 64% of transcribed SNPs are target SNPs that can modify (increase or decrease) the binding energy of putative miRNA–mRNA duplexes by more than 90%. To assess whether target SNPs are implicated in breast cancer susceptibility, we conducted a case-control population study and observed that germline occurrence of rs799917-BRCA1 and rs334348-transforming growth factor receptor 1 (TGFR1) significantly varied among populations with different risks of developing breast cancer (Fabbri et al. 2009; Nicoloso et al. 2010). Luciferase activity of target SNPs, allelic variants, and protein levels in cancer cell lines with different genotypes showed differential regulation of the target genes following overexpression of the two interacting miRNAs (miR-638 and miR-628-5p) (Nicoloso et al. 2010).

2.5 Profiling miRNAs: A Clinical Tool

The various methods of miRNA profiling have allowed the identification of miRNA signatures associated with diagnosis, staging, progression, prognosis, and treatment response of human cancers. For example, it was reported that a specific miRNA expression signature consisting of 13 miRNAs in human CLL was linked to disease progression from the time of diagnosis to the time of therapy (Calin et al. 2005). Similarly, aberrant expression of a specific miRNA expression signature consisting of 11 miRNAs was found to correlate well with the survival rate of patients with acute myelogenous leukemia (AML) (Garzon et al. 2008). In another study, researchers identified miRNA expression signatures associated with the incidence, prognosis, and therapeutic outcome of colon adenocarcinoma by using cancer-specific death as their endpoint; they reported that miRNAs were differentially expressed in adenomas and adenocarcinomas of the colon and that miRNA expression patterns were associated with survival. Those researchers concluded that miRNAs expression patterns are systematically altered in colon adenocarcinomas (Schetter et al. 2008).

Importantly, using miRNA pattern to classify the lesions was found to be much better at establishing the correct diagnosis in cases of metastatic cancer from an unknown primary (CUP) than using mRNA criteria. Metastatic CUP is one of the ten most frequent cancer diagnoses worldwide and constitutes 3–5% of all human malignancies. Patients with CUP present with metastases (i.e., late-stage disease) without an established primary tumor (i.e., a site wherein a therapeutically curative or palliative intervention can be performed). By analyzing 17 poorly differentiated metastatic CUP tumors with nondiagnostic histological appearance, researchers showed that the miRNA pattern was much better at establishing the correct diagnosis than was the mRNA criteria (Lu et al. 2005). This result is exciting as it shows that profiling a few hundred miRNAs has a much greater predictive power for CUP diagnosis than does profiling tens of thousands of mRNAs for primary coding genes.

2.6 *Measurement of Plasma miRNAs: A New Tool for Clinicians*

Measurement of miRNAs in body fluids such as plasma and serum may represent a gold mine of noninvasive analysis of biomarkers in cancer. It has been shown consistently that serum miRNAs remain stable after being subjected to severe conditions that would normally degrade most RNAs, such as boiling, very low or high pH levels, extended storage, and ten freeze–thaw cycles (Mitchell et al. 2008). Researchers have shown a correlation between circulating miRNA expression levels and response to a given anticancer treatment, as in the case of serum *miR-21* levels that were higher in hormone-refractory prostate cancer patients whose disease was resistant to docetaxel-based chemotherapy when compared with patients whose disease was chemosensitive (Zhang et al. 2011). Recently, a high-throughput study generated miRNA signatures from plasma samples collected 12–28 months prior to detection of lung cancer and at the time of lung cancer diagnosis (Boeri et al. 2011). In this study, 21 miRNAs were identified as risk, diagnosis, and prognosis predictors and as being potentially useful in monitoring high-risk disease-free smokers. That study was one of the first to demonstrate that specific predisease signatures of miRNA expression in plasma can predict the development of lung cancer prior to diagnosis by conventional techniques and via a noninvasive technique. Furthermore, an independent study found that plasma miR-141 may represent a novel biomarker that complements carcinoembryonic antigen (CEA) in detecting colon cancer with distant metastasis and that high levels of miR-141 in plasma were associated with poor prognosis (Cheng et al. 2011).

2.7 miRNAs as Drugs or Drug Targets

Inhibiting RNA by using miRNAs could soon represent a valid option for the treatment of specific patients (Spizzo et al. 2009a, b; Rupaimoole et al. 2011). These patients should have concordant expression between a specific miRNA and the experimentally proven targets. There would be at least two advantages to using miRNAs as therapeutic tools: (1) miRNAs are "natural" products produced in human cells (unlike chemotherapy agents or antisense oligonucleotides), thereby reducing the incidence of undesired side effects, and (2) miRNAs target multiple genes from the same pathway, and therefore, the action occurs at multiple levels in the same pathway (e.g., miR-16 targets both antiapoptotic genes *BCL2* and *MCL1*) (Cimmino et al. 2005).

Two strategies that have been proven to successfully inhibit RNA expression could be implemented to treat CLL. First, the "sandwich RNA-inhibition strategy" uses multiple agents to focus on a major molecular alteration that is clearly linked to CLL pathogenesis. Given recently published studies showing the relative efficacy of oblimersen sodium in treating relapsed or refractory CLL (Cheson 2007), treatment regimens that combine anti-BCL2 oligonucleotides and miRNAs targeting BCL2, such as *miR-15* and *miR-16*, would be feasible for treating indolent CLL. Second, the "multiplex RNA-inhibition strategy" targets various molecular defects in the same pathway, such as apoptosis. With this strategy, multiple synthetic miRNAs target the overexpressed apoptosis regulators BCL2 (*miR-15* and *miR-16*) and MCL1 (the *miR-29* family and the miR-15a/16-1 cluster) and may have a better chance of consistently and robustly reducing expression of these proteins than a single-agent therapy. The potential use of miRNAs and/or their antisense inhibitors in cancer treatment has only recently been envisioned, and clinical trials to test their efficacy as anticancer agents will certainly be scheduled soon.

3 Other ncRNAs Important in Cancer

Despite the leading role of miRNAs as cancer-related ncRNAs in published research, new categories of nontranslated RNAs have recently emerged. ncRNAs are conventionally divided into categories of long and short RNAs. Long ncRNAs include those greater than 200 nt in length; they can reach up to 100 kb (Ravasi et al. 2006; Furuno et al. 2006). Establishing the function and characterization of long ncRNAs is still under way; however, it is already clear that this heterogeneous class displays important regulatory functions, as shown in developmental processes in which ncRNAs can regulate expression of homeotic genes, oncogenes, and metabolic genes (Mattick 2009). Despite their smaller size, short RNAs are equally important in development, cell biology, and disease, and their discovery triggered a general interest in ncRNAs within the scientific community. Other ncRNAs such as large intervening ncRNAs (lincRNAs) and ultraconserved genes (UCGs) were found to be abnormally expressed in cancer and to be involved in human carcinogenesis (Gupta et al. 2010; Calin et al. 2007).

As the spectrum of ncRNAs is much larger than that of miRNAs (with estimates as high as 1,000,000 ncRNA transcripts versus only 10,000 potential miRNAs), the impact on any aspect of basic and translational cancer research is expected to be huge. It was recently discovered that a specific lincRNA, called HOTAIR, is involved in cancer metastasis (Gupta et al. 2010). The lincRNAs have many more nucleotides in their sequence than do short ncRNAs. In fact, long ncRNAs are usually composed of several hundred nucleotides, as opposed to the 20 or so nucleotides that compose short ncRNAs. The involvement of long noncoding UCGs in cancer is suggested by their frequent location in CAGRs and their aberrant expression in several human cancers (Calin et al. 2007). Signatures of deregulated UCGs are cancer specific and harbor prognostic implications. Similar to miRNAs, UCGs can also act as oncogenes or TSGs, and their expression is under the control of miRNAs, suggesting the existence of a more complex regulatory layer between different groups of ncRNAs. Another group of ncRNAs, the short germline-specific PIWI-associated RNAs, which are involved in the regulation of transposable elements and mRNAs, might have implications for human carcinogenesis, although this connection remains elusive.

4 Conclusions

There is no longer any doubt that miRNAs are involved in the regulation of pathways involved in cancer initiation, development, progression, and dissemination. The question of whether miRNAs represent the “dark side” of cancer predisposition is only beginning to be answered by studies in large populations of cancer patients. Despite a few intuitive theories in the 1960s that proposed a regulatory role for RNA in controlling gene expression through base-pair complementarity (Jacob and Monod 1961; Britten and Davidson 1969), the subsequent discovery of transcription factors quenched further research in this field. As a consequence, for many years, RNA has been regarded as the exclusive intermediary molecule between DNA and protein, with the primary role in carrying the genetic information necessary for protein synthesis. Intriguingly, only about 2% of human DNA accounts for protein-coding genes, and the total number of proteins does not vary significantly between species. On the other hand, the extent of non-protein-coding DNA, regarded for a long time as junk DNA, increases proportionally with developmental complexity (Taft et al. 2007), and over 90% (Birney et al. 2007) of the genome is actually transcribed in a developmentally regulated manner to produce ncRNA that can be intergenic, intronic, or overlapping with protein-coding transcripts (Kapranov et al. 2007; Carninci et al. 2005).

miRNAs have been identified as significant new diagnostic and prognostic tools for use in the treatment of cancer patients, and miRNA-based cancer therapy is a future option for medical oncologists. In addition, ncRNAs display precise tissue-expression patterns (Mercer et al. 2008) and are differentially expressed in pathologic conditions such as cancer and immune and heart diseases (Mattick 2009; Calin et al. 2007). A progressive understanding of the implications of ncRNAs for the malignant phenotype represents the essential background needed to achieve the goal of better treatment options for cancer patients.

Acknowledgments G.A.C. is supported as a fellow by the University of Texas MD Anderson Research Trust and as a research scholar by the University of Texas System Board of Regents and by the Chronic Lymphocytic Leukemia Global Research Foundation. Work in Dr. Calin's laboratory is supported in part by the National Institutes of Health; a Department of Defense Breast Cancer Idea Award; Developmental Research Awards from the Breast Cancer, Ovarian Cancer, Brain Cancer, Multiple Myeloma, and Leukemia Specialized Programs of Research Excellence (SPORE) grants from the National Institutes of Health; a CTT/3I-TD grant; a 2009 Seena Magowitz—Pancreatic Cancer Action Network AACR Pilot Grant; and the Arnold Foundation. M.F. is supported by a 2009 Kimmel Scholar Award. We would like to thank Sue Moreau from the Department of Scientific Publications at the University of Texas MD Anderson Cancer Center for her help with the editing of this manuscript.

References

Ambros V (2008) The evolution of our thinking about microRNAs. Nat Med 14:1036–1040

Asangani IA, Rasheed SA, Nikolova DA et al (2008) MicroRNA-21 (miR-21) post-transcriptionally downregulates tumor suppressor Pdcd4 and stimulates invasion, intravasation and metastasis in colorectal cancer. Oncogene 27:2128–2136

Birney E, Stamatoyannopoulos JA, Dutta A et al (2007) Identification and analysis of functional elements in 1% of the human genome by the ENCODE pilot project. Nature 447:799–816

Boeri M, Verri C, Conte D et al (2011) MicroRNA signatures in tissues and plasma predict development and prognosis of computed tomography detected lung cancer. Proc Natl Acad Sci USA 108:3713–3718

Britten RJ, Davidson EH (1969) Gene regulation for higher cells: a theory. Science 165:349–357

Calin GA, Croce CM (2006) MicroRNA signatures in human cancers. Nat Rev Cancer 6:857–866

Calin GA, Dumitru CD, Shimizu M et al (2002) Frequent deletions and down-regulation of micro-RNA genes miR15 and miR16 at 13q14 in chronic lymphocytic leukemia. Proc Natl Acad Sci USA 99:15524–15529

Calin GA, Sevignani C, Dumitru CD et al (2004) Human microRNA genes are frequently located at fragile sites and genomic regions involved in cancers. Proc Natl Acad Sci USA 101:2999–3004

Calin GA, Ferracin M, Cimmino A et al (2005) A MicroRNA signature associated with prognosis and progression in chronic lymphocytic leukemia. N Engl J Med 353:1793–1801
Calin GA, Liu CG, Ferracin M et al (2007) Ultraconserved regions encoding ncRNAs are altered in human leukemias and carcinomas. Cancer Cell 12:215–229
Calin GA, Cimmino A, Fabbri M et al (2008) MiR-15a and miR-16-1 cluster functions in human leukemia. Proc Natl Acad Sci USA 105:5166–5171
Carninci P, Kasukawa T, Katayama S et al (2005) The transcriptional landscape of the mammalian genome. Science 309:1559–1563
Chang TC, Wentzel EA, Kent OA et al (2007) Transactivation of miR-34a by p53 broadly influences gene expression and promotes apoptosis. Mol Cell 26:745–752
Chang TC, Yu D, Lee YS et al (2008) Widespread microRNA repression by Myc contributes to tumorigenesis. Nat Genet 40:43–50
Cheng H, Zhang L, Cogdell DE et al (2011) Circulating plasma MiR-141 is a novel biomarker for metastatic colon cancer and predicts poor prognosis. PLoS One 6:e17745
Cheson BD (2007) Oblimersen for the treatment of patients with chronic lymphocytic leukemia. Ther Clin Risk Manag 3:855–870
Cimmino A, Calin GA, Fabbri M et al (2005) miR-15 and miR-16 induce apoptosis by targeting BCL2. Proc Natl Acad Sci USA 102:13944–13999
Costinean S, Zanesi N, Pekarsky Y et al (2006) Pre-B cell proliferation and lymphoblastic leukemia/high-grade lymphoma in E(mu)-miR155 transgenic mice. Proc Natl Acad Sci USA 103:7024–7029
Cully M, You H, Levine AJ, Mak TW (2006) Beyond PTEN mutations: the PI3K pathway as an integrator of multiple inputs during tumorigenesis. Nat Rev Cancer 6:184–192
Duursma AM, Kedde M, Schrier M, le Sage C, Agami R (2008) miR-148 targets human DNMT3b protein coding region. RNA 14:872–877
Edwards JK, Pasqualini R, Arap W, Calin GA (2010) MicroRNAs and ultraconserved genes as diagnostic markers and therapeutic targets in cancer and cardiovascular diseases. J Cardiovasc Transl Res 3:271–279
Eiring AM, Harb JG, Neviani P et al (2010) miR-328 functions as an RNA decoy to modulate hnRNP E2 regulation of mRNA translation in leukemic blasts. Cell 140:652–665
Esteller M (2008) Epigenetics in cancer. N Engl J Med 358:1148–1159
Fabbri M (2008) MicroRNAs and cancer epigenetics. Curr Opin Investig Drugs 9:583–590
Fabbri M, Ivan M, Cimmino A, Negrini M, Calin GA (2007a) Regulatory mechanisms of microRNAs involvement in cancer. Expert Opin Biol Ther 7:1009–1019
Fabbri M, Garzon R, Cimmino A et al (2007b) MicroRNA-29 family reverts aberrant methylation in lung cancer by targeting DNA methyltransferases 3A and 3B. Proc Natl Acad Sci USA 104:15805–15810
Fabbri M, Valeri N, Calin GA (2009) MicroRNAs and genomic variations: from Proteus tricks to Prometheus gift. Carcinogenesis 30:912–917
Fabbri M, Bottoni A, Shimizu M et al (2011) Association of a microRNA/TP53 feedback circuitry with pathogenesis and outcome of B-cell chronic lymphocytic leukemia. JAMA 305:59–67
Felli N, Fontana L, Pelosi E et al (2005) MicroRNAs 221 and 222 inhibit normal erythropoiesis and erythroleukemic cell growth via kit receptor down-modulation. Proc Natl Acad Sci USA 102:18081–18086
Filipowicz W, Bhattacharyya SN, Sonenberg N (2008) Mechanisms of post-transcriptional regulation by microRNAs: are the answers in sight? Nat Rev Cancer 9:102–114
Frankel LB, Christoffersen NR, Jacobsen A et al (2008) Programmed cell death 4 (PDCD4) is an important functional target of the microRNA miR-21 in breast cancer cells. J Biol Chem 283:1026–1033
Furuno M, Pang KC, Ninomiya N et al (2006) Clusters of internally primed transcripts reveal novel long noncoding RNAs. PLoS Genet 2:e37
Garzon R, Garofalo M, Martelli MP et al (2008) Distinctive microRNA signature of acute myeloid leukemia bearing cytoplasmic mutated nucleophosmin. Proc Natl Acad Sci USA 105:3945–3950

Garzon R, Liu S, Fabbri M et al (2009) MicroRNA-29b induces global DNA hypomethylation and tumor suppressor gene re-expression in acute myeloid leukemia by targeting directly DNMT3A and 3B and indirectly DNMT1. Blood 114:5331–5341
Gupta RA, Shah N, Wang KC et al (2010) Long non-coding RNA HOTAIR reprograms chromatin state to promote cancer metastasis. Nature 464:1071–1076
Hanahan D, Weinberg RA (2011) Hallmarks of cancer: the next generation. Cell 144:646–674
He L, Thomson JM, Hemann MT et al (2005) A microRNA polycistron as a potential human oncogene. Nature 435:828–833
He L, He X, Lowe SW, Hannon GJ (2007a) microRNAs join the p53 network—another piece in the tumour-suppression puzzle. Nat Rev Cancer 7:819–822
He L, He X, Lim LP et al (2007b) A microRNA component of the p53 tumour suppressor network. Nature 447:1130–1134
Hwang HW, Wentzel EA, Mendell JT (2007) A hexanucleotide element directs microRNA nuclear import. Science 315:97–100
Jacob F, Monod J (1961) Genetic regulatory mechanisms in the synthesis of proteins. J Mol Biol 3:318–356
Kapranov P, Willingham AT, Gingeras TR (2007) Genome-wide transcription and the implications for genomic organization. Nat Rev Genet 8:413–423
Klein U, Lia M, Crespo M et al (2010) The DLEU2/miR-15a/16-1 cluster controls B cell proliferation and its deletion leads to chronic lymphocytic leukemia. Cancer Cell 17:28–40
Liu CG, Calin GA, Volinia S, Croce CM (2008) MicroRNA expression profiling using microarrays. Nat Protoc 3:563–578
Lodygin D, Tarasov V, Epanchintsev A et al (2008) Inactivation of miR-34a by aberrant CpG methylation in multiple types of cancer. Cell Cycle 7:2591–2600
Lu J, Getz G, Miska EA et al (2005) MicroRNA expression profiles classify human cancers. Nature 435:834–838
Ma L, Teruya-Feldstein J, Weinberg RA (2007) Tumour invasion and metastasis initiated by microRNA-10b in breast cancer. Nature 449:682–688
Mattick JS (2009) The genetic signatures of noncoding RNAs. PLoS Genet 5:e1000459
Melo SA, Esteller M (2011) A precursor microRNA in a cancer cell nucleus: Get me out of here! Cell Cycle 10:922–925
Mendell JT (2008) miRiad roles for the miR-17-92 cluster in development and disease. Cell 133:217–222
Meng F, Henson R, Wehbe-Janek H et al (2007) MicroRNA-21 regulates expression of the PTEN tumor suppressor gene in human hepatocellular cancer. Gastroenterology 133:647–658
Mercer TR, Dinger ME, Sunkin SM et al (2008) Specific expression of long noncoding RNAs in the mouse brain. Proc Natl Acad Sci USA 105:716–721
Mitchell PS, Parkin RK, Kroh EM et al (2008) Circulating microRNAs as stable blood-based markers for cancer detection. Proc Natl Acad Sci USA 105:10513–10518
Negrini M, Nicoloso MS, Calin GA (2009) MicroRNAs and cancer—new paradigms in molecular oncology. Curr Opin Cell Biol 21:470–479
Nicoloso MS, Spizzo R, Shimizu M et al (2009) MicroRNAs—the micro steering wheel of tumour metastases. Nat Rev Cancer 9:293–302
Nicoloso MS, Sun H, Spizzo R, Kim H, Wickramasinghe P, Shimizu M, Wojcik SE, Ferdin J, Kunej T, Xiao L, Manoukian S, Secreto G, Ravagnani F, Wang X, Radice P, Croce CM, Davuluri RV, Calin GA (2010) Single-nucleotide polymorphisms inside microRNA target sites influence tumor susceptibility. Cancer Res 70(7):2789–98
Pelletier C, Weidhaas JB (2010) MicroRNA binding site polymorphisms as biomarkers of cancer risk. Expert Rev Mol Diagn 10:817–829
Pineau P, Volinia S, McJunkin K et al (2010) miR-221 overexpression contributes to liver tumorigenesis. Proc Natl Acad Sci USA 107:264–269
Place RF, Li LC, Pookot D et al (2008) MicroRNA-373 induces expression of genes with complementary promoter sequences. Proc Natl Acad Sci USA 105:1608–1613

Ravasi T, Suzuki H, Pang KC et al (2006) Experimental validation of the regulated expression of large numbers of non-coding RNAs from the mouse genome. Genome Res 16:11–19
Raveche ES, Salerno E, Scaglione BJ et al (2007) Abnormal microRNA-16 locus with synteny to human 13q14 linked to CLL in NZB mice. Blood 109:5079–5086
Rodriguez A, Vigorito E, Clare S et al (2007) Requirement of bic/microRNA-155 for normal immune function. Science 316:608–611
Rupaimoole R, Han HD, Lopez-Berestein G et al (2011) MicroRNA therapeutics: principles, expectations, and challenges. Chin J Cancer 30:368–370
Schetter AJ, Leung SY, Sohn JJ et al (2008) MicroRNA expression profiles associated with prognosis and therapeutic outcome in colon adenocarcinoma. JAMA 299:425–436
Shen J, Ambrosone CB, Zhao H (2009) Novel genetic variants in microRNA genes and familial breast cancer. Int J Cancer 124:1178–1182
Small EM, Frost RJ, Olson EN (2010) MicroRNAs add a new dimension to cardiovascular disease. Circulation 121:1022–1032
Song FJ, Chen KX (2011) Single-nucleotide polymorphisms among microRNA: big effects on cancer. Chin J Cancer 30:381–391
Spizzo R, Rushworth D, Guerrero M et al (2009a) RNA inhibition, microRNAs, and new therapeutic agents for cancer treatment. Clin Lymphoma Myeloma 9:S313–S318
Spizzo R, Nicoloso MS, Croce CM et al (2009b) SnapShot: MicroRNAs in cancer. Cell 137:586–e1
Taft RJ, Pheasant M, Mattick JS (2007) The relationship between non-protein-coding DNA and eukaryotic complexity. Bioessays 29:288–299
Tay Y, Zhang J, Thomson AM et al (2008) MicroRNAs to Nanog, Oct4 and Sox2 coding regions modulate embryonic stem cell differentiation. Nature 455:1124–1128
Thai TH, Calado DP, Casola S et al (2007) Regulation of the germinal center response by microRNA-155. Science 316:604–608
Tili E, Michaille JJ, Cimino A et al (2007) Modulation of miR-155 and miR-125b levels following lipopolysaccharide/TNF-alpha stimulation and their possible roles in regulating the response to endotoxin shock. J Immunol 179:5082–5089
Vasudevan S, Tong Y, Steitz JA (2007) Switching from repression to activation: microRNAs can up-regulate translation. Science 318:1931–1934
Vatolin S, Navaratne K, Weil RJ (2006) A novel method to detect functional microRNA targets. J Mol Biol 358:983–996
Wei JS, Song YK, Durinck S et al (2008) The MYCN oncogene is a direct target of miR-34a. Oncogene 27:5204–5213
Zenz T, Mohr J, Eldering E et al (2009) miR-34a as part of the resistance network in chronic lymphocytic leukemia. Blood 113:3801–3808
Zhang HL, Yang LF, Zhu Y et al (2011) Serum miRNA-21: elevated levels in patients with metastatic hormone-refractory prostate cancer and potential predictive factor for the efficacy of docetaxel-based chemotherapy. Prostate 71:326–331

Targeting Non-coding RNAs for Cancer Therapy

Satya K. Kota and Savithri Balasubramanian

Contents

Abstract Cancer remains one of the leading causes of morbidity and mortality, with limited progress in the development of cancer cell-specific, targeted, long-term therapeutic strategies. In the past decade, non-coding RNAs have emerged as novel and key modulators of cellular gene expression influencing both the transcriptional and post-transcriptional output. Seminal contributions of non-coding RNAs during development

S.K. Kota (✉)
Institute of Molecular Genetics, CNRS, UMR5535, 1919 route de Mende, 34293 Montpellier, Cedex 5, France
e-mail: satya.kota@igmm.cnrs.fr; keerthires@gmail.com

S. Balasubramanian
Department of Medicine, The Transplant Institute, Beth Israel Deaconess Medical Center and Harvard Medical School, Boston, MA 02115, USA

V.A. Erdmann and J. Barciszewski (eds.), *From Nucleic Acids Sequences to Molecular Medicine*, RNA Technologies, DOI 10.1007/978-3-642-27426-8_24,

and disease have been described in many eukaryotic species including humans. To date, several different classes of non-coding RNA have been identified and can be broadly categorized mainly based on their size, structural characteristics, and proteins involved in their biogenesis. Recently, microRNAs, a small non-coding RNA species, and various long non-coding RNAs have been implicated and studied widely in multiple aspects of cancer initiation, progression, and metastasis. Understanding the mechanistic aspects of non-coding RNAs might hold the key, not only to decipher what goes awry in cancer but also most importantly to devise novel targeted therapeutic strategies in the coming years. In this chapter, we will provide a brief overview of non-coding RNAs, with emphasis on those RNAs whose levels are modulated during carcinogenesis.

Keywords Non-coding RNA • Small non-coding RNA • Long non-coding RNA • Cancer • MicroRNA • piwiRNA • RNA interference • Chromatin modification • Translational repression • mRNA degradation • Antisense oligonucleotide • miRNA mimics • Polycomb proteins

1 Introduction

Our understanding of RNA as a cellular mediator of gene function has undergone profound changes in the past couple of decades with the discovery of a large number of small and long non-coding RNA species, now collectively called the non-coding RNA (ncRNA). The ncRNA family is comprised of a number of different small RNA species such as siRNA, miRNA, tasiRNA, natsiRNA, tncRNA, scnRNA, rasiRNA, and piRNA that are between 21 and 200 nucleotides in length (Farazi et al. 2008) and a growing list of long noncoding RNAs (lncRNA) of size more than 200 nucleotides, including T-UCR and lincRNA (Esteller 2011), which differ widely in their size, expression pattern, biogenesis, effector mechanisms, etc. (Brosnan and Voinnet 2009). Of these ncRNAs, the role of small ncRNAs in fine-tuning gene expression by mRNA destabilization and translational and transcriptional repression has been well established (Valencia-Sanchez et al. 2006), while the understanding of lncRNA is still evolving with a growing consensus that ncRNAs play a vital role in development and disease.

Among the small ncRNAs, the naturally occurring ~22-nt-long small RNA species, namely, microRNAs (miRNA), have gathered widespread attention of late, because of their prevalence of expression, differential expression patterns in normal and disease conditions such as cancer and the anticancer therapeutic potential they offer (Kota and Balasubramanian 2010; Kasinski and Slack 2011). In fact, the first evidence for the existence and role of small RNAs came from the elegant study by Lee et al. who reported that the *Caenorhabditis elegans* gene, lin-4, coded for a small antisense RNA complementary to a developmentally regulated protein-coding gene lin-14 (Lee et al. 1993). The aforementioned report of microRNA preceded the actual discovery of the phenomenon of RNA interference later in 1999 (Fire et al. 1998). Since then, several miRNA encoding genes have been cloned and shown to be evolutionarily conserved from plants to humans and as negative regulators of gene expression (Pasquinelli et al. 2000). miRNAs negatively regulate gene expression by binding to perfect or nearly perfect complementary sequences present at the 3′ UTR of target genes, leading to translational repression and/or mRNA destabilization (Guo et al. 2010b). While the initial studies reported the blockade of protein translation as the primary mechanism of action of miRNA, subsequent high-throughput studies indeed identified a significant reduction in the target mRNA levels as well.

The other well-studied class of small ncRNA is siRNA which are ~21 nt long and can be endogenous as in the case of *Drosophila melanogaster* (Czech et al. 2008), mouse oocytes (Watanabe et al. 2008), and embryonic stem cells (Babiarz et al. 2008) or virally derived. siRNAs are the most utilized in experimental down-regulation of candidate gene expression, delivered either directly as mature siRNAs or as shRNA plasmids encoding the hairpin loop of the precursor small RNA. By and large, siRNAs regulate gene expression by target mRNA degradation by binding to perfect complementary sequence in the coding region of the target mRNA (Farazi et al. 2008).

tasiRNAs (trans-acting siRNA) are ~21–22 nt long and are generated from the noncoding regions of the genome. tasiRNAs are found predominantly in plants and are implicated in development. Originating from double stranded RNA precursors (dsRNA), tasiRNAs are 2′O methylated at the 3′ end and mediate target mRNA cleavage (Farazi et al. 2008).

natsiRNAs (natural antisense siRNA) are ~21–22-nt-long endogenous small RNAs and are generated from overlapping sense and antisense transcripts (Farazi et al. 2008).

hcRNAs (heterochromatic small RNA) are endogenous small RNAs, ~24 nt long, and are derived from repeat-associated genomic regions. They are implicated in heterochromatic regulation in plants and fungi. hcRNA association with Ago1, 4, and 6 as well as with chromatin modifiers such as DNA methyltransferase and SNF2-like chromatin remodeling proteins has been found (Farazi et al. 2008).

piRNAs (piwi-interacting small RNA) are 28–33-nt-long small RNAs that are evolutionarily conserved and involve the piwi clade of Argonaute proteins in their effector complexes. piRNAs are processed from single stranded primary transcripts, and mammalian piRNAs are involved in transposon regulation. In mammals, piRNAs are thought to silence the selfish DNA elements. piRNAs carry 2′-*O*-methyl modification at their 3′ end. piRNAs are widely reported in flies and mouse germ cells.

Figure 1 depicts the schematic of various noncoding RNA classes that are implicated in cancer progression.

The biology of various classes of lncRNAs and their role in cancer are discussed in Sect. 5.

2 Molecular Mechanisms of Small RNA Biogenesis and Action

The biogenesis of miRNAs and siRNAs shares considerable similarities in terms of the enzymes and compartments involved (Kim et al. 2009). Around 50% of mammalian miRNAs are found to be in clusters and therefore transcribed as single

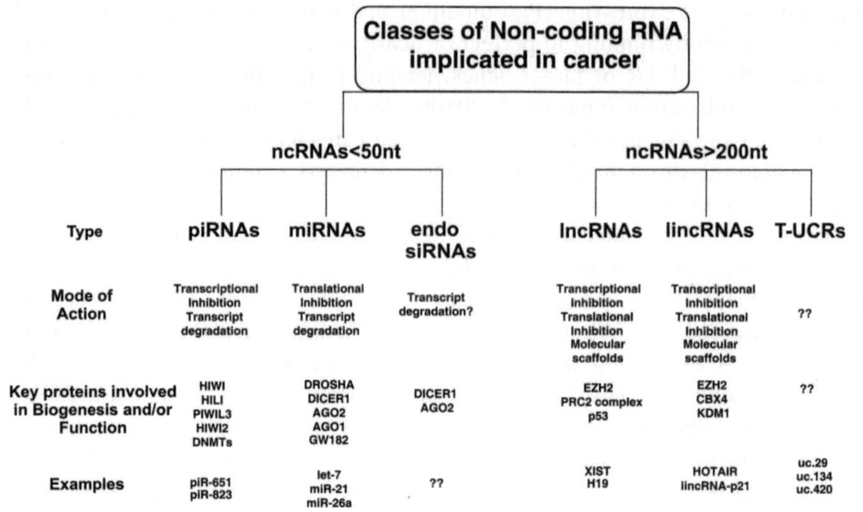

Fig. 1 Depiction of various classes of ncRNA implicated in cancer classified on the basis of their size, biogenesis, mechanisms of action, etc.

polycistronic transcription units. The transcription units that give rise to miRNAs could be either protein-coding or noncoding units. The majority of the miRNA coding sequences are found in the intronic regions both in the protein-coding and non-coding transcription units, whereas only ~10% are encoded within the exonic region of the noncoding transcription units. miRNAs are transcribed by RNA polymerase II (Pol II) as capped, polyadenylated large RNA precursors, whereas a minor set of miRNAs that are associated with certain repeat elements are transcribed by RNA polymerase III. The product of Pol II is several kilobases long containing stem loop structures and is called primary miRNA (pri-miRNA). The pri-miRNA is processed in the nucleus by a RNase III enzyme called Drosha, resulting in the formation of a small hairpin RNA termed precursor miRNA (pre-miRNA). Drosha requires the cofactor protein Pasha in Drosophila or DGCR8 in humans together forming a large microprocessor complex. Pre-miRNA that arises from Drosha processing is 70 nt long which forms imperfect stem–loop structures. Following the nuclear processing, these pre-miRNAs are transported into the cytoplasm via the exportin-5 complex, which are subsequently processed by another RNase III enzyme namely, Dicer, resulting in the formation of ~22-nt-long mature miRNA. The action of miRNA occurs in a multiprotein complex named miRNA-associated RNA-induced silencing complex (miRISC), wherein the double stranded mature miRNAs are bound by a protein complex including Ago and dsRNA binding proteins such as TRBP. In the miRISC, one strand is selected to be the guiding strand or the miRNA, and the other strand is degraded; this is determined purely based on the relative thermodynamic stability of the miRNA strands. The guiding strand is selectively retained on the RISC complex, which then directs the binding of miRNA to its complementary sequence at the 3′

UTR of the target transcript. It is important to mention the regulation of miRNA biogenesis. There are several rate-limiting steps involved in the generation and maintenance of cellular levels of miRNA. Transcription factors play a critical role in the induction of miRNA transcription as in the case of p53-induced miR-34 family and MyoD-mediated induction of miR-1 and miR-133. Post-transcriptional regulation of miRNA occurs at the level of Drosha- and Dicer-mediated miRNA maturation as well as factors altering the turnover of miRNA.

The siRNA biogenesis pathway involves certain unique processes, proteins, and RNA intermediates resulting in the generation of ~21-nt single-stranded RNA. Numerous endogenously produced siRNAs are derived from transposon transcripts or sense–antisense transcripts or long stem–loop structures, and they also arise from convergent transcription from the protein-coding regions. Endo-siRNAs are shown to associate with Ago2. In Drosophila, involvement of the RNA-dependent RNA polymerase (RdRP) enzyme has been reported in the generation of the precursor molecule. Similar to miRNAs, the RNAse III enzyme Dicer plays an important role in the generation of the effector siRNAs, but instead of Dicer 1, Dicer 2 is the major isoform involved in both exo- and endo-siRNA generation. Ago2 protein has been implicated in exogenously derived siRNAs (exo-siRNA) that are processed from long double stranded RNA (dsRNA). Interestingly, some endo-siRNAs are shown to arise from piRNA-generating loci and are 2′-*O*-methyl modified at the 3′ end (Kim et al. 2009).

piRNAs are generated from intergenic repetitive elements called piRNA clusters in flies, and piwi-associated piRNAs are preferentially transcribed from the antisense strand. piRNA production in flies does not require Dicer but instead depends on Aubergine or piwi proteins to generate the 5′ end of the sense strand of piRNA. Further, Ago3 cleaves the 5′ end of the antisense transcript, while factors required for the generation of the 3′ end are still not clear. In mouse, a piRNA generation is reported during prepachytene and pachytene stages of meiosis and is derived from repeat and transposon-rich clusters. The biogenesis of mammalian piRNA shares some similarities with that of flies and is shown to associate with the piwi proteins such as MILI, MIWI2, and MIWI (Kim et al. 2009).

3 MicroRNA and Cancer: Lessons Learned from Model Organisms to Humans- From Diagnosis to Therapy

Past several years have seen leaps and bounds in the identification, validation, and targeting of a number of cancer-specific miRNAs. These cancer-specific miRNAs, widely referred to as OncomiRs, uniquely identify cancer tissues from their normal counterparts (Esquela-Kerscher and Slack 2006). Several studies point that there is a global down-regulation of miRNAs (tumor suppressor) in cancer cells compared to the normal counterparts, while the upregulation of a number of oncogenic miRNAs has also been reported. Several families of miRNAs such as let-7, miR-34, miR-17-92, and miR-21 are modulated up or down in various cancer types.

The distinctive modulation of level of these miRNAs in cancer cells led to reliable profiling of cancer tissues, thus emerging as valuable tool for cancer diagnosis and prognosis (Kota and Balasubramanian 2010; Kasinski and Slack 2011 and references therein).

Detailed study of the biology of various human OncomiRs has enabled devising strategies to modulate or correct their levels in order to stall or reverse cancer progression in vivo in mice models. The results from small animal models targeting miRNA levels promise an immediate translation into a powerful modality of treating human cancer when coupled with research to enhance biosafety, bioavailability, and pharmacokinetics of miRNA and/or -pathway targeting molecules. The miRNA targeting approach is comprised of (1) miRNA inhibition therapy using antisense oligonucleotides against oncogenic miRNAs (Ma et al. 2010a) and (2) miRNA replacement therapy by supplementation of tumor suppressor miRNAs (Kota et al. 2009; Ibrahim et al. 2011). As many cancer tissues are marked by specific up- or down-regulation of certain miRNAs compared to the normal tissues, anti-miRNA oligonucleotides (AntagomiRs) against oncogenic miRNAs that are upregulated in cancers and miRNA replacement therapy for tumor suppressor miRNAs that are down-regulated specifically in tumors are therefore to be well tolerated by normal tissues without any adverse effect. As more details emerge from individual miRNAs and their role in oncogenesis, progression, or suppression as well as their interaction network in individual cancer type, miRNA-based cancer therapy is likely to be a far-reaching modality for cancer therapy in the near future.

Not only the levels of miRNAs reflect the oncogenic status of a cell, but the levels of small RNA processing enzymes Dicer and Drosha are also of diagnostic value (Karube et al. 2005; Merritt et al. 2008). Similar to the protein-coding oncogenes and tumor suppressors, genetic mutations and polymorphisms have been detected in the coding sequences of OncomiRs as well as the binding sites in the target genes and have revealed prognostic details in various cancer types (Hu et al. 2009; Yang et al. 2010). Thus, expression levels of miRNA pathway genes and allelic information about miRNAs and their targets are emerging as useful tools to predict cancer risk, cancer grade, therapeutic responsiveness, and overall prognosis of the disease. In the futuristic approach, validated miRNAs and their targets could serve as indicators in the cancer preventive screening measures.

4 Emerging Topics in OncomiR Study

A new concept, namely, miRNA sponges, is gaining momentum as an effective way to quench the higher levels of oncogenic miRNAs (Ebert and Sharp 2010; Ma et al. 2010b). Several reports suggest that natural miRNA sponges do exist and are derivatives of viral genome (Cazalla et al. 2010) or pseudogenes of tumor suppressors (Poliseno et al. 2010). For targeting cancer-specific miRNAs using sponges that serve as decoys, specific miRNA target sequences are cloned in multiple copies in tandem at the 3′ UTR of a transcript that is driven by a strong promoter. Essentially, such a construct upon delivered into a tumor cell should act as a sponge for the respective miRNA that is targeted and relieve its natural target(s) from repression.

Along the lines of miRNA sponges, a recently emerging concept in the study of miRNAs in cancer biology is the competitive endogenous mRNAs (ceRNA) (Salmena et al. 2011). It is a novel phenomenon wherein endogenous mRNAs compete for the miRNAs which also target tumor suppressors (Karreth et al. 2011; Tay et al. 2011). These studies were based on PTEN, a commonly mutated tumor suppressor gene implicated in a number of cancers including prostate cancer and melanoma. Noncoding RNAs derived from pseudogenes of PTEN as well as endogenous protein-coding transcripts such as ZEB2 acted as decoys of miR-181, miR-200b, miR-25, and miR-92a. These miRNAs were found to target both PTEN and ZEB2. Silencing of ZEB2 by RNA interference resulted in the augmented repression of PTEN by the aforementioned miRNAs. Thus, the transcript levels of protein-coding genes could modulate the transcript and protein levels of other genes in a cell via modulating the activity of miRNA(s) common to both (Fig. 2). In principle, ceRNAs regulating oncogenes might also exist. These studies provide yet another handle for cancer therapy and prognosis.

These reports on ceRNA trace back to the initial hypothesis, namely, "target mimicry," first discovered in plants (Franco-Zorrilla et al. 2007). In this study, endogenous mRNA levels of IPS1 (induced by phosphate starvation 1) altered the protein levels of PHO2 (phosphate2), by modulating the effects of miR399 on the stability and translation of PHO2. Messenger RNAs like IPS1 were termed as target mimics which are the equivalents of ceRNAs. These findings reiterate that (1) miRNAs play a critical role in various aspects of cellular functions through complex regulatory mechanisms and (2) the biological pathways involving miRNA are highly conserved.

5 Long Non-coding RNAs and Their Roles in Cancer

High-throughput studies on transcriptomes have revealed a striking feature that majority of the mammalian genome is transcriptionally active and yet only $<2\%$ of it is protein coding. Only recently, the other spurious or regulated transcripts including several long noncoding RNAs (lncRNA) (>200 bases in length) have been described as crucial regulators of a myriad of physiological processes including cell cycle. In animal models, these lncRNAs have been shown to play critical roles during embryonic development, and misregulation of lncRNA levels has been reported in several types of cancers. In majority of the cases, lncRNA are species-specific and appear to have regulatory roles both *cis* and *trans* of their sites of origin. lncRNAs are known to regulate gene expression at both the transcriptional and post-transcriptional levels, and detailed functional roles of these transcripts are yet to be deduced. More interestingly, lncRNA-mediated gene regulation could either be repressive or activating. In the following section, we will discuss briefly a few examples of lncRNAs, their mode of action, and additional protein players

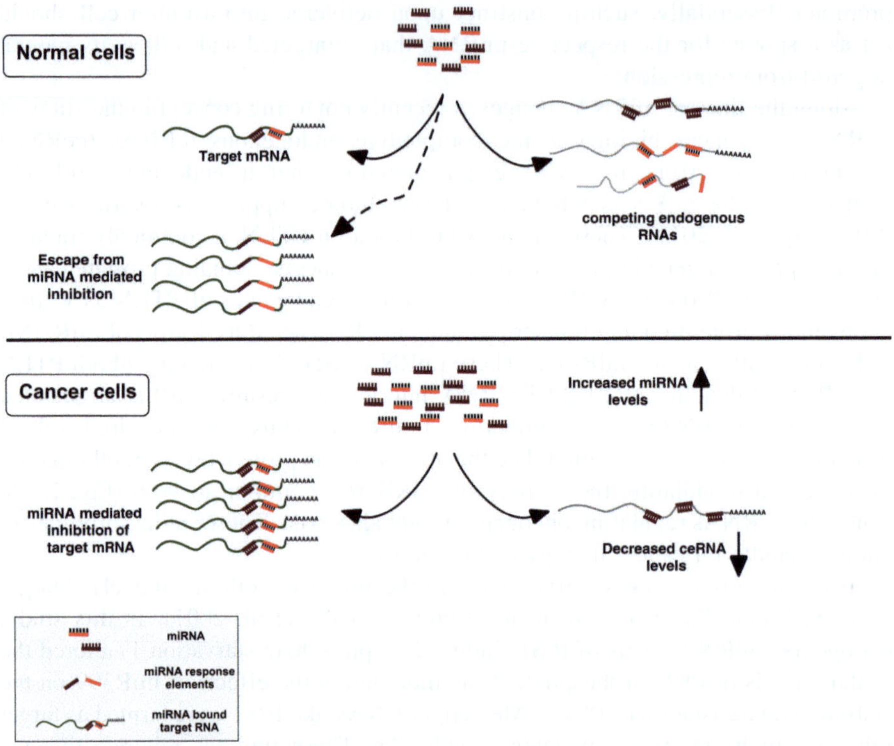

Fig. 2 MicroRNA-mediated transcriptional and translational inhibition of protein-coding genes is regulated by competing endogenous RNAs (ceRNAs) that share homologous miRNA binding sequence/s or microRNA response elements (MRE) forming ceRNA network/s. Increased levels of specific microRNAs or decreased levels of ceRNAs or both in cancer cells might lead to increased targeting of important transcripts including tumor suppressors by miRNAs leading to decreased protein levels of miRNA targets

involved, with emphasis on their role in human cancers and potential uses of these ncRNAs both as diagnostic as well as therapeutic targets.

5.1 X Inactive-Specific Transcript

X inactive-specific transcript (XIST) is one of the earliest and well-studied lncRNA in higher mammals in the context of gene dosage compensation and of late implicated in the pathology of breast and ovarian cancers. Compared to one X chromosome in male mammals, the presence of two X chromosomes in females poses a serious imbalance of X-inked gene dosage and is countered by silencing of one of the X chromosomes in all the cells of female mammals (Avner and Heard 2001). In mice and humans, this process of X chromosome silencing is known as X inactivation and is orchestrated by a region within the X chromosome known as X inactivation center (Augui et al. 2011). This region is known to be transcriptionally active, giving rise to several noncoding transcripts, and among them the functionally well-studied and crucial RNA is XIST (X inactive-specific transcript). This lncRNA of ~17 kb in length with certain conserved repeat elements is transcribed only from the inactive X chromosome and localizes to that chromosome in *cis* (Brown et al. 1992). Several chromatin modifiers/modifications, like polycomb group proteins, H3-K9 methylation, incorporation of histone variant, macroH2A, and DNA methylation, accumulate on the promoters of X-linked genes on the Xist-coated X chromosome, eventually leading to the repression of transcription.

Unlike most of the genes on Y chromosome, X-linked genes have roles in several physiological processes in multiple tissues, and therefore, defects in silencing of X-linked genes or alteration of gene expression might result in changes in cellular proliferation leading to cancer. Moreover, X chromosome encodes several proteins that have direct critical roles in regulating cellular proliferation. Skewing of X inactivation, wherein one of the X chromosomes is preferentially inactivated in adult tissues, is also found in cancers. Preferential inactivation of the X chromosome harboring wild-type copy of FOXP3, an X-linked tumor suppressor gene, is frequently observed in breast cancer (Medema and Burgering 2007).

Though there is a lack of direct functional link between the XIST RNA and occurrence of cancers, several studies have reported misregulation of XIST expression and/or its localization especially in breast, ovarian, cervical, and testicular cancer cell lines (Sirchia et al. 2005, 2009). In ovarian cancers, relationship between XIST down-regulation and sensitivity to chemotherapy was reported (Huang et al. 2002). Several protein factors whose expression or function is altered in cancers such as BRCA1 were shown to be regulating the XIST levels or its function. BRCA1, a protein frequently mutated in breast and ovarian cancers, has been shown to localize onto inactive X chromosome, associate with and aid in the localization of XIST RNA without affecting its levels (Ganesan et al. 2002; Silver et al. 2007). Loss and gain-of-function analyses showed that BRCA1 is a protein factor that aids in the *cis*-localization of XIST and the ensuing histone modifications that are associated with the inactive X chromosome such as H3K27 trimethylation and macro-H2A. Consistent with this finding, major Xi-associated chromatin markers were conspicuously reduced or all together lacking in a significant number of basal-like human breast cancer samples (Ganesan et al. 2005). Similarly, upregulation of XIST transcript levels by BRCA1 was also reported (Welcsh et al. 2002). Contrary to these findings, subsequent studies have been unable to find a link between the BRCA1 mutation/s or its overexpression on XIST localization as well as its localization onto inactive X chromosome (Xiao et al. 2007). Altogether, these studies indicate that much of the phenotype observed in BRCA1-defective breast cancer samples may not be due to the direct action of BRCA1 on XIST localization but due to aberrant genetic and epigenetic changes that accompany cancer progression (Pageau et al. 2007).

In human breast cancer cell lines, miR-103/107-mediated down-regulation of DICER-1, the RNase III enzyme (Martello et al. 2010), was found to be a key event in increasing the metastasizing potential of non-metastatic breast cancer cells, which is also correlated with high epithelial to mesenchymal transition capacity (Grelier et al. 2009). Considering the impact of XIST on the expression of X-linked genes and DNA stability (Diaz-Perez et al. 2006), it will be interesting to see whether the breast cancer cells that show reduced levels or mutations of DICER-1 also show defects in XIST localization; if so, the subsequent activation of silenced X-linked genes might compound the pathology and offer XIST as a diagnostic or therapeutic target in metastatic breast cancers. DICER-1 levels were found to be reduced in ovarian cancers (Pampalakis et al. 2010), leaving the door open to probe the associated lncRNA modulation, specifically XIST. Also, very little is studied about the existence and relevance of other newly described ncRNAs from mouse XIC such as repeat A (Zhao et al. 2008) that helps in the recruitment of polycomb group proteins during initial phases of Xi establishment and its levels and functional relevance in the above-described human cancers.

5.2 *HOTAIR*

HOX genes are master regulators of spatiotemporal patterning in mammals. In humans, there are 39 HOX genes that are found in four distinct clusters, HOX A–D (Pearson et al. 2005). Several long and small ncRNAs from the HOX clusters have been described, majority of them being identified using array-based expression analysis of RNA from human fibroblasts (Rinn et al. 2007). Among them, HOTAIR, a large intergenic noncoding RNA (lincRNA) of >2 kb in length is of intense interest. HOTAIR is a peculiar ncRNA; it is transcribed from HOX-C locus, and its major action is on the HOX-D locus in *trans*, regulating the positional identity during organismal development. HOTAIR interacts with many protein complexes including polycomb repressor complex 2 (PRC2) (Rinn et al. 2007). PRC2 is a functional complex of polycomb group family of proteins that coordinate a plethora of chromatin regulatory functions including the important role in maintenance of gene expression patterns during development (Schuettengruber and Cavalli 2009). HOTAIR mediates PRC2 recruitment to the target chromatin, facilitating methylation on histone H3 lysine 27 by the PRC2 member EZH2, a histone lysine methyltransferase, resulting in the repression of gene expression (Rinn et al. 2007). The exact molecular mechanisms of HOTAIR functions vis-à-vis in PRC2-mediated silencing have yet to be deciphered, while novel strategies deciphering whole-genome binding sites of chromatin modifiers have provided some interesting insights (Tsai et al. 2010; Chu et al. 2011). Interestingly, HOTAIR simultaneously interacts with histone H3 lysine 4 demethylase LSD1/CoREST/REST complex as well, leading to recruitment of multiple gene repressive complexes to target loci (Tsai et al. 2010).

Of particular interest for human cancer are the recent studies that have shown a remarkable link between HOTAIR and the metastatic potential of breast cancer cells, poor prognosis in colorectal cancer, and recurrence in liver tumors (Gupta et al. 2010; Kogo et al. 2011; Yang et al. 2011b). In breast cancer cells, HOTAIR

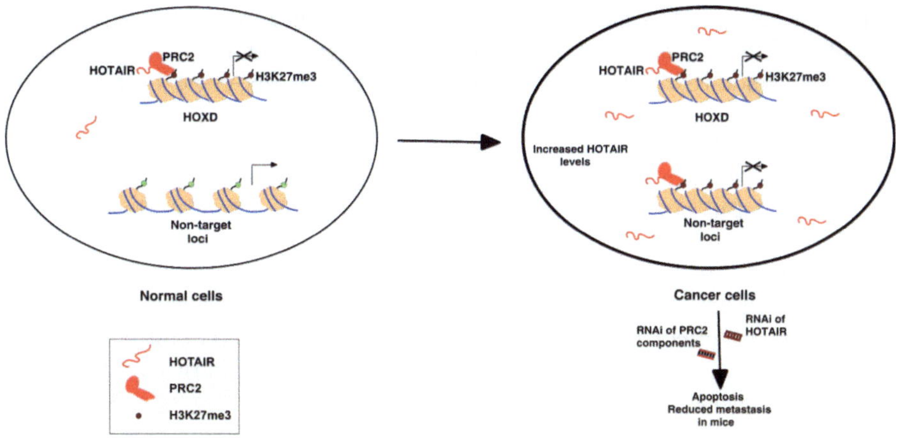

Fig. 3 Higher levels of HOTAIR transcript are found in many cancer cell types compared to normal cells; the HOTAIR ncRNA brings about repression of gene expression through trimethylation of histone H3 lysine 27, via interaction with polycomb repressive complex 2. Silencing of HOTAIR or components of PRC2 via RNA interference in hepatic and breast cancer cells resulted in increased apoptosis and decreased metastasis in vivo in mice

RNA levels were highly upregulated and correlated well with the metastatic potential and mortality. Enforced HOTAIR overexpression in non-metastatic cancer cells led to metastasis of those cells when injected in vivo in mice. Also, increased HOTAIR expression led to ectopic occupancy of PRC2 subunits onto several new genes and increased H3K27 trimethylation levels (Fig. 3). The polycomb binding profiles of HOTAIR overexpressing breast cancer cell line closely resembled those of embryonic fibroblasts, indicating a possible critical role of HOTAIR in breast cancer pathology (Gupta et al. 2010). Notably, the metastatic potential of HOTAIR expressing breast cancer cells was mitigated via reduction in the levels of PRC2 components (Gupta et al. 2010). Apart from HOTAIR, the PRC2 complex protein EZH2 was notably overexpressed in several cancers including breast cancer (Kleer et al. 2003). In another study, increased expression of HOTAIR was found in colorectal cancer cells, with an associated increase in metastasis to liver (Kogo et al. 2011). Lack of conservation of HOTAIR function(s) between mice and humans (Schorderet and Duboule 2011) is also an indication of specialized functional roles for this RNA in humans and possibly other primates.

5.3 *H19*

Monoallelic gene expression or genomic imprinting is a phenomenon wherein despite inheriting two alleles one from each parent, gene expression predominantly originates either from the maternally or paternally inherited allele. One such example of such parent-of-origin specific gene expression is the monoallelically expressed RNA, H19, which is predominantly transcribed from the maternal allele in both mice and humans. H19 RNA is also a precursor for an evolutionarily conserved microRNA, miR-675. Levels of both H19 RNA and miR-675 were found to be deregulated in human cancers (Ariel et al. 1995, 1998; Hibi et al. 1996; Fellig et al. 2005; Matouk et al. 2007; Schmitz et al. 2011). There is a lack of consensus on the exact functional role of H19 RNA as a tumor suppressor or an oncogene, and its targets in human cancers are yet to be established (Yoshimizu et al. 2008). H19 expression was found to be negatively controlled by P53 in vivo (Matouk et al. 2010). Also, miR-675 is reported to regulate the levels of pRb in colorectal cancer cells (Tsang et al. 2010). In contrast, tumor suppressor RB also has a repressive effect on H19 promoter, and E2F1 augments H19 mRNA levels in cancer (Berteaux et al. 2005). All these results indicate an intricate relationship between oncogenic pathways and H19 RNA.

5.4 MALAT1

Metastasis-Associated Lung Adenocarcinoma Transcript-1 (MALAT1), also known as Nuclear Enriched Abundant Transcript-2 (NEAT2), is another example of lncRNA that is highly expressed in several cancers. This RNA is evolutionarily highly conserved and was identified in a subtractive hybridization screen as a RNA that is abundant in metastasized non-small cell lung cancer samples (Ji et al. 2003). Several studies have characterized the misexpression of MALAT1 in a wide range of cancers and its expression also highly correlated with the metastatic potential of several cancers (Tano et al. 2010; Chang et al. 2011; Schmidt et al. 2011; Xu et al. 2011). In lung adenocarcinomas, RNA interference-mediated inhibition of MALAT1 expression impaired the in vitro motility of those cells (Tano et al. 2010). Reduced MALAT1 levels resulted in altered levels of pre-mRNA of some motility-related protein-coding transcripts. These results indicated an important nuclear role for MALAT1 in regulating the expression of other genes. Consistent with this finding, MALAT1 was found associated with nuclear speckles, a nuclear territory associated with RNA transcription, processing, and storage (Hutchinson et al. 2007). Though MALAT1 is not directly involved in the organization of nuclear speckles, it is required for recruitment and accumulation of certain proteins that are involved in RNA splicing (Tripathi et al. 2010). Antisense-mediated inhibition of MALAT1 revealed an important role for MALAT1 in RNA processing (Lin et al. 2011). Significantly, mutations in the 3′ end of MALAT1 were found in colorectal cancer cell lines as well as in clinical cancer samples, the domain harboring 3′ end being implicated in cell growth and invasion (Xu et al. 2011). Similarly, in cervical cancer cells, inhibition of MALAT1 transcript by RNAi led to cell cycle arrest in G1, reduced tumor growth in mice, as well as increased expression of proapoptotic genes (Guo et al. 2010a).

5.5 *piwi-Interacting RNAs*

piRNAs are relatively new players that are found to have significant roles in certain human cancers. Majority of piRNAs and the proteins involved in their biogenesis, namely, the piwi clad of Argonaute family proteins, are predominantly expressed during male and female germline development in several species (Siomi et al. 2011). During germline development, transposon- and repeat-derived piRNAs guide chromatin remodeling complexes to homologous sequences and thereby play important roles in silencing of transposable elements and other selfish genetic elements (Saito and Siomi 2010). There are four human homologues of Drosophila piwi family, namely, HIWI, HILI, PIWIL3, and HIWI2 (Carmell et al. 2002). One of the family member, HIWI, is highly expressed in several human cancers (Qiao et al. 2002; Liu et al. 2006; Grochola et al. 2008; Garzon et al. 2009; Jiang et al. 2011; Sun et al. 2011). Recently, certain piRNAs have been shown to play important roles in gastric carcinogenesis (Cheng et al. 2011; Cui et al. 2011). piR-651 was found to be highly expressed in human gastric cancers, and inhibition of this piRNA in those cells resulted in cell cycle arrest (Cheng et al. 2011). Intriguingly, levels of another piRNA, namely, piR-823, were found to be lower in gastric cancer samples compared to normal tissues, and restoring its levels using a piR-823 mimic led to growth suppression of cancer cells both in vitro and in vivo (Cheng et al. 2012). However, it is not clear as how HIWI or certain piRNAs function during carcinogenesis. It is plausible that the aberrant activation of piRNA pathway in cancers might result in sequence-independent chromatin remodeling and silencing of genes that have key regulatory roles in cellular growth and proliferation. It will be interesting to investigate whether the pathway is activated in cancer stem-like cells and confers any growth advantages. Modulation of the piRNA levels seems to be a promising cancer therapeutic strategy.

5.6 *Ultraconserved Regions*

Ultraconserved regions are noncoding DNA sequences of ~200–800 nt long and comprised of ~481 such DNA sequences, both inter- and intragenic, that are conserved 100% among several mammalian species. These UCRs are presumed to be functional regions due to their sequence retention during evolution (Bejerano et al. 2004; Katzman et al. 2007). Extensive transcription across these regions has been described in normal tissues, and levels of the transcripts originating from UCR (T-UCR) are significantly varied in leukemia, carcinomas (Calin et al. 2007), and neuroblastoma (Scaruffi et al. 2009). Targets and functional details for many of these RNAs both under physiological conditions as well as during malignancy have not been yet discovered. It will be interesting to understand the functional parallels and differences in mechanisms and targets between the transcripts (RNAs) from these ultraconserved regions and other noncoding RNAs.

6 ncRNA, Polycomb Group Proteins

Polycomb proteins were first indentified in Drosophila as regulators of Hox gene expression and are evolutionarily conserved epigenetic modifiers in several eukaryotic species including humans. PcG proteins are known to maintain long-term gene repression at several gene loci including those involved in development (Schuettengruber and Cavalli 2009). PcG proteins form two distinct complexes: polycomb repressive complex 1 (PRC1) and polycomb repressive complex 2 (PRC2). One of the major protein components of PRC2 complex is EZH2. EZH2 has histone methyltransferase activity and shown to catalyze histone 3 lysine 27 di- and trimethylation. Methylated H3K27 serves as a binding site for PRC1 complex to stably propagate the gene repression memory (Margueron and Reinberg 2011). Several ncRNAs such as XIST and HOTAIR have been shown to bind with EZH2 in mouse and human cell lines and cancer cells and mediate H3K27 methylation and recruitment of PRC complex proteins to target loci. Analysis of EZH2-bound RNAs by RNA immunoprecipitation followed by sequencing (RIP-Seq) experiments in mouse ESC has identified a large repertoire of bound RNAs including several noncoding RNAs that directly or indirectly interact with EZH2 (Zhao et al. 2010). Along with recruitment of methyltransferases such as EZH2 to target loci, ncRNAs also function as molecular scaffolds that help in bringing multiple protein complexes together. Recently, it has been shown that methylation status of polycomb group protein-2, PC2 (CBX4), a component of polycomb repressive complexes, brings about changes in intranuclear localization of PC2-bound growth control gene loci that largely depend on their binding specificity to two distinct nuclear ncRNAs (Yang et al. 2011a). In this study, the authors described that PC2 is a substrate for SUVAR3-9H1 (KMT1A), a histone lysine 9 (H3K9) methyltransferase that specifically catalyzes dimethylation of K191 residue (PC2K191me2). This process might be dynamic, as methylation in vivo is also susceptible for demethylation by lysine-specific demethylase KDM4C. Most importantly, it was observed that methylated PC2 (PC2K191me2) was localized in polycomb bodies, a nuclear entity with an enrichment for transcriptional repressive factors, whereas unmethylated PC2 was found in interchromatin granules (ICG), a nuclear milieu with enrichment for transcriptional permissive environment. Under conditions of serum starvation, a significant percentage of PC2-bound E2F1 target growth control genes were found in PC bodies, whereas the same loci were found in interchromatin granules upon serum supplementation (Fig. 4). These localization changes in the target loci were brought about by the specific interaction between the methylated PC2 with ncRNA TUG1 localized in the PC bodies or the binding of unmethylated PC2 to ncRNA NEAT2 in interchromatin granules. These RNAs serve as catalysts for assembling several transcriptional corepressors or coactivators in PC bodies or ICG depending on external growth stimuli. NEAT2 interacting unmethylated PC2 mediates SUMOylation of E2F1 resulting in transcriptional activation of growth control genes.

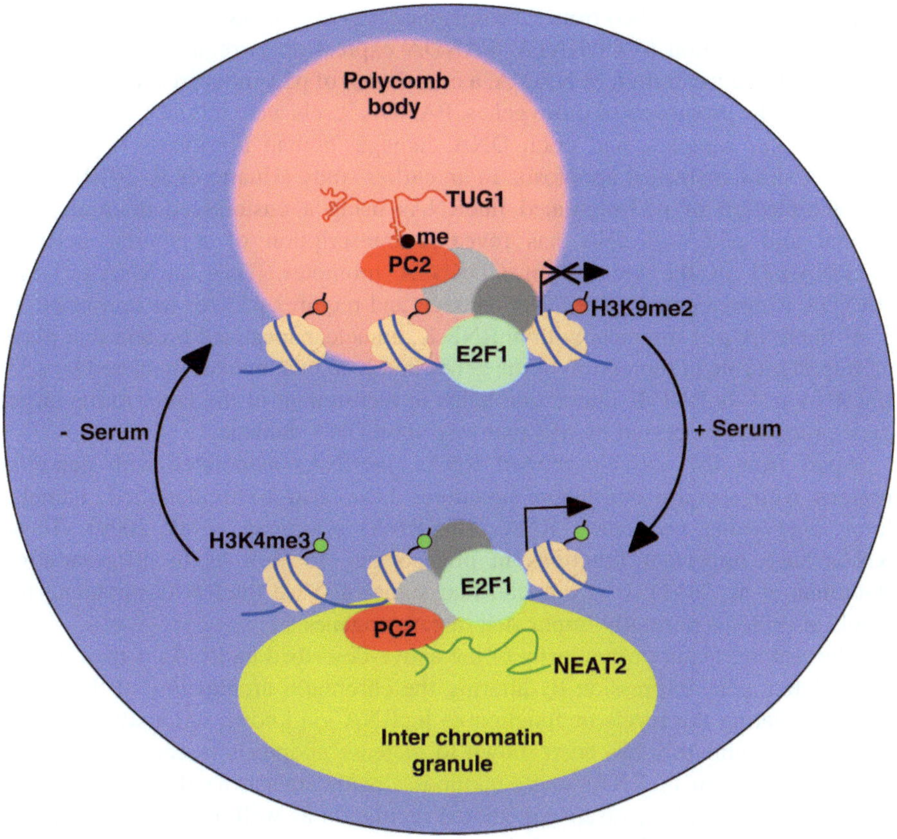

Fig. 4 Methylation of polycomb 2 protein changes its binding specificity to polycomb body localized ncRNA TUG1, and growth stimulation of cells via serum increases demethylation of PC2 and binding of unmethylated PC2 to interchromatin granule localized ncRNA, NEAT2. This dynamic modulation of PC2 localization also leads to relocalization of PC2-bound growth control genes between intranuclear bodies leading to expression of ICG-localized growth control genes upon supplementation with serum(adapted from Yang et al. 2011a)

Search for transcription surrounding 56 cell cycle genes using a customized ultrahigh-resolution microarray has revealed more than 200 noncoding transcripts (Hung et al. 2011). The majority of these transcripts were >200 nt, and a higher percentage of these were found transcribed from the upstream regions of the known cell cycle genes compared to intronic, or those that are transcribed from downstream sequences. Importantly, similar to the cell cycle genes, significant expression changes in a few of these ncRNAs were documented during cell cycle, distinct cell types, growth conditions, and DNA damage. Functional relevance of these lncRNAs especially in terms of their effect on cellular growth during normalcy or malignancy is yet to be ascertained. In the same study, the authors have described PANDA (P21 associated ncRNA DNA damage activated) a DNA damage-induced lncRNA that is transcribed from 5 kb upstream region of CDKN1A promoter in the antisense orientation to CDKN1A. PANDA expression is regulated by p53 and mediates the sequestration of NF-YA, a coactivator of p53-induced apoptotic genes away from the promoters of those genes. PANDA levels were induced in metastatic breast cancer samples, and upon DNA damage, human fibroblasts depleted of PANDA show increased apoptosis. In an earlier study (Huarte et al. 2010) aimed at identification of p53-regulated lincRNAs using a customized microarray in normal and p53 null cells has revealed transcription of a similar ncRNA, lincRNA-p21 in the vicinity of CDKN1A promoter. This lincRNA-p21-like PANDA is upregulated upon DNA damage and requires p53 for its transcription. 5′ of lincRNA-p21 interacts with hnRNP-K, a nuclear-localized protein that plays an important role in repression of the genes involved in apoptosis mediated by p53. lincRNA-p21–hnRNP-K interaction helps in recruitment of the latter to the target gene promoters repressed by p53 pathway upon DNA damage.

Apart from the above-described RNAs, lncRNAs associated with genomic regions with specific chromatin signatures have also been identified, namely, large intervening noncoding RNAs (lincRNA) (Guttman et al. 2009). These RNAs have important functions in pluripotency as well as in differentiation (Guttman et al. 2011). It is notable, that some of these lincRNAs modulate the levels of proteins normally implicated in carcinogenesis.

The striking feature of majority of the above-described lncRNAs is their ability to modulate gene expression by altering the chromatin architecture at the target loci. Regulating the levels or function of lncRNA via known chromatin or DNA methylation inhibitors has been shown to increase apoptosis in cancer cell lines. Direct targeting of lncRNAs using methods already standardized for silencing of messenger RNAs like antisense-mediated regulation as well as RNA interference via siRNAs and siRNA analogs with modified bases for extended silencing have been employed in cancer cells with considerable success. The intricate link between lncRNA, cancer epigenome, and transcriptome forms an additional layer of complexity in understanding the biology of cancer as well as offers a unique opportunity to design cancer-specific therapeutic modalities. The ease and efficacy of in vivo silencing of target RNAs, including mRNA targets (Soutschek et al. 2004; Sato et al. 2008), miRNAs (Krutzfeldt et al. 2005; Elmen et al. 2008), and viruses (Bitko et al. 2005; Palliser et al. 2006; Lanford et al. 2010), observed in mice models and primates (Zimmermann et al. 2006) can be extended to modulate the levels of lncRNA targets whose levels are highly upregulated in cancers.

7 Conclusion and Future Prospects

The current decade has seen exciting discoveries in the field of noncoding RNAs that fine-tune various biological processes as well as their perturbation in several disease conditions, especially cellular proliferation and cancer. The wealth of knowledge acquired using high-end, high-throughput sequencing technologies such as deep sequencing of transcriptome of individual cancer type will lead to discovery of novel diagnostic and therapeutic targets in a variety of cancer tissues including cancer types with limitations to obtain large amount of clinical samples. Simultaneously, the interactome analyses of individual ncRNA and its targets at the level of chromatin and transcriptome, under the condition of carcinogenesis and metastasis, will contribute tremendously towards the overall understanding of cancer biology. These advances offer a unique opportunity to develop highly efficient and effective ncRNA-based cancer-specific therapeutic modalities in the near future.

References

Ariel I, Lustig O, Schneider T et al (1995) The imprinted H19 gene as a tumor marker in bladder carcinoma. Urology 45:335–338

Ariel I, Miao HQ, Ji XR et al (1998) Imprinted H19 oncofetal RNA is a candidate tumour marker for hepatocellular carcinoma. Mol Pathol 51:21–25

Augui S, Nora EP, Heard E (2011) Regulation of X-chromosome inactivation by the X-inactivation centre. Nat Rev Genet 12:429–442

Avner P, Heard E (2001) X-chromosome inactivation: counting, choice and initiation. Nat Rev Genet 2:59–67

Babiarz JE, Ruby JG, Wang Y et al (2008) Mouse ES cells express endogenous shRNAs, siRNAs, and other Microprocessor-independent, Dicer-dependent small RNAs. Genes Dev 22:2773–2785

Bejerano G, Pheasant M, Makunin I et al (2004) Ultraconserved elements in the human genome. Science 304:1321–1325

Berteaux N, Lottin S, Monte D et al (2005) H19 mRNA-like noncoding RNA promotes breast cancer cell proliferation through positive control by E2F1. J Biol Chem 280:29625–29636

Bitko V, Musiyenko A, Shulyayeva O et al (2005) Inhibition of respiratory viruses by nasally administered siRNA. Nat Med 11:50–55

Brosnan CA, Voinnet O (2009) The long and the short of noncoding RNAs. Curr Opin Cell Biol 21:416–425

Brown CJ, Hendrich BD, Rupert JL et al (1992) The human XIST gene: analysis of a 17 kb inactive X-specific RNA that contains conserved repeats and is highly localized within the nucleus. Cell 71:527–542

Calin GA, Liu CG, Ferracin M et al (2007) Ultraconserved regions encoding ncRNAs are altered in human leukemias and carcinomas. Cancer Cell 12:215–229

Carmell MA, Xuan Z, Zhang MQ et al (2002) The Argonaute family: tentacles that reach into RNAi, developmental control, stem cell maintenance, and tumorigenesis. Genes Dev 16:2733–2742

Cazalla D, Yario T, Steitz JA (2010) Down-regulation of a host microRNA by a Herpesvirus saimiri noncoding RNA. Science 328:1563–1566

Chang HY, Yang YL, Jensen MP et al (2011) The experience of and coping with lumbopelvic pain among pregnant women in Taiwan. Pain Med 12:846–853

Cheng J, Guo JM, Xiao BX et al (2011) piRNA, the new non-coding RNA, is aberrantly expressed in human cancer cells. Clin Chim Acta 412:1621–1625

Cheng J, Deng H, Xiao B et al (2012) piR-823, a novel non-coding small RNA, demonstrates in vitro and in vivo tumor suppressive activity in human gastric cancer cells. Cancer Lett 315:12–17

Chu C, Qu K, Zhong FL et al (2011) Genomic maps of long noncoding RNA occupancy reveal principles of RNA-chromatin interactions. Mol Cell 44:667–678

Cui L, Lou Y, Zhang X et al (2011) Detection of circulating tumor cells in peripheral blood from patients with gastric cancer using piRNAs as markers. Clin Biochem 44:1050–1057

Czech B, Malone CD, Zhou R et al (2008) An endogenous small interfering RNA pathway in Drosophila. Nature 453:798–802

Diaz-Perez SV, Ferguson DO, Wang C et al (2006) A deletion at the mouse Xist gene exposes trans-effects that alter the heterochromatin of the inactive X chromosome and the replication time and DNA stability of both X chromosomes. Genetics 174:1115–1133

Ebert MS, Sharp PA (2010) MicroRNA sponges: progress and possibilities. RNA 16:2043–2050

Elmen J, Lindow M, Schutz S et al (2008) LNA-mediated microRNA silencing in non-human primates. Nature 452:896–899

Esquela-Kerscher A, Slack FJ (2006) Oncomirs—microRNAs with a role in cancer. Nat Rev Cancer 6:259–269

Esteller M (2011) Non-coding RNAs in human disease. Nat Rev Genet 12:861–874

Farazi TA, Juranek SA, Tuschl T (2008) The growing catalog of small RNAs and their association with distinct Argonaute/Piwi family members. Development 135:1201–1214

Fellig Y, Ariel I, Ohana P et al (2005) H19 expression in hepatic metastases from a range of human carcinomas. J Clin Pathol 58:1064–1068

Fire A, Xu S, Montgomery MK et al (1998) Potent and specific genetic interference by double-stranded RNA in Caenorhabditis elegans. Nature 391:806–811

Franco-Zorrilla JM, Valli A, Todesco M et al (2007) Target mimicry provides a new mechanism for regulation of microRNA activity. Nat Genet 39:1033–1037

Ganesan S, Silver DP, Greenberg RA et al (2002) BRCA1 supports XIST RNA concentration on the inactive X chromosome. Cell 111:393–405

Ganesan S, Richardson AL, Wang ZC et al (2005) Abnormalities of the inactive X chromosome are a common feature of BRCA1 mutant and sporadic basal-like breast cancer. Cold Spring Harb Symp Quant Biol 70:93–97

Garzon R, Heaphy CE, Havelange V et al (2009) MicroRNA 29b functions in acute myeloid leukemia. Blood 114:5331–5341

Grelier G, Voirin N, Ay AS et al (2009) Prognostic value of Dicer expression in human breast cancers and association with the mesenchymal phenotype. Br J Cancer 101:673–683

Grochola LF, Greither T, Taubert H et al (2008) The stem cell-associated Hiwi gene in human adenocarcinoma of the pancreas: expression and risk of tumour-related death. Br J Cancer 99:1083–1088

Guo F, Li Y, Liu Y et al (2010a) Inhibition of metastasis-associated lung adenocarcinoma transcript 1 in CaSki human cervical cancer cells suppresses cell proliferation and invasion. Acta Biochim Biophys Sin (Shanghai) 42:224–229

Guo H, Ingolia NT, Weissman JS et al (2010b) Mammalian microRNAs predominantly act to decrease target mRNA levels. Nature 466:835–840

Gupta RA, Shah N, Wang KC et al (2010) Long non-coding RNA HOTAIR reprograms chromatin state to promote cancer metastasis. Nature 464:1071–1076

Guttman M, Amit I, Garber M et al (2009) Chromatin signature reveals over a thousand highly conserved large non-coding RNAs in mammals. Nature 458:223–227

Guttman M, Donaghey J, Carey BW et al (2011) lincRNAs act in the circuitry controlling pluripotency and differentiation. Nature 477:295–300

Hibi K, Nakamura H, Hirai A et al (1996) Loss of H19 imprinting in esophageal cancer. Cancer Res 56:480–482

Hu Z, Liang J, Wang Z et al (2009) Common genetic variants in pre-microRNAs were associated with increased risk of breast cancer in Chinese women. Hum Mutat 30:79–84

Huang KC, Rao PH, Lau CC et al (2002) Relationship of XIST expression and responses of ovarian cancer to chemotherapy. Mol Cancer Ther 1:769–776

Huarte M, Guttman M, Feldser D et al (2010) A large intergenic noncoding RNA induced by p53 mediates global gene repression in the p53 response. Cell 142:409–419

Hung T, Wang Y, Lin MF et al (2011) Extensive and coordinated transcription of noncoding RNAs within cell-cycle promoters. Nat Genet 43:621–629

Hutchinson JN, Ensminger AW, Clemson CM et al (2007) A screen for nuclear transcripts identifies two linked noncoding RNAs associated with SC35 splicing domains. BMC Genomics 8:39

Ibrahim AF, Weirauch U, Thomas M et al (2011) MicroRNA replacement therapy for miR-145 and miR-33a is efficacious in a model of colon carcinoma. Cancer Res 71:5214–5224

Ji P, Diederichs S, Wang W et al (2003) MALAT-1, a novel noncoding RNA, and thymosin beta4 predict metastasis and survival in early-stage non-small cell lung cancer. Oncogene 22:8031–8041

Jiang J, Zhang H, Tang Q et al (2011) Expression of HIWI in human hepatocellular carcinoma. Cell Biochem Biophys 61:53–58

Karreth FA, Tay Y, Perna D et al (2011) In vivo identification of tumor- suppressive PTEN ceRNAs in an oncogenic BRAF-induced mouse model of melanoma. Cell 147:382–395

Karube Y, Tanaka H, Osada H et al (2005) Reduced expression of Dicer associated with poor prognosis in lung cancer patients. Cancer Sci 96:111–115

Kasinski AL, Slack FJ (2011) MicroRNAs en route to the clinic: progress in validating and targeting microRNAs for cancer therapy. Nat Rev Cancer 11:849–864

Katzman S, Kern AD, Bejerano G et al (2007) Human genome ultraconserved elements are ultraselected. Science 317:915

Kim VN, Han J, Siomi MC (2009) Biogenesis of small RNAs in animals. Nat Rev Mol Cell Biol 10:126–139

Kleer CG, Cao Q, Varambally S et al (2003) EZH2 is a marker of aggressive breast cancer and promotes neoplastic transformation of breast epithelial cells. Proc Natl Acad Sci USA 100:11606–11611

Kogo R, Shimamura T, Mimori K et al (2011) Long noncoding RNA HOTAIR regulates polycomb-dependent chromatin modification and is associated with poor prognosis in colorectal cancers. Cancer Res 71:6320–6326

Kota SK, Balasubramanian S (2010) Cancer therapy via modulation of micro RNA levels: a promising future. Drug Discov Today 15:733–740

Kota J, Chivukula RR, O'Donnell KA et al (2009) Therapeutic microRNA delivery suppresses tumorigenesis in a murine liver cancer model. Cell 137:1005–1017

Krutzfeldt J, Rajewsky N, Braich R et al (2005) Silencing of microRNAs in vivo with 'antagomirs'. Nature 438:685–689

Lanford RE, Hildebrandt-Eriksen ES, Petri A et al (2010) Therapeutic silencing of microRNA-122 in primates with chronic hepatitis C virus infection. Science 327:198–201

Lee RC, Feinbaum RL, Ambros V (1993) The *C. elegans* heterochronic gene lin-4 encodes small RNAs with antisense complementarity to lin-14. Cell 75:843–854

Lin R, Roychowdhury-Saha M, Black C et al (2011) Control of RNA processing by a large non-coding RNA over-expressed in carcinomas. FEBS Lett 585:671–676

Liu X, Sun Y, Guo J et al (2006) Expression of hiwi gene in human gastric cancer was associated with proliferation of cancer cells. Int J Cancer 118:1922–1929

Ma L, Reinhardt F, Pan E et al (2010a) Therapeutic silencing of miR-10b inhibits metastasis in a mouse mammary tumor model. Nat Biotechnol 28:341–347

Ma L, Young J, Prabhala H et al (2010b) miR-9, a MYC/MYCN-activated microRNA, regulates E-cadherin and cancer metastasis. Nat Cell Biol 12:247–256

Margueron R, Reinberg D (2011) The polycomb complex PRC2 and its mark in life. Nature 469:343–349

Martello G, Rosato A, Ferrari F et al (2010) A MicroRNA targeting dicer for metastasis control. Cell 141:1195–1207

Matouk IJ, DeGroot N, Mezan S et al (2007) The H19 non-coding RNA is essential for human tumor growth. PLoS One 2:e845

Matouk IJ, Mezan S, Mizrahi A et al (2010) The oncofetal H19 RNA connection: hypoxia, p53 and cancer. Biochim Biophys Acta 1803:443–451
Medema RH, Burgering BM (2007) The X factor: skewing X inactivation towards cancer. Cell 129:1253–1254
Merritt WM, Lin YG, Han LY et al (2008) Dicer, Drosha, and outcomes in patients with ovarian cancer. N Engl J Med 359:2641–2650
Pageau GJ, Hall LL, Lawrence JB (2007) BRCA1 does not paint the inactive X to localize XIST RNA but may contribute to broad changes in cancer that impact XIST and Xi heterochromatin. J Cell Biochem 100:835–850
Palliser D, Chowdhury D, Wang QY et al (2006) An siRNA-based microbicide protects mice from lethal herpes simplex virus 2 infection. Nature 439:89–94
Pampalakis G, Diamandis EP, Katsaros D et al (2010) Down-regulation of dicer expression in ovarian cancer tissues. Clin Biochem 43:324–327
Pasquinelli AE, Reinhart BJ, Slack F et al (2000) Conservation of the sequence and temporal expression of let-7 heterochronic regulatory RNA. Nature 408:86–89
Pearson JC, Lemons D, McGinnis W (2005) Modulating Hox gene functions during animal body patterning. Nat Rev Genet 6:893–904
Poliseno L, Salmena L, Zhang J et al (2010) A coding-independent function of gene and pseudogene mRNAs regulates tumour biology. Nature 465:1033–1038
Qiao D, Zeeman AM, Deng W et al (2002) Molecular characterization of hiwi, a human member of the piwi gene family whose overexpression is correlated to seminomas. Oncogene 21:3988–3999
Rinn JL, Kertesz M, Wang JK et al (2007) Functional demarcation of active and silent chromatin domains in human HOX loci by noncoding RNAs. Cell 129:1311–1323
Saito K, Siomi MC (2010) Small RNA-mediated quiescence of transposable elements in animals. Dev Cell 19:687–697
Salmena L, Poliseno L, Tay Y et al (2011) A ceRNA hypothesis: the Rosetta Stone of a hidden RNA language? Cell 146:353–358
Sato Y, Murase K, Kato J et al (2008) Resolution of liver cirrhosis using vitamin A-coupled liposomes to deliver siRNA against a collagen-specific chaperone. Nat Biotechnol 26:431–442
Scaruffi P, Stigliani S, Moretti S et al (2009) Transcribed-ultra conserved region expression is associated with outcome in high-risk neuroblastoma. BMC Cancer 9:441
Schmidt LH, Spieker T, Koschmieder S et al (2011) The long noncoding MALAT-1 RNA indicates a poor prognosis in non-small cell lung cancer and induces migration and tumor growth. J Thorac Oncol 6:1984–1992
Schmitz KJ, Helwig J, Bertram S et al (2011) Differential expression of microRNA-675, microRNA-139-3p and microRNA-335 in benign and malignant adrenocortical tumours. J Clin Pathol 64:529–535
Schorderet P, Duboule D (2011) Structural and functional differences in the long non-coding RNA hotair in mouse and human. PLoS Genet 7:e1002071
Schuettengruber B, Cavalli G (2009) Recruitment of polycomb group complexes and their role in the dynamic regulation of cell fate choice. Development 136:3531–3542
Silver DP, Dimitrov SD, Feunteun J et al (2007) Further evidence for BRCA1 communication with the inactive X chromosome. Cell 128:991–1002
Siomi MC, Sato K, Pezic D et al (2011) PIWI-interacting small RNAs: the vanguard of genome defence. Nat Rev Mol Cell Biol 12:246–258
Sirchia SM, Ramoscelli L, Grati FR et al (2005) Loss of the inactive X chromosome and replication of the active X in BRCA1-defective and wild-type breast cancer cells. Cancer Res 65:2139–2146
Sirchia SM, Tabano S, Monti L et al (2009) Misbehaviour of XIST RNA in breast cancer cells. PLoS One 4:e5559
Soutschek J, Akinc A, Bramlage B et al (2004) Therapeutic silencing of an endogenous gene by systemic administration of modified siRNAs. Nature 432:173–178

Sun G, Wang Y, Sun L et al (2011) Clinical significance of Hiwi gene expression in gliomas. Brain Res 1373:183–188

Tano K, Mizuno R, Okada T et al (2010) MALAT-1 enhances cell motility of lung adenocarcinoma cells by influencing the expression of motility-related genes. FEBS Lett 584:4575–4580

Tay Y, Kats L, Salmena L et al (2011) Coding-independent regulation of the tumor suppressor PTEN by competing endogenous mRNAs. Cell 147:344–357

Tripathi V, Ellis JD, Shen Z et al (2010) The nuclear-retained noncoding RNA MALAT1 regulates alternative splicing by modulating SR splicing factor phosphorylation. Mol Cell 39:925–938

Tsai MC, Manor O, Wan Y et al (2010) Long noncoding RNA as modular scaffold of histone modification complexes. Science 329:689–693

Tsang WP, Ng EK, Ng SS et al (2010) Oncofetal H19-derived miR-675 regulates tumor suppressor RB in human colorectal cancer. Carcinogenesis 31:350–358

Valencia-Sanchez MA, Liu J, Hannon GJ et al (2006) Control of translation and mRNA degradation by miRNAs and siRNAs. Genes Dev 20:515–524

Watanabe T, Totoki Y, Toyoda A et al (2008) Endogenous siRNAs from naturally formed dsRNAs regulate transcripts in mouse oocytes. Nature 453:539–543

Welcsh PL, Lee MK, Gonzalez-Hernandez RM et al (2002) BRCA1 transcriptionally regulates genes involved in breast tumorigenesis. Proc Natl Acad Sci USA 99:7560–7565

Xiao C, Sharp JA, Kawahara M et al (2007) The XIST noncoding RNA functions independently of BRCA1 in X inactivation. Cell 128:977–989

Xu C, Yang M, Tian J et al (2011) MALAT-1: a long non-coding RNA and its important 3′ end functional motif in colorectal cancer metastasis. Int J Oncol 39:169–175

Yang R, Schlehe B, Hemminki K et al (2010) A genetic variant in the pre-miR-27a oncogene is associated with a reduced familial breast cancer risk. Breast Cancer Res Treat 121:693–702

Yang L, Lin C, Liu W et al (2011a) ncRNA- and Pc2 methylation-dependent gene relocation between nuclear structures mediates gene activation programs. Cell 147:773–788

Yang Z, Zhou L, Wu LM et al (2011b) Overexpression of long non-coding RNA HOTAIR predicts tumor recurrence in hepatocellular carcinoma patients following liver transplantation. Ann Surg Oncol 18:1243–1250

Yoshimizu T, Miroglio A, Ripoche MA et al (2008) The H19 locus acts in vivo as a tumor suppressor. Proc Natl Acad Sci USA 105:12417–12422

Zhao J, Sun BK, Erwin JA et al (2008) Polycomb proteins targeted by a short repeat RNA to the mouse X chromosome. Science 322:750–756

Zhao J, Ohsumi TK, Kung JT et al (2010) Genome-wide identification of polycomb-associated RNAs by RIP-seq. Mol Cell 40:939–953

Zimmermann TS, Lee AC, Akinc A et al (2006) RNAi-mediated gene silencing in non-human primates. Nature 441:111–114

MicroRNA Pathways in *Drosophila*

Geetanjali Chawla and Nicholas S. Sokol

Contents

Abstract MicroRNAs belong to a class of 20–25 nucleotide noncoding RNAs that posttranscriptionally regulate gene expression in a sequence-specific manner. Members of the microRNA family play roles in the normal physiology and development of diverse organisms. In this review, we focus on the recent progress made in understanding the roles of microRNAs in *Drosophila* biology. The topics discussed include biogenesis, discovery, mechanism of action, and function of microRNAs in diverse aspects of fly development.

Keywords *Drosophila* • Development • MicroRNA • miRNA

G. Chawla (✉) • N.S. Sokol (✉)
Department of Biology, Indiana University, Jordan Hall A502, Bloomington, IN 47405, USA
e-mail: chawlag@indiana.edu; nsokol@indiana.edu

V.A. Erdmann and J. Barciszewski (eds.), *From Nucleic Acids Sequences to Molecular Medicine*, RNA Technologies, DOI 10.1007/978-3-642-27426-8_25,

1 Introduction

MicroRNAs (miRNAs) are small endogenous RNAs that silence target messenger RNAs (mRNAs) by inhibiting translation or by facilitating mRNA decay. They are abundantly expressed and comprise 1–2% of animal genes (Ruby et al. 2007; Bartel 2009). A single miRNA can regulate a large fraction of genes (Brennecke et al. 2005; Krek et al. 2005). Moreover, the regulatory potential of miRNAs is further increased due to the presence of target sites for several miRNAs within many genes, allowing for a combinatorial control of gene expression that is reminiscent of transcriptional control (Hobert 2004). Finally, a high level of conservation of some miRNAs suggests an ancient role for miRNAs similar to transcription factors (Pasquinelli et al. 2000).

Genetic screens for regulators that control timing of developmental events in *Caenorhabditis elegans* identified two small temporal RNAs (stRNAs), *lin-4* and *let-7* (Lee et al. 1993; Reinhart et al. 2000). Subsequently, these two stRNAs were reported to represent a very widespread class of small endogenous RNAs found in worms, flies, and mammals that were named microRNAs (Lagos-Quintana et al. 2001; Lau et al. 2001; Lee and Ambros 2001). Genetic and genomic approaches have contributed toward the currently annotated list of fly miRNAs. Gain-of-function screens undertaken in the 1990s uncovered genetic loci that were recognized to encode miRNAs several years later (Li and Carthew 2005; Brennecke et al. 2003; Teleman et al. 2006). In addition, analysis of various hypermorphic alleles led to the discovery of regulatory motifs in the 3′-UTRs of several genes that were subsequently identified as miRNA binding sites (Lai and Posakony 1997; Lai 2002).

Subsequently, sequencing of *Drosophila pseudoobscura* facilitated comparative genomic studies that led to the identification of novel miRNAs (Lai et al. 2003) and the prediction of miRNA target genes (Enright et al. 2003; Stark et al. 2003; Grun et al. 2005). These were followed up with studies leading to refinement of rules for miRNA-target recognition (Brennecke et al. 2005) and the description of global patterns of miRNA regulatory networks (Stark et al. 2005). In this review, we discuss the contribution that the *Drosophila* model system has made toward the current understanding of different aspects of miRNA biology.

2 Biogenesis of miRNAs

Analysis of genomic positions of miRNA sequences revealed that miRNAs are either encoded as single units (monocistronic) or as clusters of sequences generated from a single locus (polycistronic). Many miRNA loci are present in intergenic regions and are transcribed from their own promoters (Lagos-Quintana et al. 2001; Lau et al. 2001; Lee et al. 2002; Reinhart et al. 2002). miRNAs have also been found within the introns or 3′-UTRs of protein-coding genes or within introns or

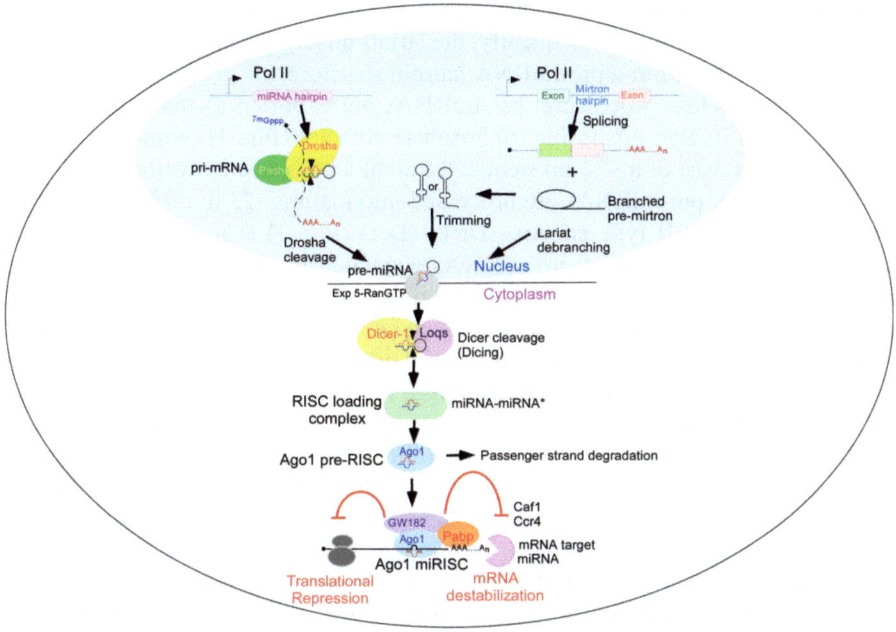

Fig. 1 miRNA biogenesis pathway. miRNA genes are transcribed by RNA polymerase II to generate primary transcripts whose hairpin structures are cleaved by the Drosha–Pasha complex to release ~65 nt pre-miRNAs. Mirtrons are derived from small introns that are spliced and then debranched, yielding pre-miRNA. Some introns have tails at either the 5′ end or 3′ end, so they need to be trimmed before pre-miRNA export. Pre-miRNAs are exported by Exportin 5–RanGTP from the nucleus to the cytoplasm, where they are processed by Dicer-1/Loquacious complex to generate a duplex containing two strands, termed the miRNA and miRNA*. The passenger strand of the duplex is discarded and undergoes degradation. The functional strand of the miRNA guides RISC to silence target mRNAs through translational repression or mRNA destabilization. Deadenylation of target RNAs is mediated by GW182 proteins. The N-terminal region of GW182 interacts with Ago proteins, and the C-terminal region interacts with Pabp and recruits the deadenylases Ccr4 and Caf1

exons of noncoding RNAs. More than 25% of the canonical miRNAs in flies are intronic and are expressed coordinately with their host gene (Ruby et al. 2007). Transcription of the miRNA gene in most cases is catalyzed by RNA polymerase II and results in the generation of primary miRNAs (pri-miRNAs). These are often several kilobases long and contain local stem-loop structures (Lee et al. 2004) (Fig. 1). A nuclear ribonuclease III (RNAse III) enzyme known as Drosha catalyzes an endonucleolytic cleavage at the stem of the hairpin structure, releasing a 60–70-nucleotide (nt) hairpin that is termed the precursor miRNA (pre-miRNA) (Fig. 1) (Lee et al. 2002). A double-stranded RNA-binding protein named Partner of Drosha (Pasha) functions as a Drosha cofactor during this step. Drosha and Pasha form a ~500-kDa complex referred to as the "Microprocessor complex" (Lee et al. 2003).

Another distinct class of miRNAs referred to as mirtrons have been identified through deep sequencing analysis (Fig. 1) (Okamura et al. 2007). These miRNAs are located within short introns and are processed by the spliceosomal machinery to generate lariat structures. Subsequently, the lariats undergo debranching to generate linear mirtrons that form a pre-miRNA hairpin structure.

Following nuclear processing, pre-miRNAs are exported to the cytoplasm by Exportin 5 and its Ran-guanosine triphosphate cofactor (Fig. 1) (Kim et al. 2009). Pre-miRNAs consist of a ~22-bp stem, a terminal loop, and a 3′ overhang of ~2 nt. In the cytoplasm, pre-miRNAs are processed into mature ~22 nt miRNA duplexes by another RNAse III type enzyme, Dicer (Dcr) (Fig. 1) (Kim et al. 2009). Dcr anchors the 3′ and 5′ end of pre-miRNAs generated by Drosha through its PAZ domain and cleaves the double-stranded region ~22 nt away from the 5′ end (Park et al. 2011). This cleavage results in the generation of a miRNA duplex (miRNA: miRNA*). Dcr proteins associate with several other proteins such as R2D2, Loquacious, and FMR1 in flies and Argonaute (Ago) family proteins in various organisms (Liu et al. 2003; Ishizuka et al. 2002; Jiang et al. 2005; Saito et al. 2005). These Dcr-interacting proteins play various roles in miRNA stability and effector complex formation and action. They enhance the affinity of Dcr for RNAs and participate in the selection of mature miRNA strands as well as the transfer of miRNAs to Ago proteins.

Dcr-mediated cleavage is followed by the formation of an effector complex, termed the miRNA-containing RNA-induced silencing complex (miRISC, see Fig. 1) (Kawamata and Tomari 2010). The Ago family proteins form the core of the miRISC, and the association of a small RNA with a specific Ago protein determines its function. In flies, Ago1 functions in the miRNA pathway and Ago2 functions in the RNAi pathway (Forstemann et al. 2007). In the first step of miRISC assembly, the miRNA: miRNA* duplexes are inserted into Ago proteins to form a pre-RISC complex. The second step involves strand dissociation or unwinding of the two strands of the miRNA duplex to form the active miRISC complex. The RNA strand retained in the RISC is called the "guide strand" and the other discarded strand is referred to as the "passenger strand."

3 Regulation of miRNA Biogenesis

miRNA biogenesis is stringently controlled at several levels and is often subject to feedback regulation. Transcriptional regulation is a major level of control responsible for the tissue-specific or development-specific expression of miRNAs. For example, myogenic transcription factors Twist and Mef2 are involved in the muscle-specific expression of *Drosophila miR-1* (Sokol and Ambros 2005).

miRNAs and transcription factors are frequently components of regulatory networks, where they regulate each other's expression (Herranz and Cohen 2010). Some of the fly miRNAs that participate in feedback and feed-forward loops are *miR-9a, miR-14*, and *miR-7* (Herranz and Cohen 2010; Li et al. 2009).

A number of proteins have been shown to regulate miRNA processing either by interacting with Drosha or Dcr, or by binding directly to miRNA precursors (Siomi and Siomi 2010). Most of these studies have been carried out in other model systems, so these regulatory relationships still need to be verified in flies. The best-studied negative regulator of miRNA biogenesis is the Lin-28, RNA-binding protein (Viswanathan and Daley 2010). Repression by Lin-28 affects members of the *let-7* family. Lin-28 binds to the terminal loop of *pri-let-7* transcripts, and this binding interferes with its Drosha cleavage and blocks its processing by Dicer (Viswanathan and Daley 2010). Interestingly, *lin-28* mRNA is targeted by *let-7*, thus resulting in double-negative feedback regulatory motif (Rybak et al. 2008). Though both *let-7* and Lin-28 protein are conserved in flies, so far no study has demonstrated such a regulatory relationship in this system.

4 Mechanism of miRNA Action

Most animal miRNAs imperfectly base-pair with sequences in the 3′-UTR of target mRNAs and inhibit protein synthesis by either repressing translation or promoting mRNA deadenylation and decay (Fabian et al. 2010). Efficient mRNA targeting requires continuous base-pairing of the mRNA with the second to eight nt of the miRNA, a region known as the "seed" (Brennecke et al. 2005). The mechanism of miRNA-mediated translational repression is not as clearly understood as the mechanisms of miRNA-mediated mRNA deadenylation (Fabian et al. 2010). Deadenylation of mRNAs is mediated by GW182, a protein that physically interacts with Ago proteins. The amino-terminal region of GW182 interacts with Ago proteins, whereas its carboxy-terminus (C terminus) interacts with the poly (A)-binding protein (Pabp) and recruits the deadenylases Ccr4 and Caf1 (Fabian et al. 2010). In addition to the C terminus, GW182 contains two additional domains that function in translational repression. These three repressive domains may be differentially regulated or may target distinct sets of mRNAs (Fabian et al. 2010). It is also now clear that miRNAs lead to the degradation of target mRNAs. When RISC containing Ago2 in mammals or flies encounters mRNAs bearing sites containing near perfectly complementary to a miRNA, these mRNAs are cleaved endonucleolytically and degraded (Fabian et al. 2010). Finally, although most reports have focused on the repression of mRNAs by targeting miRNAs, some reports have indicated that miRNAs also activate translation of mRNA targets (Vasudevan et al. 2007).

5 Discovery of *Drosophila* miRNAs

The current tally of *Drosophila* miRNAs is 240. Many of these miRNA genes were discovered by high-throughput cloning and sequencing of small RNA libraries (Griffiths-Jones 2004; Lagos-Quintana et al. 2001; Lee and Ambros 2001; Aravin et al. 2003). Genomic sequences flanking these cloned fragments were then retrieved to confirm that they folded into the hairpin structures that are typical of pre-miRNAs (Bentwich 2005). Recent advances in high-throughput sequencing technology have greatly increased the sensitivity of these approaches, leading to discovery of many novel miRNAs that had escaped previous detection due to low abundance (Stark et al. 2007; Berezikov et al. 2010; Ruby et al. 2007). These studies used computational methods to accurately predict miRNAs conserved among *Drosophila* species and high-throughput sequencing of libraries of small RNAs from various tissues or stages to experimentally confirm and complement these predictions (Berezikov et al. 2010; Ruby et al. 2007; Stark et al. 2007). Moreover, structural and evolutionary patterns of known miRNAs were employed to derive rules for discriminating miRNAs from the nonfunctional hairpins in the genome. Some of the structural features utilized for prediction included length of miRNA hairpins, length of the arms and the hairpin loop, secondary structures generated by folding of precursor sequences, and loop structures in the miRNA hairpins. The characteristic evolutionary profile of miRNAs in *Drosophila* genomes was determined by aligning previously cloned miRNAs and their flanking regions across all 12 *Drosophila* species (Stark et al. 2007). These approaches have vastly increased the list of annotated miRNAs in flies. In addition to identifying novel miRNAs, these comparative genomic studies have provided new insights on miRNA function, biogenesis, and evolution (Stark et al. 2007; Berezikov et al. 2010).

6 Identification of miRNA Targets

The identification of miRNA targets is a critical step towards the functional characterization of miRNAs. The mRNA targets of the founding members, *lin-4* and *let-7*, were identified through genetic approaches. Similar genetic approaches aided in the functional analysis of other miRNAs. An example was the identification of proapoptotic gene *hid* as a biologically relevant target of the fly miRNA *bantam* (Brennecke et al. 2003). Since miRNAs are short, and animal miRNAs have limited sequence complementarity with their targets, accurate prediction of target mRNAs has been difficult. Moreover, due to the laborious nature of experiments and the absence of high-throughput experimental approaches to identify targets, a number of sophisticated bioinformatic methods have been developed to allow reliable and testable predictions of miRNA targets (Bentwich 2005; Min and Yoon 2010; Yue et al. 2009). The most commonly used parameter in target prediction is complementarity (Stark et al. 2003; Brennecke et al. 2005). Other features include evolutionary conservation and mRNA secondary structure (Kertesz et al. 2007; Long et al. 2007). Some of the methods used to predict fly miRNAs are discussed below.

The first large-scale prediction of fly miRNA targets utilized a conserved database generated by comparing 3′-UTR regions in *Drosophila melanogaster* and *Drosophila pseudoobscura* (Stark et al. 2003). This approach combined a sequence database search with an RNA folding algorithm to evaluate the quality of predicted miRNA–target duplexes. The 3′-UTRs obtained from the prediction algorithm were filtered for conservation in *D. pseudoobscura* and *Anopheles gambiae*. The identified sequences were evaluated for their ability to form thermodynamically favorable duplexes with the miRNA. This method was able to detect all previously known targets, which ranked as top-scoring candidates. Many new target sites were also predicted (e.g., *miR-7*: *hairy*, *HLHm3*; *miR-2*: *reaper*, *grim*, *sickle*; *miR-277*: enzymes involved in the amino acid metabolic pathway).

Miranda is a prediction method initially designed for miRNA targets in flies (Enright et al. 2003). It uses sequence complementarity to identify potential target sites in 3′-UTRs. The scoring matrix is designed so that complementary bases at the 5′ end of the miRNA are rewarded more than those at 3′ end. Hence, binding sites that exhibit a near perfect match at the seed region display a better score. Subsequently, the resulting potential binding sites are evaluated thermodynamically.

PicTar is a prediction method that relies on comparative data from several species to identify common targets for miRNAs and has also been applied to target identification in flies (Grun et al. 2005). In addition, this algorithm also computes the maximum likelihood that a given sequence is bound by one or more miRNAs. The initial parameters used for target prediction include optimal binding free energy, and the resulting target set is tested statistically using genome-wide alignment. PicTar was the first method that utilized the criteria of co-expression of miRNA and its target. Cross-species comparison of predicted targets suggested that, on average, 54 genes are regulated by a given miRNA and that fly miRNAs regulate expression of target genes coordinately (Grun et al. 2005).

7 Functions of miRNAs in *Drosophila*

The *Drosophila* system has provided great insights into the function of numerous miRNAs (Table 1). miRNAs can function in regulating developmental transitions (*let-7*) (Sokol et al. 2008). Some miRNAs function by "fine-tuning" mRNA targets that are active in their expression domain (*miR-9*/*Senseless*) (Li et al. 2006). Yet other miRNAs function by repressing key targets: phenotypes resulting from the removal of these miRNAs can be suppressed by partially reducing the expression of a single target gene (*miR-14*/*EcR*) (Varghese and Cohen 2007).

7.1 Regulation of Metamorphosis

The *let-7-Complex* (*let-7-C*) encodes three ancient miRNAs: *let-7*, *miR-100*, and *miR-125* (Sokol et al. 2008). The sequences and developmentally regulated expression profiles of *let-7* and *lin-4* are conserved among diverse bilaterians (Pasquinelli et al. 2000; Sempere et al. 2003). *let-7-C* knockout flies appear normal morphologically but display defects in adult behavior as well as juvenile features in their neuromusculature (Sokol et al. 2008). In addition, *let-7* and *miR-125* loss-of-

Table 1 *Drosophila* miRNAs with assigned functional roles

Function	miRNA	Target	Transcriptional regulator	References
Apoptosis/cell proliferation				
Apoptosis/cell proliferation	*bantam*	*Head involution defective*	Hippo pathway	Nolo et al. (2006), Thompson and Cohen (2006), Brennecke et al. (2003)
Apoptosis	*miR-2*	*reaper, grim, sickle*	–	Leaman et al. (2005), Stark et al. (2003)
Apoptosis	*miR-13*	*reaper, grim, sickle*	–	Leaman et al. (2005), Stark et al. (2003)
Apoptosis	*miR-11*	*reaper, grim, dE2F1*	–	Leaman et al. (2005), Stark et al. (2003), Truscott et al. (2011)
Apoptosis	*miR-6*	*hid, reaper, sickle*	–	Leaman et al. (2005), Stark et al. (2003)
Apoptosis	*miR-308*	*grim, sickle*	–	Leaman et al. (2005), Stark et al. (2003)
Developmental timing, metamorphosis				
Remodeling of neuro-musculature	*let-7*	*abrupt*	–	Caygill and Johnston (2008), Sokol et al. (2008), Sempere et al. (2003)
Spiracle eversion, physiology and life span	*miR-14*	*Ecdysone receptor*	Ecdysone receptor	Varghese and Cohen (2007)
Neuronal functions				
Prevention of neuro-degeneration	*miR-8*	*atrophin*	–	Karres et al. (2007)
Sensory organ specification	*miR-9a*	*senseless*	–	Li et al. (2006)
Photoreceptor differentiation	*miR-7*	*Yan*	Yan, Pointed-P1, Tramtrack69	Li et al. (2009), Li and Carthew (2005)
Sensory neuron development	*bantam*	–	–	Parrish et al. (2009)
Neurotransmission in presynaptic neurons	*miR-310 cluster*	*khc-73*	–	Tsurudome et al. (2010)
Olfactory neuron fate determination	*miR-279*	*nerfin*	–	Cayirlioglu et al. (2008)
Morphogenesis and function of neuromuscular junction	*miR-8*	*enabled*	–	Loya et al. (2009)
Patterning of retina	*miR263a/b*	*Head involution defective*	–	Hilgers et al. (2010)
Metabolism and insulin signaling				
Insulin signaling	*miR-278*	*expanded*	–	Teleman et al. (2006)
Insulin production and metabolism	*miR-14*	*Sugarbabe*	–	Varghese et al. (2010)
Insulin signaling and growth	*miR-8*	*u-shaped*	–	Hyun et al. (2009)
Germ line stem cell differentiation				
Germ line stem cell differentiation	*miR-184*	*saxophone*	–	Iovino et al. (2009)

(continued)

Table 1 (continued)

Function	miRNA	Target	Transcriptional regulator	References
Dorsoventral patterning of the egg shell	*miR-184*	*K10*	–	Iovino et al. (2009)
Anteroposterior blastoderm patterning	*miR-184*	*tramtrack69*	–	Iovino et al. (2009)
Differentiation of germ line stem cells	*miR-7*	*bag-of-marbles*	–	Pek et al. (2009)
Notch signaling/wing development				
Notch signaling/wing development chordotonal organ development	*miR-7*	*enhancer of split*	Atonal	Li et al. (2009), Stark et al. (2003)
Wing development	*miR-9a*	*Lim-only*	–	Biryukova et al. (2009), Bejarano et al. (2010)
Regulation of the wingless pathway				
Upregulates wingless pathway	*miR-315*	*axin, notum*	–	Silver et al. (2007)
Repression of wingless pathway	*miR-8*	*wntless, CG32767, TCF*	–	Kennell et al. (2008)
Hox gene regulation				
Wing/haltere specification	*miR-iab-4-5p*	*ultrabithorax*	–	Ronshaugen et al. (2005)
Wing/haltere specification	*miR-iab-4AS*	*ultrabithorax*	–	Stark et al. (2008)
Wing/haltere specification	*mir-iab-8*	*ultrabithorax, abdominal-A*	–	Tyler et al. (2008)
Circadian regulation				
Circadian rhythm	*bantam*	*Clock*	Hippo pathway/ Yorkie	Kadener et al. (2009)
Regulation of muscle function				
Muscle development and maintenance	*miR-1*	*delta*	Mef-2, Twist	Kwon et al. (2005), Sokol and Ambros (2005)

function mutants have smaller wings, which was attributed to a significant reduction in cell size in spite of the presence of more cells than wild type (Caygill and Johnston 2008). The perdurance of juvenile features in *Drosophila let-7* mutants is analogous to the reiteration of larval cell fates in adult *C. elegans let-7* mutants (Reinhart et al. 2000) and is consistent with the idea that *let-7* might control developmental transitions in diverse bilateria (Pasquinelli et al. 2000; Sokol et al. 2008; Caygill and Johnston 2008). Furthermore, Abrupt(Ab) was identified as a target for *let-7* (Caygill and Johnston 2008). The 3′-UTR of *ab* mRNA contains five predicted *let-7* binding sites, and ectopic *let-7* expression in wing disc cells lead to a downregulation of Ab protein levels. However, *abrupt* activity does not seem to account for the cell-cycle defect or shortened life span of *let-7* mutants, so other important targets might be deregulated in these settings.

A defect in anterior spiracle eversion during the larval-to-pupal transition suggested a role for *miR-14* in metamorphosis (Varghese and Cohen 2007).

miR-14 was shown to modulate a positive autoregulatory loop controlling steroid hormone signaling in flies. The steroid hormone ecdysone and its receptor (EcR) play crucial roles during development and metamorphosis. *miR-14* participates in a mutual antagonistic feedback loop, wherein *miR-14* inhibits EcR-positive autoregulation, and ecdysone receptor reciprocally inhibits *miR-14* expression (Varghese and Cohen 2007).

7.2 Regulation of Neuronal Function

A group of miRNAs including *miR-8*, *miR-7*, *miR-9*, *bantam*, and *miR-310* cluster regulate different aspects of neuronal function. *miR-8* is a highly conserved miRNA present in single copy in flies. The *miR-8* null mutants display reduced survival, morphological defects in legs and wings, and progressive neurodegeneration (Karres et al. 2007). A combination of computational approaches and expression profiling identified *atrophin* as one of the targets of *miR-8*. Notably, overexpression of *atrophin* in *miR-8* cells resulted in a spectrum of defects resembling those in *miR-8* null mutants, while removal of a single copy of *atrophin* reduced the severity of the *miR-8* phenotype. These data indicated that misregulation of *atrophin* contributed substantially to the defects associated with loss of *miR-8* function (Karres et al. 2007).

Another role for *miR-8* is the regulation of morphogenesis and function of neuromuscular junctions (NMJ) (Loya et al. 2009). Third instar *miR-8* null mutants display a decrease in the number of synaptic boutons, reduction in the presynaptic terminal arbors, and fewer axonal branches. Comparable NMJ phenotypes were observed in transgenic flies ubiquitously expressing *miR-8* sponges. *miR-8* has been shown to directly repress the expression of *enabled* (*ena*) in thc NMJ. Protein levels of Ena were significantly increased in *miR-8* mutants, and Ena overexpression elicited the same NMJ phenotype as *miR-8* removal. In addition, reducing levels of Ena protein resulted in rescue of all *miR-8* NMJ phenotypes, indicating that *miR-8* promotes presynaptic growth by postsynaptically limiting Ena expression (Loya et al. 2009).

miR-7 is expressed in photoreceptor cells of eye imaginal discs at the time when neuronal differentiation is first detected. Analysis of *miR-7* loss-of-function mutants revealed a reciprocal negative regulation between Yan protein and *miR-7* in retinal cells (Li and Carthew 2005). Yan is a neural repressor expressed in undifferentiated progenitor cells, where it inhibits *miR-7* transcription. Cells undergoing photoreceptor differentiation activate *miR-7*, which inhibits Yan protein synthesis (Li and Carthew 2005). The degradation of Yan allows activation of its competitor Ptd-P1, which further upregulates *miR-7* transcription (Li and Carthew 2005). In addition, *miR-7* and Yan both participate in coherent feed-forward loops (Li et al. 2009). In one of the feed-forward loops, Yan affects *miR-7* transcription by repressing the expression of *phyllopod*, which encodes the ubiquitin E3 ligase subunit that degrades the Ttk69 repressor of *miR-7* transcription

(Li and Carthew 2005). In a second feed-forward loop, Pnt-P1activates *miR-7*, which in turn represses Yan. The two coherent feed-forward loops are interlocked together to generate a double-negative feedback loop between *miR-7* and Yan that determines the fate of photoreceptor cells (Li et al. 2009).

miR-7 also participates in regulatory networks that determine the fates of proprioceptor organs and olfactory organs (Li et al. 2009). *yan* and *E (spl)* are direct targets of *miR-7*, and these factors are essential for the development of insect sensory organs (Li and Carthew 2005; Lai et al. 2005). *miR-7* transcription is activated by Atonal and causes the repression of *E (spl)*. Since Ato also directly activates transcription of *E (spl)*, it participates in an incoherent feed-forward loop where it directly activates and indirectly represses *E (spl)* and thereby imparts an accelerated and transient pulse of downstream gene expression. In addition, *E (spl)* feedbacks on Ato to create a double-negative feedback loop that is interconnected with the feed-forward loop. In this network, fluctuating peaks of Atonal result in transient pulses of *Ato* repression by *E (spl)*, but sustained increase of Atonal results in sustained repression of *E (spl)* by *miR-7* and stabilization of Atonal (Li et al. 2009).

Though *miR-7* is expressed in developing sensory organs, loss-of-function mutants of *miR-7* display little detectable impact on the development of these organs under laboratory conditions. Moreover, *miR-7* mutants had only minor defects in the protein expression of their target Yan and displayed normal Ato expression levels (Li and Carthew 2005). Thus, it was proposed that *miR-7* participates in feed-forward and feedback loops to buffer developmental programs against variation and imparts robustness to diverse networks. Consistent with this hypothesis, *miR-7* mutant subjected to environmental perturbation using temperature cycling displayed a reduction in Ato signal and a strong increase in Yan expression in the eyes, suggesting that under these conditions the mutant failed to activate Ato and repress Yan (Li et al. 2009).

Another miRNA involved in neural specification is *miR-9a*. Flies lacking *miR-9a* exhibit mild ectopic sensory organ formation in the larva as well as the adult anterior wing margin and thorax. These phenotypes are consistent with an inability to repress neural identity (Li et al. 2006). These flies also exhibited a completely penetrant loss of posterior wing margin. The proneural gene *senseless* (*sens*) was reported as a critical target of *miR-9a*. Notably, heterozygosity for *sens* rescued the phenotype on the anterior wing margin (Li et al. 2006). Based on the sensory organ phenotype of *miR-9a*, it was proposed that this miRNA helps prevent inappropriate induction of sensory organ precursors by keeping *sens* levels below a certain threshold (Cohen et al. 2006).

A recent study established a role for the *miR-310–313* cluster in synaptic transmission at the larval NMJ through its direct regulation of kinesin family member Khc-73. Khc-73 expression increased in *miR-310–313* mutants, and synaptic transmission defects were corrected by heterozygosity of Khc-73 (Tsurudome et al. 2010).

7.3 Regulation of Muscle Differentiation and Maintenance

miR-1 is a highly conserved miRNA that is specifically expressed in muscle tissue (Sokol and Ambros 2005). The biological role of *miR-1* in flies has been deciphered by analysis of null mutants of *miR-1* (Sokol and Ambros 2005; Kwon et al. 2005). Loss of *miR-1* during development resulted in a highly penetrant phenotype; few *miR-1* mutants died as first instars, but none progressed beyond second instar (Sokol and Ambros 2005). *miR-1* was shown to play a key role in maintaining muscle integrity during the dramatic, postmitotic growth in muscle mass that occurs during larval growth (Sokol and Ambros 2005). The Notch ligand Delta was identified as a target for *miR-1* (Kwon et al. 2005). A common theme that has emerged from a number of computational prediction studies is that mRNAs targeted by *miR-1* are predominantly expressed in non-muscle tissues. These observations suggest that *miR-1* could downregulate mRNAs that are specific for non-muscle tissues and whose products need to be stringently excluded from muscle tissue (Nguyen and Frasch 2006; Sokol and Ambros 2005).

7.4 Regulation of Cell Growth and Metabolism

The miRNAs *bantam*, *miR-278*, *miR-8*, and *miR-14* have all been implicated in regulating growth and metabolism. *Bantam* is an example of a *Drosophila*-specific miRNA with multiple biological roles. The *bantam* locus was initially identified in a gain-of-function screen for genes that affect tissue growth (Hipfner et al. 2002) and was later shown to encode a miRNA involved in cell growth and apoptosis (Brennecke et al. 2003). The complete elimination of *bantam* results in lethality at early pupal stage, whereas flies homozygous for hypomorphic alleles of *bantam* survive to adulthood but are small and display female fertility defects. Conversely, ectopic expression of *bantam* promotes growth of various tissues such as eyes and wings, due to an increase in the number of cells, with no increase in cell size. The mRNA targets through which *bantam* regulates growth are still unknown. However, *bantam* was shown to be a downstream target of Yorkie (Nolo et al. 2006; Thompson and Cohen 2006), which is a component of the Hippo signaling pathway that restricts tissue size by promoting apoptosis and cell-cycle arrest. Consistently, *hippo* mutant cells had elevated levels of *bantam*, and *bantam* was shown to be required for Yorkie-driven outgrowth. Additionally, overexpression of *bantam* rescued the growth defects of *yorkie* mutant cells (Nolo et al. 2006; Thompson and Cohen 2006). Thus, *bantam* forms a crucial component of the growth signaling pathways operating in diverse cellular as well as developmental contexts.

miR-278 is another miRNA that promotes growth and inhibits apoptosis, and has also been shown to play a role in controlling energy homeostasis. *miR-278* was initially identified in a gain-of-function screen for genes that affect growth during fly development (Teleman et al. 2006). *miR-278* mutants have elevated insulin levels and are correspondingly lean (Teleman et al. 2006), likely due to misregulation of the expanded gene: *expanded* mRNA levels increased severalfold in *miR-278* mutant tissue and overexpression of *expanded* in *miR-278* cells caused elevated expression of *Insulin like peptide* genes and produced the lean phenotype (Teleman et al. 2006).

Phenotypic analysis of *miR-8* null flies revealed a significant growth defect at larval stages, slightly delayed adult eclosion, and a smaller size and mass of adults which was a result of a reduced cell number and not cell size (Karres et al. 2007). *miR-8* was found abundantly expressed in the larval fat body, an important organ involved in the control of growth and metabolism. Fat body-specific expression of *miR-8* was sufficient to rescue the small body phenotype of *mir-8* null mutants, suggesting that *miR-8* in the fat body was important for systemic growth. The most dramatic rescue of *miR-8* null flies was observed in the depletion of *u-shaped* (*ush*), and *ush* RNA and protein levels were elevated in *miR-8* flies. Furthermore, insulin signaling was defective in the fat body of *miR-8* null flies, and Ush and its ortholog FOG2 regulate the insulin signaling pathway by directly interacting with the regulatory subunit of PI3 kinase (Hyun et al. 2009).

miR-14 has been recently shown to act in neurosecretory cells in the adult brain to control metabolism through its direct target *sugarbabe* (Varghese et al. 2010). Sugarbabe is a zinc finger protein that regulates insulin gene expression in neurosecretory cells. Regulation of *sugarbabe* by nutrients and *miR-14* allows it to function as a regulator of energy balance in a nutritionally variable environment (Varghese et al. 2010).

7.5 Regulation of Hox *Genes*

Homeobox (*hox*) genes encode homeodomain-containing transcription factors that specify the anterior–posterior axis and segment identity of organisms. *miR-10* and *miR-iab-4*/*mir-196* are encoded by analogous *hox* gene clusters in flies and vertebrates (Lagos-Quintana et al. 2001; Aravin et al. 2003). The *miR-iab-4* locus in flies encodes two miRNAs, *miR-iab-4-5p* and *miR-iab-4-3p*, that are generated from the two arms of the same hairpin (Aravin et al. 2003). In vivo analysis of *miR-iab-4-5p* showed that it inhibits the activity of one of its predicted targets *ultrabithorax* (*ubx*) (Grun et al. 2005; Stark et al. 2003): ectopic expression of *miR-iab-4-5p* prevented the accumulation of endogenous Ubx protein and induced a classical homeotic mutant phenotype that transformed halteres to wings (Ronshaugen et al. 2005). A subsequent study showed that the *iab-4* locus produced a second miRNA from the opposite DNA strand (*miR-iab-4AS*) that regulates neighboring *hox* genes (Stark et al. 2008). Computational analysis and luciferase reporter assays confirmed the presence of highly conserved binding sites for *miR-iab-4AS* in the 3′-UTRs of *abdominal-A*, *ubx*, and *antennapedia* (Stark et al. 2007). Moreover, a clear homeotic transformation of halteres to wings was observed when *miR-iab-4AS* was misexpressed in the haltere imaginal disc (Stark et al. 2008).

8 Concluding Remarks

Functional studies in *Drosophila* suggest that miRNAs play critical roles in maintaining the fidelity and precision of cellular and developmental programs. Moreover, these studies also suggest that miRNAs employ multiple modes of action to regulate different target mRNAs. Aspects of miRNA biology that have not received much attention include their role as reversible posttranscriptional regulators in neurons, mechanisms that control miRNA decay, and posttranscriptional control of miRNAs. As the annotation of miRNAs is approaching saturation, the stage is now set for analysis of the diverse miRNA–target interactions. Genetic analysis of miRNA loss-of-function mutants in *Drosophila* and their regulatory relationships with their target genes will continue to reveal new insights into the regulation and functioning of miRNAs.

References

Aravin AA, Lagos-Quintana M, Yalcin A et al (2003) The small RNA profile during *Drosophila melanogaster* development. Dev Cell 5:337–350

Bartel DP (2009) MicroRNAs: target recognition and regulatory functions. Cell 136:215–233

Bejarano F, Smibert P, Lai EC (2010) miR-9a prevents apoptosis during wing development by repressing *Drosophila* LIM-only. Dev Biol 338:63–73

Bentwich I (2005) Prediction and validation of microRNAs and their targets. FEBS Lett 579:5904–5910

Berezikov E, Robine N, Samsonova A et al (2010) Deep annotation of *Drosophila melanogaster* microRNAs yields insights into their processing, modification, and emergence. Genome Res 21:203–215

Biryukova I, Asmar J, Abdesselem H et al (2009) *Drosophila* mir-9a regulates wing development via fine-tuning expression of the LIM only factor, dLMO. Dev Biol 327:487–496

Brennecke J, Hipfner DR, Stark A et al (2003) Bantam encodes a developmentally regulated microRNA that controls cell proliferation and regulates the proapoptotic gene hid in *Drosophila*. Cell 113:25–36

Brennecke J, Stark A, Russell RB et al (2005) Principles of microRNA-target recognition. PLoS Biol 3:e85

Caygill EE, Johnston LA (2008) Temporal regulation of metamorphic processes in *Drosophila* by the let-7 and miR-125 heterochronic microRNAs. Curr Biol 18:943–950

Cayirlioglu P, Kadow IG, Zhan X et al (2008) Hybrid neurons in a microRNA mutant are putative evolutionary intermediates in insect CO2 sensory systems. Science 319:1256–1260

Cohen SM, Brennecke J, Stark A (2006) Denoising feedback loops by thresholding—a new role for microRNAs. Genes Dev 20:2769–2772

Enright AJ, John B, Gaul U et al (2003) MicroRNA targets in *Drosophila*. Genome Biol 5:R1

Fabian MR, Sonenberg N, Filipowicz W (2010) Regulation of mRNA translation and stability by microRNAs. Annu Rev Biochem 79:351–379

Forstemann K, Horwich MD, Wee L et al (2007) *Drosophila* microRNAs are sorted into functionally distinct argonaute complexes after production by dicer-1. Cell 130:287–297

Griffiths-Jones S (2004) The microRNA registry. Nucleic Acids Res 32(Database issue): D109–D111

Grun D, Wang YL, Langenberger D et al (2005) MicroRNA target predictions across seven *Drosophila* species and comparison to mammalian targets. PLoS Comput Biol 1:e13
Herranz H, Cohen SM (2010) MicroRNAs and gene regulatory networks: managing the impact of noise in biological systems. Genes Dev 24:1339–1344
Hilgers V, Bushati N, Cohen SM (2010) *Drosophila* microRNAs 263a/b confer robustness during development by protecting nascent sense organs from apoptosis. PLoS Biol 8:e1000396
Hipfner DR, Weigmann K, Cohen SM (2002) The bantam gene regulates *Drosophila* growth. Genetics 161:1527–1537
Hobert O (2004) Common logic of transcription factor and microRNA action. Trends Biochem Sci 29:462–468
Hyun S, Lee JH, Jin H et al (2009) Conserved microRNA miR-8/miR-200 and its target USH/FOG2 control growth by regulating PI3K. Cell 139:1096–1108
Iovino N, Pane A, Gaul U (2009) miR-184 has multiple roles in *Drosophila* female germline development. Dev Cell 17:123–133
Ishizuka A, Siomi MC, Siomi H (2002) A *Drosophila* fragile X protein interacts with components of RNAi and ribosomal proteins. Genes Dev 16:2497–2508
Jiang F, Ye X, Liu X et al (2005) Dicer-1 and R3D1-L catalyze microRNA maturation in *Drosophila*. Genes Dev 19:1674–1679
Kadener S, Menet JS, Sugino K et al (2009) A role for microRNAs in the *Drosophila* circadian clock. Genes Dev 23:2179–2191
Karres JS, Hilgers V, Carrera I et al (2007) The conserved microRNA miR-8 tunes atrophin levels to prevent neurodegeneration in *Drosophila*. Cell 131:136–145
Kawamata T, Tomari Y (2010) Making RISC. Trends Biochem Sci 35:368–376
Kennell JA, Gerin I, MacDougald OA et al (2008) The microRNA miR-8 is a conserved negative regulator of Wnt signaling. Proc Natl Acad Sci USA 105:15417–15422
Kertesz M, Iovino N, Unnerstall U et al (2007) The role of site accessibility in microRNA target recognition. Nat Genet 39:1278–1284
Kim VN, Han J, Siomi MC (2009) Biogenesis of small RNAs in animals. Nat Rev Mol Cell Biol 10:126–139
Krek A, Grun D, Poy MN et al (2005) Combinatorial microRNA target predictions. Nat Genet 37:495–500
Kwon C, Han Z, Olson EN et al (2005) MicroRNA1 influences cardiac differentiation in *Drosophila* and regulates Notch signaling. Proc Natl Acad Sci USA 102:18986–18991
Lagos-Quintana M, Rauhut R, Lendeckel W et al (2001) Identification of novel genes coding for small expressed RNAs. Science 294:853–858
Lai EC (2002) Micro RNAs are complementary to 3′UTR sequence motifs that mediate negative post-transcriptional regulation. Nat Genet 30:363–364
Lai EC, Posakony JW (1997) The bearded box, a novel 3′UTR sequence motif, mediates negative post-transcriptional regulation of bearded and enhancer of split complex gene expression. Development 124:4847–4856
Lai EC, Tam B, Rubin GM (2005) Pervasive regulation of *Drosophila* Notch target genes by GY-box-, Brd-box-, and K-box-class microRNAs. Genes Dev 19:1067–1080
Lai EC, Tomancak P, Williams RW et al (2003) Computational identification of *Drosophila* microRNA genes. Genome Biol 4:R42
Lau NC, Lim LP, Weinstein EG et al (2001) An abundant class of tiny RNAs with probable regulatory roles in *Caenorhabditis elegans*. Science 294:858–862
Leaman D, Chen PY, Fak J et al (2005) Antisense-mediated depletion reveals essential and specific functions of microRNAs in *Drosophila* development. Cell 121:1097–1108
Lee RC, Ambros V (2001) An extensive class of small RNAs in *Caenorhabditis elegans*. Science 294:862–864
Lee RC, Feinbaum RL, Ambros V (1993) The *C. elegans* heterochronic gene lin-4 encodes small RNAs with antisense complementarity to lin-14. Cell 75:843–854
Lee Y, Ahn C, Han J et al (2003) The nuclear RNase III Drosha initiates microRNA processing. Nature 425:415–419

Lee Y, Jeon K, Lee JT et al (2002) MicroRNA maturation: stepwise processing and subcellular localization. EMBO J 21:4663–4670
Lee Y, Kim M, Han J et al (2004) MicroRNA genes are transcribed by RNA polymerase II. EMBO J 23:4051–4060
Li X, Carthew RW (2005) A microRNA mediates EGF receptor signaling and promotes photoreceptor differentiation in the *Drosophila* eye. Cell 123:1267–1277
Li X, Cassidy JJ, Reinke CA et al (2009) A microRNA imparts robustness against environmental fluctuation during development. Cell 137:273–282
Li Y, Wang F, Lee JA et al (2006) MicroRNA-9a ensures the precise specification of sensory organ precursors in *Drosophila*. Genes Dev 20:2793–2805
Liu Q, Rand TA, Kalidas S et al (2003) R2D2, a bridge between the initiation and effector steps of the *Drosophila* RNAi pathway. Science 301:1921–1925
Long D, Lee R, Williams P et al (2007) Potent effect of target structure on microRNA function. Nat Struct Mol Biol 14:287–294
Loya CM, Lu CS, Van Vactor D et al (2009) Transgenic microRNA inhibition with spatiotemporal specificity in intact organisms. Nat Methods 6:897–903
Min H, Yoon S (2010) Got target? Computational methods for microRNA target prediction and their extension. Exp Mol Med 42:233–244
Nguyen HT, Frasch M (2006) MicroRNAs in muscle differentiation: lessons from *Drosophila* and beyond. Curr Opin Genet Dev 16:533–539
Nolo R, Morrison CM, Tao C et al (2006) The bantam microRNA is a target of the hippo tumor-suppressor pathway. Curr Biol 16:1895–1904
Okamura K, Hagen JW, Duan H et al (2007) The mirtron pathway generates microRNA-class regulatory RNAs in *Drosophila*. Cell 130:89–100
Park JE, Heo I, Tian Y et al (2011) Dicer recognizes the 5′end of RNA for efficient and accurate processing. Nature 475:201–205
Parrish JZ, Xu P, Kim CC et al (2009) The microRNA bantam functions in epithelial cells to regulate scaling growth of dendrite arbors in *Drosophila* sensory neurons. Neuron 63:788–802
Pasquinelli AE, Reinhart BJ, Slack F et al (2000) Conservation of the sequence and temporal expression of let-7 heterochronic regulatory RNA. Nature 408:86–89
Pek JW, Lim AK, Kai T (2009) *Drosophila* maelstrom ensures proper germline stem cell lineage differentiation by repressing microRNA-7. Dev Cell 17:417–424
Reinhart BJ, Slack FJ, Basson M et al (2000) The 21-nucleotide let-7 RNA regulates developmental timing in *Caenorhabditis elegans*. Nature 403:901–906
Reinhart BJ, Weinstein EG, Rhoades MW et al (2002) MicroRNAs in plants. Genes Dev 16:1616–1626
Ronshaugen M, Biemar F, Piel J et al (2005) The *Drosophila* microRNA iab-4 causes a dominant homeotic transformation of halteres to wings. Genes Dev 19:2947–2952
Ruby JG, Stark A, Johnston WK et al (2007) Evolution, biogenesis, expression, and target predictions of a substantially expanded set of *Drosophila* microRNAs. Genome Res 17:1850–1864
Rybak A, Fuchs H, Smirnova L et al (2008) A feedback loop comprising lin-28 and let-7 controls pre-let-7 maturation during neural stem-cell commitment. Nat Cell Biol 10:987–993
Saito K, Ishizuka A, Siomi H et al (2005) Processing of pre-microRNAs by the Dicer-1-Loquacious complex in *Drosophila* cells. PLoS Biol 3:e235
Sempere LF, Sokol NS, Dubrovsky EB et al (2003) Temporal regulation of microRNA expression in *Drosophila melanogaster* mediated by hormonal signals and broad-complex gene activity. Dev Biol 259:9–18
Silver SJ, Hagen JW, Okamura K et al (2007) Functional screening identifies miR-315 as a potent activator of Wingless signaling. Proc Natl Acad Sci USA 104:18151–18156
Siomi H, Siomi MC (2010) Posttranscriptional regulation of microRNA biogenesis in animals. Mol Cell 38:323–332
Sokol NS, Ambros V (2005) Mesodermally expressed *Drosophila* microRNA-1 is regulated by Twist and is required in muscles during larval growth. Genes Dev 19:2343–2354

Sokol NS, Xu P, Jan YN et al (2008) *Drosophila* let-7 microRNA is required for remodeling of the neuromusculature during metamorphosis. Genes Dev 22:1591–1596
Stark A, Brennecke J, Bushati N et al (2005) Animal MicroRNAs confer robustness to gene expression and have a significant impact on 3′UTR evolution. Cell 123:1133–1146
Stark A, Brennecke J, Russell RB et al (2003) Identification of *Drosophila* microRNA targets. PLoS Biol 1:E60
Stark A, Bushati N, Jan CH et al (2008) A single Hox locus in *Drosophila* produces functional microRNAs from opposite DNA strands. Genes Dev 22:8–13
Stark A, Kheradpour P, Parts L et al (2007) Systematic discovery and characterization of fly microRNAs using 12 *Drosophila* genomes. Genome Res 17:1865–1879
Teleman AA, Maitra S, Cohen SM (2006) *Drosophila* lacking microRNA miR-278 are defective in energy homeostasis. Genes Dev 20:417–422
Thompson BJ, Cohen SM (2006) The Hippo pathway regulates the bantam microRNA to control cell proliferation and apoptosis in *Drosophila*. Cell 126:767–774
Truscott M, Islam AB, Lopez-Bigas N et al (2011) mir-11 limits the proapoptotic function of its host gene, dE2f1. Genes Dev 25:1820–1834
Tsurudome K, Tsang K, Liao EH et al (2010) The *Drosophila* miR-310 cluster negatively regulates synaptic strength at the neuromuscular junction. Neuron 68:879–893
Tyler DM, Okamura K, Chung WJ et al (2008) Functionally distinct regulatory RNAs generated by bidirectional transcription and processing of microRNA loci. Genes Dev 22:26–36
Varghese J, Cohen SM (2007) MicroRNA miR-14 acts to modulate a positive autoregulatory loop controlling steroid hormone signaling in *Drosophila*. Genes Dev 21:2277–2282
Varghese J, Lim SF, Cohen SM (2010) *Drosophila* miR-14 regulates insulin production and metabolism through its target, sugarbabe. Genes Dev 24:2748–2753
Vasudevan S, Tong Y, Steitz JA (2007) Switching from repression to activation: microRNAs can up-regulate translation. Science 318:1931–1934
Viswanathan SR, Daley GQ (2010) Lin28: a microRNA regulator with a macro role. Cell 140:445–449
Yue D, Liu H, Huang Y (2009) Survey of computational algorithms for MicroRNA target prediction. Curr Genomics 10:478–492

Viroids: The Smallest Known Infectious Agents Cause Accumulation of Viroid-Specific Small RNAs

Jaroslav Matoušek, Detlev Riesner, and Gerhard Steger

Contents

Abstract Viroids are plant-infectious, noncoding, unencapsidated, circular RNAs ranging in size from 250 to 400 nucleotides that are transcribed in a rolling-circle mechanism either in nuclei (*Pospiviroidae*) or in chloroplasts (*Avsunviroidae*) of plant hosts. The pathogenic effect caused by viroids is still an enigma: Potato spindle tuber viroid (PSTVd), the type strain of *Pospiviroidae*, causes typical symptoms in tomato plants, but the severity of symptoms depends on the tomato cultivar; different strains of PSTVd, which vary in sequence by a few mutations

J. Matoušek
Institute of Plant Molecular Biology, Biology Centre v.v.i., Academy of Science CR, 37005 České Budějovice, Czech Republic
e-mail: jmat@umbr.cas.cz

D. Riesner (✉) • G. Steger
Institut für Physikalische Biologie, Heinrich-Heine-Universität Düsseldorf, 40204 Düsseldorf, Germany
e-mail: riesner@biophys.uni-duesseldorf.de; steger@biophys.uni-duesseldorf.de

V.A. Erdmann and J. Barciszewski (eds.), *From Nucleic Acids Sequences to Molecular Medicine*, RNA Technologies, DOI 10.1007/978-3-642-27426-8_26,

from each other, induce symptoms from very mild up to necrosis upon infection of a cultivar. According to recent findings, viroids cause the accumulation of viroid-specific small RNAs (vsRNA) similar in size to small interfering (siRNA) and miRNAs, but they do escape the cytoplasmic silencing mechanism. In this chapter, we will discuss these findings and hypotheses on the biogenesis of viroid-specific small RNAs and connections to symptom induction.

Keywords Pathways of vsRNA production • Potential targets of vsRNA • Transcription factors

1 Introduction

Viroids are single-stranded, circular RNA molecules that replicate autonomously in infected host plants. Their length is 250–400 nucleotides, depending on the viroid species. As noncoding RNA, viroids do not code for any protein but possess the biological activity and function of a minimal parasite replicating autonomously in the host cell and promoting specific disease of the host including significant losses in the yield of some important crops. The viroid RNA does encode signals within its sequence and its thermodynamically stable or metastable structures that allow for them to exploit the host machinery for their own replication, processing, transport, and induction of pathogenesis [for recent reviews, see Hadidi et al. (2003), Tabler and Tsagris (2004), Flores et al. (2005), Daròs et al. (2006), Schmitz and Steger (2007), and Ding (2010)].

1.1 Taxonomy of Viroids

Viroids are classified into two families (Flores et al. 1998), *Pospiviroidae* with 27 species and *Avsunviroidae* with four species (see the "Subviral RNA Database"[1] Rocheleau and Pelchat 2006). *Pospiviroidae* is named for *potato spindle tuber viroid* (PSTVd); further species in this family are *citrus exocortis viroid* (CEVd), *hop stunt viroid* (HSVd), and *coconut cadang-cadang viroid* (CCCVd). Species of this family possess a rod-like secondary structure in their thermodynamically optimal state [for an example, see Fig. 1a; for review, see Riesner and Gross (1985) and Steger and Riesner (2003)], are localized primarily in the nucleolus (Schumacher et al. 1983; Harders et al. 1989; Qi and Ding 2003), are replicated in an asymmetric rolling-circle mechanism (Branch and Robertson 1984) by DNA-dependent RNA polymerase II of the host (Schindler and Mühlbach 1992; Warrilow and Symons 1999; Kolonko et al. 2006), and only (+)-stranded

[1] http://subviral.med.uottawa.ca.

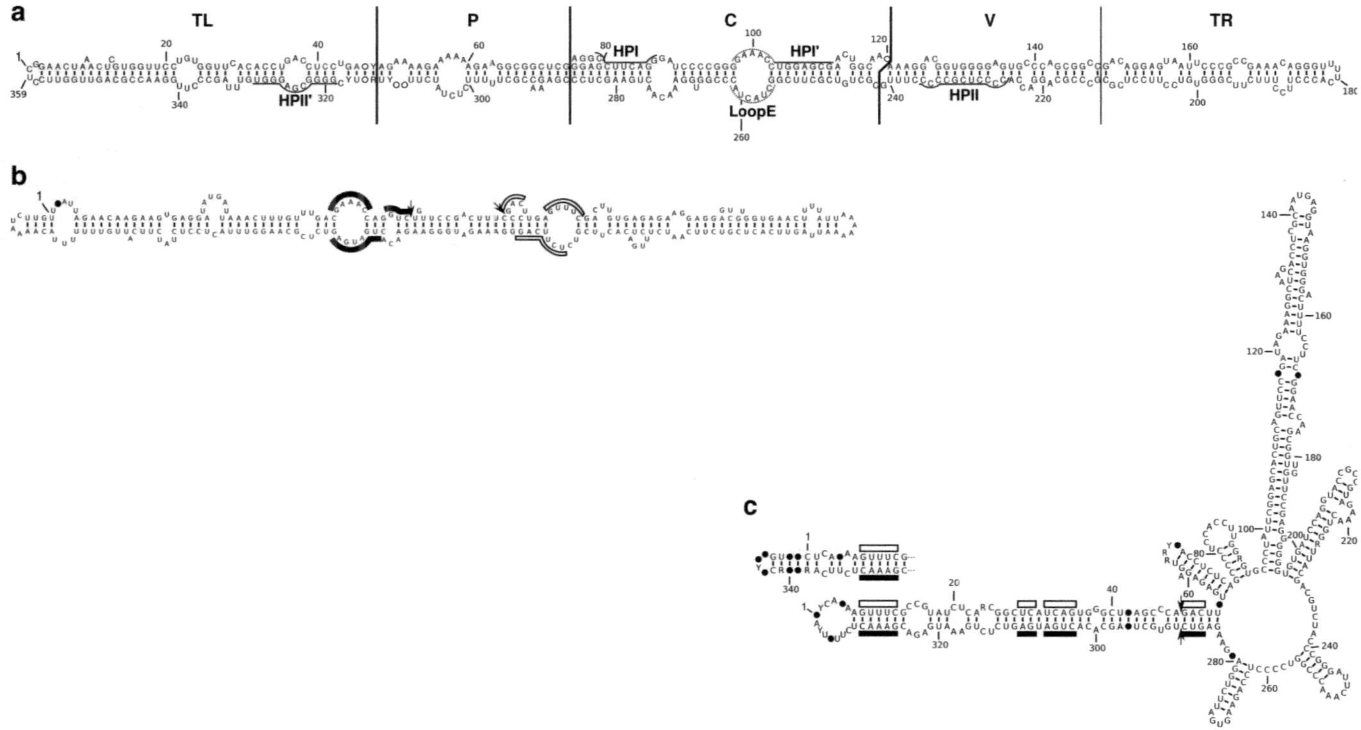

Fig. 1 Secondary structure of viroids. The shown consensus structures and sequences are based on 131, 90, and 331 different sequences of PSTVd (**a**), ASBVd (**b**), and PLMVd (**c**), respectively, obtained from the "Subviral RNA Database" (Rocheleau and Pelchat 2006). Structures were predicted by ConStruct (Wilm et al. 2008) and drawn by R2R (Weinberg and Breaker 2011); non-conserved nucleotides are marked by *filled circles*. (**a**) The five homology domains of pospiviroids are marked as proposed by Keese and Symons (1985): terminal left and right (TL, TR), pathogenicity-modulating (P), central conserved (C), and variable (V) domain; the P domain overlaps with the thermodynamically most unstable "premelting" region (Steger et al. 1984).

replication intermediates are processed by host enzymes to monomeric circles (Baumstark et al. 1997; Schrader et al. 2003). *Avsunviroidae* is named for *avocado sun blotch viroid* (ASBVd); a further species is *peach latent mosaic viroid* (PLMVd). Species of this family differ from *Pospiviroidae* by localization in the chloroplast (Lima et al. 1994; Pelchat et al. 2001), replication in a symmetric rolling-circle mechanism (Daròs et al. 1994) by a chloroplastic RNA polymerase (Navarro et al. 2000; Navarro and Flores 2000), and self-cleavage of replication intermediates via hammerhead ribozymes in the (+) as well as in the (−) strand (Daròs et al. 1994; Daròs and Flores 2002). For predicted structure of ASBVd and PLMVd, see Fig. 1b, c, respectively.

1.2 Movement of Viroids

During systemic infection of a host plant, a viroid has to exit the nucleus or chloroplast, move from cell to cell to invade neighboring cells, to enter the phloem for long-distance trafficking into upper leaves and the root (Palukaitis 1987; Zhu et al. 2001); from the phloem, it has to enter nonvascular cells to establish infection in further tissue [for recent reviews, see Ding and Wang (2009), Ding (2010), and Wang and Ding (2010)]. Several loops of the native structure of PSTVd seem to be critical for trafficking (Zhong et al. 2008). Host proteins have to be involved by complex formation with the viroid RNA; possible candidates are the viroid RNA-binding protein 1 (Virp1; Martínez de Alba et al. 2003; Kalantidis et al. 2007), the phloem protein 2 of *Cucumis sativus* (Owens et al. 2001; Gómez and Pallás 2001, 2004), ribosomal protein L5, and/or transcription factor IIIA (Eiras et al. 2010). To identify further host factors, a system based on *Arabidopsis thaliana* would be desirable. However, no viroid occurs naturally in this model plant, in which replication and processing of viroids are supported to a certain extent, but trafficking seems to be the limiting step (Daròs and Flores 2004; Matoušek et al. 2004).

Fig. 1 (continued) The C domain contains the loop E motif (Branch et al. 1985), which possesses structural similarity to loop E of eukaryotic 5S RNA (Zhong et al. 2006) and is involved in processing of (+)-intermediates to mature circles (Baumstark et al. 1997). Regions forming the thermodynamically extrastable hairpins (HP) I and II in metastable structures are marked (Steger and Riesner 2003). (**b**) and (**c**) The three regions most conserved in hammerhead ribozymes (Ruffner et al. 1990) are marked by *bars* [*filled*, (+)-strand; *open*, (−)-strand]; the cleavage sites are marked by an *arrow*. (**c**) The full structure is shown for "short" PLMVd (based on 270 sequences); certain RNAs possess an elongated left terminal hairpin (61 sequences). For the right, branched structure part, different alternatives are thermodynamically probable [see, e.g., Dubé et al. (2011)]

1.3 Host Range and Symptoms of Viroids

Viroids do infect a wide range of food, industrial, and ornamental plants and can produce severe crop losses [for review, see Randles (2003)]; for example, CCCVd leads to a ceased nut production and is finally lethal to coconut palms, CEVd causes bark shelling and scaling on trifoliate orange, and HSVd causes plant stunting and reduces hop cone weight and alpha acid content by about 50%. PLMVd generally shows only rarely symptoms on leaves but affects fruit quality, whereas certain PLMVd variants, which differ by an insertion of 11–13 nt (see Fig. 1c; Rodio et al. 2006), cause white patterns covering most or all of the leaf area, a disease called peach calico.

1.4 Severity of Viroid Symptoms

The symptoms elicited by a certain viroid species depend on both the host and the viroid strain. For example, a certain PSTVd strain (named "Diener" or "intermediate"; GenBank ACM88678) produces symptoms of stunting and leaf epinasty in the PSTVd-sensitive tomato (*Solanum lycopersicum*) cultivar (cv) Rutgers, whereas cultivars Goldkugel and Moneymaker are tolerant and practically symptomless (Stark-Lorenzen et al. 1997; Matoušek et al. 2007). A PSTVd strain like AS1 (AC AY518939) causes strong symptoms such as dwarfing and necrosis upon infection of tomato plants cv Rutgers, whereas a mild strain like QFA (AC U23059) shows nearly no symptoms in cv Rutgers. The sequence differences between the PSTVd strains of different severity are located mostly [but not exclusively: see, e.g., Sano et al. (1992), Rodriguez and Randles (1993), and Reanwarakorn and Semancik (1998)] in the so-called pathogenicity-modulating (P) domain of the PSTVd structure (cf. Fig. 1a). These sequence differences give rise to different thermodynamic stability of this domain (Schnölzer et al. 1985); the stability is, however, not (inversely) correlated with pathogenicity (Owens et al. 1996), as originally proposed based on only a few PSTVd strains (Schnölzer et al. 1985). A model that tried to correlate the three-dimensional bending of the P domain (Schmitz and Riesner 1998; Owens et al. 1996) does also not hold for all strains.

1.5 Symptom Induction

Over the years, several hypotheses were presented that tried to correlate a viroid feature with symptom expression. Viroids activate a general plant defense system based on pathogenesis-related (PR) proteins, which is also induced by other pathogens including viruses, bacteria, and fungi (Camacho Henriquez and Sänger 1982; Gadea et al. 1996; López-Gresa et al. 2010). However, PSTVd infection is also followed by suppression or expression of unique host genes (Itaya et al. 2002;

Hammond and Zhao 2009). A few viroid-interacting proteins like histones (Wolff et al. 1985), dsRNA-activated protein kinase (Diener et al. 1993), ribosomal proteins S25 (Werner et al. 1995) and L5 (Eiras et al. 2010), phloem protein 2 (Gómez and Pallás 2001; Owens et al. 2001), Virp1 (Martínez de Alba et al. 2003; Kalantidis et al. 2007), and transcription factor IIIA (Eiras et al. 2010) have been reported; their involvement in symptom induction is mostly unknown. Direct base-pairing interaction of host RNAs with viroids was proposed for 7S RNA (Haas et al. 1988), U1 RNA/mRNA splicing (Dickson 1981), and ribosomal RNA (Rodio et al. 2007), but could not be confirmed experimentally, at least for full-length viroid RNA with these host RNAs. Viroid–5S rRNA complexes could be identified inside cells by a cross-linking methodology (Aschermann 2001; Stroeks 2008).

After the finding that a PSTVd infection leads to de novo methylation of PSTVd cDNA (Wassenegger et al. 1994) and the detection of small RNAs with viroid sequence in plants infected with viroids from *Pospiviroidae* (Papaefthimiou et al. 2001; Itaya et al. 2001) or *Avsunviroidae* (Markarian et al. 2003; Martínez de Alba et al. 2002), ongoing research concentrates on a possible connection of these viroid-specific small RNAs to (mis)regulation of the host's gene expression via transcriptional (TGS) or posttranscriptional gene silencing (PTGS).

2 Viroid-Specific Small RNAs

RNAs with a length of 21–24 nt and viroid sequence of both polarities accumulate in viroid-infected plants (Papaefthimiou et al. 2001; Itaya et al. 2001; Markarian et al. 2004; Matoušek et al. 2007; Machida et al. 2007; Martín et al. 2007; Gómez et al. 2008; Martinez et al. 2010; Wang et al. 2011). Such RNAs, called viroid-derived small RNAs (vd-sRNA) or viroid-specific RNAs (vsRNA), are similar in size to other small RNAs like silencing RNAs (siRNA), microRNAs (miRNA), or trans-acting silencing RNAs (ta-siRNA) [for reviews, see Ramachandran and Chen (2008), Gómez et al. (2009), and Jamalkandi and Masoudi-Nejad (2009)]. In the following, we will summarize findings and hypotheses on the vsRNAs.

2.1 Pathway of vsRNA Production

At present, the pathway responsible to generate the vsRNAs is still a matter of debate; almost none of the possible pathways for genesis of small RNAs can be excluded. Common to all pathways in plants is the involvement of a protein from the Dicer-like (DCL) family acting either on double-stranded RNA (dsRNA, leading to siRNA and ta-siRNA, respectively) or on partially complementary single-stranded RNA [primary (pri-miRNA) and precursor miRNA (pre-miRNA) leading to miRNA]. The dsRNA could be a product of, for example, RNA-virus replication (in an siRNA pathway) or of host-encoded RNA-dependent RNA polymerase (RDR, in a ta-siRNA pathway). DCL proteins have been described to localize in the nucleus and in the cytoplasm, but not in chloroplasts; that is, vsRNAs from members of families *Pospiviroidae* and *Avsunviroidae* might be generated in the cytoplasm during cell-to-cell movement, whereas vsRNAs from members of *Pospiviroidae* might be generated in the nucleus and transported to the cytoplasm. In fact, PSTVd-specific small RNAs are found in the cytoplasm as most other miRNAs and siRNAs, but not in the nucleus (Denti et al. 2004), which is the place of pospiviroid transcription and storage.

Several authors [e.g., see Denti et al. (2004) and Martinez et al. (2010)] favor an siRNA-like mechanism including a viroid dsRNA. A viroid dsRNA is not involved in the replication cycle of viroids, but might be (accidentally) formed by (+)- and (−)-stranded replication intermediates. Such a viroid dsRNA could, however, never be shown in vivo, for example, after psoralen cross-linking in plants.

Gómez et al. (2008, 2009) favor a ta-siRNA-like generation in which RDR6 activity is required for producing the double-stranded precursors of secondary siRNAs. This is supported by the finding that RDR6 activity is required for HSVd-induced symptoms in *Nicotiana benthamiana*. In contrast, Di Serio et al. (2009a) showed that symptom expression in comparable PSTVd-infected plants is not dependent on RDR6.

The single-stranded viroid RNA in its typical, highly base-paired, at least partially rod-shaped structures (cf. Fig. 1) suggests that viroid RNA might mimic a pri-miRNA or pre-miRNA. Indeed, crude DCL preparations do release small RNAs of 21–24 nt from transcripts of PSTVd (Itaya et al. 2007) and from partial transcripts of PLMVd encompassing the hammerhead-containing hairpin (see Fig. 1c; Landry and Perreault 2005). Furthermore, a computer tool, able to predict pre-miRNAs and miRNA/miRNA* complexes in plant genomes, identifies PSTVd (and the PLMVd hairpin) as pre-miRNAs (Teune and Steger 2010).

2.2 Sequence of vsRNAs

By small- and large-scale sequencing of vsRNA isolated from different host–viroid combinations, their sequences were determined: PSTVd-infected tomato cv Rutgers and Moneymaker (Itaya et al. 2007; Machida et al. 2007; Diermann et al. 2010; Wang et al. 2011), PSTVd-infected *N. benthamiana* (Di Serio et al. 2009a), HSVd- and *grapevine yellow speckle viroid* 1 (GYSVd1)-infected grapevine (Navarro et al. 2009), HSVd-infected cucumber (Martinez et al. 2010), and PLMVd-infected peach (Di Serio et al. 2009b; Bolduc et al. 2010). Common to all analyses is the detection of "hotspots" of vsRNA (cf. Fig. 2); that is, vsRNA sequences are not evenly distributed over the respective viroid genome.

The hotspots for (+)- and (−)-stranded vsRNAs might have overlap but do not coincide. Furthermore, the vsRNA's distributions from stems versus leafs and from the two tomato cultivars are clearly different (Wang et al. 2011); for example, more (−)-stranded vsRNAs were detected in Moneymaker than in Rutgers plants. One hotspot of PSTVd-specific small RNAs is located in the P domain (Diermann et al.

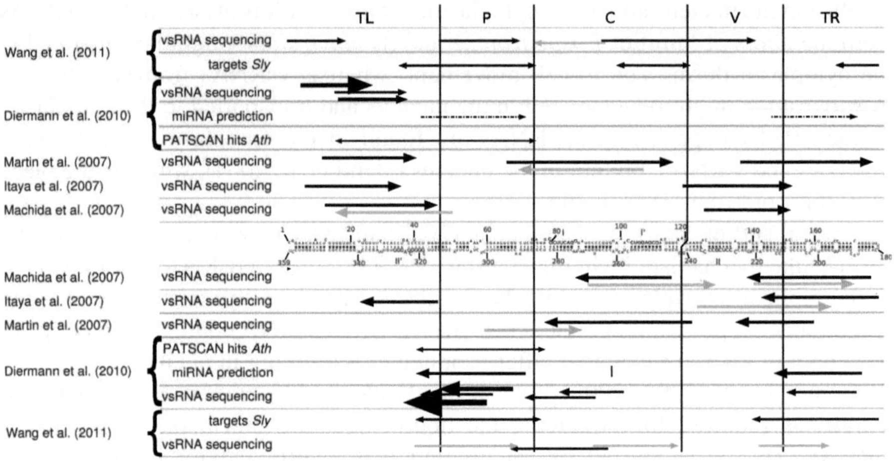

Fig. 2 Mapping of vsRNAs in circular PSTVd. In the *center* is shown a consensus secondary structure of (+)-stranded circular PSTVd (see Fig. 1a). In the *boxes* above and below the structure are summarized mappings of (+)-stranded (*black arrows*) and (−)-stranded (*gray arrows*) vsRNAs. The *boxes* marked "Wang et al. (2011)," "Diermann et al. (2010)," "Machida et al. (2007)," and "Itaya et al. (2007)" are based on PSTVd infection; the *box* marked "Martín et al. (2007)" shows a mapping for CEVd. The *solid and dotted arrow* pairs in *boxes* marked "miRNA prediction" denote the miRNA/miRNA* complexes predicted by NOVOMIR (Teune and Steger 2010)

2010; Wang et al. 2011), which would allow for a correlation of vsRNAs with symptoms; vsRNAs of PSTVd strains differing in pathogenicity were not analyzed up to now. PLMVd-specific small RNAs from the left terminus of the hammerhead-containing hairpin, which differs in length between mosaic- and calico-inducing PLMVd strains (cf. Fig. 1c and Sect. 1.3), are underrepresented (Di Serio et al. 2009b).

2.3 *Influence of vsRNA on Viroid Accumulation*

Up to now, the level of vsRNA accumulation and viroid titer was analyzed only by hybridization, which does not allow conclusions on vsRNA levels from different sequence regions or domains. Anyway, the results described in the literature are quite divergent, which might be due to the use of different experimental methods and different time points of analysis. For example, for PSTVd, no difference in vsRNA levels could be detected for strains of different virulence (Papaefthimiou et al. 2001; Itaya et al. 2001); in a different study, the ratio of vsRNA to full-length viroid RNA was found to be higher for a lethal than for a mild strain (Matoušek et al. 2007). For CEVd, the vsRNA level was directly correlated with symptom expression, but not with the viroid level (Markarian et al. 2003). The levels of ASBVd-specific sRNA were described either as proportional to the ASBVd level (Markarian et al. 2003) or

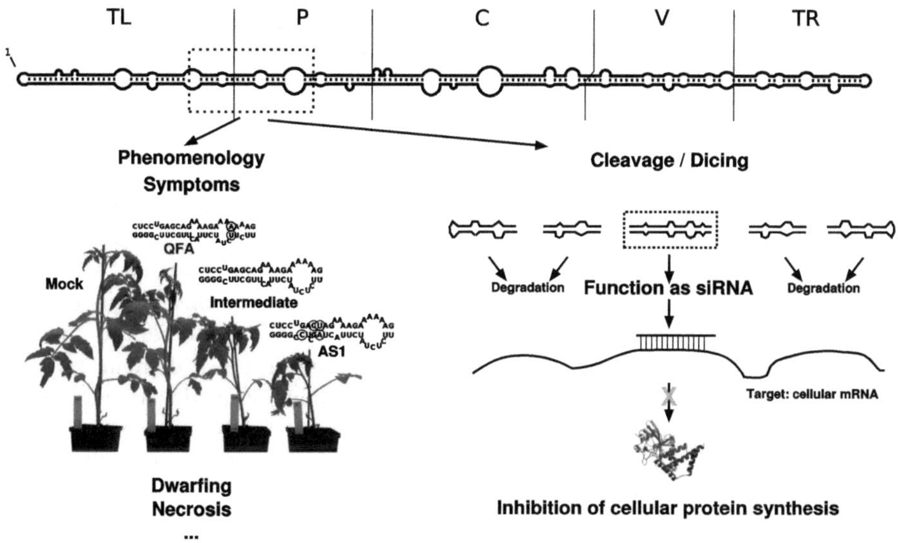

Fig. 3 Model for pathogenesis of viroid infection. Small sequence differences in or close to the P domain (*boxed* in the viroid structure, *top*) lead to drastic differences in symptoms. An example is tomato plants either mock-inoculated or infected with PSTVd strains QFA (mild), intermediate, and AS1 (lethal), respectively (*bottom left*); the few mutations (*circled*) between these strains are located in the P domain. Cleavage of PSTVd (*right*) leads to vsRNA complementary to a host mRNA followed by misregulation of cellular protein synthesis and symptoms

as undetectable (Markarian et al. 2003). For two variants of *chrysanthemum chlorotic mottle viroid* (CChMVd), belonging to *Avsunviroidae*, no dependence in the level of vsRNA on symptoms was detected (Martínez de Alba et al. 2002).

3 Potential Targets of vsRNAs

In general, RNA silencing refers to eukaryotic sequence-specific RNA-based pathways downregulating gene expression either by transcriptional repression via DNA methylation of promoter regions or by posttranscriptional regulation via inactivation or repression of target RNAs. Small RNAs bind to Argonaute (AGO) proteins and guide the resulting RNA-induced silencing complex (RISC) to the partially or fully complementary targets.

If vsRNA(s) is involved in symptom expression via (m)RNA silencing (cf. Fig. 3), one expects to find such targets either by bioinformatic means in the host genome sequence and/or by experimental detection of differential RNA expression. Several authors have published shorter or longer lists of differentially regulated mRNAs (e.g., Itaya et al. 2002; Wang et al. 2011). Up to now, however, no direct or indirect link between a vsRNA and a target could be established. Searches for targets, either by simple scans for (partial) complementarities between viroid and host genome sequences or by more refined methods, found a zoo of mRNAs, introns, and non-protein-coding regions in *A. thaliana* and *S. lycopersicum*, respectively. Many of the complementarities were found in both upper and lower P domain of PSTVd (see Fig. 2, lines marked as "targets *Sly*" and "PATSCAN hits *Ath*"), which is, however, not conclusive because of the biased nucleotide composition of PSTVd's P domain.

Lethal PSTVd strains like AS1 (AC AY518939), KF440-2 (AC X58388), and C3 (AC HE575349) promote very similar disease symptoms in solanaceous species, suggesting an involvement of common regulatory network(s) in symptom expression, but these strains differ by multiple point mutations within their P domains. For instance, the pathogenicity of PSTVd C3 is similar to that of PSTVd AS1, but both strains differ by five mutations in the P domain. From a thermodynamic point of view and in terms of gene silencing-mediated pathogenesis, one has to assume a high probability of different vsRNA sequence(s) and consequently different vsRNA targets to initiate pathogenesis within the same network. In addition, besides several potential targets, one has to consider another factor important in the development of viroid disease: tissue and cellular specificity of potential targeting. The tissue-specific effects were, for instance, detected for the PSTVd AS1-activated apoptotic nuclease TBN1 in tomato vascular system, whereas essentially no changes in nuclease activity were detected in leaf blades (Matoušek et al. 2007). The part of the network connected to vascular system degeneration involves, for instance, the oligofamily of Vsf-1 transcription factors (TFs) (Torres-Schumann et al. 1996) and VPE proteases with plant metacaspase activity initiating activation of lytic vacuoles and apoptotic changes (e.g., Hatsugai et al. 2006). TCP3 TFs organizing tomato leaf morphology (Ori et al. 2007), one of the computer-predicted PSTVd targets established in our work, is another potential factor involved in part of the network. According to our unpublished results (Matoušek et al., unpublished), PSTVd infection with lethal AS1 and C3 strains suppress activation of several promoters in infiltrated leaves of *N. benthamiana*, suggesting targeting of some TF(s); HSVd infection of hops led to imbalance of the expression of several transcription factors from Myb, WD40, and bHLH families connected to production of lupulin metabolome. Besides tissue specificity, the hypothesis about possible involvement of targeting of some TF(s) by vsRNA is further supported by the following observations: (1) Symptom expression is strongly dependent on the stages of plant development, (2) pathogenesis is very complex on the transcriptome level, and (3) phytohormone networking is involved in the process of pathogenesis. Further studies are necessary to specify PSTVd targets within potentially misregulated networks.

4 Conclusions

According to the present knowledge, viroid infections cause accumulation of viroid-specific small RNAs with the potential of targeting or "multitargeting" genetic information of the host on various levels of its expression. From the recent point of view, the detailed research on viroid-caused disease involves analysis of regulatory networks and their misregulation by a combination of bioinformatics and experimental approaches. Such analysis will obviously require a more "structured" approach including analysis of target expression on the level of specific tissues or individual cells and using combinations of genetically distinct viroid strains, mutants, and host genotypes.

Acknowledgments Our work on viroids is supported by joint grants from the German Research Foundation to GS (465/7) and from the Czech Science Foundation to MJ (GACR P501/10/J018).

References

Aschermann M (2001) Identifikation eines zellulären RNA-Wechselwirkungspartners von PSTVd in vitro und in vivo. Thesis, Heinrich-Heine-Universität Düsseldorf, http://docserv.uni-duesseldorf.de/servlets/DocumentServlet?id=2094

Baumstark T, Schröder A, Riesner D (1997) Viroid processing: switch from cleavage to ligation is driven by a change from a tetraloop to a loop E conformation. EMBO J 16:599–610, http://dx.doi.org/10.1093/emboj/16.3.599

Bolduc F, Hoareau C, St-Pierre P, Perreault J (2010) In-depth sequencing of the siRNAs associated with peach latent mosaic viroid infection. BMC Mol Biol 11:16, http://dx.doi.org/10.1186/1471-2199-11-16

Branch A, Robertson H (1984) A replication cycle for viroids and other small infectious RNA's. Science 223:450–455

Branch A, Benenfeld B, Robertson H (1985) Ultraviolet light-induced crosslinking reveals a unique region of local tertiary structure in potato spindle tuber viroid and HeLa 5S RNA. Proc Nat Acad Sci USA 82:6590–6594, http://www.pnas.org/content/82/19/6590.full.pdf+html?sid=bd413635-93ac-4ab7-b4a6-a55a78ae6f34

Camacho Henriquez A, Sänger H (1982) Analysis of acid-extractable tomato leaf proteins after infection with a viroid, two viruses and a fungus and partial purification of the "pathogenesis-related" protein p14. Arch Virol 74:181–196

Daròs J, Flores R (2002) A chloroplast protein binds a viroid RNA in vivo and facilitates its hammerhead-mediated self-cleavage. EMBO J 21:749–759, http://dx.doi.org/10.1093/emboj/21.4.749

Daròs J, Flores R (2004) *Arabidopsis thaliana* has the enzymatic machinery for replicating representative viroid species of the family Pospiviroidae. Proc Nat Acad Sci USA 101:6792–6797, http://dx.doi.org/10.1073/pnas.0401090101

Daròs J, Marcos J, Hernández C, Flores R (1994) Replication of avocado sunblotch viroid: evidence for a symmetric pathway with two rolling circles and hammerhead ribozyme processing. Proc Nat Acad Sci USA 91:12813–12817, http://dx.doi.org/

Daròs J, Elena S, Flores R (2006) Viroids: an Ariadne's thread into the RNA labyrinth. EMBO Rep 7:593–598, http://dx.doi.org/10.1038/sj.embor.7400706

Denti M, Boutla A, Tsagris M, Tabler M (2004) Short interfering RNAs specific for potato spindle tuber viroid are found in the cytoplasm but not in the nucleus. Plant J 37:762–769, http://dx.doi.org/10.1111/j.1365-313X.2004.02001.x

Di Serio F, Martínez de Alba A, Navarro B, Gisel A, Flores R (2009a) RNA-dependent RNA polymerase 6 delays accumulation and precludes meristem invasion of a viroid that replicates in the nucleus. J Virol 84:2477–2489, http://dx.doi.org/10.1128/JVI.02336-09

Di Serio F, Gisel A, Navarro B, Delgado S, Martínez de Alba A, Donvito G, Flores R (2009b) Deep sequencing of the small RNAs derived from two symptomatic variants of a chloroplastic viroid: implications for their genesis and for pathogenesis. PLoS ONE 4:e7539, http://dx.doi.org/10.1371/journal.pone.0007539

Dickson E (1981) A model for the involvement of viroids in RNA splicing. Virology 115:216–221, http://dx.doi.org/10.1016/0042-6822(81)90104-5

Diener T, Hammond R, Black T, Katze M (1993) Mechanism of viroid pathogenesis: differential activation of the interferon-induced, double-stranded RNA-activated, Mr 68,000 protein kinase by viroid strains of varying pathogenicity. Biochimie 75:533–538, http://dx.doi.org/10.1016/0300-9084(93)90058-Z

Diermann N, Matoušek J, Junge M, Riesner D, Steger G (2010) Characterization of plant-miRNAs and small RNAs from potato spindle tuber viroid (PSTVd) in infected tomato. Biol Chem 391:1379–1390, http://dx.doi.org/10.1515/BC.2010.148

Ding B (2010) Viroids: self-replicating, mobile, and fast-evolving noncoding regulatory RNAs. Wiley Interdiscip Rev RNA 1:362–375, http://dx.doi.org/10.1002/wrna.22

Ding B, Wang Y (2009) Viroids: uniquely simple and tractable models to elucidate regulation of cell-to-cell trafficking of RNA. DNA Cell Biol 28:51–56, http://dx.doi.org/10.1089/dna.2008.0811

Dubé A, Bolduc F, Bisaillon M, Perreault J (2011) Mapping studies of the Peach latent mosaic viroid reveal novel structural features. Mol Plant Pathol 12:688–701, http://dx.doi.org/10.1111/j.1364-3703.2010.00703.x

Eiras M, Nohales M, Kitajima E, Flores R, Daròs J (2010) Ribosomal protein L5 and transcription factor IIIA from *Arabidopsis thaliana* bind in vitro specifically *Potato spindle tuber viroid* RNA. Arch Virol 156:529–533, http://dx.doi.org/10.1007/s00705-010-0867-x

Flores R, Randles J, Bar-Joseph M, Diener T (1998) A proposed scheme for viroid classification and nomenclature. Arch Virol 143:623–629, http://dx.doi.org/10.1007/s007050050318

Flores R, Hernández C, Martínez de Alba A, Daròs J, Di Serio F (2005) Viroids and viroid-host interactions. Annu Rev Phytopathol 43:117–139, http://dx.doi.org/10.1146/annurev.phyto.43.040204.140243

Gadea J, Mayda M, Conejero V, Vera P (1996) Characterization of defense-related genes ectopically expressed in viroid-infected tomato plants. Mol Plant Microbe Interact 9:409–415, http://dx.doi.org/10.1094/MPMI-9-0409

Gómez G, Pallás V (2001) Identification of an in vitro ribonucleoprotein complex between a viroid RNA and a phloem protein from cucumber plants. Mol Plant Microbe Interact 14:910–913, http://dx.doi.org/10.1094/MPMI.2001.14.7.910

Gómez G, Pallás V (2004) A long-distance translocatable phloem protein from cucumber forms a ribonucleoprotein complex in vivo with *Hop stunt viroid* RNA. J Virol 78:10104–10110, http://jvi.asm.org/cgi/reprint/78/18/10104

Gómez G, Martínez G, Pallás V (2008) Viroid-induced symptoms in *Nicotiana benthamiana* plants are dependent of RDR6 activity. Plant Physiol 148:414–423, http://dx.doi.org/10.1104/pp.108.120808

Gómez G, Martínez G, Pallás V (2009) Interplay between viroid-induced pathogenesis and RNA silencing pathways. Trends Plant Sci 14:264–269, http://dx.doi.org/10.1016/j.tplants.2009.03.002

Haas B, Klanner A, Ramm K, Sänger H (1988) The 7S RNA from tomato leaf tissue resembles a signal recognition particle RNA and exhibits a remarkable sequence complementarity to viroids. EMBO J 7:4063–4074

Hadidi A, Flores R, Randles J, Semancik J (eds) (2003) Viroids. CSIRO, Australia

Hammond R, Zhao Y (2009) Modification of tobacco plant development by sense and antisense expression of the tomato viroid-induced AGC VIIIa protein kinase PKV suggests involvement in gibberellin signaling. BMC Plant Biol 9:108, http://dx.doi.org/10.1186/1471-2229-9-108

Harders J, Lukács N, Robert-Nicoud M, Jovin T, Riesner D (1989) Imaging of viroids in nuclei from tomato leaf tissue by in situ hybridization and confocal laser scanning microscopy. EMBO J 8:3941–3949

Hatsugai N, Kuroyanagi M, Nishimura M, Hara-Nishimura I (2006) A cellular suicide strategy of plants: vacuole-mediated cell death. Apoptosis 11:905–911, http://dx.doi.org/10.1007/s10495-006-6601-1

Itaya A, Folimonov A, Matsuda Y, Nelson R, Ding B (2001) *Potato spindle tuber viroid* as inducer of RNA silencing in infected tomato. Mol Plant Microbe Interact 14:1332–1334, http://dx.doi.org/10.1094/MPMI.2001.14.11.1332

Itaya A, Matsuda Y, Gonzales R, Nelson R, Ding B (2002) *Potato spindle tuber viroid* strains of different pathogenicity induces and suppresses expression of common and unique genes in infected tomato. Mol Plant Microbe Interact 15:990–999, http://dx.doi.org/10.1094/MPMI.2002.15.10.990

Itaya A, Zhong X, Bundschuh R, Qi Y, Wang Y, Takeda R, Harris A, Molina C, Nelson R, Biao Ding B (2007) A structured viroid RNA is substrate for Dicer-like cleavage to produce biologically active small RNAs but is resistant to RISC-mediated degradation. J Virol 81:2980–2994, http://dx.doi.org/10.1128/JVI.02339-06

Jamalkandi S, Masoudi-Nejad A (2009) Reconstruction of *Arabidopsis thaliana* fully integrated small RNA pathway. Funct Integr Genomics 9:419–432, http://dx.doi.org/10.1007/s10142-009-0141-z

Kalantidis K, Denti M, Tzortzakaki S, Marinou E, Tabler M, Tsagris M (2007) Virp1 is a host protein with a major role in Potato spindle tuber viroid infection in Nicotiana plants. J Virol 81:12872–12880, http://dx.doi.org/10.1128/JVI.00974-07

Keese P, Symons R (1985) Domains in viroids: evidence of intermolecular RNA rearrangement and their contribution to viroid evolution. Proc Nat Acad Sci USA 82:4582–4586, http://www.pnas.org/cgi/reprint/82/14/4582.pdf

Kolonko N, Bannach O, Aschermann K, Hu KH, Moors M, Schmitz M, Steger G, Riesner D (2006) Transcription of potato spindle tuber viroid by RNA polymerase II starts in the left terminal loop. Virology 347:392–404, http://dx.doi.org/10.1016/j.virol.2005.11.039

Landry P, Perreault J (2005) Identification of a peach latent mosaic viroid hairpin able to act as a Dicer-like substrate. J Virol 79:6540–6543, http://dx.doi.org/10.1128/JVI.79.10.6540-6543.2005

Lima M, Fonseca M, Flores R, Kitajima E (1994) Detection of avocado sunblotch viroid in chloroplasts of avocado leaves by in situ hybridization. Arch Virol 138:385–390

López-Gresa M, Maltese F, Bellés J, Conejero V, Kim H, Choi Y, Verpoorte R (2010) Metabolic response of tomato leaves upon different plant-pathogen interactions. Phytochem Anal 21:89–94, http://dx.doi.org/10.1002/pca.1179

Machida S, Yamahata N, Watanuki H, Owens R, Sano T (2007) Successive accumulation of two size classes of viroid-specific small RNA in potato spindle tuber viroid-infected tomato plants. J Gen Virol 88:3452–3457, http://dx.doi.org/10.1099/vir.0.83228-0

Markarian N, Li H, Ding S, Semancik J (2003) RNA silencing as related to viroid induced symptom expression. Arch Virol 149:397–406, http://dx.doi.org/10.1007/s00705-003-0215-5

Markarian N, Li H, Ding S, Semancik J (2004) RNA silencing as related to viroid induced symptom expression. Arch Virol 149:397–406, http://dx.doi.org/10.1007/s00705-003-0215-5

Martín R, Arenas C, Daròs J, Covarrubias A, Reyes J, Chua N (2007) Characterization of small RNAs derived from *Citrus exocortis viroid* (CEVd) in infected tomato plants. Virology 367:135–146, http://dx.doi.org/10.1016/j.virol.2007.05.011

Martínez de Alba A, Flores R, Hernández C (2002) Two chloroplastic viroids induce the accumulation of small RNAs associated with posttranscriptional gene silencing. J Virol 76:13094–13096, http://dx.doi.org/10.1128/JVI.76.24.13094-13096.2002

Martínez de Alba A, Sägesser R, Tabler M, Tsagris M (2003) A bromodomain-containing protein from tomato specifically binds potato spindle tuber viroid RNA in vitro and in vivo. J Virol 77:9685–9694, http://dx.doi.org/10.1128/JVI.77.17.9685-9694.2003

Martinez G, Donaire L, Llave C, Pallas V, Gomez G (2010) High-throughput sequencing of *Hop stunt viroid*-derived small RNAs from cucumber leaves and phloem. Mol Plant Pathol 11:347–359, http://dx.doi.org/10.1111/j.1364-3703.2009.00608.x

Matoušek J, Orctová L, Steger G, Škopek J, Moors M, Dědič P, Riesner D (2004) Analysis of thermal stress-mediated PSTVd variation and biolistic inoculation of progeny of viroid "thermomutants" to tomato and *Brassica* species. Virology 323:67–78, http://dx.doi.org/10.1016/j.virol.2004.02.010

Matoušek J, Kozlová P, Orctová L, Schmitz A, Pešina K, Bannach O, Diermann D, Steger G, Riesner D (2007) Accumulation of viroid-specific small RNAs and increase of nucleolytic activities linked to viroid-caused pathogenesis. Biol Chem 388:1–13, http://dx.doi.org/10.1515/BC.2007.001

Navarro J, Flores R (2000) Characterization of the initiation sites of both polarity strands of a viroid RNA reveals a motif conserved in sequence and structure. EMBO J 19:2662–2670, http://dx.doi.org/10.1093/emboj/19.11.2662

Navarro J, Vera A, Flores R (2000) A chloroplastic RNA polymerase resistant to tagetitoxin is involved in replication of avocado sunblotch viroid. Virology 268:218–225, http://dx.doi.org/10.1006/viro.1999.0161

Navarro B, Pantaleo V, Gisel A, Moxon S, Dalmay T, Bisztray G, Di Serio F, Burgyán J (2009) Deep sequencing of viroid-derived small RNAs from grapevine provides new insights on the role of RNA silencing in plant-viroid interaction. PLoS ONE 4:e7686, http://dx.doi.org/10.1371/journal.pone.0007686

Ori N, Cohen A, Etzioni A, Brand A, Yanai O, Shleizer S, Menda N, Amsellem Z, Efroni I, Pekker I, Alvarez J, Blum E, Zamir D, Eshed Y (2007) Regulation of *LANCEOLATE* by *miR319* is required for compound-leaf development in tomato. Nat Genet 39:787–791, http://dx.doi.org/10.1038/ng2036

Owens R, Steger G, Hu Y, Fels A, Hammond R, Riesner D (1996) RNA structural features responsible for potato spindle tuber viroid pathogenicity. Virology 222:144–158, http://dx.doi.org/10.1006/viro.1996.0405

Owens R, Blackburn M, Ding B (2001) Possible involvement of the phloem lectin in long-distance viroid movement. Mol Plant Microbe Interact 14:905–909, http://dx.doi.org/10.1094/MPMI.2001.14.7.905

Palukaitis P (1987) Potato spindle tuber viroid: investigation of the long-distance, intra-plant transport route. Virology 158:239–241, http://dx.doi.org/10.1016/0042-6822(87)90260-1

Papaefthimiou I, Hamilton A, Denti M, Baulcombe D, Tsagris M, Tabler M (2001) Replicating potato spindle tuber viroid RNA is accompanied by short RNA fragments that are characteristic of post-transcriptional gene silencing. Nucleic Acids Res 29:2395–2400, http://nar.oxfordjournals.org/cgi/reprint/29/11/2395

Pelchat M, Coté F, Perreault JP (2001) Study of the polymerization step of the rolling circle replication of peach latent mosaic viroid. Arch Virol 146:1753–1763, http://dx.doi.org/10.1007/s007050170061

Qi Y, Ding B (2003) Differential subnuclear localization of RNA strands of opposite polarity derived from an autonomously replicating viroid. Plant Cell 15:2566–2577, http://dx.doi.org/10.1105/tpc.016576

Ramachandran V, Chen X (2008) Small RNA metabolism in *Arabidopsis*. Trends Plant Sci 13:368–374, http://dx.doi.org/10.1016/j.tplants.2008.03.008

Randles J (2003) Economic impact of viroid diseases. In: Flores R, Randles J, Semancik J, Hadidi A (eds) Viroids. CSIRO, Australia, pp 3–11

Reanwarakorn K, Semancik J (1998) Regulation of pathogenicity in hop stunt viroid-related group II citrus viroids. J Gen Virol 79(Pt 12):3163–3171

Riesner D, Gross H (1985) Viroids. Annu Rev Biochem 54:531–564

Rocheleau L, Pelchat M (2006) The Subviral RNA Database: a toolbox for viroids, the hepatitis delta virus and satellite RNAs research. BMC Microbiol 6:24, http://dx.doi.org/10.1186/1471-2180-6-24

Rodio M, Delgado S, Flores R, Di Serio F (2006) Variants of Peach latent mosaic viroid inducing peach calico: uneven distribution in infected plants and requirements of the insertion

containing the pathogenicity determinant. J Gen Virol 87:231–240, http://dx.doi.org/10.1099/vir.0.81356-0
Rodio M, Delgado S, De Stradis A, Gómez M, Flores R, Di Serio F (2007) A viroid RNA with a specific structural motif inhibits chloroplast development. Plant Cell 19:3610–3626, http://dx.doi.org/10.1105/tpc.106.049775
Rodriguez M, Randles J (1993) Coconut cadang-cadang viroid (CCCVd) mutants associated with severe disease vary in both the pathogenicity domain and the central conserved region. Nucleic Acids Res 21:2771, http://dx.doi.org/10.1093/nar/21.11.2771
Ruffner D, Stormo G, Uhlenbeck O (1990) Sequence requirements of the hammerhead RNA self-cleavage reaction. Biochemistry 29:10695–10702, http://dx.doi.org/10.1021/bi00499a018
Sano T, Candresse T, Hammond R, Diener T, Owens R (1992) Identification of multiple structural domains regulating viroid pathogenicity. Proc Nat Acad Sci USA 89:10104–10108
Schindler IM, Mühlbach HP (1992) Involvement of nuclear DNA-dependent RNA polymerases in potato spindle tuber viroid replication: a reevaluation. Plant Sci 84:221–229
Schmitz A, Riesner D (1998) Correlation between bending of the VM region and pathogenicity of different potato spindle tuber viroid strains. RNA 4:1295–1303, http://www.rnajournal.org/cgi/reprint/4/10/1295.pdf
Schmitz M, Steger G (2007) Potato spindle tuber viroid (PSTVd). Plant Viruses 1:106–115, http://globalsciencebooks.info/JournalsSup/07PV_1_1.html
Schnölzer M, Haas B, Ramm K, Hofmann H, Sänger H (1985) Correlation between structure and pathogenicity of potato spindle tuber viroid (PSTV). EMBO J 4:2181–2190, http://www.pubmedcentral.gov/articlerender.fcgi?artid=554484
Schrader O, Baumstark T, Riesner D (2003) A Mini-RNA containing the tetraloop, wobble-pair and loop E motifs of the central conserved region of potato spindle tuber viroid is processed into a minicircle. Nucleic Acids Res 31:988–998, http://nar.oxfordjournals.org/cgi/reprint/31/3/988.pdf
Schumacher J, Sänger H, Riesner D (1983) Subcellular localization of viroids in highly purified nuclei from tomato leaf tissue. EMBO J 2:1549–1555
Stark-Lorenzen P, Guitton M, Werner R, Mühlbach H (1997) Detection and tissue distribution of potato spindle tuber viroid in infected tomato plants by tissue print hybridization. Arch Virol 142:1289–1296, http://dx.doi.org/10.1007/s007050050160
Steger G, Riesner D (2003) Properties of viroids: molecular characteristics. In: Flores R, Randles J, Semancik J, Hadidi A (eds) Viroids. CSIRO, Australia, pp 15–29
Steger G, Hofmann H, Förtsch J, Gross H, Randles J, Sänger H, Riesner D (1984) Conformational transitions in viroids and virusoids: comparison of results from energy minimization algorithm and from experimental data. J Biomol Struct Dyn 2:543–571
Stroeks P (2008) Kontaktstellen zwischen dem Viroid PSTVd und der ribosomalen 5S-RNA. Thesis, Heinrich-Heine-Universität Düsseldorf, http://docserv.uni-duesseldorf.de/servlets/DocumentServlet?id=8473
Tabler M, Tsagris M (2004) Viroids: petite RNA pathogens with distinguished talents. Trends Plant Sci 9:339–348, http://dx.doi.org/10.1016/j.tplants.2004.05.007
Teune J, Steger G (2010) NOVOMIR: de novo prediction of microRNA-coding regions in a single plant genome. J Nucleic Acids 2010, http://dx.doi.org/10.4061/2010/495904
Torres-Schumann S, Ringli C, Heierli D, Amrhein N, Keller B (1996) In vitro binding of the tomato bZIP transcriptional activator VSF-1 to a regulatory element that controls xylem-specific gene expression. Plant J 9:283–296, http://dx.doi.org/10.1046/j.1365-313X.1996.09030283.x
Wang Y, Ding B (2010) Viroids: small probes for exploring the vast universe of RNA trafficking in plants. J Integr Plant Biol 52:28–39, http://dx.doi.org/10.1111/j.1744-7909.2010.00900.x
Wang Y, Shibuya M, Taneda A, Kurauchi T, Senda M, Owens R, Sano T (2011) Accumulation of Potato spindle tuber viroid-specific small RNAs is accompanied by specific changes in gene expression in two tomato cultivars. Virology 413:72–83, http://dx.doi.org/10.1016/j.virol.2011.01.021

Warrilow D, Symons R (1999) Citrus exocortis viroid RNA is associated with the largest subunit of RNA polymerase II in tomato in vivo. Arch Virol 144:2367–2375

Wassenegger M, Heimes S, Riedel L, Sänger H (1994) RNA-directed *de novo* methylation of genomic sequences in plants. Cell 76:567–576, http://dx.doi.org/10.1016/0092-8674(94)90119-8

Weinberg Z, Breaker R (2011) R2R–software to speed the depiction of aesthetic consensus RNA secondary structures. BMC Bioinformatics 12:3, http://dx.doi.org/10.1186/1471-2105-12-3

Werner R, Mühlbach H, Guitton M (1995) Isolation of viroid-RNA-binding proteins from an expression library with nonradioactive-labeled RNA probes. Biotechniques 19:218–222

Wilm A, Linnenbrink K, Steger G (2008) ConStruct: improved construction of RNA consensus structures. BMC Bioinformatics 9:219, http://dx.doi.org/10.1186/1471-2105-9-219

Wolff P, Gilz R, Schumacher J, Riesner D (1985) Complexes of viroids with histones and other proteins. Nucleic Acids Res 13:355–367

Zhong X, Leontis N, Qian S, Itaya A, Qi Y, Boris-Lawrie K, Ding B (2006) Tertiary structural and functional analyses of a viroid RNA motif by isostericity matrix and mutagenesis reveal its essential role in replication. J Virol 80:8566–8581, http://jvi.asm.org/cgi/reprint/80/17/8566

Zhong X, Archual A, Amin A, Ding B (2008) A genomic map of viroid RNA motifs critical for replication and systemic trafficking. Plant Cell 20:35–47

Zhu Y, Green L, Woo Y, Owens R, Ding B (2001) Cellular basis of potato spindle tuber viroid systemic movement. Virology 279:69–77, http://dx.doi.org/10.1006/viro.2000.0724

Index

V.A. Erdmann and J. Barciszewski (eds.), *From Nucleic Acids Sequences to Molecular Medicine*, RNA Technologies, DOI 10.1007/978-3-642-27426-8,

N

O

P

R